Proceedings

of the

Second International Tampere Conference in Statistics

PROCEEDINGS

of the

Second International
Tampere Conference in Statistics

University of Tampere, Tampere, Finland

1 - 4 June 1987

Edited by

Tarmo Pukkila and Simo Puntanen

Published by

Department of Mathematical Sciences/Statistics, University of Tampere
P.O. Box 607, SF-33101 Tampere, Finland

1987

Report No. A 184

ISBN 951-44-2168-X
ISSN 0356-4231

Printed by Vammalan Kirjapaino Oy, Vammala

PREFACE

The Second International Tampere Conference in Statistics was held at the University of Tampere, Tampere, Finland, from 1-4 June 1987. The conference brought together more than 200 researchers – from 25 different countries – in linear models, multivariate analysis, statistical computing, time series analysis and related themes. This conference continued the spirit kindled by the First International Tampere Seminar on Linear Statistical Models and their Applications that was held at the University of Tampere from August 30 - September 2, 1983.

The conference comprised Plenary Sessions (of keynote addresses with discussion), Invited Talks, and Contributed Papers. The keynote speakers were Professors T.W. Anderson (Stanford, U.S.A.), George E.P. Box (Madison, U.S.A.), E.J. Hannan (Canberra, Australia) and C. Radhakrishna Rao (Pittsburgh, U.S.A. and New Delhi, India). These four statisticians do not need any introduction to the statistical community – their contributions to the development of modern statistical science have indeed been outstanding. It is not only a pleasure to publish their excellent papers in these *Proceedings* but it was also thrilling to meet and visit with these keynote speakers in the various conference activities.

The thanks of the organizers also go to the discussants of the keynote addresses, respectively: Professors Michael D. Perlman (U.S.A.), Stratis Kounias (Greece), Knut Conradsen (Denmark) and C. G. Khatri (India). The invited talks were given by Professors A.C. Atkinson (U.K.), Jerzy K. Baksalary (Poland), Knut Conradsen (Denmark), Takeaki Kariya (Japan), C.G. Khatri (India), Seppo Mustonen (Finland), Michael D. Perlman (U.S.A.), John W. Pratt (U.S.A.), Daryl Pregibon (U.S.A.), Friedrich Pukelsheim (F.R. Germany), Tarmo Pukkila (Finland), Jorma Rissanen (U.S.A), Alastair J. Scott (New Zealand), Bimal K. Sinha (U.S.A.), Terry P. Speed (Australia) and George P.H. Styan (Canada). Special thanks are due to these speakers for their stimulating talks and for preparing such authoritative papers for these *Proceedings*. Professor P.R. Krishnaiah of the University of Pittsburgh, U.S.A., was also an invited speaker but due to serious illness was unable to come: we were deeply saddened to learn that Professor Krishnaiah passed away on 1 August 1987.

Dr. Timo Teräsvirta of the Research Institute of the Finnish Economy (Helsinki) gave the Opening Address of the conference. I am also grateful to Elina Mustonen and University Rector Jarmo Visakorpi for their contributions in the Opening Session.

There were more than 70 speakers in the contributed paper sessions. I wish to thank each of them for their kind cooperation as well as the chairmen of these sessions.

All papers in this volume have been refereed, including all the keynote addresses and invited papers. Exceptionally warm thanks should go to the 140 referees (whose names appear on page 701); the quality of these *Proceedings* is very much due to their unselfish efforts. Of the 48 contributed papers submitted for publication, 30 were accepted for publication and appear in these *Proceedings*.

Many of the papers have been typeset in our Department (using a Xerox Star) and thanks for this typesetting go to Leena Kaunisto, Elisa Lahtinen, Pirjo Larima, Virpi Mäntylä, Tuija Nummi and Marita Raita. We are also grateful to Arto Kallinen for his help with several technical arrangements while preparing this volume.

The conference was organized by a local committee within the Statistics Group of the Department of Mathematical Sciences, consisting of Paula Hietala, Simo Puntanen, George P.H. Styan and myself. Yes, indeed, there is Professor Styan of McGill University in this "local group" – he participated in several long, enthusiastic, and very helpful discussions, taking place in various parts of the world, concerning both this conference and these *Proceedings*.

Also Pentti Huuhtanen, Raija Leppälä and Tapio Nummi very kindly helped with various arrangements. Pirjo Larima deserves very special thanks for her considerable work for the conference during the summer 1987. Thanks also go to Professor Sergio G. Koreisha (University of Oregon) for assistance in planning the sessions during his visit to Tampere in March 1987.

I wish to thank Kalevi Kankaala for preparing the drawings of the four keynote speakers and the associated poster for the conference; this poster is the basis of the dust jacket of these *Proceedings*.

The conference was financially supported by the Ministry of Education of Finland, the University of Tampere and various associations and companies whose names are listed on page VIII. All deserve our warmest thanks for their invaluable support. We are also indebted to the City of Tampere for inviting all participants to a Civic Reception in the Town Hall.

The organizers were also glad to see the active interest shown in the social programme of the conference – particularly by experiencing the sweat in the sauna party in the traditional Finnish way. There was talk while planning for this conference that the organizers were all of a sweat: as it turned out the winter preceding the conference was the coldest in Finland this century.

TARMO PUKKILA
Conference Director

2 November 1987

E.J. Hannan, George E.P. Box, T.W. Anderson

Kalevi Kankaala, Tarmo Pukkila, C. Radhakrishna Rao

The conference was sponsored by

The Ministry of Education of Finland

The University of Tampere

The Foundation of the University of Tampere

The City of Tampere

The Finnish Cultural Foundation

Finnair Oy

Oy International Business Machines Ab

Rank Xerox Oy

Industry Mutual

Kansallis-Osake-Pankki

The Finnish Metalworkers' Union

Pension Insurance Company Ilmarinen

Hotel Ilves

Okobank

Sampo Insurance Company

The British Council

CONTENTS

Contributed Papers

Proc. Second International Tampere Conference in Statistics
(Tampere, Finland, 1-4 June 1987)
Tarmo Pukkila and Simo Puntanen, *Editors*

pp. 1 - 7

OPENING ADDRESS

How we got the data

Timo TERÄSVIRTA
Research Institute of the Finnish Economy, Helsinki, Finland

Key words and phrases: Data collection, history of statistics, population statistics, Tabellverket.

ABSTRACT

This opening address tells how Sweden-Finland as the first country in the world started collecting demographic data on a regular basis in 1749.

The Story

"*As should be well-known to most people in this audience, Finland shares with Sweden the honour of having been the first country to establish (in 1749) regular population statistics, actually based on records of births and deaths submitted by the clergy.*" These words may be well-known at least to a few of you. They are the first words of the opening address of the First International Tampere Seminar in 1983 (Elfving, 1985). From there, the late Professor Gustav Elfving went on to present some of the early representatives of Finnish mathematical statistics and their statistical contributions.

If Professor Elfving deemed those words good enough for the beginning of his opening address here in Tampere, I see no reason why they could not be successfully recycled and used again for the same purpose today. However, I do not intend to tell you anything about the past, present or future of Finnish mathematical statistics. Instead, I shall relate how the Finnish and Swedish statisticians got what keeps quite a few of them busy even today but may sometimes even scare the rest: the data. The story is about how collecting population data on a regular basis was started in Sweden-Finland. It will basically be a story of two memoranda, one written by a scientist and the other by an officer (and, as it seems, a well-educated gentleman).

Collecting demographic data before 1749

Collecting demographic data in Sweden-Finland had started already in the 17th century. Many parishes made lists on baptisms and marriages. The new ecclesiastic law in 1686 obliged priests in every parish to record their members, births, deaths, marriages and people entering and leaving the congregation. This produced a wealth of data, but the government did not try to make systematic use of this information (Arosenius, 1928). A suggestion in 1728 to arrange a general census in the country did not receive any positive response in the parliament. The attitude of the government started changing a few years later. Responding to a request of the parliament, King Adolf Friedrich in 1735 decreed that the governors of the provinces should, for each parliamentary session, write a report on the state of the province. The report also had to contain information on changes in the number of inhabitants in the area. Some governors did provide this information, others did not, and as a whole the numbers were obviously presented in as many ways as there were reporting governors. As we shall see, this impractical and annoying diversity was an important reason for one of the two memoranda I am going to talk about.

Sanitary Commission needs data

The next important step in developing population statistics in Sweden-Finland was the establishment of the so-called Sanitary Commission in 1737. The task of the commission was to keep watch on unusual illnesses and epidemics so that precautions against them could be taken as quickly as possible. The first measure of the commission was to propose a system for collecting statistics on mortality and illnesses. This proposal was successful in that the King obliged the dioceses to collect such data from their parishes and send them to the governors. In practice, the data were reported in a haphazard way; the material obtained was incomplete and partially incorrect and thus not very useful to the decision-makers of the day (Hjelt, 1900).

Academy of Science sends a memorandum

Although this effort of compiling statistics on mortality and illnesses did not yet have great practical significance, it must have attracted keen attention in the Royal Academy of Science, established in 1739. It seems that the members of the new academy were quick in grasping the importance of reliable population statistics to the country. The secretary of the academy since 1744, Pehr Elvius, was the most active member in these matters: in 1744 he published a scientific paper in demography, the first one of its kind in Sweden-Finland, entitled "*Förtekning uppå Barnens årliga antal, som äro födde uti U... Stad under de sistförflutne 50 åren. Jemte anledning til anmärkningar deröfver.*" [1)]

Two years later, the Royal Academy of Science decided that something had to be done to improve the system of compiling vital statistics in the country. It sent a memorandum written by its secretary (Elvius, 1746) to the Secret Committee, the most important standing committee of the parliament.

The memorandum did not adopt a hard-sell approach to the problem. Instead of insisting on how useful and profitable it would be for the government to have reliable data on the population, it merely reported results of a study carried out using such data. In fact, the main result was an estimate of the population of the whole country and that of each province separately. The figures must have been a major piece of news to the Secret Committee. The reason was that the total was alarmingly small, only 2,097,000 people, whereas the other estimates available at the time put the figure near 3 million, see Westergaard (1932, pp. 44, 53).

The estimates were obtained starting from the assumption of a stationary population, the annual number of deaths in the country (70000), and mortality tables based on data from certain parts of the kingdom. Elvius indicated that he would publish his estimation technique in the Proceedings of the Academy, but the paper never appeared. The reason for this is unclear, in particular as no referee's reports have survived. However, Westergaard (1932, pp. 54-55) has reconstructed the method of Elvius and given detailed comments to it.

---

1) "Report on the annual number of babies born in the town of U... during the last 50 years. And reason for remarks thereupon." U... means Uppsala.

Elvius (1746) concluded:

"Fast det eij kan lofvas, det denna Uträkning skall gifva antalet af Rikets menighet så noga, som en värkelig Mantalsskrifning öfver hela Riket det skulle förmå, hvilken också deremot kostade oförlikneligen mera möda och arbete, så lärer den likväl förmodeligen vara af en tilräklig noghet för Regeringen uti hvarjehanda förslagers uprättande som på något sätt fordra Invånarnes antal til grund. Och i hvad mål som desse uträkningar ännu kunna förbättras igenom en Samling af fullkomligare och flere års Förtekningar öfver de i Riket födda och döda, hvaruppå de böra grundas, deruppå lärer Academien eij underlåta at använda sin flit, enär tid och tilfälle det medgifva."

Thus, the Academy did not end up suggesting anything specific but its strategy seemed clear: let the numbers speak for themselves. The memorandum did contain a small bait: if the estimates in the memorandum could be improved upon through more complete and longer time series, the Academy, under proper conditions, would not save their effort in order to achieve such improvements.

The General joins the battle

When Pehr Elvius was drafting the Academy memorandum, another important thing happened: Brigadier General Jakob Albrecht von Lantingshausen returned to his home country after an illustrious career in the French army. During those years he had also been a private tutor of two German princes and supervised their studies at the University of Leyden (Malmström, 1886). He assumed at once a seat in the parliament and became a member of a section of the Secret Committee. As a member he had to read through the reports on the state of the provinces the governors were sending to the parliament. The lack of rules in reporting the numbers greatly distressed the general accustomed to order and military precision. This may have spurred him to give a deeper thought to the problem of obtaining reliable population data for decision makers. The result of his labour was a memorandum in French, the Swedish translation of which (von Lantingshausen, 1746) appeared in Hjelt (1900).[2)]

After first noting the serious disorder in governors' reporting the memorandum pointed out that it was absolutely necessary for an administration to have accurate information about the size of the population. One had to know the strength of the country in agriculture, trade and defence. It was rather natural for the old pro to stress the last sector, but any economist could have enthusiastically agreed with him on the first two ones.

---

2) Hjelt (1900) mentions that the French original has disappeared from the archives and only the Swedish translation can be found.

Von Lantingshausen did not just content himself to underline the advantages of population statistics in decision-making. He also made a detailed proposal, complete with drafts for tables, on what kind of data were needed and explained why. In suggesting that the statistics should include the number of marriages he motivated his idea as follows:

"Så kan en god uträkning af de nuvarande gifftermåls afföda hafwa för framtiden sin stora nytta. Som alting är uträkning underkastadt; så kunna naturens alster eij derifrån undantagas: Han går sin jämna wäg i alla sina wärkningar och hans Fölgder swara så mathematiskt emot hans Physiska orsaker, hwarifrån de herröra, at man med temmelig wisshet kan uträkna hans bekanta wärkan, utan fruktan at mycket bedraga sig och utan at det lönar mödan, at hålla räkning öfwer skiljemonen."

This is indeed a strong statement in favour of mathematical modelling and model forecasts: it is possible to model "laws of nature" so accurately that the error component can safely be ignored in practice.

Although the author liked to emphasize the practical usefulness of the population data, he did not overlook their use for more abstract scientific purposes either. He was aware of birth statistics in London and Amsterdam showing that the number of male births was higher than that of female births. The memorandum proposed research to find out what the situation was like in Sweden-Finland and what the causes of the phenomenon might be.

Welfare arguments

After discussing the need for statistics on births and deaths, von Lantingshausen noted the importance of accurate information on the causes of death. This he considered crucial to any administration caring about the welfare of their citizens:

"Ännu en stor förmån finner jag heraf, på hwilken jag fruktar at wj lika så litet anwändt wår upmärksamhet som på de förra: om dödzförtekningarne blefwo stälte på det sätt, som föreslagit är, skulle Hans Maij:st få tilräckeliga tilfällen, at låta lysa sin ömhet om sina undersåtare genom faderlige och kloka anstalter emot alt det, som kan förorsaka deras förtidiga afgång. Man märker wäl heraf, at min mening är, det Herrar Landshöfdingarnes Berättelser borde äfwen innehålla de hwarjehanda slags Siukdomar, som mäst upfylla dödzlistorne; ty man kan eij neka, at ju brist på läkedom, illa lämpade hus-medel, elak föda, hunger och fattigdom giöra på landsbygderne stort elände och offta förorsaka Smittosamma Siukdomar, som föröda mycket folk. En försiktig regering, som är mohn om undersåtarnas wältrefnad, biuder til på alt möijeligt wjs, at afböija sådane almänna olyckor."

Those were obviously not yet the magic words of a modern welfare society but remarkable enough to be repeated here. At this point the author demonstrated considerable diplomatic talent. He suggested that the statistics on the causes of death would, when properly used, give His Majesty an excellent opportunity to let his tenderness shine upon his subjects. On the other hand, he carefully avoided all mention of how much such loving care would cost. It was only two centuries later that the subjects themselves fully discovered how dreadfully expensive it could be to pursue the idea of letting the crown show extensive concern about their welfare. Nonetheless, I am convinced Brigadier General von Lantingshausen would be pleased to know that although the people of today keep grumbling they also keep paying. The memorandum ended with a strong plea to the King to organize the data collection from the provinces in a practical and useful way so as to help His Majesty to make his subjects happier.

Epilogue

Those two memoranda did not remain without response. In 1747 the Secret Committee sent a proposal to the King[3], urging His Majesty to set up a system for collecting demographic data. The proposal was very detailed and included all the tables to be filled in by the clergy; it even specified which kind of paper was to be used for different tables, and who would print them and at which cost (Hjelt, 1900).

The King agreed to the proposal and the rest, as they say, was history. Tabellverket was born and started functioning in the beginning of 1749, not yet as a statistical office but at least as a well-organized data-collecting network. The first report based on data collected by Tabellverket was prepared in 1755 (Carleson *et al.*, 1755). It disclosed, among other things, that the population of Sweden-Finland was 2,132,619 people, not far from the estimate of Elvius (1746). This figure was so low that it, together with other results, could not be published; see Westergaard (1932, p. 59). It would inevitably have revealed to the enemy how sparsely populated and thereby weak this 18th century declining superpower actually was. However, preparing the first demographic tables and reporting results is already another story which I am quite happy to spare for future speakers in these conferences.

---

3) This proposal has been published in Hjelt (1900, pp. 87-94).

Acknowledgements

I wish to thank Lars-Erik Öller and Eine Perhonen for their help in preparing this address. Financial support from the Yrjö Jahnsson Foundation is gratefully acknowledged.

REFERENCES

Arosenius, E. (1928). *Bidrag till det svenska tabellverkets historia* (avec un resumé en français). Norstedt & Söner, Stockholm.

Carleson, Edv., U. Rudenschöld, J. Faggot, P. Wargentin and Jac. J. Faggot (1755). Ödmjukt betänkande om Tabellwärket. Unpublished memorandum; later published in Hjelt (1899).

Elfving, G. (1985). Finnish mathematical statistics in the past. *Proceedings of the First International Tampere Seminar on Linear Statistical Models and their Applications* (T. Pukkila and S. Puntanen, *eds.*). Dept. of Mathematical Sciences, University of Tampere, Finland, 3-8.

Elvius, P. (1746). Svenska Vetenskapsakademins betänkande angående folkmängden i Sverige och Finland. Originally unpublished memorandum; later published in Hjelt (1900).

Hjelt, A. (1899). De första officiela relationerna om svenska Tabellverket åren 1749-1757. Några bidrag till den svensk-finska befolkningsstatistikens historia (mit deutscher Zusammenfassung). *Fennia* 16 (3), 1-136.

Hjelt, A. (1900). Det svenska Tabellverkets uppkomst, organisation och tidigare verksamhet. Några minnesblad ur den svensk-finska befolkningsstatistikens historia (mit deutscher Zusammenfassung). *Fennia* 16 (2), 1-109.

Lantingshausen, J.A. von (1746). Nödvändigheten af närmare underrättelsers inhemtande om Rikets styrka i anseende till dess Inbyggares antal, tillväxt och afgång med mera. Swedish translation of the original unpublished memorandum in French; later published in Hjelt (1900).

Malmström, C. G. (1886). Minne af öfverståthållaren m.m. friherre Jakob Albrecht von Lantingshausen. *Svenska Akademiens handlingar* 62, 85-215.

Westergaard, H. (1932). *Contributions to the History of Statistics.* King & Son, London.

Received 13 February 1987
Revised 24 June 1987

Research Institute of the Finnish Economy
Lönnrotinkatu 4 B
SF - 00120 Helsinki
Finland

Proc. Second International Tampere Conference in Statistics
(Tampere, Finland, 1-4 June 1987)
Tarmo Pukkila and Simo Puntanen, *Editors*

pp. 9 - 36

KEYNOTE ADDRESS

Multivariate linear relations

T.W. ANDERSON
Stanford University, Stanford, CA, U.S.A.

Key words and phrases: Multivariate linear relations, factor analysis, covariance structures, asymptotic robustness, maximum likelihood estimates.

ABSTRACT

Linear functional and structural relations can be defined by $\boldsymbol{B}\boldsymbol{y} = \boldsymbol{\beta}$ in a model $\boldsymbol{x} = \boldsymbol{y} + \boldsymbol{u}$, where $\boldsymbol{x}$ is the observed vector, $\boldsymbol{y}$ and $\boldsymbol{u}$ are unobservable vectors, and $\boldsymbol{B}$ and $\boldsymbol{\beta}$ are a matrix and a vector of parameters, respectively. Such a model is equivalent to a factor analysis model $\boldsymbol{x} = \boldsymbol{\mu} + \boldsymbol{\Lambda}\boldsymbol{f} + \boldsymbol{u}$ if $\boldsymbol{B}\boldsymbol{\Lambda} = \boldsymbol{0}$ and $\boldsymbol{B}\boldsymbol{\mu} = \boldsymbol{\beta}$. Usually the error vector $\boldsymbol{u}$ is assumed to have a diagonal covariance matrix. These models are generalized to $\boldsymbol{x} = \boldsymbol{\mu} + \sum_{g=0}^{G} \boldsymbol{\Lambda}_g(\boldsymbol{\lambda})\boldsymbol{z}^{(g)}$, where the unobservable vectors $\boldsymbol{z}^{(0)}, \ldots, \boldsymbol{z}^{(G)}$ are independent with arbitrary covariance matrices. The maximum likelihood estimates of the vector of parameters $\boldsymbol{\lambda}$ and of the covariance matrices of the $\boldsymbol{z}^{(g)}$'s are obtained under assumed normality of the $\boldsymbol{z}^{(g)}$'s. The asymptotic normal distribution of the estimates is derived. The asymptotic distribution of the estimate of $\boldsymbol{\lambda}$ is shown to be valid for nonnormal distributions of the $\boldsymbol{z}^{(g)}$'s if some mild conditions are satisfied. Several applications of the results are made.

1. INTRODUCTION

A general kind of multivariate model represents an observable p–component vector $\boldsymbol{x}_\alpha$ in terms of two unobservable vectors $\boldsymbol{y}_\alpha$, called the systematic part, and $\boldsymbol{u}_\alpha$, called the error, as

$$\boldsymbol{x}_\alpha = \boldsymbol{y}_\alpha + \boldsymbol{u}_\alpha, \qquad \alpha = 1, \ldots, N. \tag{1.1}$$

Linear structural relations are defined as linear conditions on the systematic parts when they are random vectors:

$$\boldsymbol{B}\boldsymbol{y}_\alpha = \boldsymbol{\beta}. \tag{1.2}$$

If the systematic parts are nonstochastic, the relations are termed functional. The factor analysis model fits this pattern with $\boldsymbol{y}_\alpha$ replaced by $\boldsymbol{\mu} + \boldsymbol{\Lambda}\boldsymbol{f}_\alpha$; that is,

$$\boldsymbol{x}_\alpha = \boldsymbol{\mu} + \boldsymbol{\Lambda}\boldsymbol{f}_\alpha + \boldsymbol{u}_\alpha. \tag{1.3}$$

Here $\boldsymbol{\mu}$ is a parameter vector, $\boldsymbol{\Lambda}$ is the matrix of factor loadings, and $\boldsymbol{f}_\alpha$ is the vector of factor scores. If

$$\boldsymbol{B}\boldsymbol{\Lambda} = \boldsymbol{0}, \qquad \boldsymbol{B}\boldsymbol{\mu} = \boldsymbol{\beta}, \tag{1.4}$$

the two models are equivalent. Usually in factor analysis the mean vector of the random error is $\mathcal{E}\,\boldsymbol{u}_\alpha = \boldsymbol{0}$ and the covariance matrix of the random error vector, $\mathcal{E}\,\boldsymbol{u}_\alpha\boldsymbol{u}'_\alpha = \boldsymbol{\Psi}$, say, is diagonal. Then the factor scores "explain" the dependence between components of the observed vector. The systematic parts and errors are uncorrelated: $\mathcal{E}\,\boldsymbol{f}_\alpha\boldsymbol{u}'_\alpha = \boldsymbol{0}$. Anderson (1984a) has discussed these models in considerable detail.

If $\boldsymbol{f}_\alpha$ is random with mean $\mathcal{E}\,\boldsymbol{f}_\alpha = \boldsymbol{0}$, the mean vector of $\boldsymbol{x}_\alpha$ is $\boldsymbol{\mu}$. If the covariance matrix of $\boldsymbol{f}_\alpha$ is $\mathcal{E}\,\boldsymbol{f}_\alpha\boldsymbol{f}'_\alpha = \boldsymbol{\Phi}$, the covariance matrix of the observable vector is

$$\begin{aligned}\boldsymbol{\Sigma} &= \mathcal{E}(\boldsymbol{x}_\alpha - \boldsymbol{\mu})(\boldsymbol{x}_\alpha - \boldsymbol{\mu})' \\ &= \mathcal{E}(\boldsymbol{\Lambda}\boldsymbol{f}_\alpha + \boldsymbol{u}_\alpha)(\boldsymbol{\Lambda}\boldsymbol{f}_\alpha + \boldsymbol{u}_\alpha)' \\ &= \boldsymbol{\Lambda}\boldsymbol{\Phi}\boldsymbol{\Lambda}' + \boldsymbol{\Psi}.\end{aligned} \tag{1.5}$$

This is an example of a *covariance structure*. The observable covariance matrix $\boldsymbol{\Sigma}$ is a function of the components of $\boldsymbol{\Lambda}, \boldsymbol{\Phi}$, and $\boldsymbol{\Psi}$.

In (1.3) there is the indeterminacy of replacing $\boldsymbol{\Lambda}$ by $\boldsymbol{\Lambda C}$ and $\boldsymbol{f}_\alpha$ by $\boldsymbol{C}^{-1}\boldsymbol{f}_\alpha$, where $\boldsymbol{C}$ is an arbitrary nonsingular matrix, or equivalently in (1.5) replacing $\boldsymbol{\Lambda}$ by $\boldsymbol{\Lambda C}$ and $\boldsymbol{\Phi}$ by $\boldsymbol{C}^{-1}\boldsymbol{\Phi}\boldsymbol{C}^{-1\prime}$. To eliminate this indeterminacy some conditions may be put on $\boldsymbol{\Lambda}$ and $\boldsymbol{\Phi}$. For example, one may require

$$\boldsymbol{\Lambda} = \begin{pmatrix} \boldsymbol{\Lambda}^* \\ \boldsymbol{I} \end{pmatrix}. \tag{1.6}$$

Here certain elements of $\boldsymbol{\Lambda}$ are specified as 0 and others as 1; $\boldsymbol{\Lambda}$ is a function of the components of $\boldsymbol{\Lambda}^*$. We can consider the factor loading matrix in general as a function of some parameter vector $\boldsymbol{\lambda}$; that is, $\boldsymbol{\Lambda} = \boldsymbol{\Lambda}(\boldsymbol{\lambda})$. Then $\boldsymbol{\Sigma}$ is a function of $\boldsymbol{\lambda}$, the components of $\boldsymbol{\Phi}$, and the diagonal elements of $\boldsymbol{\Psi}$.

There is ambiguity in the linear structural/functional relations given by (1.2), also. The matrix $\boldsymbol{B}$ and vector $\boldsymbol{\beta}$ can be replaced by $\boldsymbol{AB}$ and $\boldsymbol{A\beta}$, respectively, where $\boldsymbol{A}$ is an arbitrary nonsingular matrix. One way of ruling out this indeterminacy is to specify

$$\boldsymbol{B} = (\boldsymbol{I} \;\; \boldsymbol{B}^*). \tag{1.7}$$

If (1.6) and (1.7) hold, then (1.4) implies

$$\boldsymbol{B}^* = -\boldsymbol{\Lambda}^*. \tag{1.8}$$

We can more generally consider $\boldsymbol{B}$ as a function of a parameter vector $\boldsymbol{B} = \boldsymbol{B}(\boldsymbol{\lambda})$.

In this paper we shall study a generalization of (1.3) to

$$\boldsymbol{x}_\alpha = \boldsymbol{\mu} + \sum_{g=0}^{G} \boldsymbol{\Lambda}_g(\boldsymbol{\lambda})\boldsymbol{z}_\alpha^{(g)}, \tag{1.9}$$

where

$$\mathcal{E}\, \boldsymbol{z}_\alpha^{(g)} = \boldsymbol{0}, \quad \mathcal{E}\, \boldsymbol{z}_\alpha^{(g)}\boldsymbol{z}_\alpha^{(g)\prime} = \boldsymbol{\Phi}_g, \tag{1.10}$$

$$\mathcal{E}\, \boldsymbol{z}_\alpha^{(g)}\boldsymbol{z}_\alpha^{(h)\prime} = \boldsymbol{0}, \quad g \neq h. \tag{1.11}$$

Then

$$\boldsymbol{\Sigma} = \mathcal{E}(\boldsymbol{x}_\alpha - \boldsymbol{\mu})(\boldsymbol{x}_\alpha - \boldsymbol{\mu})' = \sum_{g=0}^{G} \boldsymbol{\Lambda}_g(\boldsymbol{\lambda})\boldsymbol{\Phi}_g\boldsymbol{\Lambda}_g'(\boldsymbol{\lambda}). \tag{1.12}$$

A purpose of this notation is to emphasize the zero correlation of the components of $\boldsymbol{z}_\alpha^{(g)}$ and $\boldsymbol{z}_\alpha^{(h)}$ for $g \neq h$. For the sake of generality we let $\boldsymbol{\Phi}_0$ be a function of a parameter vector $\boldsymbol{\tau}$; that is, $\boldsymbol{\Phi}_0 = \boldsymbol{\Phi}_0(\boldsymbol{\tau})$. Many covariance structures of interest are special cases of this model. Examples will be given later. Browne and Shapiro (1986) introduced this kind of model; it is a special case of the general linear model of Bentler (1983).

We shall obtain the maximum likelihood estimators of $\boldsymbol{\lambda}, \boldsymbol{\tau}$, and $\boldsymbol{\Phi}_1, \dots, \boldsymbol{\Phi}_G$ when the $\boldsymbol{z}_\alpha^{(g)}$'s are assumed to be normally distributed and derive the asymptotic distribution of the estimates under normality. Then the asymptotic distribution is obtained when $\boldsymbol{z}_\alpha^{(0)}, \dots, \boldsymbol{z}_\alpha^{(G)}$ are not necessarily normal, but their fourth–order moments exist. It will be shown that the asymptotic distribution of the estimate of $\boldsymbol{\lambda}$ obtained for normally distributed $\boldsymbol{z}_\alpha^{(g)}$'s holds when they are not normally distributed; we require only the existence of second–order moments of those $\boldsymbol{z}_\alpha^{(g)}$'s. In fact, one vector can be nonstochastic.

Anderson and Rubin (1956) stated that the asymptotic distribution of the estimate of $\boldsymbol{\lambda}$ in the factor analysis model is normal when 0's and 1's are specified in $\boldsymbol{\Lambda}$ and some mild conditions on the $\boldsymbol{f}_\alpha$'s and $\boldsymbol{u}_\alpha$'s are met. Anderson and Amemiya (1985) stated more precise theorems about the factor analysis model; roughly speaking, if the components of $\boldsymbol{u}_\alpha$ are mutually independent and independent of $\boldsymbol{f}_\alpha$, the asymptotic distribution of the estimate of $\boldsymbol{\lambda}$ does not depend on the distributions of $\boldsymbol{f}_\alpha$ and $\boldsymbol{u}_\alpha$ except for the second–order moments. Browne (1987) treated the factor analysis model with $\boldsymbol{\Lambda}$ and $\boldsymbol{\Psi}$ functions of $\boldsymbol{\lambda}$ assuming $\boldsymbol{f}_\alpha$ has finite eighth–order moments and $\boldsymbol{u}_\alpha$ is normal. In terms of linear structural/functional relations asymptotic normality when $\boldsymbol{u}_\alpha$ is normal has been studied by several authors, including Gleser (1983) and Amemiya and Fuller (1984).

Browne and Shapiro (1986) have studied inference in the model (1.9) to (1.11) and have found the asymptotic distribution of the normal maximum likelihood estimators for general distributions of the $\boldsymbol{z}_\alpha^{(g)}$'s. Their treatment is based on the asymptotic normal distribution of the sample covariance matrix. Since second–order moments of the sample covariance matrix are needed, that is, fourth–order moments of the $\boldsymbol{x}_\alpha$'s, they assume that the $\boldsymbol{z}^{(g)}$'s have finite fourth–order moments. The approach in this paper is different from that of Browne and Shapiro. Much of the asymptotic theory is developed here on the assumption that $\boldsymbol{z}_\alpha^{(1)}, \ldots, \boldsymbol{z}_\alpha^{(G)}$ have only finite second–order moments. It may be surprising that the asymptotic distribution of the estimate of $\boldsymbol{\lambda}$ depends only on the second–order moments although the estimate of $\boldsymbol{\lambda}$ is a function of the sample covariance matrix. We separate out some effects of the covariances of the different $\boldsymbol{z}_\alpha^{(g)}$'s in the estimation of $\boldsymbol{\lambda}$.

The behavior of normal maximum likelihood estimates (as well as other estimates) for more general covariance structures has been investigated; see, for example, Browne (1984) and Mooijart and Bentler (1985). More recently Satorra and Bentler (1986) and Mooijart and Bentler (1987) have also studied analysis of covariance structures when the observed vector or part of it is composed of systematic parts and errors. In these studies the emphasis is on the validity of the large–sample theory when normality of the constituents is not assumed. The relaxation of the assumptions of normality is known as *asymptotic robustness.*

Computer programs, for example, LISREL, are available for computing the maximum likelihood estimates based on the normal likelihood function. Such a program may also give the asymptotic covariance matrix of the estimates by evaluation or approximation of the Hessian (the matrix of second derivatives of the logarithm of the normal likelihood function). That the estimates are asymptotically robust implies that this asymptotic covariance matrix is appropriate for a wide class of distributions of the constituent $\boldsymbol{z}_\alpha^{(g)}$'s.

In the first part of the paper we treat $\boldsymbol{z}_\alpha^{(0)}, \boldsymbol{z}_\alpha^{(1)}, \ldots, \boldsymbol{z}_\alpha^{(G)}$ as random, in fact, independently identically distributed, but later we permit one of $\boldsymbol{z}_\alpha^{(0)}, \boldsymbol{z}_\alpha^{(1)}, \ldots, \boldsymbol{z}_\alpha^{(G)}$ to be nonstochastic. In Section 2 we develop maximum likelihood estimation for a general covariance structure, summarizing earlier results; see Browne (1984) or Shapiro (1983), for example. In Section 3 we use these results to obtain asymptotic distributions under minimal conditions. Section 4 provides examples, and Section 5 extends some of the limit theorems.

2. THE GENERAL COVARIANCE STRUCTURE

2.1 The model

The inference procedures considered in this paper are based on the sample covariance matrix

$$\boldsymbol{S} = \frac{1}{n} \sum_{\alpha=1}^{N} (\boldsymbol{x}_\alpha - \bar{\boldsymbol{x}})(\boldsymbol{x}_\alpha - \bar{\boldsymbol{x}})', \tag{2.1}$$

where $n = N - 1$, $\bar{\boldsymbol{x}} = (1/N) \sum_{\alpha=1}^{N} \boldsymbol{x}_\alpha$, and $\boldsymbol{x}_1, \ldots, \boldsymbol{x}_N$ are the p–component observation vectors. If the observations are normally distributed with mean $\boldsymbol{\mu}$ and covariance matrix $\boldsymbol{\Sigma}$, then $\boldsymbol{S}$ is the sufficient statistic invariant with respect to translations $\boldsymbol{x}_\alpha \to \boldsymbol{x}_\alpha + \boldsymbol{a}$, and $n\boldsymbol{S}$ is distributed according to the Wishart distribution $W(\boldsymbol{\Sigma}, n)$, which has the density

$$\text{const}\, |\boldsymbol{\Sigma}|^{-\frac{1}{2}n} |\boldsymbol{S}|^{\frac{1}{2}(n-p-1)} \exp\left(-\frac{1}{2} n \text{ tr } \boldsymbol{S}\boldsymbol{\Sigma}^{-1}\right). \tag{2.2}$$

In this section we shall assume that the population covariance matrix is a function of an r–component parameter vector $\boldsymbol{\theta}$; that is, $\boldsymbol{\Sigma} = \boldsymbol{\Sigma}(\boldsymbol{\theta})$. In the case of normality the Wishart maximum likelihood estimator of $\boldsymbol{\theta}$ based on $\boldsymbol{S}$, written $\hat{\boldsymbol{\theta}}$, is the value of $\boldsymbol{\theta}$ minimizing

$$\ell(\boldsymbol{\theta}, \boldsymbol{S}) = \log |\boldsymbol{\Sigma}(\boldsymbol{\theta})| + \text{tr } \boldsymbol{S}\boldsymbol{\Sigma}^{-1}(\boldsymbol{\theta}) \tag{2.3}$$

over the set $\Theta = \{\boldsymbol{\theta} | \boldsymbol{\Sigma}(\boldsymbol{\theta}) \text{ is positive definite}\}$, which we assume to be an open set. Note that $\ell(\boldsymbol{\theta}, \boldsymbol{S})$ differs from $-2/n$ times the logarithm of the Wishart likelihood only by terms not depending on $\boldsymbol{\theta}$. The normal likelihood function based on the observations $\boldsymbol{x}_1, \ldots, \boldsymbol{x}_N$ is $\text{const}\, |\boldsymbol{\Sigma}|^{-\frac{1}{2}N} \exp\left(-\frac{1}{2} n \text{ tr } \boldsymbol{S}\boldsymbol{\Sigma}^{-1}\right)$ after replacing $\boldsymbol{\mu}$ by its maximum likelihood estimate $\bar{\boldsymbol{x}}$; if n in this last expression is replaced by N, its maximization is equivalent to maximizing (2.2). Asymptotic results hold, of course, for such estimators.

The covariance matrix $\boldsymbol{\Sigma}$ is a (single–valued) function of the parameter $\boldsymbol{\theta}$. Conversely, we require that $\boldsymbol{\Sigma}$ uniquely determine $\boldsymbol{\theta}$. We shall ask that the inverse of the function $\boldsymbol{\Sigma} = \boldsymbol{\Sigma}(\boldsymbol{\theta})$ be a continuous function of $\boldsymbol{\Sigma}$ (for values of $\boldsymbol{\Sigma}$ that can be generated by the model) at $\boldsymbol{\theta}_0$, the value of the parameter of the distribution actually sampled (the "true value"). This is called the strong identification condition. [See Browne and Shapiro (1986) or Rao (1973), Section 5e.2.] We use norms $\| \boldsymbol{\theta} \| = \sqrt{\boldsymbol{\theta}'\boldsymbol{\theta}}$ and $\| \boldsymbol{\Sigma} \| = \sqrt{\text{tr } \boldsymbol{\Sigma}^2}$.

ASSUMPTION 1. *For any* $\varepsilon > 0$ *there exists* $\delta > 0$ *such that if* $\| \boldsymbol{\Sigma}(\boldsymbol{\theta}) - \boldsymbol{\Sigma}(\boldsymbol{\theta}_0) \| < \delta$, *then* $\| \boldsymbol{\theta} - \boldsymbol{\theta}_0 \| < \varepsilon$.

For positive definite $\boldsymbol{\Sigma}$ and $\boldsymbol{\Sigma}^*$ we define the loss function [Sec. 7.8, Anderson (1984b)] or discrepancy function [Browne (1984)]

$$L(\boldsymbol{\Sigma}, \boldsymbol{\Sigma}^*) = \log|\boldsymbol{\Sigma}| - \log|\boldsymbol{\Sigma}^*| + \operatorname{tr} \boldsymbol{\Sigma}^{-1}\boldsymbol{\Sigma}^* - p, \qquad (2.4)$$
$$= \operatorname{tr} \boldsymbol{\Sigma}^{-1}\boldsymbol{\Sigma}^* - \log|\boldsymbol{\Sigma}^{-1}\boldsymbol{\Sigma}^*| - p.$$

Note that

$$\ell(\boldsymbol{\theta}, \boldsymbol{S}) = L[\boldsymbol{\Sigma}(\boldsymbol{\theta}), \boldsymbol{S}] + \log|\boldsymbol{S}| + p. \qquad (2.5)$$

We see that

$$L(\boldsymbol{\Sigma}, \boldsymbol{\Sigma}) = 0, \;\; L(\boldsymbol{\Sigma}, \boldsymbol{\Sigma}^*) > 0 \quad \text{if} \quad \boldsymbol{\Sigma} \neq \boldsymbol{\Sigma}^*, \qquad (2.6)$$

and $L[\cdot,\cdot]$ is continuous in both arguments. Maximizing the Wishart likelihood is equivalent to minimizing $L[\boldsymbol{\Sigma}(\boldsymbol{\theta}), \boldsymbol{S}]$ with respect to $\boldsymbol{\theta}$. However, $L[\boldsymbol{\Sigma}(\boldsymbol{\theta}), \boldsymbol{S}]$ can be considered an objective function without reference to the normal distribution.

2.2 Consistency

We shall now show that if $\boldsymbol{S}$ is a consistent estimator of $\boldsymbol{\Sigma}$, then $\hat{\boldsymbol{\theta}}$ is a consistent estimator of $\boldsymbol{\theta}$. In particular, this is the case if the $\boldsymbol{x}_\alpha$ are independently identically distributed with mean $\boldsymbol{\mu}$ and covariance matrix $\boldsymbol{\Sigma}$ (not necessarily normal). We assume that as $n \to \infty$

$$\boldsymbol{S} \xrightarrow{\text{P}} \boldsymbol{\Sigma}(\boldsymbol{\theta}_0). \qquad (2.7)$$

THEOREM 2.1. *If Assumption* 1 *and* (2.7) *hold, then as* $n \to \infty$

$$\hat{\boldsymbol{\theta}} \xrightarrow{\text{P}} \boldsymbol{\theta}_0. \qquad (2.8)$$

Proof. From (2.7) and the continuity of $L(\cdot,\cdot)$ we obtain

$$L[\boldsymbol{\Sigma}(\boldsymbol{\theta}_0), \boldsymbol{S}] \xrightarrow{\text{P}} L[\boldsymbol{\Sigma}(\boldsymbol{\theta}_0), \boldsymbol{\Sigma}(\boldsymbol{\theta}_0)] = 0. \qquad (2.9)$$

Then

$$0 \leq L[\boldsymbol{\Sigma}(\hat{\boldsymbol{\theta}}), \boldsymbol{S}] \leq L[\boldsymbol{\Sigma}(\boldsymbol{\theta}_0), \boldsymbol{S}] \xrightarrow{\text{P}} 0; \tag{2.10}$$

that is,

$$L[\boldsymbol{\Sigma}(\hat{\boldsymbol{\theta}}), \boldsymbol{S}] \xrightarrow{\text{P}} 0. \tag{2.11}$$

LEMMA 2.1. *For positive definite* $\boldsymbol{A}$

$$tr\ \boldsymbol{A} - \log |\boldsymbol{A}| - p \to 0 \tag{2.12}$$

implies $\boldsymbol{A} \to \boldsymbol{I}$.

Proof of Lemma 2.1. Let $\boldsymbol{A} = \boldsymbol{PDP}'$, where $\boldsymbol{D}$ is diagonal with diagonal elements $d_1, \ldots, d_p$ and $\boldsymbol{P}$ is orthogonal. Then

$$\text{tr}\ \boldsymbol{A} - \log |\boldsymbol{A}| - p = \sum_{i=1}^{p} (d_i - \log d_i - 1). \tag{2.13}$$

The assumption (2.12) implies $d_i - \log d_i - 1 \to 0$, which in turn implies $d_i \to 1$ because $x - \log x - 1$ is nonnegative for $x > 0$, has a unique minimum at $x = 1$, is continuous, and approaches ∞ as $x \to 0$ or $x \to \infty$. Thus $\boldsymbol{D} \to \boldsymbol{I}$ and $\boldsymbol{A} \to \boldsymbol{I}$. ∎

Thus (2.11) implies

$$\boldsymbol{\Sigma}^{-\frac{1}{2}}(\hat{\boldsymbol{\theta}}) \boldsymbol{S} \boldsymbol{\Sigma}^{-\frac{1}{2}}(\hat{\boldsymbol{\theta}}) \xrightarrow{\text{P}} \boldsymbol{I}. \tag{2.14}$$

Combining (2.14) and (2.7) we obtain

$$\boldsymbol{\Sigma}(\hat{\boldsymbol{\theta}}) \xrightarrow{\text{P}} \boldsymbol{\Sigma}(\boldsymbol{\theta}_0), \tag{2.15}$$

and the theorem follows by use of Assumption 1. ∎

Consistency of a larger class of estimates has been given by Shapiro (1984) and Kano (1986), for example.

2.3 Asymptotic normality

Now we turn to the asymptotic normality of the estimator $\hat{\boldsymbol{\theta}}$. We define the vec operator by

$$\text{vec}\ \boldsymbol{A} = \begin{pmatrix} \boldsymbol{a}_1 \\ \vdots \\ \boldsymbol{a}_m \end{pmatrix}, \qquad \boldsymbol{A} = (\boldsymbol{a}_1, \ldots, \boldsymbol{a}_m). \tag{2.16}$$

ASSUMPTION 2. $\boldsymbol{\Sigma}(\boldsymbol{\theta})$ *is twice continuously differentiable in a neighborhood of* $\boldsymbol{\theta}_0$, *and* $\partial \operatorname{vec} \boldsymbol{\Sigma}(\boldsymbol{\theta})/\partial\boldsymbol{\theta}'$ *has full column rank at* $\boldsymbol{\theta} = \boldsymbol{\theta}_0$.

The (Wishart) maximum likelihood estimates may be obtained by setting to 0 the derivatives

$$\begin{aligned}\frac{\partial \ell(\boldsymbol{\theta}, \boldsymbol{S})}{\partial \theta_i} &= \operatorname{tr} \boldsymbol{\Sigma}^{-1}(\boldsymbol{\theta}) \frac{\partial \boldsymbol{\Sigma}(\boldsymbol{\theta})}{\partial \theta_i} - \operatorname{tr} \boldsymbol{\Sigma}^{-1}(\boldsymbol{\theta}) \boldsymbol{S} \boldsymbol{\Sigma}^{-1}(\boldsymbol{\theta}) \frac{\partial \boldsymbol{\Sigma}(\boldsymbol{\theta})}{\partial \theta_i} \\ &= -\operatorname{tr} \boldsymbol{\Sigma}^{-1}(\boldsymbol{\theta})[\boldsymbol{S} - \boldsymbol{\Sigma}(\boldsymbol{\theta})]\boldsymbol{\Sigma}^{-1}(\boldsymbol{\theta}) \frac{\partial \boldsymbol{\Sigma}(\boldsymbol{\theta})}{\partial \theta_i}, \qquad i = 1, \ldots, r.\end{aligned} \tag{2.17}$$

With probability approaching 1 as $n \to \infty$, the minimum of $\ell(\boldsymbol{\theta}, \boldsymbol{S})$ will occur in a neighborhood of $\boldsymbol{\theta}_0$ and hence the MLE will satisfy the derivative equations. Using the relation [Magnus and Neudecker (1986)]

$$\operatorname{tr} \boldsymbol{ABCD} = (\operatorname{vec} \boldsymbol{D}')'(\boldsymbol{C}' \otimes \boldsymbol{A}) \operatorname{vec} \boldsymbol{B}, \tag{2.18}$$

we obtain from (2.17)

$$\frac{\partial \ell(\boldsymbol{\theta}, \boldsymbol{S})}{\partial \boldsymbol{\theta}} = -\left[\frac{\partial \operatorname{vec} \boldsymbol{\Sigma}(\boldsymbol{\theta})}{\partial \boldsymbol{\theta}'}\right]' [\boldsymbol{\Sigma}^{-1}(\boldsymbol{\theta}) \otimes \boldsymbol{\Sigma}^{-1}(\boldsymbol{\theta})] \operatorname{vec}[\boldsymbol{S} - \boldsymbol{\Sigma}(\boldsymbol{\theta})]. \tag{2.19}$$

To investigate the asymptotic normality of $\hat{\boldsymbol{\theta}}$ we need the second partial derivatives, which are

$$\begin{aligned}\frac{\partial^2 \ell(\boldsymbol{\theta}, \boldsymbol{S})}{\partial \theta_i \partial \theta_j} = &-\operatorname{tr} \boldsymbol{\Sigma}^{-1}(\boldsymbol{\theta}) \frac{\partial \boldsymbol{\Sigma}(\boldsymbol{\theta})}{\partial \theta_i} \boldsymbol{\Sigma}^{-1}(\boldsymbol{\theta}) \frac{\partial \boldsymbol{\Sigma}(\boldsymbol{\theta})}{\partial \theta_j} \\ &+ \operatorname{tr} \boldsymbol{\Sigma}^{-1}(\boldsymbol{\theta}) \frac{\partial^2 \boldsymbol{\Sigma}(\boldsymbol{\theta})}{\partial \theta_i \partial \theta_j} \\ &+ 2 \operatorname{tr} \boldsymbol{\Sigma}^{-1}(\boldsymbol{\theta}) \boldsymbol{S} \boldsymbol{\Sigma}^{-1}(\boldsymbol{\theta}) \frac{\partial \boldsymbol{\Sigma}(\boldsymbol{\theta})}{\partial \theta_i} \boldsymbol{\Sigma}^{-1}(\boldsymbol{\theta}) \frac{\partial \boldsymbol{\Sigma}(\boldsymbol{\theta})}{\partial \theta_j} \\ &- \operatorname{tr} \boldsymbol{\Sigma}^{-1}(\boldsymbol{\theta}) \boldsymbol{S} \boldsymbol{\Sigma}^{-1}(\boldsymbol{\theta}) \frac{\partial^2 \boldsymbol{\Sigma}(\boldsymbol{\theta})}{\partial \theta_i \partial \theta_j}, \qquad i, j = 1, \ldots, r.\end{aligned} \tag{2.20}$$

Since $\mathcal{E}\boldsymbol{S} = \boldsymbol{\Sigma}(\boldsymbol{\theta})$,

$$\mathcal{E} \left.\frac{\partial^2 \ell(\boldsymbol{\theta}, \boldsymbol{S})}{\partial \theta_i \partial \theta_j}\right|_{\boldsymbol{\theta} = \boldsymbol{\theta}_0} = \operatorname{tr} \boldsymbol{\Sigma}^{-1}(\boldsymbol{\theta}_0) \left.\frac{\partial \boldsymbol{\Sigma}(\boldsymbol{\theta})}{\partial \theta_i}\right|_{\boldsymbol{\theta} = \boldsymbol{\theta}_0} \boldsymbol{\Sigma}^{-1}(\boldsymbol{\theta}_0) \left.\frac{\partial \boldsymbol{\Sigma}(\boldsymbol{\theta})}{\partial \theta_j}\right|_{\boldsymbol{\theta} = \boldsymbol{\theta}_0}. \tag{2.21}$$

Using (2.18), we write the matrix whose elements are (2.21) as

$$H(\theta_0) = \left[\frac{\partial \text{ vec } \Sigma(\theta)}{\partial \theta'}\bigg|_{\theta = \theta_0}\right]' [\Sigma^{-1}(\theta_0) \otimes \Sigma^{-1}(\theta_0)] \left[\frac{\partial \text{ vec } \Sigma(\theta)}{\partial \theta'}\bigg|_{\theta = \theta_0}\right] \tag{2.22}$$

This matrix is positive definite by Assumption 2. The derivative equations can be expanded as

$$\begin{aligned} 0 &= \frac{\partial \ell(\theta, S)}{\partial \theta}\bigg|_{\theta = \hat{\theta}} \\ &= \frac{\partial \ell(\theta, S)}{\partial \theta}\bigg|_{\theta = \theta_0} + \frac{\partial^2 \ell(\theta, S)}{\partial \theta \partial \theta'}\bigg|_{\theta = \theta^*} (\hat{\theta} - \theta_0), \end{aligned} \tag{2.23}$$

where θ^* is on the line segment joining $\hat{\theta}$ and θ_0. If $S \xrightarrow{P} \Sigma(\theta_0)$, then $\hat{\theta} \xrightarrow{P} \theta_0, \theta^* \xrightarrow{P} \theta_0$, and

$$\frac{\partial^2 \ell(\theta, S)}{\partial \theta \partial \theta'}\bigg|_{\theta = \theta^*} \xrightarrow{P} H(\theta_0). \tag{2.24}$$

We have

$$\hat{\theta} - \theta_0 = H^{-1}(\theta_0) \left[\frac{\partial \text{ vec } \Sigma(\theta)}{\partial \theta'}\bigg|_{\theta = \theta_0}\right]' [\Sigma^{-1}(\theta_0) \otimes \Sigma^{-1}(\theta_0)] \text{ vec} [S - \Sigma(\theta_0)] + o_p\left(\frac{1}{\sqrt{n}}\right) \tag{2.25}$$

if $S - \Sigma(\theta_0) = O_p(1/\sqrt{n})$. As $n \to \infty$, $\sqrt{n}(\hat{\theta} - \theta_0)$ has the limiting distribution of a linear function of $\sqrt{n}$ vec $[S - \Sigma(\theta_0)]$.

Now suppose

$$\sqrt{n} \text{ vec}[S - \Sigma(\theta_0)] \xrightarrow{L} N(\mathbf{0}, \Gamma). \tag{2.26}$$

THEOREM 2.2. *If Assumptions* 1 *and* 2 *and* (2.29) *hold, then*

$$\sqrt{n}(\hat{\theta} - \theta_0) \xrightarrow{L} N\left\{\mathbf{0}, H^{-1}(\theta_0) \left[\frac{\partial \text{ vec } \Sigma(\theta)}{\partial \theta'}\bigg|_{\theta = \theta_0}\right]' [\Sigma^{-1}(\theta_0) \otimes \Sigma^{-1}(\theta_0)]\Gamma \right.$$
$$\left. [\Sigma^{-1}(\theta_0) \otimes \Sigma^{-1}(\theta_0)] \left[\frac{\partial \text{ vec } \Sigma(\theta)}{\partial \theta'}\bigg|_{\theta = \theta_0}\right] H^{-1}(\theta_0)\right\}. \tag{2.27}$$

This theorem has also been given by Browne (1984) in Proposition 2.

COROLLARY 2.1. *If Assumptions* 1 *and* 2 *hold and the* $\boldsymbol{x}_\alpha$ *are independently identically distributed with* $\mathcal{E}(\boldsymbol{x}_\alpha'\boldsymbol{x}_\alpha)^2 < \infty$, *then* (2.27) *holds with*

$$\begin{aligned}\boldsymbol{\Gamma} = \mathcal{E}\ &\text{vec}\ [(\boldsymbol{x}_\alpha - \boldsymbol{\mu})(\boldsymbol{x}_\alpha - \boldsymbol{\mu})']\,\{\text{vec}\,[(\boldsymbol{x}_\alpha - \boldsymbol{\mu})(\boldsymbol{x}_\alpha - \boldsymbol{\mu})']\}' \\ &- \text{vec}\ \boldsymbol{\Sigma}(\boldsymbol{\theta}_0)[\text{vec}\ \boldsymbol{\Sigma}(\boldsymbol{\theta}_0)]'. \end{aligned} \tag{2.28}$$

The condition $\mathcal{E}(\boldsymbol{x}_\alpha'\boldsymbol{x}_\alpha)^2 < \infty$ is simply that the fourth–order moments of the components of $\boldsymbol{x}_\alpha$ are finite. Then (2.26) holds with $\boldsymbol{\Gamma}$ given by (2.28).

COROLLARY 2.2. *If Assumptions* 1 *and* 2 *hold and the* $\boldsymbol{x}_\alpha$*'s are independently identically normally distributed, then*

$$\sqrt{n}(\hat{\boldsymbol{\theta}} - \boldsymbol{\theta}_0) \xrightarrow{\text{L}} N(\boldsymbol{0}, 2\boldsymbol{H}_0^{-1}). \tag{2.29}$$

This result follows from the standard asymptotic theory of maximum likelihood estimates or by calculation of (2.28). When $\boldsymbol{x}_\alpha = (x_{1\alpha}, \ldots, x_{p\alpha})'$ is normally distributed with mean $\boldsymbol{\mu}$ and covariance matrix $\boldsymbol{\Sigma} = (\sigma_{ij})$, then

$$\mathcal{E}(x_{i\alpha} - \mu_i)(x_{j\alpha} - \mu_j)(x_{k\alpha} - \mu_k)(x_{\ell\alpha} - \mu_\ell) = \sigma_{ij}\sigma_{k\ell} + \sigma_{ik}\sigma_{j\ell} + \sigma_{i\ell}\sigma_{jk}. \tag{2.30}$$

[See Sec. 2.6, Anderson (1984b), for example.] Then

$$\text{Cov}\,[(x_{i\alpha} - \mu_i)(x_{j\alpha} - \mu_j), (x_{k\alpha} - \mu_k)(x_{\ell\alpha} - \mu_\ell)] = \sigma_{ik}\sigma_{j\ell} + \sigma_{i\ell}\sigma_{jk}. \tag{2.31}$$

We can write the limiting covariance matrix of vec $\boldsymbol{S}$ in the case of normality as

$$(\boldsymbol{I} + \boldsymbol{K}_{pp})[\boldsymbol{\Sigma}(\boldsymbol{\theta}_0) \otimes \boldsymbol{\Sigma}(\boldsymbol{\theta}_0)]. \tag{2.32}$$

Here $\boldsymbol{K}_{pp}$ is the commutation matrix; that is,

$$\text{vec}\ \boldsymbol{A}' = \boldsymbol{K}_{pp}\ \text{vec}\ \boldsymbol{A} \tag{2.33}$$

for an arbitrary $p \times p$ matrix $\boldsymbol{A}$. [Nel (1980) uses the notation $\boldsymbol{I}_{(p,p)}$ for $\boldsymbol{K}_{pp}$.]

The difference between (2.28) in general and (2.32) consists of fourth–order cumulants of $\boldsymbol{x}_\alpha$. Because the limiting distribution of $\sqrt{n}[\boldsymbol{S} - \boldsymbol{\Sigma}(\boldsymbol{\theta}_0)]$ depends only on the second– and fourth–order moments of $\boldsymbol{x}_\alpha$, the limiting distribution of $\sqrt{n}[\boldsymbol{S} - \boldsymbol{\Sigma}(\boldsymbol{\theta})]$ and hence of $\sqrt{n}(\hat{\boldsymbol{\theta}} - \boldsymbol{\theta}_0)$ will be that of the case of $\boldsymbol{x}_\alpha$ normal if the fourth–order cumulants of $\boldsymbol{x}_\alpha$ are 0.

COROLLARY 2.3. *If Assumptions* 1 *and* 2 *hold and the fourth–order cumulants of* $\boldsymbol{x}_\alpha$ *are* 0, *then* (2.29) *holds.*

3. A COVARIANCE STRUCTURE BASED ON A LINEAR MODEL

3.1 The model

We now consider the model (1.9), (1.10), and (1.11) with $\boldsymbol{z}_\alpha = (\boldsymbol{z}_\alpha^{(0)\prime}, \boldsymbol{z}_\alpha^{(1)\prime}, \ldots, \boldsymbol{z}_\alpha^{(G)\prime})', \alpha = 1, \ldots, N$, independently identically distributed. The mean of an observed vector is $\mathcal{E}\boldsymbol{x}_\alpha = \boldsymbol{\mu}$, and the covariance matrix is given by (1.12). A particular case of this model is the factor analysis model

$$\begin{aligned}
\boldsymbol{x}_\alpha &= \boldsymbol{\mu} + \boldsymbol{\Lambda}\boldsymbol{f}_\alpha + \boldsymbol{u}_\alpha \\
&= \boldsymbol{\mu} + \boldsymbol{\Lambda}\boldsymbol{f}_\alpha + \begin{pmatrix} u_{1\alpha} \\ u_{2\alpha} \\ \vdots \\ u_{p\alpha} \end{pmatrix} \\
&= \boldsymbol{\mu} + \boldsymbol{\Lambda}\boldsymbol{f}_\alpha + \begin{pmatrix} 1 \\ 0 \\ \vdots \\ 0 \end{pmatrix} u_{1\alpha} + \begin{pmatrix} 0 \\ 1 \\ \vdots \\ 0 \end{pmatrix} u_{2\alpha} + \ldots + \begin{pmatrix} 0 \\ 0 \\ \vdots \\ 1 \end{pmatrix} u_{p\alpha} \\
&= \boldsymbol{\mu} + \boldsymbol{\Lambda}_0(\boldsymbol{\lambda})\boldsymbol{z}_\alpha^{(0)} + \boldsymbol{\Lambda}_1(\boldsymbol{\lambda})\boldsymbol{z}_\alpha^{(1)} + \boldsymbol{\Lambda}_2(\boldsymbol{\lambda})\boldsymbol{z}_\alpha^{(2)} + \ldots + \boldsymbol{\Lambda}_G(\boldsymbol{\lambda})\boldsymbol{z}_\alpha^{(G)}.
\end{aligned} \tag{3.1}$$

That is, $\boldsymbol{\Lambda}_0(\boldsymbol{\lambda}) = \boldsymbol{\Lambda}$, $\boldsymbol{z}_\alpha^{(0)} = \boldsymbol{f}$, $\boldsymbol{\Phi}_0 = \boldsymbol{\Phi}$, $\boldsymbol{\Lambda}_g(\boldsymbol{\lambda})$ is the g–th column of $\boldsymbol{I}$, $\boldsymbol{z}_\alpha^{(g)} = u_{g\alpha}$, and $\boldsymbol{\Phi}_g = \psi_{gg}$, $g = 1, \ldots, G = p$. If the specification (1.6) is made, $\boldsymbol{\lambda} = \text{vec } \boldsymbol{\Lambda}^*$ and $\boldsymbol{\Phi}_0 = \boldsymbol{\Phi}$ is unrestricted.

In the general model the coefficient matrices $\boldsymbol{\Lambda}_g$ are functions of a vector $\boldsymbol{\lambda}$ of dimension q, $\boldsymbol{\Lambda}_g = \boldsymbol{\Lambda}_g(\boldsymbol{\lambda})$, $g = 0, 1, \ldots, G$. The moment matrix $\boldsymbol{\Phi}_0$ will be treated as a function of a vector $\boldsymbol{\tau}$ of dimension t, $\boldsymbol{\Phi}_0 = \boldsymbol{\Phi}_0(\boldsymbol{\tau})$. Then in addition to the mean $\boldsymbol{\mu}$ the vector of parameters is

$$\boldsymbol{\theta} = \begin{pmatrix} \boldsymbol{\lambda} \\ \boldsymbol{\tau} \\ \boldsymbol{\phi}_1 \\ \vdots \\ \boldsymbol{\phi}_G \end{pmatrix}, \tag{3.2}$$

where

$$\boldsymbol{\phi}_g = \operatorname{vech} \boldsymbol{\Phi}_g, \tag{3.3}$$

the vector of dimensionality $k_g(k_g+1)/2$ consisting of the first column of $\boldsymbol{\Phi}_g$ followed by the part of the second column on or below the diagonal, etc. [Browne writes vecs for vech.] The vector $\boldsymbol{\phi}_g$ is restricted by the requirement that $\boldsymbol{\Phi}_g$ is positive definite. We shall write

$$\boldsymbol{\Sigma} = \boldsymbol{\Sigma}(\boldsymbol{\theta}) = \boldsymbol{\Lambda}_0(\boldsymbol{\lambda})\boldsymbol{\Phi}_0(\boldsymbol{\tau})\boldsymbol{\Lambda}_0'(\boldsymbol{\lambda}) + \sum_{g=1}^{G} \boldsymbol{\Lambda}_g(\boldsymbol{\lambda})\boldsymbol{\Phi}_g\boldsymbol{\Lambda}_g'(\boldsymbol{\lambda}). \tag{3.4}$$

In a sense the model (1.9) with $\boldsymbol{\Phi}_0$ an arbitrary function of $\boldsymbol{\tau}$ is too general; in fact, with $\boldsymbol{\Lambda}_0 = \boldsymbol{I}$ we could dispense with $\boldsymbol{z}_\alpha^{(1)}, \ldots, \boldsymbol{z}_\alpha^{(G)}$ and have the general model $\boldsymbol{\Sigma}(\boldsymbol{\theta})$. The interesting cases are where the class of $\boldsymbol{\Phi}_0(\boldsymbol{\tau})$ is restricted or where $\boldsymbol{z}_\alpha^{(0)}$ is omitted. The model of Browne and Shapiro (1986) has $\sum_{g=G+1}^{H} \boldsymbol{\Lambda}_g(\boldsymbol{\lambda})\boldsymbol{\Phi}_g(\boldsymbol{\lambda})\boldsymbol{\Lambda}_g'(\boldsymbol{\lambda}$ in place of $\boldsymbol{\Lambda}_0(\boldsymbol{\lambda})\boldsymbol{\Phi}_0(\boldsymbol{\tau})\boldsymbol{\Lambda}_0'(\boldsymbol{\lambda})$ Their model is more general in that the covariance matrices and loading matrices depend on the same vector $\boldsymbol{\lambda}$, but the sum is no more general than the single product because

$$\sum_{g=G+1}^{H} \boldsymbol{\Lambda}_g(\boldsymbol{\lambda})\boldsymbol{\Phi}_g(\boldsymbol{\lambda})\boldsymbol{\Lambda}_g'(\boldsymbol{\lambda}) = \tag{3.5}$$

$$[\boldsymbol{\Lambda}_{G+1}(\boldsymbol{\lambda}), \ldots, \boldsymbol{\Lambda}_H(\boldsymbol{\lambda})] \begin{pmatrix} \boldsymbol{\Phi}_{G+1}(\boldsymbol{\lambda}) & \ldots & \mathbf{0} \\ \vdots & & \vdots \\ \mathbf{0} & \ldots & \boldsymbol{\Phi}_H(\boldsymbol{\lambda}) \end{pmatrix} \begin{pmatrix} \boldsymbol{\Lambda}_{G+1}'(\boldsymbol{\lambda}) \\ \vdots \\ \boldsymbol{\Lambda}_H'(\boldsymbol{\lambda}) \end{pmatrix},$$

which is a special form of $\boldsymbol{\Lambda}_0(\boldsymbol{\lambda})\boldsymbol{\Phi}_0(\boldsymbol{\lambda})\boldsymbol{\Lambda}_0'(\boldsymbol{\lambda})$. The moment matrix in (3.5) is a block diagonal version of $\boldsymbol{\Phi}_0(\boldsymbol{\lambda})$. The reason for writing out

$\boldsymbol{z}_\alpha^{(0)}, \boldsymbol{z}_\alpha^{(1)}, \ldots, \boldsymbol{z}_\alpha^{(G)}$ here is to indicate the lack of correlation among the subvectors and the unrestrictedness of $\boldsymbol{\Phi}_1, \ldots, \boldsymbol{\Phi}_G$.

3.2 Consistency

The properties of the estimators depend on the distribution of the unobservable vectors $\boldsymbol{z}_\alpha$ (the latent variables or errors). We define the unobservable sample covariances

$$\boldsymbol{\Phi}_{gh}(n) = \frac{1}{n} \sum_{\alpha=1}^{p} (\boldsymbol{z}_\alpha^{(g)} - \bar{\boldsymbol{z}}^{(g)})(\boldsymbol{z}_\alpha^{(h)} - \bar{\boldsymbol{z}}^{(h)})', \quad g, h = 0, 1, \ldots, G, \tag{3.6}$$

where $\bar{\boldsymbol{z}}^{(g)} = (1/N) \sum_{\alpha=1}^{N} \boldsymbol{z}_\alpha^{(g)}$. Furthermore, we define $\boldsymbol{\Phi}_{gg} = \boldsymbol{\Phi}_g$ and $\boldsymbol{\Phi}_{gh} = \mathbf{0}, g \neq h$. Let $\boldsymbol{\Phi}_{00}^0 = \boldsymbol{\Phi}_0(\boldsymbol{\tau}_0)$ and $\boldsymbol{\Phi}_{gg}^0$, $g = 1, \ldots, G$, be the values in the distributions sampled and $\boldsymbol{\Phi}_{gh}^0 = \mathbf{0}$ for $g \neq h$. If the $\boldsymbol{z}_\alpha$'s are independently identically distributed,

$$\operatorname*{plim}_{n \to \infty} \boldsymbol{\Phi}_{gh}(n) = \boldsymbol{\Phi}_{gh}^0, \quad g, h = 0, 1, \ldots, G. \tag{3.7}$$

THEOREM 3.1. *If Assumption* 1 *and* (3.7) *hold, then*

$$\operatorname*{plim}_{n \to \infty} \hat{\boldsymbol{\theta}} = \boldsymbol{\theta}_0 = \begin{pmatrix} \boldsymbol{\lambda}_0 \\ \boldsymbol{\tau}_0 \\ \boldsymbol{\phi}_1^0 \\ \vdots \\ \boldsymbol{\phi}_G^0 \end{pmatrix}, \tag{3.8}$$

where $\boldsymbol{\phi}_g^0 = \operatorname{vech} \boldsymbol{\Phi}_{gg}^0$.

Proof. We have

$$\begin{aligned} \boldsymbol{S} &= \frac{1}{n} \sum_{\alpha=1}^{N} \left[\sum_{g=0}^{G} \boldsymbol{\Lambda}_g(\boldsymbol{\lambda}_0)(\boldsymbol{z}_\alpha^{(g)} - \bar{\boldsymbol{z}}^{(g)}) \right] \left[\sum_{h=0}^{G} \boldsymbol{\Lambda}_h(\boldsymbol{\lambda}_0)(\boldsymbol{z}_\alpha^{(h)} - \bar{\boldsymbol{z}}^{(h)}) \right]' \qquad (3.9) \\ &= \frac{1}{n} \sum_{\alpha=1}^{N} \sum_{g,h=0}^{G} \boldsymbol{\Lambda}_g(\boldsymbol{\lambda}_0)(\boldsymbol{z}_\alpha^{(g)} - \bar{\boldsymbol{z}}^{(g)})(\boldsymbol{z}_\alpha^{(h)} - \bar{\boldsymbol{z}}^{(h)})' \boldsymbol{\Lambda}_h'(\boldsymbol{\lambda}_0) \\ &= \sum_{g,h=0}^{G} \boldsymbol{\Lambda}_g(\boldsymbol{\lambda}_0) \boldsymbol{\Phi}_{gh}(n) \boldsymbol{\Lambda}_h'(\boldsymbol{\lambda}_0). \end{aligned}$$

Then (3.7) implies

$$S \xrightarrow{\mathrm{p}} \sum_{g,h=0}^{G} \boldsymbol{\Lambda}_g(\boldsymbol{\lambda}_0)\boldsymbol{\Phi}_{gh}^0\boldsymbol{\Lambda}_h'(\boldsymbol{\lambda}_0) = \sum_{g=0}^{G} \boldsymbol{\Lambda}_g(\boldsymbol{\lambda}_0)\boldsymbol{\Phi}_{gg}^0\boldsymbol{\Lambda}_g(\boldsymbol{\lambda}_0) = \boldsymbol{\Sigma}(\boldsymbol{\theta}_0), \tag{3.10}$$

and Theorem 3.1 follows from Theorem 2.1. ∎

3.3 Asymptotic normality

Now we turn to the asymptotic normality of $\hat{\boldsymbol{\theta}}$. The following assumption implies Assumption 2.

ASSUMPTION 3. *$\boldsymbol{\Lambda}_g(\boldsymbol{\lambda})$ is twice continuously differentiable in a neighborhood of $\boldsymbol{\lambda}_0$, $g = 0, 1, \ldots, G$, and $\boldsymbol{\Phi}_0(\boldsymbol{\tau})$ is twice continuously differentiable in a neighborhood of $\boldsymbol{\tau}_0$. The partial derivative matrix ∂ vec $\boldsymbol{\Sigma}(\boldsymbol{\theta})/\partial\boldsymbol{\theta}'$ has full column rank at $\boldsymbol{\theta} = \boldsymbol{\theta}_0$.*

The procedures given in Section 2 hold, and Theorem 2.2 holds. However, the linear structure of (1.9) permits relaxing conditions for the asymptotic normality of the estimates of $\boldsymbol{\lambda}$ and $\boldsymbol{\tau}$. We define the hybrid vector, consisting of "true values" and (unobservable) sample quantities,

$$\boldsymbol{\theta}(n) = \begin{pmatrix} \boldsymbol{\lambda}_0 \\ \boldsymbol{\tau}_0 \\ \boldsymbol{\phi}_1(n) \\ \vdots \\ \boldsymbol{\phi}_G(n) \end{pmatrix}, \tag{3.11}$$

where

$$\boldsymbol{\phi}_g(n) = \text{vech } \boldsymbol{\Phi}_{gg}(n). \tag{3.12}$$

We consider the limiting distribution of $\sqrt{n}[\hat{\boldsymbol{\theta}} - \boldsymbol{\theta}(n)]$. We have

$$\boldsymbol{\Sigma}[\boldsymbol{\theta}(n)] = \boldsymbol{\Lambda}_0(\boldsymbol{\lambda}_0)\boldsymbol{\Phi}_0(\boldsymbol{\tau}_0)\boldsymbol{\Lambda}_0'(\boldsymbol{\lambda}_0) + \sum_{g=1}^{G} \boldsymbol{\Lambda}_g(\boldsymbol{\lambda}_0)\boldsymbol{\Phi}_{gg}(n)\boldsymbol{\Lambda}_g'(\boldsymbol{\lambda}_0). \tag{3.13}$$

Then

$$\begin{aligned} S - \boldsymbol{\Sigma}[\boldsymbol{\theta}(n)] = &\sum_{\substack{g,h=0 \\ g\neq h}}^{G} \boldsymbol{\Lambda}_g(\boldsymbol{\lambda}_0)\boldsymbol{\Phi}_{gh}(n)\boldsymbol{\Lambda}_h'(\boldsymbol{\lambda}_0) \\ &+ \boldsymbol{\Lambda}_0(\boldsymbol{\lambda}_0)[\boldsymbol{\Phi}_{00}(n) - \boldsymbol{\Phi}_0(\boldsymbol{\tau}_0)]\boldsymbol{\Lambda}_0'(\boldsymbol{\lambda}_0). \end{aligned} \tag{3.14}$$

The likelihood equations can be expanded as

$$0 = \left.\frac{\partial \ell(\boldsymbol{\theta}, \boldsymbol{S})}{\partial \boldsymbol{\theta}}\right|_{\boldsymbol{\theta}=\hat{\boldsymbol{\theta}}} = \left.\frac{\partial \ell(\boldsymbol{\theta}, \boldsymbol{S})}{\partial \boldsymbol{\theta}}\right|_{\boldsymbol{\theta} = \boldsymbol{\theta}(n)} + \left.\frac{\partial^2 \ell(\boldsymbol{\theta}, \boldsymbol{S})}{\partial \boldsymbol{\theta}\partial \boldsymbol{\theta}'}\right|_{\boldsymbol{\theta} = \boldsymbol{\theta}^*} [\hat{\boldsymbol{\theta}} - \boldsymbol{\theta}(n)], \tag{3.15}$$

where $\boldsymbol{\theta}^*$ is on the line segment joining $\boldsymbol{\theta}(n)$ and $\hat{\boldsymbol{\theta}}$. If (3.7) holds, $\boldsymbol{\theta}(n) \xrightarrow{\mathrm{P}} \boldsymbol{\theta}_0$ as well as $\hat{\boldsymbol{\theta}} \xrightarrow{\mathrm{P}} \boldsymbol{\theta}_0$, and

$$\left.\frac{\partial^2 \ell(\boldsymbol{\theta}, \boldsymbol{S})}{\partial \boldsymbol{\theta}\partial \boldsymbol{\theta}'}\right|_{\boldsymbol{\theta} = \boldsymbol{\theta}^*} \xrightarrow{\mathrm{P}} \left.\frac{\partial^2 \ell[\boldsymbol{\theta}, \boldsymbol{\Sigma}(\boldsymbol{\theta}_0)]}{\partial \boldsymbol{\theta}\partial \boldsymbol{\theta}'}\right|_{\boldsymbol{\theta} = \boldsymbol{\theta}_0} = \boldsymbol{H}(\boldsymbol{\theta}_0). \tag{3.16}$$

Then if $\boldsymbol{\Phi}_{gh}(n) = O_p(1/\sqrt{n})$, $g \neq h$, and $\boldsymbol{\Phi}_{00}(n) - \boldsymbol{\Phi}_0(\boldsymbol{\tau}_0) = O_p(1/\sqrt{n})$

$$\hat{\boldsymbol{\theta}} - \boldsymbol{\theta}(n) = \boldsymbol{H}^{-1}(\boldsymbol{\theta}_0) \left[\left.\frac{\mathrm{vec}\ \boldsymbol{\Sigma}(\boldsymbol{\theta})}{\partial \boldsymbol{\theta}'}\right|_{\boldsymbol{\theta} = \boldsymbol{\theta}_0}\right]' \left[\boldsymbol{\Sigma}^{-1}(\boldsymbol{\theta}_0) \otimes \boldsymbol{\Sigma}^{-1}(\boldsymbol{\theta}_0)\right]$$
$$\left[\sum_{\substack{g,h=0 \\ g\neq h}}^{G} \boldsymbol{\Lambda}_g(\boldsymbol{\lambda}_0)\boldsymbol{\Phi}_{gh}(n)\boldsymbol{\Lambda}_h'(\boldsymbol{\lambda}_0) + \boldsymbol{\Lambda}_0(\boldsymbol{\lambda}_0)[\boldsymbol{\Phi}_{00}(n)\right.$$
$$\left. - \boldsymbol{\Phi}_0(\boldsymbol{\tau}_0)]\boldsymbol{\Lambda}_0'(\boldsymbol{\lambda}_0)\right] + o_p\left(\frac{1}{\sqrt{n}}\right). \tag{3.17}$$

Note that the sum in (3.17) does not include $\boldsymbol{\Phi}_{gg}(n)$, $g = 1, \ldots, G$.

LEMMA 3.1. *If the $\boldsymbol{z}_\alpha$'s are independently identically distributed and if $\boldsymbol{z}_\alpha^{(0)}, \boldsymbol{z}_\alpha^{(1)}, \ldots, \boldsymbol{z}_\alpha^{(G)}$ are independent, then $\sqrt{n}\,\boldsymbol{\Phi}_{01}(n), \ldots, \sqrt{n}\,\boldsymbol{\Phi}_{0G}(n)$, $\sqrt{n}\,\boldsymbol{\Phi}_{12}(n), \ldots, \sqrt{n}\,\boldsymbol{\Phi}_{G-1,G}(n)$ have a limiting normal distribution in which the matrices are independent, $\mathcal{E}\boldsymbol{\Phi}_{gh}(n) = \mathbf{0}$, and*

$$\lim_{n\to\infty} n\mathcal{E}\ \mathrm{vec}\ \boldsymbol{\Phi}_{gh}(n)[\mathrm{vec}\ \boldsymbol{\Phi}_{gh}(n)]' = \boldsymbol{\Phi}_{hh}^0 \otimes \boldsymbol{\Phi}_{gg}^0, \qquad g \neq h. \tag{3.18}$$

Proof. The matrix $\sqrt{n}\,\boldsymbol{\Phi}_{gh}(n)$ has the limiting distribution of

$$\frac{1}{\sqrt{n}} \sum_{\alpha=1}^{N} \boldsymbol{z}_\alpha^{(g)} \boldsymbol{z}_\alpha^{(h)'} \tag{3.19}$$

since $\bar{z}^{(g)} \xrightarrow{p} \mathbf{0}$ The vectors of all components of $z_\alpha^{(g)} z_\alpha^{(h)\prime}$ for $g < h, g, h = 0, 1, \ldots, G$ are independently identically distributed. The covariance of an element of $z_\alpha^{(g)} z_\alpha^{(h)\prime}$ with an element of $z_\alpha^{(k)} z_\alpha^{(\ell)\prime}$ is 0 unless $(g,h) = (k,\ell)$ or $(g,h) = (\ell,k)$. Then (3.18) follows from

$$
\begin{aligned}
\mathcal{E} \text{ vec } z_\alpha^{(g)} z_\alpha^{(h)\prime} \left(\text{vec } z_\alpha^{(g)} z_\alpha^{(h)\prime} \right)' &= \mathcal{E}\left[z_\alpha^{(h)} \otimes z_\alpha^{(g)} \right] \left[z_\alpha^{(h)} \otimes z_\alpha^{(g)} \right]' \\
&= \mathcal{E}\left[(z_\alpha^{(h)} z_\alpha^{(h)\prime}) \otimes (z_\alpha^{(g)} z_\alpha^{(g)\prime}) \right] \\
&= \boldsymbol{\Phi}_{hh} \otimes \boldsymbol{\Phi}_{gg}, \quad g \neq h,
\end{aligned} \tag{3.20}
$$

the independence of $z_\alpha^{(g)}$ and $z_\alpha^{(h)}$, and the fact that vec $\boldsymbol{a}\boldsymbol{b}' = \boldsymbol{b} \otimes \boldsymbol{a}$. By the multivariate central limit theorem [Theorem 3.4.3 of Anderson (1984b), for example] the vector of all components of (3.19) has a joint limiting normal distribution with mean $\mathbf{0}$. ∎

THEOREM 3.2. *If Assumptions 1 and 3 hold, if the z_α's are independently identically distributed, if $z_\alpha^{(0)}, z_\alpha^{(1)}, \ldots, z_\alpha^{(G)}$ are independent, and if $\mathcal{E}(z_\alpha^{(0)\prime} z_\alpha^{(0)})^2 < \infty$, then $\sqrt{n}[\hat{\boldsymbol{\theta}} - \boldsymbol{\theta}(n)]$ has a limiting normal distribution with mean $\mathbf{0}$ and covariance matrix depending on $\boldsymbol{\theta}_0$ and possibly other parameters of the distribution of $z_\alpha^{(0)}$.*

Note that $\sqrt{n}(\hat{\boldsymbol{\lambda}} - \boldsymbol{\lambda}_0)$ has a limiting normal distribution if $z_\alpha^{(0)} \equiv 0$ since $\boldsymbol{\Phi}_1, \ldots, \boldsymbol{\Phi}_G$ are unrestricted; the only conditions on the $z_\alpha^{(g)}$'s, $g = 1, \ldots, G$, is that they are independent and that their second-order moments exist. If $z_\alpha^{(0)}$ is nontrivial and Φ_0 is restricted, then the fourth-order moments of $z_\alpha^{(0)}$ must exist to obtain asymptotic normality. In the factor analysis model where the identification conditions are imposed on the factor loadings and the second-order moment matrix of the factor scores is unrestricted, the limiting normal distribution of $\sqrt{n}(\hat{\boldsymbol{\lambda}} - \boldsymbol{\lambda}_0)$ depends only on second-order moments and the population loadings. When the moment matrices are unrestricted, the asymptotic covariances of $\sqrt{n}[\hat{\boldsymbol{\theta}} - \boldsymbol{\theta}(n)]$ obtained for the normal distribution hold for any arbitrary distribution of the z_α's. If $\boldsymbol{\Phi}_0$ is restricted to be $\boldsymbol{\Phi}_0(\boldsymbol{\tau})$, the limiting distributions of $\sqrt{n}[\hat{\boldsymbol{\theta}} - \boldsymbol{\theta}(n)]$ may depend on the fourth-order moments of $z_\alpha^{(0)}$, but only on the second-order moments of $z_\alpha^{(1)}, \ldots, z_\alpha^{(G)}$.

Now consider

$$\boldsymbol{\theta}(n) - \boldsymbol{\theta}_0 = \begin{pmatrix} \mathbf{0} \\ \mathbf{0} \\ \boldsymbol{\phi}_1(n) - \boldsymbol{\phi}_1^0 \\ \vdots \\ \boldsymbol{\phi}_G(n) - \boldsymbol{\phi}_G^0 \end{pmatrix}. \tag{3.21}$$

The subvectors $\boldsymbol{\phi}_1(n) - \boldsymbol{\phi}_1^0, \ldots, \boldsymbol{\phi}_G(n) - \boldsymbol{\phi}_G^0$ have zero means and are independently distributed if $\boldsymbol{z}_\alpha^{(1)}, \ldots, \boldsymbol{z}_\alpha^{(G)}$ are independent. Then

$$\begin{aligned} \lim_{n\to\infty} \; & n\,\mathcal{E}\left\{\mathrm{vec}\left[\boldsymbol{\Phi}_{gg}(n) - \boldsymbol{\Phi}_{gg}^0\right]\right\}\left\{\mathrm{vec}\left[\boldsymbol{\Phi}_{gg}(n) - \boldsymbol{\Phi}_{gg}^0\right]\right\}' \\ &= \mathcal{E}\ \mathrm{vec}\ \boldsymbol{z}_\alpha^{(g)}\boldsymbol{z}_\alpha^{(g)\prime}\left[\mathrm{vec}\ \boldsymbol{z}_\alpha^{(g)}\boldsymbol{z}_\alpha^{(g)\prime}\right]' - \mathrm{vec}\ \boldsymbol{\Phi}_{gg}^0\left(\mathrm{vec}\ \boldsymbol{\Phi}_{gg}^0\right)' \end{aligned} \tag{3.22}$$

if $\mathcal{E}[\boldsymbol{z}_\alpha^{(g)\prime}\boldsymbol{z}_\alpha^{(g)}]^2 < \infty$; that is, if the fourth-order moments of the components of $\boldsymbol{z}_\alpha^{(g)}$ exist. In particular, if $\boldsymbol{z}_\alpha^{(g)}$ is normally distributed, (3.22) is

$$\left(\boldsymbol{I} + \boldsymbol{K}_{k_g k_g}\right)\left(\boldsymbol{\Phi}_{gg}^0 \otimes \boldsymbol{\Phi}_{gg}^0\right). \tag{3.23}$$

If the fourth-order moments of $\boldsymbol{z}_\alpha^{(g)}$ exist, then $\sqrt{n}[\boldsymbol{\phi}_g(n) - \boldsymbol{\phi}_g^0]$ has a limiting normal distribution and in the limiting distribution the set $\{\sqrt{n}[\boldsymbol{\phi}_g(n) - \boldsymbol{\phi}_g^0]\}$ is independent of the set $\{\mathrm{vec}\ \boldsymbol{\Phi}_{gh}(n) - \mathrm{vec}\ \boldsymbol{\Phi}_{gh}^0,\ g \neq h\}$.

THEOREM 3.3. *If Assumptions* 1 *and* 3 *hold, if* $\mathcal{E}(\boldsymbol{z}_\alpha'\boldsymbol{z}_\alpha)^2 < \infty$, *if the* $\boldsymbol{z}_\alpha$*'s are independently identically distributed, and if* $\boldsymbol{z}_\alpha^{(0)}, \boldsymbol{z}_\alpha^{(1)}, \ldots, \boldsymbol{z}_\alpha^{(G)}$ *are independent, then* $\sqrt{n}[\hat{\boldsymbol{\theta}} - \boldsymbol{\theta}(n)]$ *and* $\sqrt{n}[\boldsymbol{\theta}(n) - \boldsymbol{\theta}_0]$ *have a limiting normal distribution in which the two vectors are independent.*

Thus under the conditions of Theorem 3.3 the limiting distribution of $\sqrt{n}(\hat{\boldsymbol{\theta}} - \boldsymbol{\theta}_0)$ is normal with a covariance matrix that is the sum of the covariance matrices of the limiting distributions of $\sqrt{n}[\hat{\boldsymbol{\theta}} - \boldsymbol{\theta}(n)]$ and $\sqrt{n}\,[\boldsymbol{\theta}(n) - \boldsymbol{\theta}_0]$. If the $\boldsymbol{z}_\alpha$'s are normal, $\hat{\boldsymbol{\theta}}$ is the maximum likelihood estimate of $\boldsymbol{\theta}$ and by the usual theory for maximum likelihood estimates

$$\sqrt{n}(\hat{\boldsymbol{\theta}} - \boldsymbol{\theta}_0) \xrightarrow{\mathrm{L}} N\left[\mathbf{0}, 2\boldsymbol{H}^{-1}(\boldsymbol{\theta}_0)\right]. \tag{3.24}$$

To give the limiting distribution when the $\boldsymbol{z}_\alpha$'s are not necessarily normally distributed, we define the cumulant matrix as (3.22) minus (3.23), namely

$$\boldsymbol{C}_g = \mathcal{E}\ \text{vec}\ \boldsymbol{z}_\alpha^{(g)}\boldsymbol{z}_\alpha^{(g)\prime}[\text{vec}\ \boldsymbol{z}_\alpha^{(g)}\boldsymbol{z}_\alpha^{(g)\prime}]' - \text{vec}\ \boldsymbol{\Phi}_{gg}^0(\text{vec}\ \boldsymbol{\Phi}_{gg}^0)' - (\boldsymbol{I}+\boldsymbol{K}_{k_g k_g})(\boldsymbol{\Phi}_{gg}^0 \otimes \boldsymbol{\Phi}_{gg}^0). \tag{3.25}$$

Then

$$\begin{aligned}\lim_{n\to\infty} \mathcal{E}\,[\boldsymbol{\phi}_g(n) - \boldsymbol{\phi}_g^0]\,[\boldsymbol{\phi}_g(n) - \boldsymbol{\phi}_g^0]' &\\ = \boldsymbol{D}_{k_g}^+\ \left[(\boldsymbol{I}+\boldsymbol{K}_{k_g k_g})\left(\boldsymbol{\Phi}_{gg}^0 \otimes \boldsymbol{\Phi}_{gg}^0\right) + \boldsymbol{C}_g\right] \boldsymbol{D}_{k_g}^{+\prime}&\\ = \boldsymbol{D}_{k_g}^+\ \left[2(\boldsymbol{\Phi}_{gg}^0 \otimes \boldsymbol{\Phi}_{gg}^0) + \boldsymbol{C}_g\right] \boldsymbol{D}_{k_g}^{+\prime},&\end{aligned} \tag{3.26}$$

where

$$\boldsymbol{D}_k^+ = (\boldsymbol{D}_k'\boldsymbol{D}_k)^{-1}\boldsymbol{D}_k', \tag{3.27}$$

the Moore–Penrose inverse of $\boldsymbol{D}_k$, has the property

$$\boldsymbol{D}_k^+\ \text{vec}\ \boldsymbol{A} = \text{vech}\ \boldsymbol{A}, \tag{3.28}$$

and $\boldsymbol{D}_k$ has the property

$$\text{vec}\ \boldsymbol{A} = \boldsymbol{D}_k\ \text{vech}\ \boldsymbol{A} \tag{3.29}$$

for an arbitrary symmetric $k \times k$ matrix $\boldsymbol{A}$. Note that

$$\boldsymbol{D}_k^+(\boldsymbol{I}+\boldsymbol{K}_{kk})\boldsymbol{A} = 2\boldsymbol{D}_k^+\boldsymbol{A}. \tag{3.30}$$

[Browne (1987) writes $\boldsymbol{K}_k'$ for $\boldsymbol{D}_k^+$ and $\boldsymbol{K}_k^{-\prime}$ for $\boldsymbol{D}_k$; see also Nel (1980).] In conformity with the partitioning of $\boldsymbol{\theta}$ define

$$\boldsymbol{C} = \begin{pmatrix} \boldsymbol{0} & \boldsymbol{0} & \boldsymbol{0} & \cdots & \boldsymbol{0} \\ \boldsymbol{0} & \boldsymbol{0} & \boldsymbol{0} & \cdots & \boldsymbol{0} \\ \boldsymbol{0} & \boldsymbol{0} & \boldsymbol{D}_{k_1}^+\boldsymbol{C}_1\boldsymbol{D}_{k_1}^{+\prime} & \cdots & \boldsymbol{0} \\ \vdots & \vdots & \vdots & & \vdots \\ \boldsymbol{0} & \boldsymbol{0} & \boldsymbol{0} & \cdots & \boldsymbol{D}_{k_G}^+\boldsymbol{C}_G\boldsymbol{D}_{k_G}^{+\prime} \end{pmatrix}. \tag{3.31}$$

THEOREM 3.4. *If Assumptions* 1 *and* 3 *hold, if* $\mathcal{E}(\boldsymbol{z}_\alpha'\boldsymbol{z}_\alpha)^2 < \infty$, *if the* $\boldsymbol{z}_\alpha$*'s are independently identically distributed, and if* $\boldsymbol{z}_\alpha^{(0)}, \boldsymbol{z}_\alpha^{(1)}, \ldots, \boldsymbol{z}_\alpha^{(G)}$

are independent, $\sqrt{n}[\boldsymbol{\theta}(n)-\boldsymbol{\theta}_0]$ *has a limiting normal distribution with mean* $\mathbf{0}$ *and covariance matrix*

$$\begin{pmatrix} \mathbf{0} & \mathbf{0} & \mathbf{0} & \cdots & \mathbf{0} \\ \mathbf{0} & \mathbf{0} & \mathbf{0} & \cdots & \mathbf{0} \\ \mathbf{0} & \mathbf{0} & 2\boldsymbol{D}_{k_1}^{+}(\boldsymbol{\Phi}_{11}^{0}\otimes\boldsymbol{\Phi}_{11}^{0})\boldsymbol{D}_{k_1}^{+'} & \cdots & \mathbf{0} \\ \vdots & \vdots & \vdots & & \vdots \\ \mathbf{0} & \mathbf{0} & \mathbf{0} & \cdots & 2\boldsymbol{D}_{k_G}^{+}(\boldsymbol{\Phi}_{GG}^{0}\otimes\boldsymbol{\Phi}_{GG}^{0})\boldsymbol{D}_{k_G}^{+'} \end{pmatrix} + \boldsymbol{C}\,. \tag{3.32}$$

THEOREM 3.5. *If Assumptions 1 and 3 hold, if the fourth–order cumulants of* $\boldsymbol{z}_\alpha^{(0)}$ *are* $\mathbf{0}$, *if* $\mathcal{E}\left(\boldsymbol{z}'_\alpha\boldsymbol{z}_\alpha\right)^2<\infty$, *if the* $\boldsymbol{z}_\alpha$*'s are independently identically distributed, and if* $\boldsymbol{z}_\alpha^{(0)},\boldsymbol{z}_\alpha^{(1)},\ldots,\boldsymbol{z}_\alpha^{(G)}$ *are independent, then*

$$\sqrt{n}\left(\hat{\boldsymbol{\theta}}-\boldsymbol{\theta}_0\right)\xrightarrow{\text{L}}N\left(\mathbf{0},2\boldsymbol{H}^{-1}(\boldsymbol{\theta}_0)+\boldsymbol{C}\right). \tag{3.33}$$

Proof. If the $\boldsymbol{z}_\alpha^{(g)}$'s are normally distributed, the covariance matrix of the limiting distribution of $\sqrt{n}(\hat{\boldsymbol{\theta}}-\boldsymbol{\theta}_0)$ is $2\boldsymbol{H}^{-1}(\boldsymbol{\theta}_0)$. The covariance matrix of the limiting distribution of $\sqrt{n}[\hat{\boldsymbol{\theta}}-\boldsymbol{\theta}(n)]$ does not depend on normality; the effect of nonnormality on the covariance matrix of the limiting distribution of $\sqrt{n}[\boldsymbol{\theta}(n)-\boldsymbol{\theta}_0]$ is to add $\boldsymbol{C}$. Hence, the effect of nonnormality on the asymptotic covariance matrix of $\hat{\boldsymbol{\theta}}-\boldsymbol{\theta}_0$ is to add $\boldsymbol{C}$. ∎

This theorem essentially coincides with Proposition 3.1 of Browne and Shapiro (1986).

From (2.29) and Theorems 3.3 and 3.5 we can now obtain the covariance matrix of the limiting distribution of $\sqrt{n}[\hat{\boldsymbol{\theta}}-\boldsymbol{\theta}(n)]$ when the fourth-order cumulants of $\boldsymbol{z}_\alpha^{(0)}$ are $\mathbf{0}$ as

$$2\boldsymbol{H}^{-1}(\boldsymbol{\theta}_0)-\begin{pmatrix} \mathbf{0} & \mathbf{0} & \mathbf{0} & \cdots & \mathbf{0} \\ \mathbf{0} & \mathbf{0} & \mathbf{0} & \cdots & \mathbf{0} \\ \mathbf{0} & \mathbf{0} & 2\boldsymbol{D}_{k_1}^{+}(\boldsymbol{\Phi}_{11}^{0}\otimes\boldsymbol{\Phi}_{11}^{0})\boldsymbol{D}_{k_1}^{+'} & \cdots & \mathbf{0} \\ \vdots & \vdots & \vdots & & \vdots \\ \mathbf{0} & \mathbf{0} & \mathbf{0} & \cdots & 2\boldsymbol{D}_{k_G}^{+}(\boldsymbol{\Phi}_{GG}^{0}\otimes\boldsymbol{\Phi}_{GG}^{0})\boldsymbol{D}_{k_G}^{+'} \end{pmatrix} \tag{3.34}$$

We can make the derivatives more explicit by using

$$\text{vec}\, \boldsymbol{ABC} = (\boldsymbol{C}' \otimes \boldsymbol{A})\, \text{vec}\, \boldsymbol{B}, \tag{3.35}$$

and vec $\boldsymbol{ABC} = \boldsymbol{K}_{pp}\, \text{vec}(\boldsymbol{ABC})'$. From (1.12) we obtain

$$\frac{\partial \boldsymbol{\Sigma}(\boldsymbol{\theta})}{\partial \lambda_i} = \sum_{g=0}^{G} \left(\frac{\partial \boldsymbol{\Lambda}_g(\boldsymbol{\lambda})}{\partial \lambda_i} \boldsymbol{\Phi}_g \boldsymbol{\Lambda}_g'(\boldsymbol{\lambda}) + \boldsymbol{\Lambda}_g(\boldsymbol{\lambda}) \boldsymbol{\Phi}_g \frac{\partial \boldsymbol{\Lambda}_g'(\boldsymbol{\lambda})}{\partial \lambda_i} \right), \tag{3.36}$$

$$\begin{aligned} \frac{\partial\, \text{vec}\, \boldsymbol{\Sigma}(\boldsymbol{\theta})}{\partial \boldsymbol{\lambda}'} &= \sum_{g=0}^{G} \Bigg\{ [\boldsymbol{I} \otimes \boldsymbol{\Lambda}_g(\boldsymbol{\lambda}) \boldsymbol{\Phi}_g] \frac{\partial\, \text{vec}\, \boldsymbol{\Lambda}_g'(\boldsymbol{\lambda})}{\partial \boldsymbol{\lambda}'} \\ &\qquad + [\boldsymbol{\Lambda}_g(\boldsymbol{\lambda}) \boldsymbol{\Phi}_g \otimes \boldsymbol{I}] \frac{\partial\, \text{vec}\, \boldsymbol{\Lambda}_g(\boldsymbol{\lambda})}{\partial \boldsymbol{\lambda}'} \Bigg\} \\ &= (\boldsymbol{I} + \boldsymbol{K}_{pp}) \sum_{g=1}^{G} [\boldsymbol{\Lambda}_g(\boldsymbol{\lambda}) \boldsymbol{\Phi}_g \otimes \boldsymbol{I}] \frac{\partial\, \text{vec}\, \boldsymbol{\Lambda}_g(\boldsymbol{\lambda})}{\partial \boldsymbol{\lambda}'}. \end{aligned} \tag{3.37}$$

Further we have

$$\begin{aligned} \frac{\partial\, \text{vec}\, \boldsymbol{\Sigma}(\boldsymbol{\theta})}{\partial \boldsymbol{\phi}_h'} &= [\boldsymbol{\Lambda}_h(\boldsymbol{\lambda}) \otimes \boldsymbol{\Lambda}_h(\boldsymbol{\lambda})] \frac{\partial\, \text{vec}\, \boldsymbol{\Phi}_h}{\partial \boldsymbol{\phi}_h'} \\ &= [\boldsymbol{\Lambda}_h(\boldsymbol{\lambda}) \otimes \boldsymbol{\Lambda}_h(\boldsymbol{\lambda})]\, \boldsymbol{D}_{k_h}, \end{aligned} \tag{3.38}$$

$$\frac{\partial\, \text{vec}\, \boldsymbol{\Sigma}(\boldsymbol{\theta})}{\partial \boldsymbol{\tau}'} = [\boldsymbol{\Lambda}_0(\boldsymbol{\lambda}) \otimes \boldsymbol{\Lambda}_0(\boldsymbol{\lambda})] \frac{\partial\, \text{vec}\, \boldsymbol{\Phi}_0(\boldsymbol{\tau})}{\partial \boldsymbol{\tau}'}. \tag{3.39}$$

4. EXAMPLES

4.1 Multiple battery factor analysis

Suppose $\boldsymbol{x}_\alpha$ is partitioned into G subvectors of $k_1, \ldots, k_G$ components, respectively, so

$$\boldsymbol{x}_\alpha = \begin{pmatrix} \boldsymbol{x}_\alpha^{(1)} \\ \boldsymbol{x}_\alpha^{(2)} \\ \vdots \\ \boldsymbol{x}_\alpha^{(G)} \end{pmatrix} = \boldsymbol{\mu} + \boldsymbol{\Lambda}_0\, \boldsymbol{z}_\alpha^{(0)} + \begin{pmatrix} \boldsymbol{z}_\alpha^{(1)} \\ \boldsymbol{z}_\alpha^{(2)} \\ \vdots \\ \boldsymbol{z}_\alpha^{(G)} \end{pmatrix}, \tag{4.1}$$

where $\boldsymbol{z}_\alpha^{(0)}, \boldsymbol{z}_\alpha^{(1)}, \ldots, \boldsymbol{z}_\alpha^{(G)}$ are independent. If $\boldsymbol{z}_\alpha^{(1)}, \ldots, \boldsymbol{z}_\alpha^{(G)}$ are scalars, the model is that of factor analysis as seen in Section 3. If some or all of $\boldsymbol{z}_\alpha^{(1)}, \ldots, \boldsymbol{z}_\alpha^{(G)}$ are vectors, the model is that of multiple battery factor analysis. It can be put in the form of (1.9) by letting

$$(\boldsymbol{\Lambda}_1, \boldsymbol{\Lambda}_2, \ldots, \boldsymbol{\Lambda}_G) = \begin{pmatrix} \boldsymbol{I}_{k_1} & \mathbf{0} & \ldots & \mathbf{0} \\ \mathbf{0} & \boldsymbol{I}_{k_2} & \ldots & \mathbf{0} \\ \vdots & \vdots & & \vdots \\ \mathbf{0} & \mathbf{0} & \ldots & \boldsymbol{I}_{k_G} \end{pmatrix}. \tag{4.2}$$

In this model

$$\begin{aligned} \boldsymbol{S} - \boldsymbol{\Sigma}[\boldsymbol{\theta}(n)] = {} & \boldsymbol{\Lambda}_0(\boldsymbol{\lambda}_0)[\boldsymbol{\Phi}_{00}(n) - \boldsymbol{\Phi}_0(\boldsymbol{\tau}_0)]\boldsymbol{\Lambda}_0'(\boldsymbol{\lambda}_0) \\ & + \begin{pmatrix} \boldsymbol{\Phi}_{10}(n) \\ \boldsymbol{\Phi}_{20}(n) \\ \vdots \\ \boldsymbol{\Phi}_{G0}(n) \end{pmatrix} \boldsymbol{\Lambda}_0'(\boldsymbol{\lambda}_0) + \boldsymbol{\Lambda}_0(\boldsymbol{\lambda}_0)[\boldsymbol{\Phi}_{01}(n), \boldsymbol{\Phi}_{02}(n), \ldots, \boldsymbol{\Phi}_{0G}(n)] \\ & + \begin{pmatrix} \mathbf{0} & \boldsymbol{\Phi}_{12}(n) & \ldots & \boldsymbol{\Phi}_{1G}(n) \\ \boldsymbol{\Phi}_{21}(n) & \mathbf{0} & \ldots & \boldsymbol{\Phi}_{2G}(n) \\ \vdots & \vdots & & \vdots \\ \boldsymbol{\Phi}_{G1}(n) & \boldsymbol{\Phi}_{G2}(n) & \ldots & \mathbf{0} \end{pmatrix}. \end{aligned} \tag{4.3}$$

The asymptotic distribution of $\hat{\boldsymbol{\lambda}}$ and $\hat{\boldsymbol{\tau}}$ depends on $\boldsymbol{\Phi}_{00}(n)$ and $\boldsymbol{\Phi}_{gh}(n)$ for $g \neq h$. If $\boldsymbol{\Phi}_0$ is unrestricted (that is, $\boldsymbol{\tau} = \phi_0$), then the limiting distribution of $\sqrt{n}(\hat{\boldsymbol{\lambda}} - \boldsymbol{\lambda}_0)$ depends on $\boldsymbol{\Phi}_{gh}(n)$, $g \neq h$, and hence is normal with mean $\mathbf{0}$ and covariance matrix the same as if the $\boldsymbol{z}_\alpha$'s were normally distributed. Browne (1979), (1980) has found the maximum likelihood estimates of the parameters under normality.

If $\boldsymbol{\Lambda}_0 \boldsymbol{z}_\alpha^{(0)}$ is omitted, the model is that of independence of subvectors. Then the parameters are the components of $\boldsymbol{\Phi}_g = \boldsymbol{\Sigma}_{gg}$, the g–th diagonal block of $\boldsymbol{\Sigma}$, and their estimates are the components of $\boldsymbol{\Phi}_{gg}(n) = \boldsymbol{S}_{gg}$, the g–th diagonal block of $\boldsymbol{S}$, $g = 1, \ldots, G$.

4.2 Linear structural relations

The set of linear structural relations described in Section 1 can be related to the general linear model of Section 3 as

$$\boldsymbol{y}_\alpha = \boldsymbol{\Lambda}_0 \boldsymbol{z}_\alpha^{(0)}, \qquad \boldsymbol{u}_\alpha = \sum_{g=1}^{G} \boldsymbol{\Lambda}_g \boldsymbol{z}_\alpha^{(g)}, \tag{4.4}$$

with $\boldsymbol{B\Lambda}_0 = \mathbf{0}$. The second equation of (4.4) specifies the structure of the error covariance matrix. If $\boldsymbol{\Lambda}_0$ is of full rank, namely k_0, then the $(p-k_0)\times p$ matrix $\boldsymbol{B}$ is determined by $\boldsymbol{B\Lambda}_0 = \mathbf{0}$ except for multiplication of $\boldsymbol{B}$ on the left by an arbitrary $(p-k_0)\times(p-k_0)$ nonsingular matrix. That indeterminacy can be eliminated by the imposition of (1.7), for instance. Then $\boldsymbol{B}^* = -\boldsymbol{\Lambda}_{10}\boldsymbol{\Lambda}_{20}^{-1}$ and the maximum likelihood estimate of $\boldsymbol{B}^*$ is $\hat{\boldsymbol{B}}^* = -\hat{\boldsymbol{\Lambda}}_{10}\hat{\boldsymbol{\Lambda}}_{20}^{-1}$, where

$$\boldsymbol{\Lambda}_0 = \begin{pmatrix}\boldsymbol{\Lambda}_{10}\\ \boldsymbol{\Lambda}_{20}\end{pmatrix}, \qquad \hat{\boldsymbol{\Lambda}}_0 = \begin{pmatrix}\hat{\boldsymbol{\Lambda}}_{10}\\ \hat{\boldsymbol{\Lambda}}_{20}\end{pmatrix}. \tag{4.5}$$

Then the limiting distribution of $\sqrt{n}(\hat{\boldsymbol{B}}^* - \boldsymbol{B}_0^*)$ is that of

$$-\sqrt{n}(\hat{\boldsymbol{\Lambda}}_{10} - \boldsymbol{\Lambda}_{10})\boldsymbol{\Lambda}_{20}^{-1} + \boldsymbol{\Lambda}_{10}\boldsymbol{\Lambda}_{20}^{-1}\sqrt{n}(\hat{\boldsymbol{\Lambda}}_{20} - \boldsymbol{\Lambda}_{20})\boldsymbol{\Lambda}_{20}^{-1}. \tag{4.6}$$

4.3 Panel studies

An autoregressive process with error is often used to model repeated time series. It also has the form of a Kalman filter. Let the unobservable m–component vector $\boldsymbol{y}_t$ be generated by the first–order autoregressive process

$$\boldsymbol{y}_t = \boldsymbol{B}\boldsymbol{y}_{t-1} + \boldsymbol{u}_t, \quad t = 2,\ldots,T, \tag{4.7}$$

where

$$\begin{aligned} &\mathcal{E}\,\boldsymbol{y}_1 = \mathbf{0}, \quad \mathcal{E}\,\boldsymbol{u}_t = \mathbf{0}, \quad \mathcal{E}\,\boldsymbol{y}_1\boldsymbol{y}_1' = \boldsymbol{\Phi}_1, \\ &\mathcal{E}\,\boldsymbol{u}_t\boldsymbol{u}_t' = \boldsymbol{\Phi}_t, \quad \mathcal{E}\,\boldsymbol{y}_1\boldsymbol{u}_t' = \mathbf{0}, \quad t = 2,\ldots,T, \\ &\mathcal{E}\,\boldsymbol{u}_t\boldsymbol{u}_s' = \mathbf{0}, \quad t \neq s. \end{aligned} \tag{4.8}$$

The observable vector at time t is

$$\begin{aligned} \boldsymbol{x}^{(t)} &= \boldsymbol{\mu}_t + \boldsymbol{y}_t + \boldsymbol{v}_t, \qquad t = 1,\ldots,T-1, \\ \boldsymbol{x}^{(T)} &= \boldsymbol{\mu}_T + \boldsymbol{y}_T, \end{aligned} \tag{4.9}$$

where

$$\begin{aligned} &\mathcal{E}\,\boldsymbol{v}_t = \mathbf{0}, \quad \mathcal{E}\,\boldsymbol{v}_t\boldsymbol{v}_t' = \boldsymbol{\Phi}_{T+t}, \\ &\mathcal{E}\,\boldsymbol{v}_t\boldsymbol{v}_s' = \mathbf{0}, \quad t \neq s, \quad \mathcal{E}\,\boldsymbol{v}_t\boldsymbol{u}_s' = \mathbf{0}, \\ &\mathcal{E}\,\boldsymbol{v}_t\boldsymbol{y}_1' = \mathbf{0}. \end{aligned} \tag{4.10}$$

The "error" term $\boldsymbol{v}_T$ is dropped from $\boldsymbol{x}^{(T)}$ because $\boldsymbol{u}_T$ and $\boldsymbol{v}_T$ could not be distinguished. We assemble (4.9) into a vector

$$\boldsymbol{x} = \begin{pmatrix} \boldsymbol{x}^{(1)} \\ \boldsymbol{x}^{(2)} \\ \boldsymbol{x}^{(3)} \\ \vdots \\ \boldsymbol{x}^{(T-1)} \\ \boldsymbol{x}^{(T)} \end{pmatrix} = \begin{pmatrix} \boldsymbol{\mu}_1 \\ \boldsymbol{\mu}_2 \\ \boldsymbol{\mu}_3 \\ \vdots \\ \boldsymbol{\mu}_{T-1} \\ \boldsymbol{\mu}_T \end{pmatrix} + \begin{pmatrix} \boldsymbol{I} \\ \boldsymbol{B} \\ \boldsymbol{B}^2 \\ \vdots \\ \boldsymbol{B}^{T-2} \\ \boldsymbol{B}^{T-1} \end{pmatrix} \boldsymbol{y}_1 \tag{4.11}$$

$$+ \begin{pmatrix} \boldsymbol{0} \\ \boldsymbol{I} \\ \boldsymbol{B} \\ \vdots \\ \boldsymbol{B}^{T-3} \\ \boldsymbol{B}^{T-2} \end{pmatrix} \boldsymbol{u}_2 + \begin{pmatrix} \boldsymbol{0} \\ \boldsymbol{0} \\ \boldsymbol{I} \\ \vdots \\ \boldsymbol{B}^{T-4} \\ \boldsymbol{B}^{T-3} \end{pmatrix} \boldsymbol{u}_3 + \dots + \begin{pmatrix} \boldsymbol{0} \\ \boldsymbol{0} \\ \boldsymbol{0} \\ \vdots \\ \boldsymbol{0} \\ \boldsymbol{I} \end{pmatrix} \boldsymbol{u}_T$$

$$+ \begin{pmatrix} \boldsymbol{I} \\ \boldsymbol{0} \\ \boldsymbol{0} \\ \vdots \\ \boldsymbol{0} \\ \boldsymbol{0} \end{pmatrix} \boldsymbol{v}_1 + \begin{pmatrix} \boldsymbol{0} \\ \boldsymbol{I} \\ \boldsymbol{0} \\ \vdots \\ \boldsymbol{0} \\ \boldsymbol{0} \end{pmatrix} \boldsymbol{v}_2 + \dots + \begin{pmatrix} \boldsymbol{0} \\ \boldsymbol{0} \\ \boldsymbol{0} \\ \vdots \\ \boldsymbol{I} \\ \boldsymbol{0} \end{pmatrix} \boldsymbol{v}_{T-1}.$$

The covariance matrix of $\boldsymbol{x}$ of Tm components is composed of T^2 blocks of $m \times m$ submatrices $\boldsymbol{\Sigma}_{ts}$, where

$$\boldsymbol{\Sigma}_{ts} = \sum_{r=1}^{s} \boldsymbol{B}^{t-r} \boldsymbol{\Phi}_r (\boldsymbol{B}')^{s-r}, \quad t \geq s, \tag{4.12}$$

and $\boldsymbol{\Sigma}_{st} = \boldsymbol{\Sigma}'_{ts}$. Note that $\boldsymbol{\Sigma}_{21} = \boldsymbol{B}\boldsymbol{\Phi}_1$, $\boldsymbol{\Sigma}_{31} = \boldsymbol{B}^2\boldsymbol{\Phi}_1 = \boldsymbol{B}\boldsymbol{\Sigma}_{21}$. If $|\boldsymbol{B}_0| \neq 0$, then $|\boldsymbol{\Sigma}^0_{21}| \neq 0$, $\boldsymbol{\Sigma}^0_{31}(\boldsymbol{\Sigma}^0_{21})^{-1} = \boldsymbol{B}_0$, $\boldsymbol{\Phi}^0_1 = \boldsymbol{B}_0^{-1}\boldsymbol{\Sigma}^0_2$, and $\boldsymbol{\Phi}^0_{T+1} = \boldsymbol{\Sigma}^0_{11} - \boldsymbol{\Phi}^0_1$. Similarly $\boldsymbol{\Phi}^0_2$ and $\boldsymbol{\Phi}^0_{T+2}$ can be obtained from $\boldsymbol{\Sigma}^0_{22}$ and $\boldsymbol{\Sigma}^0_{23}$, etc. We see that if $|\boldsymbol{B}_0| \neq 0$ and if $T \geq 3$ the parameters are identified. In this case $\boldsymbol{\lambda} = \text{vec } \boldsymbol{B}$.

Consider N observations on $\boldsymbol{x}$, say $\boldsymbol{x}_1, \dots, \boldsymbol{x}_N$, and find the maximum likelihood estimates of $\boldsymbol{\theta}' = \left[(\text{vec } \boldsymbol{B})', (\text{vech } \boldsymbol{\Phi}_1)', \dots, (\text{vech } \boldsymbol{\Phi}_{2T-1})'\right]$ from the Wishart likelihood of $\boldsymbol{S}$ equivalent to (2.5). Then the limiting distribution of $\sqrt{n}(\hat{\boldsymbol{B}} - \boldsymbol{B}_0)$ as $N \to \infty$ does not depend on the distribution of $\boldsymbol{y}_{1\alpha}, \boldsymbol{u}_{2\alpha}, \dots, \boldsymbol{u}_{T\alpha}, \boldsymbol{v}_{1\alpha}, \dots, \boldsymbol{v}_{T-1,\alpha}$ (except for the moment matrices $\Phi_1, \ldots, \Phi_{2T-1}$) as long as these vectors are mutually independent.

The underlying autoregressive process can be of higher order. For example, (4.7) can be replaced by

$$y_t = B_1 y_{t-1} + B_2 y_{t-2} + u_t, \quad t = 3, \ldots, T, \tag{4.13}$$

with y_1 and y_2 being independent with covariance matrices Φ_1 and Φ_2. If some aspects of stationarity are imposed on the model, for example, that $\Phi_2 = \ldots = \Phi_T$ in the first-order model, the version of (4.7) includes a moment matrix that is a function of τ.

5. EXTENSIONS OF LIMIT THEOREMS

In the previous sections we have assumed that the z_α's were independently identically distributed. However, to obtain limiting distributions we have only used the fact that $\sqrt{n}\,\Phi_{gh}(n), g \neq h$, have limiting normal distributions and that $\sqrt{n}\left[\Phi_{00}(n) - \Phi_0(\tau_0)\right]$ has a limiting distribution in the case that Φ_0 is restricted. The assumption that the z_α's (that is, the x_α's) are independently identically distributed can be relaxed considerably. As one important enlargment of the scope of the asymptotic theory, we shall treat one set of $z_\alpha^{(g)}$'s as arbitrary; they can be nonstochastic or random.

Suppose one subvector, say $z_\alpha^{(0)}$, consists of nonstochastic variables such that as $n \to \infty$

$$\Phi_{00}(n) \to \Phi_0^0, \tag{5.1}$$

where now (5.1) defines Φ_0^0; since the $z_\alpha^{(0)}$'s are nonstochastic, Φ_0^0 is not a "true value" (of the covariance of $z_\alpha^{(0)}$). Then

$$\text{vec}\ \Phi_{0g}(n) \xrightarrow{\text{L}} N\left(\mathbf{0}, \Phi_g^0 \otimes \Phi_0^0\right),\ g = 1, \ldots, G\ ; \tag{5.2}$$

see Section 2.6 of Anderson (1971), for example. Browne (1987) has used this result in the factor analysis model.

In a more general situation the components of $z_\alpha^{(0)}$ may be both stochastic and nonstochastic. In that case if (5.1) holds almost surely, then (5.2) holds. See Gleser (1983), Amemiya, Fuller, and Pantula

(1987), and Anderson and Amemiya (1985). These authors have applied the theorem to the factor analysis model where $\boldsymbol{z}_\alpha^{(0)} = \boldsymbol{f}_\alpha$.

A more general theorem states that (5.2) holds if $\boldsymbol{\Phi}_{00}(n) \xrightarrow{\text{P}} \boldsymbol{\Phi}_0^0$ [Anderson (1987)]. These theorems could be further generalized so that the $\{\boldsymbol{z}_\alpha^{(0)}\}$ sequence consists of martingale differences.

THEOREM 5.1. *Suppose* $\{\boldsymbol{z}_\alpha^{(0)}\}, \{\boldsymbol{z}_\alpha^{(1)}\}, \ldots, \{\boldsymbol{z}_\alpha^{(G)}\}$ *are independent sequences of vectors such that*

$$\frac{1}{n}\sum_{\alpha=1}^{n} \boldsymbol{z}_\alpha^{(0)} \boldsymbol{z}_\alpha^{(0)\prime} \xrightarrow{\text{P}} \boldsymbol{\Phi}_0^0 \tag{5.3}$$

and $\boldsymbol{z}_\alpha^{(g)}$ *are independently identically distributed with* $\mathcal{E}\,\boldsymbol{z}_\alpha^{(g)} = \mathbf{0}$ *and* $\mathcal{E}\,\boldsymbol{z}_\alpha^{(g)} \boldsymbol{z}_\alpha^{(g)\prime} = \boldsymbol{\Phi}_g^0$, $g = 1, \ldots, G$. *Then the limiting distribution of* $\sqrt{n}\,\mathrm{vec}\,\boldsymbol{\Phi}_{gh}(n)$, $g \neq h$ *, is normal with mean* $\mathbf{0}$ *and covariance matrix* $\boldsymbol{\Phi}_{hh}^0 \otimes \boldsymbol{\Phi}_{gg}^0$. *In the limiting distribution* $\sqrt{n}\Phi_{01}(n), \ldots, \sqrt{n}\boldsymbol{\Phi}_{G-1,G}(n)$ *are independent.*

In the theorems concerning the limiting distributions of $\sqrt{n}[\hat{\boldsymbol{\theta}} - \boldsymbol{\theta}(n)]$ the assumptions on $\{\boldsymbol{z}_\alpha\}$ can be replaced by those of Theorem 5.1 if $\boldsymbol{\Phi}_0$ is unrestricted.

ACKNOWLEDGEMENTS

Research sponsored by National Science Foundation Grant No. DMS-86-03779 at Stanford University and conducted at the U.S. Naval Post-graduate School. The author is indebted to Yasuo Amemiya, Naoto Kunitomo, T. L. Lai, and anonymous reviewers for helpful suggestions.

REFERENCES

Amemiya, Yasuo, and Fuller, Wayne A. (1984). Estimation for the multivariate errors–in–variables model with estimated error covariance matrix. *Annals of Statistics*, **12**, 497-509.

Amemiya, Yasuo, Fuller, Wayne A., and Pantula, S. G. (1987). The asymptotic distributions of some estimators for a factor analysis model. *Journal of Multivariate Analysis*, **22**, 51-64.

Anderson, T. W. (1971). *The Statistical Analysis of Time Series*. John Wiley and Sons, New York.

Anderson, T. W. (1984a). Estimating linear statistical relationships (1982 Abraham Wald Memorial Lectures). *Annals of Statistics,* **12**, 1-45.

Anderson, T. W. (1984b). *An Introduction to Multivariate Statistical Analysis*, Second Edition. John Wiley & Sons, New York.

Anderson, T. W. (1987). Linear models and covariance structures. Technical Report No. 27. Econometric Workshop, Stanford University.

Anderson, T. W., and Amemiya, Y. (1985). The asymptotic normal distribution of estimators in factor analysis under general conditions. Technical Report No. 12. Econometric Workshop, Stanford University. *Annals of Statistics,* **16**, in press.

Anderson, T. W., and Rubin, H. (1956). Statistical inference in factor analysis. *Proceedings of the Third Berkeley Symposium on Mathematical Statistics and Probability*, Vol. 5, University of California Press, Berkeley, 111-150.

Bentler, P. M. (1983). Some contributions to efficient statistics in structural models: specification and estimation of moment structures. *Psychometrika,* **48**, 493-517.

Browne, M. W. (1979). The maximum likelihood solution in inter–battery factor analysis. *British Journal of Mathematical and Statistical Psychology,* **32**, 75-86.

Browne, M. W. (1980). Factor analysis of multiple batteries by maximum likelihood. *British Journal of Mathematical and Statistical Psychology,* **33**, 184-199.

Browne, M. W. (1984). Asymptotic distribution–free methods for the analysis of covariance structures. *British Journal of Mathematical and Statistical Psychology*, **37**, 62-83.

Browne, M. W. (1987). Robustness of statistical inference in factor analysis and related models. *Biometrika*, **74**, 375-384.

Browne, M. W., and Shapiro, A (1986). Robustness of normal theory methods in the analysis of linear latent variate models. Research Report No. 86/5. Department of Statistics, University of South Africa.

Gleser, L. J. (1983). Functional, structural and ultrastructural errors–in–variables models. *Proceedings of the Business and Economics Statistics Section*, American Statistical Association, 57-66.

Kano, Yutaka (1986). Conditions on consistency of estimators in covariance structure model. *Journal of the Japan Statistical Society*, **16**, 75-80.

Magnus, Jan R., and H. Neudecker (1986). Symmetry, $0-1$ matrices and Jacobians: A review. *Econometric Theory,* **2**, 157-190.

Mooijart, Ab, and Bentler, P. M. (1985). The weight matrix in asymptotic distribution-free methods. *British Journal of Mathematical and Statistical Psychology*, **38**, 190-196.

Mooijart, Ab, and Bentler, Peter (1987). Robustness of normal theory statistics in structural equation models. Technical Report. Leiden University.

Nel, D. G. (1980). On matrix differentiation in statistics. *South African Statistical Journal,* **14**, 137-193.

Rao, C. R. (1973). *Linear Statistical Inference and Its Applications.* Second Edition. John Wiley and Sons, Inc., New York.

Satorra, A., and Bentler, P. M. (1986). Robustness properties of ML statistics in covariance structure analysis. Unpublished.

Shapiro, Alexander (1983). Asymptotic distribution theory in the analysis of covariance structures (A unified approach). *South African Statistical Journal,* **17**, 33-81.

Shapiro, Alexander (1984). A note on the consistency of estimators in the analysis of moment structures. *British Journal of Mathematical and Statistical Psychology*, **37**, 84-88.

Received 17 August 1987
Revised 25 September and 29 October 1987

Department of Statistics
Stanford University
Sequoia Hall
Stanford, CA 94305
U.S.A.

Proc. Second International Tampere Conference in Statistics
(Tampere, Finland, 1-4 June 1987)
Tarmo Pukkila and Simo Puntanen, *Editors*

pp. 37 - 52

KEYNOTE ADDRESS

Some aspects of statistical design in quality improvement

George E.P. BOX and R. Daniel MEYER
University of Wisconsin-Madison, Madison, WI, U.S.A.
and The Lubrizol Corporation, Wickliffe, OH, U.S.A.

Key words and phrases: Fractional factorial designs, factor sparsity, dispersion effects, location effects, screening designs.

ABSTRACT

Consideration of certain aspects of scientific method leads to discussion of recent research on the role of screening designs in the improvement of quality. A projective rationale for the use of these designs in the circumstances of *factor sparsity* is advanced. In this circumstance the possibility of identification of sparse *dispersion* effects as well as sparse *location* effects is considered. A new method for the *analysis of fractional factorial designs* is advanced.

1. INTRODUCTION

Humans beings differ from other animals most remarkably in their ability to learn. It is clear that, although throughout the history of mankind technological learning has taken place, until three or four hundred years ago change occurred very slowly. One reason for this was that in order to learn something - for example, how to make fire or champagne - two *rare events* needed to coincide: (a) an informative event had to *occur*, and (b) a person able to draw logical conclusions and to act on them had to be *aware* of that informative event.

Observation by an informed observer is a way of increasing the probability that the rare informative event will be constructively taken note of and is exemplified by quality control charting methods. Thus a Shewhart chart is a means to ensure that possibly informative events are brought to the attention of those who may be able to discover in them an "assignable cause" [Shewhart (1931)] and to act appropriately.

Active intervention by experimentation aims, in addition, to increase the probability of an informative event *actually occurring*. A designed experiment conducted by a qualified experimenter can dramatically increase the probability of learning because it increases simultaneously the probability of an informative event occurring and also the probability of the event being constructively witnessed. Recently there has been much use of experimental design in Japanese industry particularly by Genichi Taguchi [Taguchi and Wu (1980)] and his followers. In off-line experimentation he has in particular emphasized the use of highly fractionated designs and orthogonal arrays and the minimization of variance.

In the remainder of this paper we briefly outline some recent research on such uses of screening designs.

2. USE OF SCREENING DESIGNS TO IDENTIFY "ACTIVE" FACTORING

Table 1 shows in summary a highly fractionated two-level factorial design employed as a screening design in an off-line welding experiment performed by the National Railway Corporation of Japan [Taguchi and Wu (1980)]. To facilitate later discussion we have set out the design and labelled the levels somewhat differently from Taguchi. In the column to the right of the table is shown the observed tensile strength of the weld, one of several quality characteristics measured.

The design was chosen on the assumption that in addition to main effects only the two-factor interactions AC, AG, AH, and GH were expected to be present. On that supposition, all nine main effects and the four selected two-factor interactions can be separately estimated by appropriate orthogonal contrasts and the two remaining contrasts corresponding to the columns labelled e_1 and e_2 measure only experimental error. When, using a valuable procedure for analysis due to Cuthbert Daniel (1959, 1976), the effects are plotted on normal probability paper, thirteen of them plot roughly as a straight line but the remaining two, corresponding to the main effects for factors B and C, fall markedly off

A: Kind of Welding Rods
B: Period of Drying
C: Welded Material
D: Thickness
E: Angle
F: Opening
G: Current
H: Welding Method
J: Preheating

Factor			D	H	e_1	G	F	GH	AC	A	E	AH	e_1	AG	J	B	C	Tensile strength kg/mm^2
Column Number		0	1	2	3	4	5	6	7	8	9	10	11	12	13	14	15	
	1	+	−	−	+	−	+	+	−	−	+	+	−	+	−	−	+	43.7
	2	+	+	−	−	−	−	+	+	−	−	+	+	+	+	−	−	40.2
	3	+	−	+	−	−	+	−	+	−	+	−	+	+	−	+	−	42.4
	4	+	+	+	+	−	−	−	−	−	−	−	−	+	+	+	+	44.7
	5	+	−	−	+	+	−	−	+	−	+	+	−	−	+	+	−	42.4
	6	+	+	−	−	+	+	−	−	−	−	+	+	−	−	+	+	45.9
	7	+	−	+	−	+	−	+	−	−	+	−	+	−	+	−	+	42.2
Run	8	+	+	+	+	+	+	+	+	−	−	−	−	−	−	−	−	40.6
	9	+	−	−	+	−	+	+	−	+	−	−	+	−	+	+	−	42.4
	10	+	+	−	−	−	−	+	+	+	+	−	−	−	−	+	+	45.5
	11	+	−	+	−	−	+	−	+	+	−	+	−	−	+	−	+	43.6
	12	+	+	+	+	−	−	−	−	+	+	+	+	−	−	−	−	40.6
	13	+	−	−	+	+	−	−	+	+	−	−	+	+	−	−	+	44.0
	14	+	+	−	−	+	+	−	−	+	+	−	−	+	+	−	−	40.2
	15	+	−	+	−	+	−	+	−	+	−	+	−	+	−	+	−	42.5
	16	+	+	+	+	+	+	+	+	+	+	+	+	+	+	+	+	46.5
Effect		43.0	.13	−.15	+.30	+.15	.40	−.03	.38	.40	−.05	.43	.13	.13	−.38	2.15	3.10	

TABLE 1. A fractional two-level design used in a welding experiment showing observed tensile strength and effects.

the line, suggesting that over the ranges studied, only factors B and C affect tensile strength location by amounts not readily attributed to noise.

If this conjecture is true, at least approximately, the sixteen runs could be regarded as four replications of a 2^2 factorial design in factors B and C only. However, when the results are plotted in Figure 1 so as to reflect this, inspection suggests the existence of a dramatic effect of a different kind - when factor C is at its plus level the spread of the data appears much larger than when it is at its minus level. (Data of this kind might be accounted for by the effect of one or more variables other than B that affected tensile strength only at the "plus level" of C. Analysis of the eight runs made at the plus level of C does not support this possibility, however). Thus, in addition to detecting shifts in location due to B and C, the experiment may also have detected what we will call a *dispersion effect* due to C. The example raises the general possibility of analyzing unreplicated designs for dispersion effects as well as for the more usual location effects.

3. RATIONALES FOR USING SCREENING DESIGNS

Before proceeding we need to consider the question, "In what situations are screening designs, such as highly fractionated factorials, useful?"

3.1 Effect sparsity

A common industrial problem is to find from a rather large number of factors those few that are responsible for *large effects*. The idea is comparable to that which motivates the use in quality control studies of the "Pareto diagram". [See, for example, Ishikawa (1976)]. The situation is approximated by postulating that only a small proportion of effects will be "*active*" and the rest "*inert*". We call this the postulate of *effect sparsity*. For studying such situations, highly fractionated designs and other orthogonal arrays [Finney (1945), Plackett and Burman (1946), Rao (1947), Taguchi and Wu (1980)], which can screen moderately large numbers of variables in rather few runs, are of great interest. Two main rationalizations have been suggested for the use of these designs; both ideas rely on the postulate of effect sparsity but in somewhat different ways.

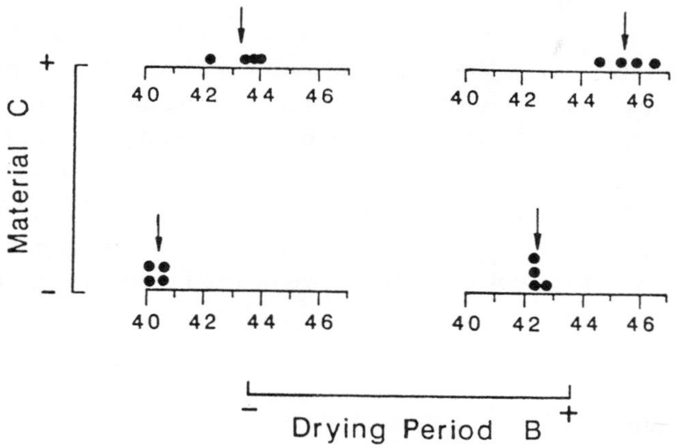

FIGURE 1. Welding experiment data presented as four replicates of a 2^2 factorial design in factors B and C only. Arrows indicate sample averages.

3.2 Rationale based on prior selection of important interactions

It is argued [see for example Davies (1954)] that in some circumstances physical knowledge of the process will make only a few interactions likely and that the remainder may be assumed negligible. For example, in the welding experiment described above there were 36 possible two-factor interactions between the nine factors, but only four were regarded as likely, leaving 32 such interactions assumed negligible. The difficulty with this idea is that in many applications the picking out of a few "likely" interactions is difficult if not impossible. Indeed the investigator might justifiably protest that, in the circumstance where an experiment is needed to determine which *first order* (main) effects are important, it is illogical that he expected to guess in advance which effects of *second order* (interactions) are important.

3.3 Projective rationale and factor sparsity

A somewhat different notion is that of *factor sparsity*. Thus suppose that, of the k factors considered, only a small subset of vaguely known size d, *whose identity is, however, unknown*, will be active in providing main effects and interactions within that subset. Arguing as in Box and Hunter (1961) a two-level design enabling us to study such a system is a fraction of resolution $R=d+1$ [or in the terminology of Rao (1947) an array of strength d] which produces complete factorials (possibly replicated) in every one of the $\binom{k}{d}$ subspaces of $d=R-1$ dimensions. For example, we have seen that on the assumption that only factors B and C are important, the welding design could be regarded as four replicates of a 2^2 factorial in just those two factors. But because the design is of resolution $R=\text{III}$ the same would have been true for any of the 36 choices of two out of the nine factors tested. Thus the design would be appropriate if it were believed that not more than two of the factors were likely to be "active".

	Columns	1	2	3	4	5	6	7	8	9	10	11	12	13	14	15
(a)	2^{15-11}_{III}	.	.	.	.	.	.	.	.	.	.	.	.	.	.	.
(b)	2^{8-4}_{IV}	.	.		.			.	.			.		.	.	
(c)	2^{5-1}_{V}	.	.		.				.							.
(d)	2^4	.	.		.				.							

TABLE 2. Some alternative uses of the orthogonal array of Table 2.

For further illustration we consider again the sixteen-run orthogonal array of Table 1 and, adopting a roman subscript to denote the resolution R of the design, we indicate in Table 2 various ways in which that array might be used. It may be shown that

(a) If we associated the fifteen contrast columns of the design with fifteen factors, we would generate a 2^{15-11}_{III} design providing four-fold replication of 2^2 factorials in every one of the 105 two-dimensional projections.

(b) If we associated only columns 1, 2, 4, 7, 8, 11, 13, and 14 with eight factors we would generate a 2^{8-4}_{IV} design providing two-fold replication of 2^3 factorials in every one of the 56 three-dimensional projections.

(c) If we associated only columns 1, 2, 4, 8, and 15 with five factors we would generate a 2^{5-1}_{V} design providing a 2^4 factorial in every one of the four-dimensional projections.

(d) If we associated only columns 1, 2, 4 and 8 with four factors we would obtain the complete 2^4 design from which this orthogonal array was in fact generated.

Designs (a), (b) and (c) would thus be appropriate for situations where we believed respectively that not more than 2, 3, or 4 factors would be active. The designs also give partial coverage for a larger number of factors. For example [Box and Hunter (1961)], 56 of the 70 four-dimensional projections of the 2^{8-4}_{IV} yield a full factorial in four variables. Notice that intermediate values of k could be accommodated by suitably omitting certain columns. Thus the welding design is a 2^{9-5}_{III} arrangement which can be obtained by omitting 6 columns from the complete 2^{15-11}_{III}. Notice finally that for intermediate designs we can take advantage of both rationales by arranging, as was done for the welding design, that particular interactions are isolated.

A discussion of the iterative model building process by Box and Jenkins (1970) characterized three steps in the iterative data analysis cycle indicated below

→ identification → fitting → diagnostic checking ↩

Most of the present paper is concerned with model *identification* - the search for a model worthy to be formally entertained and subsequently fitted by an efficient procedure such as maximum likelihood. The situation we address concerns the analysis of fractional designs such as the welding design. It is supposed that only a few of the factors are likely to have effects which may include dispersion effects as well as location effects and we are looking for clues to identify these. The clues will be checked with later experimentation.

4. DISPERSION EFFECTS

We again use the design of Table 1 for illustration. There are 16 runs from which 16 quantities -- the average and 15 effect contrasts -- have been calculated. Now if we were also interested in possible dispersion effects we could also calculate 15 variance ratios. For example, in column 1 we can compute the sample variance s^2_{1-} for those observations associated with a minus sign and compare it with the sample variance s^2_{1+} for observations associated with a plus sign to provide the ratio $F_1 = s^2_{1-}/s^2_{1+}$. If this is done for the welding data we obtain values for lnF_i given in Figure 2(a). In this figure familiar normal theory significance levels are also shown. Obviously the necessary assumptions are not satisfied in this case, but these percentages provide a rough indication of magnitude. It will be recalled that in the earlier analysis a large dispersion effect associated with factor C (column 15) was found, but in Figure 2(a) the effect for factor C is not especially extreme, instead the dispersion effect for factor D (column 1) stands out from all the rest. This misleading indication occurs because we have not so far taken account of the aliasing of location with dispersion effects. Since sixteen linearly independent location effects have already been calculated for the original data, calculated dispersion effects must be functions of these. Recently [Box and Meyer (1986a)] a general theory of location-dispersion aliasing has been obtained for factorials and fractional factorials at two levels. For illustration, in this particular example it turns out that the following identity exists for the dispersion effect F_1, that is the F ratio associated with factor D and hence for column 1 of the design.

$$F_1 = \frac{(\hat{2}+\hat{3})^2+(\hat{4}+\hat{5})^2+(\hat{6}+\hat{7})^2+(\hat{8}+\hat{9})^2+(\hat{10}+\hat{11})^2+(\hat{12}+\hat{13})^2+(\hat{14}+\hat{15})^2}{(\hat{2}-\hat{3})^2+(\hat{4}-\hat{5})^2+(\hat{6}-\hat{7})^2+(\hat{8}-\hat{9})^2+(\hat{10}-\hat{11})^2+(\hat{12}-\hat{13})^2+(\hat{14}-\hat{15})^2} \qquad (1)$$

Now (see Table 1) $\hat{14} = \hat{B} = 2.15$ and $\hat{15} = \hat{C} = 3.10$ are the two largest location effects, standing out from all the others. The extreme value of F_1 associated with an apparent dispersion effect of factor $D(1)$ is largely accounted for by the squared sum and squared difference of the location effects B and C which appear respectively as the last terms in the denominator and numerator of equation (1). A natural way to proceed is to compute variances from the residuals obtained after eliminating large location effects. After such elimination the alias relations of equation (1) remain the same except that location effects from eliminated variables drop out, that is zeros are substituted for eliminated variables. Variance analysis for the residuals after eliminating effects of B and C are shown in Figure 2(b). The dispersion effect associated with C (factor 15) is now correctly indicated as extreme. It is shown in the paper referenced above how, more generally, under circumstances of effect sparsity a location-dispersion model may be correctly identified when a few effects of both kinds are present.

5. ANALYSIS OF UNREPLICATED FRACTIONAL DESIGNS

Another important problem in the analysis of unreplicated fractional designs and other orthogonal arrays concerns the picking out of "active" factors. A serious difficulty is that with unreplicated fractional designs no simple estimate of the experimental error variance against which to judge the effects is available.

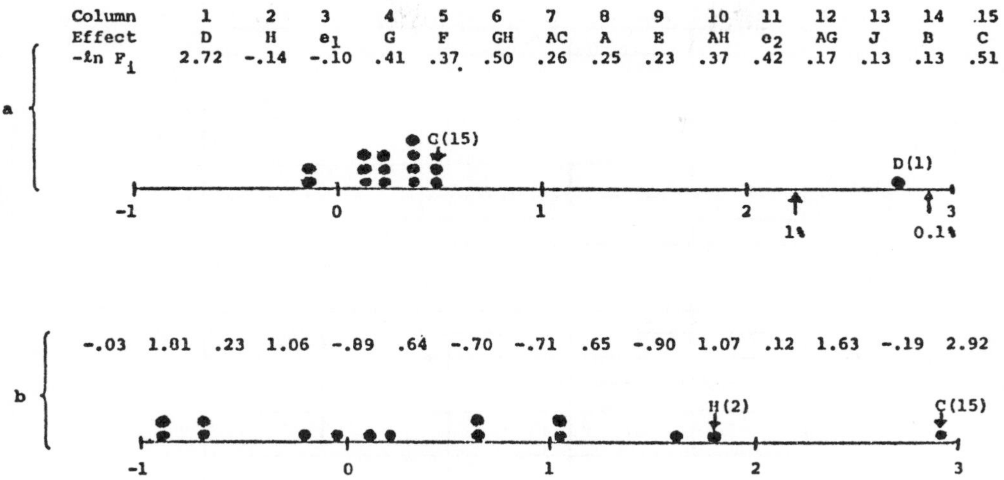

FIGURE 2. Welding experiment log dispersion effects (a) before, and (b) after elimination of location effects for B and C.

For illustration Table 3 shows the calculated effects from a 2^{8-4}_{IV} design used in an experiment on injection molding [Box, Hunter and Hunter (1978, p. 399)]. These effects are plotted on normal probability paper in Figure 3.

T_1	= - 0.7		$\rightarrow 1$	mold temperature
T_2	= - 0.1		$\rightarrow 2$	moisture content
T_3	= 5.5		$\rightarrow 3$	holding pressure
T_4	= - 0.3		$\rightarrow 4$	cavity thickness
T_5	= - 3.8		$\rightarrow 5$	booster pressure
T_6	= - 0.1		$\rightarrow 6$	cycle time
T_7	= 0.6		$\rightarrow 7$	gate size
T_8	= 1.2		$\rightarrow 8$	screw speed
T_9	$= T_{1.2}$	= - 0.6	$\rightarrow 1.2+3.7+4.8+5.6$	
T_{10}	$= T_{1.3}$	= 0.9	$\rightarrow 1.3+2.7+4.6+5.8$	
T_{11}	$= T_{1.4}$	= - 0.4	$\rightarrow 1.4+2.8+3.6+5.7$	
T_{12}	$= T_{1.5}$	= 4.6	$\rightarrow 1.5+2.6+3.8+4.7$	
T_{13}	$= T_{1.6}$	= - 0.3	$\rightarrow 1.6+2.5+3.4+7.8$	
T_{14}	$= T_{1.7}$	= - 0.2	$\rightarrow 1.7+2.3+6.8+4.5$	
T_{15}	$= T_{1.8}$	= - 0.6	$\rightarrow 1.8+2.4+3.5+6.7$	

TABLE 3. Calculated effects from a 2^{8-4}_{IV} design showing alias structure assuming three-factor and higher order interactions negligible.

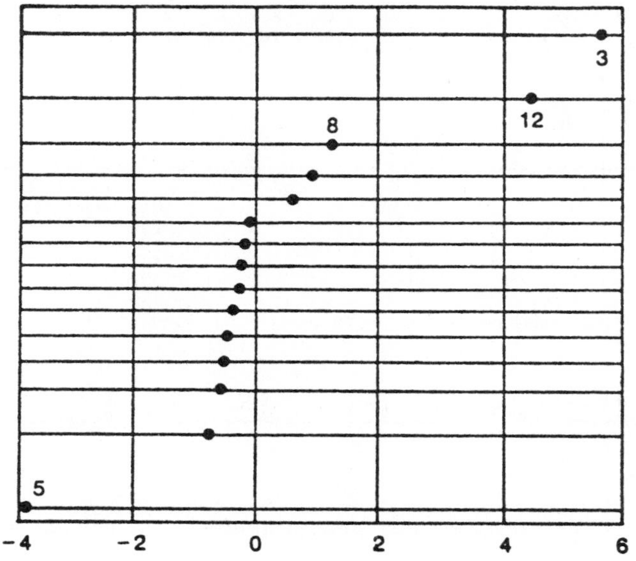

FIGURE 3. Normal plot of effects.

An alternative Bayesian approach [Box and Meyer (1986b)] is as follows: Let $T_1, T_2, \ldots, T_v$ be standardized effects (that is, effects scaled so that they all have equal variances) with

$$T_i = e_i \quad \text{if effect inert,}$$

$$T_i = e_i + \tau_i \quad \text{if effect active,}$$

$$e_i \to N(0,\sigma^2), \quad \tau_i \to N(0,\sigma_\tau^2), \quad k^2 = \frac{\sigma^2 + \sigma_\tau^2}{\sigma^2}.$$

Suppose the probability that an effect is active is α.

Let $a_{(r)}$ be the event that a particular set of r of the v factors are active, and let $\tilde{T}_{(r)}$ be the vector of estimated effects corresponding to active factors of $a_{(r)}$. Then, [Box and Tiao (1968)] with $p(\sigma) \propto 1/\sigma$ the posterior probability that $\tilde{T}_{(r)}$ are the only active effects is

$$P[a_{(r)}|\tilde{T}, \alpha, k] \propto \left[\frac{\alpha k^{-1}}{1-\alpha}\right]^r \left\{1-(1-\frac{1}{k^2})\ \frac{S_{(r)}}{S}\right\}^{-\nu/2},$$

where $S_{(r)} = \tilde{T}'_{(r)}\tilde{T}_{(r)}$ and $S = \tilde{T}'\tilde{T}$. In particular the marginal probability that an effect i is active given T, α and k is proportional to

$$\sum_{\substack{a_{(r)} \\ i \text{ active}}} \left[\frac{\alpha k^{-1}}{1-\alpha}\right]^r \left\{1-(1-\frac{1}{k^2})\ \frac{S_{(r)}}{S}\right\}^{-\nu/2}.$$

A study of the fractional factorials appearing in Davies (1954), Daniel (1976) and Box, Hunter and Hunter (1978) suggested that α might range from 0.15 to 0.45 while k might range from 5 to 15. The posterior probabilities computed with the (roughly average) values $\alpha=0.30$ and $k=10$ are shown in Figure 4(a) in which N denotes the probability (negligible for this example) that there are no active effects. The results from a sensitivity analysis in which α and k were altered to vary over the ranges mentioned above are shown in Figure 4(b).

It will be seen that Figure 4(a) points to the conclusion that active effects are associated with columns 3, 5 and 12 of the design and that column 8 might possibly also be associated with an active factor. Figure 4(b) suggests that this conclusion is very little affected by widely different choices for α and k. Further research has been conducted concerning marginalization with respect to k, different choices of the error distribution, and calculation of distributions of the active effects (see Meyer (1985)).

6. ALLOWANCE FOR FAULTY OBSERVATIONS

Recent work [Box and Meyer (1987)] has shown how a double application of the scale-contamination model (both to the observations themselves as well as to the effects) can make it possible to allow for faulty observations in the analysis of unreplicated factorials or fractional factorials.

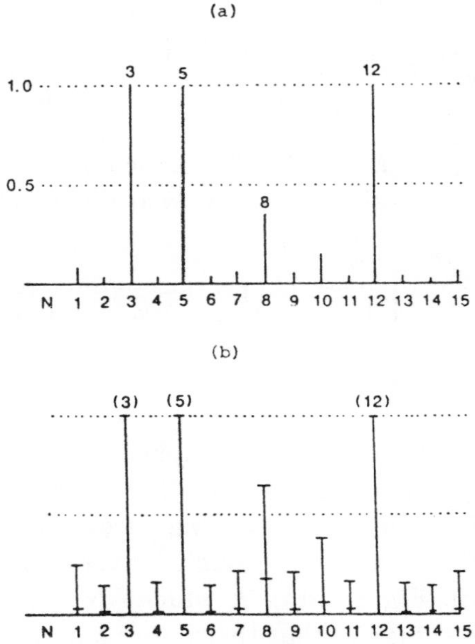

FIGURE 4. (a) Welding experiment. Posterior probability that factor i is active (α = 0.30, k = 10). (b) Sensitivity analysis for posterior probability (α = 0.15 - 0.45, k = 5-15).

REFERENCES

Box, G.E.P. and Hunter, J.S. (1961). The 2^{k-p} fractional factorial designs. *Technometrics*, 3, 311-351, 449-458.

Box, G.E.P. and Tiao, G.C. (1968). A Bayesian approach to some outlier problems. *Biometrika*, 55, 119.

Box, G.E.P. and Jenkins, G.M. (1970). *Time Series Analysis, Forecasting and Control.* Holden-Day, San Francisco.

Box, G.E.P., Hunter, W.G. and Hunter, J.S. (1978). *Statistics for Experimenters.* Wiley, New York.

Box, G.E.P. and Meyer, R.D. (1986a). Dispersion effects from fractional designs. *Technometrics*, 28, 19-27.

Box, G.E.P. and Meyer, R.D. (1986b). An analysis for unreplicated fractional factorials. *Technometrics*, 28, 11-18.

Box, G.E.P. and Meyer, R.D. (1987). Analysis of unreplicated factorials allowing for possibly faulty observations. To appear in a Festshrift Volume for Cuthbert Daniel. To be published by Wiley, New York.

Daniel, C. (1959). Use of half-normal plots in interpreting factorial two-level experiments. *Technometrics*, 1, 311-341.

Daniel, C. (1976). *Application of Statistics to Industrial Experimentation.* Wiley, New York.

Davies, O.L., Editor (1954). *The Design and Analysis of Industrial Experiments.* Oliver and Boyd, London.
Finney, D.J. (1945). The fractional replication of factorial arrangements. *Annals of Eugenics*, 12, 291-301.
Ishikawa, K. (1976). *Guide to Quality Control.* Asian Productivity Organization, Tokyo.
Meyer, R.D. (1985). Analysis of factorial experiments. University of Wisconsin Mathematics Research Center Technical Summary Report #2865.
Plackett, R.L. and Burman, J.P. (1946). Design of optimal multifactorial experiments. *Biometrika*, 23, 305-325.
Rao, C.R. (1947). Factorial experiments derivable from combinatorial arrangements of arrays. *J. Roy. Statist. Soc., B9*, 128-140.
Taguchi, G. and Wu, Y. (1980). *Introduction to Off-Line Quality Control.* Central Japan Quality Control Association, Nagoya, Japan.

Received 17 March 1987

Center for Quality and Productivity Improvement
University of Wisconsin-Madison
610 Walnut Street
Madison, WI 53705
U.S.A.

The Lubrizol Corporation
29400 Lakeland Blvd.
Wickliffe, OH 44092
U.S.A.

DISCUSSION

Stratis KOUNIAS
Aristotle University of Thessaloniki, Thessaloniki, Greece

The interesting paper by George E.P. Box and R. Daniel Meyer gives a new insight in studying and analyzing factorial experiments. Professor Box is one of the few statisticians who has a genuine interest and is motivated from applied problems. He enjoys a wide experience in theory and practice.

The interest in studying factorial experiments is to isolate those factors which do not have negligible location and dispersion effects. The difficulty in these problems is that the number of experimental runs is small due to limited budget and time. The revival of quality control in industry makes it more important to have "good" or "efficient" designs with a small number of experimental runs.

I will not go into the analysis of factorial experiments for which Professor Box gave an expert's view. I will rather discuss the optimality of factorial experiments which is related to their efficiency and I will restrict my comments on two-level factorials.

If we consider that the second and higher order interactions are negligible i.e. absorbed into the error, then we have the so-called main effect plans or resolution III designs.

The linear model is:

$$\mathbf{Y} = \mathbf{X}\mathbf{B} + \boldsymbol{\varepsilon},$$

where the design matrix **X** is $n \times k$, n is the number of observations and k is the number of factors. The elements of **X** here are $x_{ij} = \pm 1$ denoting the high and low level of each factor. The information matrix of the design is $\mathbf{M} = \mathbf{X}'\mathbf{X}$, $\mathbf{M}$: $k \times k$.

An "efficient" or "optimal" design minimizes some function of the information matrix. The optimality criteria usually employed require the minimization of:

(i) The generalized variance, i.e., max det **M**, called D-optimality.

(ii) The sum of the variances of the main effect estimators, i.e., min trace($\mathbf{M}^{-1}$), called A-optimality.

(iii) The maximum variance among all linear combinations of the parameters, i.e., max $\lambda_{min}(\mathbf{M})$, called E-optimality.

(iv) The maximum variance of the estimated response surface, i.e.,

$$\min_{d} \max_{x} (\mathrm{var} \sum_{i=1}^{k} x_i \hat{\beta}_i) \text{ with } x_i = \pm 1, \text{ called G-optimality.}$$

In many cases the experimenter does not know what type of optimality is preferable, although he wants the design to be efficient.

When the number of runs (observations) is a multiple of 4 i.e. $n \equiv 0$ mod 4, the orthogonal plans of Plackett and Burman are universally optimal i.e. optimal with respect to all the criteria. Such orthogonal plans are known for most values of n and are widely used ($n = 428$ is the smallest unknown orthogonal plan). In this case $\mathbf{M}^{-1}$ is diagonal and the estimates are easily calculated. It is like Yate's algorithm.

When the number of runs is not multiple of 4, i.e., $n \neq 0$ mod 4 it is not so easy to construct the most efficient design.

For D-optimality it is known that:

(i) If $n = 4m + 1$ and the number of factors $k < n$, the optimal design is constructed by adding one arbitrary observation to an orthogonal design with $n - 1$ observations.

These designs are balanced, i.e., all main effect estimators are equally correlated and there are $n - k$ degrees of freedom left to estimate the error variance.

If the design is saturated i.e. $k = n$, no degrees of freedom are left for estimating the error variance but all main effects are estimable. In this case the efficient (D-optimal) designs are known for only a few values of $n = 5, 9, 13, 17, 21, 25, 41, 61$. These are sufficient for our needs. For $n = 9$, 17, 21 the D-optimal designs are not balanced and all the estimators do not have the same variance.

If we insist on balanced designs for $k = n$, $n = 9$, 17, 21, then we will lose in efficiency sometimes more than 20% if a balanced design exists.

For the saturated case, the optimal designs for $n = 4m + 1$ are not derived from the orthogonal ones.

(ii) If $n = 4m + 2$ and the number of factors is $k \leq n - 2$, then the D-optimal designs are constructed by adding two appropriate runs to an orthogonal plan with $n - 2$ runs.

The optimal saturated $n = k$ or almost saturated $k = n - 1$ are *not* constructed from the orthogonal ones by adding 1 or 2 observations. For $n \leq 50$ the only unknown designs are for $n = 22, 34$.

For $n = 4m + 2$ the factors in the optimal design are divided in two groups. Every two estimators within each group are equally correlated and between groups are uncorrelated. Again the balance is lost here for all estimators.

(iii) If $n = 4m + 3$ and the number of factors $k \leq (n + 5)/2$, the D-optimal design is found by subtracting one observation from an orthogonal design with k factors and $n + 1$ observations. These are balanced.

If $(n + 5)/2 < k \leq n$, the construction is unknown for many values of n and k. For example if $k = 13, 14, 15$ and $n = 15$ the optimal designs are unknown. If all or some of the interactions are not negligible, then we have to go to resolution IV or resolution V designs. Again here if n is a multiple of 16 i.e. $n = 16, 32, 48, \ldots$, the orthogonal designs derived from orthogonal arrays are universally optimal. So if we have up to 5 factors with their interactions we need at least $1 + 5 + \binom{5}{2} = 16$ runs. If $k = 6$ we need 22 runs and if $k = 7$ we need 29 runs.

Since the optimal designs for 22 or 29 runs are not known we use a 32 run experiment.

The question is:

(i) If we have 35 observations, then shall we perform 32 for which the design is known and lose in efficiency?

(ii) Shall we insist on balanced designs although we can have more efficient not balance designs?

For $n \neq 0 \bmod 16$ very little is known about the D, G, A or E-optimal designs.

In today's paper of Box and Meyer, some of the main effects and some of the interactions are of interest and performing 16 or 32 runs we are on safe grounds with respect to efficiency. This is the influence of Plackett and Burman and sometimes the experimenter must have in mind that even with $n < 16$ observations can have good designs. So for $k = 4$ we have $1 + 4 + \binom{4}{2} = 11$ unknown parameters, if we can perform $n =$

16 runs, then we have 5 d.f. for the error variance, otherwise we have to look for the optimal design with $n < 16$ observations.

In conclusion:

Professor Box gave a meaningful use of 2-level factorials in industrial quality control and at the same time he suggests challenging problems for the construction of optimal designs for these problems.

Received 12 July 1987

Department of Mathematics
Aristotle University of Thessaloniki
54006 Thessaloniki
Greece

REJOINDER

George E.P. Box

I am grateful to Professor Kounias for his excellent discussion of the properties of two level orthogonal arrays insofar as various characteristics of their information matrices are concerned. These measures of what I have called "alphabetic optimality" (Box, 1982; Box and Draper, 1987) do not, of course, by themselves, tell us what is the best design for a particular purpose. A list of some fourteen such properties is, for example, given by Box and Draper (1975; 1987). Although each one of these can be important in the choice of a design, none use directly any of the alphabetic optimality measures. I believe this is an area where we must work harder to bring the mathematical theory of design and the real design of experiments closer together.

ACKNOWLEDGEMENT

This research was sponsored by the National Science Foundation Grant No. DMS-8420968.

REFERENCES

Box, G.E.P. (1982). Choice of response surface design and alphabetic optimality. *Utilitias Mathematica*, 21B, 11-55.

Box, G.E.P. and Draper, N.R. (1987). *Empirical Model Building and Response Surfaces.* Wiley, New York.

Box, G.E.P. and Draper, N.R. (1975). Robust designs. *Biometrika*, 62, 347-352.

Proc. Second International Tampere Conference in Statistics
(Tampere, Finland, 1-4 June 1987)
Tarmo Pukkila and Simo Puntanen, *Editors*

pp. 53 - 72

KEYNOTE ADDRESS

The statistical theory of linear systems

E.J. HANNAN
Australian National University, Canberra, Australia

Key words and phrases: Hankel matrix, ARMAX, description length, martingale, recursive calculation, lattice algorithm.

ABSTRACT

The fitting of ARMAX models is viewed as a process of approximation to a true structure rather than of estimation. The structure of the set of all ARMAX systems is described. Criteria for choosing a good approximant are considered. Some theory relating to these criteria and to approximants is presented. Finally, recursive, adaptive, on line calculations are discussed and algorithms for these are described.

1. APPROXIMATION OF LINEAR SYSTEMS

It will generally be agreed that most time series problems are not problems of estimating a finite number of parameters in some known, true, structure but are rather problems of finding a suitable approximation to some aspect of the structure of the stochastic process generating the data. Of course much good method has come out of the paradigm of estimation but here approximation is to be emphasised.

In Brillinger (1984, pp. 1143-1153) J.W. Tukey discusses the dichotomy between these two approaches in relation to spectral estimation and an "overt", model free, approach is favoured. In that context the approximation process is usually founded upon an appropriate form of Fourier transformation of the data. This method has been very successful and widely used. The nature of a good approximation process depends, however, on the end purpose of the investigation and there are many situations where that purpose is special rather than investigative. Examples are prediction, control, encoding data for transmission and speech recognition. In such situations one may need models, the model set being sufficiently wide to permit a good approximation. In the present paper the model set will be that of rational transfer function linear systems. The main problem is the linearity. One way of handling this is to view the linear structure as local in time and to seek to adapt that structure to the changing local situation. This last aspect will be discussed in Section 5.

Some consequences of the approach outlined in the previous paragraph are as follows:

(i) Some appropriate criterion is needed on which to base a choice of model. This will be dealt with in Section 3.

(ii) The number of parameters for the models in the model set will vary and this will have to be determined from the data.

(iii) Partly because of (ii), above, the mathematical theory will be complex since the number of parameters fitted will depend on the amount of data available. Some such theory will be dealt with in Section 4.

(iv) Theorems will often have only a suggestive value since they will be proved under unreal conditions. This is always so but is more evident in the present context. Simulations can play an important part in conjunction with theory.

Before going on to discuss the problems mentioned above we need to discuss the structure of linear systems and this is done in the next section.

2. LINEAR SYSTEMS

Consider an output $y(t)$ of n components and an input $z(t)$ of m components related linearly by

$$y(t) = \sum_{1}^{\infty} L(j)\, z(t-j) + \zeta(t). \tag{2.1}$$

We shall take $\zeta(t)$ as stationary and linearly purely non deterministic with finite variance, so that

$$\zeta(t) = \sum_{0}^{\infty} K(j)\varepsilon(t-j)\,, \quad K(0) = I_n, \quad E\{\varepsilon(s)\,\varepsilon(t)'\} = \delta_{st}\Sigma, \quad \Sigma > 0. \tag{2.2}$$

The $\varepsilon(t)$ are the linear innovations for $\zeta(t)$. Evidently $\Sigma \| K(j) \|^2 < \infty$ and we shall also at least assume $\Sigma \| L(j) \|^2 < \infty$. Here $\| X \|^2 = tr(XX')$.

Let

$$H = [K(i+j-1),\ L(i+j-1)], \qquad i, j = 1, 2, \ldots,$$

where the (i, j)th block, of n rows and $n+m$ columns, in the infinite matrix, H, is shown. H is the Hankel matrix of the system. Let H_0 be a set of rows of H spanning all rows of H (which lie in ℓ_2, the Hilbert space of all square summable sequences). Put $H_0 = [B, L, H_2]$ where $[B, L]$ is the block composed of the first $n+m$ columns of H_0. It is easily seen that H_2 is composed of rows of H so that $H_2 = AH_0$. The first n rows of H, H_1 say, can also be written as $H_1 = CH_0$. Put

$$x(t) = H_0(\varepsilon(t)', z(t)', \varepsilon(t-1)', z(t-1)', \ldots)'.$$

Then

$$x(t+1) = Ax(t) + Lz(t) + B\,\varepsilon(t),\ y(t) = Cx(t) + \varepsilon(t). \qquad (2.3)$$

This is the general prediction error, state space representation.

It is well known that H is of finite rank if and only if the matrix functions

$$k(z) = \sum_{0}^{\infty} K(j) z^{-j}, \qquad \ell(z) = \sum_{1}^{\infty} L(j) z^{-j}$$

are rational, i.e., composed of rational functions.

Thus given a system with a linear input-output structure and purely non-deterministic, stationary error structure it is natural to approximate by a system of finite rank for H. (This rank is called the McMillan degree.) We need now to describe the set of all finite rank structures, i.e., all rational $[k(z), \ell(z)]$. We may do this via (2.3) for the only arbitrary element in that construction was the choice of H_0. We avoid this arbitrariness by choosing H_0 to be composed of the first linearly independent set found as the rows of H are examined from top to bottom. Let $r(u, j)$ be the jth row in the uth block of n rows. Then the basic set of rows discovered in this way is of the form

$$r(u, j),\ u = 1, \ldots, d_j, \quad j = 1, \ldots, n. \qquad (2.4)$$

Thus $d = \Sigma d_j$ is the rank of H. The d_j are called Kronecker indices. Now A, B, C, L in (2.3) are uniquely determined. The freely varying elements in these, i.e., those not identically 0 or 1, serve, along with the on and above diagonal elements of Σ to parametrize the set of all systems with given Kronecker indices. We speak of the parameters in A, B, C, L as system parameters, distinct from variances and covariances. The number of system parameters is

$$d(n+m+1) + \sum_{j<i} \{\min(d_i, d_j) + \min(d_i+1, d_j)\}. \tag{2.5}$$

For proofs of this and related results quoted below the reader may consult Hannan and Deistler (1987). We can arrange the sets $\{d_i\}$ of Kronecker indices in dictionary order. If α is the index of a typical set in that order we call V_α the set of all systems, for arbitrary Σ, having the Kronecker indices indexed by α. Of course there are other restrictions on the system parameters consequent on the fact that $k(z)$, $\ell(z)$ are analytic, $|z| \geq 1$, and det $k(z) \neq 0$, $|z| > 1$. However we always assume det $k(z) \neq 0$, $|z| \geq 1$. Then the system parameters map V_α, in a one to one fashion, into an open set in Euclidean space of dimension (2.5).

For statistical purposes a more useful parameterisation of V_α may be via the canonical form for $k(z)$, $\ell(z)$. We may write this as $(k, \ell) = \tilde{a}^{-1}(\tilde{b}, \tilde{d})$, where $\tilde{a}, \tilde{b}, \tilde{d}$ are matrices of polynomials, with $\tilde{a}$ having a monic element of degree d_j in the jth place in the main diagonal. (Monic here means that z^{d_j} has unity as coefficient.) Then this matrix fraction description of (k, ℓ) is uniquely determined if the degree relationships in the following diagram hold. (All inequalities are for the degrees of elements in the indicated rows and columns relative to the diagonal element of $\tilde{a}$.)

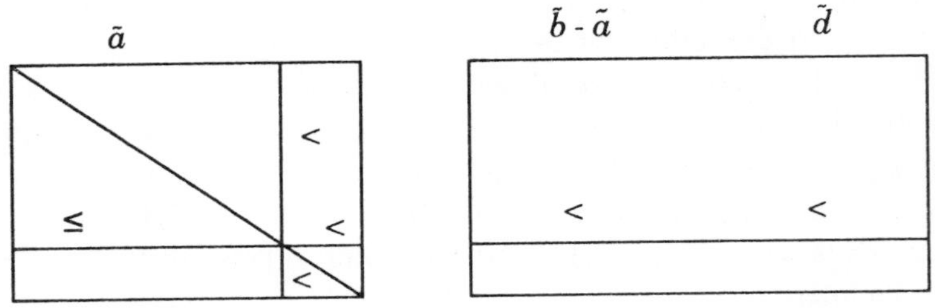

This diagram uniquely determines the 0,1 coefficients in coefficient matrices of $\tilde{a}$, $\tilde{b}$, $\tilde{d}$ and hence a set of system parameters, (2.3) in number. To obtain the corresponding ARMAX form for the system, i.e.,

$$\Sigma A(j)y(t-j) = \Sigma D(j)z(t-j) + \Sigma B(j)\varepsilon(t-j)\,,\; A(0) = B(0), \qquad (2.6)$$

we introduce the diagonal matrix, Z, with z^{dj} in the jth place and put

$$[a(z), b(z), d(z)] = Z\,[\tilde{a}(z^{-1}), \tilde{b}(z^{-1}), \tilde{d}(z^{-1})]$$

then

$$[a, b, d] = \Sigma\,[A(j), B(j), D(j)]\,z^j\,.$$

Again the fully varying elements of the $A(j), B(j), D(j)$, recalling that $A(0) = B(0)$, are of number (2.3) and constitute a vector, τ let us say, of system parameters.

The union of all V_α for $\Sigma d_j = d$ we call $M(d)$. This set of all systems of McMillan degree d may be topologised as an analytic manifold and is, indeed, a kind of algebraic variety. It is of dimension $d(2n+m)$. $M(d)$ may be covered by at most $\binom{d+n-1}{d}$ neighbourhoods each open and dense in $M(d)$. Only one of the V_α constitutes such a neighbourhood, some others being submanifolds of $M(d)$ of lower dimension. This is the V_α for which (2.4) constitutes the first d rows of H. Then if $d = np + q$ $0 \le q < n$ we have $d_1 = d_2 = \ldots = d_q = p + 1$, $d_{q+1} = \ldots = d_n = p$. The $A(j), B(j), D(j)$ in (2.6) are now in the following form where all partitions are after row or column q and an asterisk indicates a submatrix of freely varying elements.

$$A(0) = B(0) = \begin{pmatrix} I_q & 0 \\ * & I_{n-q} \end{pmatrix} \qquad D(0) = 0,$$

$$A(1) = \begin{pmatrix} * & 0 \\ * & * \end{pmatrix} \qquad A(p+1), B(p+1), D(p+1) = \begin{pmatrix} * \\ 0 \end{pmatrix}.$$

All other $A(j), B(j), D(j)$, $j < p+1$ are unconstrained. We do *not* mean that $A(p+1), B(p+1), D(p+1)$ are equal.

There has been little explicit use of this structure theory, in statistics, except for $n = 1$. For $n = 1$ there is only one neighbourhood with $d = p$, $q = 0$ and $a(z), b(z), d(z)$ are of degree p. Then $d(2n+m) = p(m+2)$. Of course often the degrees of $a(z)$, $b(z)$, $d(z)$, are not taken equal but that can be regarded as imposing prior constraints. All systems for $n = 1$ are uniquely listed by the single integer $d = p$ and the $p(m+2)$ continuously varying coefficients in $a(z)$, $b(z)$, $d(z)$, with $\alpha(0) = \beta(0) = 1$. [*In the scalar case,* $n = 1$, *we shall customarily use lower case letters* $\alpha(j)$, $\beta(j)$, $\delta(j)$, σ^2 *in place of* $A(j)$, $B(j)$, $D(j)$, Σ.] The reasons for the lack of use of the theory for $n > 1$ include lack of familiarity, the complexity of the general description and the fact that often there is physical knowledge leading to prior constraints that uniquely determine a representation, say in the form (2.3), and such that the constraints would be very difficult to determine for the canonical forms.

There is an extensive literature relating to approximation of a general H by one of finite rank. See Hannan (1987) for some details and references. The theory is deterministic, i.e., assumes H known, and seeks to determine an approximating $\hat{H}$ so as to make the Euclidean norm, i.e., the largest singular value of $H - \hat{H}$, as small as possible. However we need approximations to k, ℓ and that is a more complicated matter. We refer the reader to Glover (1984) for an account of the theory and some references. We proceed to discuss statistical criteria for choosing a good approximant.

3. APPROXIMATION CRITERIA

All approximations commence from a likelihood function for the data. Let us call $L_T(\theta)$ the log likelihood, where θ stands for the stochastic structure. For clarity, but only in this paragraph, let us follow Findley (1985) and write $E_T(\theta) = \mathcal{E}\{L_t(\theta)\}$ where $\mathcal{E}$ denotes expectation with respect to the true structure, θ_0. Since $E_T(\theta)$ is maximised at θ_0 it is natural to prefer θ_1 before θ_2 if $E_T(\theta_1) > E_T(\theta_2)$. However $E_T(\theta)$ is unknown. Under suitable conditions (see Findley, 1985 and references therein) it may be shown that $L_T(\theta_1) - L_T(\theta_2) + \delta(\theta_2) - \delta(\theta_1)$ is an asymptotically unbiased estimate of $E_T(\theta_1) - E_T(\theta_2)$, where $\delta(\theta)$ is the dimension of the vector θ. Thus we may reasonably choose to maximise $L_T(\theta) - \delta(\theta)$ and thus to choose δ to maximise $L_T(\hat{\theta}) - \delta(\hat{\theta})$ where $\hat{\theta}$ is the maximum likelihood estimate given δ. If we restrict ourselves to Gaussian models, ignore constants and lower order terms and multiply by $-T/2$ we arrive at the criterion, for the choice of δ, which is to be minimised,

$$\text{AIC: } \log \det \hat{\Sigma}_\delta + 2\delta/T .$$

Here $\hat{\Sigma}_\delta$ is the *ML* estimator of Σ for $\delta(\theta)$ fixed at δ. We may also take δ as the dimension of the vector of system parameters since the number of variances and covariances is fixed. Though AIC is derived on Gaussian assumptions the criterion may be more generally used and analysed.

An alternative approach is due to Rissanen (1978, 1983, 1986). He makes no assumption, initially about the data and assumes only a prescribed model set, the choice of which embodies all prior knowledge. The model is to be chosen which minimise a measure of the complexity of the data. The measure of complexity is the minimum number of bits for an encoding of the data given a model. Let $P(. \mid \theta)$ be the probability density of the (discrete) data, recorded to a fixed accuracy. It is known that one may encode with a number of bits as near to $-\log_2 P(y_T|\theta)$ but no smaller and that this encoding gives, effectively, a minimal average code length. Here $y_T' = (y(1)',\ldots, y(T)')$. We discuss only the ARMA case for brevity but the final formula is general. However the code must be decodable by someone knowing only the rules and thus θ must also be encoded. This will take $f(\theta)$ further bits. We shall not here go to the complex considerations to determine $f(\theta)$ (see Rissanen, 1983) but observe that there is little use in recording θ to more accuracy than can be known from the data. Since the standard deviation will be $0(T^{-\frac{1}{2}})$ then each element of θ needs about $\frac{1}{2}\log_2 T$ bits and thus $(\delta/2)\log_2 T$ bits are needed in all. Again we adopt Gaussian assumptions, ignore constants and lower order terms, multiply by $-2T^{-1}$ and change to natural logarithms and are led to the criterion

$$\text{BIC: } \log\det \hat{\Sigma}_\delta + \delta \log T/T.$$

Partly to avoid the argument evaluating $f(\theta)$ Rissanen (1986) also proceeds to derive a further criterion. Let $f_{\theta,\delta}(y(t+1)|y_t)$ be the conditional density of $y(t+1)$ given the previous data. Though δ is implicit in θ we separately indicate it. Then

$$-\log_2 P(y_T|\,\theta) = -\sum_{0}^{T-1} \log_2 f_{\theta,\delta}\,(y(t+1)|\,y_t).$$

Given δ and data to time t we choose θ to be $\hat{\theta}(t)$, the *ML* estimator. At time t, Rissanen (1986) suggests, we choose $\hat{\delta}(t)$ to minimise

$$-\sum_{0}^{t-1} \log_2 f_{\hat{\theta}(s),\delta}\,(y(s+1)|\,y_s).$$

Thus the overall measure of complexity is

$$- \sum_{0}^{T-1} \log_2 f_{\hat{\theta}(t),\hat{\delta}(t)} \left(y(t+1) | y_t \right).$$

Making the same Gaussian assumptions, etc., as for BIC we arrive at

$$\text{PMDL: } \frac{1}{t} \sum_{0}^{t-1} \{\log \det \hat{\Sigma}_\delta(s) + e_\delta(s+1)' \hat{\Sigma}_\delta(s)^{-1} e_\delta(s+1)\}.$$

The $\hat{\Sigma}_\delta(s)$ is the *ML* estimator of Σ for a model of dimension δ and data to time s. The vector $e_\delta(s+1)$ is the estimate of $\varepsilon(s+1)$ using data to time s to estimate the system parameters. The dimension $\hat{\delta}(t)$ is to be obtained so as to minimise PMDL. Note that the corresponding encoding is valid, for at time t the decoder will know y_{t-1} and hence $\hat{\theta}(t-1)$, $\hat{\delta}(t-1)$ and hence be able to decode $y(t)$. He, in fact, runs precisely the same algorithm as that run by the encoder. Note that

$$\hat{\Sigma}_\delta(s) = \frac{1}{s} \sum_{1}^{T} \hat{\varepsilon}_s(u) \hat{\varepsilon}_s(u)',$$

where $\hat{\varepsilon}_s(u)$ is the estimate of $\varepsilon(u)$, using all data to time s. Thus $e_\delta(s+1) = \hat{\varepsilon}_s(s+1)$ but we have suppressed the subscript s, for $u = s+1$, so as to be able to indicate the dimension δ by a subscript.

Rissanen (1986) shows that his procedures do provide a measure of complexity, i.e., to a suitable approximation and in an asymptotic sense they give a minimal encoding. The remaining thing to do is to investigate the statistical properties of estimates got by these procedures, AIC, BIC, PMDL. This, to some extent, we do in the next section.

4. SOME ASYMPTOTIC THEORY

For the construction of an asymptotic theory more has to be said about $\varepsilon(t)$ and $z(t)$. For the latter it could be assumed that, in addition to (2.1), (2.2)

$$z(t) = \sum_{0}^{\infty} M(j) y(t-j) + \sum_{0}^{\infty} N(j) \eta(t-j), \tag{4.1}$$

with the $\eta(t)$ being of the same nature as $\varepsilon(t)$. However, then $y(t)$ and $z(t)$ are on so much the same footing with the vector composed of $y(t)$, $z(t)$

being linearly represented in terms of $\varepsilon(t-j)$, $\eta(t-j)$, $j \geq 0$. To avoid a profusion of notation we therefore consider only the situation where

$$y(t) = \sum_0^\infty K(j)\,\varepsilon(t-j), \quad \sum_0^\infty \|K(j)\|^2 < \infty, \quad K(0) = I_n. \tag{4.2}$$

There are reasons why one might not wish to do that because $z(t)$ is special in relation to (4.1). For some treatment of that kind of situation see Hannan and Kavalieris (1986), Hannan and Deistler (1987). So far as $\varepsilon(t)$ is concerned we require that it be ergodic and using F_t for the σ - algebra determined by $\varepsilon(s)$, $s \leq t$,

$$E\{\varepsilon(t)|\, F_{t-1}\} = 0, \quad E\{\varepsilon(t)\,\varepsilon(t)' \,|\, F_{-\infty}\} = \Sigma, \quad E\{\varepsilon_j(t)^4\} < \infty, \quad j = 1,\ldots,n. \tag{4.3}$$

If variances have to be evaluated, for example, we need

$$E\{\varepsilon(t)\,\varepsilon(t)' \,|\, F_{t-1}\} = \Sigma \tag{4.4}$$

which is much stronger than the middle part of (4.3). There are many contexts where the conditional variance in (4.4) is not time invariant and thus (4.4) is to be avoided if possible. The first part of (4.3) is natural and the other parts of (4.3) are fairly innocuous, as also is

$$E\{\varepsilon_j(t)^4 \log_+|\,\varepsilon_j(t)\,|\} < \infty, \quad j = 1,\ldots, n. \tag{4.5}$$

Let us consider the situation where $k(z) \in M(d_0)$, $d_0 < \infty$. This is not the situation basically to be considered here but has some relevance in the sense of point (iv) in Section 1. We need a further condition, namely, using a zero subscript for a true value,

$$\det\{k_0(z)\} \neq 0, \quad |z| \geq 1 - \eta, \quad \eta > 0; \quad |\tau_0| \leq \rho < \infty. \tag{4.6}$$

It is assumed that η, ρ are known but they are arbitrary, subject to the constraints in (4.6). To encompass AIC, BIC introduce the criterion, with $\delta_0 \leq \Delta$,

$$\log\det \hat{\Sigma}_\delta + \delta C_T/T, \quad C_T > 0,\ C_T/T \to 0, \quad \delta \leq \Delta. \tag{4.7}$$

Then we have the following

THEOREM 4.1. *If* $y(t)$ *is generated by* (4.2) *with* $k_0(z) \in M(d_0)$ *and* $\hat{\delta}$ *is chosen to minimise* (4.7) *then if* (4.3), (4.6) *hold there are* $0 < c_0 < c_1 < \infty$ *so that:*

(a) *If*

$$\liminf_{T\to\infty} C_T/(2 \operatorname{loglog} T) > c_1$$

then $\hat{\delta} \to \delta_0$, *a.s. If* (4.4) *also holds we may take* $c_1 = 1$.

(b) *If*

$$\limsup_{T\to\infty} C_T/(2 \operatorname{loglog} T) < c_0$$

then a.s. convergence of $\hat{\delta}$ *to* δ_0 *does not hold. If also* (4.4) *holds we may take* $c_0 = 1$.

(c) *If*

$$C_T \to \infty \quad \text{then } \hat{\delta} \to \delta_0 \text{ in probability}.$$

(d) *If*

$$\limsup C_T < \infty \tag{4.8}$$

then $\hat{\delta}$ *does not converge to* δ_0 *in probability*.

(e) *If* (4.4) *also holds and* (4.8) *then*

$$\lim_{\eta\to 0} \lim_{T\to\infty} P\{\hat{\delta} > \delta_0\} = 1.$$

There are analogous results for the d_j. See Hannan and Deistler (1987). The theorem favours BIC. However, it may be argued that the assumption that $d_0 < \infty$ is unreasonable. Part (*e*) depends critically on the use of ML.

The case of PMDL is more complex. We discuss $n = 1$ and an autoregression (AR), namely

$$\sum_{j=0}^{p} \phi(j) y(t-j) = \varepsilon(t), \ \phi(0) = 1, \quad \sum_{j=0}^{p} \phi(j) z^j \neq 0, \ |z| \leq 1. \tag{4.9}$$

Then if p_0 is the true value, under (4.3), (4.4), taking $\delta = p$ in PMDL

$$\frac{1}{t}\sum_{1}^{t} e_p(s)^2 = \frac{1}{t}\sum_{1}^{t} \varepsilon(s)^2 + \sigma^2 \frac{p \log t}{t} \{1 + o(1)\}, \qquad p \geq p_0, \tag{4.10}$$

where the $o(1)$ quantities are of that order a.s. For t sufficiently large it is evident that (4.10) is minimised at p_0. The result (4.10) probably extends to $p_0 = \infty$, under suitable conditions. The proof will be published in another place. (The work is joint with Kavaliers, McDougall and Poskitt.) It may be shown that (4.10) is the essential component in PMDL so that PMDL, in these circumstances, consistently estimates p_0.

For the general case a key part is played by the autocovariances

$$\Gamma(t) = E\{y(s)\,y(s+t)'\},\ \hat{\Gamma}(t) = \frac{1}{T}\sum_{1}^{T-1} y(s)\,y(s+t)',\ \hat{\Gamma}(t) = 0,\ t \geq T.$$

It is convenient to introduce

$$\dot{\Gamma}(t) = \sum_{0}^{\infty} K(j)\,\dot{\Sigma}\,K(j+t),\quad \dot{\Sigma} = \frac{1}{T}\sum_{1}^{T} \varepsilon(t)\,\varepsilon(t)'. \tag{4.11}$$

If $\dot{\Sigma}$ is replaced by Σ then $\dot{\Gamma}(t)$ becomes $\Gamma(t)$.
We also need

$$\Sigma\, j^{\frac{1}{2}} \| K(j) \| < \infty,\ \sup j \| K(j) \| < \infty,\ \det\{k_0(z)\} \neq 0,\ |z| \geq 1. \tag{4.12}$$

THEOREM 4.2.

(a) *If* (4.2) *holds then*

$$\lim_{T\to 0}\ \sup_{0\leq t<\infty} \| \hat{\Gamma}(t) - \Gamma(t) \| = 0, \text{ a.s.}$$

(b) *If additionally* (4.3), (4.12) *hold then*

$$\sup_{0\leq t<\infty} \| \hat{\Gamma}(t) - \dot{\Gamma}(t) \| = 0\,\{(\log T/T)^{\frac{1}{2}}\}.$$

If (4.4) *holds* $\Gamma(t)$ *may replace* $\dot{\Gamma}(t)$ *in the above.*

In any case $\dot{\Gamma}(t)$ may be replaced by $\Gamma(t)$ to the order of accuracy with which $\dot{\Sigma}$ may replace Σ. Since $\dot{\sigma}^2$, σ^2 may be taken out as factors in (4.11), when $n = 1$, then since all calculations of system parameters are scale free then replacing $\hat{\gamma}(t)$ by $\hat{\rho}(t) = \hat{\gamma}(t)/\hat{\gamma}(0)$ and $\gamma(t)$ by $\rho(t) = \gamma(t)/\gamma(0)$ then the following is relevant:

$$\sup_{0 \le t < \infty} |\hat{\rho}(t) - \rho(t)| = 0\{(\log T/T)^{\frac{1}{2}}\}.$$

Let $\Phi_p(j)$ minimise

$$E\{\| \sum_0^p \Phi_p(j) y(t-j) \|^2\}, \quad \Phi_p(0) = I_n$$

and $\hat{\Phi}_p(j)$ satisfy

$$\sum_0^p \hat{\Phi}_p(j)\, \hat{\Gamma}(j-k) = \delta_{ok}\, \hat{\Sigma}_p\,, \; k = 0, 1, \ldots, p. \tag{4.13}$$

Put

$$k(z)^{-1} = \sum_0^\infty \Phi(j) z^{-j}, \qquad \Phi(0) = I_n.$$

THEOREM 4.3. *Under* (4.2), (4.3), (4.12), *uniformly in* $p = 0\{(T/\log T)^{\frac{1}{2}}\}$,

$$\max \|\hat{\Phi}_p(j) - \Phi_p(j)\| = 0\{(\log T/T)^{\frac{1}{2}}\} + o(1) \sum_{p+1}^\infty \|\Phi(j)\|. \tag{4.14}$$

When $n = 1$ *or when* (4.4) *holds the* $o(1)$ *term is* $0\{(\log T/T)^{\frac{1}{2}}\}$.

The second factor in the last term in (4.14) is $o(p^{-\frac{1}{2}})$. (See Hannan and Kavalieris, 1986.) The $o(1)$ term is $0\{(\log T)^{\delta}/T)^{\frac{1}{2}}\}$, $\delta > \frac{1}{2}$, under rather general circumstances. Since p will be $0(\log T)$, at least when chosen by BIC unless $y(t)$ is an AR then it is the first term that matters, on the right in (4.14).

We conclude with a result concerning

$$\hat{\varepsilon}_p(t) = \sum_0^p \hat{\Phi}_p(j)\, y(t-j).$$

As we shall see

$$\liminf_{T \to \infty} p_T/(\log\log T)^2 = \infty \tag{4.15}$$

seems no constraint in relation to BIC nor would

$$\liminf_{T\to\infty} p_T/(\log T)^2 > 0 \tag{4.16}$$

unless $y(t)$ is ARMA. We write $o_p(\cdot)$ for an order relation holding in probability. We write $\hat{\varepsilon}_T(t)$ for $\hat{\varepsilon}_{p_T}(t)$ when p depends on T.

THEOREM 4.4. *If* (4.2), (4.3), (4.4), (4.12), (4.15) *hold*

$$\frac{1}{T}\sum_1^T \{\hat{\varepsilon}_T(t)-\varepsilon(t)\}\{\hat{\varepsilon}_T(t)-\varepsilon(t)\}'$$

$$= (\mathbf{\Sigma}_{p_T} - \mathbf{\Sigma} - \frac{n\,p_T}{T}\mathbf{\Sigma})\{1+o(1)\} + o_p(p_T/T).$$

If (4.16) *holds then* $o_p(p_T/T)$ *becomes* $o(p_T/T)$.

This result may be used to show, for example, that if

$$tr\{\mathbf{\Sigma}^{-1}(\mathbf{\Sigma}_p - \mathbf{\Sigma})\} \sim c_0 p^{-\beta},\ \beta > 1$$

then if p_T is obtained by BIC

$$p_T = \{c\beta T/(n^2 \log T)\}^{1/(1+\beta)}\{1+o(1)\}.$$

See Shibata (1980) for similar considerations.

No doubt much or all of theorems (4.3), (4.4) can be extended to ARMAX approximation.

5. RECURSIVE CALCULATIONS

One of the important developments in time series in recent years is that of recursive methods for estimating ARMAX systems, the work being mainly due to people in systems and control. Two recent books are Young (1984), Ljung and Söderström (1983). For brevity we deal only with $n=1$ and the ARMA case. The extension to the ARMAX case, and to the case $n > 1$ to a large extent, is not very complex.

All procedures effectively proceed via autoregressions, possibly a number run in parallel. For example consider

$$y(t) = -\sum_1^p \alpha(j)y(t-j) + \sum_1^p \beta(j)\,\varepsilon(t-j) + \varepsilon(t). \tag{5.1}$$

This may be converted to a regression by replacing $\varepsilon(t-j)$, $j > 0$, by an estimate. This could be got from a parallel calculation, using an AR, or from estimates $\alpha_s(j)$, $\beta_s(j)$ at the earlier true point, $s = t-j$.

The autoregression may be calculated recursively as follows, putting τ_t for the estimate of the vector of coefficient τ:

$$\tau_t = \tau_{t-1} + P(t)\,x(t)\,e(t), \tag{5.2}$$

$$P(t)^{-1} = \sum_1^t x(s)x(s)', \quad e(t) = y(t) - \tau'_{t-1}x(t), \tag{5.3}$$

$$P(t) = P(t-1) - \{1 + x(t)'P(t-1)x(t)\}^{-1}P(t-1)x(t)x(t)'P(t-1). \tag{5.4}$$

Here $x(t)$, τ must be prescribed. For an AR (4.9), $x(t)' = (-y(t-1),\dots,-y(t-p))$, $\tau' = (\phi(1),\dots,\phi(p))$. For (5.1) with $\hat{\varepsilon}(t-j)$ replacing $\varepsilon(t-j)$ we have $x(t)' = (-y(t-1),\dots,-y(t-p),\ \hat{\varepsilon}(t-1),\dots,\hat{\varepsilon}(t-p))$, $\tau' = (\alpha(1),\dots,\alpha(p),\beta(1),\dots,\beta(p))$. Here we could take $\hat{\varepsilon}(t-j) = e(t-j)$, from (5.3) or, slightly better, $\hat{\varepsilon}(t-j) = y(t-j) - \tau(t-j)'x(t-j)$. A third case is that where τ is as in the last example but $x(t)' = (\eta(t-1),\dots,\eta(t-p),\ \xi(t-1),\dots,\xi(t-q))$

$$\left.\begin{aligned}
\eta(t) &= -\sum_1^q \beta_t(j)\eta(t-j) + y(t),\quad y(t), \eta(t) = 0, t \le 0\\
\xi(t) &= -\sum_1^q \beta_t(j)\xi(t-j) + \hat{\varepsilon}(t),\quad \xi(t), \hat{\varepsilon}(t) = 0, t \le 0\\
\hat{\varepsilon}(t) &= -\sum_1^q \beta_t(j)\hat{\varepsilon}(t-j) + \sum_0^p \alpha_t(j)y(t-j),\quad \hat{\varepsilon}(t), y(t) = 0, t \le 0.
\end{aligned}\right\} \tag{5.5}$$

In this last case the dependent variable in the regression is actually $\eta(t) - \xi(t) + \hat{\varepsilon}(t)$, but does not appear explicitly. It would also be possible to find $\eta(t)$, $\xi(t)$, $\hat{\varepsilon}(t)$ from a parallel calculation, e.g. that corresponding to (5.1) described as the second example.

There is a large literature concerning the convergence of these algorithms. We refer to the literature cited above for details. One technique is to first calculate an AR, of order determined by BIC at time t. This is used to provide $\hat{\varepsilon}(t)$ for a parallel estimate of (5.1). Then the output of this is used to form (5.5) which is used in a final form of (5.2), (5.3), (5.4), again done in parallel, to produce the final estimates. If $y(t)$ is generated by an ARMA process and (4.3), (4.4) hold it is shown in Hannan, Kavalieris and Mackisack (1986) that the τ_t from the final stage converges a.s. to τ_0, the true value and $t^{\frac{1}{2}}(\tau_t - \tau_0)$ is asymptotically normal with zero mean vector and covariance matrix that for the central limit result for the ML estimator on Gaussian assumptions.

Again this theorem is of only suggestive value. For one thing it may be shown, as at the end of Section 4, that BIC will give an order for the AR that is $c_0 \log T\{1+o(1)\}$, where $c_0^{-1} = -2 \log \rho_0$ and ρ_0 is the modulus of the zero of $k_0(z)$ nearest to $|z| = 1$. Thus eventually the algorithm cannot run in real time. However that is largely irrelevant because in almost any case where the algorithms of this section are used the far past will be forgotten so that the algorithm will more readily adapt to a changing situation. Thus the data vector will be multiplied by $\ell_t(s)^{\frac{1}{2}}$, $s = 1, 2, \ldots, t$, when

$$\ell_t(s) = a_s \prod_{s+1}^{t} \lambda(n), \qquad \ell_t(t) = a_t.$$

In practice the $\lambda(n)$ might eventually settle down at some chosen value, $\lambda(n) = \lambda$, $0 < \lambda < 1$. The net effect on (5.2) to (5.3) is to replace (5.4) by

$$P(t) = \frac{1}{\lambda(t)}\left[P(t-1) - \{\lambda(t) + x(t)' P(t-1)x(t)\}^{-1} P(t-1)x(t)x(t)' P(t-1)\right]$$

and to insert a factor a_t in the second term in (5.2). When using BIC etc. we need to redefine t used in defining $\hat{\Sigma}_\delta(t)$, (see section 3), (4.10), and in the term $\log t/t$ in BIC, for example. We replace t by $f(t)$

$$f(t+1) = \sum_{1}^{t+1} \ell_t(s) = \lambda(t+1)f(t) + a_{t+1}, \qquad f(0) = 0.$$

However, when PMDL is used with forgetting, simulations and theory suggest the $\log f(t)$ term disappears from the right in (4.10) and it seems better to use BIC, adjusted to replace t by $f(t)$, since the PMDL formula seems too adaptive.

One can of course use these procedures to choose p as well as the AR order by criteria such as BIC. An algorithm for that purpose is described in Hannan, Kavalieris and Mackisack (1986), p being bounded above by some fixed value P. Again, with forgetting, this bounding does not seem important.

We conclude by pointing out that an autoregression may also be calculated by some very fast algorithms, of great importance, called lattice algorithms. The regression (5.1), with $\hat{\varepsilon}(t)$ replacing $\varepsilon(t)$, may be viewed as the first row of a vector autoregression, of

$$\begin{pmatrix} y(t) \\ \hat{\varepsilon}(t) \end{pmatrix}$$

on lagged values of itself. The third stage of the algorithm in Hannan, Kavalieris and Mackisack (1986) is not quite of the same nature since it is $\eta(t) + \xi(t) - \hat{\varepsilon}(t)$ that is being regressed on lagged values of

$$\begin{pmatrix} \eta(t) \\ \xi(t) \end{pmatrix}. \tag{5.6}$$

However, if the vector autoregression of this upon lagged values of itself is effected then the backwards residuals (see below) allow a recursive calculation of the regression of $\eta(t) + \xi(t) - \hat{\varepsilon}(t)$ on lagged values of (5.6) to be carried out by the obvious device of making the last vector, $(\eta(t-p)$, $\xi(t-p))'$ in the regression orthogonal to earlier ones. We omit details and refer the reader to Hannan and Deistler (1987). Thus in all of the calculations above a key part is played by a vector autoregression. The so-called Toeplitz calculation (4.13) leads to simple procedures for increasing the order p. (See Whittle, 1963.) This follows by considering the array

$$\begin{array}{c|cccc}
y(1) & 0 & \cdot & \cdot & 0 \\
y(2) & y(1) & \cdot & \cdot & 0 \\
\vdots & \vdots & & & \vdots \\
y(p) & y(p-1) & \cdot & \cdot & 0 \\
\vdots & \vdots & & & \vdots \\
y(t) & y(t-1) & \cdot & \cdot & y(t-p) \\
0 & y(t) & & & y(t-p+1) \\
\vdots & \vdots & & & \vdots \\
0 & 0 & & & y(t)
\end{array}$$

When p is increased by 1 then another column is added and previous columns do not change. Moreover the covariance matrix of the vectors to the right of the vertical line is Toeplitz, i.e., has all elements the same down any diagonal. However when t increases by unity not only is another row added but the $p-1$ previous rows also change. The standard lattice algorithms in Friedlander (1982) use only the first t rows of the array at time t. This new array is said to be pre-windowed but not post windowed, the terminology referring to a metaphorical window edge that prevents us from seeing before $t = 1$ and forces $y(t)=0$, $t \leq 0$. Though the order increase is thereby made a little more complex the time update is much easier.

We shall not describe these lattice calculations in detail. (See Friedlander, 1982, for example, and references therein.) They proceed by directly calculating forwards and backwards prediction errors, $\varepsilon_p(t)$, $r_p(t)$ according to the formulae

$$\varepsilon_{p+1}(t) = \varepsilon_p(t) - K^r_{p+1}(t) r_p(t-1), \quad r_{p+1}(t) = r_p(t-1) - \{K^\varepsilon_{p+1}(t)\}' \varepsilon_p(t) \qquad (5.7)$$

with $\varepsilon_0(t) = r_0(t) = y(t)$, $r_p(0) \equiv 0$. The $K^r_p(t)$ and $K^\varepsilon_p(t)$ are called backwards and forwards matrices of reflection coefficients. However $K^r_{p+1}(t)$ corresponds to $\phi_{p+1}(p+1)$, in the notation of (4.13), and would be regarded as a forwards coefficient by statisticians. $\{K^\varepsilon_{p+1}(t)\}'$ would be the backwards coefficient. The algorithms are easily adjusted to allow for forgetting. The $r_p(t)$ are the "prediction" errors for the time reversed process and thus constitute residuals from the regression of $y(t-p-1)$ on $y(t-1),\ldots,y(t-p)$. [See the discussion below (5.6) to see how they may be useful.] The calculations do not necessarily involve computing the autoregressive coefficient matrices. [The $\hat{\delta}_p(j)$ are the forwards matrices for the calculation in (4.14).] These are called the predictor coefficients in the jargon of the subject. If these are not computed the number of operations to time t for $p \leq P$ is cPt, i.e., linear in P and t. If the predictor coefficients are needed another c_1P^2 calculations must be done at each time from t for which they are calculated. (This is effectively a back-substitution.) The algorithms are very fast and very important. It may be noted however that for (5.5), for example, inverse filterings are required, i.e., $\eta(t)$, for example, is got from $y(t)$ by a filter with transfer function $(\Sigma\beta_t(j)z^{-j})^{-1}$. This can be done by reversing part of the basic lattice calculation (see Friedlander, 1982) but requires p to be fixed. If the calculation is to be done for all p, $0 \leq p \leq P$, then the predictor coefficients are needed.

REFERENCES

Brillinger, D. (1984). *The Collected Works of John Tukey, Vol. II Time Series, 1965-1984*. Ed. D.R. Brillinger, Wadsworth, Monterey.

Findley, D.F. (1985). On the unbiasedness property of AIC for exact or approximating linear stochastic time series models. *J. Time Series Anal.*, 6, 229-252.

Friedlander, B. (1982). Lattice filters for adaptive processing. *Proc. I.E.E.E.*, 70, 830-867.

Glover, K. (1984). All optimal Hankel-norm approximations of linear multivariate systems and their L^∞ error bounds. *Int. J. Control*, 39, 1115-1193.

Hannan, E.J. (1987). Rational transfer function approximation, *Statistical Science*, 2.

Hannan, E.J. and Deistler, M. (1987). *The Statistical Theory of Linear Systems*. Wiley, New York. (To be published.)

Hannan, E.J. and Kavalieris, L. (1986). Regression, autoregression models. *J. Time Series Anal.*, 7, 27-50.

Hannan, E.J., Kavalieris, L. and Mackisack, M.(1986). Recursive estimation of linear systems. *Biometrika*, 73, 119-134.

Ljung, L. and Söderström, T. (1983). *Theory and Practice of Recursive Identification*. The MIT Press, Cambridge, Massachusetts.

Rissanen, J.(1978). Modeling by shortest data description. *Automatica*, 14,465-471.

Rissanen, J. (1983). A universal prior for integers and estimation by minimum description length. *Ann. Statist.*, 11, 416-431.

Rissanen, J. (1986). Stochastic complexity and modeling. *Ann. Statist.*, 14, 1080-1100.

Shibata, R. (1980). Asymptotically efficient estimation of the order of the models for estimating parameters of a linear process. *Ann. Statist.*, 8, 147-164.

Young, P.C. (1984). *Recursive Estimation and Time Series Analysis*. Springer-Verlag, Berlin.

Received 10 February 1987

Department of Statistics
Faculty of Economics and Commerce
Australian National University
GPO Box 4, Canberra, A.C.T. 2601
Australia

DISCUSSION

Knut CONRADSEN

Technical University of Denmark, Lyngby, Denmark

I am very pleased to have been asked to discuss Professor Hannan's very interesting paper. In my discussion I should like to concentrate on some aspects of different parametrizations and relate this to some recent results obtained by other authors, some of which were pointed out to me by Professor Jan Holst.

As Professor Hannan states, time series problems are not problems of estimating a finite number of parameters in some known, true structure, but are rather problems of finding a suitable approximation to some aspect of the stochastic process generating the data. A main problem is often the assumed linearity. It is suggested to view the linear structure as local in time and seek to adapt that structure to the changing local situation. This is an important issue and my comments are dependent on this.

The model set used by Professor Hannan is that of rational transfer functions. The description is given via the Hankel matrix and by selecting a set of rows that are spanning all rows we obtain the so-called canonical representation. Now, this approach is not without problems. A disadvantage of using the canonical representation is that the Kronecker invariants must be determined (from the data) and the estimation of these invariants is critical, since the parameter estimates are not consistent if the structural invariants have been wrongly assessed (Caines and Rissanen, 1974).

The parametrization problem is equivalent to cover a set (related to the set of systems of order n) with coordinate systems that may or may not overlap. The non-overlapping case corresponds to the canonical forms. An alternative to this parametrization would be to use the overlapping systems. Now, as Gevers and Wertz (1984) point out with respect to the overlapping forms, each local coordinate system will in general cover a dense subset, wherefore almost any choice of structure indices will normally produce a parametrization allowing an exact description of the process. This again enables an on-line structure selection for the multivariable state space models (van Overbeek and Ljung, 1982).

In Section 2 Professor Hannan later mentions that the matrix fraction description is a more useful parametrization for statistical purposes. The state space models and the matrix fraction description models are most widely used in control. In econometrics and in system identification it is often more natural to use ARMA or ARMAX models. These models are connected to MFD models in a very obvious way, and one should expect that the properties of canonical MFD models should apply to canonical ARMA models. This, however, is not so. Gevers (1986) points out some problems with canonical ARMA models obtained from a canonical matrix fraction description. These problems include a lack of connection between certain parameters and the Kronecker indices. Furthermore it will generally not be possible to construct a monic ARMA model such that the number of free parameters corresponds to the dimension of the analytic manifold of matrix functions with the given Kronecker indices. This sums up to drawbacks that could strongly limit the potential use of ARMA-models in identification and control. Now, the canonical matrix fraction description involves a basis selection rule that corresponds to the idea of constructing the present state from the future outputs. As Bokor and Keviczky (1987) mention, the use of the alternative constructibility invariants corresponds to constructing the present state of a system from the most recent past outputs. This inversion requires that there are no poles at the origin, but on the other hand, the disadvantages mentioned by Gevers (1986) disappear.

The main objective of these remarks have been to emphasize that canonical parametrizations treated in this paper are not canonical in the sense that they represent the only natural (= useful) parametrizations when one is analysing linear systems. I am sure that Professor Hannan's paper will give inspiration to a lot of work and investigations in this important field. Before one starts, a good idea may be to read the forthcoming book by Professors Deistler and Hannan (1987) which I am sure will give the solution to many of the problems that can not be dealt with in the limited space that a single paper provides. In conclusion, it gives me very great pleasure to propose a strong vote of thanks to Professor Hannan for his very stimulating and interesting paper.

REFERENCES

Bokor, J. and Keviczky, L. (1987). ARMA canonical forms obtained from constructibility invariants. *International Journal of Control*, 45, 861-873.

Caines, P.E. and Rissanen, J. (1974). Maximum likelihood estimation of parameters in multivariate Gaussian stochastic processes. *IEEE Transactions on Information Theory*, IT-20, 102-104.

Deistler, M. and Hannan, J. (1987). *The Statistical Theory of Linear Systems.* Wiley, New York. To appear.

Gevers, M. (1986). ARMA models, their Kronecker indices and their McMillan degree. *International Journal of Control*, 43, 1745-1761.

Gevers, M. and Wertz, V. (1984). Uniquely identifiable state-space and ARMA parametrizations for multivariable linear systems. *Automatica*, 20, 331-347.

van Overbeek, A.J.M. and Ljung, L. (1982). On-line structure selection for multivariable state-space models. *Automatica*, 18, 529-543.

Received 17 August 1987

IMSOR
The Technical University of Denmark
DK-2800 Lyngby
Denmark

Proc. Second International Tampere Conference in Statistics
(Tampere, Finland, 1-4 June 1987)
Tarmo Pukkila and Simo Puntanen, *Editors*

pp. 73 - 98

KEYNOTE ADDRESS

Estimation in linear models with mixed effects: a unified theory

C. Radhakrishna RAO
University of Pittsburgh, Pittsburgh, PA, U.S.A.

Key words and phrases: Inverse partitioned matrix method, genetic selection, MINQE, mixed Gauss-Markoff model, simultaneous estimation, simultaneous prediction, variance and covariance components.

ABSTRACT

A unified theory without any assumptions on the ranks of the matrices involved is developed for the estimation of a linear function of unknown parameters ($\boldsymbol{\beta}$) and hypothetical variables ($\boldsymbol{\xi}$) in a mixed Gauss-Markoff linear model, $\mathbf{Y} = \mathbf{X}\boldsymbol{\beta} + \mathbf{U}\boldsymbol{\xi} + \boldsymbol{\varepsilon}$. The expressions for the estimates and the mean square errors of the estimates depend on the elements of an inverse partitioned matrix as in the case of the Gauss-Markoff model with fixed effects developed earlier by the author (Rao 1971, 1985).

The general theory is applied to several special problems that arise in social and biological sciences. In particular, explicit expressions are obtained for simultaneous estimation or prediction in several similar and dissimilar linear models with or without concomitant variables and in special univariate and multivariate linear models in genetic selection.

1. INTRODUCTION

The Gauss-Markoff model with some fixed and some random effects, called the mixed effects model, is written in the form

$$\mathbf{Y} = \mathbf{X}\boldsymbol{\beta} + \mathbf{U}\boldsymbol{\xi} + \boldsymbol{\varepsilon}, \tag{1.1}$$

where $\mathbf{Y}$ is an n-vector of observations, $\mathbf{X}$ is a given $n \times m$ matrix, $\boldsymbol{\beta}$ is an m-vector of unknown fixed parameters, $\mathbf{U}$ is a given $n \times p$ matrix, $\boldsymbol{\xi}$ is a p-vector of hypothetical random variables and $\boldsymbol{\varepsilon}$ is an n-vector of unknown error variables. We make the following assumptions on the first and second order moments of $\boldsymbol{\xi}$ and $\boldsymbol{\varepsilon}$

$$E(\boldsymbol{\xi}) = \mathbf{A}\boldsymbol{\gamma},\ \ E(\boldsymbol{\varepsilon}) = \mathbf{0},\ \ D(\boldsymbol{\xi}) = \boldsymbol{\Gamma},\ \ D(\boldsymbol{\varepsilon}) = \boldsymbol{\Sigma},\ \ cov(\boldsymbol{\xi}, \boldsymbol{\varepsilon}) = \mathbf{0}, \tag{1.2}$$

where $\mathbf{A}$ is a $p \times r$ given matrix, E stands for the expectation and D for the dispersion (variance-covariance) operators. Such a model has been considered by various authors. For a historical account of the model, methods of estimation of the unknown parameters and prediction of unobserved variables, the reader is referred to the preface in a forthcoming book by Kleffe and Rao (1987). In this paper, we refer to (1.1) - (1.2) as the GM(M) model where M within the brackets stands for mixed effects. The corresponding model with fixed effects will be referred to as the GM(F) when a distinction has to be made or simply as the GM model.

We develop a unified theory for the estimation of $\boldsymbol{\beta}$ and the prediction of $\boldsymbol{\xi}$ when the other parameters $\boldsymbol{\gamma}$, $\boldsymbol{\Gamma}$ and $\boldsymbol{\Sigma}$ in the GM(M) model, (1.1)-(1.2), are partly known or completely unknown. The theory is similar to the one developed by the author for the GM(F) model without making any assumptions on the ranks of the matrices involved (Rao, 1971, 1972, 1973, pp. 298-300). We consider special cases which arise in practice and provide satisfactory solutions.

2. UNIFIED THEORY FOR THE GM(M) MODEL

The GM(M) model we consider is

$$\mathbf{Y} = \mathbf{X}\boldsymbol{\beta} + \mathbf{U}\boldsymbol{\xi} + \boldsymbol{\varepsilon} \tag{2.1}$$

with the first and second order moments

$$E(\boldsymbol{\varepsilon}) = \mathbf{0},\ \ D(\boldsymbol{\varepsilon}) = \boldsymbol{\Sigma},\ \ E(\boldsymbol{\xi}) = \mathbf{A}\boldsymbol{\gamma},\ \ D(\boldsymbol{\xi}) = \boldsymbol{\Gamma},\ \ cov(\boldsymbol{\varepsilon}, \boldsymbol{\xi}) = \mathbf{0}, \tag{2.2}$$

where $\boldsymbol{\gamma}$ and $\boldsymbol{\beta}$ are unknown parameters. The problem is that of estimating a linear function

$$\mathbf{p}'\boldsymbol{\beta} + \mathbf{q}'\boldsymbol{\xi} \tag{2.3}$$

of the fixed parameter $\boldsymbol{\beta}$ and the random variable $\boldsymbol{\xi}$ by a linear function $a+\mathbf{L}'\mathbf{Y}$ of $\mathbf{Y}$ such that

$$E(a+\mathbf{L}'\mathbf{Y}-\mathbf{p}'\boldsymbol{\beta}-\mathbf{q}'\boldsymbol{\xi}) = 0, \text{ independently of } \boldsymbol{\gamma} \text{ and } \boldsymbol{\beta}, \tag{2.4}$$

and

$$E(a+\mathbf{L}'\mathbf{Y}-\mathbf{p}'\boldsymbol{\beta}-\mathbf{q}'\boldsymbol{\xi})^2 \quad \text{is a minimum.} \tag{2.5}$$

The condition (2.4) leads to the equation

$$a+\mathbf{L}'(\mathbf{X}\boldsymbol{\beta}+\mathbf{U}\mathbf{A}\boldsymbol{\gamma}) = \mathbf{p}'\boldsymbol{\beta}+\mathbf{q}'\mathbf{A}\boldsymbol{\gamma} \qquad \forall\, \boldsymbol{\beta}, \boldsymbol{\gamma}$$

which is equivalent to

$$a = 0, \quad \mathbf{X}'\mathbf{L} = \mathbf{p}, \quad \mathbf{A}'\mathbf{U}'\mathbf{L} = \mathbf{A}'\mathbf{q}. \tag{2.6}$$

Thus the unbiasedness condition (2.4) holds iff there exist a vector $\mathbf{L}$ satisfying the equations (2.6), or the vector $\mathbf{s} = (\mathbf{p}' : \mathbf{q}'\mathbf{A})'$ belongs to $R(\mathbf{S}')$, the column space of $\mathbf{S}' = (\mathbf{X} : \mathbf{U}\mathbf{A})'$. When (2.6) holds

$$E(a+\mathbf{L}'\mathbf{Y}-\mathbf{p}'\boldsymbol{\beta}-\mathbf{q}'\boldsymbol{\xi})^2 = \mathbf{L}'\mathbf{G}\mathbf{L}-2\mathbf{q}'\boldsymbol{\Gamma}\mathbf{U}'\mathbf{L}+\mathbf{q}'\boldsymbol{\Gamma}\mathbf{q} \tag{2.7}$$

where $\mathbf{G} = \boldsymbol{\Sigma}+\mathbf{U}\boldsymbol{\Gamma}\mathbf{U}'$. The minimum of (2.7) subject to (2.6) is attained when $\mathbf{L}$ satisfies the equations

$$\begin{pmatrix} \mathbf{G} & \mathbf{S} \\ \mathbf{S}' & \mathbf{0} \end{pmatrix}\begin{pmatrix} \mathbf{L} \\ \boldsymbol{\lambda} \end{pmatrix} = \begin{pmatrix} \mathbf{U}\boldsymbol{\Gamma}\mathbf{q} \\ \mathbf{s} \end{pmatrix} \tag{2.8}$$

which is the same type of equation which occurs in the unified theory of linear estimation in the GM(F) model given in Rao (1971; 1973, pp. 298-300; 1985, p. 24).

Let

$$\begin{pmatrix} \mathbf{G} & \mathbf{S} \\ \mathbf{S}' & \mathbf{0} \end{pmatrix}^{-} = \begin{pmatrix} \mathbf{C}_1 & \mathbf{C}_2 \\ \mathbf{C}_3 & -\mathbf{C}_4 \end{pmatrix} \tag{2.9}$$

for any choice of the g-inverse. Then one solution for $\mathbf{L}$ is

$$\mathbf{L}_* = \mathbf{C}_1\mathbf{U}\boldsymbol{\Gamma}\mathbf{q}+\mathbf{C}_2\mathbf{s} \tag{2.10}$$

leading to the optimum estimate of $\mathbf{p}'\boldsymbol{\beta}+\mathbf{q}'\boldsymbol{\xi}$

$$\mathbf{s}'\mathbf{C}_2{}'\mathbf{Y}+\mathbf{q}'\boldsymbol{\Gamma}\mathbf{U}'\mathbf{C}_1{}'\mathbf{Y}. \tag{2.11}$$

Writing $\mathbf{C}_2 = (\mathbf{C}_\beta : \mathbf{C}_\gamma)$, the expression (2.11), which is the estimate of $\mathbf{p}'\boldsymbol{\beta}+\mathbf{q}'\boldsymbol{\xi}$, can be written as

$$\mathbf{p}'\hat{\boldsymbol{\beta}}+\mathbf{q}'\hat{\boldsymbol{\xi}} = \mathbf{p}'\hat{\boldsymbol{\beta}}+\mathbf{q}'\mathbf{A}\hat{\boldsymbol{\gamma}}+\mathbf{q}'\hat{\boldsymbol{\eta}}, \tag{2.12}$$

where

$$\hat{\boldsymbol{\beta}} = \mathbf{C}_\beta{}'\mathbf{Y}, \quad \hat{\boldsymbol{\gamma}} = \mathbf{C}_\gamma{}'\mathbf{Y}, \quad \hat{\boldsymbol{\eta}} = \boldsymbol{\Gamma}\mathbf{U}'\mathbf{C}_1{}'\mathbf{Y}, \quad \hat{\boldsymbol{\xi}} = \mathbf{A}\hat{\boldsymbol{\gamma}}+\hat{\boldsymbol{\eta}}. \tag{2.13}$$

Let

$$\mathbf{C}_4 = \begin{pmatrix} \mathbf{C}_{\beta\beta} & \mathbf{C}_{\beta\gamma} \\ \mathbf{C}_{\gamma\beta} & \mathbf{C}_{\gamma\gamma} \end{pmatrix}. \tag{2.14}$$

Then we have the following results analogous to those in the unified theory for the GM(F) model as discussed in Rao (1971; 1973, pp. 298-300; 1985, p. 24).

(1) The best estimate of $\mathbf{p}'\boldsymbol{\beta}$ alone when estimable, i.e., when (2.6) holds with $\mathbf{q} = \mathbf{0}$, is $\mathbf{p}'\hat{\boldsymbol{\beta}}$ where $\hat{\boldsymbol{\beta}}$ is as in (2.13) and

$$E(\mathbf{p}'\hat{\boldsymbol{\beta}} - \mathbf{p}'\boldsymbol{\beta})^2 = \mathbf{p}'\mathbf{C}_{\beta\beta}\mathbf{p}. \tag{2.15}$$

(2) The best estimate of $\mathbf{r}'\boldsymbol{\gamma}$ when estimable, i.e., when (2.6) holds with $\mathbf{p} = \mathbf{0}$ and $\mathbf{q} = \mathbf{r}$, is $\mathbf{r}'\hat{\boldsymbol{\gamma}}$ where $\hat{\boldsymbol{\gamma}}$ is as in (2.13) and

$$E(\mathbf{r}'\hat{\boldsymbol{\gamma}} - \mathbf{r}'\boldsymbol{\gamma})^2 = \mathbf{r}'\mathbf{C}_{\gamma\gamma}\mathbf{r}. \tag{2.16}$$

(3) The best estimate of $\mathbf{p}'\boldsymbol{\beta}+\mathbf{r}'\boldsymbol{\gamma}$ is $\mathbf{p}'\hat{\boldsymbol{\beta}}+\mathbf{r}'\hat{\boldsymbol{\gamma}}$ and

$$E(\mathbf{p}'\hat{\boldsymbol{\beta}}+\mathbf{r}'\hat{\boldsymbol{\gamma}} - \mathbf{p}'\boldsymbol{\beta} - \mathbf{r}'\boldsymbol{\gamma})^2 = (\mathbf{p}' : \mathbf{r}')\mathbf{C}_4(\mathbf{p}' : \mathbf{r}')'. \tag{2.17}$$

(4) The best predictor of $\mathbf{q}'\boldsymbol{\xi}$ is $\mathbf{q}'\hat{\boldsymbol{\xi}}$ where $\hat{\boldsymbol{\xi}}$ is as in (2.13) and

$$E(\mathbf{q}'\hat{\boldsymbol{\xi}} - \mathbf{q}'\boldsymbol{\xi})^2 = \mathbf{q}'(\mathbf{A}\mathbf{C}_{\gamma\gamma}\mathbf{A}' + \boldsymbol{\Gamma} - \boldsymbol{\Gamma}\mathbf{U}'\mathbf{C}_1{}'\mathbf{U}\boldsymbol{\Gamma} - 2\mathbf{A}\mathbf{C}_\gamma{}'\mathbf{U}\boldsymbol{\Gamma})\mathbf{q}. \tag{2.18}$$

(5) The best estimate of $\mathbf{p}'\boldsymbol{\beta}+\mathbf{q}'\boldsymbol{\xi}$ is

$$\begin{aligned}\mathbf{p}'\hat{\boldsymbol{\beta}}+\mathbf{q}'\hat{\boldsymbol{\xi}} &= \mathbf{p}'\hat{\boldsymbol{\beta}}+\mathbf{q}'\mathbf{A}\hat{\boldsymbol{\gamma}}+\mathbf{q}'\hat{\boldsymbol{\eta}} \\ &= \mathbf{s}'\begin{pmatrix}\hat{\boldsymbol{\beta}} \\ \hat{\boldsymbol{\gamma}}\end{pmatrix}+\mathbf{q}'\hat{\boldsymbol{\eta}},\end{aligned} \tag{2.19}$$

where $\mathbf{s}' = (\mathbf{p}' : \mathbf{q}'\mathbf{A})$, and

$$E(\mathbf{p}'\hat{\boldsymbol{\beta}}+\mathbf{q}'\hat{\boldsymbol{\xi}}-\mathbf{p}'\boldsymbol{\beta}-\mathbf{q}'\boldsymbol{\xi})^2 = \mathbf{s}'\mathbf{C}_4\mathbf{s}+\mathbf{q}'(\boldsymbol{\Gamma}-\boldsymbol{\Gamma}\mathbf{U}'\mathbf{C}_1{}'\mathbf{U}\boldsymbol{\Gamma})\mathbf{q}-2\mathbf{s}'\mathbf{C}_2{}'\mathbf{U}\boldsymbol{\Gamma}\mathbf{q}. \tag{2.20}$$

The results (2.15) - (2.20) depend only on the elements of the inverse partitioned matrix in (2.9) and are derived in the same way as in the unified theory for the GM(F) model (see Rao 1973, pp. 298-300). Thus we have a complete theory for the GM(M) model in the most general case without making any assumptions on the ranks of the matrices involved.

NOTE 1. The estimates given in (2.13) can be obtained directly by solving the equations

$$\begin{pmatrix} \mathbf{G} & \mathbf{X} & \mathbf{UA} \\ \mathbf{X}' & \mathbf{0} & \mathbf{0} \\ \mathbf{A}'\mathbf{U}' & \mathbf{0} & \mathbf{0} \end{pmatrix} \begin{pmatrix} \hat{\boldsymbol{\alpha}} \\ \hat{\boldsymbol{\beta}} \\ \hat{\boldsymbol{\gamma}} \end{pmatrix} = \begin{pmatrix} \mathbf{Y} \\ \mathbf{0} \\ \mathbf{0} \end{pmatrix}, \tag{2.21}$$

where $\hat{\boldsymbol{\beta}}$ and $\hat{\boldsymbol{\gamma}}$ are the estimates of $\boldsymbol{\beta}$ and $\boldsymbol{\gamma}$ as given in (2.13) and $\hat{\boldsymbol{\eta}} = \boldsymbol{\Gamma}\mathbf{U}'\hat{\boldsymbol{\alpha}}$.

The equations (2.21) are the most general and fundamental in the analysis of mixed linear models. An important feature is that they involve only the given matrices $\mathbf{G} = \boldsymbol{\Gamma}\mathbf{U}\boldsymbol{\Gamma}'+\boldsymbol{\Sigma}$, $\mathbf{X}$, $\mathbf{A}$ and $\mathbf{Y}$ without making further computations on them.

In the fixed effects case, the corresponding equations are

$$\begin{pmatrix} \mathbf{G} & \mathbf{X} \\ \mathbf{X}' & \mathbf{0} \end{pmatrix} \begin{pmatrix} \hat{\boldsymbol{\alpha}} \\ \hat{\boldsymbol{\beta}} \end{pmatrix} = \begin{pmatrix} \mathbf{Y} \\ \mathbf{0} \end{pmatrix}, \tag{2.21$'$}$$

where $\mathbf{G} = \boldsymbol{\Sigma}$, which is obtained by putting $\mathbf{U} = \mathbf{0}$ in (2.21). The solution for $\hat{\boldsymbol{\beta}}$ provides the BLUE of $\boldsymbol{\beta}$ in the fixed effects case.

The method of solving (2.21) depends on the complexity of the matrices $\mathbf{U}, \boldsymbol{\Gamma}$ and $\boldsymbol{\Sigma}$. When $\boldsymbol{\Gamma}$ and $\boldsymbol{\Sigma}$ are non-singular, the equations (2.21) can be written in an alternative form such as that of Henderson's mixed model equations. [In Henderson's model $\mathbf{A} = \mathbf{0}$ and we get exactly the same equations as those given in Henderson (1984, p. 19).] The equations (2.21) and its reduced form (2.21)′ for the fixed effects model cover the most general cases when nothing is assumed about the ranks of the matrices and have the simplicity of being expressed directly in terms of given matrices. Further, in the process of solving the equations we also obtain the expressions for the mean square errors of the estimates as in the equations (2.15) - (2.20) of this paper and (2.59) - (2.60) of the previous paper, Rao (1985) on fixed effects.

It is interesting to note that in the fixed effects case, once we obtain the solutions $\hat{\boldsymbol{\alpha}}$ and $\hat{\boldsymbol{\beta}}$ from the equations (2.21)′, the estimates of the unknowns $\mathbf{X}\boldsymbol{\beta}$ and $\boldsymbol{\varepsilon}$ in the Gauss-Markoff model $\mathbf{Y} = \mathbf{X}\boldsymbol{\beta} + \boldsymbol{\varepsilon}$ are given by $\mathbf{X}\hat{\boldsymbol{\beta}}$ and $\mathbf{G}\hat{\boldsymbol{\alpha}}$ respectively. In this sense (2.21)′ may be more appropriately called normal equations, and not the one derived from (2.21)′ by eliminating $\boldsymbol{\alpha}$. Thus, the appropriate normal equations in the mixed effects case are (2.21).

NOTE 2. Harville (1976) and Henderson (1950) considered the case where $\boldsymbol{\gamma} = \mathbf{0}$. The expressions for the estimates and mean square errors in such a case are obtained by putting $\mathbf{A} = \mathbf{0}$ in all the expressions (2.15) - (2.20). In theory we may assume that $\boldsymbol{\gamma} = \mathbf{0}$ without loss of generality by taking $\boldsymbol{\gamma}$ as a part of the fixed parameter $\boldsymbol{\beta}$. However, in practice it would be useful to keep the distinction between the $\boldsymbol{\gamma}$ and $\boldsymbol{\beta}$ parameters.

NOTE 3. If $\boldsymbol{\gamma}$ is known, then we take $\mathbf{Y}_* = \mathbf{Y} - \mathbf{UA}\boldsymbol{\gamma}$ and start with the model

$$\mathbf{Y}_* = \mathbf{X}\boldsymbol{\beta} + \mathbf{U}\boldsymbol{\xi} + \boldsymbol{\varepsilon}, \tag{2.22}$$

$$E(\boldsymbol{\varepsilon}) = \mathbf{0},\quad E(\boldsymbol{\xi}) = \mathbf{0},\quad D(\boldsymbol{\xi}) = \boldsymbol{\Gamma},\quad cov(\boldsymbol{\xi}, \boldsymbol{\varepsilon}) = \mathbf{0},\quad D(\boldsymbol{\varepsilon}) = \boldsymbol{\Sigma}, \tag{2.23}$$

which is in the form mentioned in Note 1.

NOTE 4. It is shown in Rao (1973, p. 296) that one choice of $\mathbf{C}_1$, $\mathbf{C}_2$, $\mathbf{C}_3$ and $\mathbf{C}_4$ in (2.9), when no assumptions are made on the ranks of the matrices involved in the model, is

$$\begin{aligned} \mathbf{C}_1 &= \mathbf{T}^-(\mathbf{I} - \mathbf{S}\mathbf{C}_3), & \mathbf{C}_3 &= (\mathbf{S}'\mathbf{T}^-\mathbf{S})^-\mathbf{S}'\mathbf{T}^-, \\ \mathbf{C}_2 &= \mathbf{C}_3', & \mathbf{C}_4 &= -\mathbf{K} + (\mathbf{S}'\mathbf{T}^-\mathbf{S})^-, \end{aligned} \tag{2.24}$$

where $\mathbf{T} = \mathbf{G} + \mathbf{SKS}'$, with $\mathbf{K}$ as any matrix such that $R(\mathbf{S}) \subseteq R(\mathbf{T})$, i.e., the column space of $\mathbf{S}$ is contained in the column space of $\mathbf{T}$. We may always choose $\mathbf{K} = \mathbf{I}$. It is seen that

$$(\hat{\boldsymbol{\beta}}' : \hat{\boldsymbol{\gamma}}')' = (\mathbf{S}'\mathbf{T}^-\mathbf{S})^-\mathbf{S}'\mathbf{T}^-\mathbf{Y}, \tag{2.25}$$

which shows that $(\hat{\boldsymbol{\beta}}' : \hat{\boldsymbol{\gamma}}')$ is the Gauss-Markoff estimate of $(\boldsymbol{\beta}' : \boldsymbol{\gamma}')$ in the GM(F) model

$$\mathbf{Y} = \mathbf{X\beta} + \mathbf{UA\gamma} + \boldsymbol{\varepsilon}, \ D(\boldsymbol{\varepsilon}) = \mathbf{U\Gamma U'} + \boldsymbol{\Sigma} = \mathbf{G}. \tag{2.26}$$

Further

$$\hat{\boldsymbol{\xi}} = \mathbf{A}\hat{\boldsymbol{\gamma}} + \boldsymbol{\Gamma}\mathbf{U'T^-}(\mathbf{Y} - \mathbf{X}\hat{\boldsymbol{\beta}} - \mathbf{UA}\hat{\boldsymbol{\gamma}}), \tag{2.27}$$

where we could also use any g-inverse of $\mathbf{G}$ instead of $\mathbf{T}^-$.

NOTE 5. If $\mathbf{G}$ is non-singular, we can write the equations for obtaining $\boldsymbol{\beta}$, $\boldsymbol{\gamma}$ and $\boldsymbol{\xi}$ as

$$(\mathbf{S'G^{-1}S})(\hat{\boldsymbol{\beta}}' : \hat{\boldsymbol{\gamma}}')' = \mathbf{S'G^{-1}Y},$$

$$\hat{\boldsymbol{\xi}} = \mathbf{A}\hat{\boldsymbol{\gamma}} + \boldsymbol{\Gamma}\mathbf{U'G^-}(\mathbf{Y} - \mathbf{X}\hat{\boldsymbol{\beta}} - \mathbf{UA}\hat{\boldsymbol{\gamma}}), \tag{2.28}$$

in which the only large matrix to be inverted is $\mathbf{G} = \mathbf{U\Gamma U'} + \boldsymbol{\Sigma}$. This may be done when $\boldsymbol{\Sigma}^{-1}$ exists in a simple way by using the formula

$$(\mathbf{U\Gamma U'} + \boldsymbol{\Sigma})^{-1} = \boldsymbol{\Sigma}^{-1} - \boldsymbol{\Sigma}^{-1}\mathbf{U}(\mathbf{U'}\boldsymbol{\Sigma}^{-1}\mathbf{U} + \boldsymbol{\Gamma}^{-1})^{-1}\mathbf{U'}\boldsymbol{\Sigma}^{-1}. \tag{2.29}$$

The reader is also referred to Chapter 8 in Kleffe and Rao (1987) for computational algorithms for inverting large matrices.

NOTE 6. If $R(\mathbf{S}) \subseteq R(\mathbf{G})$, then one choice of $\mathbf{C}_1$, $\mathbf{C}_2$, $\mathbf{C}_3$ and $\mathbf{C}_4$ is

$$\mathbf{C}_1 = \mathbf{G}^- - \mathbf{G}^-\mathbf{S}\mathbf{C}_3, \quad \mathbf{C}_3 = (\mathbf{S'G^-S})^-\mathbf{S'G^-},$$

$$\mathbf{C}_2 = \mathbf{C}_3', \qquad \mathbf{C}_4 = (\mathbf{S'G^-S})^-, \tag{2.30}$$

using any g-inverses of $\mathbf{G}$ and $\mathbf{S'G^{-1}S}$.

NOTE 7. It was assumed in deriving the expressions (2.15) - (2.20) that $\boldsymbol{\Gamma}$ and $\boldsymbol{\Sigma}$, the dispersion matrices of $\boldsymbol{\xi}$ and $\boldsymbol{\varepsilon}$ respectively, are known. In practice, they may be known only as functions of some unknown parameters. In such a case the unknown parameters could be estimated using an appropriate method such as MINQE (minimum norm quadratic estimation) or MLE (maximum likelihood estimation) as described in Kleffe and Rao (1987). The estimates of $\boldsymbol{\Gamma}$ and $\boldsymbol{\Sigma}$ could be substituted in (2.15) - (2.20). We consider some special cases in the following sections, where some improvements are made by making some adjustments in the estimates of $\boldsymbol{\Gamma}$ and $\boldsymbol{\Sigma}$.

3. SIMULTANEOUS PREDICTION IN SIMILAR LINEAR MODELS

We consider k linear models

$$\mathbf{Y}_i = \mathbf{U}\boldsymbol{\xi}_i + \boldsymbol{\varepsilon}_i, \quad i = 1,\ldots, k, \tag{3.1}$$

associated with k individuals drawn at random from a population, where each $\mathbf{Y}_i$ is an n-vector of observations, $\mathbf{U}$ is a $n \times m$ known matrix, and $\boldsymbol{\xi}_i$ and $\boldsymbol{\varepsilon}_i$ are m- and n-vector variables. We assume that $\boldsymbol{\xi}_i$, $i = 1,\ldots, k$, are i.i.d. with

$$E(\boldsymbol{\xi}_i) = \boldsymbol{\gamma} \quad \text{and} \quad D(\boldsymbol{\xi}_i) = \boldsymbol{\Gamma} \tag{3.2}$$

and $\boldsymbol{\varepsilon}_1,\ldots, \boldsymbol{\varepsilon}_k$ are i.i.d. with

$$E(\boldsymbol{\varepsilon}_i) = \mathbf{0}, \quad cov\,(\boldsymbol{\xi}_i, \boldsymbol{\varepsilon}_j) = \mathbf{0}, \quad D(\boldsymbol{\varepsilon}_i) = \sigma^2\mathbf{V}. \tag{3.3}$$

In the model (3.1), $\boldsymbol{\xi}_i$ is a hypothetical (unobservable) measurement specific to the i-th individual and $\boldsymbol{\varepsilon}_i$ is in the nature of error. The problem is that of simultaneous prediction of $\mathbf{q}'\boldsymbol{\xi}_i$, $i = 1,\ldots, k$, for given $\mathbf{q}$. We assume for simplicity that $\boldsymbol{\Gamma}$, $\mathbf{V}$ and $\mathbf{U}'\mathbf{V}^{-1}\mathbf{U}$ are all full rank matrices.

CASE 1: $\boldsymbol{\gamma}, \boldsymbol{\Gamma}, \sigma^2$ *known*

Let

$$\boldsymbol{\xi}_i^{(\ell)} = (\mathbf{U}'\mathbf{V}^{-1}\mathbf{U})^{-1}\mathbf{U}'\mathbf{Y}_i \qquad \text{(least squares estimator)},$$

$$\boldsymbol{\xi}_i^{(r)} = (\sigma^2\boldsymbol{\Gamma}^{-1} + \mathbf{U}'\mathbf{V}^{-1}\mathbf{U})^{-1}\mathbf{U}'\mathbf{V}^{-1}\mathbf{Y}_i \qquad \text{(ridge regression estimator)},$$

$$\mathbf{W} = (\mathbf{U}'\mathbf{V}^{-1}\mathbf{U})^{-1}. \tag{3.4}$$

From the general theory, the best linear predictor of $\mathbf{q}'\boldsymbol{\xi}_i$ is $\mathbf{q}'\boldsymbol{\xi}_i^{(b)}$ where $\boldsymbol{\xi}_i^{(b)}$ can be represented in several alternative forms given below:

$$\begin{aligned}\boldsymbol{\xi}_i^{(b)} &= \boldsymbol{\gamma} + \boldsymbol{\Gamma}\mathbf{U}'(\mathbf{U}\boldsymbol{\Gamma}\mathbf{U}' + \sigma^2\mathbf{V})^{-1}(\mathbf{Y}_i - \mathbf{U}\boldsymbol{\gamma}) \\ &= \boldsymbol{\gamma} + (\sigma^2\boldsymbol{\Gamma}^{-1} + \mathbf{W}^{-1})^{-1}\mathbf{U}'\mathbf{V}^{-1}(\mathbf{Y}_i - \mathbf{U}\boldsymbol{\gamma})\end{aligned}$$

$$= (\sigma^2\boldsymbol{\Gamma}^{-1} + \mathbf{W}^{-1})^{-1}\sigma^2\boldsymbol{\Gamma}^{-1}\boldsymbol{\gamma} + \boldsymbol{\xi}_i^{(r)}$$

$$= \boldsymbol{\gamma} + \boldsymbol{\Gamma}(\boldsymbol{\Gamma} + \sigma^2\mathbf{W})^{-1}(\boldsymbol{\xi}_i^{(\ell)} - \boldsymbol{\gamma})$$

$$= \sigma^2\mathbf{W}(\boldsymbol{\Gamma} + \sigma^2\mathbf{W})^{-1}\boldsymbol{\gamma} + \boldsymbol{\Gamma}(\boldsymbol{\Gamma} + \sigma^2\mathbf{W})^{-1}\boldsymbol{\xi}_i^{(\ell)}$$

$$= \boldsymbol{\xi}_i^{(\ell)} - \sigma^2\mathbf{W}(\boldsymbol{\Gamma} + \sigma^2\mathbf{W})^{-1}(\boldsymbol{\xi}_i^{(\ell)} - \boldsymbol{\gamma}). \tag{3.5}$$

We call $\boldsymbol{\xi}_i^{(b)}$ as the Bayes predictor of $\boldsymbol{\xi}_i$. The mean square error of the Bayes predictor of $\mathbf{q}'\boldsymbol{\xi}_i$ is

$$E(\mathbf{q}'\boldsymbol{\xi}_i^{(b)} - \mathbf{q}'\boldsymbol{\xi}_i)^2 = \sigma^2\mathbf{q}'\mathbf{Q}\mathbf{q}, \tag{3.6}$$

where

$$\sigma^2\mathbf{Q} = \sigma^2\mathbf{W} - \sigma^2\mathbf{W}(\boldsymbol{\Gamma} + \sigma^2\mathbf{W})^{-1}\mathbf{W}\sigma^2$$

$$= \sigma^2\boldsymbol{\Gamma}(\boldsymbol{\Gamma} + \sigma^2\mathbf{W})^{-1}\mathbf{W} = \boldsymbol{\Gamma} - \boldsymbol{\Gamma}(\boldsymbol{\Gamma} + \sigma^2\mathbf{W})^{-1}\boldsymbol{\Gamma}$$

$$= (\sigma^2\mathbf{W}^{-2} + \boldsymbol{\Gamma}^{-1})^{-1} \quad \text{if } \boldsymbol{\Gamma}^{-1} \text{ exists.} \tag{3.7}$$

All these results follow from the general theory developed in Section 2.

NOTE 1. An alternative predictor of $\mathbf{q}'\boldsymbol{\xi}_i$ is $\mathbf{q}'\boldsymbol{\xi}_i^{(\ell)}$ using the least squares estimator $\boldsymbol{\xi}_i^{(\ell)}$ defined in (3.4). Such a predictor does not make use of the stochastic nature of $\boldsymbol{\xi}_i$. The mean square error of the least squares predictor is

$$E(\mathbf{q}'\boldsymbol{\xi}_i^{(\ell)} - \mathbf{q}'\boldsymbol{\xi})^2 = \sigma^2\mathbf{q}'\mathbf{W}\mathbf{q}, \tag{3.8}$$

which is greater than (3.6), the mean square error for the Bayes predictor $\mathbf{q}'\boldsymbol{\xi}_i^{(b)}$.

NOTE 2. Although the overall prediction error is smaller for the Bayes predictor $\mathbf{q}'\boldsymbol{\xi}_i^{(b)}$ compared to the least squares predictor $\mathbf{q}'\boldsymbol{\xi}_i^{(\ell)}$, it is not true that conditionally for given $\boldsymbol{\xi}_i$

$$E[(\mathbf{q}'\boldsymbol{\xi}_i^{(b)} - \mathbf{q}'\boldsymbol{\xi}_i)^2 \mid \boldsymbol{\xi}_i] < E[\mathbf{q}'\boldsymbol{\xi}_i^{(\ell)} - \mathbf{q}'\boldsymbol{\xi}_i)^2 \mid \boldsymbol{\xi}_i]. \tag{3.9}$$

This inequality may be reversed if $|\mathbf{q}'\boldsymbol{\xi}_i - \mathbf{q}'\boldsymbol{\gamma}|$ is large or small. The inequality (3.9) holds only when a further expectation w.r.t. $\boldsymbol{\xi}_i$ is taken.

CASE 2: γ, Γ *and* σ^2 *unknown*

When γ, Γ and σ^2 are unknown, we substitute for them suitable estimates in the formula (3.5) to obtain what is called an empirical Bayes predictor of ξ_i denoted by $\xi_i^{(e)}$. The best linear unbiased estimator of γ and the minimum norm quadratic unbiased and invariant estimators [MINQE(U, I)] of σ^2 and $\Gamma+\sigma^2\mathbf{W}$ are as follows:

$$k\hat{\gamma} = \sum_{i=1}^{k} \xi_i^{(\ell)}, \tag{3.10}$$

$$k(n-m)\sigma^2 = \sum_{i=1}^{k} \mathbf{Y}_i'\mathbf{V}^{-1}(\mathbf{Y}_i - \mathbf{U}\xi_i^{(\ell)}) = S \tag{3.11}$$

$$(k-1)(\Gamma+\sigma^2\mathbf{W}) = \sum_{i=1}^{k} (\xi_i^{(\ell)} - \hat{\gamma})(\xi_i^{(\ell)} - \hat{\gamma})' = \mathbf{B}. \tag{3.12}$$

Using the last expression in the formula (3.5) and substituting the estimates (3.10) - (3.12) we can write the empirical predictor of ξ_i as

$$\xi_i^{(e)} = \xi_i^{(\ell)} - c\,S\mathbf{W}\mathbf{B}^{-1}(\xi_i^{(\ell)} - \hat{\gamma}),\ \ i = 1,\ldots, k, \tag{3.13}$$

where c is a constant to be chosen to minimize the total expected dispersion error

$$E\big[\sum_{i=1}^{k} (\xi_i^{(e)} - \xi_i)(\xi_i^{(e)} - \xi_i)'\big]. \tag{3.14}$$

If ξ_i and ε_i have multivariate normal distributions, the expression (3.14) can be computed to be

$$k\sigma^2\mathbf{W} + g\sigma^2\mathbf{W}(\Gamma+\sigma^2\mathbf{W})^{-1}\mathbf{W}\sigma^2 \tag{3.15}$$

as shown in Rao (1975), where

$$g = \frac{c^2k(n-m)(kn-km+2)}{k-m-2} - 2ck(n-m). \tag{3.16}$$

Minimizing (3.15) with respect to c, which is equivalent to minimizing g with respect to c, the optimum c is found to be

$$c = \frac{k-m-2}{kn-km+2}. \tag{3.17}$$

With the choice of c as in (3.17), the empirical Bayes predictor of $\mathbf{q}'\boldsymbol{\xi}_i$ is $\mathbf{q}'\boldsymbol{\xi}_i^{(e)}$ where

$$\boldsymbol{\xi}_i^{(e)} = \boldsymbol{\xi}_i^{(\ell)} - \frac{k-m-2}{kn-km+2} S\mathbf{W}\mathbf{B}^{-1}(\boldsymbol{\xi}_i^{(\ell)} - \hat{\boldsymbol{\gamma}}). \tag{3.18}$$

and the mean dispersion error in simultaneous prediction of $\boldsymbol{\xi}_1,\ldots,\boldsymbol{\xi}_k$ is

$$k^{-1}E\sum_{i=1}^{k}(\boldsymbol{\xi}_i^{(e)}-\boldsymbol{\xi}_i)(\boldsymbol{\xi}_i^{(e)}-\boldsymbol{\xi}_i)' = \sigma^2\mathbf{W} - \frac{\sigma^4(n-m)(k-m-2)}{k(n-m)+2}\mathbf{W}(\boldsymbol{\Gamma}+\sigma^2\mathbf{W})^{-1}\mathbf{W}. \tag{3.19}$$

We still have the inequality

$$E\left[\sum_{i=1}^{k}(\mathbf{q}'\boldsymbol{\xi}_i^{(e)}-\mathbf{q}'\boldsymbol{\xi}_i)^2\right] \le E\left[\sum_{i=1}^{k}(\mathbf{q}'\boldsymbol{\xi}_i^{(\ell)}-\mathbf{q}'\boldsymbol{\xi}_i)^2\right] \tag{3.20}$$

so that the compound mean square error in the simultaneous prediction of $\boldsymbol{\xi}_1,\ldots,\boldsymbol{\xi}_k$ by the empirical Bayes predictors, $\boldsymbol{\xi}_1^{(e)},\ldots,\boldsymbol{\xi}_k^{(e)}$, is smaller than that by the least squares predictors, $\boldsymbol{\xi}_1^{(\ell)},\ldots,\boldsymbol{\xi}_k^{(\ell)}$. It may be noted that

$$E\left[\sum_{i=1}^{k}(\mathbf{q}'\boldsymbol{\xi}_i^{(e)}-\mathbf{q}'\boldsymbol{\xi}_i)^2\right] \ge E\left[\sum_{i=1}^{k}(\mathbf{q}'\boldsymbol{\xi}_i^{(b)}-\mathbf{q}'\boldsymbol{\xi}_i)^2\right] \tag{3.21}$$

in view of the additional loss due to the estimation of the unknown parameters $\boldsymbol{\gamma}$, $\boldsymbol{\Gamma}$ and σ^2.

It is interesting to note that

$$E\left(\left[\sum_{i=1}^{k}(\boldsymbol{\xi}_i^{(\ell)}-\boldsymbol{\xi}_i)(\boldsymbol{\xi}_i^{(\ell)}-\boldsymbol{\xi}_i)'\right]\Big|\,\boldsymbol{\xi}_1,\ldots,\boldsymbol{\xi}_k\right)$$
$$\le E\left(\left[\sum_{i=1}^{k}(\boldsymbol{\xi}_i^{(e)}-\boldsymbol{\xi}_i)(\boldsymbol{\xi}_i^{(e)}-\boldsymbol{\xi}_i)'\right]\Big|\,\boldsymbol{\xi}_1,\ldots,\boldsymbol{\xi}_k\right) \tag{3.22}$$

so that the empirical Bayes estimators have a smaller total dispersion error even when $\boldsymbol{\xi}_1,\ldots,\boldsymbol{\xi}_k$ are considered as fixed parameters.

4. SIMULTANEOUS PREDICTION IN DISSIMILAR LINEAR MODELS

We consider k linear models

$$\mathbf{Y}_i = \mathbf{U}_i \boldsymbol{\xi}_i + \boldsymbol{\varepsilon}_i, \quad i = 1, \ldots, k, \tag{4.1}$$

where $\mathbf{Y}_i$ is an n_i-vector of observations, $\mathbf{U}_i$ is a $n_i \times m$ known matrix, and $\boldsymbol{\xi}_i$ and $\boldsymbol{\varepsilon}_i$ are hypothetical vector variables of order m and n_i respectively. We assume that $\boldsymbol{\xi}_i$, $i = 1, \ldots, k$, are i.i.d. observations with

$$E(\boldsymbol{\xi}_i) = \boldsymbol{\gamma}, \quad D(\boldsymbol{\xi}_i) = \boldsymbol{\Gamma},$$

and $\boldsymbol{\varepsilon}_i$, $i = 1, \ldots, k$, are independent with

$$E(\boldsymbol{\varepsilon}_i) = \mathbf{0}, \; D(\boldsymbol{\varepsilon}_i) = \sigma^2 \mathbf{V}_i, \; cov(\boldsymbol{\xi}_i, \boldsymbol{\varepsilon}_j) = \mathbf{0}. \tag{4.2}$$

If $\boldsymbol{\gamma}$, $\boldsymbol{\Gamma}$ and σ^2 are known, the least squares and Bayes predictors of $\boldsymbol{\xi}_i$ are, using the expressions (3.4) and (3.5),

$$\boldsymbol{\xi}_i^{(\ell)} = \mathbf{W}_i \mathbf{U}_i' \mathbf{V}_i^{-1} \mathbf{Y}_i,$$

$$\boldsymbol{\xi}_i^{(b)} = \boldsymbol{\xi}_i^{(\ell)} - \sigma^2 \mathbf{W}_i (\boldsymbol{\Gamma} + \sigma^2 \mathbf{W}_i)^{-1} (\boldsymbol{\xi}_i^{(\ell)} - \boldsymbol{\gamma}), \tag{4.3}$$

where $\mathbf{W}_i = (\mathbf{U}_i' \mathbf{V}_i^{-1} \mathbf{U}_i)^{-1}$. When $\boldsymbol{\gamma}$, $\boldsymbol{\Gamma}$ and σ^2 are unknown, we can obtain empirical Bayes predictors by substituting suitable estimates for the unknown $\boldsymbol{\gamma}$, $\boldsymbol{\Gamma}$ and σ^2 in (4.3).

Methods for estimating $\boldsymbol{\gamma}$, $\boldsymbol{\Gamma}$ and σ^2 are described in Chapter 5 of Kleffe and Rao (1987). However, simpler estimates of $\boldsymbol{\gamma}$, $\boldsymbol{\Gamma}$ and σ^2 such as those obtained from the following equations may provide fairly good empirical Bayes predictors of $\boldsymbol{\xi}_i$, provided n_i are not very much different:

$$\left[\sum_{i=1}^{k} (n_i - m)\right] \hat{\sigma}^2 = \sum_{i=1}^{k} \mathbf{Y}_i' \mathbf{V}_i^{-1} (\mathbf{Y}_i - \mathbf{U}_i \boldsymbol{\xi}_i^{(\ell)}), \tag{4.4}$$

$$(k-1)(\hat{\boldsymbol{\Gamma}} + \sigma^2 \overline{\mathbf{W}}) = \sum_{i=1}^{k} (\boldsymbol{\xi}_i^{(\ell)} - \bar{\boldsymbol{\xi}}^{(\ell)})(\boldsymbol{\xi}_i^{(\ell)} - \bar{\boldsymbol{\xi}}^{(\ell)})', \tag{4.5}$$

$$[\sum_{i=1}^{k}(\hat{\mathbf{\Gamma}}+\hat{\sigma}^2\mathbf{W}_i)^{-1}]\hat{\boldsymbol{\gamma}}=\sum_{i=1}^{k}(\hat{\mathbf{\Gamma}}+\hat{\sigma}^2\mathbf{W}_i)^{-1}\boldsymbol{\xi}_i^{(\ell)}, \tag{4.6}$$

where

$$k\bar{\boldsymbol{\xi}}^{(\ell)} = \boldsymbol{\xi}_1^{(\ell)}+\ldots+\boldsymbol{\xi}_k^{(\ell)}, \tag{4.7}$$

$$k\overline{\mathbf{W}} = \mathbf{W}_1+\ldots+\mathbf{W}_k. \tag{4.8}$$

The equation (4.4) provides a direct estimate $\hat{\sigma}^2$ of σ^2. Then $\mathbf{\Gamma}$ is obtained from the equation (4.5) substituting the value of $\hat{\sigma}^2$ given by (4.4). Finally the value of $\hat{\boldsymbol{\gamma}}$ is computed from the equation (4.6). It may be noted that the estimate $\hat{\mathbf{\Gamma}}$ derived from (4.4) and (4.5) may not be positive definite. In such a case we may alter the diagonal elements of $\hat{\mathbf{\Gamma}}$ slightly by adding small positive quantities to ensure positive definiteness of the resulting estimate of $\mathbf{\Gamma}$.

5. SIMULTANEOUS ESTIMATION IN GENERAL DISSIMILAR LINEAR MODELS

Consider k linear models

$$\mathbf{Y}_i = \mathbf{X}_i\boldsymbol{\beta}+\mathbf{U}_i\boldsymbol{\xi}_i+\boldsymbol{\varepsilon}_i,\quad i = 1,\ldots, k, \tag{5.1}$$

where $\mathbf{U}_i$, $\boldsymbol{\xi}_i$ and $\boldsymbol{\varepsilon}_i$ are as defined in (4.1) and $\mathbf{X}_i$ is a $n_i\times p$ known matrix and $\boldsymbol{\beta}$ is a p-vector of fixed unknown parameters. We can rewrite the model (5.1) in the form

$$\mathbf{Y}_i = (\mathbf{X}_i : \mathbf{U}_i)\begin{pmatrix}\boldsymbol{\beta}\\ \boldsymbol{\gamma}\end{pmatrix}+\mathbf{U}_i\boldsymbol{\eta}_i+\boldsymbol{\varepsilon}_i,\quad i = 1,\ldots, k, \tag{5.2}$$

$$E(\boldsymbol{\eta}_i) = \mathbf{0},\ D(\boldsymbol{\eta}_i) = \mathbf{\Gamma},\ E(\boldsymbol{\varepsilon}_i) = \mathbf{0},\ D(\boldsymbol{\varepsilon}_i) = \sigma^2\mathbf{V}_i,\ cov(\boldsymbol{\varepsilon}_i, \boldsymbol{\eta}_j) = \mathbf{0}. \tag{5.3}$$

The model (5.1) with $\boldsymbol{\gamma} = \mathbf{0}$ has been considered by Laird and Ware (1982). If $\mathbf{\Gamma}$ and σ^2 are known, the model for estimating $(\boldsymbol{\beta}' : \boldsymbol{\gamma}')'$ can be written in the form

$$\mathbf{Y}_i = (\mathbf{X}_i : \mathbf{U}_i)\begin{pmatrix}\boldsymbol{\beta}\\ \boldsymbol{\gamma}\end{pmatrix}+\boldsymbol{\zeta}_i,\quad i = 1,\ldots, k,$$

$$E(\boldsymbol{\zeta}_i) = \mathbf{0}\ \text{ and }\ D(\boldsymbol{\zeta}_i) = \mathbf{U}_i\mathbf{\Gamma}\mathbf{U}_i'+\sigma^2\mathbf{V}_i = \mathbf{W}_i^{-1}+\sigma^2\mathbf{V}_i. \tag{5.4}$$

The best linear unbiased estimate of $(\boldsymbol{\beta}' : \boldsymbol{\gamma}')'$ is the solution of the equation

$$[\sum_{i=1}^{k} (\mathbf{X}_i : \mathbf{U}_i)'\mathbf{T}_i^{-1}(\mathbf{X}_i : \mathbf{U}_i)]\binom{\boldsymbol{\beta}}{\boldsymbol{\gamma}} = \sum_{i=1}^{k} (\mathbf{X}_i : \mathbf{U}_i)'\mathbf{T}_i^{-1}\mathbf{Y}_i. \tag{5.5}$$

where $\mathbf{T}_i = (\mathbf{W}_i^{-1} + \sigma^2\mathbf{V}_i)$. Then the Bayes predictor of $\boldsymbol{\eta}_i$ when $\boldsymbol{\gamma}$ is unknown, but $\boldsymbol{\Gamma}$ and σ^2 are known is

$$\boldsymbol{\eta}_i^{(b)} = \boldsymbol{\Gamma}\mathbf{U}_i'\mathbf{T}_i^{-1}(\mathbf{Y}_i - \mathbf{X}_i\hat{\boldsymbol{\beta}} - \mathbf{U}_i\hat{\boldsymbol{\gamma}}), \tag{5.6}$$

and the predictor of $\mathbf{q}'\boldsymbol{\xi}_i$ can be written in the form

$$\mathbf{q}'\boldsymbol{\xi}_i^{(b)} = \mathbf{q}'\hat{\boldsymbol{\gamma}} + \mathbf{q}'\boldsymbol{\eta}_i^{(b)}. \tag{5.7}$$

If $\boldsymbol{\Gamma}$ and σ^2 are unknown, we have to estimate them and substitute the estimates for the uknowns in (5.7) to obtain an empirical predictor of $\mathbf{q}'\boldsymbol{\xi}_i$. Some methods of estimating $\boldsymbol{\Gamma}$ and σ^2 are discussed in Chapter 5 of Kleffe and Rao (1987). However, we suggest simpler estimates of $\boldsymbol{\Gamma}$ and σ^2 which may provide reasonably good empirical predictors of $\boldsymbol{\xi}_i$, provided $\mathbf{W}_i$ and $\mathbf{V}_i$ are not very much different.

To estimate σ^2, we consider $\boldsymbol{\xi}_1, \ldots, \boldsymbol{\xi}_k$ as fixed parameters and write the k models as a single linear model

$$\mathbf{Y} = \mathbf{X}\boldsymbol{\delta} + \boldsymbol{\varepsilon}, \quad E(\boldsymbol{\varepsilon}) = \mathbf{0}, \quad D(\boldsymbol{\varepsilon}) = \sigma^2\mathbf{V}, \tag{5.8}$$

where

$$\mathbf{Y}' = (\mathbf{Y}_1' : \ldots : \mathbf{Y}_k'),$$

$$\mathbf{X} = \begin{pmatrix} \mathbf{X}_1 & : & \mathbf{U}_1 & : & \mathbf{0} & : & \ldots & : & \mathbf{0} \\ \mathbf{X}_2 & : & \mathbf{0} & : & \mathbf{U}_2 & : & \ldots & : & \mathbf{0} \\ . & & . & & . & & \ldots & & . \\ \mathbf{X}_k & : & \mathbf{0} & : & \mathbf{0} & : & \ldots & : & \mathbf{U}_k \end{pmatrix},$$

$$\boldsymbol{\delta}' = (\boldsymbol{\beta}' : \boldsymbol{\xi}_1' : \ldots : \boldsymbol{\xi}_k') \tag{5.9}$$

and $\mathbf{V}$ is a block diagonal matrix with the matrices $\mathbf{V}_1, \ldots, \mathbf{V}_k$ in the diagonal and null matrices elsewhere. We can then estimate σ^2 in the usual way from the model (5.8) - (5.9) assuming $\boldsymbol{\xi}_i$ as fixed parameters,

$$[n_1 + \ldots + n_k - \rho(\mathbf{X})]\hat{\sigma}^2 = \mathbf{Y}'[\mathbf{V}^{-1} - \mathbf{V}^{-1}\mathbf{X}(\mathbf{X}'\mathbf{V}^{-1}\mathbf{X})^{-1}\mathbf{X}'\mathbf{V}^{-1}]\mathbf{Y}, \tag{5.10}$$

where $\rho(\mathbf{X})$ denotes the rank of $\mathbf{X}$. In this process, we obtain the least squares estimators

$$\boldsymbol{\beta}^{(\ell)},\ \boldsymbol{\xi}_1^{(\ell)},\dots,\ \boldsymbol{\xi}_k^{(\ell)} \tag{5.11}$$

of $\boldsymbol{\beta}, \boldsymbol{\xi}_1,\dots, \boldsymbol{\xi}_k$ considered as fixed parameters with the variance-covariance matrix

$$\sigma^2(\mathbf{X}'\mathbf{V}^{-1}\mathbf{X})^{-1} = \sigma^2 \begin{pmatrix} \mathbf{D}_{00} & \mathbf{D}_{01} & \cdots & \mathbf{D}_{0k} \\ \mathbf{D}_{10} & \mathbf{D}_{11} & \cdots & \mathbf{D}_{1k} \\ \cdot & \cdot & \cdots & \cdot \\ \mathbf{D}_{k0} & \mathbf{D}_{k1} & \cdots & \mathbf{D}_{kk} \end{pmatrix}. \tag{5.12}$$

From (5.11) and (5.12), we can set up the estimating equation

$$\sum_{i=1}^{k} (\boldsymbol{\xi}_i^{(\ell)} - \bar{\boldsymbol{\xi}}^{(\ell)})(\boldsymbol{\xi}_i^{(\ell)} - \bar{\boldsymbol{\xi}}^{(\ell)})' = (k-1)[\hat{\boldsymbol{\Gamma}} + \frac{\hat{\sigma}^2}{k}(\mathbf{D}_{11} + \dots + \mathbf{D}_{kk})]. \tag{5.13}$$

The two equations (5.10) and (5.13) provide the estimates $\hat{\boldsymbol{\Gamma}}$ and $\hat{\sigma}^2$,which can be substituted for the unknown $\boldsymbol{\Gamma}$ and σ^2 in (5.7) to obtain empirical Bayes estimators.

6. A SPECIAL MODEL IN GENETIC SELECTION

6.1 Univariate case

An interesting case of simultaneous prediction arises in animal breeding programs considered by Henderson (1984). We illustrate the method for one of the many models considered by Henderson,

$$Y_{ij} = x_{i1}\beta_1 + \dots + x_{im}\beta_m + \xi_i + \varepsilon_{ij},\ \ j=1,\dots, n_i,\ \ i=1,\dots, k, \tag{6.1.1}$$

where $Y_{ij}, j=1,\dots, n_i$, are repeated measurements on individual i, $\beta_1,\dots, \beta_m$ are fixed parameters, $x_{i1},\dots, x_{im}$ are concomitant observations on the i-th individual, ξ_i is the hypothetical breeding quality of the i-th individual and ε_{ij} is a random error. The problem is the simultaneous prediction of $\xi_1,\dots, \xi_k$ for purposes of comparison and selection. The second order moments associated with the model (6.1.1) are usually of the form

$$E(\xi_i) = \gamma,\ \ V(\xi_i) = \sigma_i^2,\ \ E(\varepsilon_{ij}) = 0,\ \ V(\varepsilon_{ij}) = \sigma_0^2,$$

$$cov(\varepsilon_{ij}, \xi_i) = 0,\ \ cov\,(\varepsilon_{ij}, \varepsilon_{rs}) = 0 \text{ if } (i,j) \neq (r, s). \tag{6.1.2}$$

Note that Y_{ij}, ξ_i, ε_{ij} in (6.1.1) are all scalar variables. Writing

$$Y_i = n_i^{-1}(Y_{i1}+\ldots+Y_{in_i}),\quad \mathbf{Y}' = (Y_1,\ldots, Y_k)',\quad \mathbf{X} = (\mathbf{u} : (x_{ij})),$$

$$\boldsymbol{\eta}' = (\eta_1,\ldots,\eta_m),\quad \varepsilon_i = n_i^{-1}(\varepsilon_{i1}+\ldots+\varepsilon_{in_i}),$$

$$\boldsymbol{\delta}' = (\gamma, \boldsymbol{\beta}'),$$

where $\mathbf{u}$ is a k-vector of unities and $\eta_i = \xi_i - \gamma$, the model for estimating $\boldsymbol{\delta}$ and $\boldsymbol{\eta}$ can be written as

$$\mathbf{Y} = \mathbf{X}\boldsymbol{\delta}+\boldsymbol{\eta}+\boldsymbol{\varepsilon} \tag{6.1.3}$$

with the first and second order moments

$$E(\boldsymbol{\eta})=\mathbf{0},\ \ D(\boldsymbol{\eta})=\sigma_1^{\,2}\mathbf{I}_k,\ \ E(\boldsymbol{\varepsilon})=\mathbf{0},\ \ D(\boldsymbol{\varepsilon})=\sigma_0^{\,2}\boldsymbol{\Delta},$$

where $\boldsymbol{\Delta}=(\delta_{ij})$ is a diagonal matrix with $\delta_{ii}=n_i^{-1}$ and $\delta_{ij}=0$ for $i\neq j$. From the general theory of Section 2, the predictor of η_i is

$$\hat{\eta}_i = (1+n_i^{-1}\phi)^{-1}(Y_i-\mathbf{X}_i'\hat{\boldsymbol{\delta}}), \tag{6.1.4}$$

where

$$\mathbf{X}_i' = (1, x_{i1},\ldots, x_{im}),\quad \phi = \sigma_0^{\,2}/\sigma_1^{\,2},$$

$$\hat{\boldsymbol{\delta}} = (\mathbf{X}'\mathbf{V}^{-1}\mathbf{X})^{-1}\mathbf{X}'\mathbf{V}^{-1}\mathbf{Y} = (\hat{\gamma}, \hat{\beta}_1,\ldots, \hat{\beta}_k)',$$

$$\mathbf{V} = (\delta_{ij}),\ \ \delta_{ii} = (1+n_i^{-1}\phi)^{-1},\ \ \delta_{ij} = 0 \text{ for } i\neq j. \tag{6.1.5}$$

Combining (6.1.4) and (6.1.5), the predictor of ξ_i is

$$\hat{\xi}_i = \hat{\gamma}+\hat{\eta}_i = \hat{\gamma}+(1+n_i^{-1}\phi)^{-1}(Y_i-\mathbf{X}_i'\hat{\boldsymbol{\delta}}). \tag{6.1.6}$$

The formula (6.1.6) involves the parameter ϕ which is the ratio of the variances $\sigma_0^{\,2}$ and $\sigma_1^{\,2}$. If these variance components are not known, they could be estimated by the MINQE(U, I) method and substituted for the unkown values in (6.1.6) to obtain an empirical Bayes predictor of ξ_i.

It is interesting to note that when all n_i are equal to n say, then we have simple formulas for the estimate of the variance components

$$k(n-1)\hat{\sigma}_0^2 = W = \sum_{i=1}^{k}\sum_{j=1}^{n_i}(Y_{ij}-\bar{Y}_i)^2, \tag{6.1.7}$$

$$[n-\rho(\mathbf{X})](n^{-1}\hat{\sigma}_0^{\,2}+\hat{\sigma}_1^{\,2}) = B = (\mathbf{Y}-\mathbf{X}\hat{\boldsymbol{\delta}})'(\mathbf{Y}-\mathbf{X}\hat{\boldsymbol{\delta}}), \tag{6.1.8}$$

where $\hat{\boldsymbol{\delta}} = (\mathbf{X}'\mathbf{X})^{-1}\mathbf{X}'\mathbf{Y}$ is the best unbiased estimator of $\boldsymbol{\delta} = (\gamma, \beta_1,\ldots, \beta_m)'$. We consider an empirical estimator of $\boldsymbol{\eta} = (\eta_1,\ldots, \eta_k)'$ of the form

$$\boldsymbol{\eta}^{(e)} = (\mathbf{Y}-\mathbf{X}\hat{\boldsymbol{\delta}}) - cWB^{-1}(\mathbf{Y}-\mathbf{X}\hat{\boldsymbol{\delta}}) \tag{6.1.9}$$

and choose c to minimize the compound mean square error of prediction,

$$E(\boldsymbol{\eta}^{(e)}-\boldsymbol{\eta})'(\boldsymbol{\eta}^{(e)}-\boldsymbol{\eta}). \tag{6.1.10}$$

The minimum value of c, under the assumption of normality of the variables involved, can be computed to be

$$c = \frac{k-m-2}{n(kn-k+2)} \tag{6.1.11}$$

giving the compound mean square of prediction with the optimum choice of c as

$$\sigma_0^2\frac{k}{n}[1-\frac{(n-1)(k-m-2)}{kn-k+2}\frac{\phi}{n+\phi}]. \tag{6.1.12}$$

6.2 Multivariate case

In Section 6.1, we considered the univariate model (6.1.1) for the estimation of breeding quality of an individual based on a single response. In general, multiple measurements are taken on an individual which enables better estimation of breeding quality. For instance, in selecting varieties of wheat for high yield, Fairfield Smith (1936) used the observed yield as well as some other plant characteristics to construct the selection index. Such a possibility was originally suggested by R. A. Fisher and an appropriate theory for this purpose was developed by the author in Rao (1953).

We consider the following multivariate version of the Henderson model (6.1.1):

$$\mathbf{Y}_{ij} = \mathbf{x}_i\mathbf{B}+\boldsymbol{\eta}_i+\boldsymbol{\varepsilon}_{ij},\ j = 1,\ldots, n_i,\ \ i = 1,\ldots, k, \tag{6.2.1}$$

where $\mathbf{Y}_{ij}'$ is a p-vector of responses, $\mathbf{x}_i'$ is an m-vector of concomitant measurements, $\boldsymbol{\eta}_i'$ is a p-vector of hypothetical breeding qualities of the i-th individual and $\mathbf{B}$ is an $m\times p$ matrix of unknown regression coefficients. We assume without loss of generality that

$$E(\boldsymbol{\eta}_i) = \mathbf{0},\ E(\boldsymbol{\varepsilon}_{ij}) = \mathbf{0},\ D(\boldsymbol{\eta}_i) = \boldsymbol{\Gamma},\ D(\boldsymbol{\varepsilon}_{ij}) = \boldsymbol{\Sigma} \tag{6.2.2}$$

and all the covariances are zero. Taking the average over j for fixed i in the model (6.2.1), we arrive at the model

$$\mathbf{Y}_i = \mathbf{x}_i\mathbf{B}+\boldsymbol{\eta}_i+\boldsymbol{\varepsilon}_i,\ \ i = 1,\ldots, k, \tag{6.2.3}$$

analogous to (6.1.3) in the univariate case, where

$$\mathbf{Y}_i = n_i^{-1}(\mathbf{Y}_{i1}+\ldots+\mathbf{Y}_{in_i}) \text{ and } \boldsymbol{\varepsilon}_i = n_i^{-1}(\boldsymbol{\varepsilon}_{i1}+\ldots+\boldsymbol{\varepsilon}_{in_i})$$

$$E(\boldsymbol{\eta}_i) = \mathbf{0},\ E(\boldsymbol{\varepsilon}_i) = \mathbf{0},\ D(\boldsymbol{\eta}_i) = \boldsymbol{\Gamma},\ D(\boldsymbol{\varepsilon}_i) = \boldsymbol{\Sigma}_i = n_i^{-1}\boldsymbol{\Sigma}.$$

Let $\mathbf{G}_i = \boldsymbol{\Gamma}+\boldsymbol{\Sigma}_i$, $\mathbf{G}_i^{-1} = (g_i^{rs})$ and

$$\mathbf{X}' = (\mathbf{x}_1' : \ldots : \mathbf{x}_k'),\ \mathbf{B} = (\boldsymbol{\beta}_1 : \boldsymbol{\beta}_2 : \ldots : \boldsymbol{\beta}_p),$$

$$\mathbf{Y}' = (\mathbf{Y}_1' : \ldots : \mathbf{Y}_k') = (\mathbf{W}_1 : \ldots : \mathbf{W}_p)'.$$

Then the least squares estimates of $\boldsymbol{\beta}_1,\ldots, \boldsymbol{\beta}_p$ are the solutions of the equations

$$\begin{aligned} \mathbf{A}_{11}\boldsymbol{\beta}_1+\ldots+\mathbf{A}_{1p}\boldsymbol{\beta}_p &= \mathbf{C}_{11}+\ldots+\mathbf{C}_{1p} \\ \cdot \quad \cdots \quad \cdot &\quad \cdot \quad \cdots \quad \cdot \\ \mathbf{A}_{p1}\boldsymbol{\beta}_1+\ldots+\mathbf{A}_{pp}\boldsymbol{\beta}_p &= \mathbf{C}_{p1}+\ldots+\mathbf{C}_{pp} \end{aligned} \tag{6.2.4}$$

where

$$\mathbf{A}_{ij} = \mathbf{X}'\boldsymbol{\Delta}_{ij}\mathbf{X},\ \ \mathbf{C}_{ij} = \mathbf{X}'\boldsymbol{\Delta}_{ij}\mathbf{W}_j, \tag{6.2.5}$$

and $\boldsymbol{\Delta}_{ij}$ is a diagonal matrix with $g_1^{ij},\ldots, g_k^{ij}$ in the diagonal. Let $\hat{\boldsymbol{\beta}}_1, \ldots, \hat{\boldsymbol{\beta}}_p$ be a solution of (6.2.4) and denote $\hat{\mathbf{B}} = (\hat{\boldsymbol{\beta}}_1 : \ldots : \hat{\boldsymbol{\beta}}_p)$. Then using the general theory of Section 2, an estimate of $\boldsymbol{\eta}_i$ is

$$\hat{\boldsymbol{\eta}}_i = (\mathbf{Y}_i - \mathbf{x}_i\hat{\mathbf{B}})\mathbf{G}_i^{-1}\boldsymbol{\Gamma}. \tag{6.2.6}$$

The estimate (6.2.6) depends on $\mathbf{\Gamma}$ and $\mathbf{\Sigma}$. If they are unknown, we substitute their estimates in (6.2.6) to obtain empirical Bayes estimates.

We suggest the following estimating equations for $\mathbf{\Sigma}$ and $\mathbf{\Gamma}$ which are simple to compute:

$$(n.-k)\hat{\mathbf{\Sigma}} = \sum_{i=1}^{k}\sum_{j=1}^{n_i} (\mathbf{Y}_{ij}-\mathbf{Y}_i)'(\mathbf{Y}_{ij}-\mathbf{Y}_i), \tag{6.2.7}$$

$$(p-k)\hat{\mathbf{\Sigma}} + (\text{tr}\,\mathbf{Q})\hat{\mathbf{\Gamma}} = \mathbf{Y}'\mathbf{Q}\mathbf{Y}, \tag{6.2.8}$$

where

$$\mathbf{Q} = \mathbf{N} - \mathbf{N}\mathbf{X}(\mathbf{X}'\mathbf{N}\mathbf{X})^{-1}\mathbf{X}'\mathbf{N}', \quad n. = n_1 + \ldots + n_k, \tag{6.2.9}$$

and $\mathbf{N}$ is the diagonal matrix with $n_1,\ldots, n_k$ as the diagonal elements. Other methods for the estimation of $\mathbf{\Sigma}$ and $\mathbf{\Gamma}$ are discussed in Chapter 5 of Kleffe and Rao (1987).

We illustrate the computations in special cases where certain improvements could be made in the estimation of $\mathbf{\Gamma}$ and $\mathbf{\Sigma}$. When there are no concomitant variables, the model (6.2.3) assumes the form

$$\mathbf{Y}_i = \boldsymbol{\gamma} + \boldsymbol{\eta}_i + \boldsymbol{\varepsilon}_i, \quad i = 1,\ldots, k, \tag{6.2.10}$$

where $\mathbf{Y}_i$, $\boldsymbol{\eta}_i$ and $\boldsymbol{\varepsilon}_i$ are as in (6.2.3) and $\boldsymbol{\gamma}'$ is unknown p-vector representing the overall mean of $\mathbf{Y}_i$. If $\boldsymbol{\gamma}'$, $\mathbf{\Gamma}$ and $\mathbf{\Sigma}$ are known, then the Bayes estimate of $\boldsymbol{\eta}_i$ is

$$\hat{\boldsymbol{\eta}}_i = (\mathbf{Y}_i - \boldsymbol{\gamma})(\mathbf{I} - n_i^{-1}\mathbf{G}_i^{-1}\mathbf{\Sigma}), \quad i = 1,\ldots, k, \tag{6.2.11}$$

where $\mathbf{G}_i = \mathbf{\Gamma} + n_i^{-1}\mathbf{\Sigma}$. If $\boldsymbol{\gamma}$ is not known but $\mathbf{\Gamma}$ and $\mathbf{\Sigma}$ are known, then the least squares estimate of $\boldsymbol{\gamma}$,

$$\hat{\boldsymbol{\gamma}} = \Big(\sum_{i=1}^{k} \mathbf{Y}_i\mathbf{G}_i^{-1}\Big)\Big(\sum_{i=1}^{k} \mathbf{G}_i^{-1}\Big)^{-1}, \tag{6.2.12}$$

is substituted in (6.2.11) to obtain an empirical Bayes estimate. If $\mathbf{\Gamma}$ and $\mathbf{\Sigma}$ are also not known then they can be estimated using the formulas (6.2.7) and (6.2.8) with $\mathbf{X}' = (1,\ldots,1)$. Some simplifications are possible when all the n_i's are equal.

Let $n_i = n$ for each i and $\boldsymbol{\gamma}$ and $\mathbf{\Gamma}$ be unknown but $\mathbf{\Sigma}$ is known. Then an estimate of $\boldsymbol{\eta}_i$ can be written in the form

$$\hat{\boldsymbol{\eta}}_i = (\mathbf{Y}_i - \hat{\boldsymbol{\gamma}})(\mathbf{I} - cn^{-1}\mathbf{B}^{-1}\mathbf{\Sigma}), \quad i = 1,\ldots, k, \tag{6.2.13}$$

where c is a constant to be suitably chosen and

$$\hat{\mathbf{Y}} = k^{-1} \sum_{i=1}^{k} \mathbf{Y}_i, \quad \mathbf{B} = \sum_{i=1}^{k} (\mathbf{Y}_i - \hat{\mathbf{Y}})'(\mathbf{Y}_i - \hat{\mathbf{Y}}). \tag{6.2.14}$$

The total expected dispersion error (EDE) of the estimates (6.2.13), when $\boldsymbol{\eta}_i$ and $\boldsymbol{\varepsilon}_i$ have multivariate normal distributions is

$$E \sum_{i=1}^{k} (\hat{\boldsymbol{\eta}}_i - \boldsymbol{\eta}_i)'(\hat{\boldsymbol{\eta}}_i - \boldsymbol{\eta}_i)$$
$$= kn^{-1}\boldsymbol{\Sigma} + [c^2(k-p-2) - 2c]n^{-2}\boldsymbol{\Sigma}(\boldsymbol{\Gamma} + n^{-1}\boldsymbol{\Sigma})^{-1}\boldsymbol{\Sigma}. \tag{6.2.15}$$

Choosing $c = k - p - 2$ to minimize (6.2.15), the estimates (6.2.13) can be written as

$$\hat{\boldsymbol{\eta}}_i = (\mathbf{Y}_i - \hat{\mathbf{Y}})[\mathbf{I} - (k-p-2)n^{-1}\mathbf{B}^{-1}\boldsymbol{\Sigma}] \tag{6.2.16}$$

with the total EDE

$$kn^{-1}\boldsymbol{\Sigma} - (k-p-2)n^{-2}\boldsymbol{\Sigma}(\boldsymbol{\Gamma} + n^{-1}\boldsymbol{\Sigma})^{-1}\boldsymbol{\Sigma}. \tag{6.2.17}$$

The total EDE when $\boldsymbol{\gamma}$, $\boldsymbol{\Gamma}$ and $\boldsymbol{\Sigma}$ are known is

$$kn^{-1}\boldsymbol{\Sigma} - kn^{-2}\boldsymbol{\Sigma}(\boldsymbol{\Gamma} + n^{-1}\boldsymbol{\Sigma})^{-1}\boldsymbol{\Sigma} \tag{6.2.18}$$

and the difference between (6.2.17) and (6.2.18) is the additional loss due to the estimation of $\boldsymbol{\gamma}$ and $\boldsymbol{\Gamma}$. If $\boldsymbol{\gamma}$, $\boldsymbol{\Gamma}$ and $\boldsymbol{\Sigma}$ are all unknown, we have to estimate $\boldsymbol{\Sigma}$ in addition. This is easily done by computing the pooled within dispersion matrix

$$\mathbf{S} = \sum_{i=1}^{k} \sum_{j=1}^{n} (\mathbf{Y}_{ij} - \mathbf{Y}_i)'(\mathbf{Y}_{ij} - \mathbf{Y}_i), \tag{6.2.19}$$

which has the Wishart distribution on $s = k(n-1)$ degrees of freedom, so that $E(\mathbf{S}) = s\boldsymbol{\Sigma}$. We may write the empirical estimators in the form

$$\hat{\boldsymbol{\eta}}_i = (\mathbf{Y}_i - \hat{\mathbf{Y}})(\mathbf{I} - cn^{-1}\mathbf{B}^{-1}\mathbf{S}), \quad i = 1, \dots, k, \tag{6.2.20}$$

where c in a constant to be suitably chosen. The total EDE in this case is

$$kn^{-1}\Sigma + (k-p-2)^{-1}c^2\{(s^2+s)n^{-2}\Sigma(\Gamma+n^{-1}\Sigma)^{-1}\Sigma$$
$$+ n^{-2}s[\mathrm{tr}(\Gamma+n^{-1}\Sigma)^{-1}\Sigma]\Sigma\} - 2csn^{-2}\Sigma(\Gamma+n^{-1}\Sigma)^{-1}\Sigma. \qquad (6.2.21)$$

Unfortunately, there is no value of c independent of the unknown parameters which minimizes (6.2.21). However, we may choose $c = (k - p-2)/(s+p+1)$ and obtain the empirical estimates of $\boldsymbol{\eta}_i$ as

$$\hat{\boldsymbol{\eta}}_i = (\mathbf{Y}_i - \hat{\mathbf{Y}})(\mathbf{I} - \frac{k-p-2}{s+p+1} n^{-1}\mathbf{B}^{-1}\mathbf{S}), \quad i = 1,\ldots, k. \qquad (6.2.22)$$

The total EDE associated with these estimators is

$$kn^{-1}\Sigma - \frac{(k-p-2)s}{(s+p+1)} n^{-2}\Sigma(\Gamma+n^{-1}\Sigma)^{-1}\Sigma$$
$$+ \frac{(k-p-2)s}{(s+p+1)^2}\{[\mathrm{tr}(\Gamma+n^{-1}\Sigma)^{-1}\Sigma]n^{-2}\Sigma - pn^{-2}\Sigma(\Gamma+n^{-1}\Sigma)^{-1}\Sigma\}. \qquad (6.2.23)$$

The third term in (6.2.23) arises due to the estimation of Σ, but its magnitude is likely to be small.

ACKNOWLEDGEMENT

This work is supported by the Air Force Office of Scientific Research under contract C 49620-85-C-0008. The United States Government is authorised to reproduce and distribute reprints for governmental purposes notwithstanding any copyright notation hereon.

REFERENCES

Fairfield Smith, H. (1936). A discriminant function for plant selection. *Annals of Eugenics* (London) 7, 240-260.

Harville, D. A. (1976). Extension of the Gauss-Markoff theorem to include the estimation of random effects. *Annals of Statistics*, 4, 384-395.

Henderson, C. R. (1950). Estimation of genetic parameters, *Annals of Mathematical Statistics*, 21, 309-310 (abstract).

Henderson, C. R. (1984). *Applications of Linear Models in Animal Breeding*, Univ. of Guelph.

Kleffe, Jürgen and Rao, C. Radhakrishna (1987). *Estimation of Variance Components and Applications*. Wiley, New York (in press).

Laird, N. R. and Ware, J. H. (1982). Random-effects models for longitudinal data, *Biometrics*, 38, 963-974.

Rao, C. Radhakrishna (1953). Discriminant function for genetic differentiation and selection. *Sankhyā*, 12, 229-246.

Rao, C. Radhakrishna (1971). Unified theory of linear estimation. *Sankhyā A*, 33, 370-396 and *Sankhyā A*, 34, 477.

Rao, C. Radhakrishna (1972). A note on the IPM method in the unified theory of linear estimation. *Sankhyā A*, 34, 285-288.

Rao, C. Radhakrishna (1973). *Linear Statistical Inference and its Applications*, Second Edition. Wiley, New York.

Rao, C. Radhakrishna (1975). Simultaneous estimation of parameters in different linear models and applications to biometric problems. *Biometrics*, 31, 545-554.

Rao, C. Radhakrishna (1977). Simultaneous estimation of parameters - a compound decision problem. *Statistical Decision Theory and Related Topics* (S. S. Gupta and D. S. Moore, *eds.*). Academic Press, New York, 327-350.

Rao, C. Radhakrishna (1985). A unified approach to inference from linear models. *Proceedings of the First International Tampere Seminar on Linear Statistical Models and their Applications* (Tarmo Pukkila and Simo Puntanen, *eds.*). Department of Mathematical Sciences, University of Tampere, Finland, 9-36.

Received 5 May 1987
Revised 5 October 1987

Center for Multivariate Analysis
University of Pittsburgh
Pittsburgh, PA 15260
U.S.A.

DISCUSSION

C.G. KHATRI

Gujarat University, Ahmedabad, India

(1) Professor Rao (1985) has given a unified theory of estimation in linear models with fixed effects, and has given Bayes and empirical Bayes estimates of fixed effects under the assumption of prior distribution of fixed effects. In the mixed effect models, the estimation of random effects has natural prior distribution and the Bayes and empirical Bayes estimates of random effects naturally arise. This is appropriately investigated by Professor Rao in the present paper.

(2) Professor Rao (1985) has given the method of minimum dispersion unbiased estimates of fixed effects' parameters. Here, too, this method can be adopted in the following way:

$\begin{pmatrix} \mathbf{L}_1\mathbf{y} \\ \mathbf{L}_2\mathbf{y} \end{pmatrix}$ will be the minimum dispersion unbiased estimate of $\begin{pmatrix} \mathbf{X}\boldsymbol{\beta} \\ \mathbf{U}\boldsymbol{\xi} \end{pmatrix}$ if and only if

(i) $E\begin{pmatrix} \mathbf{L}_1\mathbf{y} - \mathbf{X}\boldsymbol{\beta} \\ \mathbf{L}_2\mathbf{y} - \mathbf{U}\boldsymbol{\xi} \end{pmatrix} = \mathbf{0}$ for all $\boldsymbol{\beta}$ and $\boldsymbol{\gamma}$ and

(ii) $Var\begin{pmatrix} \mathbf{L}_1\mathbf{y} - \mathbf{X}\boldsymbol{\beta} \\ \mathbf{L}_2\mathbf{y} - \mathbf{U}\boldsymbol{\xi} \end{pmatrix} \leq Var\begin{pmatrix} \mathbf{P}_1\mathbf{y} - \mathbf{X}\boldsymbol{\beta} \\ \mathbf{P}_2\mathbf{y} - \mathbf{U}\boldsymbol{\xi} \end{pmatrix}$ for all $\begin{pmatrix} \mathbf{P}_1\mathbf{y} \\ \mathbf{P}_2\mathbf{y} \end{pmatrix}$ for which

$$E\begin{pmatrix} \mathbf{P}_1\mathbf{y} - \mathbf{X}\boldsymbol{\beta} \\ \mathbf{P}_2\mathbf{y} - \mathbf{U}\boldsymbol{\xi} \end{pmatrix} = \mathbf{0}.$$

The above inequality (ii) is in the Löwner's sense [see, for example, Hartwig and Styan (1987)].

Notice that the the condition (i) implies

$$\begin{pmatrix} \mathbf{L}_1 \\ \mathbf{L}_2 \end{pmatrix}(\mathbf{X} : \mathbf{UA}) = \begin{pmatrix} \mathbf{X} & \mathbf{0} \\ \mathbf{0} & \mathbf{UA} \end{pmatrix},$$

and (ii) implies that for every vector $\mathbf{p}$ and $\mathbf{q}$

$$Var\,[(\mathbf{p}'\mathbf{L}_1 + \mathbf{q}'\mathbf{L}_2)\mathbf{y} - \mathbf{p}'\mathbf{X}\boldsymbol{\beta} - \mathbf{q}'\mathbf{U}\boldsymbol{\xi}]$$

is minimum for all $\mathbf{L}_1$ and $\mathbf{L}_2$ satisfying the above conditions for $\mathbf{L}_1$ and $\mathbf{L}_2$. Thus, we have to minimize

$$\phi = (\mathbf{p}'\mathbf{L}_1 + \mathbf{q}'\mathbf{L}_2)\,\mathbf{G}(\mathbf{L}_1'\mathbf{p} + \mathbf{L}_2'\mathbf{q}) - 2\mathbf{q}'(\mathbf{U}\boldsymbol{\Gamma}\mathbf{U}')(\mathbf{L}_1'\mathbf{p} + \mathbf{L}_2'\mathbf{q})$$
$$+ \mathbf{q}'\mathbf{L}_2\mathbf{U}\boldsymbol{\Gamma}\mathbf{U}'\mathbf{L}_2\mathbf{q} + 2[(\mathbf{p}'\mathbf{L}_1 + \mathbf{q}'\mathbf{L}_2)(\mathbf{X}\boldsymbol{\lambda}_1 + \mathbf{UA}\boldsymbol{\lambda}_2) - \mathbf{p}'\mathbf{X}\boldsymbol{\lambda}_1 - \mathbf{q}'\mathbf{UA}\boldsymbol{\lambda}_2]$$

with respect to $\mathbf{L}_1$, $\mathbf{L}_2$, $\boldsymbol{\lambda}_1$ and $\boldsymbol{\lambda}_2$ for all $\mathbf{p}$ and $\mathbf{q}$. For given $\mathbf{p}$ and $\mathbf{q}$, ϕ will be minimum if $\boldsymbol{\ell} = \mathbf{L}_1'\mathbf{p} + \mathbf{L}_2'\mathbf{q}$ is a solution of

$$\begin{pmatrix} \mathbf{G} & \mathbf{S} \\ \mathbf{S}' & \mathbf{0} \end{pmatrix}\begin{pmatrix} \boldsymbol{\ell} \\ \boldsymbol{\lambda} \end{pmatrix} = \begin{pmatrix} \mathbf{U}\boldsymbol{\Gamma}\mathbf{U}'\mathbf{q} \\ \mathbf{X}'\mathbf{p} \\ \mathbf{A}'\mathbf{U}'\mathbf{q} \end{pmatrix} \text{ with } \boldsymbol{\lambda} = \begin{pmatrix} \boldsymbol{\lambda}_1 \\ \boldsymbol{\lambda}_2 \end{pmatrix} \text{ and } \mathbf{S} = (\mathbf{X} : \mathbf{UA})$$

which is the same as (2.8) of Rao. From this, we get $\mathbf{L}_1$ and $\mathbf{L}_2$ by equating the coefficients of $\mathbf{p}$ and $\mathbf{q}$. Taking $\boldsymbol{\lambda}_1 = \boldsymbol{\Lambda}_1\mathbf{p}$ and $\boldsymbol{\lambda}_2 = \boldsymbol{\Lambda}_2\mathbf{q}$, the above equation is equivalent to

$$\begin{pmatrix} \mathbf{G} & \mathbf{S} \\ \mathbf{S}' & \mathbf{0} \end{pmatrix} \begin{pmatrix} \mathbf{L}_1' & \mathbf{L}_2' \\ \boldsymbol{\Lambda}_1 & \mathbf{0} \\ \mathbf{0} & \boldsymbol{\Lambda}_2 \end{pmatrix} = \begin{pmatrix} \mathbf{0} & \mathbf{U\Gamma U}' \\ \mathbf{X}' & \mathbf{0} \\ \mathbf{0} & \mathbf{A}'\mathbf{U}' \end{pmatrix}$$

and if

$$\begin{pmatrix} \mathbf{G} & \mathbf{S} \\ \mathbf{S}' & \mathbf{0} \end{pmatrix}^- = \begin{pmatrix} \mathbf{C}_1 & \mathbf{C}_{21} & \mathbf{C}_{22} \\ \mathbf{C}_{21}' & -\mathbf{C}_{41} & -\mathbf{C}_{42} \\ \mathbf{C}_{22}' & -\mathbf{C}_{42}' & -\mathbf{C}_{43} \end{pmatrix}$$

is symmetric, then $\mathbf{L}_1' = \mathbf{C}_{21}\mathbf{X}'$ and $\mathbf{L}_2' = \mathbf{C}_1\mathbf{U\Gamma U}' + \mathbf{C}_{22}\mathbf{A}'\mathbf{U}'$. Further,

$$Var\begin{pmatrix} \mathbf{X}'\hat{\boldsymbol{\beta}} \\ \mathbf{U}'\hat{\boldsymbol{\xi}} \end{pmatrix} = \begin{pmatrix} \mathbf{C}_{41} & \mathbf{C}_{42} \\ \mathbf{C}_{42}' & \mathbf{C}_{43} \end{pmatrix} \text{with } \hat{\boldsymbol{\beta}} = \mathbf{C}_{21}'\mathbf{y} \text{ and } \hat{\boldsymbol{\xi}} = \mathbf{A}\hat{\boldsymbol{\gamma}} + \boldsymbol{\Gamma}\mathbf{U}'\hat{\boldsymbol{\varepsilon}};$$

$\hat{\boldsymbol{\gamma}} = \mathbf{C}_{22}'\mathbf{y}$ and $\hat{\boldsymbol{\varepsilon}} = \mathbf{C}_1\mathbf{y}$. Notice that the linear restrictions

$$\begin{pmatrix} \mathbf{L}_1 \\ \mathbf{L}_2 \end{pmatrix}\mathbf{S} = \begin{pmatrix} \mathbf{X} & \mathbf{0} \\ \mathbf{0} & \mathbf{UA} \end{pmatrix}$$

obtained above on account of condition (i) may be inconsistent, because

$$\text{rank}\begin{pmatrix} \mathbf{S}' & \vdots & \mathbf{X}' & \mathbf{0} \\ & \vdots & \mathbf{0} & \mathbf{A}'\mathbf{U}' \end{pmatrix} = \text{rank}\,\mathbf{X} + \text{rank}\,\mathbf{UA} \geq \text{rank}\,\mathbf{S}.$$

Hence, we shall reconsider the above method by rewriting the model as

$$\mathbf{y} = \mathbf{S}\boldsymbol{\delta} + \mathbf{U}\boldsymbol{\xi}_1 + \boldsymbol{\varepsilon} \text{ with } \boldsymbol{\xi}_1 = \boldsymbol{\xi} - \mathbf{A}\boldsymbol{\gamma} \text{ and } \boldsymbol{\delta}' = (\boldsymbol{\beta}', \boldsymbol{\gamma}').$$

Then, $\begin{pmatrix} \mathbf{L}_{(1)}\mathbf{y} \\ \mathbf{L}_{(2)}\mathbf{y} \end{pmatrix}$ will be an unbiased estimate of $\begin{pmatrix} \mathbf{S}\boldsymbol{\delta} \\ \boldsymbol{\xi}_1 \end{pmatrix}$ in the sense

$$E\begin{pmatrix} \mathbf{L}_{(1)}\mathbf{y} - \mathbf{S}\boldsymbol{\delta} \\ \mathbf{L}_{(2)}\mathbf{y} - \boldsymbol{\xi}_1 \end{pmatrix} = \mathbf{0} \text{ for all } \boldsymbol{\delta}, \text{ which gives } \begin{pmatrix} \mathbf{L}_{(1)} \\ \mathbf{L}_{(2)} \end{pmatrix}\mathbf{S} = \begin{pmatrix} \mathbf{S} \\ \mathbf{0} \end{pmatrix} \text{ or}$$

$$\begin{pmatrix} \mathbf{L}_{(1)} \\ \mathbf{L}_{(2)} \end{pmatrix} = \begin{pmatrix} \mathbf{P}_\mathbf{S} \\ \mathbf{0} \end{pmatrix} + \mathbf{L}_{(.)}(\mathbf{I} - \mathbf{P}_\mathbf{S}) \text{ and } \mathbf{P}_\mathbf{S} = \mathbf{S}(\mathbf{S}'\mathbf{S})^-\mathbf{S}'.$$

Taking $\mathbf{R}_{11} = (\mathbf{I} - \mathbf{P}_\mathbf{S})\mathbf{G}(\mathbf{I} - \mathbf{P}_\mathbf{S})$, $\mathbf{R}_{12} = (\mathbf{I} - \mathbf{P}_\mathbf{S})(\mathbf{G}\mathbf{P}_\mathbf{S} : -\mathbf{U}\boldsymbol{\Gamma})$ and

$$\mathbf{R}_{22} = \begin{pmatrix} \mathbf{P_S G P_S} & -\mathbf{P_S U \Gamma} \\ -\mathbf{\Gamma U' P_S} & \mathbf{\Gamma} \end{pmatrix},$$

it can be shown that

$$Var\begin{pmatrix} \mathbf{L}_{(1)}\mathbf{y} - \mathbf{S\delta} \\ \mathbf{L}_{(2)}\mathbf{y} - \boldsymbol{\xi}_1 \end{pmatrix} = \mathbf{L}_{(.)}\mathbf{R}_{11}\mathbf{L}_{(.)}' + \mathbf{L}_{(.)}\mathbf{R}_{12} + \mathbf{R}_{12}'\mathbf{L}_{(.)}' + \mathbf{R}_{22}$$
$$\geq \mathbf{R}_{22} - \mathbf{R}_{12}'\mathbf{R}_{11}^{-}\mathbf{R}_{12}$$

and the equality holds if and only if

$$\mathbf{L}_{(\cdot)} = -\mathbf{R}_{12}'\mathbf{R}_{11}^{-} + \mathbf{L}_3(\mathbf{I} - \mathbf{R}_{11}\mathbf{R}_{11}^{-}) \text{ for any arbitrary } \mathbf{L}_3.$$

Then, the minimum dispersion unbiased estimate of $\begin{pmatrix} \mathbf{X\beta} + \mathbf{UA\gamma} \\ \boldsymbol{\xi} - \mathbf{A\gamma} \end{pmatrix}$ is

$$\begin{pmatrix} \mathbf{P_S y} - \mathbf{P_S G}(\mathbf{I} - \mathbf{P_S})\mathbf{R}_{11}^{-}(\mathbf{I} - \mathbf{P_S})\mathbf{y} \\ \mathbf{\Gamma U'}(\mathbf{I} - \mathbf{P_S})\mathbf{R}_{11}^{-}(\mathbf{I} - \mathbf{P_S})\mathbf{y} \end{pmatrix}$$

because $(\mathbf{I} - \mathbf{R}_{11}\mathbf{R}_{11}^{-})(\mathbf{I} - \mathbf{P_S})\mathbf{y} = \mathbf{0}$ almost surely. These can be shown to be the same as those obtained above (or obtained by Professor Rao).

(3) Suppose the mixed model is

$$\mathbf{y} = \mathbf{X\beta} + \mathbf{U}\boldsymbol{\xi} + \boldsymbol{\varepsilon} \text{ with } E(\boldsymbol{\xi}) = \mathbf{A\gamma}, E\,\boldsymbol{\varepsilon} = \mathbf{0}, Var\begin{pmatrix} \boldsymbol{\xi} \\ \boldsymbol{\varepsilon} \end{pmatrix} = \begin{pmatrix} \mathbf{\Gamma} & \mathbf{\Delta} \\ \mathbf{\Delta}' & \mathbf{\Sigma} \end{pmatrix},$$

where $\mathbf{\Delta} = cov(\boldsymbol{\xi}, \boldsymbol{\varepsilon})$ is not zero. If $\mathbf{\Delta} = \mathbf{0}$, we get (1.1) and (1.2). We observe that if $\boldsymbol{\varepsilon}_1 = \boldsymbol{\varepsilon} - \mathbf{\Delta}'\mathbf{\Gamma}^{-}(\boldsymbol{\xi} - \mathbf{A\gamma})$, then

$$Var\begin{pmatrix} \boldsymbol{\xi} \\ \boldsymbol{\varepsilon}_1 \end{pmatrix} = \begin{pmatrix} \mathbf{\Gamma} & \mathbf{0} \\ \mathbf{0} & \mathbf{\Sigma}_1 \end{pmatrix}, \quad \mathbf{\Sigma}_1 = \mathbf{\Sigma} - \mathbf{\Delta}'\mathbf{\Gamma}^{-}\mathbf{\Delta},$$

and

$$\mathbf{y} = \mathbf{S}\begin{pmatrix} \boldsymbol{\beta} \\ \boldsymbol{\gamma} \end{pmatrix} + \mathbf{U}_1\boldsymbol{\eta} + \boldsymbol{\varepsilon}_1 \text{ with } \mathbf{U}_1 = \mathbf{U} + \mathbf{\Delta}'\mathbf{\Gamma}^{-} \text{ and } \boldsymbol{\eta} = \boldsymbol{\xi} - \mathbf{A\gamma}.$$

This is the model considered by Professor Rao and note that $\mathbf{U}_1$ must be known for applications; that is, some structure on $\mathbf{\Delta}$ must be known.

REFERENCES

Hartwig, R. E. and Styan, G.P.H. (1987). Partially ordered idempotent matrices. *Proceedings of the Second International Tampere Conference in Statistics* (Tarmo Pukkila and Simo Puntanen, *eds.*). Department of Mathematical Sciences, University of Tampere, Finland, 361-383.

Rao, C.R. (1985). A unified approach to inference from linear models. *Proceedings of the First International Tampere Seminar on Linear Models (1983)* (Tarmo Pukkila and Simo Puntanen, *eds.*). Department of Mathematical Sciences, University of Tampere, Finland, 9-36.

Received 16 July 1987
Revised 26 August and 23 October 1987

Department of Statistics
Gujarat University
Ahmedabad - 380 009
India

Proc. Second International Tampere Conference in Statistics
(Tampere, Finland, 1-4 June 1987)
Tarmo Pukkila and Simo Puntanen, *Editors*

pp. 99 - 112

INVITED PAPER

Robust regression and unmasking transformations

A.C. ATKINSON
Imperial College, London, U.K.

Key words and phrases: Graphical methods, least median of squares regression, outliers, power transformation, profile loglikelihood, regression diagnostics.

ABSTRACT

Evidence for parametric transformation of the response in a regression model may sometimes depend crucially on one or a few observations. The paper describes the exploratory use of least median of squares regression at several values of the transformation parameter. The structure of the residuals from this robust analysis indicates outliers and influential observations. The confirmatory stage of the analysis is based on plots of profile loglikelihoods for the transformation parameter as observations are removed from the analysis.

1. INTRODUCTION

The paper is concerned with inference about transformation of the response in a multiple regression model. The purpose is to determine the effect of groups of observations on the estimated transformation. Diagnostic methods based on the deletion of single observations are well established for this purpose. Multiple deletion methods are, on the other hand, rarely used because of the combinatorial explosion of possibilities when the deletion of all m tuples is considered, even for fixed m. In those examples where the importance of observations is not evident unless several are deleted at once, masking is said to occur. In the presence of masking, diagnostic methods based on the deletion of single observations fail to reveal outlying and influential observations.

For a regression model without transformation, Atkinson (1986b) suggested a two-stage procedure for overcoming masking. The first, exploratory, stage consists of least median of squares regression, a robust method which resists nearly 50% of contamination in the data (Rousseeuw, 1984; Rousseeuw and Leroy, 1987). In the second, confirmatory, stage the diagnostic methods of least squares regression are used to check the indications of the robust method. The extension of the procedure to transformations of the response indexed by a parameter λ is given by Atkinson (1988). The strategy is to perform the robust analysis at a series of values of λ. Features of the analysis which vary in a regular way with λ are subjected to confirmatory analysis and so indicate a suitable transformation in the presence of masking.

A brief outline of the diagnostic use of least median of squares regression is given in Section 2. Section 3 reviews likelihood procedures for transformation of the response in a regression model. In Atkinson (1988) the emphasis is on the development of addition diagnostics for the approximate score test for transformations and the use of these, together with deletion diagnostics, in the analysis of data. In the present paper the emphasis is instead on the use of profile loglikelihoods to confirm the status of the m outliers and $n - m$ "good" observations indicated by the robust analysis. In Section 4 the output of least median of squares regression is considered as the transformation parameter λ varies. A plot of standardized residuals from the robust regression as a function of λ is found, for one example, to be a useful indicator of the structure of the data, even in the presence of masking. Three further examples are analysed in Section 5.

2. EXPLORATORY ROBUST ANALYSIS

In this section a brief summary is given of the exploratory use of least median of squares regression. Discussion of transformations of the response is left until Section 3.

For the present suppose that the majority of the observations follow the standard linear regression model

$$E(Y) = X\beta. \tag{1}$$

where the $n \times p$ matrix X consists of the known values of the p carriers which are functions of the explanatory variables. The errors are assumed additive and independent with constant variance σ^2.

For the parameter value b let the residual $r_i = y_i - x_i^{\mathrm{T}}b$. Then two criteria for the choice of b are:

Least Sum of Squares Regression: $\min_b \Sigma\, r_i^2$

Least Median of Squares Regression: $\min_b \text{ median } r_i^2$.

The intention of least median of squares regression in the presence of outliers is to fit a line to the "good" observations whilst revealing the "bad" observations as such.

The numerical method used to find b is a form of random search due to Rousseeuw (1984). If the rank of the regression model is p, samples of p observations are taken, to each of which, except for singular samples which are abandoned, the regression model can be fitted exactly. Hawkins, Bradu and Kass (1984), who use the method for the direct detection of outliers, call these samples "elemental sets". Sampling with calculation of the median of the squares of the non-zero residuals continues until either a stable pattern of residuals emerges, or until there is a specified probability, for a given level of contamination, of obtaining at least one elemental set which consists solely of "good" observations.

For the Jth elemental set let the residuals be r_{iJ}, at least p of which will be zero. If the elemental set giving rise to the minimum median squared residual is denoted by T, then

$$\tilde{r}^2_T = \underset{N_T}{\text{median}}\; r^2_{iT} = \min_J \underset{N_J}{\text{median}}\; r^2_{iJ} \tag{2}$$

where N_T and N_J are the number of non-zero residuals, usually $n - p$. As an estimate of σ^2 Atkinson (1986b) suggests $\tilde{s}^2 = \tilde{r}^2_T$. The estimate is used to provide standardized least median of squares residuals $\tilde{r}_i = r_{iT}/\tilde{s}$. This algorithm differs slightly from that of Rousseeuw who takes the median over all n residuals and who uses a different, but related, estimate of σ^2.

The results of Rousseeuw (1984) and of Atkinson (1986b) show that least median of squares is an excellent exploratory tool. However, the estimates of the parameters of the linear model have poor properties and a second, confirmatory, stage is required. Rousseeuw (1984) uses the least median of squares estimate as a starting point for robust regression using M estimators. Atkinson (1986b) uses extensions of the standard least-squares regression diagnostics described in the books of Belsley, Kuh and Welsch (1980), Cook and Weisberg (1982), Atkinson (1985) and Weisberg (1985, Chap. 5&6). Different diagnostic techniques are required when transformation of the response is investigated. Some methods are mentioned in the next section. A fuller description is given by Atkinson (1988).

3. AN ANALYSIS OF TRANSFORMATIONS

In this section a short review is given of likelihood-based methods for inference about transformations of the response. The hope is that there will be some value of the scalar parameter λ for which the transformed observations $y(\lambda)$ satisfy the linear model (1) to an adequate degree. As an example we use the parametric family of power transformations analysed by Box and Cox (1964). In normalized form this is

$$\begin{aligned} z(\lambda) &= (y^{\lambda} - 1)/(\lambda \dot{y}^{\lambda-1}) && (\lambda \neq 0) \\ &= \dot{y} \log y && (\lambda = 0) \end{aligned} \qquad (3)$$

where $\dot{y}$ is the geometric mean of the y_i. Some inferential consequences of data transformations are discussed by Hinkley and Runger (1984).

For the normalized transformation (3) the loglikelihood of the observations, maximized over the parameters of the linear model, is given by

$$L_{\max}(\lambda) = -(n/2)\,[1 + \log\{2\pi R(\lambda)/n\}]. \qquad (4)$$

In (4) $R(\lambda)$ is the residual sum of squares of the $z(\lambda)$ given by

$$R(\lambda) = z(\lambda)^{\mathrm{T}}(I - H)z(\lambda). \qquad (5)$$

where the hat matrix $H = X(X^{\mathrm{T}}X)^{-1}X^{\mathrm{T}}$. The maximum likelihood estimate of λ is the value $\hat{\lambda}$ for which the profile loglikelihood $L_{\max}(\lambda)$ is maximized.

Hypotheses about the value of λ can be tested using the approximate score statistic $T_p(\lambda)$ introduced by Atkinson (1973). The use of diagnostic versions of this test to confirm the exploratory robust analysis of transformations is described by Atkinson (1988). An advantage of these tests is that it is possible to investigate the influence of individual observations in a simple manner without finding the value of $\hat{\lambda}$. However, estimation of λ is computationally negligible when compared to the calculations of least median of squares regression repeated over a grid of λ values. In this paper we accordingly follow Box and Cox (1964) and use plots of $L_{max}(\lambda)$ to investigate the effect of individual observations on the estimated transformation. When a specified m observations suggested by the robust analysis are deleted, the calculations for $L_{max}(\lambda)$ in (3), (4) and (5) refer to the remaining $n - m$ observations.

4. UNMASKING TRANSFORMATIONS

This section describes the analysis of an example in which masking is present. Only after three observations have been deleted does it become apparent that the data should be transformed. As the results of Atkinson (1988) show, this structure is not revealed by single-deletion diagnostic calculations, but it is revealed by the exploratory use of least median of squares regression.

The example involves augmentation of a set of simulated data used by Cook and Wang (1983) and hinges on the presence of leverage points, that is observations at remote values of the explanatory variables. There are 13 observations with a single explanatory variable. For the first ten observations the values of x lie between 0.3 and 1.5. For these observations, which are generated from a lognormal model, the approximately standard normal score statistic $T_p(1) = -3.40$, so that there is evidence of the need for a transformation. Addition of observations 11, 12 and 13 at the leverage points $x = 2.6$, 2.8 and 3.0 with responses 6.2, 6.6 and 6.9 suppresses the evidence for a transformation with $T_p(1) = -0.23$ and $T_p(0) = 3.52$. There is thus no evidence of the need for a transformation away from 1 and the log transformation is rejected.

To unmask this information, least median of squares regression is used to provide a pattern of residuals indicative of potential outliers. As the examples in Atkinson (1988) show, the change in the robust residuals as λ varies is informative about the structure of the data. Fig. 1 is a plot, for Cook and Wang's "data" after augmentation, of the five largest standardized least median of squares residuals $\tilde{r}_i$ calculated for 21 values of λ between -1 and 1. At each value of λ, 1000 elemental sets were sampled, although evolution of the residuals for this and other examples suggests that the number of samples could be reduced if the computational burden is too heavy. For simple examples such as this, systematic enumeration of all elemental sets is a possibility which could also be exploited. The slight irregularities in Fig. 1 away from the one large jump are due to the use of medians, which do not vary smoothly with λ.

For values of λ between -1 and -0.1, the standardized residuals for observations 11, 12 and 13, with values as extreme as -19.28, clearly indicate the importance of these three. Observation 4 has a moderately large residual at $\lambda = -1$. For λ between 0 and 1 observation 10 stands out. This plot suggests that the effect of deletion of either observation 10, or of observations 11, 12 and 13 should be investigated by a confirmatory analysis.

Fig. 2 is a plot of the profile loglikelihood $L_{max}(\lambda)$ against λ for various subsets of the data. For all 13 observations the maximum of the likelihood is near one and remains near it if any one of observations 11, 12 or 13 is deleted. The effect of deletion of observation 13 is shown in the figure. Deletion of observation 10 yields a maximum slightly, but not significantly, below $\lambda = 1$. If all three influential observations are deleted, evidence for the log transformation is revealed, but the evidence depends crucially on the presence of observation 10. If observations 1 to 9 are analysed alone, there is no evidence as to whether a transformed or untransformed analysis is preferable.

The two conflicting structures of the data are revealed by the deletion of observations suggested by the plot of the robust residuals in Fig. 1. The corresponding plot of least squares residuals, Fig. 3, fails to reveal this structure. All residuals, after standardization by the mean square estimate of σ^2, lie between -1 and 2.5. The figure shows that the importance of observations 11, 12 and 13 is masked at all values of λ.

5. EXAMPLES

In this section three examples are analysed to illustrate the behaviour of the two-stage procedure when the structure of the data is not known. The first two examples, taken from Box and Cox (1964), are data sets which have often been analysed. In all 3 examples first-order models are filted, without interactions.

Example 1. *Box and Cox Worsted Data.* There are 27 observations on the number of cycles to failure of a worsted yarn, the design for which was a 3^3 factorial. The non-negative response ranges from 90 for observation 9 to 3,636 for observation 19. On prior grounds it is therefore to be expected that an untransformed model with additive errors of constant variance will be inappropriate. This feeling is supported by a value of -18.56 for $T_p(1)$.

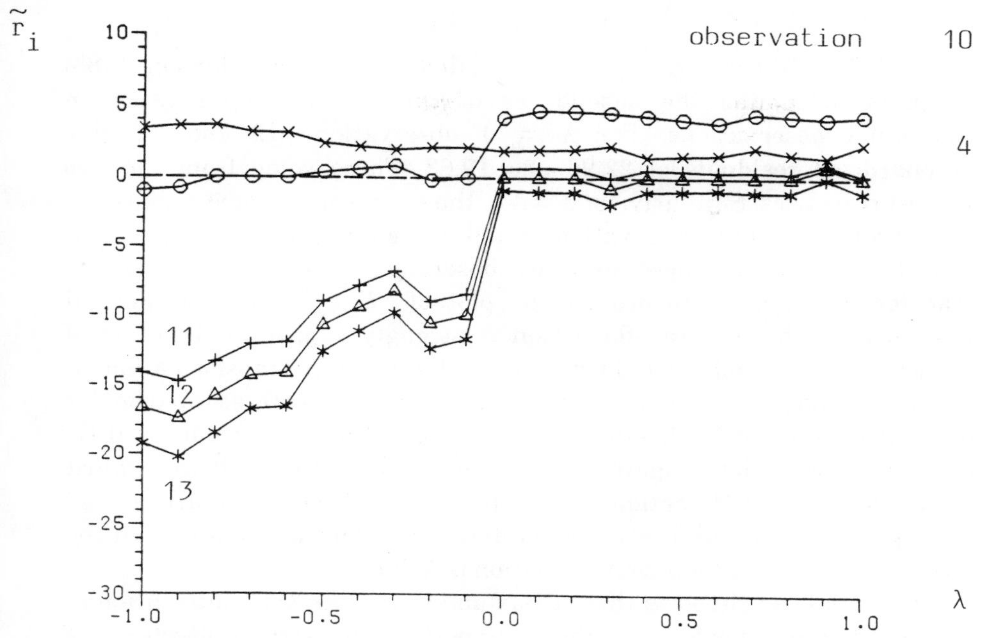

Fig. 1. Augmented Cook and Wang "data": Standardized least median of squares residuals $\tilde{r}_i$ against λ.

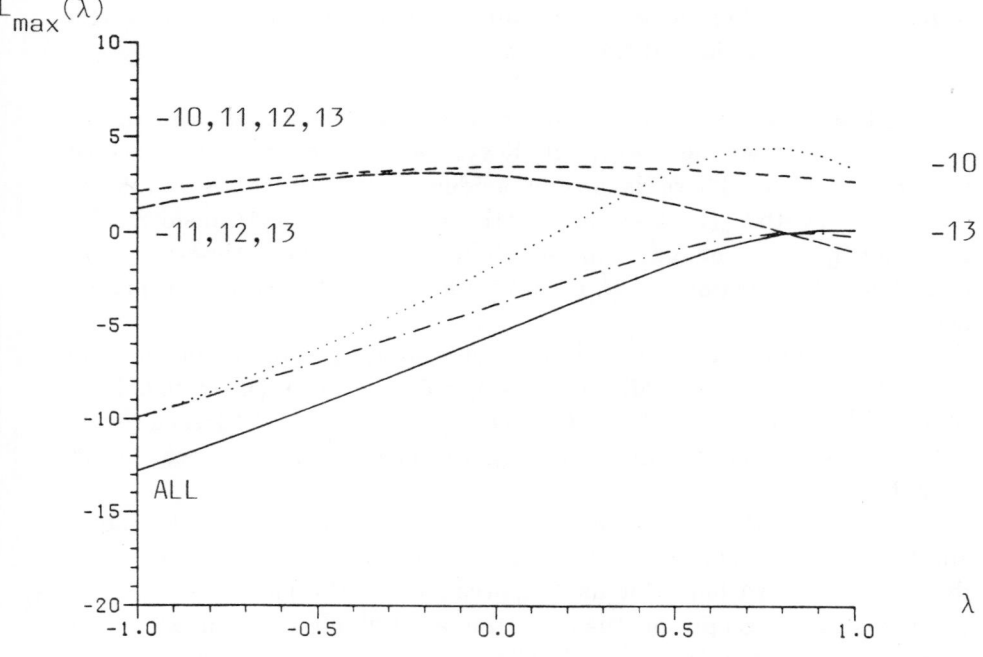

Fig. 2. Augmented Cook and Wang "data": Profile loglikelihoods $L_{max}(\lambda)$ against λ as observations are deleted.

The plot of the robust residuals $\tilde{r}_i$ against λ given by Atkinson (1988) helps to determine the dependence of this overwhelming value on individual observations. For $\lambda = 1$ observations 19 and 20 give standardized residuals of 22.96 and 19.68. These come from the two largest responses. Similarly, for $\lambda = -1$, the two most extreme values are -15.82 and -11.00 for observations 8 and 9, the two smallest values of y. The importance of these extreme observations in determining the transformation is confirmed by the plot of $L_{max}(\lambda)$, Fig. 4. For all observations the log transformation is strongly supported. Deletion of observations 19 and 20 reduces the evidence for a transformation at $\lambda = 1$, but has much less effect at $\lambda = -1$. The reverse is true for observations 8 and 9. Deletion of these small observations causes little change in the evidence against $\lambda = 1$, but weakens the evidence against the reciprocal transformation. If all four extreme observations are deleted the log transformation is still the prefered transformation, although the strength of evidence for a transformation is reduced.

This analysis supports that of Box and Cox. Here the contribution of the robust analysis is to highlight the importance of extreme observations in determining the transformation. The likelihood plots in Fig. 4 show that the transformation suggested by these influential observations is the same as that indicated by the rest of the data. The situation is thus very different from that of the previous example where a few observations were solely determining the transformation.

Example 2. *Box and Cox Poison Data.* The data consist of the results of a 3×4 factorial experiment with 4 replicates per cell, in which the response is survival time. There is strong evidence that the data should be transformed with $T_p(1) = -13.54$. For this example the log transformation does not go far enough. The maximum likelihood estimate of the transformation parameter is $\hat{\lambda} = -0.75$ and the reciprocal transformation is indicated.

Rather than subject the data to an analysis similar to that for Example 1, the effect of addition of a single outlier is demonstrated. To do this we follow Andrews (1971) and alter y_{20} from 0.23 to 0.13. As a result, the likelihood analysis indicates the log, rather than the reciprocal, transformation.

The effect of the changed observation is not apparent from the robust analysis at $\lambda = 1$. The largest least median of squares residuals belong to the largest observations. But, as λ approaches -1, the robust residuals all become small, except for that for the altered observation 20 which increases in magnitude to -17.13. The next greatest magnitude is -3.99

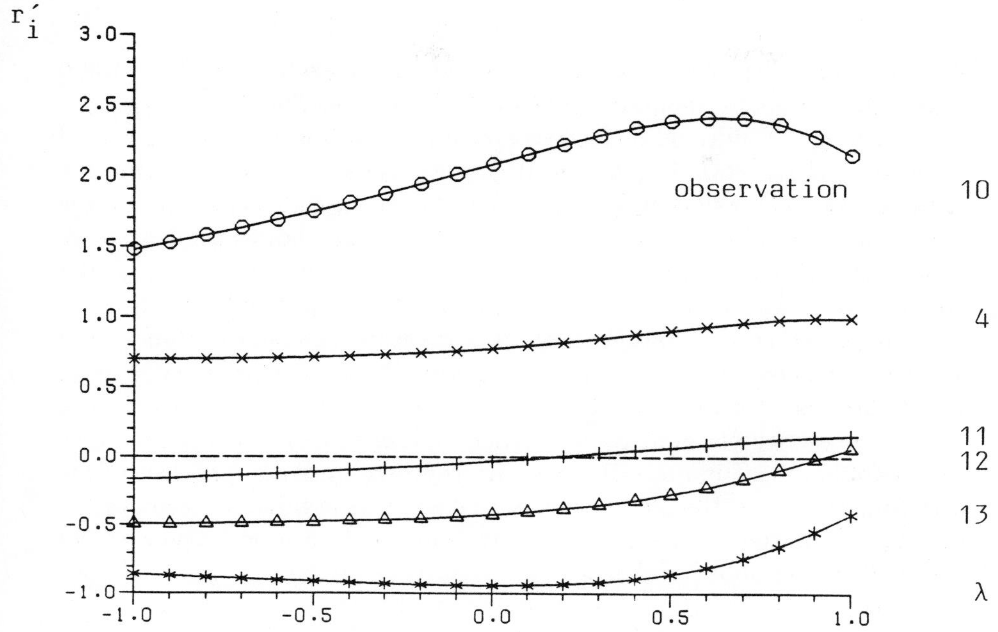

Fig. 3. Augmented Cook and Wang "data": Standardized least squares residuals r_i' against λ.

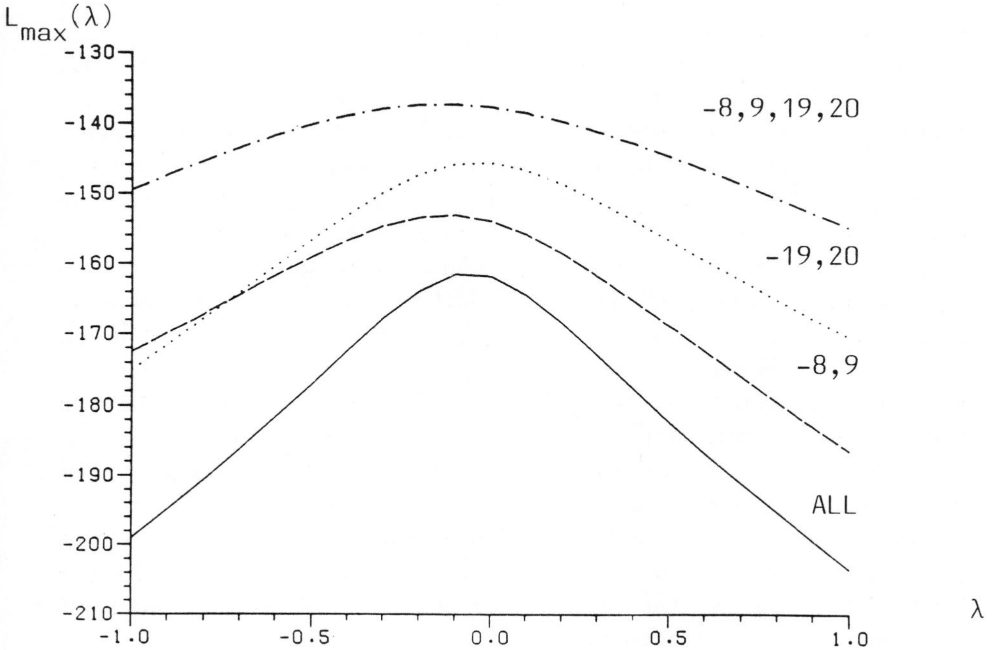

Fig. 4. Box and Cox worsted data: Profile loglikelihoods $L_{\max}(\lambda)$ against λ as observations are deleted.

for observation 35. The plot of robust residuals given by Atkinson (1988) clearly indicates the anomalous nature of observation 20.

The effect of individual observations on the estimate of λ is revealed in the plot of $L_{max}(\lambda)$, Fig. 5. For all observations the log transformation is indicated. If observation 20 is deleted, the reciprocal transformation is re-established. If observations 24 and 30, large values indicated as important by the robust analysis, are deleted, the inference is not changed: observation 20 remains in conflict with the rest of the data.

The importance of observation 20 is revealed not only by the robust analysis, but also by the plot of least squares residuals against λ, Fig. 6. Least squares here reveals the structure because the balance of the data and the presence of replication mean that there are no leverage points which could exert undue influence and there is thus no opportunity for masking. However, the standardized least squares residuals are about 1/3 the size of the robust ones. The structure of the data is more powerfully revealed by the robust analysis.

Example 3. *Hill Racing Data.* Atkinson (1986c) gives the record times for 35 hill races, together with the distance in miles and the climb in feet. The data are taken from the 1984 fixture list of the Scottish Hill Runners Association. For the calculations in this paper, the time for race 18, which is 3 miles long, has been corrected from 1 hr 18 minutes to 18 minutes.

The robust analysis on the untransformed scale shows that observations 7, 11, 33 and 35 have large positive residuals. The residuals decrease as λ decreases and score statistics for all 35 observations suggest that the square root transformation might be appropriate, as does the plot of $L_{max}(\lambda)$ for all observations in Fig. 8. However, observation 7 is revealed as important not only by the robust analysis, but also by the plot of least squares residuals in Fig. 7. The effect of deletion of observation 7 is to change the value of $T_p(1)$ from -6.24 to -3.17. If the plot of least squares residuals is repeated after observation 7 has been deleted, observation 33 is revealed as outlying. If this observation is also deleted, $T_p(1)$ becomes a non-significant -1.42 and, as Fig. 8 confirms, there is now no evidence for a transformation.

One feature of this example is that all the evidence for a transformation is due to 2 observations. A second feature is that least median of squares immediately revealed the importance of observations 7 and 33 (as well as the unimportant observations 11 and 35). But, when least squares was used, the importance of observation 33 was not apparent until observation 7 had been deleted.

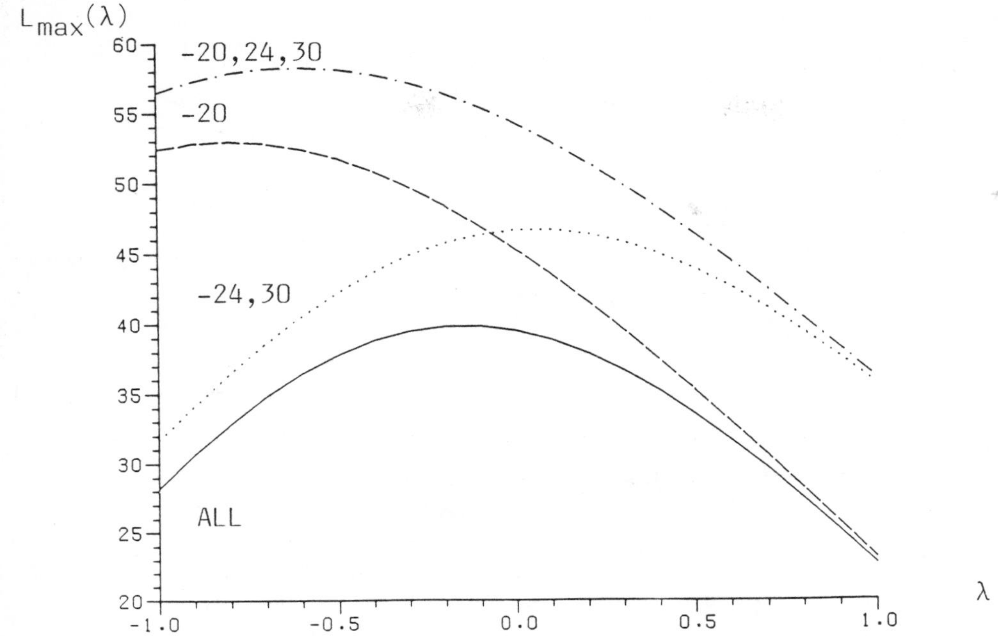

Fig. 5. Altered Box and Cox poison data: Profile loglikelihoods $L_{\max}(\lambda)$ against λ as observations are deleted.

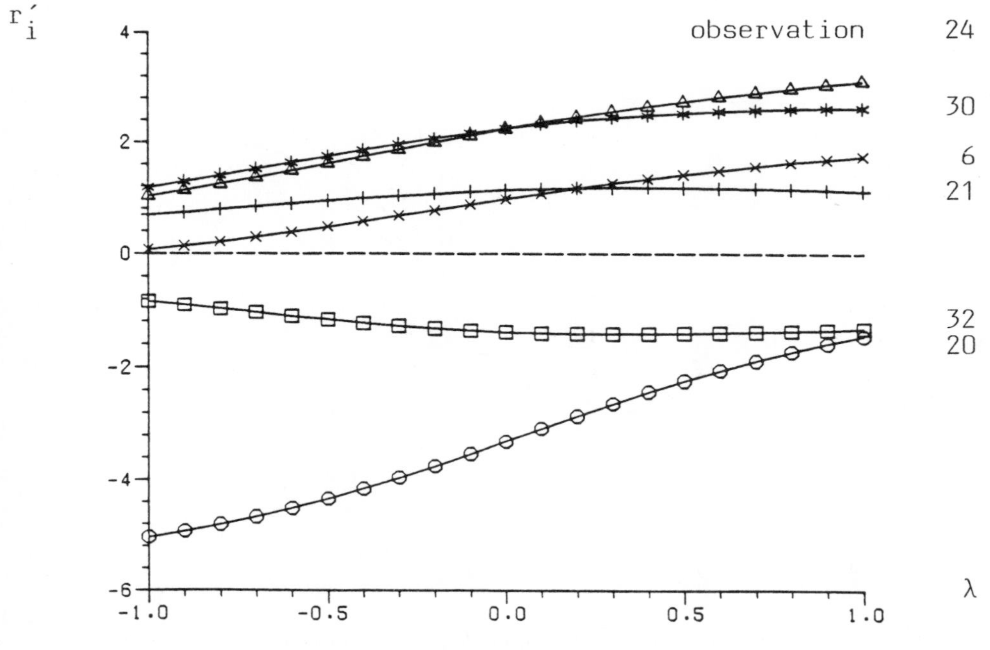

Fig. 6. Altered Box and Cox poison data: Standardized least squares residuals r_i' against λ.

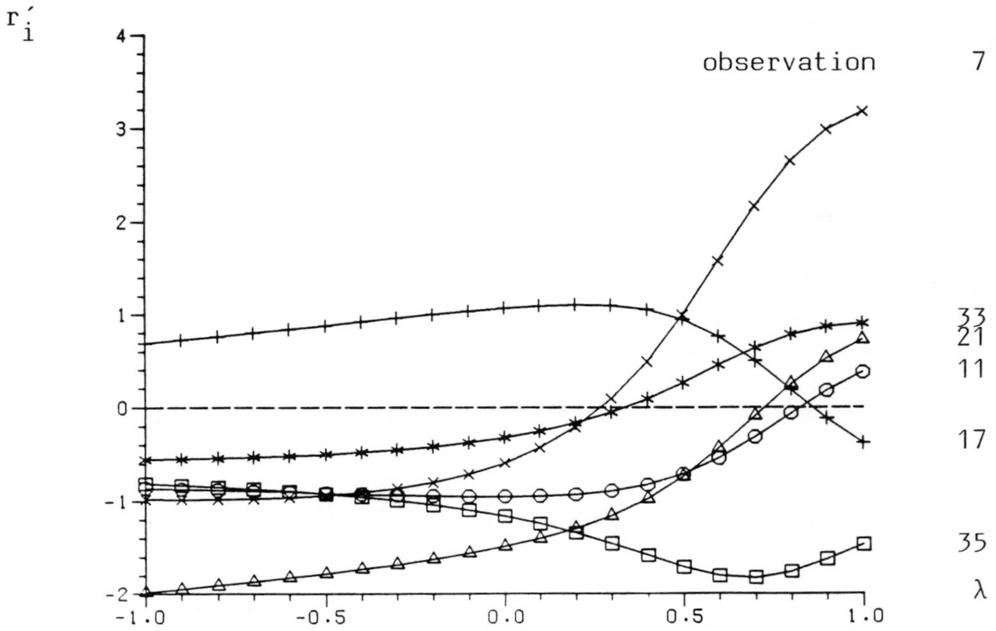

Fig. 7. Hill racing data: Standardized least squares residuals r_i' against λ.

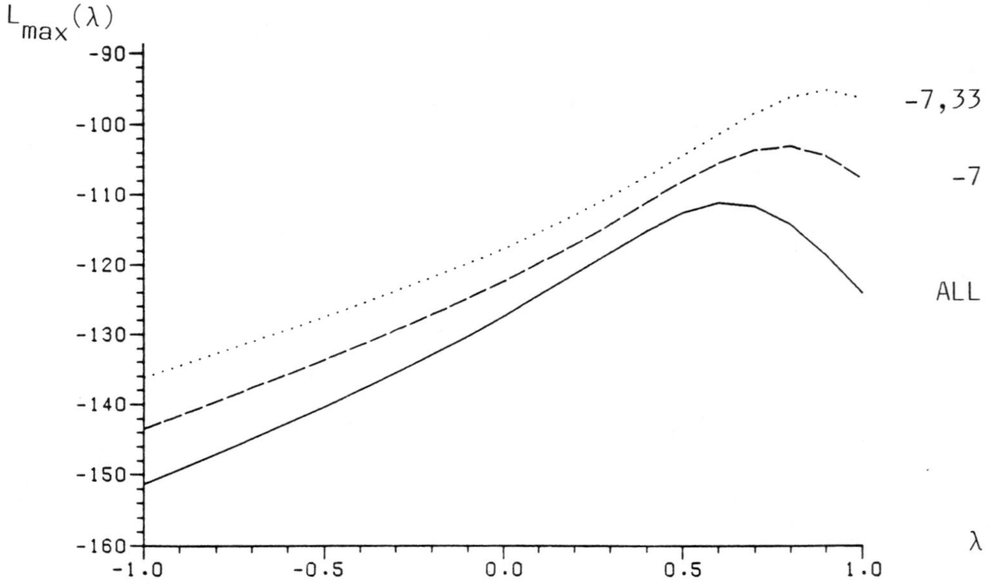

Fig. 8. Hill racing data: Profile loglikelihoods $L_{\max}(\lambda)$ as observations are deleted.

6. COMMENT

The robust analyses described by Atkinson (1988) combine plots of least median of squares residuals with diagnostic procedures based on the approximate score statistic for transformations. The analyses described in this paper show the extra information to be gained by coupling least median of squares with plots of profile loglikelihoods. Detailed properties of $L_{max}(\lambda)$ for the transformation (3) are discussed by Cox and Reid (1987). Both sets of analyses reinforce the necessity of following the exploratory robust methods with a confirmatory analysis. An instance is the information about observations 11 and 35 in the hill racing data. The examples also show that exploratory information on outlying and potentially influential observations is more clearly revealed by plots of least median of squares residuals than by plots of least squares residuals.

The robust method reveals influential observations when masking is present, as in the example in Section 4, which is not revealed by least squares. For the other examples, the plot of robust residuals is more informative than the plot of least squares residuals. But, when the observations come from a balanced design, established diagnostic methods based on least squares provide the required information with appreciably reduced computation. These methods include those based on added variable plots (Cook and Weisberg, 1982, Section 2.4; Atkinson, 1985, Chapters 6-9) and on measures of the effect of deletion of single observations (Cook and Wang, 1983; Atkinson 1986a).

REFERENCES

Andrews, D.F. (1971). A note on the selection of data transformations. *Biometrika*, 58, 249-254.

Atkinson, A.C. (1973). Testing transformations to normality. *J.R. Statist.Soc. B*, 35, 473-479.

Atkinson, A.C. (1985). *Plots, Transformations, and Regression*. Oxford University Press, Oxford, UK.

Atkinson, A.C. (1986a). Diagnostic tests for transformations. *Technometrics*, 28, 29-37.

Atkinson, A.C. (1986b). Masking unmasked. *Biometrika*, 73, 533-541.

Atkinson, A.C. (1986c). Discussion of paper by Chatterjee and Hadi. *Statistical Science*, 1, 397-402.

Atkinson, A. C. (1988). Transformations unmasked. Submitted for publication.

Belsley, D.A., Kuh, E. and Welsch, R.E. (1980). *Regression Diagnostics*. Wiley, New York.

Box, G.E.P. and Cox, D.R. (1964). An analysis of transformations (with discussion). *J. R. Statist. Soc. B*, 26, 211-246.

Cook, R.D. and Wang, P.C. (1983). Transformations and influential cases in regression. *Technometrics*, 25, 337-343.

Cook, R.D. and Weisberg, S. (1982). *Residuals and Influence in Regression.* Chapman and Hall, New York and London.

Cox, D.R. and Reid, N. (1987). Parameter orthogonality and approximate conditional inference (with discussion). *J. R. Stat. Soc. B*, 49, 1-39.

Hawkins,, D.M., Bradu, D. and Kass, G.V. (1984). Location of several outliers in multiple-regression data using elemental sets. *Technometrics*, 26, 197-208.

Hinkley, D.V. and Runger, G. (1984). The analysis of transformed data (with discussion). *J. Amer. Statist. Assoc.*, 79, 302-320.

Rousseeuw, P.J. (1984). Least median of squares regression. *J. Amer. Statist. Assoc.*, 79, 871-880.

Rousseeuw, P.J. and Leroy, A. (1987). *Robust Regression and Outlier Detection.* Wiley, New York.

Weisberg, S. (1985). *Applied Linear Regression.* Second Edition. Wiley, New York.

Received 9 March 1987

Department of Mathematics
Imperial College of Science and Technology
London SW7 2BZ
U.K.

Proc. Second International Tampere Conference in Statistics
(Tampere, Finland, 1-4 June 1987)
Tarmo Pukkila and Simo Puntanen, *Editors*

pp. 113 - 142

INVITED PAPER

Algebraic characterizations and statistical implications of the commutativity of orthogonal projectors

Jerzy K. BAKSALARY

Academy of Agriculture in Poznań, Poznań, Poland

Key words and phrases: Ordinary least squares estimator, generalized least squares estimator, best linear unbiased estimator, linearly sufficient statistics, linearly minimal sufficient statistics, canonical correlations.

ABSTRACT

Let **A** and **B** be real matrices with equal numbers of rows, and let $\mathbf{P}_A$ and $\mathbf{P}_B$ denote the orthogonal projectors onto the ranges of **A** and **B**, respectively. The purpose of this paper is to collect together various algebraic characterizations of the equality $\mathbf{P}_A\mathbf{P}_B = \mathbf{P}_B\mathbf{P}_A$, and then discuss its statistical implications. In the algebraic part of the paper, fourty-five conditions equivalent to the commutativity of $\mathbf{P}_A$ and $\mathbf{P}_B$ are revealed, of which some are quotations from the literature, some others are new. The discussion of statistical aspects of the commutativity of two orthogonal projectors refers to problems concerned with canonical correlations and with comparisons between estimators and between sets of linearly sufficient statistics corresponding to different linear models. Some known solutions to the problems discussed are derived applying new approaches, some other results are new.

1. INTRODUCTION

Let $M_{n,p}$ denote the set of $n\times p$ real matrices. Given $\mathbf{L} \in M_{n,p}$, the symbols $\mathbf{L}'$, $R(\mathbf{L})$, $R^{\perp}(\mathbf{L})$, and $r(\mathbf{L})$ will stand for the transpose of $\mathbf{L}$, range of $\mathbf{L}$, the orthogonal (with respect to the standard inner product) complement to $R(\mathbf{L})$, and the rank of $\mathbf{L}$, respectively. Further, let $\mathbf{P}_{\mathbf{L}}$ denote the orthogonal projector onto $R(\mathbf{L})$. It admits representation

$$\mathbf{P}_{\mathbf{L}} = \mathbf{L}\mathbf{L}^{+}, \tag{1.1}$$

where $\mathbf{L}^{+}$ is the Moore-Penrose inverse of $\mathbf{L}$, which may be characterized as the unique solution, $\mathbf{X}$, to the four equations

$$\mathbf{LXL} = \mathbf{L},\ \mathbf{XLX} = \mathbf{X},\ \ \mathbf{LX} = (\mathbf{LX})',\ \ \mathbf{XL} = (\mathbf{XL})'. \tag{1.2}$$

Another widely used representation of $\mathbf{P}_{\mathbf{L}}$ is that originally given by Rao (1965, p. 187),

$$\mathbf{P}_{\mathbf{L}} = \mathbf{L}(\mathbf{L}'\mathbf{L})^{-}\,\mathbf{L}', \tag{1.3}$$

where $(\mathbf{L}'\mathbf{L})^{-}$ is a generalized inverse of $\mathbf{L}'\mathbf{L}$, i.e., any matrix satisfying $\mathbf{L}'\mathbf{L}(\mathbf{L}'\mathbf{L})^{-}\mathbf{L}'\mathbf{L} = \mathbf{L}'\mathbf{L}$. Consequently, if $\mathbf{Q}_{\mathbf{L}}$ denotes the orthogonal projector onto $R^{\perp}(\mathbf{L})$, then

$$\mathbf{Q}_{\mathbf{L}} = \mathbf{I}_n - \mathbf{P}_{\mathbf{L}} = \mathbf{I}_n - \mathbf{L}\mathbf{L}^{+} = \mathbf{I}_n - \mathbf{L}(\mathbf{L}'\mathbf{L})^{-}\,\mathbf{L}', \tag{1.4}$$

where $\mathbf{I}_n$ is the identity matrix of order n.

The commutativity of two projectors (not necessarily orthogonal) was established by Rao and Mitra (1971, Theorem 5.1.4) to be a sufficient condition for the product of the projectors to be a projector as well. With the restriction to orthogonal projectors, necessary and sufficient conditions for the commutativity property were derived by Afriat (1957, p. 801) and Rao and Yanai (1979, Theorem 4 and Corollary 6); see also Takeuchi, Yanai and Mukherjee (1982, Theorems 1.22 and 1.28). Some other conditions were given by Fellman (1976) and Baksalary (1984) in their discussions of a linear model of the form

$$\mathrm{M}_a = \{\mathbf{Y},\, \mathbf{W}\boldsymbol{\gamma}+\mathbf{Z}\boldsymbol{\delta},\, \sigma^2\mathbf{I}_n\}, \tag{1.5}$$

in which $\mathbf{Y}$ is an $n\times 1$ observable random vector with expectation $E(\mathbf{Y}) = \mathbf{W}\boldsymbol{\gamma}+\mathbf{Z}\boldsymbol{\delta}$ and with dispersion matrix $D(\mathbf{Y}) = \sigma^2\mathbf{I}_n$, where $\mathbf{W} \in M_{n,w}$ and $\mathbf{Z} \in M_{n,z}$ are known, while $\boldsymbol{\gamma} \in M_{w,1}$, $\boldsymbol{\delta} \in M_{z,1}$ and $\sigma^2>0$ are unknown parameters. Under the assumption that only the components of $\boldsymbol{\gamma}$ are of interest while those of $\boldsymbol{\delta}$ are nuisance parameters, a natural counterpart to (1.5) is the model

$$\mathrm{M}_0 = \{\mathbf{Y}, \mathbf{W}\boldsymbol{\gamma}, \sigma^2\mathbf{I}_n\}, \tag{1.6}$$

in which the nuisance parameters are absent. Clearly, the original model may then be viewed as an augmentation of M_0, and hence the subscript "a" in (1.5). Further conditions for the commutativity of two orthogonal projectors were derived by Puntanen (1985, 1986, 1987) in his discussion of the canonical correlations between the vectors $\mathbf{P}_{\mathbf{X}}\mathbf{Y}$ and $\mathbf{Q}_{\mathbf{X}}\mathbf{Y}$ under a general linear model of the form

$$\mathrm{M} = \{\mathbf{Y}, \mathbf{X}\boldsymbol{\beta}, \sigma^2\mathbf{V}\}, \tag{1.7}$$

in which each of the known matrices $\mathbf{X}$ and $\mathbf{V}$ may be deficient in rank.

The purpose of this paper is to collect together various algebraic characterizations of the equality $\mathbf{P}_{\mathbf{A}}\mathbf{P}_{\mathbf{B}} = \mathbf{P}_{\mathbf{B}}\mathbf{P}_{\mathbf{A}}$, and then discuss its statistical implications. In the algebraic part of the paper, fourty-five conditions equivalent to the commutativity of $\mathbf{P}_{\mathbf{A}}$ and $\mathbf{P}_{\mathbf{B}}$ are revealed. Some of them are quotations from the literature, several other are new. The discussion of statistical aspects of the commutativity of two orthogonal projectors includes certain problems concerned with: (i) canonical correlations between the vectors $\mathbf{W}'\mathbf{Y}$ and $\mathbf{Z}'\mathbf{Y}$ under the model M_a, (ii) comparisons of the ordinary least squares estimator (OLSE) of $\mathbf{X}\boldsymbol{\beta}$ and the generalized least squares estimator (GLSE) of $\mathbf{X}\boldsymbol{\beta}$ to the best linear unbiased estimator (BLUE) of $\mathbf{X}\boldsymbol{\beta}$ under the model M, (iii) comparisons between the BLUEs of linear functions of $\boldsymbol{\gamma}$ under the model M_a and its reduced counterpart M_0, and (iv) comparisons between the sets of linearly sufficient statistics for $\mathbf{W}'\mathbf{Q}_{\mathbf{Z}}\mathbf{W}\boldsymbol{\gamma}$ and between the sets of linearly minimal sufficient statistics for $\mathbf{W}'\mathbf{Q}_{\mathbf{Z}}\mathbf{W}\boldsymbol{\gamma}$ under the models M_a and M_0. Some known solutions to these problems are derived applying new approaches, some other results are new.

2. ALGEBRAIC CHARACTERIZATIONS

Fourty-five various characterizations of the commutativity of orthogonal projectors are collected together in Theorem 1. These characterizations are expressed in terms of matrices and their ranks, as well as in terms of subspaces and their dimensions. The proof of Theorem 1 is given in the final section of the paper.

THEOREM 1. *Let* $\mathbf{A} \in M_{n,a}$ *and* $\mathbf{B} \in M_{n,b}$, *and let* $\mathbf{S}$ *and* $\mathbf{T}$ *be any matrices such that*

$$R(\mathbf{S}) = R(\mathbf{A}) + R(\mathbf{B}) = R(\mathbf{A} : \mathbf{B}), \tag{2.1}$$

$$R(\mathbf{T}) = R(\mathbf{A}) \cap R(\mathbf{B}). \tag{2.2}$$

Further, let $(\mathbf{A}'\mathbf{A})^-$ *and* $(\mathbf{B}'\mathbf{B})^-$ *be any generalized inverses of* $\mathbf{A}'\mathbf{A}$ *and* $\mathbf{B}'\mathbf{B}$, *let*

$$\mathbf{G} = (\mathbf{B}'\mathbf{B})^- \, \mathbf{B}'\mathbf{A}(\mathbf{A}'\mathbf{A})^-, \tag{2.3}$$

and let the symbol $\boxplus$ *denote the orthogonal sum of subspaces. Then the following fourty-six statements are equivalent:*

(A1) $\mathbf{P}_\mathbf{A}\mathbf{P}_\mathbf{B} = \mathbf{P}_\mathbf{B}\mathbf{P}_\mathbf{A}$,
(A2) $\mathbf{A}'\mathbf{B} = \mathbf{A}'\mathbf{P}_\mathbf{B}\mathbf{P}_\mathbf{A}\mathbf{B}$,
(A3) $\mathbf{G}$ *is a generalized inverse of* $\mathbf{A}'\mathbf{B}$,
(A4) $\mathbf{A}'\mathbf{B}$ *is a generalized inverse of* $\mathbf{G}$,
(A5) $\mathbf{A}'\mathbf{B}\mathbf{G} = (\mathbf{A}'\mathbf{B}\mathbf{G})^2$,
(A6) $\mathbf{G}\mathbf{A}'\mathbf{B} = (\mathbf{G}\mathbf{A}'\mathbf{B})^2$,
(A7) $\mathbf{P}_\mathbf{A}\mathbf{P}_\mathbf{B} = (\mathbf{P}_\mathbf{A}\mathbf{P}_\mathbf{B})^2$,
(A8) $\mathbf{B}'\mathbf{P}_\mathbf{A}\mathbf{B} = \mathbf{B}'\mathbf{P}_\mathbf{A}\mathbf{P}_\mathbf{B}\mathbf{P}_\mathbf{A}\mathbf{B}$,
(A9) $\mathbf{B}'\mathbf{Q}_\mathbf{A}\mathbf{B} = \mathbf{B}'\mathbf{Q}_\mathbf{A}\mathbf{P}_\mathbf{B}\mathbf{Q}_\mathbf{A}\mathbf{B}$,
(A10) $\mathbf{Q}_\mathbf{B}\mathbf{P}_\mathbf{A}\mathbf{B} = \mathbf{0}$,
(A11) $\mathbf{P}_\mathbf{A}\mathbf{P}_\mathbf{B} = \mathbf{P}_\mathbf{T}$,
(A12) $\mathbf{A}'\mathbf{B} = \mathbf{A}'\mathbf{P}_\mathbf{T}\mathbf{B}$,
(A13) $\mathbf{P}_\mathbf{A}\mathbf{Q}_\mathbf{B}$ *is the orthogonal projector onto* $R(\mathbf{A}) \cap R^\perp(\mathbf{T})$,
(A14) $\mathbf{P}_\mathbf{A}\mathbf{Q}_\mathbf{B}$ *is the orthogonal projector onto* $R(\mathbf{A}) \cap R^\perp(\mathbf{B})$,
(A15) $\mathbf{P}_\mathbf{A}\mathbf{Q}_\mathbf{B}$ *is the orthogonal projector onto* $R(\mathbf{S}) \cap R^\perp(\mathbf{B})$,
(A16) $\mathbf{P}_\mathbf{S} = \mathbf{P}_\mathbf{A} + \mathbf{P}_\mathbf{B} - \mathbf{P}_\mathbf{A}\mathbf{P}_\mathbf{B}$,
(A17) $\mathbf{P}_\mathbf{A} + \mathbf{P}_\mathbf{B} - {}^3\!/_2\mathbf{P}_\mathbf{A}\mathbf{P}_\mathbf{B} = (\mathbf{P}_\mathbf{A} + \mathbf{P}_\mathbf{B})^+$,
(A18) $\mathbf{P}_\mathbf{A} + \mathbf{P}_\mathbf{B} - {}^3\!/_2\mathbf{P}_\mathbf{A}\mathbf{P}_\mathbf{B}$ *is a generalized inverse of* $\mathbf{P}_\mathbf{A} + \mathbf{P}_\mathbf{B}$,
(A19) $R(\mathbf{Q}_\mathbf{T}\mathbf{A})$ *and* $R(\mathbf{Q}_\mathbf{T}\mathbf{B})$ *are orthogonal,*
(A20) $R(\mathbf{A}) \cap R^\perp(\mathbf{T})$ *and* $R(\mathbf{B}) \cap R^\perp(\mathbf{T})$ *are orthogonal,*
(A21) $R(\mathbf{A}) \cap R^\perp(\mathbf{T})$ *and* $R(\mathbf{B})$ *are orthogonal,*
(A22) $R(\mathbf{P}_\mathbf{A}\mathbf{B}) \subseteq R(\mathbf{B})$,
(A23) $R(\mathbf{P}_\mathbf{A}\mathbf{B}) \subseteq R(\mathbf{T})$,
(A24) $R(\mathbf{P}_\mathbf{A}\mathbf{B}) = R(\mathbf{T})$,
(A25) $R(\mathbf{A}'\mathbf{B}) = R(\mathbf{A}'\mathbf{T})$,
(A26) $R(\mathbf{Q}_\mathbf{B}\mathbf{A}) = R(\mathbf{A}) \cap R^\perp(\mathbf{T})$,
(A27) $R(\mathbf{Q}_\mathbf{B}\mathbf{A}) = R(\mathbf{A}) \cap R^\perp(\mathbf{B})$,
(A28) $R(\mathbf{S}) \cap R^\perp(\mathbf{B}) = R(\mathbf{A}) \cap R^\perp(\mathbf{T})$,

(A29) $R(\mathbf{S}) \cap R^{\perp}(\mathbf{B}) = R(\mathbf{A}) \cap R^{\perp}(\mathbf{B})$,
(A30) $R(\mathbf{A}) = R(\mathbf{T}) \boxplus R(\mathbf{S}) \cap R^{\perp}(\mathbf{B})$,
(A31) $R(\mathbf{A}) = R(\mathbf{T}) \boxplus R(\mathbf{Q}_{\mathbf{B}}\mathbf{A})$,
(A32) $R(\mathbf{S}) = R(\mathbf{B}) \boxplus R(\mathbf{A}) \cap R^{\perp}(\mathbf{T})$,
(A33) $R(\mathbf{S}) = R(\mathbf{B}) \boxplus R(\mathbf{A}) \cap R^{\perp}(\mathbf{B})$,
(A34) $\mathrm{r}(\mathbf{A}'\mathbf{B}) = \mathrm{r}(\mathbf{T})$,
(A35) $\dim R(\mathbf{A}) \cap R^{\perp}(\mathbf{T}) = \mathrm{r}(\mathbf{A}) - \mathrm{r}(\mathbf{A}'\mathbf{B})$,
(A36) $\dim R(\mathbf{S}) \cap R^{\perp}(\mathbf{B}) = \mathrm{r}(\mathbf{A}) - \mathrm{r}(\mathbf{A}'\mathbf{B})$,
(A37) $\mathrm{r}(\mathbf{Q}_{\mathbf{B}}\mathbf{A}) = \mathrm{r}(\mathbf{A}) - \mathrm{r}(\mathbf{A}'\mathbf{B})$,
(A38) $\mathrm{r}(\mathbf{S}) = \mathrm{r}(\mathbf{A}) + \mathrm{r}(\mathbf{B}) - \mathrm{r}(\mathbf{A}'\mathbf{B})$,
(A39) $\dim R(\mathbf{A}) \cap R^{\perp}(\mathbf{B}) = \mathrm{r}(\mathbf{Q}_{\mathbf{B}}\mathbf{A})$,
(A40) $\dim R(\mathbf{A}) \cap R^{\perp}(\mathbf{B}) = \mathrm{r}(\mathbf{S}) - \mathrm{r}(\mathbf{B})$,
(A41) $\dim R(\mathbf{A}) \cap R^{\perp}(\mathbf{B}) = \mathrm{r}(\mathbf{A}) - \mathrm{r}(\mathbf{T})$,
(A42) $\dim R(\mathbf{A}) \cap R^{\perp}(\mathbf{B}) = \dim R(\mathbf{A}) \cap R^{\perp}(\mathbf{T})$,
(A43) $\dim R(\mathbf{A}) \cap R^{\perp}(\mathbf{B}) = \dim R(\mathbf{S}) \cap R^{\perp}(\mathbf{B})$,
(A44) $\mathrm{r}(\mathbf{I}_a - \mathbf{A}'\mathbf{B}\mathbf{G}) = a - \mathrm{r}(\mathbf{A}'\mathbf{B})$,
(A45) $\mathrm{r}(\mathbf{I}_b - \mathbf{G}\mathbf{A}'\mathbf{B}) = b - \mathrm{r}(\mathbf{A}'\mathbf{B})$,
(A46) $\mathbf{P}_{\mathbf{A}}$ *and* $\mathbf{P}_{\mathbf{B}}$ *are simultaneously reducible to diagonal matrices by an orthogonal transformation.*

Among the conditions collected in Theorem 1, some are quotations from the literature. The statement (A46) is a particular case of the well known characterization of the commutativity of symmetric matrices; see e.g., Ben-Israel and Greville (1974, pp. 250-251) who trace back a criterion for a simultaneous diagonalization to Eckart and Young (1939). The equivalence of (A1) to (A20) has been noted by Afriat (1957, p. 801). The conditions (A2), (A3), (A44), and (A45) were established by Fellman (1976). The conditions (A11) and (A16) are due to Rao and Yanai (1979). Combining Lemmas 1 and 2 in Puntanen (1985) yields (A34), (A37), (A41), and two other conditions, not listed in Theorem 1. Moreover, Lemma 1 in Puntanen (1986) and Lemma 3.4.1 in Puntanen (1987, Part 1) include (A22), (A24), and a version of (A31).

The commutativity of orthogonal projectors is closely related to the concept of commutative linear subspaces which has been introduced in a more general context by Birkhoff and von Neumann (1936). For the definition of this concept and its fundamental properties the reader is referred to a recent survey by Nordström and von Rosen (1987, Section 3). Adapted to the particular case considered in this paper, their Theorem 3.1 states that the subspaces $R(\mathbf{A})$ and $R(\mathbf{B})$ are commutative if and only if any of the conditions (A1), (A11), (A16), (A20) holds. In view of the representations $R(\mathbf{A}) = R(\mathbf{T}) \boxplus R(\mathbf{A}) \cap R^{\perp}(\mathbf{T})$ and $R(\mathbf{B}) = R(\mathbf{T}) \boxplus R(\mathbf{B}) \cap R^{\perp}(\mathbf{T})$, the last of these four conditions reveals a simple geometric meaning of the commutativity property. The subspaces $R(\mathbf{A})$ and $R(\mathbf{B})$ satisfying (A20) are called "orthogonally incident", cf. Afriat (1957, p. 801), or "geometrically orthogonal", cf. Tjur (1984, p. 40).

The following two corollaries to Theorem 1 will be useful from the point of view of statistical applications considered in the next section.

COROLLARY 1. *Let* $\mathbf{A} \in M_{n,a}$ and $\mathbf{B} \in M_{n,b}$ *be such that* $R(\mathbf{A}) \cap R(\mathbf{B}) = \{\mathbf{0}\}$. *Then the projectors* $\mathbf{P}_{\mathbf{A}}$ *and* $\mathbf{P}_{\mathbf{B}}$ *commute if and only if* $\mathbf{A}'\mathbf{B} = \mathbf{0}$, *i.e., if and only if the subspaces* $R(\mathbf{A})$ *and* $R(\mathbf{B})$ *are orthogonal.*

COROLLARY 2. *Let* $\mathbf{A} \in M_{n,a}$ *and* $\mathbf{B} \in M_{n,b}$ *be such that* $R(\mathbf{A}) \cap R(\mathbf{B}) = R(\mathbf{1}_n)$, *where* $\mathbf{1}_n \in M_{n,1}$ *has all its components equal to one. Further, let* $\mathbf{c}_{\mathbf{A}} \in M_{a,1}$ *and* $\mathbf{c}_{\mathbf{B}} \in M_{b,1}$ *be the vectors of the column sums of* $\mathbf{A}$ *and* $\mathbf{B}$, *respectively, i.e.,* $\mathbf{c}_{\mathbf{A}} = \mathbf{A}'\mathbf{1}_n$ *and* $\mathbf{c}_{\mathbf{B}} = \mathbf{B}'\mathbf{1}_n$. *Then the projectors* $\mathbf{P}_{\mathbf{A}}$ *and* $\mathbf{P}_{\mathbf{B}}$ *commute if and only if* $\mathbf{A}'\mathbf{B} = \mathbf{c}_{\mathbf{A}}\mathbf{c}_{\mathbf{B}}'/n$.

Notice that Corollary 1 is an immediate consequence of the part "(A1) $\Leftrightarrow$ (A12)" of Theorem 1. It may also be viewed as a particular case of a result given by Rao and Yanai (1979, p. 4). On the other hand, Corollary 2, quoted from Rao and Yanai (1979, pp. 12-13), is an immediate consequence of the part "(A1) $\Leftrightarrow$ (A11)" of Theorem 1 and the fact that, according to (1.3), the orthogonal projector onto $R(\mathbf{1}_n)$ is of the form $\mathbf{1}_n\mathbf{1}_n'/n$.

3. STATISTICAL IMPLICATIONS

3.1 Canonical correlations

This subsection is concerned with the canonical correlations between the vectors $\mathbf{W}'\mathbf{Y}$ and $\mathbf{Z}'\mathbf{Y}$ under the model $\mathrm{M}_a = \{\mathbf{Y}, \mathbf{W}\boldsymbol{\gamma} + \mathbf{Z}\boldsymbol{\delta}, \sigma^2\mathbf{I}_n\}$ given in (1.5). Since the dispersion matrix

$$D\begin{pmatrix} \mathbf{W}'\mathbf{Y} \\ \mathbf{Z}'\mathbf{Y} \end{pmatrix} = \sigma^2\begin{pmatrix} \mathbf{W}'\mathbf{W} & \mathbf{W}'\mathbf{Z} \\ \mathbf{Z}'\mathbf{W} & \mathbf{Z}'\mathbf{Z} \end{pmatrix}$$

may be singular, the general theory developed by Khatri (1976), Seshadri and Styan (1980), and Rao (1981) is to be adopted for discussing these correlations. They were thoroughly studied by Styan (1985, Section 2), and the purpose here is to supplement his considerations by some comments referring to the commutativity of orthogonal projectors.

According to Styan's (1985) Theorem 2.1, the nonzero eigenvalues of the matrix

$$(\mathbf{W}'\mathbf{W})^{-}\mathbf{W}'\mathbf{Z}(\mathbf{Z}'\mathbf{Z})^{-}\mathbf{Z}'\mathbf{W}$$

are the squares of the canonical correlations between $\mathbf{W}'\mathbf{Y}$ and $\mathbf{Z}'\mathbf{Y}$, irrespective of the choice of generalized inverses $(\mathbf{W}'\mathbf{W})^{-}$ and $(\mathbf{Z}'\mathbf{Z})^{-}$. Consequently, the number of nonzero canonical correlations between $\mathbf{W}'\mathbf{Y}$ and $\mathbf{Z}'\mathbf{Y}$ is

$$s = \mathrm{r}(\mathbf{W}'\mathbf{Z}), \tag{3.1}$$

including the number

$$s_1 = \mathrm{r}(\mathbf{W}) + \mathrm{r}(\mathbf{Z}) - \mathrm{r}(\mathbf{W} : \mathbf{Z}) \tag{3.2}$$

of those which are equal to one.

One extreme situation is when all the nonzero canonical correlations between $\mathbf{W}'\mathbf{Y}$ and $\mathbf{Z}'\mathbf{Y}$ are less than one. In view of (3.2), this is the case if and only if

$$\mathrm{r}(\mathbf{W} : \mathbf{Z}) = \mathrm{r}(\mathbf{W}) + \mathrm{r}(\mathbf{Z}). \tag{3.3}$$

But in general,

$$\mathrm{r}(\mathbf{W} : \mathbf{Z}) = \mathrm{r}(\mathbf{W}) + \mathrm{r}(\mathbf{Z}) - \dim R(\mathbf{W}) \cap R(\mathbf{Z}); \tag{3.4}$$

cf. (2.17) in Marsaglia and Styan (1974). Consequently, comparing (3.3) with (3.4) shows that

$$s_1 = 0 \Leftrightarrow R(\mathbf{W}) \cap R(\mathbf{Z}) = \{0\}. \tag{3.5}$$

The other extreme situation is when all the nonzero canonical correlations between $\mathbf{W}'\mathbf{Y}$ and $\mathbf{Z}'\mathbf{Y}$ are equal to one. In view of (3.1) and (3.2), this is the case if and only if

$$\mathrm{r}(\mathbf{W} : \mathbf{Z}) = \mathrm{r}(\mathbf{W}) + \mathrm{r}(\mathbf{Z}) - \mathrm{r}(\mathbf{W}'\mathbf{Z}). \tag{3.6}$$

But (3.6) corresponds to (A38) in Theorem 1, and thus it follows that

$$s = s_1 \Leftrightarrow \mathbf{P}_{\mathbf{W}}\mathbf{P}_{\mathbf{Z}} = \mathbf{P}_{\mathbf{Z}}\mathbf{P}_{\mathbf{W}}. \tag{3.7}$$

An alternative proof of (3.7) is obtainable by utilizing Theorem 2.2 of Styan (1985), which states that

$$s - s_1 = r(\mathbf{W}'\mathbf{Q}_Z\mathbf{W} - \mathbf{W}'\mathbf{Q}_Z\mathbf{P}_W\mathbf{Q}_Z\mathbf{W}).$$

Hence $s = s_1$ if and only if $\mathbf{W}'\mathbf{Q}_Z\mathbf{W} = \mathbf{W}'\mathbf{Q}_Z\mathbf{P}_W\mathbf{Q}_Z\mathbf{W}$, which corresponds to (A9) in Theorem 1.

Another Styan's (1985) result (Theorem 2.5) asserts that the canonical correlations between the vectors $\mathbf{W}'\mathbf{Q}_Z\mathbf{Y}$ and $\mathbf{Z}'\mathbf{Q}_W\mathbf{Y}$ are all less than one, and are precisely the $s - s_1$ positive canonical correlations between the vectors $\mathbf{W}'\mathbf{Y}$ and $\mathbf{Z}'\mathbf{Y}$ that are not equal to one. An immediate corollary to this result is that $s - s_1 = 0$ if and only if the vectors $\mathbf{W}'\mathbf{Q}_Z\mathbf{Y}$ and $\mathbf{Z}'\mathbf{Q}_W\mathbf{Y}$ are uncorrelated. Hence, in view of (3.7),

$$Cov(\mathbf{W}'\mathbf{Q}_Z\mathbf{Y}, \mathbf{Z}'\mathbf{Q}_W\mathbf{Y}) = \mathbf{0} \Leftrightarrow \mathbf{P}_W\mathbf{P}_Z = \mathbf{P}_Z\mathbf{P}_W. \tag{3.8}$$

A direct proof of (3.8) is obtainable by noting that

$$Cov(\mathbf{W}'\mathbf{Q}_Z\mathbf{Y}, \mathbf{Z}'\mathbf{Q}_W\mathbf{Y}) = \sigma^2\mathbf{W}'\mathbf{Q}_Z\mathbf{Q}_W\mathbf{Z} = \sigma^2(\mathbf{W}'\mathbf{P}_Z\mathbf{P}_W\mathbf{Z} - \mathbf{W}'\mathbf{Z}),$$

and then referring to the part "(A1) $\Leftrightarrow$ (A2)" of Theorem 1.

The canonical correlations between the vectors $\mathbf{W}'\mathbf{Y}$ and $\mathbf{Z}'\mathbf{Y}$ were applied by Styan (1985) to compare the quadratic forms

$$S_H = \mathbf{Y}'\mathbf{Q}_Z\mathbf{W}(\mathbf{W}'\mathbf{Q}_Z\mathbf{W})^-\mathbf{W}'\mathbf{Q}_Z\mathbf{Y} \tag{3.9}$$

and

$$S_H^* = \mathbf{Y}'\mathbf{Q}_Z\mathbf{W}(\mathbf{W}'\mathbf{W})^-\mathbf{W}'\mathbf{Q}_Z\mathbf{Y}. \tag{3.10}$$

The former of these forms is the usual numerator sum of squares in the F-test of the testable part of the hypothesis H: $\boldsymbol{\delta} = \mathbf{0}$ in the model $\mathrm{M}_a = \{\mathbf{Y}, \mathbf{W}\boldsymbol{\gamma} + \mathbf{Z}\boldsymbol{\delta}, \sigma^2\mathbf{I}_n\}$, while the latter is formed from S_H by omitting the $\mathbf{Q}_Z$ in the middle. Styan's Theorem 2.4 asserts that $S_H^* = S_H$ if and only if $s - s_1 = 0$, and thus, on account of (3.7),

$$S_H^* = S_H \Leftrightarrow \mathbf{P}_W\mathbf{P}_Z = \mathbf{P}_Z\mathbf{P}_W. \tag{3.11}$$

A direct proof of (3.11) follows by observing that, in view of (3.9), (3.10) and Marsaglia and Styan's (1974) Theorem 2,

$$\begin{aligned} S_H^* = S_H &\Leftrightarrow \mathbf{Q}_Z\mathbf{W}(\mathbf{W}'\mathbf{Q}_Z\mathbf{W})^-\mathbf{W}'\mathbf{Q}_Z = \mathbf{Q}_Z\mathbf{P}_W\mathbf{Q}_Z \\ &\Leftrightarrow \mathbf{W}'\mathbf{Q}_Z\mathbf{W} = \mathbf{W}'\mathbf{Q}_Z\mathbf{P}_W\mathbf{Q}_Z\mathbf{W}, \end{aligned}$$

which corresponds to (A9) in Theorem 1.

The considerations of this subsection are concluded by assembling the results (3.7), (3.8) and (3.11).

THEOREM 2. *Let* $\mathbf{W} \in M_{n,w}$ *and* $\mathbf{Z} \in M_{n,z}$, *and let* $D(\mathbf{Y}) = \sigma^2\mathbf{I}_n$. *Then the following statements are equivalent*:

(A1) $\mathbf{P_W P_Z} = \mathbf{P_Z P_W}$,

(S1) *the nonzero canonical correlations between* $\mathbf{W'Y}$ *and* $\mathbf{Z'Y}$ *are all equal to one*,

(S2) $Cov(\mathbf{W'Q_Z Y}, \mathbf{Z'Q_W Y}) = \mathbf{0}$,

(S3) *the quadratic forms* S_H *and* $S_H{}^*$, *specified in* (3.9) *and* (3.10), *are equal.*

3.2 Comparisons between three estimators

The second part of the discussion on statistical implications of the commutativity of orthogonal projectors is concerned with the equalities

$$\text{OLSE}(\mathbf{X}\boldsymbol{\beta}) = \text{BLUE}(\mathbf{X}\boldsymbol{\beta}) \tag{3.12}$$

and

$$\text{GLSE}(\mathbf{X}\boldsymbol{\beta}) = \text{BLUE}(\mathbf{X}\boldsymbol{\beta}), \tag{3.13}$$

where

$$\text{OLSE}(\mathbf{X}\boldsymbol{\beta}) = \mathbf{P_X Y} \tag{3.14}$$

is the orthogonal projection of $\mathbf{Y}$ onto $R(\mathbf{X})$,

$$\text{GLSE}(\mathbf{X}\boldsymbol{\beta}) = \mathbf{X}(\mathbf{X}'\mathbf{V}^+\mathbf{X})^+\mathbf{X}'\mathbf{V}^+\mathbf{Y} \tag{3.15}$$

is the projection of $\mathbf{Y}$ onto $R(\mathbf{X})$ with respect to the seminorm specified by the Moore-Penrose inverse of $\mathbf{V}$, cf. Rao (1974, Lemma 2.9), and where, according to Rao (1978),

$$\text{BLUE}(\mathbf{X}\boldsymbol{\beta}) = \mathbf{AY} \Leftrightarrow \mathbf{A}(\mathbf{X} : \mathbf{VQ_X}) = (\mathbf{X} : \mathbf{0}). \tag{3.16}$$

The equalities (3.12) and (3.13) are understood to hold almost surely under the model $\mathrm{M} = \{\mathbf{Y}, \mathbf{X}\boldsymbol{\beta}, \sigma^2\mathbf{V}\}$ given in (1.7), i.e., for every

$$\mathbf{Y} \in R(\mathbf{X} : \mathbf{V}) = R(\mathbf{X} : \mathbf{VQ_X}). \tag{3.17}$$

The relation (3.17) is known as the consistency condition of the model M; cf. Rao (1973, p. 278) and Feuerverger and Fraser (1980, p. 44).

Since the seminal paper by Anderson (1948), numerous criteria for the validity of (3.12) have been devised; cf. recent surveys by Alalouf and Styan (1983), Puntanen (1987), and Puntanen and Styan (1987). In the general case, where no assumptions are made on the ranks of $\mathbf{X}$ and/or $\mathbf{V}$, the problem was first solved by Rao (1967). His criterion may be expressed as

$$R(\mathbf{VX}) \subseteq R(\mathbf{X}). \tag{3.18}$$

An alternative form of (3.18) is the equality

$$\mathbf{P_X V} = \mathbf{V P_X}, \tag{3.19}$$

originally given by Zyskind (1967). On the other hand, a characterization of the validity of (3.13) is due to Zyskind and Martin (1969). Their criterion is

$$R(\mathbf{X}) \subseteq R(\mathbf{V}), \tag{3.20}$$

or, equivalently,

$$\mathbf{P_V P_X} = \mathbf{P_X}. \tag{3.21}$$

Our discussion of the equalities (3.12) and (3.13) will follow that by Puntanen (1985; 1986; 1987, Part 1) in emphasizing their connections with certain characteristics of the canonical correlations between the vectors $\mathbf{P_X Y}$ and $\mathbf{Q_X Y}$. Since the dispersion matrix

$$D\begin{pmatrix} \mathbf{P_X Y} \\ \mathbf{Q_X Y} \end{pmatrix} = \sigma^2 \begin{pmatrix} \mathbf{P_X V P_X} & \mathbf{P_X V Q_X} \\ \mathbf{Q_X V P_X} & \mathbf{Q_X V Q_X} \end{pmatrix}$$

is singular, again the general theory of Khatri (1976), Seshadri and Styan (1980) and Rao (1981) is to be adopted for considering these correlations. Their squares are equal to the nonzero eigenvalues of the matrix

$$(\mathbf{P_X V P_X})^- \mathbf{P_X V Q_X} (\mathbf{Q_X V Q_X})^- \mathbf{Q_X V P_X},$$

irrespective of the choice of $(\mathbf{P_X V P_X})^-$ and $(\mathbf{Q_X V Q_X})^-$. Consequently, the number of nonzero canonical correlations between $\mathbf{P_X Y}$ and $\mathbf{Q_X Y}$ is

$$t = \mathrm{r}(\mathbf{Q_X V X}), \tag{3.22}$$

including the number

$$t_1 = \mathrm{r}(\mathbf{VX}) - \dim R(\mathbf{X}) \cap R(\mathbf{V}) \tag{3.23}$$

of those which are equal to one; cf. Puntanen (1985, Lemma 1). On account of the part "(A1) $\Leftrightarrow$ (A34)" of Theorem 1, it follows at once from (3.23) that

$$t_1 = 0 \Leftrightarrow \mathbf{P_X P_V} = \mathbf{P_V P_X}, \tag{3.24}$$

as originally stated in Puntanen's (1985) Lemma 2. Commenting these results, Puntanen (1985) observed an interesting fact that $t_1 = 0$ is a necessary condition for both (3.12) and (3.13). In view of (3.24), it is therefore natural to ask about conditions that should be added to the commutativity property

$$\mathbf{P_X P_V} = \mathbf{P_V P_X} \tag{3.25}$$

for assuring the validity of (3.12) and (3.13), respectively.

THEOREM 3. *Let* $\mathrm{M} = \{\mathbf{Y}, \mathbf{X}\boldsymbol{\beta}, \sigma^2\mathbf{V}\}$ *be the general linear model, and let* $\mathbf{F}$ *be any matrix such that* $R(\mathbf{F}) = R(\mathbf{X}) \cap R(\mathbf{V})$. *Then* $\mathrm{OLSE}(\mathbf{X}\boldsymbol{\beta}) = \mathrm{BLUE}(\mathbf{X}\boldsymbol{\beta})$ *if and only if* $\mathbf{P_X P_V} = \mathbf{P_V P_X}$ *holds along with*

$$R(\mathbf{VF}) \subseteq R(\mathbf{X}), \tag{3.26}$$

while $\mathrm{GLSE}(\mathbf{X}\boldsymbol{\beta}) = \mathrm{BLUE}(\mathbf{X}\boldsymbol{\beta})$ *if and only if* $\mathbf{P_X P_V} = \mathbf{P_V P_X}$ *holds along with*

$$R(\mathbf{X}) \cap R^{\perp}(\mathbf{V}) = \{\mathbf{0}\}. \tag{3.27}$$

Proof. It is easily seen that (3.19) implies $\mathbf{VP_X P_V X} = \mathbf{VX}$, and hence (3.25) follows directly from the part "(A2) $\Rightarrow$ (A1)" of Theorem 1. Moreover, since $R(\mathbf{VF}) \subseteq R(\mathbf{VX})$ holds trivially, it is clear that (3.18) entails (3.26). Conversely, according to the equivalence of (A1) to (A25) in Theorem 1, (3.25) may be reexpressed as $R(\mathbf{VX}) = R(\mathbf{VF})$. Combining this equality with (3.26) yields (3.18), thus concluding the proof of the first part. Further, (3.20) and (3.21) obviously imply (3.27) and (3.25), respectively. Conversely, it is clear that (3.25) is equivalent to $\mathbf{P_X Q_V} = \mathbf{Q_V P_X}$. In view of Corollary 1, combining this commutativity property with (3.27) yields $\mathbf{Q_V X} = \mathbf{0}$, and hence (3.20). □

It is obvious that if $\mathbf{V}$ is idempotent, then $\mathbf{VF} = \mathbf{F}$. Consequently, the condition (3.26) becomes trivial, and the first part of Theorem 3 leads to the conclusion that when $\mathbf{V}$ is idempotent, then $\mathrm{OLSE}(\mathbf{X}\boldsymbol{\beta}) = \mathrm{BLUE}(\mathbf{X}\boldsymbol{\beta})$

if and only if there are no unit canonical correlations between $\mathbf{P}_X\mathbf{Y}$ and $\mathbf{Q}_X\mathbf{Y}$; cf. (3.24) in Puntanen (1985). Another particular case in which (3.26) holds trivially is that where $R(\mathbf{X}) \cap R(\mathbf{V}) = \{\mathbf{0}\}$. Combining then the first part of Theorem 3 with Corollary 1 leads to the conclusion that if $R(\mathbf{X}) \cap R(\mathbf{V}) = \{\mathbf{0}\}$, then a necessary and sufficient condition for OLSE($\mathbf{X\beta}$) = BLUE($\mathbf{X\beta}$) is the orthogonality of $R(\mathbf{X})$ and $R(\mathbf{V})$. A straightforward proof of this statement follows from the fact that $R(\mathbf{X}) \cap R(\mathbf{V}) = \{\mathbf{0}\}$ is a necessary and sufficient condition for D[BLUE($\mathbf{X\beta}$)] = $\mathbf{0}$; cf. (3.33) in Puntanen (1986).

The second part of Theorem 3, in fact, becomes trivial when expressed in the geometric language mentioned in a comment to Theorem 1: the condition (3.20) is clearly equivalent to $R(\mathbf{X}) \cap R^{\perp}(\mathbf{F}) = \{\mathbf{0}\}$, whereas the conditions (3.25) and (3.27) mean that $R(\mathbf{X}) \cap R^{\perp}(\mathbf{F})$ is orthogonal to $R(\mathbf{V}) \cap R^{\perp}(\mathbf{F})$ and is disjoint with the orthogonal complement to $R(\mathbf{V}) \cap R^{\perp}(\mathbf{F})$, respectively. On the other hand, the result (3.14) in Marsaglia and Styan (1974) asserts that, for any matrices $\mathbf{K}$ and $\mathbf{L}$ for which the product $\mathbf{KL}$ exists,

$$\mathrm{r}(\mathbf{KL}) = \mathrm{r}(\mathbf{K}) - \dim R(\mathbf{K}') \cap R^{\perp}(\mathbf{L}). \tag{3.28}$$

Hence (3.27) may be reexpressed as $\mathrm{r}(\mathbf{V}^{+}\mathbf{X}) = \mathrm{r}(\mathbf{X})$. In view of Lemma 2.2.6(c) of Rao and Mitra (1971), this shows that (3.27) is actually equivalent to

$$\mathbf{X}(\mathbf{X}'\mathbf{V}^{+}\mathbf{X})^{+}\mathbf{X}'\mathbf{V}^{+}\mathbf{X} = \mathbf{X}, \tag{3.29}$$

the condition for the unbiasedness of GLSE($\mathbf{X\beta}$). The second part of Theorem 3 asserts, therefore, that GLSE($\mathbf{X\beta}$) = BLUE($\mathbf{X\beta}$) if and only if GLSE($\mathbf{X\beta}$) is unbiased and there are no unit canonical correlations between $\mathbf{P}_X\mathbf{Y}$ and $\mathbf{Q}_X\mathbf{Y}$. This statement has originally been established by Puntanen (1987, Part 1, Lemma 3.7.1).

Since GLSE($\mathbf{X\beta}$) is not unbiased for $\mathbf{X\beta}$ unless the condition (3.27) holds,

$$D[\mathrm{GLSE}(\mathbf{X\beta})] = \sigma^2\mathbf{X}(\mathbf{X}'\mathbf{V}^{+}\mathbf{X})^{+}\mathbf{X}' \tag{3.30}$$

may coincide with the dispersion matrix of BLUE($\mathbf{X\beta}$) even if these two estimators are not equal for all $\mathbf{Y} \in R(\mathbf{X}:\mathbf{V})$. One such case was characterized in Lemma 1 of Puntanen (1987, Part 4) by the equalities $\mathbf{X}'\mathbf{X} = \mathbf{I}_p$ and $\mathbf{P}_X\mathbf{P}_V = \mathbf{P}_V\mathbf{P}_X$. Theorem 4 below, being a restatement of Theorems 3.6.2, 3.6.3 and 3.6.4 of Puntanen (1987, Part 1) supplemented by the condition (e), contains a general solution to this problem. It appears that

$$D[\mathrm{GLSE}(\mathbf{X\beta})] = D[\mathrm{BLUE}(\mathbf{X\beta})] \tag{3.31}$$

is equivalent to the conditions derived by Alalouf (1975, p. 101) and Baksalary and Kala (1980, p. 19) for the linear estimator devised by Ahlers and Lewis (1971) to represent BLUE($\mathbf{X\beta}$).

THEOREM 4. *Let* $M = \{\mathbf{Y}, \mathbf{X\beta}, \sigma^2\mathbf{V}\}$ *be the general linear model, and let* $\mathbf{F}$ *be any matrix such that* $R(\mathbf{F}) = R(\mathbf{X}) \cap R(\mathbf{V})$. *Then the following statements are equivalent*:

(a) $D[\mathrm{BLUE}(\mathbf{X\beta})] = \sigma^2\mathbf{X}(\mathbf{X}'\mathbf{V}^+\mathbf{X})^+\mathbf{X}'$,

(b) $R(\mathbf{XX}'\mathbf{V}) \subseteq R(\mathbf{V})$,

(c) $\mathbf{XX}'\mathbf{P}_V = \mathbf{P}_V\mathbf{XX}'$,

(d) $\mathbf{P}_X\mathbf{P}_V = \mathbf{P}_V\mathbf{P}_X$ *and* $R(\mathbf{X}'\mathbf{XX}'\mathbf{V}) = R(\mathbf{X}'\mathbf{V})$,

(e) $\mathbf{P}_X\mathbf{P}_V = \mathbf{P}_V\mathbf{P}_X$ *and* $R(\mathbf{XX}'\mathbf{F}) \subseteq R(\mathbf{V})$,

(f) $\mathrm{BLUE}(\mathbf{X\beta}) = \mathbf{X}(\mathbf{X}'\mathbf{V}^+\mathbf{X})^+\mathbf{X}'\mathbf{V}^+\mathbf{Y} + \mathbf{X}(\mathbf{X}'\mathbf{Q}_V\mathbf{X})^+\mathbf{X}'\mathbf{Q}_V\mathbf{Y}$.

Proof. One of the points of the inverse partitioned matrix (IPM) method, introduced by Rao (1971, 1972), states that $D[\mathrm{BLUE}(\mathbf{X\beta})]$ is expressible as $\sigma^2\mathbf{XC}_4\mathbf{X}'$ if and only if $-\mathbf{C}_4$ is the $p \times p$ south-east submatrix of a generalized inverse

$$\begin{pmatrix} \mathbf{C}_1 & \mathbf{C}_3 \\ \mathbf{C}_2 & -\mathbf{C}_4 \end{pmatrix}$$

of the bordered matrix

$$\begin{pmatrix} \mathbf{V} & \mathbf{X} \\ \mathbf{X}' & \mathbf{0} \end{pmatrix}.$$

Hall and Meyer (1975) and Baksalary and Kala (1980b) showed that $\mathbf{C}_4$ has this property if and only if

$$\mathbf{XC}_4\mathbf{X}' = \mathbf{V} - \mathbf{VC}_1\mathbf{V}, \tag{3.32}$$

where the right-hand side of (3.32) is invariant upon the choice of $\mathbf{C}_1$ satisfying $\mathbf{Q}_X(\mathbf{V} - \mathbf{VC}_1\mathbf{V}) = \mathbf{0}$, $\mathbf{X}'\mathbf{C}_1\mathbf{V} = \mathbf{0}$, $\mathbf{VC}_1\mathbf{X} = \mathbf{0}$, and $\mathbf{X}'\mathbf{C}_1\mathbf{X} = \mathbf{0}$. Hence it is clear that (a) implies

$$R[\mathbf{X}(\mathbf{X}'\mathbf{V}^+\mathbf{X})^+\mathbf{X}'] \subseteq R(\mathbf{V}), \tag{3.33}$$

and (b) follows from (3.33) by noting that

$$R[\mathbf{X}(\mathbf{X}'\mathbf{V}^+\mathbf{X})^+\mathbf{X}'] = R(\mathbf{X}\mathbf{X}'\mathbf{V}^+\mathbf{X}) = R(\mathbf{X}\mathbf{X}'\mathbf{V}).$$

The part "(b) $\Leftrightarrow$ (f)" was established by Baksalary and Kala (1980a), and the part "(f) $\Rightarrow$ (a)" follows straightforwardly. It remains to prove, therefore, that (b) $\Leftrightarrow$ (c) $\Leftrightarrow$ (d) $\Leftrightarrow$ (e). If (b) holds, then $\mathbf{P}_\mathbf{V}\mathbf{X}\mathbf{X}'\mathbf{P}_\mathbf{V} = \mathbf{X}\mathbf{X}'\mathbf{P}_\mathbf{V}$, and hence (c). A consequence of (c) is that $R(\mathbf{P}_\mathbf{V}\mathbf{X}) \subseteq R(\mathbf{X})$. On account of the part "(A22) $\Rightarrow$ (A1)" of Theorem 1, this entails $\mathbf{P}_\mathbf{X}\mathbf{P}_\mathbf{V} = \mathbf{P}_\mathbf{V}\mathbf{P}_\mathbf{X}$. Moreover,

$$R(\mathbf{X}'\mathbf{X}\mathbf{X}'\mathbf{V}) = R(\mathbf{X}'\mathbf{X}\mathbf{X}'\mathbf{P}_\mathbf{V}) = R(\mathbf{X}'\mathbf{P}_\mathbf{V}\mathbf{X}\mathbf{X}') = R(\mathbf{X}'\mathbf{V}).$$

Hence $R(\mathbf{X}\mathbf{X}'\mathbf{V}) = R(\mathbf{P}_\mathbf{X}\mathbf{V})$, and applying the relations (A25) and (A22) of Theorem 1 leads to the latter condition in (e). Using again (A25) yields (b), thus concluding the proof. □

Notice that the conditions $R(\mathbf{X}'\mathbf{X}\mathbf{X}'\mathbf{V}) = R(\mathbf{X}'\mathbf{V})$ and $R(\mathbf{X}\mathbf{X}'\mathbf{F}) \subseteq R(\mathbf{V})$, supplementing the commutativity condition in (d) and (e) of Theorem 4, are not related one to the other. For example, if

$$\mathbf{X} = \begin{pmatrix} 1 & 0 \\ 1 & 1 \\ 0 & 1 \end{pmatrix} \quad \text{and} \quad \mathbf{V} = \begin{pmatrix} 1 & \upsilon & 0 \\ \upsilon & \upsilon & 0 \\ 0 & 0 & \upsilon \end{pmatrix},$$

then $\upsilon = 0$ yields $R(\mathbf{X}\mathbf{X}'\mathbf{F}) \subseteq R(\mathbf{V})$ together with $R(\mathbf{X}'\mathbf{X}\mathbf{X}'\mathbf{V}) \neq R(\mathbf{X}'\mathbf{V})$, while $\upsilon = 1$ yields $R(\mathbf{X}'\mathbf{X}\mathbf{X}'\mathbf{V}) = R(\mathbf{X}'\mathbf{V})$ together with $R(\mathbf{X}\mathbf{X}'\mathbf{F}) \not\subseteq R(\mathbf{V})$.

It is clear that if $\mathbf{X}$ is a partial isometry, i.e., if $\mathbf{X}\mathbf{X}' = \mathbf{P}_\mathbf{X}$ [cf. Ben-Israel and Greville (1974, p. 252) or Rao and Mitra (1971, p. 13)], then $\mathbf{X}'\mathbf{X}\mathbf{X}'\mathbf{V} = \mathbf{X}'\mathbf{V}$ and $\mathbf{X}\mathbf{X}'\mathbf{F} = \mathbf{F}$. Consequently, the latter conditions in (d) and (e) are trivially fulfilled and $\mathbf{P}_\mathbf{X}\mathbf{P}_\mathbf{V} = \mathbf{P}_\mathbf{V}\mathbf{P}_\mathbf{X}$ is then necessary and sufficient for the statements (a) and (f); cf. Theorem 1 of Puntanen (1986) pertaining to the particular case $\mathbf{X}'\mathbf{X} = \mathbf{I}_p$.

We may point out that the commutativity of $\mathbf{P}_\mathbf{X}$ and $\mathbf{P}_\mathbf{V}$ is a necessary condition also for

$$\text{GLSE}(\mathbf{X}\boldsymbol{\beta}) = \text{OLSE}(\mathbf{X}\boldsymbol{\beta}). \tag{3.34}$$

In fact, in view of (3.14), (3.15) and (3.17), the equality (3.34) means that

$$\mathbf{X}(\mathbf{X}'\mathbf{V}^+\mathbf{X})^+\mathbf{X}'\mathbf{V}^+(\mathbf{X} : \mathbf{V}) = \mathbf{P}_\mathbf{X}(\mathbf{X} : \mathbf{V}). \tag{3.35}$$

Postmultiplying the right submatrices on each side of (3.35) by $\mathbf{V}^+$ gives

$$\mathbf{X}(\mathbf{X}'\mathbf{V}^+\mathbf{X})^+\mathbf{X}'\mathbf{V}^+ = \mathbf{P}_\mathbf{X}\mathbf{P}_\mathbf{V}.$$

Hence $\mathbf{P}_\mathbf{X}\mathbf{P}_\mathbf{V}$ is idempotent, and on account of the part "(A1) $\Leftrightarrow$ (A7)" of Theorem 1 it follows that $\mathbf{P}_\mathbf{X}\mathbf{P}_\mathbf{V} = \mathbf{P}_\mathbf{V}\mathbf{P}_\mathbf{X}$. Moreover, comparing the left submatrices on each side of (3.35) yields (3.29), which is equivalent to (3.27). Consequently, the second part of Theorem 3 leads to the conclusion that if GLSE($\mathbf{X}\boldsymbol{\beta}$) = OLSE($\mathbf{X}\boldsymbol{\beta}$), then GLSE($\mathbf{X}\boldsymbol{\beta}$) = BLUE($\mathbf{X}\boldsymbol{\beta}$) and OLSE($\mathbf{X}\boldsymbol{\beta}$) = BLUE($\mathbf{X}\boldsymbol{\beta}$), as originally established by Baksalary and Kala (1983).

With the attention focused on the commutativity of orthogonal projectors, the following results may be extracted from the above discussion on relationships between OLSE($\mathbf{X}\boldsymbol{\beta}$), GLSE($\mathbf{X}\boldsymbol{\beta}$) and BLUE($\mathbf{X}\boldsymbol{\beta}$).

THEOREM 5. *Let* $\mathrm{M} = \{\mathbf{Y}, \mathbf{X}\boldsymbol{\beta}, \sigma^2\mathbf{V}\}$ *be the general linear model, and let* $\mathbf{F}$ *be any matrix such that* $R(\mathbf{F}) = R(\mathbf{X}) \cap R(\mathbf{V})$. *Then the following statements are equivalent*:

(A1) $\mathbf{P}_\mathbf{X}\mathbf{P}_\mathbf{V} = \mathbf{P}_\mathbf{V}\mathbf{P}_\mathbf{X}$,

(S4) *there are no unit canonical correlations between* $\mathbf{P}_\mathbf{X}\mathbf{Y}$ *and* $\mathbf{Q}_\mathbf{X}\mathbf{Y}$,

(S5) OLSE($\mathbf{X}\boldsymbol{\beta}$) = BLUE($\mathbf{X}\boldsymbol{\beta}$), *whenever* $R(\mathbf{VF}) \subseteq R(\mathbf{X})$,

(S6) GLSE($\mathbf{X}\boldsymbol{\beta}$) = BLUE($\mathbf{X}\boldsymbol{\beta}$), *whenever* $R(\mathbf{X}) \cap R^{\perp}(\mathbf{V}) = \{\mathbf{0}\}$,

(S7) BLUE($\mathbf{X}\boldsymbol{\beta}$) $= \mathbf{X}(\mathbf{X}'\mathbf{V}^+\mathbf{X})^+\mathbf{X}'\mathbf{V}^+\mathbf{Y} + \mathbf{X}(\mathbf{X}'\mathbf{Q}_\mathbf{V}\mathbf{X})^+\mathbf{X}'\mathbf{Q}_\mathbf{V}\mathbf{Y}$,
whenever $R(\mathbf{X}'\mathbf{X}\mathbf{X}'\mathbf{V}) = R(\mathbf{X}'\mathbf{V})$ *or* $R(\mathbf{X}\mathbf{X}'\mathbf{F}) \subseteq R(\mathbf{V})$,

(S8) D[BLUE($\mathbf{X}\boldsymbol{\beta}$)] $= \sigma^2\mathbf{X}(\mathbf{X}'\mathbf{V}^+\mathbf{X})^+\mathbf{X}'$,
whenever $R(\mathbf{X}'\mathbf{X}\mathbf{X}'\mathbf{V}) = R(\mathbf{X}'\mathbf{V})$ *or* $R(\mathbf{X}\mathbf{X}'\mathbf{F}) \subseteq R(\mathbf{V})$.

3.3 Comparisons between two models

The third part of Section 3 is concerned with certain comparisons between the models $\mathrm{M}_a = \{\mathbf{Y}, \mathbf{W}\boldsymbol{\gamma} + \mathbf{Z}\boldsymbol{\delta}, \sigma^2\mathbf{I}_n\}$ and $\mathrm{M}_0 = \{\mathbf{Y}, \mathbf{W}\boldsymbol{\gamma}, \sigma^2\mathbf{I}_n\}$ given in (1.5) and (1.6), respectively. It is well known that the set of all parametric functions $\mathbf{p}'\boldsymbol{\gamma}$ that are (unbiasedly) estimable in M_0 may be characterized as

$$E_0 = \{\mathbf{p}'\boldsymbol{\gamma} : \mathbf{p} \in R(\mathbf{W}')\}. \tag{3.36}$$

Further, among various characterizations of the estimability of $\mathbf{p}'\boldsymbol{\gamma}$ in M_a [cf. Fellman (1976, 1985), Seely and Birkes (1980), Baksalary (1984, 1985)], an analogue to (3.36) is

$$E_a = \{\mathbf{p}'\mathbf{\gamma} : \mathbf{p} \in R(\mathbf{W}'\mathbf{Q}_z)\}. \tag{3.37}$$

From (3.36) and (3.37) it is clear that $E_a \subseteq E_0$, with the equality if and only if $R(\mathbf{W}'\mathbf{Q}_z) = R(\mathbf{W}')$. But $R(\mathbf{W}'\mathbf{Q}_z) = R(\mathbf{W}')$ is equivalent to $r(\mathbf{W}'\mathbf{Q}_z) = r(\mathbf{W})$, and thus, in view of (3.28),

$$E_a = E_0 \Leftrightarrow R(\mathbf{W}) \cap R(\mathbf{Z}) = \{\mathbf{0}\}; \tag{3.38}$$

cf. Baksalary (1984, Theorem 1.3). On account of (3.5), an alternative interpretation of (3.38) is that the presence of the nuisance parameters $\boldsymbol{\delta}$ in M_a does not reduce the set of linear functions of $\boldsymbol{\gamma}$ estimable in M_0 if and only if there are no unit canonical correlations between the vectors $\mathbf{W}'\mathbf{Y}$ and $\mathbf{Z}'\mathbf{Y}$.

From (3.36) and (3.37) it is seen that

$$R(\mathbf{C}') = R(\mathbf{W}'\mathbf{Q}_z) \tag{3.39}$$

is a necessary and sufficient condition for the components of $\mathbf{C}\boldsymbol{\gamma}$ to generate all possible linear functions of $\boldsymbol{\gamma}$ that are estimable in both M_0 and M_a. A simple choice of $\mathbf{C}$ satisfying (3.39) is the matrix $\mathbf{W}'\mathbf{Q}_z\mathbf{W}$, with the corresponding vector of parametric functions

$$\boldsymbol{\eta} = \mathbf{W}'\mathbf{Q}_z\mathbf{W}\boldsymbol{\gamma}. \tag{3.40}$$

To avoid triviality in the specification (3.40) it will henceforth be assumed that $\mathbf{Q}_z\mathbf{W} \neq \mathbf{0}$, or, equivalently, that $R(\mathbf{W}) \not\subseteq R(\mathbf{Z})$.

It is known [cf. Theorem 2.2 in Baksalary (1984)] that the BLUEs of $\boldsymbol{\eta}$ under the models M_0 and M_a are

$$\hat{\boldsymbol{\eta}}_0 = \mathbf{W}'\mathbf{Q}_z\mathbf{P}_w\mathbf{Y} \tag{3.41}$$

and

$$\hat{\boldsymbol{\eta}}_a = \mathbf{W}'\mathbf{Q}_z\mathbf{Y}, \tag{3.42}$$

respectively, with the corresponding dispersion matrices

$$D(\hat{\boldsymbol{\eta}}_0) = \sigma^2\mathbf{W}'\mathbf{Q}_z\mathbf{P}_w\mathbf{Q}_z\mathbf{W} \tag{3.43}$$

and

$$D(\hat{\boldsymbol{\eta}}_a) = \sigma^2\mathbf{W}'\mathbf{Q}_z\mathbf{W}. \tag{3.44}$$

From (3.41) and (3.42) it is seen that the equality

$$\hat{\eta}_a = \hat{\eta}_0 \tag{3.45}$$

holds for all $\mathbf{Y} \in M_{n,1}$ if and only if

$$\mathbf{W}'\mathbf{Q}_\mathbf{Z} = \mathbf{W}'\mathbf{Q}_\mathbf{Z}\mathbf{P}_\mathbf{W}. \tag{3.46}$$

But (3.46) is equivalent to $\mathbf{W}'\mathbf{P}_\mathbf{Z}\mathbf{Q}_\mathbf{W} = \mathbf{0}$, which corresponds to the condition (A10) in Theorem 1. Consequently, it follows that

$$\hat{\eta}_a = \hat{\eta}_0 \Leftrightarrow \mathbf{P}_\mathbf{W}\mathbf{P}_\mathbf{Z} = \mathbf{P}_\mathbf{Z}\mathbf{P}_\mathbf{W}. \tag{3.47}$$

A necessary and sufficient condition for the BLUE of every parametric function $\mathbf{p}'\boldsymbol{\gamma}$, which is estimable in M_a, to be the same under M_a and under the reduced model M_0 has originally been established by Fellman (1976, Theorem 2.3) in the form

$$\mathbf{W}'\mathbf{Z} = \mathbf{W}'\mathbf{Z}(\mathbf{Z}'\mathbf{Z})^-\mathbf{Z}'\mathbf{W}(\mathbf{W}'\mathbf{W})^-\mathbf{W}'\mathbf{Z}. \tag{3.48}$$

The equivalence of (3.48) to $\mathbf{P}_\mathbf{W}\mathbf{P}_\mathbf{Z} = \mathbf{P}_\mathbf{Z}\mathbf{P}_\mathbf{W}$, which is actually the part "(A1) $\Leftrightarrow$ (A2)" of Theorem 1, was pointed out by Baksalary (1984, p. 10).

It is clear that if $\hat{\eta}_a = \hat{\eta}_0$, then $\hat{\eta}_0$ (being, by the definition, an unbiased estimator of $\boldsymbol{\eta}$ under M_0) is an unbiased estimator of $\boldsymbol{\eta}$ also under the model M_a and the dispersion matrices of $\hat{\eta}_a$ and $\hat{\eta}_0$ are equal. In general, if $\boldsymbol{\kappa} = \mathbf{K}\boldsymbol{\beta}$ is a vector of parametric functions estimable in both $\mathrm{M}_1 = \{\mathbf{Y}, \mathbf{X}_1\boldsymbol{\beta}, \sigma^2\mathbf{I}_n\}$ and $\mathrm{M}_2 = \{\mathbf{Y}, \mathbf{X}_2\boldsymbol{\beta}, \sigma^2\mathbf{I}_n\}$ and if $\hat{\boldsymbol{\kappa}}_1$ and $\hat{\boldsymbol{\kappa}}_2$ are the BLUEs of $\boldsymbol{\kappa}$ under M_1 and M_2, respectively, then neither the condition that $\hat{\boldsymbol{\kappa}}_1$ is unbiased for $\boldsymbol{\kappa}$ under M_2 nor the condition that $D(\hat{\boldsymbol{\kappa}}_1) = D(\hat{\boldsymbol{\kappa}}_2)$ implies the equality $\hat{\boldsymbol{\kappa}}_1 = \hat{\boldsymbol{\kappa}}_2$; cf. Rao and Mitra (1971, Section 8.3). In the particular case considered here, however, each of such conditions appears to be sufficient. In fact, on account of (3.41) and (1.5), the unbiasedness of $\hat{\eta}_0$ under M_a is equivalent to

$$\mathbf{W}'\mathbf{Q}_\mathbf{Z}\mathbf{P}_\mathbf{W}\mathbf{Z} = \mathbf{0}. \tag{3.49}$$

But (3.49) may be reduced to the form $\mathbf{Q}_\mathbf{Z}\mathbf{P}_\mathbf{W}\mathbf{Z} = \mathbf{0}$, which corresponds to the condition (A10) in Theorem 1. Consequently, it follows that

$$\hat{\eta}_0 \text{ is unbiased for } \boldsymbol{\eta} \text{ under } \mathrm{M}_a \Leftrightarrow \mathbf{P}_\mathbf{W}\mathbf{P}_\mathbf{Z} = \mathbf{P}_\mathbf{Z}\mathbf{P}_\mathbf{W}. \tag{3.50}$$

On the other hand, in view of (3.43) and (3.44), the equality $D(\hat{\eta}_a) = D(\hat{\eta}_0)$ directly corresponds to the condition (A9) in Theorem 1, and hence

$$D(\hat{\eta}_a) = D(\hat{\eta}_0) \;\Leftrightarrow\; \mathbf{P}_w\mathbf{P}_z = \mathbf{P}_z\mathbf{P}_w. \tag{3.51}$$

Reexpressing the condition (3.19) in the form $\mathbf{P}_x\mathbf{V}\mathbf{Q}_x = \mathbf{0}$ shows that OLSE($\mathbf{X}\boldsymbol{\beta}$) $= \mathbf{P}_x\mathbf{Y}$ represents the BLUE of $\mathbf{X}\boldsymbol{\beta}$ under the model M $= \{\mathbf{Y}, \mathbf{X}\boldsymbol{\beta}, \sigma^2\mathbf{V}\}$ if and only if it is uncorrelated (under M) with the residual vector $\mathbf{Q}_x\mathbf{Y}$; cf. Puntanen (1985, p. 274). A similar result appears to be valid also in the context of comparing $\hat{\eta}_a$ with $\hat{\eta}_0$. In fact, the residual vector in M_0 is $\mathbf{Q}_w\mathbf{Y}$, and thus, in view of (3.42),

$$Cov(\hat{\eta}_a, \mathbf{Q}_w\mathbf{Y}) = \sigma^2\mathbf{W}'\mathbf{Q}_z\mathbf{Q}_w. \tag{3.52}$$

From (3.52) it is seen that the BLUE of $\boldsymbol{\eta}$ under M_a is uncorrelated with the residual vector in M_0 if and only if $\mathbf{W}'\mathbf{Q}_z\mathbf{Q}_w = \mathbf{0}$. But this is equivalent to the condition $\mathbf{W}'\mathbf{P}_z\mathbf{Q}_w = \mathbf{0}$ being a version of (A10) in Theorem 1, and hence

$$Cov(\hat{\eta}_a, \mathbf{Q}_w\mathbf{Y}) = \mathbf{0} \Leftrightarrow \mathbf{P}_w\mathbf{P}_z = \mathbf{P}_z\mathbf{P}_w. \tag{3.53}$$

A different approach to comparing the models M_a and M_0 has recently been proposed by Fellman (1985). Its main point is a characterization of the set L_a consisting of all those functions $\mathbf{p}'\boldsymbol{\gamma} \in E_a$ for which the BLUEs under M_a are as good as under M_0. According to Fellman's (1985) Theorem 3,

$$L_a = \{\mathbf{q}'\mathbf{W}\boldsymbol{\gamma} : \mathbf{q} \in R(\mathbf{W}) \cap R^{\perp}(\mathbf{Z})\}. \tag{3.54}$$

On the other hand, according to Fellman's (1985) Theorem 1 and Hedayat and Majumdar's (1985) Lemma 2.1, an alternative to the representation (3.37) is

$$E_a = \{\mathbf{q}'\mathbf{W}\boldsymbol{\gamma} : \mathbf{q} \in R(\mathbf{W} : \mathbf{Z}) \cap R^{\perp}(\mathbf{Z})\}. \tag{3.55}$$

From (3.54) and (3.55) it follows that $L_a = E_a$ if and only if

$$\dim R(\mathbf{W}) \cap R^{\perp}(\mathbf{Z}) = \dim R(\mathbf{W} : \mathbf{Z}) \cap R^{\perp}(\mathbf{Z}). \tag{3.56}$$

Noting that (3.56) corresponds to the condition (A43) in Theorem 1 yields

$$L_a = E_a \Leftrightarrow \mathbf{P}_w\mathbf{P}_z = \mathbf{P}_z\mathbf{P}_w. \tag{3.57}$$

Comparing OLSE($\mathbf{X}\boldsymbol{\beta}$) with BLUE($\mathbf{X}\boldsymbol{\beta}$) under the model M $= \{\mathbf{Y}, \mathbf{X}\boldsymbol{\beta}, \sigma^2\mathbf{V}\}$, Krämer (1980) proposed the approach that consists in characterizing all those vectors $\mathbf{Y}$ for which the value of OLSE($\mathbf{X}\boldsymbol{\beta}$) coincides with the value of BLUE($\mathbf{X}\boldsymbol{\beta}$); see also Mathew (1985). This may be viewed as an alternative to the approach that consists in characterizing all those matrices $\mathbf{V}$ for which the equality OLSE($\mathbf{X}\boldsymbol{\beta}$) = BLUE($\mathbf{X}\boldsymbol{\beta}$) holds almost surely; cf. criteria (3.18) and (3.19). Adopting Krämer's approach to the problem of comparing $\hat{\boldsymbol{\eta}}_a$ with $\hat{\boldsymbol{\eta}}_0$, let

$$Y_* = \{\mathbf{Y} \in M_{n,1}: \hat{\boldsymbol{\eta}}_a = \hat{\boldsymbol{\eta}}_0\}. \tag{3.58}$$

Then from (3.41) and (3.42) it easily follows that

$$Y_* = R^{\perp}(\mathbf{Q}_\mathbf{W}\mathbf{P}_\mathbf{Z}\mathbf{W}). \tag{3.59}$$

From (3.59) it is clear that $Y_* = M_{n,1}$ if and only if $\mathbf{Q}_\mathbf{W}\mathbf{P}_\mathbf{Z}\mathbf{W} = \mathbf{0}$, which corresponds to the condition (A10) in Theorem 1. Consequently

$$Y_* = M_{n,1} \Leftrightarrow \mathbf{P}_\mathbf{W}\mathbf{P}_\mathbf{Z} = \mathbf{P}_\mathbf{Z}\mathbf{P}_\mathbf{W}. \tag{3.60}$$

The results (3.47), (3.50), (3.51), (3.53), (3.57), and (3.60) are now collected together.

THEOREM 6. *Let* $\mathrm{M}_a = \{\mathbf{Y}, \mathbf{W}\boldsymbol{\gamma}+\mathbf{Z}\boldsymbol{\delta}, \sigma^2\mathbf{I}_n\}$ *be the linear model with nuisance parameters* $\boldsymbol{\delta}$, *and let* $\mathrm{M}_0 = \{\mathbf{Y}, \mathbf{W}\boldsymbol{\gamma}, \sigma^2\mathbf{I}_n\}$ *be the reduced form of* M_a. *Further, let* $\hat{\boldsymbol{\eta}}_a$ *and* $\hat{\boldsymbol{\eta}}_0$ *be the BLUEs of* $\boldsymbol{\eta} = \mathbf{W}'\mathbf{Q}_\mathbf{Z}\mathbf{W}\boldsymbol{\gamma}$ *under* M_a *and* M_0, *respectively. Then the following statements are equivalent*:

(A1) $\mathbf{P}_\mathbf{W}\mathbf{P}_\mathbf{Z} = \mathbf{P}_\mathbf{Z}\mathbf{P}_\mathbf{W}$,

(S9) $\hat{\boldsymbol{\eta}}_a = \hat{\boldsymbol{\eta}}_0$,

(S10) $\hat{\boldsymbol{\eta}}_0$ *is unbiased for* $\boldsymbol{\eta}$ *under* M_a,

(S11) $D(\hat{\boldsymbol{\eta}}_a) = D(\hat{\boldsymbol{\eta}}_0)$,

(S12) $Cov(\hat{\boldsymbol{\eta}}_a, \mathbf{Q}_\mathbf{W}\mathbf{Y}) = \mathbf{0}$,

(S13) *the set* L_a *of linear functions of* $\boldsymbol{\gamma}$ *having their BLUEs under* M_a *as good as under* M_0 *coincides with the set* E_a *of all linear functions of* $\boldsymbol{\gamma}$ *that are estimable in* M_a,

(S14) *the set* Y_* *of those vectors* $\mathbf{Y}$ *for which the value of* $\hat{\boldsymbol{\eta}}_a$ *is identical with the value of* $\hat{\boldsymbol{\eta}}_0$ *coincides with* $M_{n,1}$.

In view of Theorem 6, a comment is needed on the corollary following Theorem 8.3.3 of Rao and Mitra (1971), which claims that for the BLUE under M_a to be M_0- optimal for every estimable linear function of $\boldsymbol{\gamma}$, it is necessary and sufficient that

$$\mathbf{W}'\mathbf{Z} = \mathbf{0}. \tag{3.61}$$

The crucial point in this claim is the meaning of the phrase "every estimable linear function of $\mathbf{y}$". The part "(A1) $\Leftrightarrow$ (S13)" of Theorem 6 shows that if the phrase in question is understood as "every linear function of $\mathbf{y}$ that is estimable in M_a", then a necessary and sufficient condition is the commutativity of the projectors $\mathbf{P}_\mathbf{W}$ and $\mathbf{P}_\mathbf{Z}$, which is a weaker requirement than (3.61). However, the condition $\mathbf{W}'\mathbf{Z} = \mathbf{0}$ is a solution to the problem when the phrase in question is understood as "every linear function of $\mathbf{y}$ that is estimable in M_0", for then $\mathbf{P}_\mathbf{W}\mathbf{P}_\mathbf{Z} = \mathbf{P}_\mathbf{Z}\mathbf{P}_\mathbf{W}$ must hold along with the equality $E_a = E_0$, which, in view of (3.38) and Corollary 1, leads to (3.61).

The condition $\mathbf{W}'\mathbf{Z} = \mathbf{0}$ has originally been established by Ehrenfeld (1955) to be necessary and sufficient for the equality of the BLUEs of $\mathbf{y}$ in the models M_0 and M_a under the assumption that the matrix $(\mathbf{W} : \mathbf{Z})$ is of full column rank. This assumption clearly implies that $R(\mathbf{W}) \cap R(\mathbf{Z}) = \{\mathbf{0}\}$, but not other way around. Consequently, Ehrenfeld's result may be viewed as a particular case of Rao and Mitra's (1971) Theorem 8.3.3, here reworded as

$$\text{BLUE}(\mathbf{W}\mathbf{y}) \text{ under } M_0 \text{ equals BLUE}(\mathbf{W}\mathbf{y}) \text{ under } M_a \Leftrightarrow \mathbf{W}'\mathbf{Z} = \mathbf{0}. \tag{3.62}$$

It is interesting to remark that, according to Theorem 3.3 of Baksalary (1984), the statement (3.62) may be strengthened to the form

$$\text{BLUE}(\mathbf{W}\mathbf{y}) \text{ under } M_0 \text{ equals BLIMBE}(\mathbf{W}\mathbf{y}) \text{ under } M_a \Leftrightarrow \mathbf{W}'\mathbf{Z} = \mathbf{0}, \tag{3.63}$$

or even to the form

$$\text{BLIMBE}(\mathbf{y}) \text{ under } M_0 \text{ equals BLIMBE}(\mathbf{y}) \text{ under } M_a \Leftrightarrow \mathbf{W}'\mathbf{Z} = \mathbf{0}, \tag{3.64}$$

where BLIMBE in (3.63) and (3.64) stands for the best linear minimum bias estimator; cf. Chipman (1964) and Rao (1971).

Closely related to the BLUE of a given vector of parametric functions is the notion of a linearly sufficient statistic. Assuming that the vector $\mathbf{K}\boldsymbol{\beta}$ is estimable in $M = \{\mathbf{Y}, \mathbf{X}\boldsymbol{\beta}, \sigma^2\mathbf{V}\}$, $\mathbf{CY}$ is said to be linearly sufficient for $\mathbf{K}\boldsymbol{\beta}$ under M if the BLUE of $\mathbf{K}\boldsymbol{\beta}$ under M is obtainable via a linear transformation of $\mathbf{CY}$. In the case of $\mathbf{K} = \mathbf{X}$, linearly sufficient statistics were originally considered by Baksalary and Kala (1981) under the label of linear transformations preserving BLUEs, and then by Drygas (1983), who adopted the term "linearly sufficient statistics". Among his concluding remarks, Drygas (1983) mentioned that the use of this term may be traced back to Barnard (1963); see also Joshi (1985). Further, he pointed out that the linearly sufficient statistics in the sense of Barnard (1963) coincide with the linearly sufficient statistics (for $\mathbf{X\beta}$) in the sense of Baksalary and Kala (1981) and Drygas (1983) only for those models M in which $R(\mathbf{X}) \subseteq R(\mathbf{V})$. Subsequently, the linear sufficiency was considered with respect to a given vector $\mathbf{K\beta}$: in Baksalary (1984) under the additional assumption that $\mathbf{V} = \mathbf{I}_n$ and in Baksalary and Kala (1986) in the general case. The problem of comparing the sets of linearly sufficient statistics under different linear models was investigated by Baksalary (1984) and Baksalary and Mathew (1986). The discussion below follows that in the former work.

Let S_0 and S_a denote the sets of all linearly sufficient statistics for $\boldsymbol{\eta} = \mathbf{W}'\mathbf{Q}_z\mathbf{W}\boldsymbol{\gamma}$ under the models $\mathrm{M}_0 = \{\mathbf{Y}, \mathbf{W}\boldsymbol{\gamma}, \sigma^2\mathbf{I}_n\}$ and $\mathrm{M}_a = \{\mathbf{Y}, \mathbf{W}\boldsymbol{\gamma}+\mathbf{Z}\boldsymbol{\delta}, \sigma^2\mathbf{I}_n\}$, respectively. Then, according to Theorem 5.2 in Baksalary (1984),

$$S_0 = \{\mathbf{CY}\colon R(\mathbf{P}_w\mathbf{Q}_z\mathbf{W}) \subseteq R(\mathbf{C}')\} \tag{3.65}$$

and

$$S_a = \{\mathbf{CY}\colon R(\mathbf{Q}_z\mathbf{W}) \subseteq R(\mathbf{C}')\}. \tag{3.66}$$

It is clear that $S_0 \cap S_a \neq \varnothing$, and also that

$$\dim R(\mathbf{Q}_z\mathbf{W}) = \dim R(\mathbf{P}_w\mathbf{Q}_z\mathbf{W}). \tag{3.67}$$

In view of (3.65) and (3.66), the equality (3.67) means that neither of the sets S_0 and S_a can be a proper subset of each other.

An obvious consequence of the definition of the sets S_0 and S_a is that $\hat{\boldsymbol{\eta}}_0 \in S_0$ and $\hat{\boldsymbol{\eta}}_a \in S_a$, where $\hat{\boldsymbol{\eta}}_0$ and $\hat{\boldsymbol{\eta}}_a$ are the BLUEs of $\boldsymbol{\eta}$ specified in (3.41) and (3.42). Moreover, if $S_a = S_0$, then also $\hat{\boldsymbol{\eta}}_0 \in S_a$ and $\hat{\boldsymbol{\eta}}_a \in S_0$. It appears that the converse implications are true as well. In fact, from (3.41) and (3.66) it follows that $\hat{\boldsymbol{\eta}}_0 \in S_a$ if and only if

$$R(\mathbf{Q}_z\mathbf{W}) \subseteq R(\mathbf{P}_w\mathbf{Q}_z\mathbf{W}), \tag{3.68}$$

while from (3.42) and (3.65) it follows that $\hat{\boldsymbol{\eta}}_a \in S_0$ if and only if

$$R(\mathbf{P}_w\mathbf{Q}_z\mathbf{W}) \subseteq R(\mathbf{Q}_z\mathbf{W}). \tag{3.69}$$

But on account of (3.67) each of the inclusions (3.68) and (3.69) strengthens to the equality

$$R(\mathbf{Q}_z\mathbf{W}) = R(\mathbf{P}_w\mathbf{Q}_z\mathbf{W}). \tag{3.70}$$

It can easily be verified that (3.70) is equivalent to $R(\mathbf{P}_z\mathbf{W}) \subseteq R(\mathbf{W})$, which corresponds to the condition (A22) in Theorem 1. Consequently,

$$\hat{\eta}_0 \in S_a \Leftrightarrow \mathbf{P}_w\mathbf{P}_z = \mathbf{P}_z\mathbf{P}_w \tag{3.71}$$

and

$$\hat{\eta}_a \in S_0 \Leftrightarrow \mathbf{P}_w\mathbf{P}_z = \mathbf{P}_z\mathbf{P}_w. \tag{3.72}$$

Clearly, the condition (3.70) is necessary and sufficient also for $S_a = S_0$, and thus

$$S_a = S_0 \Leftrightarrow \mathbf{P}_w\mathbf{P}_z = \mathbf{P}_z\mathbf{P}_w, \tag{3.73}$$

as originally stated in Theorem 5.3 of Baksalary (1984).

A particular place within the set of all linearly sufficient statistics belongs to the subset consisting of linearly minimal sufficient statistics. Assuming that the vector $\mathbf{K\beta}$ is estimable in $\mathrm{M} = \{\mathbf{Y}, \mathbf{X\beta}, \sigma^2\mathbf{V}\}$ and denoting the set of linearly sufficient statistics for $\mathbf{K\beta}$ under M by S, $\mathbf{CY}$ is said to be linearly minimal sufficient for $\mathbf{K\beta}$ under M if $\mathbf{CY} \in S$ and if for every $\mathbf{HY} \in S$ there exists a matrix $\mathbf{L}$ such that $\mathbf{CY} = \mathbf{LHY}$ almost surely. This notion was introduced by Drygas (1983) in the case of $\mathbf{K} = \mathbf{X}$, and then extended to the general case by Baksalary (1984) under the additional assumption that $\mathbf{V} = \mathbf{I}_n$ and by Baksalary and Kala (1986) without this assumption.

Let S_0^* and S_a^* denote the sets of all linearly minimal sufficient statistics for $\eta = \mathbf{W}'\mathbf{Q}_z\mathbf{W}\gamma$ under the models $\mathrm{M}_0 = \{\mathbf{Y}, \mathbf{W}\gamma, \sigma^2\mathbf{I}_n\}$ and $\mathrm{M}_a = \{\mathbf{Y}, \mathbf{W}\gamma+\mathbf{Z}\delta, \sigma^2\mathbf{I}_n\}$ respectively. Then, according to Theorem 6.2 in Baksalary (1984),

$$S_0^* = \{\mathbf{CY}\colon\ R(\mathbf{P}_w\mathbf{Q}_z\mathbf{W}) = R(\mathbf{C}')\} \tag{3.74}$$

and

$$S_a^* = \{\mathbf{CY}\colon\ R(\mathbf{Q}_z\mathbf{W}) = R(\mathbf{C}')\}. \tag{3.75}$$

From (3.41), (3.42), (3.74), and (3.75) it is clear that the equality (3.70) is a necessary and sufficient condition for $\hat{\eta}_0 \in S_a^*$, for $\hat{\eta}_a \in S_0^*$ and for $S_a^* = S_0^*$, thus leading to the following analogues to (3.71), (3.72) and (3.73):

$$\hat{\eta}_0 \in S_a^* \Leftrightarrow \mathbf{P}_w\mathbf{P}_z = \mathbf{P}_z\mathbf{P}_w, \tag{3.76}$$

$$\hat{\eta}_a \in S_0^* \Leftrightarrow \mathbf{P}_w\mathbf{P}_z = \mathbf{P}_z\mathbf{P}_w, \tag{3.77}$$

$$S_a^* = S_0^* \Leftrightarrow \mathbf{P_W P_Z} = \mathbf{P_Z P_W}. \tag{3.78}$$

Moreover, from (3.74) and (3.75) it follows that the condition (3.70) is equivalent also to $S_a^* \cap S_0^* \neq \emptyset$, and thus

$$S_a^* \cap S_0^* \neq \emptyset \Leftrightarrow \mathbf{P_W P_Z} = \mathbf{P_Z P_W}. \tag{3.79}$$

The theorem below collects together the results (3.71), (3.72), (3.73), (3.76), (3.77), (3.78), and (3.79).

THEOREM 7. *Let* $\mathrm{M}_a = \{\mathbf{Y}, \mathbf{W}\boldsymbol{\gamma} + \mathbf{Z}\boldsymbol{\delta}, \sigma^2\mathbf{I}_n\}$ *be the linear model with nuisance parameters* $\boldsymbol{\delta}$, *and let* $\mathrm{M}_0 = \{\mathbf{Y}, \mathbf{W}\boldsymbol{\gamma}, \sigma^2\mathbf{I}_n\}$ *be the reduced form of* M_a. *Further, let* $\hat{\boldsymbol{\eta}}_a$ *and* $\hat{\boldsymbol{\eta}}_0$ *be the BLUEs of* $\boldsymbol{\eta} = \mathbf{W}'\mathbf{Q_Z W}\boldsymbol{\gamma}$, *let* S_a *and* S_0 *be the sets of all linearly sufficient statistics for* $\boldsymbol{\eta}$, *and let* S_a^* *and* S_0^* *be the sets of all linearly minimal sufficient statistics for* $\boldsymbol{\eta}$, *corresponding to the models* M_a *and* M_0, *respectively. Then the following statements are equivalent*:

(A1) $\mathbf{P_W P_Z} = \mathbf{P_Z P_W}$,

(S15) $\hat{\boldsymbol{\eta}}_0 \in S_a$,

(S16) $\hat{\boldsymbol{\eta}}_a \in S_0$,

(S17) $S_a = S_0$,

(S18) $\hat{\boldsymbol{\eta}}_0 \in S_a^*$,

(S19) $\hat{\boldsymbol{\eta}}_a \in S_0^*$,

(S20) $S_a^* = S_0^*$,

(S21) $S_a^* \cap S_0^* \neq \emptyset$.

Theorem 7 shows that comparing S_a to S_0 leads to a somewhat different conclusion than comparing S_a^* to S_0^*. A necessary and sufficient condition for the equality of the sets is in both cases identical, but, on the other hand, the set $S_a \cap S_0$ is always nonempty, while $S_a^* \cap S_0^* \neq \emptyset$ if and only if $S_a^* = S_0^*$. This corrects the remark preceding Theorem 6.2 in Baksalary (1984).

4. PROOF OF THEOREM 1

Premultiplying and postmultiplying both sides of (A1): $\mathbf{P_A P_B} = \mathbf{P_B P_A}$ by $\mathbf{A}'$ and $\mathbf{B}$, respectively, yields (A2): $\mathbf{A'B} = \mathbf{A'P_B P_A B}$. Representing then $\mathbf{P_A}$ and $\mathbf{P_B}$ as in (1.3) implies (A3) stating that $\mathbf{G} = (\mathbf{B'B})^-\mathbf{B'A}(\mathbf{A'A})^-$

is a generalized inverse of $\mathbf{A}'\mathbf{B}$. Hence (A5): $\mathbf{A}'\mathbf{B}\mathbf{G} = (\mathbf{A}'\mathbf{B}\mathbf{G})^2$ is an immediate consequence of the well known property of generalized inverses; see, e.g., Ben-Israel and Greville (1974, p. 11) or Rao and Mitra (1971, p. 21). Further, (A4): $\mathbf{G}\mathbf{A}'\mathbf{B}\mathbf{G} = \mathbf{G}$ follows by cancelling in (A5) the left $\mathbf{A}'\mathbf{B}$'s according to the rule (2.12) in Marsaglia and Styan (1974). Hence (A6): $\mathbf{G}\mathbf{A}'\mathbf{B} = (\mathbf{G}\mathbf{A}'\mathbf{B})^2$. In view of (1.3), premultiplying both sides of (A6) by $\mathbf{B}$ and cancelling the right $\mathbf{B}$'s according to the rule (2.13) in Marsaglia and Styan (1974) leads to (A7): $\mathbf{P}_{\mathrm{A}}\mathbf{P}_{\mathrm{B}} = (\mathbf{P}_{\mathrm{A}}\mathbf{P}_{\mathrm{B}})^2$. Further, premultiplying and postmultiplying both sides of (A7) by $\mathbf{B}'$ and $\mathbf{B}$, respectively, yields (A8): $\mathbf{B}'\mathbf{P}_{\mathrm{A}}\mathbf{B} = \mathbf{B}'\mathbf{P}_{\mathrm{A}}\mathbf{P}_{\mathrm{B}}\mathbf{P}_{\mathrm{A}}\mathbf{B}$. Substituting in (A8) $\mathbf{P}_{\mathrm{A}} = \mathbf{I}_n - \mathbf{Q}_{\mathrm{A}}$ gives (A9): $\mathbf{B}'\mathbf{Q}_{\mathrm{A}}\mathbf{B} = \mathbf{B}'\mathbf{Q}_{\mathrm{A}}\mathbf{P}_{\mathrm{B}}\mathbf{Q}_{\mathrm{A}}\mathbf{B}$. Replacing $\mathbf{P}_{\mathrm{B}}$ by $\mathbf{I}_n - \mathbf{Q}_{\mathrm{B}}$ transforms (A9) to the form $\mathbf{B}'\mathbf{Q}_{\mathrm{A}}\mathbf{Q}_{\mathrm{B}}\mathbf{Q}_{\mathrm{A}}\mathbf{B} = \mathbf{0}$, which is clearly equivalent to $\mathbf{Q}_{\mathrm{B}}\mathbf{Q}_{\mathrm{A}}\mathbf{B} = \mathbf{0}$, and then to (A10): $\mathbf{Q}_{\mathrm{B}}\mathbf{P}_{\mathrm{A}}\mathbf{B} = \mathbf{0}$. Hence (A22): $R(\mathbf{P}_{\mathrm{A}}\mathbf{B}) \subseteq R(\mathbf{B})$, and combining (A22) with the trivial inclusion $R(\mathbf{P}_{\mathrm{A}}\mathbf{B}) \subseteq R(\mathbf{A})$ yields (A23): $R(\mathbf{P}_{\mathrm{A}}\mathbf{B}) \subseteq R(\mathbf{T})$. An equivalent formulation of (A23) is $\mathbf{P}_{\mathrm{T}}\mathbf{P}_{\mathrm{A}}\mathbf{P}_{\mathrm{B}} = \mathbf{P}_{\mathrm{A}}\mathbf{P}_{\mathrm{B}}$. But $\mathbf{P}_{\mathrm{T}}\mathbf{P}_{\mathrm{A}} = \mathbf{P}_{\mathrm{T}} = \mathbf{P}_{\mathrm{T}}\mathbf{P}_{\mathrm{B}}$, and hence (A11): $\mathbf{P}_{\mathrm{A}}\mathbf{P}_{\mathrm{B}} = \mathbf{P}_{\mathrm{T}}$. This obviously entails (A12): $\mathbf{A}'\mathbf{B} = \mathbf{A}'\mathbf{P}_{\mathrm{T}}\mathbf{B}$.

Reexpressing (A12) as $\mathbf{A}'\mathbf{Q}_{\mathrm{T}}\mathbf{B} = \mathbf{0}$ leads to condition (A19) stating that $R(\mathbf{Q}_{\mathrm{T}}\mathbf{A})$ and $R(\mathbf{Q}_{\mathrm{T}}\mathbf{B})$ are orthogonal. But

$$R(\mathbf{Q}_{\mathrm{T}}\mathbf{A}) = R(\mathbf{A}) \cap R^{\perp}(\mathbf{T}) \quad \text{and} \quad R(\mathbf{Q}_{\mathrm{T}}\mathbf{B}) = R(\mathbf{B}) \cap R^{\perp}(\mathbf{T}), \tag{4.1}$$

and therefore (A19) is equivalent to the orthogonality of the subspaces $R(\mathbf{A}) \cap R^{\perp}(\mathbf{T})$ and $R(\mathbf{B}) \cap R^{\perp}(\mathbf{T})$, as asserted in (A20). Moreover, $R(\mathbf{A}) \cap R^{\perp}(\mathbf{T})$ is obviously orthogonal to $R(\mathbf{T})$, and thus (A20) may be strengthened to (A21) stating that $R(\mathbf{A}) \cap R^{\perp}(\mathbf{T})$ is orthogonal to $R(\mathbf{T}) \boxplus R(\mathbf{B}) \cap R^{\perp}(\mathbf{T}) = R(\mathbf{B})$. Denoting now by $\mathbf{A}^0$ any matrix such that $R(\mathbf{A}^0) = R(\mathbf{A}) \cap R^{\perp}(\mathbf{T})$ and utilizing

$$R(\mathbf{A}) = R(\mathbf{T}) \boxplus R(\mathbf{A}) \cap R^{\perp}(\mathbf{T}), \tag{4.2}$$

it follows from (A21) that

$$\mathbf{P}_{\mathrm{A}}\mathbf{Q}_{\mathrm{B}} = (\mathbf{P}_{\mathrm{T}} + \mathbf{P}_{\mathrm{A}^0})\mathbf{Q}_{\mathrm{B}} = \mathbf{P}_{\mathrm{A}^0}\mathbf{Q}_{\mathrm{B}} = \mathbf{P}_{\mathrm{A}^0},$$

which is (A13). This obviously implies (A26): $R(\mathbf{Q}_{\mathrm{B}}\mathbf{A}) = R(\mathbf{A}) \cap R^{\perp}(\mathbf{T})$. But

$$R(\mathbf{S}) = R(\mathbf{B}) \boxplus R(\mathbf{Q}_{\mathrm{B}}\mathbf{A}) = R(\mathbf{B}) \boxplus R(\mathbf{S}) \cap R^{\perp}(\mathbf{B}), \tag{4.3}$$

and hence

$$R(\mathbf{Q}_{\mathrm{B}}\mathbf{A}) = R(\mathbf{S}) \cap R^{\perp}(\mathbf{B}). \tag{4.4}$$

Consequently,

$$\begin{aligned}
\text{(A26)} &\Rightarrow \text{(A32)}: R(\mathbf{S}) = R(\mathbf{B}) \boxplus R(\mathbf{A}) \cap R^{\perp}(\mathbf{T}), \text{ by (4.3)},\\
&\Rightarrow \text{(A29)}: R(\mathbf{S}) \cap R^{\perp}(\mathbf{B}) = R(\mathbf{A}) \cap R^{\perp}(\mathbf{T}), \text{ by (4.3)},\\
&\Rightarrow \text{(A30)}: R(\mathbf{A}) = R(\mathbf{T}) \boxplus R(\mathbf{S}) \cap R^{\perp}(\mathbf{B}), \text{ by (4.2)},\\
&\Rightarrow \text{(A31)}: R(\mathbf{A}) = R(\mathbf{T}) \boxplus R(\mathbf{Q}_\mathbf{B}\mathbf{A}), \text{ by (4.4)}.
\end{aligned}$$

Further, from (A31) it is clear that $\mathbf{Q}_\mathbf{B}\mathbf{P}_\mathbf{A} = \mathbf{P}_\mathbf{A}\mathbf{Q}_\mathbf{B}$, and Theorem 5.1.4 of Rao and Mitra (1971) implies that $\mathbf{P}_\mathbf{A}\mathbf{Q}_\mathbf{B}$ is the orthogonal projector onto $R(\mathbf{A}) \cap R^{\perp}(\mathbf{B})$, as asserted in (A14). But then $\mathbf{P}_\mathbf{A}\mathbf{Q}_\mathbf{B} = \mathbf{Q}_\mathbf{B}\mathbf{P}_\mathbf{A}$ is also the orthogonal projector onto its range. In view of (4.4), this leads to (A15) stating that $\mathbf{P}_\mathbf{A}\mathbf{Q}_\mathbf{B}$ is the orthogonal projector onto $R(\mathbf{S}) \cap R^{\perp}(\mathbf{B})$. Hence, on account of (4.3),

$$\mathbf{P}_\mathbf{S} = \mathbf{P}_\mathbf{B} + \mathbf{P}_\mathbf{A}\mathbf{Q}_\mathbf{B} = \mathbf{P}_\mathbf{A} + \mathbf{P}_\mathbf{B} - \mathbf{P}_\mathbf{A}\mathbf{P}_\mathbf{B},$$

which is (A16).

Let $\mathbf{L} = \mathbf{P}_\mathbf{A} + \mathbf{P}_\mathbf{B}$ and

$$\mathbf{K} = \mathbf{P}_\mathbf{A} + \mathbf{P}_\mathbf{B} - 3/2\, \mathbf{P}_\mathbf{A}\mathbf{P}_\mathbf{B}. \tag{4.5}$$

If (A16) holds, then an alternative form of (4.5) is $\mathbf{K} = 3/2\mathbf{P}_\mathbf{S} - 1/2(\mathbf{P}_\mathbf{A} + \mathbf{P}_\mathbf{B})$. Consequently,

$$\begin{aligned}
\mathbf{LK} &= 3/2(\mathbf{P}_\mathbf{A} + \mathbf{P}_\mathbf{B}) - 1/2(\mathbf{P}_\mathbf{A} + \mathbf{P}_\mathbf{A}\mathbf{P}_\mathbf{B} + \mathbf{P}_\mathbf{B}\mathbf{P}_\mathbf{A} + \mathbf{P}_\mathbf{B})\\
&= \mathbf{P}_\mathbf{A} + \mathbf{P}_\mathbf{B} - 1/2(\mathbf{P}_\mathbf{A}\mathbf{P}_\mathbf{B} + \mathbf{P}_\mathbf{B}\mathbf{P}_\mathbf{A}) = \mathbf{P}_\mathbf{S} = \mathbf{P}_\mathbf{L}.
\end{aligned} \tag{4.6}$$

Since $\mathbf{L} = \mathbf{L}'$ and $\mathbf{K} = \mathbf{K}'$, (4.6) entails $\mathbf{KL} = \mathbf{P}_\mathbf{S}$. But $R(\mathbf{K}) \subseteq R(\mathbf{S})$ and $\mathrm{r}(\mathbf{K}) \geq \mathrm{r}(\mathbf{S})$, and hence $\mathbf{KL} = \mathbf{P}_\mathbf{K}$. Then, from (1.2) it follows that $\mathbf{K} = \mathbf{L}^+$, as asserted in (A17). A trivial consequence of (A17) is (A18) stating that $\mathbf{P}_\mathbf{A} + \mathbf{P}_\mathbf{B} - 3/2\mathbf{P}_\mathbf{A}\mathbf{P}_\mathbf{B}$ is a generalized inverse of $\mathbf{P}_\mathbf{A} + \mathbf{P}_\mathbf{B}$. Further, from Theorem 10.1.8(e) of Rao and Mitra (1971) it follows that $R[\mathbf{P}_\mathbf{A}(\mathbf{P}_\mathbf{A} + \mathbf{P}_\mathbf{B})^-\mathbf{P}_\mathbf{B}] = R(\mathbf{T})$, and hence it can easily be verified that (A18) implies (A24): $R(\mathbf{P}_\mathbf{A}\mathbf{B}) = R(\mathbf{T})$. Premultiplying in (A24) by $\mathbf{A}'$ yields (A25): $R(\mathbf{A}'\mathbf{B}) = R(\mathbf{A}'\mathbf{T})$. Hence $\mathrm{r}(\mathbf{A}'\mathbf{B}) = \mathrm{r}(\mathbf{A}'\mathbf{T})$. But $\mathrm{r}(\mathbf{A}'\mathbf{T}) = \mathrm{r}(\mathbf{P}_\mathbf{A}\mathbf{T}) = \mathrm{r}(\mathbf{T})$, thus leading to (A34): $\mathrm{r}(\mathbf{A}'\mathbf{B}) = \mathrm{r}(\mathbf{T})$.

In view of the result (3.14) in Marsaglia and Styan (1974),

$$\dim R(\mathbf{A}) \cap R^{\perp}(\mathbf{B}) = \mathrm{r}(\mathbf{A}) - \mathrm{r}(\mathbf{A}'\mathbf{B}). \tag{4.7}$$

Moreover, on account of (4.2) and (4.3),

$$\dim R(\mathbf{A}) \cap R^{\perp}(\mathbf{T}) = \mathrm{r}(\mathbf{A}) - \mathrm{r}(\mathbf{T}), \tag{4.8}$$

and

$$\dim R(\mathbf{S}) \cap R^{\perp}(\mathbf{B}) = \mathrm{r}(\mathbf{S}) - \mathrm{r}(\mathbf{B}) = \mathrm{r}(\mathbf{Q}_{\mathrm{B}}\mathbf{A}). \tag{4.9}$$

On the other hand, according to (3.4),

$$\mathrm{r}(\mathbf{S}) = \mathrm{r}(\mathbf{A}) + \mathrm{r}(\mathbf{B}) - \mathrm{r}(\mathbf{T}), \tag{4.10}$$

and hence combining (4.8) with (4.9) yields

$$\dim R(\mathbf{S}) \cap R^{\perp}(\mathbf{B}) = \dim R(\mathbf{A}) \cap R^{\perp}(\mathbf{T}). \tag{4.11}$$

Consequently,

$$
\begin{aligned}
\text{(A34)} \quad &\Rightarrow \text{(A35)}: \dim R(\mathbf{A}) \cap R^{\perp}(\mathbf{T}) = \mathrm{r}(\mathbf{A}) - \mathrm{r}(\mathbf{A}'\mathbf{B}), \ \text{by (4.8)},\\
&\Rightarrow \text{(A36)}: \dim R(\mathbf{S}) \cap R^{\perp}(\mathbf{B}) = \mathrm{r}(\mathbf{A}) - \mathrm{r}(\mathbf{A}'\mathbf{B}), \ \text{by (4.11)},\\
&\Rightarrow \text{(A38)}: \mathrm{r}(\mathbf{S}) = \mathrm{r}(\mathbf{A}) + \mathrm{r}(\mathbf{B}) - \mathrm{r}(\mathbf{A}'\mathbf{B}), \ \text{by (4.9)},\\
&\Rightarrow \text{(A37)}: \mathrm{r}(\mathbf{Q}_{\mathrm{B}}\mathbf{A}) = \mathrm{r}(\mathbf{A}) - \mathrm{r}(\mathbf{A}'\mathbf{B}), \ \text{by (4.9)},\\
&\Rightarrow \text{(A39)}: \dim R(\mathbf{A}) \cap R^{\perp}(\mathbf{B}) = \mathrm{r}(\mathbf{Q}_{\mathrm{B}}\mathbf{A}), \ \text{by (4.7)},\\
&\Rightarrow \text{(A40)}: \dim R(\mathbf{A}) \cap R^{\perp}(\mathbf{B}) = \mathrm{r}(\mathbf{S}) - \mathrm{r}(\mathbf{B}), \ \text{by (4.9)},\\
&\Rightarrow \text{(A41)}: \dim R(\mathbf{A}) \cap R^{\perp}(\mathbf{B}) = \mathrm{r}(\mathbf{A}) - \mathrm{r}(\mathbf{T}), \ \text{by (4.10)},\\
&\Rightarrow \text{(A42)}: \dim R(\mathbf{A}) \cap R^{\perp}(\mathbf{B}) = \dim R(\mathbf{A}) \cap R^{\perp}(\mathbf{T}), \ \text{by (4.8)},\\
&\Rightarrow \text{(A43)}: \dim R(\mathbf{A}) \cap R^{\perp}(\mathbf{B}) = \dim R(\mathbf{S}) \cap R^{\perp}(\mathbf{B}), \ \text{by (4.11)}.
\end{aligned}
$$

Combining (A43) with the trivial inclusion $R(\mathbf{A}) \cap R^{\perp}(\mathbf{B}) \subseteq R(\mathbf{S}) \cap R^{\perp}(\mathbf{B})$ yields (A29): $R(\mathbf{S}) \cap R^{\perp}(\mathbf{B}) = R(\mathbf{A}) \cap R^{\perp}(\mathbf{B})$. Then, in view of (4.3),

$$
\begin{aligned}
\text{(A29)} \quad &\Rightarrow \text{(A33)}: R(\mathbf{S}) = R(\mathbf{B}) \boxplus R(\mathbf{A}) \cap R^{\perp}(\mathbf{B})\\
&\Rightarrow \text{(A27)}: R(\mathbf{Q}_{\mathrm{B}}\mathbf{A}) = R(\mathbf{A}) \cap R^{\perp}(\mathbf{B}).
\end{aligned}
$$

An immediate consequence of (A27) is that $\mathbf{P}_{\mathrm{A}}\mathbf{Q}_{\mathrm{B}}\mathbf{P}_{\mathrm{A}} = \mathbf{Q}_{\mathrm{B}}\mathbf{P}_{\mathrm{A}}$. This entails (A1): $\mathbf{P}_{\mathrm{A}}\mathbf{P}_{\mathrm{B}} = \mathbf{P}_{\mathrm{B}}\mathbf{P}_{\mathrm{A}}$, thus concluding the proof that (A1) $\Leftrightarrow$ ... $\Leftrightarrow$ (A43).

To complete the proof notice that (A5): $\mathbf{A}'\mathbf{B}\mathbf{G} = (\mathbf{A}'\mathbf{B}\mathbf{G})^2$, along with the specification of $\mathbf{G}$ in (2.3), gives

$$\mathrm{r}(\mathbf{I}_a - \mathbf{A}'\mathbf{B}\mathbf{G}) = \mathrm{trace}(\mathbf{I}_a - \mathbf{A}'\mathbf{B}\mathbf{G}) = a - \mathrm{r}(\mathbf{A}'\mathbf{B}\mathbf{G}) = a - \mathrm{r}(\mathbf{A}'\mathbf{B}),$$

which is (A44). Conversely, (A44) means that the matrices $\mathbf{I}_a$ and $\mathbf{A}'\mathbf{BG}$ are rank-subtractive, and thus, according to Theorem 17 of Marsaglia and Styan (1974), $(\mathbf{A}'\mathbf{BG})\mathbf{I}_a^{-1}(\mathbf{A}'\mathbf{BG}) = \mathbf{A}'\mathbf{BG}$, which is (A5). The equivalence of (A45) to (A6) follows similarly, while the equivalence of (A46) to (A1) is well known; cf. Ben-Israel and Greville (1974, pp. 250-251).

ACKNOWLEDGEMENTS

I am extremely grateful to Tarmo Pukkila and to Simo Puntanen for their generous invitation to speak at the Second International Tampere Conference in Statistics. I am also very grateful to the four referees for their valuable comments and suggestions.

This reseach was supported in part by Grant No. CPBP 01.01.2/1.

REFERENCES

Afriat, S.N. (1957). Orthogonal and oblique projectors and the characteristics of pairs of vector spaces. *Proc. Cambridge Philos. Soc.*, 53, 800-816.

Ahlers, C. W. and Lewis, T. O. (1971). Linear estimation with a positive semidefinite covariance matrix. *Indust. Math.*, 21, 23-27.

Alalouf, I. S. (1975). Comments on a paper by Ahlers and Lewis. *Indust. Math.*, 25, 97-104.

Alalouf, I. S. and Styan, G. P. H. (1983). Characterizations of the conditions for the ordinary least squares estimator to be best linear unbiased. *Topics in Applied Statistics* (Y. P. Chaubey and T. D. Dwivedi, *eds.*). Concordia University, Montreal, pp. 331-344.

Anderson, T. W. (1948). On the theory of testing serial correlation. *Skand. Aktuarietidskr.*, 31, 88-116.

Baksalary, J. K. (1984). A study of the equivalence between a Gauss-Markoff model and its augmentation by nuisance parameters. *Math. Operationsforsch. Statist. Ser. Statist.*, 15, 3-35.

Baksalary, J. K. (1985). Milliken´s estimability criterion. *Encyclopedia of Statistical Sciences, vol. 5: Lindeberg Condition to Multitrait-Multimethod Matrices* (S. Kotz, N. L. Johnson and C. B. Read, *eds.*). Wiley, New York, pp. 503-504.

Baksalary, J. K. and Kala, R. (1980a). A note on Ahlers and Lewis´representation of the best linear unbiased estimator in the general Gauss-Markoff model. *Banach Center Publications, vol. 6: Mathematical Statistics* (R. Bartoszyński, J. Koronacki and R. Zieliński, *eds.*). Polish Scientific Publishers, Warsaw, pp. 17-21.

Baksalary, J. K. and Kala, R. (1980b). On estimation problems in a general Gauss-Markov model. *Data Analysis and Informatics* (E. Diday, L. Lebart, J. P. Pagès, and R. Tomassone, *eds.*). North-Holland, Amsterdam, pp. 163-167.

Baksalary, J. K. and Kala, R. (1981). Linear transformations preserving best linear unbiased estimators in a general Gauss-Markoff model. *Ann. Statist.*, 9, 913-916.

Baksalary, J. K. and Kala, R. (1983). On equalities between BLUEs, WLSEs, and SLSEs. *Canad. J. Statist.*, 11, 119-123.

Baksalary, J. K. and Kala, R. (1986). Linear sufficiency with respect to a given vector of parametric functions. *J. Statist. Plann. Inference*, 14, 331-338.

Baksalary, J. K. and Mathew, T. (1986). Linear sufficiency and completeness in an incorrectly specified general Gauss-Markov model. *Sankhyā Ser. A*, 48, 169-180.

Barnard, G. A. (1963). The logic of least squares. *J. Roy. Statist. Soc. Ser. B*, 25, 124-127.

Ben-Israel, A. and Greville, T. N. E. (1974). *Generalized Inverses: Theory and Applications.* Wiley, New York.

Birkhoff, G. and von Neumann, J. (1936). The logic of quantum mechanics. *Ann. Math.*, 37, 823-842.

Chipman, J. S. (1964). On least squares with insufficient observations. *J. Amer. Statist. Assoc.*, 59, 1078-1111.

Drygas, H. (1983). Sufficiency and completeness in the general Gauss-Markov model. *Sankhyā Ser. A*, 45, 88-98.

Eckart, C. and Young, G. (1939). A principal axis transformation for non-Hermitian matrices. *Bull. Amer. Math. Soc.*, 45, 118-121.

Ehrenfeld, S. (1955). On the efficiency of experimental designs. *Ann. Math. Statist.*, 26, 247-255.

Fellman, J. (1976). On the effect of "nuisance" parameters in linear models. *Sankhyā Ser. A*, 38, 197-200.

Fellman, J. (1985). Estimation in linear models with nuisance parameters. *Statistics and Decisions* (supplement issue no. 2), 161-164.

Feuerverger, A. and Fraser, D.A.S. (1980). Categorical information and the singular linear model. *Canad. J. Statist.*, 8, 41-45.

Hall, F. and Meyer, C. D. Jr. (1975). Generalized inverses of the fundamental bordered matrix used in linear estimation. *Sankhyā Ser. A*, 37, 428-438.

Hedayat, A. S. and Majumdar, D. (1985). Combining experiments under Gauss-Markov models. *J. Amer. Statist. Assoc.*, 80, 698-703.

Joshi, V. M. (1985). Linear sufficiency . *Encyclopedia of Statistical Sciences, vol. 5: Lindeberg Condition to Multitrait-Multimethod Matrices* (S. Kotz, N. L. Johnson, and C. B. Read, *eds.*). Wiley, New York, pp. 64-65.

Khatri, C. G. (1976). A note on multiple and canonical correlation for a singular covariance matrix. *Psychometrika*, 41, 465-470.

Krämer, W. (1980). A note on the equality of ordinary least squares and Gauss-Markov estimates in the general linear model. *Sankhyā Ser. A*, 42, 130-131.

Marsaglia, G. and Styan, G. P. H. (1974). Equalities and inequalities for ranks of matrices. *Linear and Multilinear Algebra*, 2, 269-292.

Mathew, T. (1985). On inference in a general linear model with an incorrect dispersion matrix. *Linear Statistical Inference* (T. Caliński and W. Klonecki, *eds.*). Springer, Berlin, pp. 200-210.

Nordström, K. and von Rosen, D. (1987). Algebra of subspaces with applications to problems in statistics. *Proceedings of the Second International Tampere Conference in Statistics* (T. Pukkila and S. Puntanen, *eds.*). University of Tampere, Tampere, pp. 603-614.

Puntanen, S. (1985). Properties of the canonical correlations between the least squares fitted values and the residuals. *Proceedings of the First International Tampere Seminar on Linear Statistical Models and Their Applications* (T. Pukkila and S. Puntanen, *eds.*). University of Tampere, Tampere, pp. 269-284 [also reprinted as Part 2 of Puntanen (1987)].

Puntanen, S. (1986). Properties of the covariance matrix of the BLUE in the general linear model. *Pacific Statistical Congress* (I. S. Francis, B. F. J. Manly, and F. C. Lam, *eds.*). Elsevier North-Holland, New York, pp. 425-430 [also reprinted as Part 3 of Puntanen (1987)].

Puntanen, S. (1987). *On the Relative Goodness of Ordinary Least Squares Estimation in the General Linear Model.* Part 1: On the relative goodness of ordinary least squares and some associated canonical correlations. Part 2: Properties of the canonical correlations between the least squares fitted values and the residuals. Part 3: Properties of the covariance matrix of the BLUE in the general linear model. Part 4: On the equality of the ordinary least squares estimator and the best linear unbiased estimator. *Acta Univ. Tamper. Ser. A*, 217.

Puntanen, S. and Styan, G.P.H. (1987). On the equality of the ordinary least squares estimator and the best linear unbiased estimator; submitted for publication.

Rao, C. R. (1965). *Linear Statistical Inference and Its Applications.* Wiley, New York.

Rao, C. R. (1967). Least squares theory using an estimated dispersion matrix and its application to measurement of signals. *Proceedings of the Fifth Berkeley Symposium on Mathematical Statistics and Probability, vol. 1* (L. LeCam and J. Neyman, *eds.*). University of California, Berkeley, pp. 355-372.

Rao, C. R. (1971). Unified theory of linear estimation. *Sankhyā Ser. A*, 33, 371-394.

Rao, C. R. (1972). A note on the IPM method in the unified theory of linear estimation. *Sankhyā Ser. A*, 34, 285-288.

Rao, C. R. (1973). Representations of the best linear unbiased estimators in the Gauss-Markoff model with a singular dispersion matrix. *J. Multivariate Anal.*, 3, 276-292.

Rao, C.R. (1974). Projectors, generalized inverses and the BLUE's. *J. Roy. Statist. Soc. Ser. B*, 36, 442-448.

Rao, C. R. (1978). Choice of best linear estimators in the Gauss-Markoff model. *Comm. Statist. A - Theory Methods*, 7, 1199-1208.

Rao, C. R. (1981). A lemma on g-inverse of a matrix and computation of correlation coefficients in the singular case. *Comm. Statist. A - Theory Methods*, 10, 1-10.

Rao, C. R. and Mitra, S. K. (1971). *Generalized Inverse of Matrices and Its Applications.* Wiley, New York.

Rao, C. R. and Yanai, H. (1979). General definition and decomposition of projectors and some applications to statistical problems. *J. Statist. Plann. Inference*, 3, 1-17.

Seely, J. and Birkes, D. (1980). Estimability in partitioned linear models. *Ann. Statist.*, 8, 399-406.

Seshadri, V. and Styan, G. P. H. (1980). Canonical correlations, rank additivity and characterizations of multivariate normality. *Colloquia Math. Soc. János Bolyai, vol. 21: Analytic Function Methods in Probability Theory.* János Bolyai, Budapest, and North-Holland, Amsterdam, pp. 331-344.

Styan, G. P. H. (1985). Schur complements and linear statistical models. *Proceedings of the First International Tampere Seminar on Linear Statistical Models and Their Applications* (T. Pukkila and S. Puntanen, *eds.*). University of Tampere, Tampere, pp. 37-75.

Takeuchi, K., Yanai, H. and Mukherjee, B.N. (1982). *The Foundations of Multivariate Analysis.* Wiley Eastern Limited, New Delhi.

Tjur, T. (1984). Analysis of variance models in orthogonal designs. *Internat. Statist. Rev.*, 52, 33-81.

Zyskind, G. (1967). On canonical forms, non-negative covariance matrices and best and simple least squares linear estimators in linear models. *Ann. Math. Statist.*, 38, 1092-1109.

Zyskind, G. and Martin, F. B. (1969). On the best linear estimation and a general Gauss-Markov theorem in linear models with arbitrary non-negative structure. *SIAM J. Appl. Math.*, 17, 1190-1202.

Received 22 April 1987
Revised 17 September 1987

Dept. of Mathematical and Statistical Methods
Academy of Agriculture in Poznań
Wojska Polskiego 28
PL-60-637 Poznań
Poland

Proc. Second International Tampere Conference in Statistics
(Tampere, Finland, 1-4 June 1987)
Tarmo Pukkila and Simo Puntanen, *Editors*

pp. 143 - 159

INVITED PAPER

Textural features useful in classification of digital images

Knut CONRADSEN and **Bjarne Kjær NIELSEN**
Technical University of Denmark, Lyngby, Denmark

Key words and phrases: Context based methods, Fourier filters, local frequency, local orientation, random fields, remote sensing.

ABSTRACT

In the present paper some problems on classification of image data are discussed. It is claimed that it often is necessary to use some measure of the spatial correlation. It is shown that estimates of local frequency and local orientation successfully may be used as additional features in classification of remotely sensed data from southern Greenland.

1. INTRODUCTION

When analyzing image data in order to decide whether particular properties are present or not in the images classical discriminant analyses have been used with considerable success e.g. in remote sensing. This approach, however, has some very serious shortcomings. In order to describe those, we must first give some basic definitions on digital images. When the imaging device is a multispectral scanner, the resulting image will be a collection of p-dimensional vectors

$$\mathbf{X}_{i,j} = \begin{pmatrix} X_1(i,j) \\ \vdots \\ X_p(i,j) \end{pmatrix} \quad i = 0,\dots, n-1; \; j = 0,\dots, m-1;$$

defined on a rectangular lattice. The lattice points are called pixels (for picture elements) and they correspond to say geographical coordinates. The values $X_k(i, j)$ represent the intensity for "colour" no. k in pixel (i, j). The X values will typically vary in a random manner and will therefore be represented as random variables. For homogeneous areas, the distribution may often be assumed to be normal, with means (and dispersions) depending on the area.

In Figure 1 is shown a one channel image with 17 grey levels. It consists of 512×512 pixels, and it is sampled by the earth observation satellite Landsat 2. The pixel values correspond to the levels of the electromagnetic reflection of the sun light in the near infrared range (~ Band 7 in the satellite scanner). The actual size of each pixel is 50×50 meters and the scene thus covers appr. 25×25 kms. The area is located in Southern Greenland. The dark parts correspond to the sea, and the remaining areas are rocks of different types, some covered with snow or vegetation, others are barren. In for instance mineral exploration, it is of great interest to map the different geological units based on data like the present, eventually combined with similar images showing the reflectance in other spectral bands. The most obvious way to achieve such an identification would be to select training areas with a known geology, estimate the distributions of the pixel values $\mathbf{X}_{i,j}$, and then determine e.g. ordinary linear discriminant functions and use those in classification of the remaining pixels. As long as different units are characterized by substantial differences in mean values this approach will work fine.

The "natural optimality" of the pixel-by-pixel rules presupposes independence. A glimpse on Figure 1 shows that this is not a reasonable assumption. There is a strong spatial continuity in the image. This may be utilized in the classification. In recent years several approaches to so-called contextual methods have been proposed. Mohn, Hjort and Storvik (1986) give a good comparison of such methods, and they show that for a range of models the error rates may be reduced considerably by using contextual methods.

The methods used in the contextual algorithms assume some kind of e.g. Markovian structure of the spatial arrangement of the populations. Some introduce a parametric, spatial dependence between error terms in the model. In many cases, however, the populations are characterized by differences in the spatial correlation. Consider e.g. the binary image shown in Figure 2. It is a realisation of a so-called second order Markovian Random Field (MRF). In order to define such a field we introduce an easy notation for the pixel values of the neighbours of a pixel x (see Figure 3).

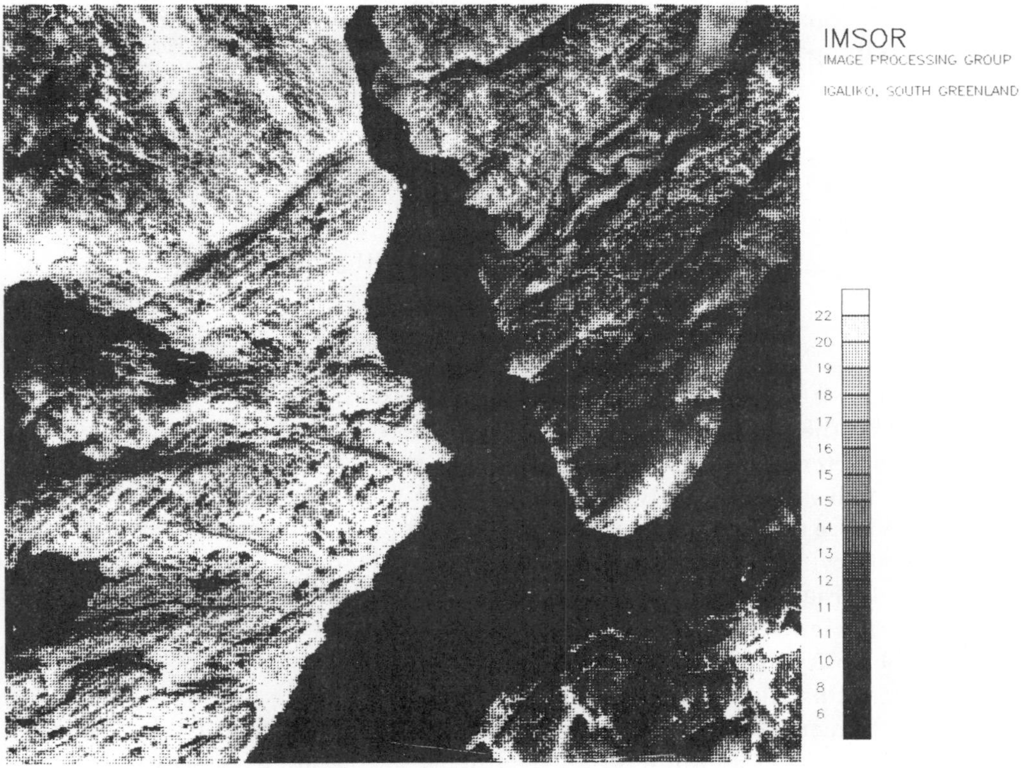

FIGURE 1: Part of Landsat 2 scene (Band 7) from South Greenland. The image consists of 512 × 512 pixels, each of size 50 × 50 ms. The original scene has 64 grey levels, but the presentation given above only shows 17 grey levels.

FIGURE 2: Inhomogeneous binary image generated by two Markovian Random Fields.

The probability structure of a second order binary MRF (Besag, 1974) is fully specified by conditional probabilities of the form

$$p\,(x \mid t, t', u, u', v, v', w, w') = \frac{\exp(xT)}{1+\exp(T)}\,, \tag{1}$$

where

$$T = a+\beta_1\,(t + t') + \beta_2\,(u + u') + \gamma_1\,(v + v') + \gamma_2\,(w + w').$$

The left part of Figure 2 was generated with parameter values

$$a= -1.3,\ \beta_1 = 1.5,\ \beta_2 = 0.5,\ \gamma_1 = \gamma_2 = -0.5.$$

by a method similar to the one described in Hassner and Sklansky (1981).

In the right part β_1 and β_2 were interchanged and the remaining parameters were unchanged. Therefore the means are the same for the two halves. In both halves approximately 40 % of the pixels are black. It is therefore obvious that a procedure based solely on mean values must fail when trying to discriminate between the two textures.

In Figure 4 we have shown the result of a discrimination using the Markovian structure assigning the class that maximizes the conditional probability (1). It follows that using the spatial dependence enables a better classification.

In more complicated situations it may, however, be difficult to obtain training sets that allow for a modelling of the spatial dependence. Instead we shall in the next sections show how information on the dependence may be extracted by suitable filters, and how these results may be used as extra features in classification.

2. ESTIMATION OF LOCAL ORIENTATION AND LOCAL FREQUENCY

In the sequel we shall study pairs of filters that may be used in estimation of local orientation and local frequency. The description will be based on a continuous Fourier representation, since the "continuous" formulas are somewhat simpler than the "discrete" ones.

We consider a function $f(\mathbf{x})$ and define its Fourier transform by

$$F(\mathbf{u}) = \int_{-\infty}^{\infty}\int_{-\infty}^{\infty} f(\mathbf{x}) \exp(-\,i2\pi\mathbf{u}'\mathbf{x})\,d\mathbf{x}.$$

The inverse transform is given by

$$f(\mathbf{x}) = \int_{-\infty}^{\infty}\int_{-\infty}^{\infty} F(\mathbf{u}) \exp(i2\pi\mathbf{u}'\mathbf{x})\, d\mathbf{u}.$$

Details on existence and properties may be found in e.g. Goodman (1968). We shall only mention that $f(\mathbf{x})$ real implies that $F(\mathbf{u})$ is Hermitian symmetric, i.e.

$$F(-\mathbf{u}) = F^*(\mathbf{u}).$$

If $f(\mathbf{x})$ is even (and real), i.e. $f(-\mathbf{x}) = f(\mathbf{x})$, the Fourier transform $F(\mathbf{u})$ is real, and if $f(\mathbf{x})$ is odd, i.e. $f(-\mathbf{x}) = -f(\mathbf{x})$, the transform is purely imaginary.

A space-invariant linear filter may be described directly as convolution with the impulse response function $h(\mathbf{x})$, i.e. the filtered "image" $\hat{f}$ is given by

$$\hat{f}(\mathbf{X}) = f * h(\mathbf{X}) = \int_{-\infty}^{\infty}\int_{-\infty}^{\infty} f(\mathbf{s})\, h(\mathbf{x} - \mathbf{s})\, d\mathbf{s}$$

or we may present $\hat{f}$ by its Fourier transform $\hat{F}$, i.e.

$$\hat{F}(\mathbf{u}) = F(\mathbf{u})\, H(\mathbf{u}),$$

where the Fourier transform H of h is the transfer function or frequency response function of the filter. We are looking for filters that give orientation estimates that should be invariant with respect to the frequency content in the estimated direction. The simplest way to obtain this is to restrict ourselves to filters that are polar separable, i.e. filters with transfer functions

$$H(\mathbf{u}) = g(\mathbf{u}'\mathbf{u})^{\frac{1}{2}}\, p(\operatorname{atan}(u_2/u_1)) = g(\delta)\, p(\phi),$$

where δ and ϕ are polar coordinates for $\mathbf{u} = (u_1, u_2)'$. Corresponding to a finite number of directions ϕ_k we consider sign functions

$$S_k(\mathbf{u}) = \operatorname{sign} \cos(\operatorname{atan}(u_2/u_1) - \phi_k).$$

The function $S_k(\mathbf{u})$ equals -1 in the shaded halfplane shown in Figure 5 and equals $+1$ in the other half plane.

Knutsson and Granlund (1983) consider filters of the form

v	t	w
u	x	u′
w′	t′	v′

FIGURE 3 : Neighbours used in definition of second order Markovian Random Fields.

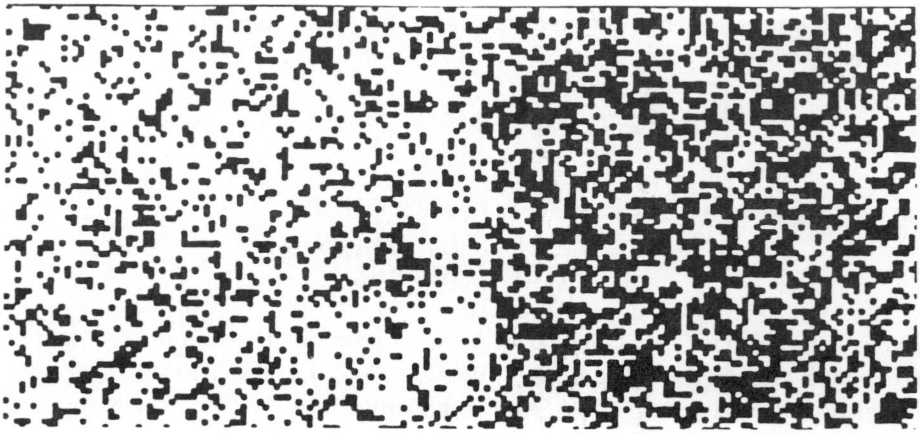

FIGURE 4 : Discrimination between the two fields shown in Figure 2.

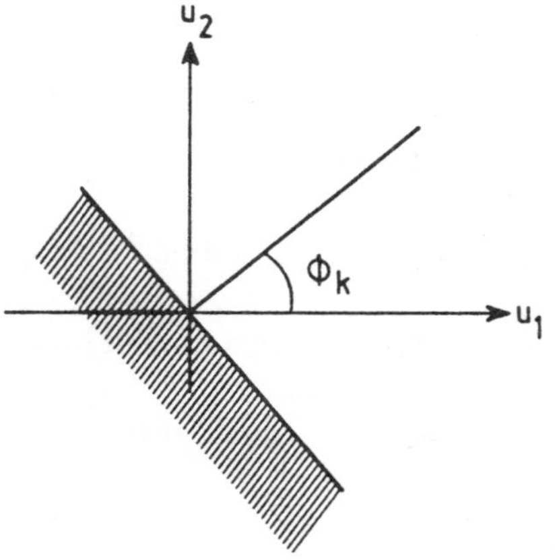

FIGURE 5 : Areas defining the sign function $S_k(\mathbf{u})$.

$$H_k^e(\mathbf{u}) = g(\delta)\cos^{2A}(\phi - \phi_k),$$

$$H_k^o(\mathbf{u}) = i\,S_k(\mathbf{u})\,H_k^e(\mathbf{u}),$$

where δ and ϕ are polar coordinates for $\mathbf{u}$ and

$$g(\delta) = \exp\left(-\frac{4}{\ln 2} B^{-2} \ln^2\left(\frac{\delta}{\delta_i}\right)\right).$$

The constants A, B, ϕ_k, and δ_i are parameters that may be used in determining the shape of the filter. A is the angle selectivity, B the bandwidth, ϕ_k the orientation and δ_i the center frequency.

Since H_k^e is real (and even) it corresponds to a zero phase filter. H_k^o is purely imaginary (and odd) and hence causes a phase shift of magnitude $\pi/2$. In the simplest (one dimensional) case a cosine would be transformed to a sine, wherefore the filter is called a quadrature filter. We shall use the term quadrature filter pairs about H_k^e and H_k^o. The corresponding impulse responses are denoted h_k^e and h_k^o. The filtered output from the two filters are

$$\hat{f}_k^e(\mathbf{x}) = f * h_k^e(\mathbf{x}) = \int\int f(\mathbf{s})\, h_k^e(\mathbf{x} - \mathbf{s})\, ds$$

$$\hat{f}_k^o(\mathbf{x}) = f * h_k^o(\mathbf{x}) = \int\int f(\mathbf{s})\, h_k^o(\mathbf{x} - \mathbf{s})\, ds$$

and we define the local orientational root mean squared (RMS) value as

$$\hat{f}_k(x) = [\,|\hat{f}_k^e(x)|^2 + |\hat{f}_k^o(x)|^2\,]^{\frac{1}{2}}.$$

If we consider K evenly distributed directions

$$\phi_k = \pi\frac{k}{K}, \quad k = 0,1,\ldots,K-1,$$

we combine the output $\hat{f}_k(\mathbf{x})$ from these K filters as

$$Z(\mathbf{x}) = \sum_{k=0}^{K-1} \hat{f}_k(\mathbf{x}) \exp\left(i2\pi\frac{k}{K}\right).$$

This corresponds to associating each $f_k(\mathbf{x})$ with an angle $2\phi_k$ and adding them as vectors. If there is a dominant direction, only that f_k will contribute, otherwise they will more or less cancel each other. A more precise argument may be found in Knutsson (1982). The direction of Z now contains information on the dominant direction in the original (i.e. $\frac{1}{2}\arg(Z)$) and the magnitude of Z is a measure of the consistency of that direction.

Estimation of local frequency is obtained by combining output from two or more filters. If a function has all its energy in a single frequency, say r, then the ratio between the output from the filter pairs with center frequencies δ_1 and δ_2 will be

$$R = \exp\left(- \frac{4}{\ln 2} B^{-2} \left[\ln^2 \frac{r}{\delta_1} - \ln^2 \frac{r}{\delta_2} \right] \right).$$

Solving this equation with respect to r yields

$$r = (\delta_1 \delta_2)^{\frac{1}{2}} R^a ,$$

where

$$a = \left[\frac{8}{\ln 2} B^{-2} \ln \frac{\delta_2}{\delta_1} \right]^{-1}$$

We see that in this case it is possible to obtain an exact assessment of the true frequency. In the general case we may use e.g. 3 sets of quadrature filters, with low, with medium, and with high center frequencies. The output from those may be combined vectorially with medium frequency corresponding to argument 0, high frequency to argument $2\pi/3$, and low frequency to $4\pi/3$. Analogously to the vectorial combination of orientation measures, this will then produce a frequency measure, where the direction corresponds to the frequency and the magnitude to the certainty of the frequency determination.

The above mentioned procedures are implemented in the GOP 300 Image Processor, Contextvision (1986). The discrete kernel weights have been determined by minimizing a weighted mean squared distance between the "theoretical" transfer funtions and the transfer functions corresponding to the digital filter. In the sequel (or_1, or_2) and (fr_1, fr_2) correspond to cartesian representations of the complex valued estimates of local orientation and local frequency. The window size actually used was 11×11 pixels. Due to the considerable size of the window we obtain less precise estimates along borders between different textures.

3. CLASSIFICATION OF DIFFERENT GEOLOGICAL UNITS

In Figure 6 is shown some training areas that are used in a multispectral classification. Area I consists of so-called Julianehaab Granites and Area II of Igaliko intrusives, and it is the objective of the study to discriminate between those two rock types. Area III, sandstone and the like, is included in order to get a better coverage of rock types, and Area IV, water, is included for similar obvious reasons. As an alternative to including classes III and IV one could have introduced a reject class. Since this would cause a merging between land and water it would then be more difficult to evaluate the classified maps visually. The granites are covered with vegetation whereas the intrusives are barren. Immediately to the west of the granite training set we also have granites, but most of those are either barren or snow covered. Due to the high reflectance of chlorophyll in the near infrared area and the absorption in the red parts of the spectrum there are substantial differences between the distributions of pixel values from barren and from vegetation covered rock. The variables used are

$B4$ = Landsat Band 4 ~ wavelength 0.5 - 0.6 µm
$B5$ = Landsat Band 5 ~ wavelength 0.6 - 0.7 µm
$B6$ = Landsat Band 6 ~ wavelength 0.7 - 0.8 µm
$B7$ = Landsat Band 7 ~ wavelength 0.8 - 1.1 µm

$or1$ = real part of smoothed local orientation based on B7
$or2$ = imaginary part of smoothed local orientation based on B7
$fr1$ = real part of smoothed local frequency based on B7
$fr2$ = imaginary part of smoothed local frequency based on B7.

Means and standard deviations of the 8 variables are shown in Table 1 for the 4 training sets. Furthermore is shown the means and standard deviations of the Landsat bands for a test area consisting of barren Julianehaab granites. It is seen that $or2$ and $fr1$ in this case are the best (individual) discriminators between the granites and the intrusives. In Table 2 is presented the correlations between the Landsat bands for the two primary training sets, the vegetation covered Julianehaab granites and the Igaliko intrusives, and for the basic test area, the barren granites. The results presented are based on all pixels in the contiguous training sets, and no attempts were made in order to avoid the biasedness (in the variances – covariances) that may result from the spatially correlated pixels.

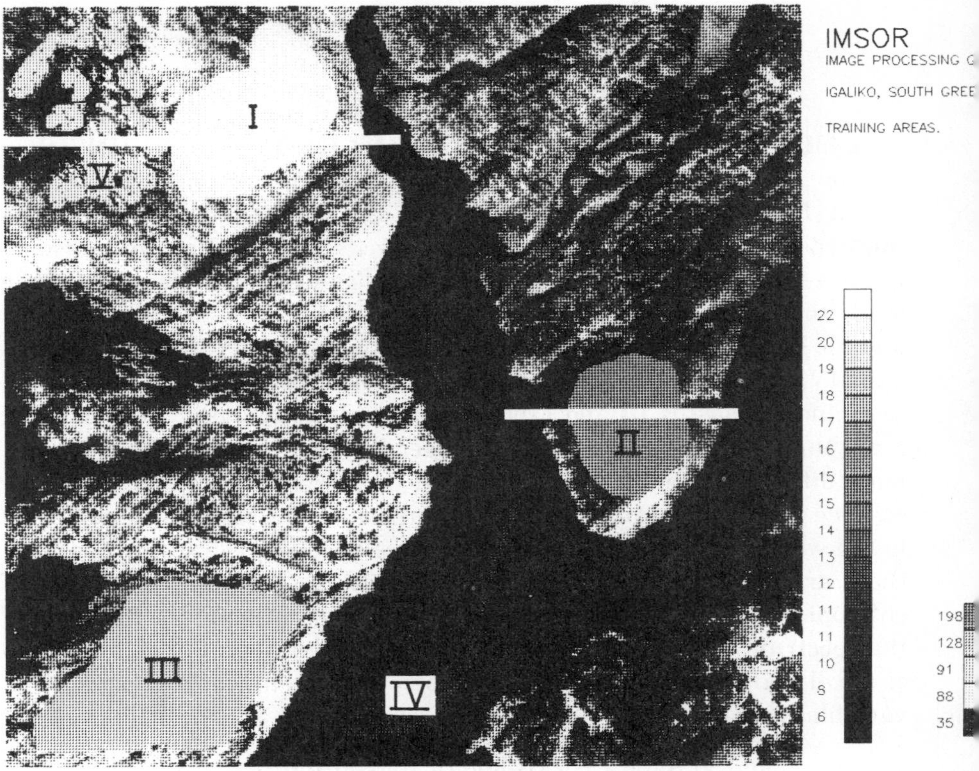

FIGURE 6: Training areas for classification of Landsat scene. The training areas are: I ~ Julianehaab Granites (covered with vegetation), II ~ Igaliko Intrusives, III ~ Sandstone etc., IV ~ water. The area V is a test area with Julianehaab Granites without vegetation.

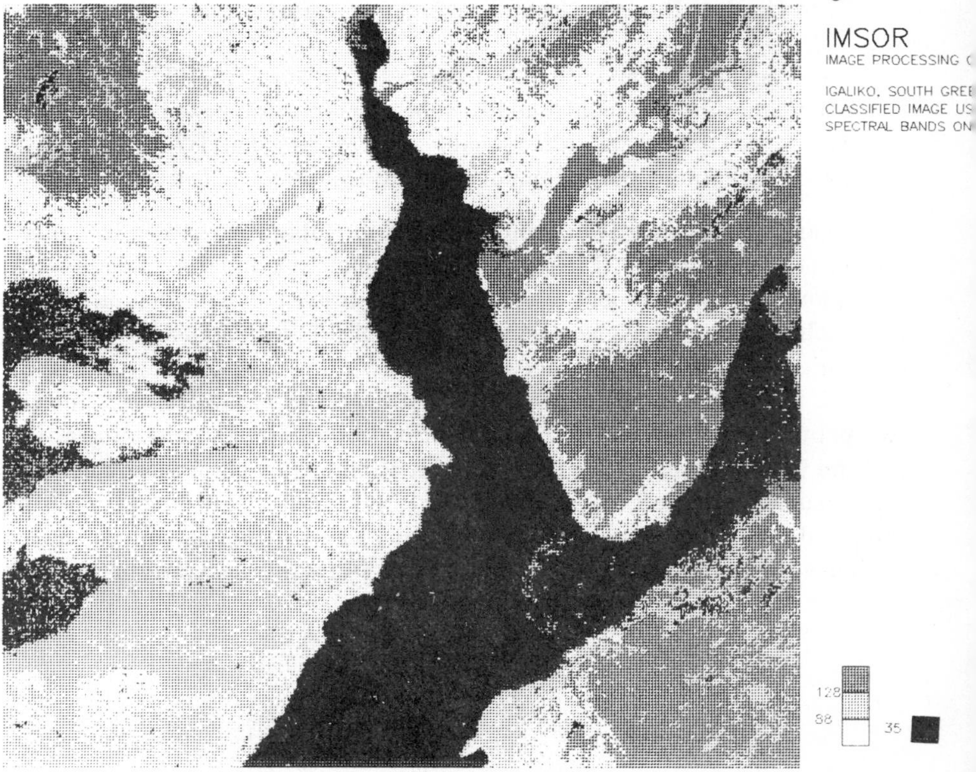

FIGURE 7 : Classification result after having used the 4 Landsat bands.

TABLE 1 : Means and Standard Deviations for the 8 variables for the 4 training sets and means and standard deviations of the 4 original bands for the barren granites.

Variable	Veg.cov.gran.		Barren gran.		Intrusives		Sandstone	
	Mean	St.dev	Mean	St.dev	Mean	St.dev	Mean	St.dev
*B*4	24.24	1.62	29.54	3.87	30.13	3.35	21.75	2.32
*B*5	52.80	5.43	67.50	10.99	68.84	9.73	44.80	8.76
*B*6	41.68	5.05	38.20	5.70	37.75	5.43	39.56	8.73
*B*7	49.26	10.86	34.37	8.19	31.54	8.53	47.67	16.77
*or*1	-2.48	5.57			0.71	4.12	-28.29	12.45
*or*2	12.17	8.47			-8.43	4.77	-19.69	12.39
*fr*1	51.54	11.79			8.59	10.03	62.11	12.74
*fr*2	-17.34	13.32			-7.48	10.77	-25.23	15.29
No. of pixels	8231		4191		5725		13277	

TABLE 2: Correlations between the 4 original bands for the two main training sets, i.e. "vegetation covered granites" and "intrusives", and for the test area "barren granites".

	Vegetation cov. gran.				Barren granites				Intrusives			
	B4	B5	B6	B7	B4	B5	B6	B7	B4	B5	B6	B7
B4	1.00				1.00				1.00			
B5	.78	1.00			.96	1.00			.93	1.00		
B6	.31	.43	1.00		.82	.87	1.00		.73	.77	1.00	
B7	.23	.34	.94	1.00	.64	.69	.89	1.00	.51	.56	.90	1.00

We assume joint normality for the variables, i.e.

$$\text{population } i \leftrightarrow N(\boldsymbol{\mu}_i, \boldsymbol{\Sigma}_i).$$

The first classifications were only based on the 4 Landsat bands. For each pixel with value $\mathbf{x}$ that must be classified we compute the 4 scores

$$S_i(\mathbf{x}) = \tfrac{1}{2}\ln(\det \hat{\boldsymbol{\Sigma}}_i) - \tfrac{1}{2}(\mathbf{x} - \hat{\boldsymbol{\mu}}_i)' \ \hat{\boldsymbol{\Sigma}}_i^{-1} (\mathbf{x} - \hat{\boldsymbol{\mu}}_i),$$

and the pixel is allocated to the population that gives the largest score. The estimated means and dispersion matrices are obtained from tables 1 and 2. This classification corresponds to ordinary quadratic discriminant analysis. The results are presented in Figure 7. It is seen that the test area with barren granites west of the granite training set is "misclassified" as intrusives. This is not strange. It was argued earlier on that the barren granites and the barren intrusives looked very similar. This is supported by the empirical moments shown in tables 1 and 2.

In order to investigate the nature of the populations further the pixel values (in band 7) along two cross sections of the granites and of the intrusives are shown in Figure 8. The two sections are shown as white bands in Figure 6. In the granites the transition from non-vegetation to vegetation occurs app. at pixel no. 100. A shift in level is observed, but the correlation structure is more or less constant, and very different from the intrusives. In figures 9, 10 and 11 are shown close-ups of granites with and without vegetation and of the intrusives. It follows that the spatial pattern looks much more similar for the two granites irrespective of differences in absolute levels than when comparing intrusives and barren granites. Therefore it seems obvious to generate new features describing the local texture in the images by means of the techniques described in section 2. The resulting values of local orientation and local frequency are then smoothed spatially in order to get values that are representative for larger areas. Such a presmoothing of variables before using them in classifications corresponds to some of the contextual methods for classifying image data that were mentioned in section 1.

The result of classifications based on the textural variables only (that in turn are based on the single band showing most local contrast, B7) is shown in Figure 12 and the result from using as well the original bands as the textural variables are shown in Figure 13. In both cases it follows that the barren granites are (practically) no longer misclassified as intrusives. Due to the fact that the granites and the intrusives are characterizable through textural measures defined by means of Fourier techniques, it is possible to distinguish between the two rocktypes irrespective of presence or absence of vegetational cover.

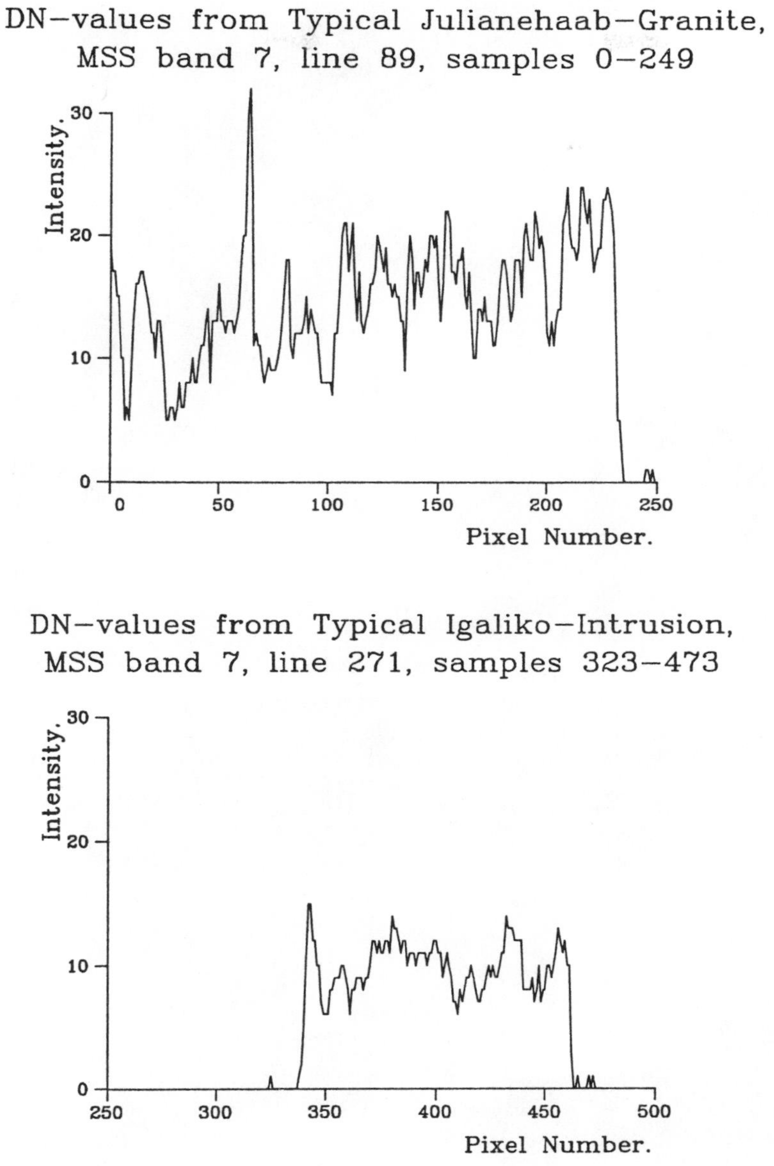

FIGURE 8: Variations of pixel values for Band 7 along the white stripes crossing the training sets in Figure 6.

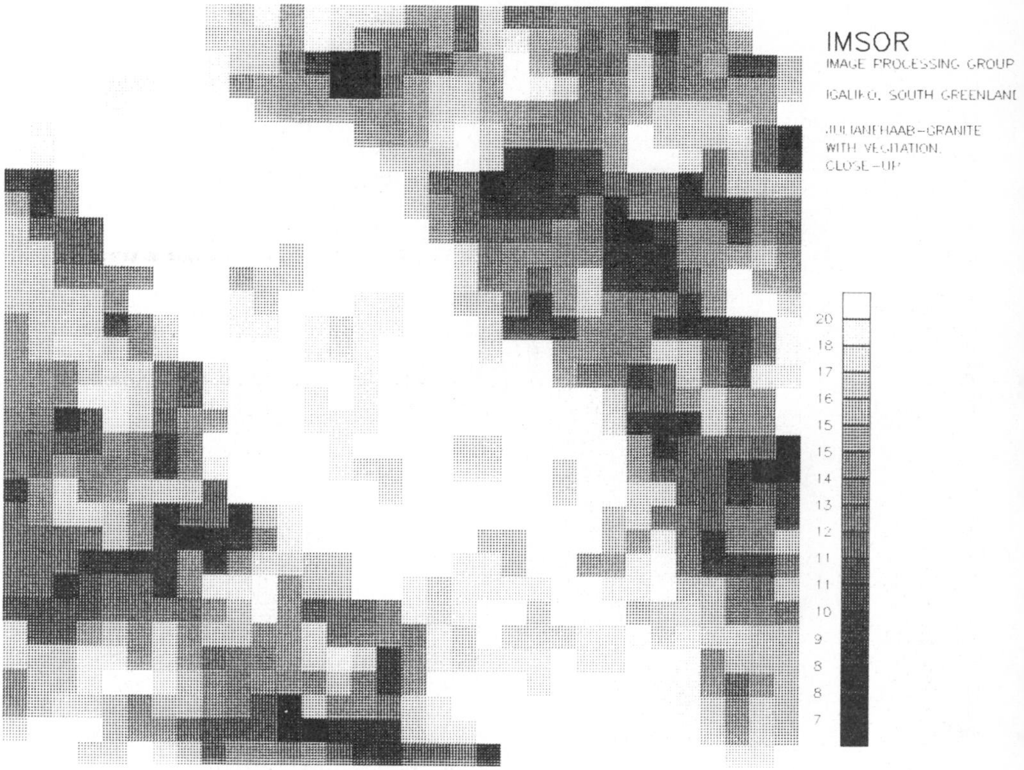

FIGURE 9: Close up of vegetation covered granites (~I).

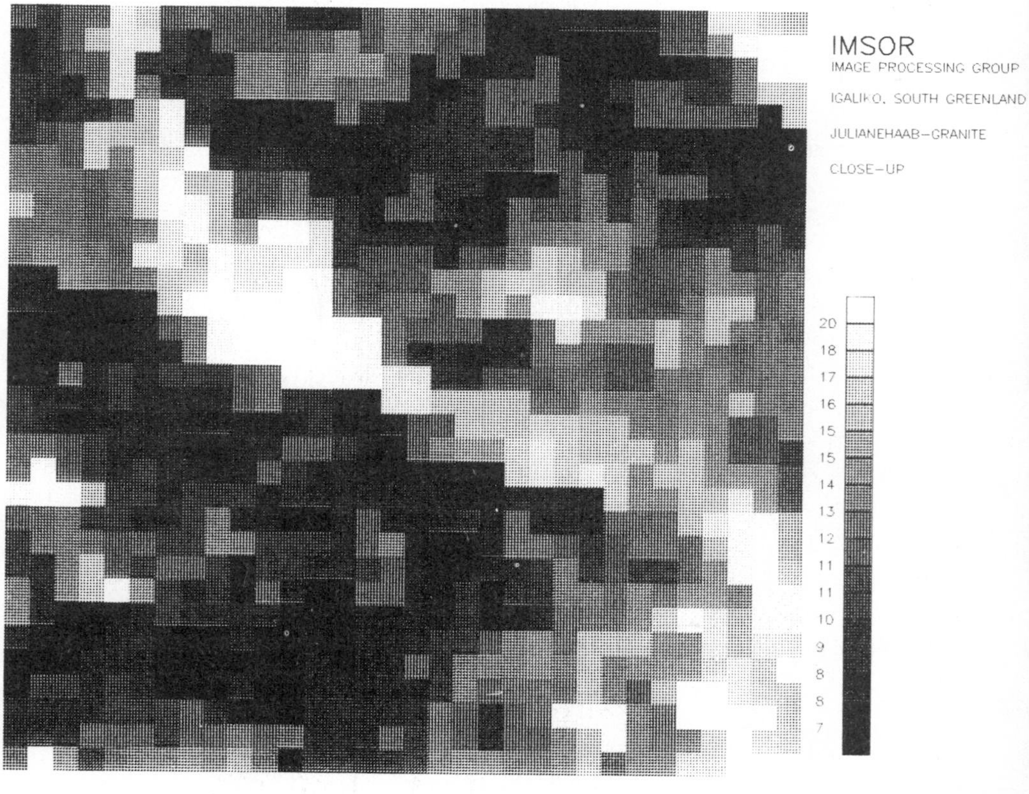

FIGURE 10: Close up of granites without vegetation (~V)

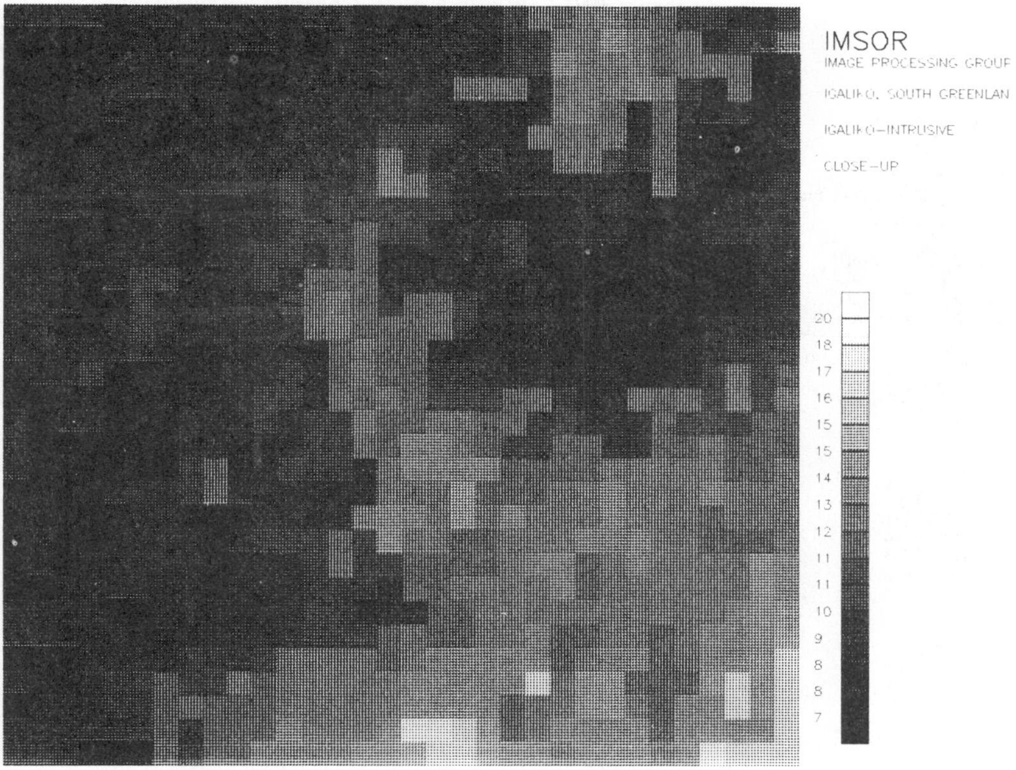

FIGURE 11: Close up of Igaliko Intrusives (~ II).

4. CONCLUSIONS

With the increasing importance of image analysis in many branches of science and technology much emphasis has recently been put on development of so called contextual methods in image classification. These methods include models e.g. of Markovian type for the spatial distribution of the populations and models for the spatial dependence in the error terms in such models. On the other hand the relevant populations in image data are often characterized not by differences in mean values and variances - covariances for the multivariate (normal) feature vector, but rather by the structure in the spatial dependence between pixels. A direct modelling of such dependences would lead to very complicated models that probably would be of no value in practical work with classification. Instead it is suggested to use quadrature filter pairs to estimate textural features that may be included in the feature vector describing each pixel. Then different discrimination methods may be applied on the augmented feature vector. The shown example on classifying satellite data demonstrates that this may be a very powerful technique.

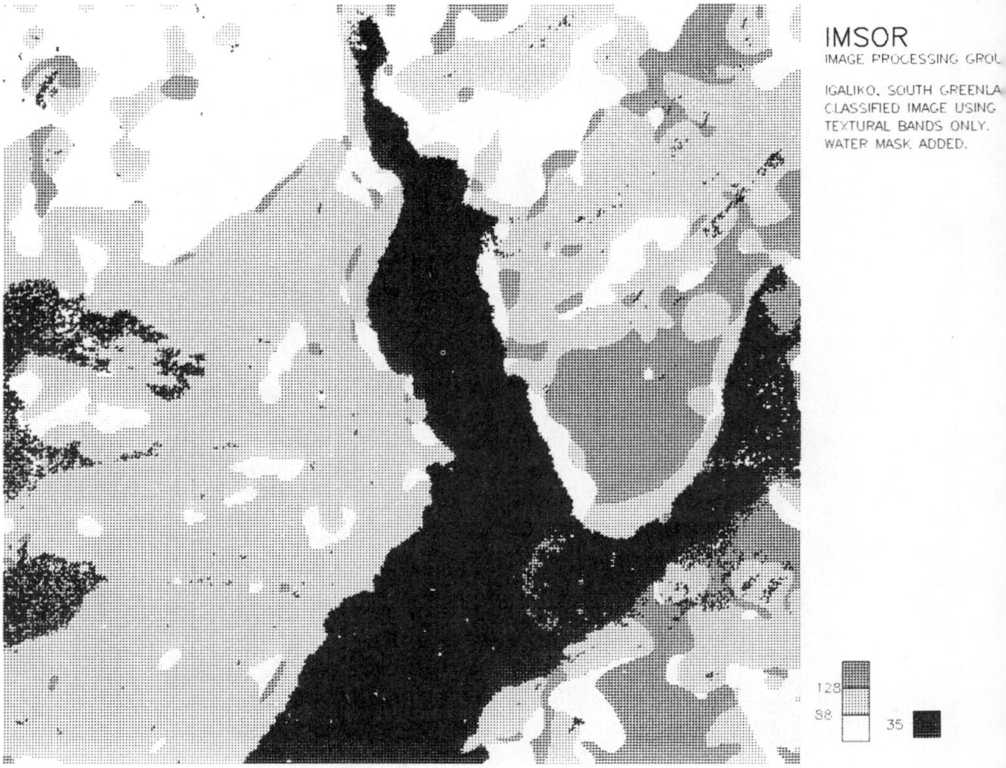
IMSOR
IMAGE PROCESSING GROU
IGALIKO, SOUTH GREENLA
CLASSIFIED IMAGE USING
TEXTURAL BANDS ONLY.
WATER MASK ADDED.
128
88
35

FIGURE 12: Classification result using textural variables only.

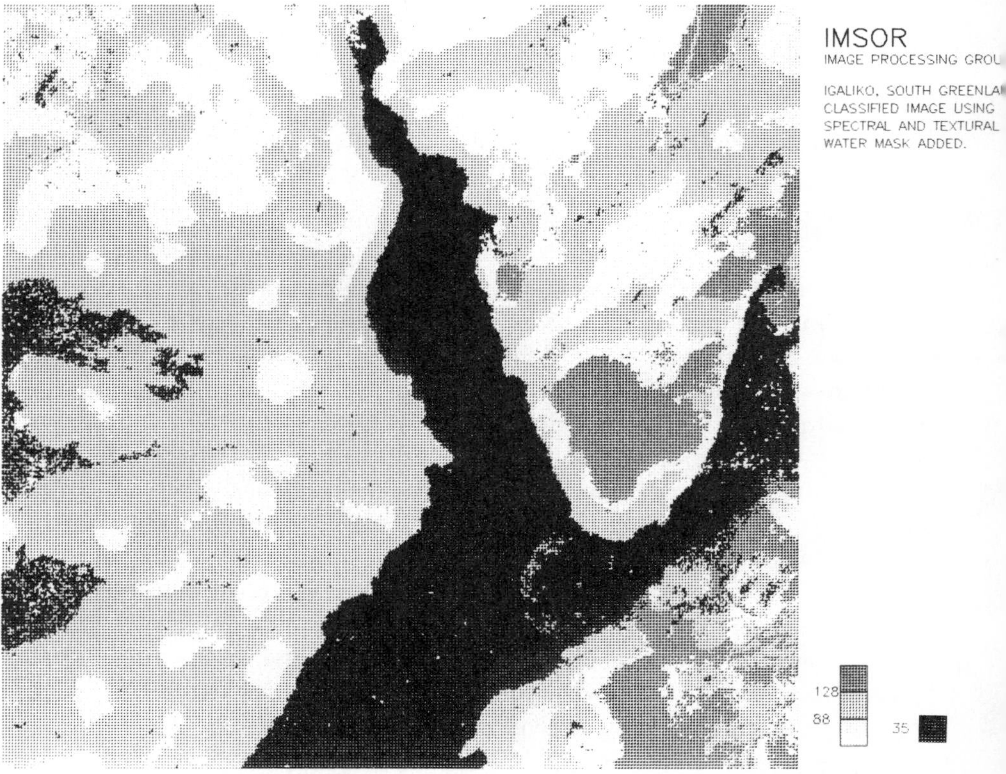
IMSOR
IMAGE PROCESSING GROU
IGALIKO, SOUTH GREENLA
CLASSIFIED IMAGE USING
SPECTRAL AND TEXTURAL
WATER MASK ADDED.
128
88
35

FIGURE 13: Classification result using as well textural variables as the original varia

5. ACKNOWLEDGEMENTS

The program for generating Figures 2 and 4 were made by Niels Anker Hansen.

REFERENCES

Besag, J. (1974). Spatial interaction and the statistical analysis of lattice systems. *Journal of the Royal Statistical Society, Series B*, vol. 36, 192-236.

ContextVision, AB (1986). Introduction to GOP-300. Linköping.

Goodman, J.W. (1986). *Introduction to Fourier Optics*. McGraw - Hill, San Francisco.

Hassner, M. and Sklansky, J. (1981). The use of Markov random fields as measure of texture. *Image Modelling* (A. Rosenfeld, *ed.*). Academic Press, New York, 185-198.

Knutsson, H. (1982). Filtering and Reconstruction in Image Processing. Linköping, Ph.D. thesis.

Knutsson, H. and Granlund, G.H. (1983). Texture analysis using two-dimensional quadrature filters. *IEEE Workshop on Computer Architecture for Pattern Analysis and Image Data Base Management*, Pasadena, California, Oct. 12-14, 206-213.

Mohn, E., Hjort, N.L. and Storvik, G. (1986). *A Comparison of Some Classification Methods in Remote Sensing by a Monte Carlo Study*, Norwegian Computing Center, Oslo.

Received 4 May 1987
Revised 19 October 1987

IMSOR
Technical University of Denmark
Building 321
DK-2800 Lyngby
Denmark

Proc. Second International Tampere Conference in Statistics
(Tampere, Finland, 1-4 June 1987)
Tarmo Pukkila and Simo Puntanen, *Editors*

pp. 161 - 176

INVITED PAPER

MTV model and its application to prediction of stock prices

Takeaki KARIYA
Hitotsubashi University, Tokyo, Japan

Key words and phrases: MTV model, principal component model, ARMA model, prediction of stock prices.

ABSTRACT

This paper proposes an MTV (Multivariate Time Series Variance Component) model as one which approximates economic and social phenomena involving great uncertainty, and shows that the model has an optimality property as such an approximation model. It may be viewed as an alternative to multivariate ARMA models. As an example, the model is applied to the prediction of the stock prices of 25 Japanese pharmaceutical companies.

1. INTRODUCTION

A multivariate time series model is often used to analyze such multidimensional complex time series phenomena as the fluctuations of stock prices, interest rates, exchange rates, etc. and to predict the future movements. A typical model is the multivariate (or vector-valued) ARMA (autoregressive moving average) model where the structure to be analyzed is treated as a black box and the stochastic regularity involved in the phenomenon is modelled to describe the structure and predict the future fluctuation. However, in the multivariate ARMA model too many parameters are introduced to specify an interdependent time series linear relation of the variables. For example, in a p-dimensional $AR(k)$ model there are $p^2 \times k$ parameters involved as the coefficients in the model, which often destabilizes its performance, especially in prediction. In this paper as an alternative model for the analysis and prediction of multidimensional complicated time series phenomena, we propose what we call an MTV (multivariate time series variance component) model.

In our MTV model, behind a p-dimensional time series process

$$\mathbf{x}_t = (x_{1t}, \dots, x_{pt}), \qquad t = 0, \pm 1, \pm 2, \dots \tag{1.1}$$

it is supposed that there are a comparatively small number of major common time series variance components through which the cross-sectional correlation structure [i.e. $Correl(x_{it}, x_{jt})$] and the serial correlation structure [i.e. $Correl(x_{it}, x_{jt-\ell})$] are formed. Formally speaking, the MTV model is regarded as a dynamic generalization of the PC (principal component) model, which assumes the serial uncorrelatedness of $\mathbf{x}_t$'s. This fact gives theoretical justification for applying the PC technique to time series data. In particular, it validates such indices as business indicators, price indices, etc. based on the PC method. On the other hand, the MTV model describes or approximates as a structural model or a prediction model such fluctuations as stock prices, interest rates, exchange rates, etc., which are too volatile to allow one to construct a causal relation model. In Section 2, it is shown that the model is well-defined (identifiable) as a model generating the $\mathbf{x}_t$'s, and that it has an optimality property as an approximation model. In Section 3, the use of the model is demonstrated and in Section 4 an estimation method of the model is given. In Section 5, as an application, the stock prices of 25 Japanese pharmaceutical companies are forecasted via the MTV model. In a univariate case a series of stock prices is often found or claimed to follow a random walk and it is said that no predictability can be elicited from the past data. However, the fact has not been tested in a multivariate situation. In fact, our result seems to show that there remains some information for predictability in the data.

2. THEORETICAL FOUNDATION OF MTV MODEL

Suppose that a p-dimensional stochastic process

$$\{\mathbf{x}_t : t = 0, \pm 1, \pm 2, \dots\} \quad \text{with} \quad \mathbf{x}_t = (x_{1t}, \dots, x_{pt})'$$

is of the structure

$$\mathbf{x}_t = \boldsymbol{\mu}_t + \boldsymbol{\alpha}_1 f_{1t} + \ldots + \boldsymbol{\alpha}_p f_{pt}, \tag{2.1}$$

where the coefficient vectors $\boldsymbol{\alpha}_j = (a_{1j},\ldots, a_{pj})' : p\times 1$ ($j = 1,\ldots, p$) are assumed to satisfy

$$\boldsymbol{\alpha}_j'\boldsymbol{\alpha}_k = \delta_{jk} \ (\delta_{jj} = 1, \delta_{jk} = 0 \text{ if } j \neq k) \tag{2.2}$$

and f_{jt}'s are variance components commonly generating the variations of x_{it}'s. The condition (2.2) is required for the model (2.1) to be identifiable. Note that from (2.2), $f_{jt} = \boldsymbol{\alpha}_j'(\mathbf{x}_t - \boldsymbol{\mu}_t)$ is obtained.

ASSUMPTION OF MTV MODEL

(A) $F_j = \{ f_{jt} : t = 0, \pm 1,\ldots\}$ is covariance stationary with mean $E(f_{jt}) = 0$. In applications we assume ARMA models for F_j's.

(B) For any n consecutive $f_{jt} \in F_j$ $(t = r + 1,\ldots, r + n)$, the covariance matrices of

$$\mathbf{f}_j(r) = (f_{jr+1},\ldots, f_{jr+n})' : n\times 1 \quad (j = 1,\ldots, p)$$

satisfies in terms of positive definiteness

$$Cov(\mathbf{f}_1(r)) > Cov(\mathbf{f}_2(r)) > \ldots > Cov(\mathbf{f}_p(r)).$$

(C) F_j and F_k are uncorrelated $(j \neq k) : E(f_{jt}f_{ks}) = 0$ for any t, s.

The MTV model can be viewed as a time series extension of a PC (principal component) model. In fact, if (A) is replaced by

(A)' F_j's are white noises ($j = 1,\ldots, p$),

then the common components f_{jt} and f_{js} of variation at time t and s are uncorrelated for $t \neq s$. Further the case of a PC model, (B) is reduced to

(B)' $\gamma_1 > \gamma_2 > \ldots > \gamma_p$, where $\gamma_j \equiv Var(f_{jt})$.

In the MTV model, F_j's follow stationary processes, through which the model can describe the time series correlation structure of the variations of $\{\mathbf{x}_t\}$ as well as the cross-sectional correlation structure. On the other hand, a factor analysis model assumes the existence of q common factors behind the variations of $\mathbf{x}_t$;

$$\mathbf{x}_t = \boldsymbol{\mu} + \boldsymbol{\beta}_1 g_{it} + \ldots + \boldsymbol{\beta}_q g_{qt} + \boldsymbol{\rho}_t \quad (\boldsymbol{\beta}_j{}'\boldsymbol{\beta}_k = 0 \text{ for } j \neq k), \tag{2.3}$$

where $G_j = \{g_{jt} : t = 0,\ t = \pm 1,\ldots\}$ is a white noise with $Var(g_{jt}) = 1$ ($j = 1,\ldots, q$), G_j and G_k are uncorrelated ($j \neq k$) and $Cov(\boldsymbol{\rho}_t) \equiv \mathbf{D}$ is diagonal independently of t. Then the covariance matrix of $\mathbf{x}_t$ is given by

$$Cov(\mathbf{x}_t) = \mathbf{BB}' + \mathbf{D} \quad \text{with} \quad \mathbf{B} = [\boldsymbol{\beta}_1,\ldots, \boldsymbol{\beta}_q] : p \times q. \tag{2.4}$$

As is well known, the factor model (2.3) is not uniquely identifiable.

Now we investigate some properties of the MTV model. It follows from (A) that $\boldsymbol{\mu}_t = E(\mathbf{x}_t)$. In the sequel, by considering $\mathbf{x}_t - \boldsymbol{\mu}_t$ we may assume $\boldsymbol{\mu}_t = \mathbf{0}$. Also by (A) the variances of f_{jt}'s are independent of t; $\gamma_j \equiv Var(f_{jt})$ and by (B) $\gamma_1 > \gamma_2 > \ldots > \gamma_p$. Hence by (C) the covariance matrix of $\mathbf{x}_t$ in (2.1) is given by

$$\boldsymbol{\Sigma} \equiv Cov(\mathbf{x}_t) = \sum_{j=1}^{p} \gamma_j \boldsymbol{\alpha}_j \boldsymbol{\alpha}_j{}' = \mathbf{A\Lambda A}', \tag{2.5}$$

which is independent of t, where

$$\mathbf{A} = [\boldsymbol{\alpha}_1,\ldots, \boldsymbol{\alpha}_p] : p \times p \quad \text{and} \quad \boldsymbol{\Lambda} = \text{diag}\{\gamma_1,\ldots, \gamma_p\}. \tag{2.6}$$

Since $\mathbf{A}'\mathbf{A} = \mathbf{I}$ from (2.2), (2.5) means that γ_j's are the latent roots of $\boldsymbol{\Sigma}$ and $\boldsymbol{\alpha}_j$'s are the latent vectors corresponding to the γ_j's.

This implies that the MTV model cross-sectionally (or for each t) corresponds to the PC model. However, our model is of the time series structure stated in the Assumption. To study the structure, set $t = 1,\ldots, n$ by the stationarity without loss of generality and let

$$\begin{cases} \mathbf{y}_i = (x_{i1},\ldots, x_{in})' : n \times 1, & \mathbf{y} = (\mathbf{y}_1{}',\ldots, \mathbf{y}_p{}')' : np \times 1, \\ \mathbf{f}_j = (f_{j1},\ldots, f_{jn})' : n \times 1, & \mathbf{f} = (\mathbf{f}_1{}',\ldots, \mathbf{f}_p{}')' : np \times 1. \end{cases} \tag{2.7}$$

Then the model (2.1) is expressed for $t = 1,\ldots, n$ as

$$\mathbf{y} = (\mathbf{A} \otimes \mathbf{I})\mathbf{f}, \tag{2.8}$$

the covariance matrix of $\mathbf{y}_i$ is given by

$$Cov(\mathbf{y}_i) = \sum_{j=1}^{p} a_{ij}^2 \boldsymbol{\Gamma}_j \quad \text{with} \quad \boldsymbol{\Gamma}_j = Cov(\mathbf{f}_j) \tag{2.9}$$

and the covariance matrix of $\mathbf{y}$ is given by

$$\boldsymbol{\Omega} \equiv Cov(\mathbf{y}) = \sum_{j=1}^{p} \boldsymbol{\alpha}_j \boldsymbol{\alpha}_j{}' \otimes \boldsymbol{\Gamma}_j . \tag{2.10}$$

This covariance matrix describes the whole structure of the model by reason of the stationarity.

Now we shall show that the MTV model formulated above is well-defined. First it is easy to see that if $\boldsymbol{\Omega} = \boldsymbol{\Omega}^*$ for two

$$\boldsymbol{\Omega} = \sum_{j=1}^{p} \boldsymbol{\alpha}_j \boldsymbol{\alpha}_j{}' \otimes \boldsymbol{\Gamma}_j \quad \text{and} \quad \boldsymbol{\Omega}^* = \sum_{j=1}^{p} \boldsymbol{\alpha}_j^* \boldsymbol{\alpha}_j^*{}' \otimes \boldsymbol{\Gamma}_j^* ,$$

satisfying the Assumption, then $\boldsymbol{\alpha}_j = \pm\, \boldsymbol{\alpha}_j^*$ and $\boldsymbol{\Gamma}_j = \boldsymbol{\Gamma}_j^*$ $(j = 1,\ldots, p)$. Second if $\mathbf{y}$ in (2.8) is expressed in two ways as

$$\mathbf{y} = (\mathbf{A} \otimes \mathbf{I})\, \mathbf{f} = (\mathbf{A}^* \otimes \mathbf{I})\, \mathbf{f}^*$$

then $\boldsymbol{\Omega} = \boldsymbol{\Omega}^*$ implies $\boldsymbol{\alpha}_j = \pm\, \boldsymbol{\alpha}_j^*$ and $\mathbf{f}_j = \pm\, \mathbf{f}_j^*$. Thus except for the simultaneous change of the signs of $\boldsymbol{\alpha}_j$ and $\mathbf{f}_j$ the model is identifiable.

Next we shall consider an optimality of the model for the case where, taking q components, we approximate the variations of $\mathbf{x}_t$ $(t = 1,\ldots, n)$ by the variations of

$$\hat{\mathbf{x}}_t = \boldsymbol{\alpha}_1 f_{1t} + \ldots + \boldsymbol{\alpha}_q f_{qt} \quad (t = 1,\ldots, n), \tag{2.11}$$

that is, we approximate $\mathbf{y}$ by

$$\hat{\mathbf{y}} = (\mathbf{A}_q \otimes \mathbf{I})\, \hat{\mathbf{f}} \tag{2.12}$$

where $\mathbf{A}_q = [\boldsymbol{\alpha}_1,\ldots, \boldsymbol{\alpha}_q]$ and $\hat{\mathbf{f}} = (\mathbf{f}_1{}',\ldots, \mathbf{f}_q{}')'$.

THEOREM 2.1. *Fix q arbitrarily. Consider the problem of approximating $\mathbf{y}$: $np \times 1$ in (2.8) by a random vector $\mathbf{y}^*$: $np \times 1$ of the form*

$$\mathbf{y}^* = (\mathbf{C}_q \otimes \mathbf{I})\, \mathbf{w}, \tag{2.13}$$

that is, the problem of approximating $\mathbf{y}_i : n \times 1$ by

$$\mathbf{y}_i^* = \sum_{j=1}^{p} c_{ij} \mathbf{w}_j \quad (i = 1,\ldots, p), \tag{2.14}$$

where $\mathbf{C}_q = (c_{ij})$ is any $p \times q$ constant matrix satisfying $\mathbf{C}_q{}'\mathbf{C}_q = \mathbf{I}_q$, and $\mathbf{w} = (\mathbf{w}_1{}', \mathbf{w}_2{}',\ldots, \mathbf{w}_q{}')' : nq \times 1$ with $\mathbf{w}_j : n \times 1$ having $\mathbf{E}_j \equiv Cov(\mathbf{w}_j)$ is any random vector satisfying

$$\mathbf{E}_1 > \mathbf{E}_2 > \ldots > \mathbf{E}_q \quad \textit{and } Cov(\mathbf{w}_j, \mathbf{w}_k) = \mathbf{0} \ (j \neq k). \tag{2.15}$$

Then the unique random vector $\mathbf{y}^*$ *that minimizes in terms of non-negative definiteness the* $n \times n$ *mean square matrix*

$$\mathbf{\Phi} = \sum_{i=1}^{p} E(\mathbf{y}_i - \mathbf{y}_i^*)(\mathbf{y}_i - \mathbf{y}_i^*)' \tag{2.16}$$

is $\hat{\mathbf{y}}$ *in* (2.12). *And the minimum value of* $\mathbf{\Phi}$ *is*

$$\sum_{j=q+1}^{p} \mathbf{\Gamma}_j .$$

The proof is deferred to the end of the paper.

COROLLARY 2.1. *In the problem of Theorem* 2.1, *the unique random vector* $\mathbf{y}^*$ *which minimizes* $E \| \mathbf{y} - \mathbf{y}^* \|^2$ *is* $\hat{\mathbf{y}}$ *in* (2.12).

COROLLARY 2.2. *In the problem of approximating* $\mathbf{x}_t$ *by a random vector of the form* $\mathbf{x}_t^* \equiv \mathbf{C}_q \mathbf{z}_t$ *where* $\mathbf{C}_q$: $n \times q$, $\mathbf{C}_q'\mathbf{C}_q = \mathbf{I}$ *and* $\mathbf{z}_t$ *is an arbitrary vector with* $Cov(\mathbf{z}_t)$, $\hat{\mathbf{x}}_t$ *in* (2.11) *minimizes* $E \| \mathbf{x}_t - \mathbf{x}_t^* \|^2$.

The result in Corollary 2.2 and Theorem 2.1 respectively correspond to those of Darroch (1965) and Okamoto and Kanazawa (1968) *in the PC model,* and our results here are regarded as generalizations of those .

Theorem 2.1 means that the best approximation to a phenomenon following an MTV structure is obtained by taking the q components f_{jt}'s with the first q largest covariance matrices in (B) and forming (2.12). The degree of the approximation may be measured by the relative ratio

$$\eta_q = \operatorname{tr} Cov(\hat{\mathbf{y}}) / \operatorname{tr} Cov(\mathbf{y}) = \sum_{j=1}^{q} \gamma_j \Big/ \sum_{j=1}^{p} \gamma_j . \tag{2.17}$$

Here note that

$$\operatorname{tr} Cov(\hat{\mathbf{y}}) = \operatorname{tr} \sum_{j=1}^{q} \boldsymbol{\alpha}_j \boldsymbol{\alpha}_j' \otimes \mathbf{\Gamma}_j = n \sum_{j=1}^{q} \gamma_j ,$$

since the diagonal elements of $\mathbf{\Gamma}_j$ are all γ_j. The measure in (2.17) is nothing but the rate of accumulative contribution in the PC model. It is noted that when $\mathbf{x}_t$'s are approximated by $\hat{\mathbf{x}}_t$'s in (2.11), then the residuals $(\mathbf{x}_t - \hat{\mathbf{x}}_t)$'s still exhibit a stationary process. Hence there remains some information to exploit in forecasting the future values of $\mathbf{x}_t$'s. However, in applications we consider the model itself an approximation to a given phenomenon to be analyzed.

The assumption of the stationary of $\{f_{jt}\}$ may be relaxed as follows. Let γ_{jt} be the variance of f_{jt} in the model (1.1). Then the covariance matrix of $\mathbf{x}_t$ is given by $\mathbf{\Sigma}_t = \mathbf{A\Lambda}_t\mathbf{A}'$ with $\mathbf{\Lambda}_t = \text{diag}\,\{\gamma_{1t}, \ldots, \gamma_{pt}\}$ where $\gamma_{1t} > \ldots > \gamma_{pt}$, and the covariance matrix of $\mathbf{y}$ in (2.7) becomes

$$\mathbf{\Omega}_t = \sum_{j=1}^{p} \boldsymbol{\alpha}_j\boldsymbol{\alpha}_j' \otimes \mathbf{\Gamma}_{jt}.$$

Then though the Theorem 2.1 no longer holds, Corollary 2.2 holds. To see the estimability of the parameter α_{ij}'s, note the fact

$$E(\mathbf{XX}') = \sum_{t=1}^{n} E(\mathbf{x}_t\mathbf{x}_t') = \mathbf{A}(\sum_{t=1}^{n} \mathbf{\Lambda}_t)\mathbf{A}',$$

where $\mathbf{X} = [\mathbf{x}_1, \ldots, \mathbf{x}_n]$. Hence from this relation an orthogonal matrix diagonalizing $\mathbf{XX}'$ may be regarded as an estimate of $\mathbf{A}$ and the i-th latent root d_j of $\mathbf{XX}'$ as an estimate of

$$\sum_{t=1}^{n} \gamma_{jt}.$$

This view facilitates or validates the PC technique applied to time series data with trends.

3. HOW TO USE THE MTV MODEL

The MTV model aims in part to find, through their time series behaviors and the sizes of their coefficients, what the time series variance components are. For this purpose, we may assume for F_j's ARMA models:

$$f_{jt} \sim \text{ARMA}\,(a_j, b_j) \quad (j = 1, \ldots, q) \tag{3.1}$$

and estimate the parameters of the ARMA models based on the scores of f_{jt}'s. Of course, the identification of the components often tends to be subjective. In the sequel we list some typical ways of using the model. For easier understanding, the exposition is made in terms of the standardized variables

$$z_{it} = (x_{it} - \mu_{it}) / \sigma_{ii}^{\frac{1}{2}}. \tag{3.2}$$

That is, we assume that z_{it}'s follow the MTV process (2.1).

[1] *Analysis of Variance.* Suppose the first q components are chosen based on the measure of the accumulative relative contribution in (2.17) as an approximation. Then just as in the PC model, the variance of each variable z_{it} is decomposed as

$$Var(z_{it}) = 1 \simeq a_{i1}^2 \gamma_1 + \dots + a_{iq}^2 \gamma_q, \tag{3.3}$$

through which the dependence of the variance of z_{it} on that of the j-th component is found out to be $a_{ij}^2 \gamma_j$.

[2] *Cross-Sectional Correlation.* For each fixed t, the cross-sectional correlation between z_{it} and z_{jt} is decomposed as

$$Correl(z_{it}, z_{jt}) \simeq a_{i1} a_{j1} \gamma_1 + \dots + a_{iq} a_{jq} \gamma_q, \tag{3.4}$$

through which it can be established how much z_{it} and z_{jt} are correlated through the k-th component; $a_{ik} a_{jk} \gamma_k$.

[3] *Time Series Correlation.* A feature of the MTV model is that the time series correlation between z_{it} and z_{jt-l} is evaluated as

$$Correl(z_{it}, z_{jt-\ell}) \simeq a_{i1} a_{j1} \gamma_1(\ell) + \dots + a_{iq} a_{jq} \gamma_q(\ell), \tag{3.5}$$

where $\gamma_k(\ell) = Cov(f_{kt}, f_{kt-\ell})$ is the autocovariance function of $\{ f_{kt} \}$ with $\gamma_k(0) \equiv \gamma_k$. Hence if a_{ij}'s and $\gamma_k(\ell)$'s are estimated, an estimate of (3.5) is obtained by substitution. When an ARMA model (3.1) is assumed for $F_k = \{ f_{kt} \}$, $\gamma_k(\ell)$ is obtained. Here it is noted that in our model the correlation between z_{it} and $z_{jt-\ell}$ is equal to the correlation between $z_{it-\ell}$ and z_{jt}. Hence a causal relation is not directly observed through the model. However, if it is required, by adding z_{jt-1}, z_{jt-2}, etc. to the set of variables z_{it}'s for the model it will be analysed.

[4] *Prediction.* Fitting the ARMA model (3.1) makes it possible to forecast the ℓ period ahead value $f_{kt+\ell}$ of f_{kt} $(k = 1,\dots, q)$. Hence from $z_{it} \simeq a_{i1} f_{1t} + \dots + a_{iq} f_{qt}$, the future values of z_{it}'s can be forecasted.

Naturally behind this predictability it is necessary for the coefficients a_{ij}'s and the ARMA model to be stable. When we wish to predict the future values of a specific variable, say z_{1t} (such as exchange rate), then it is required to choose those components which explain the variation of z_{1t} well up to a certain ratio, say 95 %.

4. ESTIMATION OF THE MODEL

It was observed in Section 2 that the coefficient vector $\mathbf{\alpha}_i$ is a latent vector of the covariance matrix $\mathbf{\Sigma}$ of $\mathbf{x}_t$. Hence $\mathbf{\alpha}_i$ can be estimated based on the sample covariance matrix

$$\mathbf{S} = \frac{1}{n}(\mathbf{X}-\mathbf{M})(\mathbf{X}-\mathbf{M})' = \frac{1}{n}\sum_{t=1}^{n}(\mathbf{x}_t-\mathbf{\mu}_t)(\mathbf{x}_t-\mathbf{\mu}_t)', \tag{4.1}$$

where $\mathbf{X} = (\mathbf{x}_1,\dots,\mathbf{x}_n)$: $p\times n$ and $\mathbf{M} = (\mathbf{\mu}_1,\dots,\mathbf{\mu}_n)$. Here the trends $\mathbf{\mu}_t$'s are assumed to be known though they must be estimated in the usual manner. As has been seen, $E(\mathbf{S}) = \mathbf{\Sigma} = \mathbf{A\Lambda A}'$ and so by a view of the moment method $\mathbf{A}$ may be estimated as an orthogonal matrix which diagonalizes $\mathbf{S}$. On the other hand, by Corollary 2.1 the best equation which approximates the variation of $\mathbf{V} \equiv \mathbf{X}-\mathbf{M}$ in terms of the MSE is given by $\mathbf{A}_q\mathbf{A}'_q\mathbf{V}$. Therefore we shall try to find an estimate $\hat{\mathbf{A}}_q$ which minimizes the squared error

$$\delta = \|\mathbf{V}-\mathbf{A}_q\mathbf{A}_q'\mathbf{V}\|^2 = \mathrm{tr}\,(\mathbf{V}-\mathbf{A}_q\mathbf{A}_q'\mathbf{V})(\mathbf{V}-\mathbf{A}_q\mathbf{A}_q'\mathbf{V})'$$

which becomes the MSE when its expectation is taken. Writing

$$\mathbf{S} = \mathbf{HDH}', \ \mathbf{H}'\mathbf{H} = \mathbf{I}_p \quad \text{and} \quad \mathbf{D} = \mathrm{diag}\,\{d_1,\dots,d_p\}$$

where $d_1 > \dots > d_p$, then

$$\frac{1}{n}\,\delta = \mathrm{tr}\,\mathbf{D} - \mathrm{tr}\mathbf{A}'_q\mathbf{HDH}'\mathbf{A}_q = \mathrm{tr}\,\mathbf{D} - \mathrm{tr}\,\mathbf{B}'\mathbf{DB} \quad \text{with} \quad \mathbf{B} = \mathbf{H}'\mathbf{A}_q.$$

Here noting $\mathbf{B}'\mathbf{B} = \mathbf{I}_q$ and arguing as in the proof of Theorem 2.1, δ is minimized when $\mathbf{B} = (\mathbf{B}_1', \mathbf{0}')'$ with $\mathbf{B}_1 = \mathrm{diag}\,\{\pm 1,\dots,\pm 1\}$ (signs are arbitrary). Hence $\hat{\mathbf{A}}_q = (\pm\mathbf{h}_1,\dots,\pm\mathbf{h}_q)$ is obtained, where $\mathbf{h}_i$ is the latent vector of $\mathbf{S}$ corresponding to d_i, and then d_i is an estimate of γ_i. Consequently f_{kt} is estimated as $\hat{f}_{kt} = \hat{\mathbf{\alpha}}_k'(\mathbf{x}_t-\mathbf{\mu}_t)$, and based on $\hat{f}_{kt}$'s an ARMA model is fitted.

5. PREDICTION OF STOCK PRICES

As an example, the MTV is applied in prediction of the stock prices of the following 25 pharmaceutical companies in Japan:

(1) Sankyo, (2) Takeda, (3) Yamanouchi, (4) Daiichi, (5) Dainippon, (6) Shionogi, (7) Tanabe, (8) Yoshitomi, (9) Fujisawa, (10) Wakamoto, (11) Banyu, (12) Nippon-shinyaku, (13) Toyama, (14) Chugai, (15) Kaken, (16) Midori, (17) Eizai, (18) Rhoto, (19) Ono, (20) Kaken, (21) Hisamitsu, (22) Tokyo tanabe, (23) Mochida, (24) Taisho, (25) Santen

The monthly data from 1971.1 to 1984.12 (12×8) adjusted for splits are used in the form of the standardized variables $z_{it} = (x_{it} - \bar{x}_i)/s_i$. No trend adjustment is made. The variations of these 25 stock prices are explained up to 95 % by the first six variance components. The variances (latent roots) γ_j of these six components ([1] ~[6]) are as follows:

[1] 17.89 (71.6%), [2] 2.48 (9.9%), [3] 2.05 (8.2%),
[4] 0.82 (3.28%), [5] 0.31 (1.24%), [6] 0.26 (1.05%)

where the numbers in the parentheses are the rate of relative contributions.

The so-called factor loading coefficients $\beta_{ij} = a_{ij}\gamma_j^{\frac{1}{2}}$ for the six components are listed in Table 5.1 and the components [1] ~ [6], i.e., f_{jt}'s are autoregressively modelled based on the AIC criterion as follows.

[1] $f_{1t} = 0.8674f_{1t-1} + 0.1767f_{1t-2}$, $\bar{R}^2 = 0.96$, $DW = 1.98$, $S = 0.79$

[2] $f_{2t} = 1.0884f_{2t-1} - 0.2995f_{2t-2} + 0.0294f_{2t-3} + 0.0665f_{2t-4} + 0.3255f_{2t-5} - 0.2781f_{2t-6}$, $\bar{R}^2 = 0.91$, $DW = 1.99$, $S = 0.45$

[3] $f_{3t} = 0.9435f_{3t-1}$, $\bar{R}^2 = 0.88$, $DW = 2.04$, $S = 0.47$

[4] $f_{4t} = 0.6049f_{4t-1} + 0.2769f_{4t-2} + \ldots - 0.5581f_{4t-20}$, $\bar{R}^2 = 0.96$, $DW = 1.77$, $S = 0.35$

[5] $f_{5t} = 0.8273f_{5t-1}$, $\bar{R}^2 = 0.66$, $DW = 1.77$, $S = 0.35$

[6] $f_{6t} = 0.7683f_{6t-1}$, $\bar{R}^2 = 0.56$, $DW = 1.88$, $S = 0.36$

By using these equations the future values of f_{jt}'s are predicted for 4 months, through which the 4 months' values of each stock are predicted. These predicted values are listed in Table 5.2 for the first 10 stocks.

Since the factor coefficient β_{ij} denotes the correlation between z_{it} (or x_{it}) and f_{jt}, using the argument in Section 3, one may interpret the cross-sectional correlation structure of the stocks and also identify the time series components [1] ~ [6] based on the AR structure of the models and the characteristics of the stocks.

In practice, for improvement it will be necessary to take into account trend adjustment, the appropriate choice of sampling period, data transformation, the appropriate combination of stocks, the simultaneous use of variables other than stocks.

TABLE 5.1

	[1]	[2]	[3]	[4]	[5]	[6]
(1)	0.914	-0.306	-0.050	-0.112	-0.106	-0.026
(2)	0.817	-0.397	-0.355	0.053	-0.031	0.018
(3)	0.943	-0.203	0.055	-0.046	-0.080	-0.064
(4)	0.929	-0.085	0.156	-0.242	-0.124	0.030
(5)	0.905	-0.020	0.379	-0.101	0.049	-0.038
(6)	0.818	-0.361	-0.395	0.074	0.068	0.006
(7)	0.873	-0.354	-0.223	0.046	-0.157	0.030
(8)	0.971	-0.045	-0.025	0.132	0.089	0.026
(9)	0.762	-0.263	-0.532	-0.142	0.108	-0.050
(10)	0.923	0.038	0.048	-0.201	0.136	0.137
(11)	0.893	0.252	0.136	0.197	-0.016	0.167
(12)	0.854	0.317	0.264	0.051	-0.064	0.151
(13)	0.394	0.601	-0.593	-0.104	0.120	0.140
(14)	0.935	-0.149	0.049	0.126	0.054	-0.155
(15)	0.861	0.260	-0.046	-0.346	0.130	-0.105
(16)	0.889	-0.109	-0.298	0.088	0.097	0.098
(17)	0.933	0.101	-0.048	0.231	-0.098	-0.150
(18)	0.839	0.131	0.389	0.275	0.033	0.124
(19)	0.876	-0.058	0.405	-0.083	0.167	0.015
(20)	0.917	0.213	-0.090	0.056	0.086	-0.162
(21)	0.131	0.944	-0.109	0.030	0.015	-0.178
(22)	0.641	0.482	-0.431	-0.059	-0.290	0.082
(23)	0.784	0.055	0.364	-0.470	-0.132	-0.017
(24)	0.917	0.062	0.242	0.179	0.044	0.014
(25)	0.925	0.113	0.080	0.239	-0.093	-0.122

TABLE 5.2

A: realized value, B: predicted value, %: error rate

		85:1	85:2	85:3	85:4
(1)	A	1,080	1,330	1,240	1,140
	B	1,019	1,047	1,041	1,042
	%	5.9	27.0	19.1	9.4
(2)	A	826	853	878	873
	B	893	919	926	918
	%	-7.6	-7.1	-5.2	-4.9
(3)	A	3,800	3.760	3,910	3,150
	B	2,038	2,084	2,086	2,096
	%	86.5	80.5	87.5	50.3
(4)	A	2,250	1,910	2,430	1,840
	B	1,985	2,061	2,050	2,089
	%	13.4	-7.3	18.5	-11.9
(5)	A	5,600	5,740	5,500	4,500
	B	5,228	5,397	5,448	5,578
	%	7.1	6.3	1.0	-19.3
(6)	A	711	711	712	722
	B	708	708	734	732
	%	0.5	0.5	-3.0	-1.4
(7)	A	1,200	1,200	1,310	1,240
	B	1,231	1,231	1,262	1,250
	%	-2.5	-2.5	3.8	-0.8
(8)	A	789	789	839	753
	B	808	808	851	868
	%	-2.3	-2.3	-1.4	-13.2
(9)	A	1,240	1,240	1,090	999
	B	1,201	1.201	1,247	1,260
	%	3.3	3.3	-12.6	-20.7
(10)	A	750	750	742	700
	B	842	842	903	937
	%	-10.9	-10.9	-17.9	-25.3

6. PROOF OF THEOREM 2.1

First consider the problem of approximating $\mathbf{y}$ by $(\mathbf{C}\otimes\mathbf{I})\mathbf{w}$ under the MSE

$$\boldsymbol{\psi} = E[\mathbf{y}-(\mathbf{C}\otimes\mathbf{I})\mathbf{w}]\,[\mathbf{y}-(\mathbf{C}\otimes\mathbf{I})\mathbf{w}]'. \tag{6.1}$$

Let

$$Cov(\mathbf{f}) = \boldsymbol{\Gamma} = \begin{pmatrix} \boldsymbol{\Gamma}_1 & & \mathbf{0} \\ & \ddots & \\ \mathbf{0} & & \boldsymbol{\Gamma}_p \end{pmatrix} : np \times np \tag{6.2}$$

and

$$Cov(\mathbf{w}) = \mathbf{E} = \begin{pmatrix} \mathbf{E}_1 & & \mathbf{0} \\ & \ddots & \\ \mathbf{0} & & \mathbf{E}_q \end{pmatrix} : nq \times nq$$

Then, since

$$Cov\begin{pmatrix} \mathbf{f} \\ \mathbf{w} \end{pmatrix} = \begin{pmatrix} \boldsymbol{\Gamma} & \mathbf{B} \\ \mathbf{B}' & \mathbf{E} \end{pmatrix} : n(p+q) \times n(p+q)$$

is nonnegative definite, $\mathbf{E}$ and $\mathbf{B}$ must satisfy

$$\boldsymbol{\Gamma} - \mathbf{B}\mathbf{E}^{-1}\mathbf{B}' \geq \mathbf{0}, \tag{6.3}$$

which is a restriction on $\mathbf{B}$ and $\mathbf{E}$. Now $\boldsymbol{\psi}$ in (6.1) is evaluated as

$$\boldsymbol{\psi} = (\mathbf{A}\otimes\mathbf{I})\,\boldsymbol{\Gamma}\,(\mathbf{A}'\otimes\mathbf{I}) - (\mathbf{A}\otimes\mathbf{I})\,\mathbf{B}\,(\mathbf{C}'\otimes\mathbf{I})$$
$$- (\mathbf{C}\otimes\mathbf{I})\mathbf{B}'\,(\mathbf{A}'\otimes\mathbf{I}) + (\mathbf{C}\otimes\mathbf{I})\,\mathbf{E}\,(\mathbf{C}'\otimes\mathbf{I}). \tag{6.4}$$

Therefore writing $\boldsymbol{\psi}_1 \equiv (\mathbf{A}'\otimes\mathbf{I})\,\boldsymbol{\psi}\,(\mathbf{A}\otimes\mathbf{I})$

$$\boldsymbol{\psi}_1 = \boldsymbol{\Gamma} - \mathbf{B}\mathbf{E}^{-1}\mathbf{B}' + [\,\tilde{\mathbf{C}}\otimes\mathbf{I} - \mathbf{B}\mathbf{E}^{-1}]\,\mathbf{E}\,[\,\tilde{\mathbf{C}}\otimes\mathbf{I} - \mathbf{B}\mathbf{E}^{-1}]' \geq \boldsymbol{\Gamma} - \mathbf{B}\mathbf{E}^{-1}\mathbf{B}' \tag{6.5}$$

where $\tilde{\mathbf{C}} = \mathbf{A}'\mathbf{C}$. Note that $\tilde{\mathbf{C}}'\tilde{\mathbf{C}} = \mathbf{I}_q$. Next, the objective function $\boldsymbol{\phi}$ in (2.16) is expressed as for any $\mathbf{a}$: $n\times 1$

$$\phi = \mathbf{a}'\boldsymbol{\phi}\mathbf{a} = \mathrm{tr}\,[\mathbf{I}\otimes\mathbf{a}']\,\boldsymbol{\psi}\,[\mathbf{I}\otimes\mathbf{a}'] = \mathrm{tr}\,[\mathbf{I}\otimes\tilde{\mathbf{a}}\,']\,\boldsymbol{\psi}_1\,[\mathbf{I}\otimes\tilde{\mathbf{a}}],$$

where $\tilde{\mathbf{a}} = \mathbf{Aa}$. Hence from (6.5)

$$\phi \geq \text{tr}\,[\mathbf{I}\otimes\tilde{\mathbf{a}}]\,(\boldsymbol{\Gamma} - \mathbf{BE}^{-1}\mathbf{B}')\,[\mathbf{I}\otimes\tilde{\mathbf{a}}] \quad \text{for all } \tilde{\mathbf{a}}$$

and the equality holds if and only if $\mathbf{BE}^{-1} = \tilde{\mathbf{C}}\otimes\mathbf{I}$. And the minimum value is

$$\phi_1 = \text{tr}\,[\mathbf{I}\otimes\tilde{\mathbf{a}}']\,[\boldsymbol{\Gamma} - (\tilde{\mathbf{C}}\otimes\mathbf{I})\,\mathbf{E}\,(\tilde{\mathbf{C}}'\otimes\mathbf{I})]\,[\mathbf{I}\otimes\tilde{\mathbf{a}}]\,. \tag{6.6}$$

Further from (6.5), $\tilde{\mathbf{C}} = \mathbf{A}'\mathbf{C}$ and (6.6)

$$[\tilde{\mathbf{C}}'\otimes\mathbf{I}]\,\boldsymbol{\Gamma}\,[\tilde{\mathbf{C}}\otimes\mathbf{I}] - \mathbf{E} \geq \mathbf{0} \tag{6.7}$$

is obtained. Thus

$$\phi_1 \geq \text{tr}\,[\mathbf{I}\otimes\tilde{\mathbf{a}}']\,[\boldsymbol{\Gamma} - (\mathbf{M}\otimes\mathbf{I})\,\boldsymbol{\Gamma}\,(\mathbf{M}\otimes\mathbf{I})]\,[\mathbf{I}\otimes\tilde{\mathbf{a}}]$$

$$= \sum_{i=1}^{p} \mathbf{a}'\boldsymbol{\Gamma}_j\mathbf{a} - \sum_{i=1}^{p}\sum_{j=1}^{p} m_{ij}^2\,\mathbf{a}'\boldsymbol{\Gamma}_j\mathbf{a} = \text{tr}\,\mathbf{D}\,[\,\mathbf{I} - \mathbf{M}\,] \equiv \phi_2\,, \tag{6.8}$$

where $\mathbf{M} = \tilde{\mathbf{C}}\tilde{\mathbf{C}}' = (m_{ij})$ and $\mathbf{D} = \text{diag}\{d_1,\ldots, d_p\}$ with $d_j = \tilde{\mathbf{a}}'\,\boldsymbol{\Gamma}_j\tilde{\mathbf{a}}$. Note that $\mathbf{M}^2 = \mathbf{M}$ and $\text{rank}(\mathbf{M}) = q$. From (6.7) the equality in (6.8) holds if and only if

$$(\mathbf{M}\otimes\mathbf{I})\,\boldsymbol{\Gamma}\,(\mathbf{M}\otimes\mathbf{I}) - (\tilde{\mathbf{C}}\otimes\mathbf{I})\,\mathbf{E}\,(\tilde{\mathbf{C}}'\otimes\mathbf{I}) = \mathbf{0}$$

or equivalently from $\tilde{\mathbf{C}}'\,\tilde{\mathbf{C}} = \mathbf{I}$

$$\mathbf{E} = (\tilde{\mathbf{C}}'\otimes\mathbf{I})\,\boldsymbol{\Gamma}\,(\tilde{\mathbf{C}}\otimes\mathbf{I}). \tag{6.9}$$

Note that for some orthogonal matrix $\mathbf{P} = (p_{ij})$, $\mathbf{M} = \mathbf{PJP}'$ with $\mathbf{J} = \text{diag}\,\{1,\ldots, 1, 0,\ldots, 0\}$. Hence ϕ_2 in (6.8) is expressed as

$$\phi_2 = \sum_{i=1}^{p} d_i - \sum_{i=1}^{p} d_i \sum_{j=1}^{q} p_{ij}^2.$$

Further from $d_1 > \ldots > d_p > 0$ for $\tilde{\mathbf{a}} \neq \mathbf{0}$ and

$$\sum_{i=1}^{p} \sum_{j=1}^{q} p_{ij}^2 = q,$$

ϕ_2 is minimized if and only if

$$\sum_{j=1}^{q} p_{ij}^2 = 1 \quad (i = 1,\dots, q)$$

and

$$\sum_{j=1}^{q} p_{ij}^2 = 0 \quad (i = q+1,\dots, p),$$

and then the minimum of ϕ_2 is

$$\sum_{i=q+1}^{p} d_i.$$

In this case $\mathbf{P}$, $\mathbf{M}$ and $\tilde{\mathbf{C}}$ are of the form

$$\mathbf{P} = \begin{pmatrix} \mathbf{P}_1 & \mathbf{0} \\ \mathbf{0} & \mathbf{P}_2 \end{pmatrix}, \quad \mathbf{M} = \begin{pmatrix} \mathbf{I} & \mathbf{0} \\ \mathbf{0} & \mathbf{0} \end{pmatrix} \text{ and } \tilde{\mathbf{C}} = \begin{pmatrix} \mathbf{T} \\ \mathbf{0} \end{pmatrix} \text{ with } \mathbf{T}'\mathbf{T} = \mathbf{I}_q \tag{6.10}$$

respectively. For this $\tilde{\mathbf{C}}$ to satisfy (6.9) with $\mathbf{E}$ in (6.2),

$$\mathbf{G} = \mathbf{T}'\mathbf{D}_1\mathbf{T} \text{ where } \mathbf{G} = \text{diag}\{g_1,\dots, g_q\} \text{ and } \mathbf{D}_1 = \text{diag}\{d_1,\dots, d_q\} \tag{6.11}$$

is necessary. Here $g_i = \tilde{\mathbf{a}}'\mathbf{E}_i\tilde{\mathbf{a}}$ and $g_1 > \dots > g_q > 0$. Therefore, it follows from $d_1 > \dots > d_p > 0$ that $\mathbf{T} = \text{diag}\{\pm 1,\dots, \pm 1\}$ with signs arbitrary. From this expression, (6.10) and $\tilde{\mathbf{C}} = \mathbf{A}'\mathbf{C}$, we obtain

$$\mathbf{C} = [\pm\boldsymbol{\alpha}_1, \pm\boldsymbol{\alpha}_2,\dots, \pm\boldsymbol{\alpha}_q]. \tag{6.12}$$

Using this $\mathbf{C}$, it follows from $\mathbf{BE}^{-1} = \tilde{\mathbf{C}}\otimes\mathbf{I}$ and (6.9) that

$$\mathbf{E} = \begin{pmatrix} \boldsymbol{\Gamma}_1 & & \mathbf{0} \\ & \ddots & \\ \mathbf{0} & & \boldsymbol{\Gamma}_q \end{pmatrix} \text{ and } \mathbf{B} = \begin{pmatrix} \pm\boldsymbol{\Gamma}_1 & & \\ & \ddots & \\ & & \pm\boldsymbol{\Gamma}_q \\ & \mathbf{0} & \end{pmatrix}. \tag{6.13}$$

For $\mathbf{E}$ and $\mathbf{B}$ in (6.13), choosing any random vector $\mathbf{w}$ satisfying

$$Cov(\mathbf{w}) = \mathbf{E} \quad \text{and} \quad Cov(\mathbf{f}, \mathbf{w}) = \mathbf{B}, \tag{6.14}$$

$(\mathbf{C}\otimes\mathbf{I})\mathbf{w}$ minimizes ϕ in (2.16) in terms of nonnegative definiteness. To show that $\mathbf{w}$ is unique, let $\mathbf{w}^* = \mathbf{B}'\mathbf{\Gamma}^{-1}\mathbf{f}$. Then for any $\mathbf{w}$ satisfying (6.14), we obtain

$$E(\mathbf{w}-\mathbf{w}^*)(\mathbf{w}-\mathbf{w}^*)' = \mathbf{E}-\mathbf{B}'\mathbf{\Gamma}^{-1}\mathbf{B}-\mathbf{B}'\mathbf{\Gamma}^{-1}\mathbf{B}+\mathbf{B}'\mathbf{\Gamma}^{-1}\mathbf{B} = \mathbf{0}$$

from which we obtain $\mathbf{w} = \mathbf{w}^*$ and $(\mathbf{C}\otimes\mathbf{I})\mathbf{w}^* = (\mathbf{A}\otimes\mathbf{I})\mathbf{f}^*$ uniquely. This completes the proof of Theorem 2.1.

REFERENCES

Darroch, J.N. (1965). An optimal property of principal components. *Annals of Mathematical Statistics*, 36, 1579-1582.

Geweke, J.F. (1977). The dynamic factor analysis of economic time series models. *Latent Variables in Socio-economic Models*, North-Holland, 365-383.

Geweke, J.F. and Singleton, K.J. (1981). Maximum likelihood "confirmatory" factor analysis of economic time series. *International Economic Review*, 22, 37-54.

King, B.F. (1966). Market and industry factors in stock price behaviour. *Journal of Business*, 28, 139-190.

Okamoto, M. and Kanazawa, M. (1968). Minimization of eigenvalues of a matrix and optimality of principal components. *Annals of Mathematical Statistics*, 39, 859-863.

Rao, C.R. and Boudreau, R. (1985). Prediction of future observations in a factor analytic type growth model. *Multivariate Analysis* VI (P.R. Krishnaiah, *ed.*). North-Holland, 449-466.

Singleton, K.J. (1980). A latent time series model of the cyclical behaviour of interest rates. *International Economic Review*, 21, 559-575.

Received 5 January 1987
Revised 1 June 1987

Institute of Economic Research
Hitotsubashi University
Kunitachi
Tokyo 186
Japan

Proc. Second International Tampere Conference in Statistics
(Tampere, Finland, 1-4 June 1987)
Tarmo Pukkila and Simo Puntanen, *Editors*

pp. 177 - 203

INVITED PAPER

Quadratic forms and null robustness for elliptical distributions

C.G. KHATRI
Gujarat University, Ahmedabad, India

Key words and phrases: Spherical distributions, matrix gamma distribution, Stieltjes manifold, quadratic forms, extension of Cochran's theorem, complex elliptical distributions.

ABSTRACT

Let $\mathbf{X}$ have a $p \times n$ matrix elliptical distribution $ME_{p\times n}(\mathbf{M}, \mathbf{V}; \psi)$ and assume that $P(\mathbf{Y}\mathbf{V}^{-}\mathbf{Y}' > \mathbf{0}) = 1$ for any generalized inverse $\mathbf{V}^{-}$ of $\mathbf{V}$ and for $\mathbf{Y} = \mathbf{X} - \mathbf{M}$. Section 2 investigates the conditions on the quadratic forms $\mathbf{Y}\mathbf{A}_i\mathbf{Y}'$ for $i = 1, 2, \ldots, k$ such that $\mathbf{Y}$ is distributed as normal. Section 3 considers the situation of testing $H_0(\mathbf{C}\boldsymbol{\xi}\mathbf{D} = \mathbf{0})$ against $H(\mathbf{C}\boldsymbol{\xi}\mathbf{D} \neq \mathbf{0})$ when $\mathbf{M} = \mathbf{B}\boldsymbol{\xi}\mathbf{A}$ and $\mathbf{V} = \mathbf{I}$, and shows that the test procedures obtained under normality conditions are all null robust. Section 4 considers univariate elliptical distribution (with $p = 1$) and obtains the necessary and sufficient conditions for the quadratic forms $\mathbf{x}'\mathbf{A}_i\mathbf{x}$ $(i = 1, 2, \ldots, k)$ to have a specified joint distribution. All these results are extended to complex elliptical situations.

1. INTRODUCTION

Let $\mathbb{R}^{p\times m}$ be a set of $p \times m$ matrices whose elements belong to the real field $\mathbb{R}$, and $\mathbb{R}^{p\times 1}$ is denoted by $\mathbb{R}^p$. Let $\mathbf{X}$ be a $p\times n$ random matrix and let its characteristic function (c.f.) be given by

$$E \exp[(-1)^{\frac{1}{2}} \operatorname{tr}(\mathbf{T}'\mathbf{X})]$$
$$= \exp[(-1)^{\frac{1}{2}} \operatorname{tr}(\mathbf{T}'\mathbf{M})]\, \psi(\mathbf{TVT}') \text{ for all } \mathbf{T} \in \mathbb{R}^{p\times n}, \quad (1.1)$$

where $\mathbf{M} \in \mathbb{R}^{p\times n}$ and $\mathbf{V}$ is an $n\times n$ positive semidefinite matrix denoted by $\mathbf{V} \geq \mathbf{0}$. We shall write $\mathbf{X} \sim ME_{p\times n}(\mathbf{M}, \mathbf{V}; \psi)$ and $\mathbf{X}$ is said to have a matrix elliptical distribution (or multivariate right elliptical distribution). If the ith row vector of $\mathbf{T}$ is $\mathbf{t}_i'$ and the ith row vector of $\mathbf{X}$ is $\mathbf{x}_i'$, then the c.f. of $\mathbf{x}_i$ is

$$E \exp[(-1)^{\frac{1}{2}} \mathbf{t}_i' \mathbf{x}_i] = E \exp[(-1)^{\frac{1}{2}} \operatorname{tr}(\mathbf{T}_1'\mathbf{X})],$$

where $\mathbf{T}_1 \in \mathbb{R}^{p\times n}$ whose ith row vector is $\mathbf{t}_i'$ and all other row vectors of $\mathbf{T}_1$ are null vectors. Then using (1.1), the c.f. of $\mathbf{x}_i$ is

$$\exp[(-1)^{\frac{1}{2}} \mathbf{t}_i' \mathbf{m}_i]\, \phi_i(\mathbf{t}_i'\mathbf{V}\mathbf{t}_i) \quad \text{for all } \mathbf{t}_i \in \mathbb{R}^n, \quad (1.2)$$

where $\mathbf{m}_i' =$ ith row vector of $\mathbf{M}$ and $\phi_i(u) = \psi(\boldsymbol{\theta}_0)$ with $\boldsymbol{\theta}_0 = (\theta_{ij})$ with $\theta_{ii} = u$ and all other θ_{ij}'s are zero for $i, j = 1, 2,\ldots, p$. This shows that $\mathbf{x}_i \sim E_n(\mathbf{m}_i, \mathbf{V}; \phi_i)$ and $\mathbf{x}_i$ is said to have an elliptical distribution. If the rank of $\mathbf{V}$ [r($\mathbf{V}$)] is m, then we can write $\mathbf{V} = \mathbf{V}_1\mathbf{V}_1'$, where $\mathbf{V}_1 \in \mathbb{R}^{n\times m}$ and $\mathrm{r}(\mathbf{V}_1) = m$. Further, we have

$$\mathbf{X} \doteq \mathbf{M} + \mathbf{Y}\mathbf{V}_1', \quad (1.3)$$

where $\mathbf{Y} \sim ME_{p\times m}(\mathbf{0}, \mathbf{I}_m; \psi)$ [$= MS_p(m; \psi)$ = a class of $p\times m$ matrix spherical distributions] and when $p = 1$, $MS_1(m; \psi)$ will be denoted as $S(m; \psi)$, a class of spherical distributions. Here $\mathbf{X} \doteq \mathbf{Z}$ means that $\mathbf{X}$ and $\mathbf{Z}$ have identical distributions. For the ith row vector $\mathbf{x}_i'$ of $\mathbf{X}$,

$$\mathbf{x}_i \doteq \mathbf{m}_i + \mathbf{V}_1\mathbf{y}_i, \quad \mathbf{y}_i \sim S(m; \phi_i) \text{ for } i = 1, 2,\ldots, p. \quad (1.4)$$

For some detailled study of $\mathbf{X}$, we shall assume throughout the paper that the distribution of $\mathbf{Y}$ is nonsingular in the sense that

$$\mathrm{P}[\mathrm{r}(\mathbf{Y}) = p] = 1 \text{ and } \mathrm{P}[\mathrm{r}(\mathbf{Y}) = k] = 0 \text{ for } k = 0, 1,\ldots, p-1. \quad (1.5)$$

If $\mathbf{S} = \mathbf{YY}'$, then on account of (1.5),

$$\mathrm{P}(\mathbf{S} > \mathbf{0}) = 1, \quad (1.6)$$

where $\mathbf{S}$ is positive definite and denoted by $\mathbf{S}>\mathbf{0}$. Given a $p\times m$ matrix $\mathbf{Y}$ of rank p, one can find a $p\times p$ nonsingular matrix $\mathbf{S}_1$ and a $p\times m$ semi-orthogonal matrix $\mathbf{U}$ such that $\mathbf{Y} = \mathbf{S}_1\mathbf{U}$ and $\mathbf{U}\mathbf{U}' = \mathbf{I}_p$. Observe that $\mathbf{S} = \mathbf{Y}\mathbf{Y}' = \mathbf{S}_1\mathbf{S}_1'$. A class of $p\times m$ semi-orthogonal matrices is denoted by $O(p, m)$ and a class of $m\times m$ orthogonal matrices by $O(m)$ [$= O(m, m)$]. Since $\mathbf{U} \in O(p, m)$, we can find an $m\times m$ orthogonal matrix $\mathbf{U}_0' = (\mathbf{U}', \mathbf{U}_1') \in O(m)$. Let $\mathbf{\Gamma}$ be a random orthogonal matrix uniformly distributed over $O(m)$ with unit invariant Haar measure $[d\mathbf{\Gamma}]$ and let it be distributed independently of $\mathbf{Y}$. Then, the c.f. of $\mathbf{Y}$ is

$$
\begin{aligned}
\psi(\mathbf{T}\mathbf{T}') &= \int_{O(m)} \psi(\mathbf{T}\mathbf{\Gamma}\mathbf{\Gamma}'\mathbf{T}')\,[d\mathbf{\Gamma}] = \int_{O(m)} E\exp[(-1)^{\frac{1}{2}}\mathrm{tr}\mathbf{T}'\mathbf{Y}\mathbf{\Gamma}'][d\mathbf{\Gamma}] \\
&= \int_{O(m)} E\exp[(-1)^{\frac{1}{2}}\,\mathrm{tr}\,\mathbf{T}'(\mathbf{S}_1, \mathbf{0})\mathbf{U}_0\mathbf{\Gamma}'][d\mathbf{\Gamma}] \\
&= \int_{O(m)} E\exp[(-1)^{\frac{1}{2}}\mathrm{tr}\mathbf{T}'(\mathbf{S}_1, \mathbf{0})\mathbf{\Gamma}'][d\mathbf{\Gamma}] \\
&= \int_{O(p,m)} E\exp[(-1)^{\frac{1}{2}}\mathrm{tr}\mathbf{T}'\mathbf{S}_1\mathbf{\Gamma}_1'][d\mathbf{\Gamma}_1]. \qquad (1.7)
\end{aligned}
$$

This shows that

$$
\mathbf{Y} \doteq \mathbf{S}_1\mathbf{U}, \qquad (1.8)
$$

where $\mathbf{S}_1$ and $\mathbf{U}$ are independently distributed, $\mathbf{U}$ is distributed uniformly over a Stieltjes manifold $O(p, m)$ and $\mathbf{S} = \mathbf{S}_1\mathbf{S}_1'$ has a nonsingular distribution. Here, $\mathbf{S}_1$ can be a lower triangular matrix with positive diagonal elements or an upper triangular matrix with positive diagonal elements or positive definite matrix and in the last case, we can write $\mathbf{S}_1 = \mathbf{S}^{\frac{1}{2}}$. Since $\mathbf{S}$ and $\mathbf{\Gamma}$ are independently distributed in (1.7), we can interchange $\int$ and E and then we use James' result [1964, p. 479, eqn. (27)], namely

$$
\int_{O(m)} \exp[(-1)^{\frac{1}{2}}\mathrm{tr}\mathbf{T}'(\mathbf{S}_1, \mathbf{0})\mathbf{\Gamma}'][d\mathbf{\Gamma}] = \sum_{k=0}^{\infty}\Sigma_K(-1)^k C_K(\mathbf{T}\mathbf{T}'\mathbf{S}/4)/k!\,(\tfrac{1}{2}m)_K, \qquad (1.9)
$$

where $K = \{k_1, k_2,..., k_p\}$, $k_1 \geq k_2 \geq ... \geq k_p \geq 0$ and $k_1 + ... + k_p = k$, is a partition of k, Σ_K denotes the summation over the partitions K of k, $C_K(.)$ denotes the zonal polynomial corresponding to the partition K [see James (1964) for more details] and $(\frac{1}{2}m)_K = \Gamma_p(\frac{1}{2}m; K)/\Gamma_p(\frac{1}{2}m)$ with

$$\Gamma_p(\tfrac{1}{2}m, K) = \pi^{\frac{1}{4}p(p-1)}\prod_{j=1}^{p}\Gamma[(m-j+1)/2+k_j)]$$

and $\Gamma_p(\frac{1}{2}m) = \Gamma_p(\frac{1}{2}m, 0)$. Then, (1.7) gives

$$\psi(\mathbf{\theta}) = E\{\sum_{k=0}^{\infty} \Sigma_K(-1)^k C_K(\mathbf{\theta S}/4)/k!\,(\tfrac{1}{2}m)_K\}, \tag{1.10}$$

where $\mathbf{\theta} \geq \mathbf{0}$ and $\mathbf{\theta} = \mathbf{TT}'$, $\mathbf{T} \in \mathbb{R}^{p\times m}$. If all the moments of $\mathbf{S}$ exist, then we can interchange E and the summation signs from (1.10); otherwise the interchange may not be permissible.

Similarly, we can define a complex matrix elliptical distribution, $CME_{p\times n}(\mathbf{M}, \mathbf{V}; \psi)$ where $\mathbf{M} \in \mathbb{C}^{p\times n}$, $\mathbb{C}$ denotes the field of complex numbers and $\mathbf{V}$ is Hermitian positive semidefinite (Hpsd). Let $\mathbf{T} \in \mathbb{C}^{p\times n}$, $(\mathbf{T})_R$ = Real part of $\mathbf{T}$ and $(\mathbf{T})_I$ = Imaginary part of $\mathbf{T}$. Let $\mathbf{T}^*$ be the conjugate transpose of $\mathbf{T}$. The c.f. of $\mathbf{X} \sim CME_{p\times n}(\mathbf{M}, \mathbf{V}; \psi)$ is

$$E \exp[(-1)^{\frac{1}{2}}\mathrm{tr}\,(\mathbf{T}^*\mathbf{X})_R] = \exp[(-1)^{\frac{1}{2}}\mathrm{tr}\,(\mathbf{T}^*\mathbf{M})_R]\,\psi(\mathbf{TVT}^*) \tag{1.11}$$

for all $\mathbf{T} \in \mathbb{C}^{p\times n}$, and the c.f. of the ith row vector of $\mathbf{X}$ is

$$\exp[(-1)^{\frac{1}{2}}(\mathbf{t}_i^*\,\mathbf{m}_i)_R]\,\phi_i(\mathbf{t}_i^*\mathbf{V}\mathbf{t}_i), \tag{1.12}$$

where $\phi_i(u) = \psi(\mathbf{\theta}_0)$ with $\mathbf{\theta}_0 = (\theta_{ij})$ being Hermitian, $\theta_{ii} = u$ and all other θ_{ij}'s being zeros. Thus, the ith row vector $\mathbf{x}_i'$ of $\mathbf{X} \sim CE_n(\mathbf{m}_i, \mathbf{V}; \phi_i)$ for $i = 1,\ldots, p$. As in the real case, we can show that

$$\mathbf{X} \doteq \mathbf{M} + \mathbf{YV}_1^*,\ \ \mathbf{Y} \doteq \mathbf{S}_1\mathbf{U}, \tag{1.13}$$

where $\mathbf{V} = \mathbf{V}_1\mathbf{V}_1^*$, $\mathbf{Y} \sim CME_{p\times m}(\mathbf{0}, \mathbf{I}_m; \psi)$, $\mathbf{S}_1$ and $\mathbf{U}$ are independently distributed, $P(\mathbf{S} = \mathbf{S}_1\mathbf{S}_1^*$ is Hpd$) = 1$ and $\mathbf{U}$ is uniformly distributed over the complex Stieltjes manifold $CO(p, m)$ = a set of $p\times m$ matrices $\mathbf{U} \in \mathbb{C}^{p\times m}$ and $\mathbf{UU}^* = \mathbf{I}_p$. Corresponding to (1.10), we have

$$\psi(\mathbf{\theta}) = E\{\sum_{k=0}^{\infty} \Sigma_K(-1)^k \tilde{C}_K(\mathbf{\theta S})/k!\,(\tilde{m})_K\}, \tag{1.14}$$

where $\tilde{C}_K(.)$ denotes the complex zonal polynomials [see for some details, James (1964, p. 487, eq. (85)], and $(\tilde{m})_K = \tilde{\Gamma}_p(m, K)/\tilde{\Gamma}_p(m)$ with

$$\tilde{\Gamma}_p(m, K) = \pi^{\frac{1}{2}p(p-1)}\prod_{j=1}^{p} \Gamma(m-j+1+k_j)$$

and $\tilde{\Gamma}_p(m) = \tilde{\Gamma}_p(m, 0)$. Here, $\boldsymbol{\theta}$ is any Hpsd, and it is assumed that the distribution of $\mathbf{Y}$ is nonsingular in the sense $\mathrm{P}[\mathrm{r}(\mathbf{Y}) = p] = 1$ and $\mathrm{P}[\mathrm{r}(\mathbf{Y}) = k] = 0$ for $k = 0, 1, \ldots, p-1$. Notice that $\mathrm{P}(\mathbf{S}$ is Hpd$) = 1$.

We shall first develop the results for the real random matrices (or vectors), and then give the corresponding results for the complex matrices and vectors. If the results are exactly parallel to the real case, then no proof will be given for the complex case and the final results will be given. For the quadratic forms, Khatri (1962, 1963, 1977, 1978, 1980a, 1980b, 1982 and 1983) has developed various types of the results for the normal random matrix $\mathbf{X}$ ($\in \mathbb{R}^{p\times n}$). In this paper, we extend the corresponding results to the elliptical distributions for real and complex variables.

We shall say that $\mathbf{S} \sim MG_p(a, \boldsymbol{\Sigma})$ (or matrix gamma distribution) with $2a > p-1$ when the density function of $\mathbf{S}$ is

$$\{\Gamma_p(a)\,|\boldsymbol{\Sigma}|^{a}\}^{-1}\,|\mathbf{S}|^{a-\frac{1}{2}(p+1)}\exp(-\mathrm{tr}\,\boldsymbol{\Sigma}^{-1}\mathbf{S}) \quad \text{for all } \mathbf{S} > \mathbf{0}, \tag{1.15}$$

while $\mathbf{S} \sim CMG_p(a, \boldsymbol{\Sigma})$ (or complex matrix gamma distribution) with $a > p-1$ when the density function of $\mathbf{S}$ is

$$\{\tilde{\Gamma}_p(a)\,|\boldsymbol{\Sigma}|^{a}\}^{-1}\,|\mathbf{S}|^{a-p}\exp(-\mathrm{tr}\,\boldsymbol{\Sigma}^{-1}\mathbf{S}) \quad \text{for all } \mathbf{S} \text{ Hpd}. \tag{1.16}$$

Further, for (1.15) and (1.16), James (1964) has given the following results, namely,

$$E\,C_K(\boldsymbol{\theta}\mathbf{S}) = C_K(\boldsymbol{\theta}\boldsymbol{\Sigma})\,(a)_K \quad \text{for the real case} \tag{1.17}$$

and

$$E\,\tilde{C}_K(\boldsymbol{\theta}\mathbf{S}) = \tilde{C}_K(\boldsymbol{\theta}\boldsymbol{\Sigma})\,(\tilde{a})_K \quad \text{for the complex case.} \tag{1.18}$$

The following James' results [1964; p. 479, eq. (23); p. 488, eq. (92)] will be useful:

$$\int_{O(n)} C_K(\mathbf{A}\boldsymbol{\Gamma}\mathbf{B}\boldsymbol{\Gamma}')\,[d\boldsymbol{\Gamma}] = C_K(\mathbf{A})C_K(\mathbf{B})/C_K(\mathbf{I}_n) \tag{1.19}$$

and for the complex case

$$\int_{CO(n)} \tilde{C}_K(\mathbf{A\Gamma B\Gamma^*})\,[d\,\Gamma] = \tilde{C}_K(\mathbf{A})\,\tilde{C}_K(\mathbf{B})/\tilde{C}_K(\mathbf{I}_n). \tag{1.20}$$

Notice that $\Sigma_K C_K(\mathbf{Q}) = (\mathrm{tr}\ \mathbf{Q})^k$ and $\Sigma_K \tilde{C}_K(\mathbf{Q}) = (\mathrm{tr}\ \mathbf{Q})^k$, where $\mathbf{Q}$ is a Hermitian (or for the real case symmetric) matrix.

2. CHARACTERIZATION OF NORMALITY

2.1 Real variables case

Let $\mathbf{Y} \sim MS_p(m; \psi)$. Then $\mathbf{y}_1 \sim S(m; \phi_1)$, where $\mathbf{y}_1 =$ the first column vector of $\mathbf{Y}'$. Hence, if $\mathbf{YAY}'$ is a quadratic form with $\mathbf{A} = \mathbf{A}'$ and $\mathbf{YAY}' \sim MG_p(a, \mathbf{\Sigma})$ with $\mathbf{\Sigma} > \mathbf{0}$ and $2a > p-1$, then $s_{11} = \mathbf{y}_1'\mathbf{A}\mathbf{y}_1 \sim G(a, \sigma_{11})$, where $G(a, \sigma_{11})$ is the Gamma distribution whose density function is

$$\{\Gamma(a)\sigma_{11}{}^{a}\}^{-1} s_{11}{}^{a-1} \exp(-s_{11}/\sigma_{11}) \quad \text{for all } s_{11} > 0.$$

Now, using the result of Khatri and Mukerjee (1987c), we must have

$$\mathbf{y}_1 \sim N(\mathbf{0}, \tfrac{1}{2}\sigma_{11}\mathbf{I}_m),\ \mathbf{A}^2 = \mathbf{A} \text{ and } \mathrm{r}(\mathbf{A}) = 2a. \tag{2.1}$$

Here, $2a\ (= a_1)$ will become a positive integer. Using these results in $\mathbf{YAY}'$, we can write

$$\mathbf{YAY}' = \mathbf{Y}_1\mathbf{Y}_1' \sim MG_p(\tfrac{1}{2}a_1; \mathbf{\Sigma}), \tag{2.2}$$

where $\mathbf{A} = \mathbf{\Gamma\Gamma}'$, $\mathbf{\Gamma} \in O(m, a_1)$ and $\mathbf{Y}_1 = \mathbf{Y\Gamma} \sim MS_p(a_1; \psi)$. Then, using (1.10), the c.f. of $\mathbf{Y}_1$ (or $\mathbf{Y}$) is given by

$$\psi(\mathbf{\theta}) = E\{\sum_{k=0}^{\infty} \Sigma_K (-1)^k C_K(\mathbf{\theta}\mathbf{Y}_1\mathbf{Y}_1'/4)/k!\,(\tfrac{1}{2}a_1)_K\} \tag{2.3}$$

for all $\mathbf{\theta} > \mathbf{0}$ and using (2.2) and (1.17), we have

$$E\,C_K(\mathbf{\theta}\mathbf{Y}_1\mathbf{Y}_1') = E\,C_K(\mathbf{\theta YAY}') = C_K(\mathbf{\theta\Sigma})\,(\tfrac{1}{2}a_1)_K.$$

Since $E\,C_K(\boldsymbol{\theta}\mathbf{Y}_1\mathbf{Y}_1')$ exists for all $\boldsymbol{\theta}$ and the series is convergent, so interchanging the two signs, we get

$$\psi(\boldsymbol{\theta}) = \exp[-\tfrac{1}{2}\mathrm{tr}(\boldsymbol{\Sigma}\boldsymbol{\theta}/2)] \quad \text{for all } \boldsymbol{\theta} \geq \mathbf{0}.$$

Taking $\boldsymbol{\theta} = \mathbf{T}\mathbf{T}'$ for any $\mathbf{T} \in \mathbb{R}^{p\times m}$, we see that $\mathbf{Y} \sim MN_{p\times m}(\mathbf{0}, \tfrac{1}{2}\boldsymbol{\Sigma}, \mathbf{I}_m)$ whose density function is given by

$$(2\pi)^{-\frac{1}{2}pm}\,|\tfrac{1}{2}\boldsymbol{\Sigma}|^{-\frac{1}{2}m}\exp[-\tfrac{1}{2}\mathrm{tr}(\tfrac{1}{2}\boldsymbol{\Sigma})^{-1}\mathbf{Y}\mathbf{Y}'] \quad \text{for all } \mathbf{Y} \in \mathbb{R}^{p\times m} \tag{2.4}$$

and $\mathbf{A}^2 = \mathbf{A}$, $\mathrm{tr}\,\mathbf{A} = a_1 = 2a$ is a positive integer. The converse result is immediate. Thus, we have established

THEOREM 1. *Let* $\mathbf{Y} \sim MS_p(m;\,\psi)$ *and let the distribution of* $\mathbf{Y}$ *be nonsingular. Let* $\mathbf{A} = \mathbf{A}'$ *be a symmetric matrix. Then,* $\mathbf{Y}\mathbf{A}\mathbf{Y}' \sim MG_p(a, \boldsymbol{\Sigma})$ *if and only if* $\mathbf{Y} \sim MN_{p\times m}(\mathbf{0}, \tfrac{1}{2}\boldsymbol{\Sigma}, \mathbf{I}_m)$, $\mathbf{A}^2 = \mathbf{A}$ *and* $\mathrm{r}(\mathbf{A}) = 2a$ *must be a positive integer* ($> p-1$).

Using (1.3), the following Corollary 1 follows from Theorem 1.

COROLLARY 1. *Let* $\mathbf{X} \sim ME_{p\times n}(\mathbf{0}, \mathbf{V};\,\psi)$ *and let* $\mathrm{P}(\mathbf{X}\mathbf{V}^{-}\mathbf{X}' > \mathbf{0}) = 1$. *Let* $\mathbf{A} = \mathbf{A}'$ *be a symmetric matrix. Then* $\mathbf{X}\mathbf{A}\mathbf{X}' \sim MG_p(a, \boldsymbol{\Sigma})$ *if and only if* $\mathbf{X} \sim MN_{p\times n}(\mathbf{0}, \tfrac{1}{2}\boldsymbol{\Sigma}, \mathbf{V})$, $\mathbf{VAVAV} = \mathbf{VAV}$ *and* $\mathrm{r}(\mathbf{VAV}) = 2a$ *must be a positive integer.*

THEOREM 2. *Let* $\mathbf{Y} \sim MS_p(m;\,\psi)$ *and let the distribution of* $\mathbf{Y}$ *be nonsingular. Assume that all the moments of* $\mathbf{Y}$ *exist. Let* $\mathbf{A}_1$ *and* $\mathbf{A}_2$ *be two symmetric matrices such that* $\mathbf{A}_1\mathbf{A}_2 = \mathbf{0}$ *and* $\mathrm{tr}\,\mathbf{A}_i \neq 0$ *for* $i = 1, 2$. *If* $\mathbf{Y}\mathbf{A}_1\mathbf{Y}'$ *and* $\mathbf{Y}\mathbf{A}_2\mathbf{Y}'$ *are independently distributed, then* $\mathbf{Y} \sim MN_{p\times m}(\boldsymbol{\theta}, \boldsymbol{\Sigma}, \mathbf{I})$ *for some* $\boldsymbol{\Sigma} > \mathbf{0}$.

Proof. Let $\mathbf{y}_1$ be the first column vector of $\mathbf{Y}'$. Then, the given conditions imply that $\mathbf{y}_1 \sim S(m;\,\phi_1)$ and $\mathbf{y}_1'\mathbf{A}_1\mathbf{y}_1$ and $\mathbf{y}_1'\mathbf{A}_2\mathbf{y}_1$ are independently distributed. Since $\mathbf{A}_1\mathbf{A}_2 = \mathbf{0}$, we can find an orthogonal matrix $\mathbf{Q} = (\mathbf{Q}_1, \mathbf{Q}_2, \mathbf{Q}_3)$ such that $\mathbf{A}_i = \mathbf{Q}_i\mathbf{D}_i\mathbf{Q}_i'$ for $i = 1, 2$ and $\mathbf{D}_i$ is a nonsingular $m_i \times m_i$ diagonal matrix. Here, $m_1 + m_2 + m_3 = m$ and $\mathrm{tr}\,\mathbf{D}_i = \mathrm{tr}\,\mathbf{A}_i \neq 0$ for $i = 1, 2$. Let $\mathbf{Q}'\mathbf{y}_1 = \mathbf{x}$ [with $\mathbf{x}_i = \mathbf{Q}_i'\mathbf{y}_1$ for $i = 1, 2, 3$ and $\mathbf{x}' = (\mathbf{x}_1', \mathbf{x}_2', \mathbf{x}_3')$] and $\mathbf{x} \sim S(m;\,\phi_1)$. Let $R_i = \mathbf{x}_i'\mathbf{x}_i$ and $u_i = \mathbf{y}_1'\mathbf{A}_i\mathbf{y}_1/R_i = \mathbf{x}_i'\mathbf{D}_i\mathbf{x}_i/R_i$ for $i = 1, 2$ because $\mathrm{P}(R_i > 0) = 1$ for all i. Then, since $\mathbf{w}_i = \mathbf{x}_i/R_i^{\frac{1}{2}}$ for $i = 1, 2, 3$ and (R_1, R_2, R_3) are independently distributed.

Further, all the moments of $\mathbf{y}_1$ exist and u_1R_1 and u_2R_2 are independently distributed. Hence,

$$(E\,u_1^{\,i})\,(E\,u_2^{\,j})\,[E(R_1^{\,i}R_2^{\,j}) - E(R_1^{\,i})\,E(R_2^{\,j})] = 0 \tag{2.5}$$

for all nonnegative integers i and j, and the c.f. of $\mathbf{x}_1$ and $\mathbf{x}_2$ is

$$\begin{aligned}\phi_1(\mathbf{t}_1'\mathbf{t}_1 + \mathbf{t}_2'\mathbf{t}_2) &= E\exp[(-1)^{\frac{1}{2}}\mathbf{t}_1'\mathbf{x}_1 + (-1)^{\frac{1}{2}}\mathbf{t}_2'\mathbf{x}_2] \\ &= E\exp[(-1)^{\frac{1}{2}}R_1\mathbf{t}_1'\mathbf{w}_1 + (-1)^{\frac{1}{2}}R_2\mathbf{t}_2'\mathbf{w}_2] \\ &= E\,[\sum_{i=0}^{\infty}\sum_{j=0}^{\infty}(-1)^{i+j}R_1^{\,i}R_2^{\,j}(\mathbf{t}_2'\mathbf{t}_2)^j(\mathbf{t}_1'\mathbf{t}_1)^i/i!\,j!\,4^{i+j}(\tfrac{1}{2}m_1)_i(\tfrac{1}{2}m_2)_j]\end{aligned}$$

for all $\mathbf{t}_1 \in \mathbb{R}^{m_1}$ and $\mathbf{t}_2 \in \mathbb{R}^{m_2}$. From this, if all the moments of $\mathbf{x}$ exist, we have

$$\phi_1^{(i+j)}(0) = (-1)^{i+j}E(R_1^{\,i}R_2^{\,j})\,/\,4^{i+j}(\tfrac{1}{2}m_1)_i(\tfrac{1}{2}m_2)_j \text{ for all } i,j = 0, 1, \ldots .$$

Hence (2.5) gives

$$E(u_1^{\,i})\,E(u_2^{\,j})\,[\phi_1^{(i+j)}(0) - \phi_1^{(i)}(0)\,\phi_1^{(j)}(0)] = 0 \tag{2.6}$$

for all $i,j = 0, 1, 2, \ldots$. Notice that $\mathbf{w}_i = \mathbf{x}_i/R_i^{\frac{1}{2}}$ is uniformly distributed over the sphere $(\mathbf{w}_i'\mathbf{w}_i = 1)$, $u_i = \mathbf{w}_i'\mathbf{D}_i\mathbf{w}_i$ for $i = 1, 2$. Hence, $E(u_i) = \operatorname{tr}\mathbf{D}_i/m_i$ $(\neq 0)$ for $i = 1, 2$ and $E\,u_i^{2a} > 0$ for all $a = 1, 2, \ldots$. Hence, taking $(i,j) = (1,1), (1,2s)$ and $(2,2s)$, we find from (2.6) that

$$\phi_1^{(r)}(0) = c_1^{\,r} \text{ for all integers } r \text{ with } c_1 = \phi_1^{(1)}(0). \tag{2.7}$$

Using this in the c.f. of $\mathbf{y}_1$, we see that

$$\phi_1(\mathbf{t}'\mathbf{t}) = \exp(-c_1\mathbf{t}'\mathbf{t})$$

and so $\mathbf{y}_1 \sim N(\mathbf{0}, 2c_1\mathbf{I}_m)$ for some $c_1 > 0$.

Now, let $\mathbf{a}$ be any given vector of $\mathbb{R}^p$. Then, $\mathbf{Y}'\mathbf{a} \sim S(m;\phi)$, where $\phi(\boldsymbol{\theta}) = \psi(\mathbf{a}\mathbf{a}'\boldsymbol{\theta})$ for all $\boldsymbol{\theta} \geq \mathbf{0}$. Then, arguing as before, $\mathbf{Y}'\mathbf{a} \sim N(\mathbf{0}, \mathbf{I}_m 2c)$, where c is a function of $\mathbf{a}$, and it is easy to see that if $E(\mathbf{S}) = E(\mathbf{Y}\mathbf{Y}') = c_1\boldsymbol{\Sigma}$, where c_1 is constant and $\boldsymbol{\Sigma} > \mathbf{0}$, then

$$2mc = c_1 \mathbf{a}' \boldsymbol{\Sigma} \mathbf{a} > 0 \text{ for all } \mathbf{a} \in \mathbb{R}^p.$$

The above results show that $\mathbf{Y}'\mathbf{a} \sim N(\mathbf{0}, \mathbf{I}_m\, c_1 \mathbf{a}' \boldsymbol{\Sigma}\mathbf{a}/m)$ for all $\mathbf{a} \in \mathbb{R}^p$ and hence $\mathbf{Y} \sim MN_{p\times m}(\mathbf{0}, c_1\boldsymbol{\Sigma}/m, \mathbf{I}_m)$ for some $c_1\boldsymbol{\Sigma}/m > \mathbf{0}$. This proves Theorem 2. □

COROLLARY 2. *Let* $\mathbf{X} \sim ME_{p\times n}(\mathbf{M}, \mathbf{V}; \psi)$ *and let* $\mathbf{X} = \mathbf{M} + \mathbf{Y}\mathbf{V}_1'$, *where* $\mathbf{V}_1 \in \mathbb{R}^{n\times m}$, $m = \mathrm{r}(\mathbf{V}) = \mathrm{r}(\mathbf{V}_1)$ *and* $\mathbf{V} = \mathbf{V}_1\mathbf{V}_1'$. *Assume that all the moments of* $\mathbf{Y}$ *exist and* $\mathbf{Y}$ *has a nonsingular distribution. Let* $\mathbf{Q}_i = \mathbf{X}\mathbf{A}_i\mathbf{X}' + \mathbf{L}_{1i}\mathbf{X}' + \mathbf{X}\mathbf{L}_{2i}' + \mathbf{C}_i$ *for* $i = 1, 2$ *such that* $\mathbf{A}_i$ $(i = 1, 2)$ *are symmetric,* $\mathbf{V}\mathbf{A}_1\mathbf{V}\mathbf{A}_2\mathbf{V} = \mathbf{0}$, $\mathbf{M}\mathbf{A}_i\mathbf{M}' + \mathbf{L}_{1i}\mathbf{M}' + \mathbf{M}\mathbf{L}_{2i}' + \mathbf{C}_i = \mathbf{0}$ and $\mathbf{V}(\mathbf{A}_i\mathbf{M}' + \mathbf{L}_{2i}) = \mathbf{0} = \mathbf{V}(\mathbf{A}_i\mathbf{M}' + \mathbf{L}_{1i})$ *for* $i = 1, 2$. *Then if* $\mathbf{Q}_1$ *and* $\mathbf{Q}_2$ *are independently distributed and* $\mathrm{tr}\,(\mathbf{A}_i\mathbf{V}) \neq 0$ *for* $i = 1, 2$, *then* $\mathbf{X} \sim ME_{p\times n}(\mathbf{M}, \boldsymbol{\Sigma}, \mathbf{V})$ *for some* $\boldsymbol{\Sigma} > \mathbf{0}$.

This follows from Theorem 2 after noting $\mathbf{Q}_i = \mathbf{Y}\mathbf{A}_{(i)}\mathbf{Y}'$ for $i = 1, 2$ and $\mathbf{A}_{(1)}\mathbf{A}_{(2)} = \mathbf{0}$ with $\mathrm{tr}\,\mathbf{A}_{(i)} \neq 0$ and $\mathbf{A}_{(i)} = \mathbf{V}_1'\mathbf{A}_i\mathbf{V}_1$ for $i = 1, 2$.

THEOREM 3. *Let* $\mathbf{Y} \sim MS_p(m; \psi)$ *and let* $\mathbf{Y}$ *have a nonsingular distribution such that* $E\,\mathbf{Y}\mathbf{Y}' = m\boldsymbol{\Sigma}\ (> \mathbf{0})$. *Let* $\mathbf{A}_1$ *and* $\mathbf{A}_2$ *be symmetric matrices such that* $\mathbf{A}_1 + \mathbf{A}_2 = \mathbf{I}_m$ *and* $\mathbf{A}_i \geq \mathbf{0}$ *for* $i = 1, 2$. *If* $\mathbf{Y}\mathbf{A}_1\mathbf{Y}'$ *and* $\mathbf{Y}\mathbf{A}_2\mathbf{Y}'$ *are independently distributed, then* $\mathbf{Y} \sim MN_{p\times m}(\mathbf{0}, \boldsymbol{\Sigma}, \mathbf{I}_m)$, *and* $\mathbf{A}_1\mathbf{A}_2 = \mathbf{0}$ *or* $\mathbf{A}_1^2 = \mathbf{A}_1$.

Proof. Let $\mathbf{a} \in \mathbb{R}^p$ be a given vector. Then, $\mathbf{Y}'\mathbf{a} \sim S(m; \phi)$, where $\phi(\boldsymbol{\theta}) = \psi(\mathbf{a}\mathbf{a}'\boldsymbol{\theta})$ for any $\boldsymbol{\theta} \geq \mathbf{0}$. If $\mathbf{Y}\mathbf{A}_1\mathbf{Y}'$ and $\mathbf{Y}\mathbf{A}_2\mathbf{Y}'$ are independently distributed and if $\mathbf{x} = \mathbf{Y}'\mathbf{a}$, then $\mathbf{x}'\mathbf{A}_1\mathbf{x}$ and $\mathbf{x}'\mathbf{A}_2\mathbf{x}$ are independently distributed and $\mathbf{x} = \mathbf{w}R^{\frac{1}{2}}$ where $\mathbf{w}$ and R are independently distributed and $\mathbf{w}$ is distributed uniformly over the sphere $(\mathbf{w}'\mathbf{w} = 1)$. Then, $\mathrm{P}(\mathbf{x}'\mathbf{A}_i\mathbf{x} > 0) = 1$ for $i = 1, 2$ and

$$R = \mathbf{x}'\mathbf{x} = \mathbf{x}'\mathbf{A}_1\mathbf{x} + \mathbf{x}'\mathbf{A}_2\mathbf{x} \quad \text{and} \quad \mathbf{x}'\mathbf{A}_2\mathbf{x}/\mathbf{x}'\mathbf{A}_1\mathbf{x} = \mathbf{w}'\mathbf{A}_2\mathbf{w}/\mathbf{w}'\mathbf{A}_1\mathbf{w}$$

are independently distributed. Hence, using Lukac´s result (1956) we see that $\mathbf{x}'\mathbf{A}_i\mathbf{x} \sim G(a_i, c)$ for some $a_i > 0$ and $c > 0$ for $i = 1, 2$. Then, using Theorem 1 for $p = 1$, we see that $\mathbf{x} = \mathbf{Y}'\mathbf{a} \sim N(\mathbf{0}, \frac{1}{2}c\mathbf{I}_m)$, $\mathbf{A}_i^2 = \mathbf{A}_i$ and $\mathrm{tr}\,\mathbf{A}_i = \mathrm{r}(\mathbf{A}_i) = 2a_i$ for $i = 1, 2$. Thus, $E\,\mathbf{x}'\mathbf{x} = \mathbf{a}'E(\mathbf{Y}\mathbf{Y}')\mathbf{a} = \mathbf{a}'\boldsymbol{\Sigma}\,\mathbf{a}\,m = m(\frac{1}{2}c)$. Hence $c = 2\mathbf{a}'\boldsymbol{\Sigma}\mathbf{a}$ for any $\mathbf{a}$. Hence $\mathbf{Y}'\mathbf{a} \sim N(\mathbf{0}, \mathbf{a}'\boldsymbol{\Sigma}\mathbf{a}\,\mathbf{I}_m)$ for all $\mathbf{a} \in \mathbb{R}^p$, which proves that $\mathbf{Y} \sim MN_{p\times n}(\mathbf{0}, \boldsymbol{\Sigma}, \mathbf{I}_m)$ for some $\boldsymbol{\Sigma} > \mathbf{0}$. Thus, Theorem 3 is established. □

NOTE 1. The result of Theorem 3 for $p=1$ was established by Anderson and Fang (1987, Theorem 1), but for $p>1$, we require the later part of the above proof, which is not necessary for $p=1$.

COROLLARY 3. *Let* $\mathbf{X} \sim ME_{p\times n}(\mathbf{M}, \mathbf{V}; \psi)$ *and let* $\mathbf{X} = \mathbf{M} + \mathbf{Y}\mathbf{V}_1{}'$ *where* $\mathbf{V}_1 \in \mathbb{R}^{n\times m}$, $m = \mathrm{r}(\mathbf{V}) = \mathrm{r}(\mathbf{V}_1)$ *and* $\mathbf{V} = \mathbf{V}_1\mathbf{V}_1{}'$ *and* $\mathbf{Y}$ *has a nonsingular distribution. Let* $\mathbf{A}_1$ *and* $\mathbf{A}_2$ *be symmetric matrices such that* $\mathbf{V}(\mathbf{A}_1 + \mathbf{A}_2)\mathbf{V} = \mathbf{V}$, $\mathbf{V}(\mathbf{A}_i\mathbf{M}' + \mathbf{L}_{1i}) = \mathbf{0} = \mathbf{V}(\mathbf{A}_i\mathbf{M}' + \mathbf{L}_{2i})$ *and* $\mathbf{M}\mathbf{A}_i\mathbf{M}' + \mathbf{M}\mathbf{L}'_{1i} + \mathbf{L}_{2i}\mathbf{M}' + \mathbf{C}_i = \mathbf{0}$ *for* $i = 1, 2$. *Let* $\mathrm{P}[\mathbf{a}'(\mathbf{X}\mathbf{A}_i\mathbf{X}' + \mathbf{X}\mathbf{L}_{1i}{}' + \mathbf{L}_{2i}\mathbf{X}' + \mathbf{C}_i)\mathbf{a} > 0] = 1$ *for all* $\mathbf{a} \in \mathbb{R}^p$ *and* $i = 1, 2$ *and* $E(\mathbf{X} - \mathbf{M})\mathbf{V}^{-}(\mathbf{X} - \mathbf{M})' = m\mathbf{\Sigma}$ $(>\mathbf{0})$. *If* $\mathbf{X}\mathbf{A}_i\mathbf{X}' + \mathbf{X}\mathbf{L}_{1i}{}' + \mathbf{L}_{2i}\mathbf{X}' + \mathbf{C}_i$ $(i = 1, 2)$ *are independently distributed, then* $\mathbf{X} \sim ME_{p\times n}(\mathbf{M}, \mathbf{\Sigma}, \mathbf{V})$ *for some* $\mathbf{\Sigma} > \mathbf{0}$.

Corollary 3 follows from Theorem 3 after noting

$$\mathbf{X}\mathbf{A}_i\mathbf{X}' + \mathbf{X}\mathbf{L}'_{1i} + \mathbf{L}_{2i}\mathbf{X}' + \mathbf{C}_i = \mathbf{Y}\mathbf{A}_{(i)}\mathbf{Y}' \text{ with } \mathbf{A}_{(i)} = \mathbf{V}_1{}'\mathbf{A}_i\mathbf{V}_1 \text{ for } i = 1, 2.$$

COROLLARY 4. *Let* $\mathbf{Y} \sim MS_p(m; \psi)$ *and let* $\mathbf{Y}$ *have a nonsingular distribution. Let* $E(\mathbf{Y}\mathbf{Y}') = m\mathbf{\Sigma}$ *exist and* $\mathbf{\Sigma} > \mathbf{0}$. *Let* $\mathbf{Y}\mathbf{A}_i\mathbf{Y}'$ *and* $\mathrm{P}(\mathbf{a}'\mathbf{Y}\mathbf{A}_i\mathbf{Y}'\mathbf{a} > 0$ for all $\mathbf{a} \in \mathbb{R}^p) = 1$, *where* $\mathbf{A}_1, \mathbf{A}_2, \ldots, \mathbf{A}_k$ *are symmetric matrices such that* $\mathbf{A}_1 + \ldots + \mathbf{A}_k = \mathbf{I}_m$ *and* $k > 2$. *If* $\mathbf{Y}\mathbf{A}_i\mathbf{Y}'$ $(i = 1, 2, \ldots, k)$ *are independently distributed, then* $\mathbf{Y} \sim ME_{p\times m}(\mathbf{0}, \mathbf{\Sigma}, \mathbf{I}_m)$ *for* $\mathbf{\Sigma} > \mathbf{0}$ *and* $\mathbf{A}_i\mathbf{A}_j = \mathbf{0}$ *for* $i \neq j$ *and* $\mathbf{A}_i{}^2 = \mathbf{A}_i$ *for* $i, j = 1, 2, \ldots, k$.

Proof. Since $\mathbf{Y}\mathbf{A}_i\mathbf{Y}'$ and $\mathbf{Y}(\mathbf{A}_2 + \ldots + \mathbf{A}_k)\mathbf{Y}'$ are independently distributed, so by Theorem 3, $\mathbf{Y} \sim MN_{p\times m}(\mathbf{0}, \mathbf{\Sigma}, \mathbf{I}_m)$ for some $\mathbf{\Sigma} > \mathbf{0}$ and $\mathbf{A}_1{}^2 = \mathbf{A}_1$ and $(\mathbf{A}_2 + \ldots + \mathbf{A}_k)^2 = \mathbf{A}_2 + \ldots + \mathbf{A}_k$. By reason of the independence of $\mathbf{Y}\mathbf{A}_i\mathbf{Y}'$ $(i = 1, 2, \ldots, k)$ and $\mathbf{Y} \sim MN_{p\times m}(\mathbf{0}, \mathbf{\Sigma}, \mathbf{I}_m)$, we get $\mathbf{A}_i\mathbf{A}_j = \mathbf{0}$ for all $i \neq j = 1, 2, \ldots, k$ and then $\mathbf{A}_i{}^2 = \mathbf{A}_i$ $(i = 1, 2, \ldots, k)$. Thus, Corollary 4 is established. □

NOTE 2. An extension of Corollary 4 similar to Corollary 3 can be given, but it is not made explicit here.

2.2 Complex variables case

We give below the results (similar to the real variables case) for the complex variables, and omit the proof in view of the similarity of expressions.

THEOREM 1c. *Let* $\mathbf{X} \sim CME_{p\times n}(\mathbf{0}, \mathbf{V}; \psi)$ *and let* $P(\mathbf{XV^-X^*}$ is Hpd$) = 1$. *Let* $\mathbf{A}$ *be a Hermitian matrix. Then* $\mathbf{XAX^*} \sim CMG_p(a, \mathbf{\Sigma})$ *if and only if* $\mathbf{X} \sim CMN_{p\times n}(\mathbf{0}, \mathbf{\Sigma}, \mathbf{V})$, $\mathbf{VAVAV} = \mathbf{VAV}$, $r(\mathbf{VAV}) = a$, *which must be a positive integer, and* $\mathbf{\Sigma}$ *is Hpd.*

THEOREM 2c. *Let* $\mathbf{X} \sim CME_{p\times n}(\mathbf{0}, \mathbf{V}; \psi)$ *and let* $P(\mathbf{XV^-X^*}$ is Hpd$) = 1$. *Let* $\mathbf{A}_1$ *and* $\mathbf{A}_2$ *be Hermitian matrices such that* $\mathbf{VA}_1\mathbf{VA}_2\mathbf{V} = \mathbf{0}$ *and* $\mathrm{tr}(\mathbf{VA}_i) \neq 0$ *for* $i = 1, 2$. *If all the moments of* $\mathbf{X}$ *exist and* $\mathbf{XA}_1\mathbf{X}^*$ *and* $\mathbf{XA}_2\mathbf{X}^*$ *are independently distributed, then* $\mathbf{X} \sim CMN_{p\times n}(\mathbf{0}, \mathbf{\Sigma}, \mathbf{V})$ *for some* $\mathbf{\Sigma} > \mathbf{0}$.

THEOREM 3c. *Let* $\mathbf{X} \sim CME_{p\times n}(\mathbf{0}, \mathbf{V}; \psi)$ *and let* $P(\mathbf{XV^-X^*}$ is Hpd$) = 1$. *Let* $E(\mathbf{XV^-X^*}) = m\mathbf{\Sigma}$ *exist with* $m = r(\mathbf{V})$ *and* $\mathbf{\Sigma}$ *being Hpd. Let* $\mathbf{A}_1, \mathbf{A}_2, \ldots, \mathbf{A}_k$ *be Hermitian matrices such that* $\mathbf{V}(\mathbf{A}_1 + \ldots + \mathbf{A}_k)\mathbf{V} = \mathbf{V}$ *and* $P(\mathbf{a^*XA}_i\mathbf{X^*a} > 0$ for all $\mathbf{a} \in \mathbb{C}^p) = 1$ *for all* $i = 1, 2, \ldots, k$. *If* $\mathbf{XA}_i\mathbf{X}^*$ *for* $i = 1, 2, \ldots, k$ *are independently distributed, then* $\mathbf{X} \sim CME_{p\times n}(\mathbf{0}, \mathbf{\Sigma}, \mathbf{V})$, $\mathbf{VA}_i\mathbf{VA}_i\mathbf{V} = \mathbf{VA}_i\mathbf{V}$ *and* $\mathbf{VA}_i\mathbf{VA}_j\mathbf{V} = \mathbf{0}$ *for all* $i \neq j$, $i, j = 1, 2, \ldots, k$.

3. NULL ROBUSTNESS FOR THE LOCATION PARAMETERS

3.1 Growth curve model (real variables)

Let $\mathbf{Y} \sim MS_p(m; \psi)$ and the distribution of $\mathbf{Y}$ be nonsingular. Further, we shall assume that the density function of $\mathbf{Y}$ exists and it is given by

$$f(\mathbf{YY}') \quad \text{for all} \quad \mathbf{Y} \in \mathbb{R}^{p\times m}. \tag{3.1}$$

Let $m \geq p$ and $\mathbf{S} = \mathbf{YY}'$. Then, the density of $\mathbf{S}$ is given by

$$\{\pi^{\frac{1}{2}pm}/\Gamma_p(\tfrac{1}{2}m)\}\,|\mathbf{S}|^{\frac{1}{2}(m-p-1)}f(\mathbf{S}) \quad \text{for } \mathbf{S} > \mathbf{0}. \tag{3.2}$$

If $\mathbf{S} = \mathbf{S}_1\mathbf{S}_1'$ and $\mathbf{U} = \mathbf{S}_1^{-1}\mathbf{Y}$, then by (1.8), $\mathbf{U}$ and $\mathbf{S}$ are independently distributed and $\mathbf{U}$ is distributed uniformly over $O(p, m)$. For some properties of the distribution of $\mathbf{U}$ one can refer to Khatri (1970).

Let $\mathbf{X} \sim ME_{p\times n}(\mathbf{B\xi A}, \mathbf{I}_n; \psi)$, where $\mathbf{B}\,(\in \mathbb{R}^{p\times q})$ and $\mathbf{A}\,(\in \mathbb{R}^{m\times n})$ are known matrices of ranks q and m respectively. We want to test $\mathbf{H}(\mathbf{C\xi D} = \mathbf{0})$ against $\mathbf{H}(\mathbf{C\xi D} \neq \mathbf{0})$, where $\mathbf{C}\,(\in \mathbb{R}^{c\times q})$ and $\mathbf{D}\,(\in \mathbb{R}^{m\times d})$ are known matrices of ranks c and d respectively. Let $\mathbf{CC}_1' = \mathbf{0}$, $\mathbf{D}_1'\mathbf{D} = \mathbf{0}$, and $\mathbf{C}_0' = (\mathbf{C}_1', \mathbf{C}')$ and $\mathbf{D}_0 = (\mathbf{D}_1, \mathbf{D})$ be nonsingular matrices. Then $\mathbf{X} \sim ME_{p\times n}(\mathbf{B}_1\boldsymbol{\eta}\mathbf{A}_1, \mathbf{I}_n; \psi)$ with $\mathbf{B}_1 = \mathbf{BC}_0^{-1}$, $\boldsymbol{\eta} = \mathbf{C}_0\boldsymbol{\xi}\mathbf{D}_0$ and $\mathbf{A}_1 = \mathbf{D}_0^{-1}\mathbf{A}$. Let $\mathbf{A}_0' = [\mathbf{A}_1'(\mathbf{A}_3')^{-1}, \mathbf{A}_2']$ be an orthogonal matrix with $\mathbf{A}_1\mathbf{A}_1' = \mathbf{A}_3\mathbf{A}_3'$

and $\mathbf{A}_3$ being a lower triangular matrix, and let $\mathbf{B}_0 = [\mathbf{B}_1(\mathbf{B}_1'\mathbf{B}_1)^{-1}, \mathbf{B}_2]$ be a nonsingular matrix with $\mathbf{B}_2'\mathbf{B}_1 = \mathbf{0}$. If $\psi_1(\boldsymbol{\theta}) = \psi(\mathbf{B}_0\boldsymbol{\theta}\mathbf{B}_0')$ for all $\boldsymbol{\theta} \geq \mathbf{0}$, then $\mathbf{Y} = \mathbf{B}_0'\mathbf{X}\mathbf{A}_0' \sim ME_{p\times n}(\mathbf{B}_0'\mathbf{B}_1\boldsymbol{\eta}\mathbf{A}_1\mathbf{A}_0', \mathbf{I}_n; \psi_1)$. Notice that $\mathbf{B}_1'\mathbf{B}_0 = (\mathbf{I}_q, \mathbf{0})$ and $\mathbf{A}_1\mathbf{A}_0' = (\mathbf{A}_3, \mathbf{0})$. Hence

$$\mathbf{M}_1 = \mathbf{B}_0'\mathbf{B}_1\boldsymbol{\eta}\mathbf{A}_1\mathbf{A}_0' = \underset{\quad m \qquad n-m}{\begin{pmatrix} \boldsymbol{\eta}\mathbf{A}_3 & \mathbf{0} \\ \mathbf{0} & \mathbf{0} \end{pmatrix}} \begin{matrix} q \\ p-q \end{matrix} ,$$

and under $\mathbf{H}_0$,

$$\boldsymbol{\eta}\mathbf{A}_3 = \underset{\quad m-d \qquad d}{\begin{pmatrix} \boldsymbol{\eta}_{11} & \boldsymbol{\eta}_{12} \\ \boldsymbol{\eta}_{21} & \mathbf{0} \end{pmatrix}} \begin{matrix} q-c \\ c \end{matrix} .$$

Let

$$\mathbf{Y} = \underset{m-d \quad d \quad n-m}{\begin{pmatrix} \mathbf{Y}_{11} & \mathbf{Y}_{12} & \mathbf{Y}_{13} \\ \mathbf{Y}_{21} & \mathbf{Y}_{22} & \mathbf{Y}_{23} \\ \mathbf{Y}_{31} & \mathbf{Y}_{32} & \mathbf{Y}_{33} \end{pmatrix}} \begin{matrix} q-c \\ c \\ p-q \end{matrix} , \quad \mathbf{R} = (\mathbf{Y}-\mathbf{M}_1)(\mathbf{Y}-\mathbf{M}_1)' ,$$

$$\Omega_0 = \{\boldsymbol{\delta}_{11} = \mathbf{Y}_{11} - \boldsymbol{\eta}_{11} \in \mathbb{R}^{(q-c)\times(m-d)}, \boldsymbol{\delta}_{21} = \mathbf{Y}_{21} - \boldsymbol{\eta}_{21} \in \mathbb{R}^{c\times(m-d)}$$

$$\text{and } \boldsymbol{\delta}_{12} = \mathbf{Y}_{12} - \boldsymbol{\eta}_{12} \in \mathbb{R}^{(q-c)\times d}\}$$

and

$$\Omega_1 = \begin{pmatrix} \mathbf{Y}_{11} & \mathbf{Y}_{12} \\ \mathbf{Y}_{21} & \mathbf{Y}_{22} \end{pmatrix} - \boldsymbol{\eta}\mathbf{A}_3 = \boldsymbol{\delta} \in \mathbb{R}^{q\times m} .$$

Then, under the distribution of $\mathbf{X}$ given in Appendix A, Note 3, the maximum likelihood ratio test statistic is given by

$$\lambda = \underset{\Omega_1}{\text{Min}}\,|\mathbf{R}| \,/\, \underset{\Omega_0}{\text{Min}}\,|\mathbf{R}| = |\mathbf{P}|\,|\mathbf{P}_{22} + \mathbf{Q}_{22}| \,/\, |\mathbf{P}_{22}||\mathbf{P}+\mathbf{Q}|, \tag{3.3}$$

(see Appendix A, Note 3) where

$$\mathbf{P} = \begin{pmatrix} \mathbf{Y}_{23} \\ \mathbf{Y}_{33} \end{pmatrix} (\mathbf{Y}'_{23}, \mathbf{Y}'_{33}), \qquad \mathbf{Q} = \begin{pmatrix} \mathbf{Y}_{22} \\ \mathbf{Y}_{32} \end{pmatrix} (\mathbf{Y}'_{22}, \mathbf{Y}'_{32}),$$

$$\mathbf{P}_{22} = \mathbf{Y}_{33}\mathbf{Y}'_{33} \quad \text{and} \quad \mathbf{Q}_{22} = \mathbf{Y}_{32}\mathbf{Y}'_{32}.$$

Let $\mathbf{Y} - \mathbf{M} = \mathbf{TU}$, where

$$\mathbf{T} = \begin{pmatrix} \mathbf{T}_{11} & \mathbf{T}_{12} & \mathbf{T}_{13} \\ \mathbf{0} & \mathbf{T}_{22} & \mathbf{T}_{23} \\ \mathbf{0} & \mathbf{0} & \mathbf{T}_{33} \end{pmatrix} \begin{matrix} q-c \\ c \\ p-q \end{matrix}$$
$$\qquad\quad q-c \qquad c \qquad p-q$$

is an upper triangular matrix, and

$$\mathbf{U} = \begin{pmatrix} \mathbf{U}_{11} & \mathbf{U}_{12} & \mathbf{U}_{13} \\ \mathbf{U}_{21} & \mathbf{U}_{22} & \mathbf{U}_{23} \\ \mathbf{U}_{31} & \mathbf{U}_{32} & \mathbf{U}_{33} \end{pmatrix} \begin{matrix} q-c \\ c \\ p-q \end{matrix} \qquad \in O(p, n).$$
$$\qquad\quad m-d \qquad d \qquad n-m$$

Then, under $\mathbf{H}_0$, Note 2 of Appendix A indicates that the statistic λ depends only on $\mathbf{U}$ and hence the test is null robust, because $\mathbf{U}$ and $\mathbf{T}$ are independently distributed. Thus, tests based on the elements of

$$\begin{pmatrix} \mathbf{U}_{22} & \mathbf{U}_{23} \\ \mathbf{U}_{32} & \mathbf{U}_{33} \end{pmatrix}$$

are null robust. Kariya and Sinha (1985) have considered a locally best invariant (LBI) test which will be null robust.

To obtain the distribution of λ under $\mathbf{H}_0$, we use Lemma 2 of Khatri (1970), and then the joint density of

$$\mathbf{U}_1 = \begin{pmatrix} \mathbf{U}_{21} & \mathbf{U}_{22} \\ \mathbf{U}_{31} & \mathbf{U}_{32} \end{pmatrix} = (\mathbf{U}_2, \mathbf{U}_3) \quad \text{with } \mathbf{U}_2 = \begin{pmatrix} \mathbf{U}_{21} \\ \mathbf{U}_{31} \end{pmatrix}$$

is given by

$$[\Gamma_{p-q+c}(\tfrac{1}{2}n)/\{\pi^{\frac{1}{2}m(p-q+c)}\Gamma_{p-q+c}[(n-m)/2]\}] \cdot$$
$$\cdot |\mathbf{I}_{p-q+c} - \mathbf{U}_1\mathbf{U}_1'|^{\frac{1}{2}(n-m-p+q-c+1)} \qquad (3.4)$$

for all $\mathbf{U}_1 \in \mathbb{R}^{(p-q+c)\times m}$ such that $\mathbf{I} - \mathbf{U}_1\mathbf{U}_1' > \mathbf{0}$. Further, λ can be rewritten as

$$|\mathbf{I}_{p-q+c} - \mathbf{U}_1\mathbf{U}_1'| \cdot |\mathbf{I}_{p-q} - \mathbf{U}_{31}\mathbf{U}_{31}'| \cdot$$
$$\cdot |\mathbf{I}_{p-q} - \mathbf{U}_{31}\mathbf{U}_{31}' - \mathbf{U}_{32}\mathbf{U}_{32}'|^{-1} \cdot |\mathbf{I}_{p-q+c} - \mathbf{U}_2\mathbf{U}_2'|^{-1}. \quad (3.5)$$

Define

$$\begin{matrix} c \\ p-q \end{matrix} \begin{pmatrix} \mathbf{W}_1 \\ \mathbf{W}_2 \end{pmatrix} = \mathbf{G}^{-1}\begin{pmatrix} \mathbf{U}_{22} \\ \mathbf{U}_{32} \end{pmatrix}, \quad (\text{columns: } d)$$

$$\mathbf{G}\mathbf{G}' = \mathbf{I}_{p-q+c} - \mathbf{U}_2\mathbf{U}_2',$$

$\mathbf{G}$ being an upper triangular matrix and $\mathbf{W}' = (\mathbf{I}_d - \mathbf{W}_2'\mathbf{W}_2)^{-\frac{1}{2}}\mathbf{W}_1'$. Note that

$$\lambda = |\mathbf{I}_d - \mathbf{W}'\mathbf{W}| \quad (3.6)$$

and the joint density of $\mathbf{W}$ is obtained from (3.4) as

$$\{\Gamma_c[(n-m-p+q+d)/2] \,/\, \pi^{\frac{1}{2}cd}\Gamma_c[(n-m-p+q)/2]\}\, |\mathbf{I}_d - \mathbf{W}'\mathbf{W}|^{\frac{1}{2}(n-m-p+q-c-1)} \quad (3.7)$$

for all $\mathbf{W} \in \mathbb{R}^{c\times d}$ such that $\mathbf{I} - \mathbf{W}'\mathbf{W} > \mathbf{0}$. Using (3.6) and (3.7), we get

$$E\lambda^h = \Gamma_c[(n-m-p+q+d)/2] \cdot \Gamma_c[(n-m-p+q)/2+h] \cdot$$
$$\cdot \{\Gamma_c[(n-m-p+q)/2]\}^{-1} \cdot \{\Gamma_c[(n-m-p+q+d)/2+h]\}^{-1}. \quad (3.8)$$

NOTE 3. In some general situations for $p=1$ [or for the density of $\mathbf{Y}$ as $|\mathbf{\Sigma}|^{-n/2}\, h(\mathrm{tr}\ \mathbf{\Sigma}^{-1}(\mathbf{Y} - \mathbf{M})(\mathbf{Y} - \mathbf{M})')$], Anderson, Fang and Hsu (1986) have established null robustness. We have considered the null robustness for the *matrix* elliptical distributions and not the particular case of the *vector* elliptical distributions as mentioned in the above reference.

3.2 Growth curve model (complex variables)

Let $\mathbf{X} \sim CME_{p\times n}(\mathbf{B}\boldsymbol{\xi}\mathbf{A}, \mathbf{I}_n; \psi)$, where $\mathbf{B}\,(\in \mathbb{C}^{p\times q})$ and $\mathbf{A}\,(\in \mathbb{C}^{m\times n})$ are known matrices. We can proceed exactly as before for testing $\mathrm{H}_0(\mathbf{C}\boldsymbol{\xi}\mathbf{D}=\mathbf{0})$

against $H(\mathbf{C\xi D} \neq \mathbf{0})$, where $\mathbf{C}$ ($\in \mathbb{C}^{c\times q}$) and $\mathbf{D}$ ($\in \mathbb{C}^{m\times d}$) are known matrices of ranks c and d respectively. The changes necessary are to change transposes to conjugate transposes, $O(p, n)$ before (3.4) to $CO(p, n)$. The density (3.4) of $\mathbf{U}_1$ is changed to

$$\{\tilde{\Gamma}_{p-q+c}(n) / \pi^{(p-q+c)d}\,\tilde{\Gamma}_{p-q+c}(n-m)\}\,|\mathbf{I}_{p-q+c} - \mathbf{U}_1\mathbf{U}_1^*|^{n-m-p+q-c}$$

for all $\mathbf{U}_1 \in \mathbb{C}^{(p-q+c)\times m}$ such that $\mathbf{I} - \mathbf{U}_1\mathbf{U}_1^*$ is Hpd, while the density of $\mathbf{W}$ of (3.7) is changed to

$$\{\tilde{\Gamma}_c(n-m-p+q+d) / \pi^{cd}\,\tilde{\Gamma}_c(n-m-p+q)\}\,|\mathbf{I}_d - \mathbf{W}^*\mathbf{W}|^{(n-m-p+q-c)}$$

for all $\mathbf{W} \in \mathbb{C}^{c\times d}$ such that $\mathbf{I} - \mathbf{W}^*\mathbf{W}$ is Hpd. Further, (3.8) becomes

$$E\lambda^h = \tilde{\Gamma}_c(n-m-p+q+d)\cdot\tilde{\Gamma}_c(n-m-p+q+h)\cdot$$
$$\cdot[\tilde{\Gamma}_c(n-m-p+q)]^{-1}\cdot[\tilde{\Gamma}_c(n-m-p+q+d+h)]^{-1}.$$

For some more results on null robustness one can refer to Khatri (1987a; 1985).

4. DISTRIBUTION OF QUADRATIC FORM

4.1 Real variables case

Let $\mathbf{X} \sim MS_p(n; \psi)$ and $\mathbf{X} = (\mathbf{X}_1, \mathbf{X}_2,\dots, \mathbf{X}_k) = \mathbf{S}^{\frac{1}{2}}(\mathbf{\Gamma}_1, \mathbf{\Gamma}_2,\dots,\mathbf{\Gamma}_k)$ with $\mathbf{\Gamma}_i$ and $\mathbf{X}_i$ being $p\times n_i$ matrices, and $n = (n_1+n_2+\dots+n_k)$. Here, if $\mathbf{X}$ has a nonsingular distribution, then $\mathbf{XX}' = \mathbf{S}$ and $\mathbf{\Gamma} = (\mathbf{\Gamma}_1,\dots,\mathbf{\Gamma}_k)$ are independently distributed by (1.8). Further, $\mathbf{\Gamma}$ has a uniform distribution over Stieltjes manifold $O(p, n)$ and the distribution of $\mathbf{S}$ is $F(\mathbf{S})$ for $\mathbf{S}>\mathbf{0}$. Let $\mathbf{V}_i = \mathbf{\Gamma}_i\mathbf{\Gamma}_i'$ *for* $i=1, 2,\dots, k$. We shall say that $\mathbf{V}_1, \mathbf{V}_2,\dots, \mathbf{V}_k$ has matrix Dirichlet's distribution $D_{p,k}(\frac{1}{2}n_1, \frac{1}{2}n_2,\dots, \frac{1}{2}n_k)$. Then, the joint distribution of $\mathbf{S}_i = \mathbf{X}_i\mathbf{X}_i' = \mathbf{S}^{\frac{1}{2}}\mathbf{V}_i\mathbf{S}^{\frac{1}{2}}$ for $i=1, 2,\dots, k-1$ is denoted by $MG_{p,k-1}(\frac{1}{2}n_1,\dots, \frac{1}{2}n_{k-1}; \frac{1}{2}n_k; \psi)$. If $n_i \geq p$ for all $i=1, 2,\dots, k$, then using Khatri (1970), the joint density of $\mathbf{V}_1, \mathbf{V}_2,\dots, \mathbf{V}_k$ is given by

$$\{\Gamma_p(\tfrac{1}{2}n) / (\prod_{i=1}^{k}\Gamma_p(\tfrac{1}{2}n_i))\}\prod_{i=1}^{k}|\mathbf{V}_i|^{\frac{1}{2}(n_i-p-1)} \text{ with } \sum_{i=1}^{k}\mathbf{V}_i = \mathbf{I}_p \tag{4.1}$$

and $\mathbf{V}_i > \mathbf{0}$ for all i, and the joint density of $\mathbf{S}_1, \mathbf{S}_2, \ldots, \mathbf{S}_{k-1}$ is

$$\{\Gamma_p(\tfrac{1}{2}n) / (\prod_{i=1}^{k} \Gamma_p(\tfrac{1}{2}n_i)\} \int_{D(\mathbf{S})} \prod_{i=1}^{k-1} |\mathbf{S}_i|^{\frac{1}{2}(n_i-p-1)} |\mathbf{S}|^{-\frac{1}{2}(n-p-1)} |\mathbf{S} - \sum_{i=1}^{k-1} \mathbf{S}_i|^{\frac{1}{2}(n_k-p-1)} \, dF(\mathbf{S}) \tag{4.2}$$

where $D(\mathbf{S}) = \{\mathbf{S} > (\mathbf{S}_1 + \ldots + \mathbf{S}_{k-1})\}$. If the density of $\mathbf{S}$ exists, then we can modify (4.2) by replacing

$$dF(\mathbf{S}) \text{ by } f(\mathbf{S})\, d\mathbf{S}$$

where $f(\mathbf{S})$ is the density of $\mathbf{S}$. This density of $MG_{p,k-1}(\frac{1}{2}n_1, \ldots, \frac{1}{2}n_{k-1}; \frac{1}{2}n_k; \psi)$ was given by Anderson and Fang (1987, Lemma 2) when $p=1$ and when $p>1$ under the existence of the density of $\mathbf{S}$ (or $\mathbf{X}$). Further, Anderson and Fang (1987, Theorem 6) have established the multivariate analog to Cochran's Theorem as

LEMMA 1. *Let* $\mathbf{X} \sim MS_p(n; \psi)$. *Let* $\mathbf{A}_1, \mathbf{A}_2, \ldots, \mathbf{A}_k$ *be symmetric matrices of ranks* n_i $(i = 1, 2, \ldots, k)$ *and* $n = n_1 + \ldots + n_{k+1}$ *with* $n_{k+1} > 0$. *Then, the joint distribution of* $\mathbf{X}\mathbf{A}_i\mathbf{X}'$ $(i = 1, 2, \ldots, k)$ *is* $MG_{p,k}(\frac{1}{2}n_1, \ldots, \frac{1}{2}n_k; \frac{1}{2}n_{k+1}; \psi)$ *if and only if* $\mathbf{A}_i\mathbf{A}_j = \mathbf{0}$ *for all* $i \neq j$ *and* $\mathbf{A}_i^2 = \mathbf{A}_i$ *for* $i, j = 1, 2, \ldots, k$ *with* $\operatorname{tr} \mathbf{A}_i = n_i$.

Now, we study some distribution aspects in the particular cases only when $\mathbf{x} \sim E_n(\boldsymbol{\mu}, \mathbf{V}; \phi)$ and $n = pm + n_1 + \ldots + n_k$. Let us denote $\mathbf{x}' = (\mathbf{x}_1', \mathbf{x}_2', \ldots, \mathbf{x}_m', \mathbf{x}'_{m+1}, \ldots, \mathbf{x}'_{m+k})$, where $\mathbf{x}_i \in \mathbb{R}^m$ for $i = 1, 2, \ldots, m$ and $\mathbf{x}_{m+j} \in \mathbb{R}^{n_j}$ for $j = 1, 2, \ldots, k$. Further, let $\mathbf{V} = \operatorname{diag}[\mathbf{I}_m \otimes \boldsymbol{\Sigma}, \mathbf{I}_{n_1}\sigma_1, \ldots, \mathbf{I}_{n_k}\sigma_k]$ be positive definite so that

$$(\mathbf{x} - \boldsymbol{\mu})' \mathbf{V}^{-1} (\mathbf{x} - \boldsymbol{\mu}) = \operatorname{tr}(\boldsymbol{\Sigma}^{-1}\mathbf{S}) + \sum_{j=1}^{k} z_j / \sigma_j,$$

where $\mathbf{S} = \sum_{i=1}^{m} (\mathbf{x}_i - \boldsymbol{\mu}_i)(\mathbf{x}_i - \boldsymbol{\mu}_i)'$ and $z_j = (\mathbf{x}_{m+j} - \boldsymbol{\mu}_{m+j})'(\mathbf{x}_{m+j} - \boldsymbol{\mu}_{m+j})$ for $j = 1, 2, \ldots, k$ and $\boldsymbol{\mu}$ is partitioned similar to $\mathbf{x}$. If the density of $\mathbf{x}$ exists and $m \geq p$, the joint density of $\mathbf{S}, z_1, \ldots, z_k$ is given by

$$\{\pi^{\frac{1}{2}pm}/\Gamma_p(\tfrac{1}{2}m)\, |\boldsymbol{\Sigma}|^{\frac{1}{2}m}\}\, |\mathbf{S}|^{\frac{1}{2}(m-p-1)}\,.$$

$$\cdot \prod_{i=1}^{k} \{\pi^{\frac{1}{2}n_i}[\Gamma(\tfrac{1}{2}n_i)\sigma_i]^{-1}(z_j/\sigma_j)^{\frac{1}{2}n_i-1}\} \cdot g(\operatorname{tr} \boldsymbol{\Sigma}^{-1}\mathbf{S} + \sum_{i=1}^{k} z_j/\sigma_j) \tag{4.3}$$

for all $\mathbf{S} > \mathbf{0}$, $z_i > 0$ $(i = 1, 2, \ldots, k)$. This distribution will be denoted by $W_{p,k}(\frac{1}{2}m, \Sigma; \frac{1}{2}n_1, \ldots, \frac{1}{2}n_k; \sigma_1, \ldots, \sigma_k; \phi)$. Let $R = \operatorname{tr} \Sigma^{-1}\mathbf{S} + (z_1/\sigma_1 + \ldots + z_k/\sigma_k)$ and $\mathbf{W} = \mathbf{S}/R$ and $w_i = z_i/R$ for $i = 1, 2, \ldots, k\ (\geq 1)$. Then, R and $(\mathbf{W}, w_1, \ldots, w_k)$ are independently distributed and the joint density of $\mathbf{W}$, $w_1, \ldots, w_k$ (for $k \geq 1$) is given by

$$\sigma_p \Gamma(\tfrac{1}{2}n) \{\Gamma_p(\tfrac{1}{2}m) [\prod_{i=1}^{k} \Gamma(\tfrac{1}{2}n_i)\, \sigma_i^{\frac{1}{2}n_i}] |\Sigma|^{\frac{1}{2}m}\}^{-1} |\mathbf{W}|^{\frac{1}{2}(m-p-1)} \prod_{i=1}^{k} \omega_i^{\frac{1}{2}n_i - 1} \tag{4.4}$$

for all $\mathbf{W} > \mathbf{0}$, $w_1 > 0, \ldots, w_k > 0$ such that $\operatorname{tr} \Sigma^{-1}\mathbf{W} + (w_1/\sigma_1 + \ldots + w_k/\sigma_k) = 1$. This distribution is denoted as $WD_{p,k}(\frac{1}{2}m, \Sigma; \frac{1}{2}n_1, \ldots, \frac{1}{2}n_k; \sigma_1, \ldots, \sigma_k)$. When ($m = 0$ or $\mathbf{S}$ does not exist and $k > 1$) or ($p = 1$ and $k \geq 1$), then (4.4) reduces to the well known Dirichlet´s distribution. When $k = 0$ (or $n_1 = \ldots = n_k = 0$ or $z_1, \ldots, z_k$ do not exist), then the density of $\mathbf{W} = \mathbf{S}/(\operatorname{tr} \Sigma^{-1}\mathbf{S})$ is given by

$$\{\Gamma(\tfrac{1}{2}pm) / \Gamma_p(\tfrac{1}{2}m) |\Sigma|^{\frac{1}{2}m}\}^{-1} |W|^{\frac{1}{2}(m-p-1)} / \sigma^{pp} \tag{4.5}$$

for all $\mathbf{W} > \mathbf{0}$ such that $\operatorname{tr} \Sigma^{-1}\mathbf{W} = 1$ and the variables are all the elements of $\mathbf{W}$ except w_{pp} where $\Sigma^{-1} = (\sigma^{ij})$. This distribution will be denoted by $WD_p(\frac{1}{2}m, \Sigma)$. When $\sigma_i = 1$ for all $i = 1, 2, \ldots, k$, we shall write

$W_{p,k}(a_0, \Sigma; \mathbf{a}; 1, \ldots, 1; \phi) = W_{p,k}(a_0, \Sigma; \mathbf{a}; \phi)$ with $\mathbf{a}' = (a_1, \ldots, a_k)$ and

$WD_{p,k}(a_0, \Sigma; \mathbf{a}; 1, \ldots, 1) = WD_{p,k}(a_0, \Sigma; \mathbf{a})$.

When $k = 0$, we shall write $W_{p,0}(a, \Sigma; ;\phi) = W_p(a, \Sigma; \phi)$, and when $p = 0$, we shall write $W_{0,k}(; \mathbf{a}; \phi) = W_{0,k}(\mathbf{a}; \phi)$.

Now, we shall establish the following.

THEOREM 4. *Let* $\mathbf{x} \sim S(n; \phi)$ *and let the density of* $\mathbf{x}$ *exist with* $P(\mathbf{x}'\mathbf{x} = R > 0) = 1$. *Let* $\mathbf{A}$ *and* $\mathbf{B}$ *be two symmetric matrices and let* $\theta_1, \theta_2, \ldots, \theta_k$ *be distinct real numbers. Then* $\mathbf{x}'\mathbf{A}\mathbf{x} \doteq \operatorname{tr}(\mathbf{BS}) + \theta_1 \mathbf{z}_1 + \ldots + \theta_k \mathbf{z}_k$, *where* $(\mathbf{S}, \mathbf{z}) \sim W_{p,k}(a_0, \Sigma; \mathbf{a}; \phi)$ *with* $\mathbf{z}' = (z_1, \ldots, z_k)$, $\mathbf{a}' = (a_1, a_2, \ldots, a_k)$ *and* $pa_0 + a_1 + \ldots + a_k = \frac{1}{2}n$, *if and only if* $2a_j$ $(j = 0, 1, \ldots, k)$ *are nonnegative integers, and there exists an orthogonal matrix* $\mathbf{T}$ *such that* $\mathbf{T}'\mathbf{A}\mathbf{T} = \operatorname{diag}(\mathbf{I}_{2a_0} \otimes \mathbf{D}_\lambda, \theta_1 \mathbf{I}_{2a_1}, \ldots, \theta_k \mathbf{I}_{2a_k})$ *and diagonal elements* λ_j*´s of* $\mathbf{D}_\lambda$ *(a diagonal matrix) are the eigenvalues of* $\Sigma\mathbf{B}$. *The same result is true for*

$\mathbf{x}'\mathbf{A}\mathbf{x}/R \doteq \operatorname{tr}(\mathbf{WB}) + \theta_1 w_1 + \ldots + \theta_k w_k$ *where* $(\mathbf{W}, \mathbf{w}) \sim WD_{p,k}(a_0, \Sigma; \mathbf{a})$.

Proof. Let $\mathbf{S}_1 = \mathbf{\Sigma}^{-\frac{1}{2}}\mathbf{S}\mathbf{\Sigma}^{-\frac{1}{2}}$ (or $\mathbf{W}_1 = \mathbf{\Sigma}^{-\frac{1}{2}}\mathbf{W}\mathbf{\Sigma}^{-\frac{1}{2}}$), $\mathbf{B}_1 = \mathbf{\Sigma}^{\frac{1}{2}}\mathbf{B}\mathbf{\Sigma}^{\frac{1}{2}}$ and $y = \mathrm{tr}(\mathbf{B}_1\mathbf{W}_1) + \theta_1 w_1 + \ldots + \theta_k w_k$. Then, $|y| \leq \lambda_0 \,\mathrm{tr}\, \mathbf{\Sigma}^{-1}\mathbf{W} + |\theta_1|w_1 + \ldots + |\theta_k|w_k \leq \lambda$, where λ_0 = maximum value of $|\lambda_j|$, $j = 1, 2, \ldots, p$, and $\lambda = \max(\lambda_0, |\theta_j|, j = 1, 2, \ldots, k)$. Hence, the c.f. of $\{y/\lambda \mid y > 0\}$ is nonzero and that of $\{-y/\lambda \mid y < 0\}$ is nonzero, because the random variables have finite nonnegative range; (see Marcinkiewicz 1939). Hence, the c.f. of (y/λ) will be nonzero. Now, let $R = \mathrm{tr}\, \mathbf{\Sigma}^{-1}\mathbf{S} + z_1 + \ldots + z_k$. Then, it is easy to see that R and $(\mathbf{W}, \mathbf{w}) = (\mathbf{S}, \mathbf{z})/R$ are independently distributed. Because $\frac{1}{2}n = pa_0 + a_1 + \ldots + a_k$, we shall write $\mathrm{tr}\, \mathbf{\Sigma}^{-1}\mathbf{S} + z_1 + \ldots + z_k = \mathbf{x}'\mathbf{x} = R$. Thus,

$$R\, y \doteq R(\mathbf{u}'\mathbf{A}\mathbf{u}),$$

where $\mathbf{u}R^{\frac{1}{2}} = \mathbf{x}$, $\mathbf{u}$ and R are independently distributed, R and y are independently distributed, c.f. of y is nonzero and $P(R > 0) = 1$. Hence, $\mathbf{u}'\mathbf{A}\mathbf{u} \doteq y$ [see, for example, Anderson and Fang (1987, Lemma 1)].

Now, let R_1 and $(\mathbf{W},\mathbf{w})$ be independently distributed, $(\mathbf{W},\mathbf{w}) \sim WD_{p,k}(a_0, \mathbf{\Sigma}; \mathbf{a})$ and $R_1 \sim G(\frac{1}{2}n, 1)$. Then

$$\mathbf{u}'\mathbf{A}\mathbf{u} \doteq y \Rightarrow R_1\, \mathbf{u}'\mathbf{A}\mathbf{u} \doteq R_1 y,$$

and $(R_1)^{\frac{1}{2}}\mathbf{u} \sim N(\mathbf{0}, \frac{1}{2}\mathbf{I}_n)$ and $R_1 y = \mathrm{tr}\, \mathbf{B}(R_1\mathbf{W}) + \theta_1(R_1 w_1) + \ldots + \theta_k(R_1 w_k)$, where $R_1\mathbf{W}$, $R_1 w_j$ $(j = 1, 2, \ldots, k)$ are independently distributed, $R_1\mathbf{W} \sim G_p(a_0, \mathbf{\Sigma})$ and $Rw_j \sim G(a_j, 1)$ for $j = 1, 2, .., k$. Then, using the c.f.´s on both sides, we get

$$|\mathbf{I}_n - (-1)^{\frac{1}{2}}\, t\mathbf{A}| = |\mathbf{I}_p - (-1)^{\frac{1}{2}}\, t\mathbf{\Sigma}\mathbf{B}|^{2a_0} \prod_{j=1}^{k} [1 - (-1)^{\frac{1}{2}}\, t\,\theta_j]^{2a_j}$$

for all $t \in \mathbb{R}$.

The solution for this equation shows that $2a_j$ (for $j = 0, 1, \ldots, k$) must be nonnegative integers, and then $\mathbf{A}$ can be represented as mentioned in Theorem 4.

COROLLARY 5. *Let* $\mathbf{x} \sim E_n(\boldsymbol{\mu}, \mathbf{V}; \phi)$, $P[(\mathbf{x} - \boldsymbol{\mu})'\mathbf{V}^{-}(\mathbf{x} - \boldsymbol{\mu}) > 0] = 1$ *and let* $\mathbf{A}$ *be a symmetric matrix such that* $\boldsymbol{\mu}'\mathbf{A}\boldsymbol{\mu} + 2\boldsymbol{\ell}'\boldsymbol{\mu} + c = 0$ *and* $\mathbf{V}(\mathbf{A}\boldsymbol{\mu} + \boldsymbol{\ell}) = \mathbf{0}$. *Let* $(\mathbf{S}, \mathbf{y}) \sim W_{p,k}(a, \mathbf{\Sigma}; \mathbf{a}; \phi)$ *and* $\mathrm{r}(\mathbf{V}) = 2pa_0 + 2a_1 + \ldots + 2a_k$. *Let* $\mathbf{B}$ *be a symmetric matrix,* θ_j*´s be distinct real numbers and* λ_j*´s be the eigenvalues of* $\mathbf{\Sigma}\mathbf{B}$. *Then,* $\mathbf{x}'\mathbf{A}\mathbf{x} + 2\boldsymbol{\ell}'\mathbf{x} + c \doteq \mathrm{tr}(\mathbf{S}\mathbf{B}) + \theta_1 y_1 + \ldots + \theta_k y_k$ *if and only if* $2a_j$ $(j = 0, 1, 2, \ldots, k)$ *must be nonnegative integers and there exists an orthogonal matrix* $\mathbf{T}$ *such that* $\mathbf{T}'(\mathbf{P}\mathbf{A}\mathbf{P}')\mathbf{T} = \mathrm{diag}(\mathbf{I}_{2a_0} \otimes \mathbf{D}_\lambda, \theta_1\mathbf{I}_{2a_1}, \ldots, \theta_k\mathbf{I}_{2a_k})$ *with* $\mathbf{V} = \mathbf{P}'\mathbf{P}$ *and* $\mathbf{P}\mathbf{P}'$ *being nonsingular.*

This follows from Theorem 4 by using (1.3) for a particular case.

NOTE 4. Particular cases of Corollary 5 (or Theorem 4):

(a) Let $n = 2pa_0$ and $\mathbf{S} \sim G_p(a_0, \boldsymbol{\Sigma}; \phi)$. Then $\mathbf{x}'\mathbf{A}\mathbf{x} + 2\boldsymbol{\ell}'\mathbf{x} + c \doteq \mathrm{tr}(\mathbf{SB})$ if and only if $2a_0$ is a positive integer ($\geq p$) and there exists an orthogonal matrix $\mathbf{T}$ such that $\mathbf{T}'(\mathbf{PAP}')\mathbf{T} = \mathbf{I}_{2a_0} \otimes \mathbf{D}_\lambda$ with $\mathbf{V} = \mathbf{P}'\mathbf{P}$ and $\mathbf{PP}'$ being nonsingular.

(b) Let $\mathbf{y} \sim G_{0,k}(\mathbf{a}; \phi)$ and $a_1 + \ldots + a_k = \frac{1}{2}n$. Then $\mathbf{x}'\mathbf{A}\mathbf{x} + 2\boldsymbol{\ell}'\mathbf{x} + c \doteq \boldsymbol{\theta}'\mathbf{y}$ with $\boldsymbol{\theta}' = (\theta_1, \ldots, \theta_k)$ if and only if $2a_j$ ($j = 1, 2, \ldots, k$) are positive integers and there exists an orthogonal matrix $\mathbf{T}$ such that $\mathbf{T}'(\mathbf{PAP}')\mathbf{T} = \mathrm{diag}(\theta_1 \mathbf{I}_{2a_1}, \ldots, \theta_k \mathbf{I}_{2a_k})$ with $\mathbf{V} = \mathbf{P}'\mathbf{P}$ and $\mathbf{PP}'$ being nonsingular.

[The result (b) was established by Anderson and Fang (1987, Lemma 3) when $2a_j$'s are assumed to be positive integers.]

THEOREM 5. *Let* $\mathbf{x} \sim E_n(\boldsymbol{\mu}, \mathbf{V}; \phi)$, $P[(\mathbf{x}-\boldsymbol{\mu})'\mathbf{V}^-(\mathbf{x}-\boldsymbol{\mu}) > 0] = 1$ *and let* $\mathbf{A}_i$ *be symmetric matrices such that* $\boldsymbol{\mu}'\mathbf{A}_i\boldsymbol{\mu} + 2\boldsymbol{\ell}_i'\boldsymbol{\mu} + c_i = 0$ *and* $\mathbf{V}(\mathbf{A}_i\boldsymbol{\mu} + \boldsymbol{\ell}_i) = \mathbf{0}$ *for all i. Let* $q_i = \mathbf{x}'\mathbf{A}_i\mathbf{x} + 2\boldsymbol{\ell}_i'\mathbf{x} + c_i$ *for all i and* $(\mathbf{S}, \mathbf{z}) \sim W_{p,k}(a_0, \boldsymbol{\Sigma}; \mathbf{a}; \phi)$ *with unit diagonal elements of* $\boldsymbol{\Sigma}$. *Then*

$$(q_1, q_2, \ldots, q_{p+k}) \doteq (s_{11}, s_{22}, \ldots, s_{pp}, z_1, z_2, \ldots, z_k)$$

if and only if $\mathrm{r}(\mathbf{V})/2 = pa_0 + a_1 + \ldots + a_k$, $\mathbf{VA}_i\mathbf{VA}_i\mathbf{V} = \mathbf{VA}_i\mathbf{V}$, $\mathbf{VA}_i\mathbf{VA}_{p+j}\mathbf{V} = \mathbf{0}$, $\mathrm{tr}(\mathbf{VA}_\alpha) = 2a_0 = \mathrm{r}(\mathbf{VA}_\alpha\mathbf{V})$, $\mathrm{r}(\mathbf{VA}_{p+j}\mathbf{V}) = \mathrm{tr}(\mathbf{VA}_{p+j}) = 2a_j$, $\mathbf{VA}_\alpha\mathbf{VA}_\beta\mathbf{VA}_\alpha\mathbf{V} = \sigma^2_{\alpha\beta}\mathbf{VA}_\alpha\mathbf{V}$ *and* $\mathbf{VA}_\alpha\mathbf{VA}_\beta\mathbf{VA}_\gamma\mathbf{VA}_\alpha\mathbf{V} = (\sigma_{\alpha\beta}\sigma_{\beta\gamma}\sigma_{\gamma\alpha})\mathbf{VA}_\alpha\mathbf{V}$ *for all* $i = 1, 2, \ldots, p+k$, $j = 1, 2, \ldots, k$ $(i \neq p+j)$, $\alpha \neq \beta \neq \gamma$, $\alpha, \beta, \gamma = 1, 2, \ldots, p$.

Proof. By (1.3) and (1.8), $\mathbf{x} \doteq \boldsymbol{\mu} + R^{\frac{1}{2}}\mathbf{P}'\mathbf{u}$ where R and $\mathbf{u}$ are independently distributed, $R = (\mathbf{x} - \boldsymbol{\mu})'\mathbf{V}^-(\mathbf{x} - \boldsymbol{\mu})$, $\mathbf{V} = \mathbf{P}'\mathbf{P}$, $\mathbf{PP}'$ is nonsingular and $\mathbf{u}$ is distributed uniformly over the sphere $\mathbf{u}'\mathbf{u} = 1$. Let $(\mathbf{W}, \mathbf{w}) \sim WD_{p,k}(a_0, \boldsymbol{\Sigma}; \mathbf{a})$ and R and $(\mathbf{W}, \mathbf{w})$ be independently distributed. Then, for any $(t_1, t_2, \ldots, t_{p+k}) \in \mathbb{R}^{p+k}$,

$$R\mathbf{u}'\left(\sum_{i=1}^{p+k} t_i \mathbf{A}_{(i)}\right)\mathbf{u} \doteq R\left(\sum_{i=1}^{p} t_i w_{ii} + \sum_{i=1}^{k} t_{p+i} w_i\right),$$

where $\mathbf{A}_{(i)} = \mathbf{PA}_i\mathbf{P}'$ for all i, $P(R > 0) = 1$. Note that $X = \mathbf{u}'(t_1\mathbf{A}_{(1)} + \ldots + t_{p+k}\mathbf{A}_{p+k})\,\mathbf{u}$ for given $t_1, \ldots, t_{p+k}$ has a finite range and a continuous distribution. Hence, the c.f.´s of $(\log X|X > 0)$ and $[\log(-X)|X < 0]$ are nonzero. Therefore, by Anderson and Fang (1987, Lemma 1),

$$X \doteq \sum_{i=1}^{p} t_i w_{ii} + \sum_{i=1}^{k} t_{p+i} w_i$$

for all $(t_1, \ldots, t_{p+k}) \in \mathbb{R}^{p+k}$. In particular

$$\mathbf{u}'\mathbf{A}_{(p+j)}\mathbf{u} \doteq w_j \sim \beta(a_j, pa_0 + \sum_{i \neq j}^{k} a_i = b \text{ (say)}).$$

Khatri (1987b) showed that this is true if and only if $pa_0 + (a_1, \ldots, a_k) = \frac{1}{2}m$ and $\mathbf{A}^2_{(p+j)} = \mathbf{A}_{(p+j)}$ with $r(\mathbf{A}_{(p+j)}) = 2a_j$ and $r(\mathbf{V}) = m$.

For all $(t_1, \ldots, t_{p+k}) \in \mathbb{R}^{p+k}$, $X \doteq t_1 w_{11} + \ldots + t_p w_{pp} + t_{p+1} w_1 + \ldots + t_{p+k} w_k \Leftrightarrow (\mathbf{u}'\mathbf{A}_{(1)}\mathbf{u}, \ldots, \mathbf{u}'\mathbf{A}_{(p+k)}\mathbf{u}) \doteq (w_{11}, \ldots, w_{pp}, w_1, \ldots, w_k)$. Now, using $R_1\mathbf{u} \sim N(\mathbf{0}, \frac{1}{2}\mathbf{I}_m)$, $m = r(\mathbf{V})$, the joint c.f. on both sides gives

$$|\mathbf{I} - (-1)^{\frac{1}{2}} \sum_{i=1}^{p+k} t_i \mathbf{A}_{(i)}| = |\mathbf{I} - (-1)^{\frac{1}{2}}\, \boldsymbol{\Sigma}\, \mathbf{D}_t|^{2a_0} \prod_{j=1}^{k} (1 - (-1)^{\frac{1}{2}} t_{p+j})^{2a_j}$$

for all real $t_1, t_2, \ldots, t_{p+k}$ and $\mathbf{D}_t$ being a $p \times p$ diagonal matrix with diagonal elements $t_1, t_2, \ldots, t_p$. This shows that $2a_0$ and $2a_j$ ($j = 1, 2, \ldots, k$) must be nonnegative integers. Upon application of Khatri´s result (1980b, Lemma 5, p. 237), the above equation gives the required result.

NOTE 5. Particular cases of Theorem 5:

(a) Let $\mathbf{S} \sim W_p(a_0, \boldsymbol{\Sigma}; \phi)$. Then $(q_1, \ldots, q_p) \doteq (s_{11}, \ldots, s_{pp})$ if and only if $2a_0 p = r(\mathbf{V})$ and $2a_0 = \text{tr}\ \mathbf{VA}_a = r(\mathbf{VA}_a\mathbf{V})$, $\mathbf{VA}_a\mathbf{VA}_a\mathbf{V} = \mathbf{VA}_a\mathbf{V}$, $\mathbf{VA}_a\mathbf{VA}_\beta\mathbf{VA}_a\mathbf{V} = \sigma_{a\beta}^2\mathbf{VA}_a\mathbf{V}$ and $\mathbf{VA}_a\mathbf{VA}_\beta\mathbf{VA}_\gamma\mathbf{VA}_a\mathbf{V} = (\sigma_{a\beta}\sigma_{\beta\gamma}\sigma_{\gamma a})\mathbf{VA}_a\mathbf{V}$ for all $a \neq \beta \neq \gamma$, $a, \beta, \gamma = 1, 2, \ldots, p$.

(b) Let $\mathbf{y} \sim G_{0,k}(\mathbf{a}; \phi)$. Then $(q_1, q_2, \ldots, q_k) \doteq (y_1, y_2, \ldots, y_k)$ if and only if $\frac{1}{2}r(\mathbf{V}) = a_1 + \ldots + a_k$, $tr\ \mathbf{VA}_j = r(\mathbf{VA}_j\mathbf{V}) = 2a_j$, $\mathbf{VA}_j\mathbf{VA}_j\mathbf{V} = \mathbf{VA}_j\mathbf{V}$ and $\mathbf{VA}_j\mathbf{VA}_{j'}\mathbf{V} = \mathbf{0}$ for all $j \neq j' = 1, 2, \ldots, k$.

[The result (b) is mentioned in Lemma 1 in different notations in a particular situation.]

NOTE 6. For results to multivariate matrix cases, refer to Khatri (1987b).

4.2 Complex variables case

Let $\mathbf{X} \sim CMS_p(n; \psi)$ and let $\mathbf{X} = (\mathbf{X}_1, \ldots, \mathbf{X}_k) = \mathbf{S}^{\frac{1}{2}}(\boldsymbol{\Gamma}_1, \boldsymbol{\Gamma}_2, \ldots, \boldsymbol{\Gamma}_k)$ have a nonsingular distribution. Then $\mathbf{S} = \mathbf{X}\mathbf{X}^*$ and $\boldsymbol{\Gamma} = (\boldsymbol{\Gamma}_1, \boldsymbol{\Gamma}_2, \ldots, \boldsymbol{\Gamma}_k)$ are independently distributed, $\boldsymbol{\Gamma}$ has a uniform distribution over the complex Stieltjes manifold $CO(p, n)$, and the distribution of $\mathbf{S}$ is $F(\mathbf{S})$ for $\mathbf{S}$ Hpd. The distribution of $\mathbf{V}_i = \boldsymbol{\Gamma}_i\boldsymbol{\Gamma}_i^*$ $(i = 1, 2, \ldots, k)$ is known as complex matrix Dirichlet's distribution and will be denoted by $CD_{p,k}(n_1, n_2, \ldots, n_k)$. The joint distribution of $\mathbf{S}_i = \mathbf{X}_i\mathbf{X}_i^*$ $(i = 1, 2, \ldots, k-1)$ is denoted by $CMD_{p,k-1}(n_1, \ldots, n_{k-1}; n_k; \psi)$. If $n_i \geq p$ for all $i = 1, 2, \ldots, k$, then the joint density of $\mathbf{S}_i = \mathbf{X}_i\mathbf{X}_i^* = \mathbf{S}^{\frac{1}{2}}\mathbf{V}_i\mathbf{S}^{\frac{1}{2}}$ for $i = 1, 2, \ldots, k-1$ is given by

$$\{\tilde{\Gamma}_p(n) / (\prod_{i=1}^{k} \tilde{\Gamma}_p(n_i))\} \prod_{i=1}^{k-1} |\mathbf{S}_i|^{n_i - p} \int_{D(\mathbf{S})} |\mathbf{S}|^{-(n-p)} |\mathbf{S} - \sum_{i=1}^{k-1} \mathbf{S}_i|^{n_k - p} \, dF(\mathbf{S})$$

where $D(\mathbf{S}) = \{\mathbf{S} - (\mathbf{S}_1 + \ldots + \mathbf{S}_{k-1})$ is Hermitian positive definite$\}$. Then, in place of Lemma 1, we have

LEMMA 1c. *Let* $\mathbf{X} \sim CMS_p(n; \psi)$. *Let* $\mathbf{A}_1, \mathbf{A}_2, \ldots, \mathbf{A}_k$ *be Hermitian matrices of ranks* n_i $(i = 1, 2, \ldots, k)$ *and* $n = (n_1 + \ldots + n_{k+1})$ with $n_{k+1} > 0$. *Then, the joint distribution of* $\mathbf{X}\mathbf{A}_i\mathbf{X}^*$ $(i = 1, 2, \ldots, k)$ *is* $CMG_{p,k}(n_1, \ldots, n_k; n_{k+1}; \psi)$ *if and only if* $\mathbf{A}_i\mathbf{A}_j = \mathbf{0}$ *for* $i \neq j$ *and* $\mathbf{A}_i^2 = \mathbf{A}_i$ *for* $i, j = 1, 2, \ldots, k$ *with* $\operatorname{tr} \mathbf{A}_i = n_i$.

The density function of the distribution $CW_{p,k}(a_0, \boldsymbol{\Sigma}; \mathbf{a}; \phi)$ is given by

$$[\pi^{a_0 p + a_1 + \ldots + a_k} / \tilde{\Gamma}_p(a_0) |\boldsymbol{\Sigma}|^{a_0}] |\mathbf{S}|^{a_0 - p} [\prod_{i=1}^{k} z_i^{a_i - 1} / \Gamma(a_i)] \, g(\operatorname{tr}\boldsymbol{\Sigma}^{-1}\mathbf{S} + \sum_{i=1}^{k} z_i)$$

for all Hpd $\mathbf{S}$ and $z_i > 0$ $(i = 1, 2, \ldots, k)$. When $k = 0$, we shall write $CW_p(a_0, \boldsymbol{\Sigma}; \phi)$, while when $p = 0$, we shall write $CW_{0,k}(\mathbf{a}; \phi)$. Then, it is easy to see that $R = \operatorname{tr} \boldsymbol{\Sigma}^{-1}\mathbf{S} + z_1 + \ldots + z_k$ and $(\mathbf{W}, \mathbf{w}) = (\mathbf{S}, \mathbf{z})/R$ are independently distributed and the distribution of $(\mathbf{W}, \mathbf{w})$ is denoted by

$CWD_{p,k}(a_0, \mathbf{\Sigma}; \mathbf{a})$. Now, the corresponding results for Corollary 5 and Theorem 5 are given by

COROLLARY 5c. *Let* $\mathbf{x} \sim CE_n(\boldsymbol{\mu}, \mathbf{V}; \phi)$, $R = (\mathbf{x}-\boldsymbol{\mu})^* \mathbf{V}^-(\mathbf{x}-\boldsymbol{\mu})$, $\mathrm{P}(R > 0) = 1$, *and let* $\mathbf{A}$ *be a Hermitian matrix such that* $\boldsymbol{\mu}^*\mathbf{A}\boldsymbol{\mu} + 2(\boldsymbol{\ell}^*\boldsymbol{\mu})_R + c = 0$ *and* $\mathbf{V}(\mathbf{A}\boldsymbol{\mu} + \boldsymbol{\ell}) = \mathbf{0}$ *(here, c is real). Let* $(\mathbf{S}, \mathbf{y}) \sim CW_{p,k}(a_0, \mathbf{\Sigma}; \mathbf{a}; \phi)$ *and* $\mathrm{r}(\mathbf{V}) = pa_0 + a_1 + \ldots + a_k$. *Let* $\mathbf{B}$ *be a Hermitian matrix and* θ_j*'s be distinct real numbers. Let* $\mathbf{D}_\lambda$ *be a diagonal matrix with diagonal elements as the eigenvalues of* $\mathbf{\Sigma B}$ *and let* $q = \mathbf{x}^*\mathbf{A}\mathbf{x} + 2(\boldsymbol{\ell}^*\mathbf{x})_R + c$. *Then* $q \doteq \mathrm{tr}(\mathbf{SB}) + \theta_1 y_1 + \ldots + \theta_k y_k$ *if and only if* a_j $(j = 0, 1,\ldots, k)$ *must be nonnegative integers and there exists a unitary matrix* $\mathbf{T}$ *such that* $\mathbf{T}^*(\mathbf{PAP}^*)\mathbf{T} = \mathrm{diag}(\mathbf{I}_{a_0}\otimes\mathbf{D}_\lambda, \theta_1\mathbf{I}_{a_1},\ldots, \theta_k\mathbf{I}_{a_k})$ *with* $\mathbf{V} = \mathbf{P}^*\mathbf{P}$ *and* $\mathbf{PP}^*$ *being nonsingular.*

THEOREM 5c. *Let* $\mathbf{x} \sim CE_n(\boldsymbol{\mu}, \mathbf{V}; \phi)$, $R = (\mathbf{x}-\boldsymbol{\mu})^* \mathbf{V}^-(\mathbf{x}-\boldsymbol{\mu})$, $\mathrm{P}(R > 0) = 1$, *and let* $\mathbf{A}_i$ $(i = 1, 2,\ldots, p+k)$ *be Hermitian matrices such that* $\boldsymbol{\mu}^*\mathbf{A}_i\boldsymbol{\mu} + 2(\boldsymbol{\ell}_i^*\boldsymbol{\mu})_R + c_i = 0$ *and* $\mathbf{V}(\mathbf{A}_i\boldsymbol{\mu} + \boldsymbol{\ell}_i) = \mathbf{0}$ *for all i. Let* $q_i = \mathbf{x}^*\mathbf{A}_i\mathbf{x} + 2(\boldsymbol{\ell}_i^*\mathbf{x})_R + c_i$ *for* $i = 1, 2,\ldots, p+k$ *and* $(\mathbf{S}, \mathbf{z}) \sim CW_{p,k}(a_0, \mathbf{\Sigma}; \mathbf{a}; \phi)$ *with unit diagonal elements of* $\mathbf{\Sigma}$. *Then*

$$(q_1, q_2,\ldots, q_{p+k}) \doteq (s_{11}, s_{22},\ldots, s_{pp}, z_1, z_2,\ldots, z_k)$$

if and only if $\mathrm{r}(\mathbf{V}) = pa_0 + a_1 + \ldots + a_k$, $\mathbf{VA}_i\mathbf{VA}_i\mathbf{V} = \mathbf{VA}_i\mathbf{V}$, $\mathbf{VA}_i\mathbf{VA}_{p+j}\mathbf{V} = \mathbf{0}$, $\mathrm{tr}(\mathbf{VA}_\alpha) = \mathrm{r}(\mathbf{VA}_\alpha\mathbf{V}) = a_0$, $\mathrm{tr}(\mathbf{VA}_{p+j}) = \mathrm{r}(\mathbf{VA}_{p+j}\mathbf{V}) = a_j$, $\mathbf{VA}_\alpha\mathbf{VA}_\beta\mathbf{VA}_\alpha\mathbf{V} = (\sigma_{\alpha\beta}\sigma_{\beta\alpha})\mathbf{VA}_\alpha\mathbf{V}$ *and* $\mathbf{VA}_\alpha\mathbf{VA}_\beta\mathbf{VA}_\gamma\mathbf{VA}_\alpha\mathbf{V} = (\sigma_{\alpha\beta}\sigma_{\beta\gamma}\sigma_{\gamma\alpha})\mathbf{VA}_\alpha\mathbf{V}$ *for all* $\alpha \neq \beta \neq \gamma$, $\alpha, \beta, \gamma = 1, 2,\ldots, p$, $i = 1, 2,\ldots, p+k$, $j = 1, 2,\ldots, k$ *and* $i \neq p+j$.

Particular cases for the corresponding results of Notes 4 and 6 can be formulated.

APPENDIX A

LEMMA A. *The minimum value of* $|\mathbf{R}|$ *over the variations of* $\boldsymbol{\eta} \in \mathbb{R}^{q\times m}$ *is* $|\mathbf{P}|\,|\mathbf{P}_{22} + \mathbf{ZZ}'|/|\mathbf{P}_{22}|$ *at* $\hat{\boldsymbol{\eta}} = \mathbf{P}_{12}\mathbf{P}_{22}^{-1}\mathbf{Z}$, *where* $\mathbf{R} = \mathbf{P} + \begin{pmatrix}\boldsymbol{\eta}\\ \mathbf{Z}\end{pmatrix}(\boldsymbol{\eta}', \mathbf{Z}')$ *and*

$$\mathbf{P} = \begin{pmatrix}\mathbf{P}_{11} & \mathbf{P}_{12}\\ \mathbf{P}_{21} & \mathbf{P}_{22}\end{pmatrix} > \mathbf{0}.$$

Proof. First we have

$$|\mathbf{R}| = |\mathbf{P}|\,|\mathbf{I} + (\boldsymbol{\eta}', \mathbf{Z}')\mathbf{P}^{-1}(\boldsymbol{\eta}', \mathbf{Z}')'|$$

$$= |\mathbf{P}|\,|\mathbf{I} + \mathbf{Z}'\mathbf{P}_{22}^{-1}\mathbf{Z} + (\boldsymbol{\eta} - \mathbf{P}_{12}\mathbf{P}_{22}^{-1}\mathbf{Z})'\,\mathbf{P}_{1\cdot 2}^{-1}(\boldsymbol{\eta} - \mathbf{P}_{12}\mathbf{P}_{22}^{-1}\mathbf{Z})|,$$

where $\mathbf{P}_{1\cdot 2} = \mathbf{P}_{11} - \mathbf{P}_{12}\mathbf{P}_{22}^{-1}\mathbf{P}_{21}$. From this, we get

$$\underset{\boldsymbol{\eta} \in \mathbb{R}^{q\times m}}{\operatorname{Min}} |\mathbf{R}| = |\mathbf{P}| \cdot |\mathbf{I} + \mathbf{Z}'\mathbf{P}_{22}^{-1}\mathbf{Z}| \text{ at } \hat{\boldsymbol{\eta}} = \mathbf{P}_{12}\mathbf{P}_{22}^{-1}\mathbf{Z}.$$

This proves Lemma A. □

LEMMA A1. *Let* $\Omega_0 = \{\boldsymbol{\eta}_1 \in \mathbb{R}^{(q-c)\times d}, \boldsymbol{\eta}_2 \in \mathbb{R}^{(q-c)\times(m-d)}, \boldsymbol{\eta}_3 \in \mathbb{R}^{c\times(m-d)}\}$,

$$\mathbf{P} = \underset{\quad q-c \quad\; c \quad\; p-q}{\begin{pmatrix} \mathbf{P}_{11} & \mathbf{P}_{12} & \mathbf{P}_{13} \\ \mathbf{P}_{21} & \mathbf{P}_{22} & \mathbf{P}_{23} \\ \mathbf{P}_{31} & \mathbf{P}_{32} & \mathbf{P}_{33} \end{pmatrix}} \begin{matrix} q-c \\ c \\ p-q \end{matrix} \qquad > \mathbf{0}$$

and $\mathbf{R} = \mathbf{P} + \mathbf{Q}\mathbf{Q}'$ *with* $\mathbf{Q} = \underset{m-d \quad d}{\begin{pmatrix} \boldsymbol{\eta}_2 & \boldsymbol{\eta}_1 \\ \boldsymbol{\eta}_3 & \mathbf{Q}_2 \\ \mathbf{Q}_1 & \mathbf{Q}_3 \end{pmatrix}} \begin{matrix} q-c \\ c \\ p-q \end{matrix}$.

Then,

$$\underset{\Omega_0}{\operatorname{Min}} |\mathbf{R}| = \frac{|\mathbf{P}| \begin{vmatrix} \mathbf{P}_{22}+\mathbf{Q}_2\mathbf{Q}_2' & \mathbf{P}_{23}+\mathbf{Q}_2\mathbf{Q}_3' \\ \mathbf{P}_{32}+\mathbf{Q}_3\mathbf{Q}_3' & \mathbf{P}_{33}+\mathbf{Q}_3\mathbf{Q}_3' \end{vmatrix} |\mathbf{P}_{33}+\mathbf{Q}_1\mathbf{Q}_1'+\mathbf{Q}_3\mathbf{Q}_3'|}{\begin{vmatrix} \mathbf{P}_{22} & \mathbf{P}_{23} \\ \mathbf{P}_{32} & \mathbf{P}_{33} \end{vmatrix} |\mathbf{P}_{33}+\mathbf{Q}_3\mathbf{Q}_3'|}$$

Proof. Using Lemma A for minimizing over $\boldsymbol{\eta}_2, \boldsymbol{\eta}_3$ given $\boldsymbol{\eta}_1$ is

$$\underset{\boldsymbol{\eta}_2, \boldsymbol{\eta}_3}{\operatorname{Min}} |\mathbf{R}| = |\mathbf{P} + \begin{pmatrix} \boldsymbol{\eta}_1 \\ \mathbf{Q}_2 \\ \mathbf{Q}_3 \end{pmatrix} (\boldsymbol{\eta}_1', \mathbf{Q}_2', \mathbf{Q}_3')| \cdot |\mathbf{P}_{33}+\mathbf{Q}_1\mathbf{Q}_1'+\mathbf{Q}_3\mathbf{Q}_3'| / |\mathbf{P}_{33}+\mathbf{Q}_3\mathbf{Q}_3'|$$

at $$\begin{pmatrix} \boldsymbol{\eta}_2 \\ \boldsymbol{\eta}_3 \end{pmatrix} = \begin{pmatrix} \mathbf{P}_{13} + \boldsymbol{\eta}_1 \mathbf{Q}_3{}' \\ \mathbf{P}_{23} + \mathbf{Q}_2 \mathbf{Q}_3{}' \end{pmatrix} (\mathbf{P}_{33} + \mathbf{Q}_3 \mathbf{Q}_3{}')^{-1} \mathbf{Q}_3.$$

Now, using Lemma A for minimizing over $\boldsymbol{\eta}_1$, we get the required result at

$$\boldsymbol{\eta}_1 = (\mathbf{P}_{12} \quad \mathbf{P}_{13}) \begin{pmatrix} \mathbf{P}_{22} & \mathbf{P}_{23} \\ \mathbf{P}_{32} & \mathbf{P}_{33} \end{pmatrix}^{-1} \begin{pmatrix} \mathbf{Q}_2 \\ \mathbf{Q}_3 \end{pmatrix}.$$

This proves Lemma A1. □

NOTE 1. Let $\mathbf{P} + \mathbf{Q}\mathbf{Q}' = \mathbf{T}\mathbf{T}'$, where

$$\mathbf{T} = \begin{pmatrix} \mathbf{T}_{11} & \mathbf{T}_{12} & \mathbf{T}_{13} \\ \cdot & \mathbf{T}_{22} & \mathbf{T}_{23} \\ \cdot & \cdot & \mathbf{T}_{33} \end{pmatrix} \begin{matrix} q-c \\ c \\ p-q \end{matrix} \qquad \text{(an upper triangular matrix)}$$
$$\qquad\quad q-c \quad c \quad p-q$$

and let $\mathbf{Q} = \mathbf{T}\mathbf{U}$ with

$$\mathbf{U}' = \begin{pmatrix} \mathbf{U}'_{11} & \mathbf{U}'_{21} & \mathbf{U}'_{31} \\ \mathbf{U}'_{12} & \mathbf{U}'_{22} & \mathbf{U}'_{32} \end{pmatrix} \begin{matrix} m-d \\ d \end{matrix}.$$
$$\qquad\quad q-c \quad c \quad p-q$$

Then, the minimum $|\mathbf{R}|$ is obtained at

$$\boldsymbol{\eta}_1 = \mathbf{T}_{12}\mathbf{U}_{22} + \mathbf{T}_{13}\mathbf{U}_{32}, \quad \begin{pmatrix} \boldsymbol{\eta}_2 \\ \boldsymbol{\eta}_3 \end{pmatrix} = \begin{pmatrix} \mathbf{T}_{11} & \mathbf{T}_{12} \\ \cdot & \mathbf{T}_{22} \end{pmatrix} \begin{pmatrix} \mathbf{U}_{11} \\ \mathbf{U}_{21} \end{pmatrix}$$

and

$$\underset{\Omega_0}{\text{Min}} |\mathbf{R}| = |\mathbf{P}| \cdot \left|\mathbf{I} - \begin{pmatrix} \mathbf{U}_{21} \\ \mathbf{U}_{31} \end{pmatrix} (\mathbf{U}'_{21}, \mathbf{U}'_{31})\right| / \left|\mathbf{I} - \begin{pmatrix} \mathbf{U}_{21} & \mathbf{U}_{22} \\ \mathbf{U}_{31} & \mathbf{U}_{32} \end{pmatrix} \begin{pmatrix} \mathbf{U}'_{21} & \mathbf{U}'_{31} \\ \mathbf{U}'_{22} & \mathbf{U}'_{32} \end{pmatrix}\right| \cdot |\mathbf{I} - \mathbf{U}_{31}\mathbf{U}'_{31}|.$$

LEMMA A2. *Let* Ω_0, $\mathbf{P}$ *and* $\mathbf{R}$ *be as defined in Lemma* A1 *and let* $\Omega = \{\Omega_0 \text{ and } \boldsymbol{\eta}_{22} = \mathbf{Q}_2 - \boldsymbol{\mu} \in R^{c \times d}\}$. *Then, the ratio statistic is*

$$\lambda = \underset{\Omega}{\text{Min}} |\mathbf{R}| / \underset{\Omega_0}{\text{Min}} |\mathbf{R}| = |\mathbf{P}_{(22)}|\, |\mathbf{P}_{33} + \mathbf{Q}_3\mathbf{Q}_3{}'| / |\mathbf{P}_{33}|\, |\mathbf{P}_{(22)} + \mathbf{Q}_{(2)}\mathbf{Q}_{(2)}{}'|,$$

where

$$\mathbf{P}_{(22)} = \begin{pmatrix} \mathbf{P}_{22} & \mathbf{P}_{23} \\ \mathbf{P}_{32} & \mathbf{P}_{33} \end{pmatrix} \text{ and } \mathbf{Q}_{(2)} = \begin{pmatrix} \mathbf{Q}_2 \\ \mathbf{Q}_3 \end{pmatrix}.$$

Lemma A2 follows from Lemma A and Lemma A1.

NOTE 2. Using the notations of Note 1 and under Ω, we can write

$$\lambda = \frac{\left|\mathbf{I} - \begin{pmatrix} \mathbf{U}_{21} & \mathbf{U}_{22} \\ \mathbf{U}_{31} & \mathbf{U}_{32} \end{pmatrix} \begin{pmatrix} \mathbf{U}'_{21} & \mathbf{U}'_{31} \\ \mathbf{U}'_{22} & \mathbf{U}'_{32} \end{pmatrix}\right| \left|\mathbf{I} - \mathbf{U}_{31}\mathbf{U}'_{31}\right|}{\left|\mathbf{I} - \begin{pmatrix} \mathbf{U}_{21} \\ \mathbf{U}_{31} \end{pmatrix} (\mathbf{U}'_{21} \quad \mathbf{U}'_{31})\right| \left|\mathbf{I} - \mathbf{U}_{31}\mathbf{U}'_{31} - \mathbf{U}_{32}\mathbf{U}'_{32}\right|}.$$

Under Ω, $\mathbf{Q}_2 - \boldsymbol{\mu}$ will have the same distribution as $\mathbf{Q}_2$ under Ω_0. Hence, under Ω, we shall write

$$\mathbf{P} + \begin{pmatrix} \boldsymbol{\eta}_2 & \boldsymbol{\eta}_1 \\ \boldsymbol{\eta}_3 & \mathbf{Q}_2 - \boldsymbol{\mu} \\ \mathbf{Q}_1 & \mathbf{Q}_3 \end{pmatrix} \begin{pmatrix} \boldsymbol{\eta}_2' & \boldsymbol{\eta}_3' & \mathbf{Q}_1' \\ \boldsymbol{\eta}_1' & (\mathbf{Q}_2 - \boldsymbol{\mu})' & \mathbf{Q}_3' \end{pmatrix} = \mathbf{T}\mathbf{T}'$$

and

$$\begin{pmatrix} \boldsymbol{\eta}_2 & \boldsymbol{\eta}_1 \\ \boldsymbol{\eta}_3 & \mathbf{Q}_2 - \boldsymbol{\mu} \\ \mathbf{Q}_1 & \mathbf{Q}_3 \end{pmatrix} = \mathbf{T} \begin{pmatrix} \mathbf{U}_{11} & \mathbf{U}_{12} \\ \mathbf{U}_{21} & \mathbf{U}_{22} \\ \mathbf{U}_{31} & \mathbf{U}_{32} \end{pmatrix}, \mathbf{T} = \begin{pmatrix} \mathbf{T}_{11} & \mathbf{T}_{12} & \mathbf{T}_{13} \\ \mathbf{0} & \mathbf{T}_{22} & \mathbf{T}_{23} \\ \mathbf{0} & \mathbf{0} & \mathbf{T}_{33} \end{pmatrix}.$$

Then $\mathbf{P}_{(22)} = \begin{pmatrix} \mathbf{T}_{22} & \mathbf{T}_{23} \\ \mathbf{0} & \mathbf{T}_{33} \end{pmatrix} [\mathbf{I} - \begin{pmatrix} \mathbf{U}_{21} & \mathbf{U}_{22} \\ \mathbf{U}_{31} & \mathbf{U}_{32} \end{pmatrix} \begin{pmatrix} \mathbf{U}'_{21} & \mathbf{U}'_{31} \\ \mathbf{U}'_{22} & \mathbf{U}'_{32} \end{pmatrix}] \begin{pmatrix} \mathbf{T}'_{22} & \mathbf{0} \\ \mathbf{T}'_{23} & \mathbf{T}'_{33} \end{pmatrix}$,

$$\mathbf{P}_{(22)} + (\mathbf{Q}_2', \quad \mathbf{Q}_3')'(\mathbf{Q}_2', \quad \mathbf{Q}_3')$$

$$= \begin{pmatrix} \mathbf{T}_{22} & \mathbf{T}_{23} \\ \cdot & \mathbf{T}_{33} \end{pmatrix} [\mathbf{I} - \begin{pmatrix} \mathbf{U}_{21} \\ \mathbf{U}_{31} \end{pmatrix} (\mathbf{U}'_{21}, \mathbf{U}'_{31}) + \begin{pmatrix} \mathbf{T}_{22}^{-1}\boldsymbol{\mu} \\ \mathbf{0} \end{pmatrix} (\mathbf{U}'_{22}, \mathbf{U}'_{32})$$

$$+ \begin{pmatrix} \mathbf{U}_{22} \\ \mathbf{U}_{32} \end{pmatrix} (\boldsymbol{\mu}'\mathbf{T}'^{-1}_{22}, \mathbf{0}) + \begin{pmatrix} \mathbf{T}'^{-1}\boldsymbol{\mu} \\ \mathbf{0} \end{pmatrix} (\boldsymbol{\mu}'\mathbf{T}'^{-1}_{22}, \mathbf{0})] \begin{pmatrix} \mathbf{T}'_{22} & \mathbf{0} \\ \mathbf{T}'_{32} & \mathbf{T}'_{33} \end{pmatrix},$$

etc. Then, though $\mathbf{T}$ and $\mathbf{U}$ are independently distributed, λ depends on $\mathbf{U}$ and $\mathbf{T}$ under Ω.

NOTE 3. Let the joint density of $\mathbf{P}$ and $\mathbf{Q}$ be given by

$$L = c\,|\boldsymbol{\Sigma}|^{-(s+m)/2}\,h[\mathrm{tr}(\boldsymbol{\Sigma}^{-1}\mathbf{R})] \text{ for all } \mathbf{P} > \mathbf{0} \text{ and } \mathbf{Q} \in \mathbb{R}^{p\times m},$$

where c is a constant depending on $\mathbf{P}$, $\mathbf{R} = \mathbf{P} + (\mathbf{Q} - \boldsymbol{\mu})(\mathbf{Q} - \boldsymbol{\mu})'$, $\boldsymbol{\mu}' = (\boldsymbol{\eta}', \mathbf{0})$, $\boldsymbol{\Sigma} > \mathbf{0}$ and $\boldsymbol{\eta} \in \mathbb{R}^{q\times m}$. Let $\mathrm{H}_0(\boldsymbol{\eta} \in \Omega_0)$ and $\mathrm{H}(\boldsymbol{\eta} \in \mathbb{R}^{q\times m} = \Omega_1)$. We can rewrite L as

$$L = c\,G^{p(s+m)/2}\,|\mathbf{R}|^{-(s+m)/2}\,h(pA),$$

where $\mathbf{T} = \mathbf{R}^{\frac{1}{2}}\boldsymbol{\Sigma}^{-1}\mathbf{R}^{\frac{1}{2}}$, $G = |\mathbf{T}|^{1/p}$ = geometric mean of $\lambda_1 \geq \lambda_2 \geq \ldots \geq \lambda_p > 0$ (eigenvalues of $\mathbf{T}$) and $\mathbf{A} = \lambda_1/p + \ldots + \lambda_p/p = (\mathrm{tr}\,\mathbf{T})/p$. Note that $G \leq A$ and the equality holds if and only if $\lambda_1 = \ldots = \lambda_p = x$ (say) or $\boldsymbol{\Sigma} = \mathbf{R}/x$ or $\mathbf{T} = x\mathbf{I}_p$. Assume that $(px)^{p(s+m)/2}\,h(px) = g(x)$ has a maximum at $x = x_h$ (> 0). Then

$$L \leq c\,|\mathbf{R}|^{-(s+m)/2}\,x_h^{\,p(s+m)/2}\,h(px_h)$$

and the equality holds if and only if $\boldsymbol{\Sigma} = \mathbf{R}/x_h$. Hence, the likelihood ratio test procedure for testing H_0 against H is to reject H_0 if

$$\lambda = \min_{\Omega}|\mathbf{R}| \,/\, \min_{\Omega_0}|\mathbf{R}| \leq c_a \text{ with } \mathrm{P}(\lambda \leq c_a \,|\, \mathrm{H}_0) = \alpha.$$

REFERENCES

Anderson, T.W. and Fang, K.-T. (1987). Cochran's Theorem for elliptically contoured distributions. [To appear in *Sankhyā Series A*.]

Anderson, T.W., Fang, K.-T. and Hsu, H. (1986). Maximum-likelihood estimates and likelihood-ratio criteria for multivariate elliptically contoured distributions. *Canadian Journal of Statistics*, 14, 55-59.

James, A.T. (1964). Distributions of matrix variates and latent roots derived from normal samples. *Annals of Mathematical Statistics*, 35, 475-501.

Kariya, T. and Sinha, B.K. (1985). Non-null and optimality robustness of some tests. *Annals of Statistics*, 13, 1182-1197.

Khatri, C.G. (1962). Conditions for Wishartness and independence of second degree polynomials in a normal vector. *Annals of Mathematical Statistics*, 33, 239-242.

Khatri, C.G. (1963). Further contributions to Wishartness and independence of second degree polynomials in normal vectors. *Journal of the Indian Statistical Association*, 1, 61-70.

Khatri, C.G. (1970). A note on Mitra's paper, "A density free approach to the matrix variates beta distribution". *Sankhyā, Series A*, 32, 311-318.

Khatri, C.G. (1977). Quadratic forms and extension of Cochran's Theorem to normal vector variables. *Multivariate Analysis IV* (P.R. Krishnaiah, *ed.*). North-Holland, Amsterdam, 79-94.

Khatri, C.G. (1980a). Quadratic forms. *Handbook of Statistics* (P.R.Krishnaiah, *ed.*). North-Holland, Amsterdam, 443-469.

Khatri, C.G. (1980b). The necessary and sufficient conditions for dependent quadratic forms to be distributed as multivariate gamma. *Journal of Multivariate Analysis*, 10, 233-242.

Khatri, C.G. (1982). A theorem on quadratic forms for normal variables. *Statistics and Probability: Essays in Honor of C.R. Rao* (G. Kallianpur, P.R. Krishnaiah and J.K. Ghosh, *eds.*). North-Holland, Amsterdam, 411-417.

Khatri, C.G. (1985). Some remarks on the spherical distributions and linear models. *Linear Statistical Inference* (Proc. International Statistical Conference on Linear Inference, Poznań, Poland, 4-9 June 1984) (T. Caliński and W. Klonecki, *eds.*). Springer-Verlag, New York.

Khatri, C.G. (1987a). Robustness study for a linear growth curve. [To appear in *Journal of Multivariate Analysis.*]

Khatri, C. G. (1987b). Quadratic forms to have a specified distribution. *K.C.S. Pillai's Memorial Volume* (A.K. Gupta, *ed.*). North-Holland, Amsterdam.

Khatri, C.G. and Mukerjee, R. (1987). Characterization of normality within the class of elliptical distributions. *Statistics and Probability Letters*, 5, 187-190.

Lukacs, E. (1956). Characterization of populations by properties of suitable statistics. *Proceedings of the Third Berkeley Symposium on Mathematical Statistics and Probability*, Vol. 2, University of California, 195-214.

Marcinkiewics, J. (1938). Sur les functions independantes III. *Fundamenta Mathematical*, 31, 66-102.

Received 15 October 1986
Revised 26 August 1987

Department of Statistics
Gujarat University
Ahmedabad - 380 009
India

Proc. Second International Tampere Conference in Statistics
(Tampere, Finland, 1-4 June 1987)
Tarmo Pukkila and Simo Puntanen, *Editors*

pp. 205 - 224

INVITED PAPER

Editorial approach in statistical computing

Seppo MUSTONEN
University of Helsinki, Finland

Key words and phrases: Interactive analysis, work station, operating system, SURVO 84.

ABSTRACT

This paper describes an environment for statistical computing, data management, graphics and report generating on a micro computer. The main approach is *editorial mode* which permits the statistician to control all stages of the work by a general text editor. Even the statistical data sets can be written in the edit field which is a visible work sheet on the screen. For large data sets special data files are provided. Information from other sources, like text files, can be easily processed.

The editorial approach is characterized mostly by examples taken from the SURVO 84C system which is entirely based on the editorial approach.

1. INTRODUCTION

SURVO 84 is an integrated interactive system for statistical analysis, computing, graphics and report generation. It also includes features related to spread sheet computing, matrix algebra and computer aided teaching. It provides tools for making application programs in various special areas. All functions are based on the **editorial approach** developed by the author in 1979. The center of the activities is an **edit field** that at all times is partially visible on the screen. The edit field is maintained by the **SURVO 84 Editor.**

The user works with SURVO 84 by typing text in the edit field and by activating various operations and commands written among the text. In many applications it is convenient to create **work schemes** including several extra **specifications,** also written in the text and in arbitrary order.

The data and the results of various operations and application schemes (like plotting schemes and matrix programs) are displayed in the same edit field when required. For more extensive data sets and tables of results SURVO 84 provides its own file representations. SURVO 84 can also communicate with text (ASCII) files.

From the user's point of view SURVO 84 is one huge program which is controlled along certain general principles. The truth is, however, that SURVO 84 is a collection of several technically independent programs (modules) which are called by the SURVO 84 editor according to the user's activations. The user hardly notices the shifting of programs, but sees the system as one integrated world without any need to know its internal structure.

As a collection of programs, SURVO 84 is open for additional modules made by experienced users according to certain rules. These rules and different tools for making modules are described in a separate document *"Programming SURVO 84 modules in C"*. After a new module has been programmed and compiled, the commands and operations defined in it can be used as any standard SURVO 84 operation.

The open structure of SURVO 84 allows calling any other program and using it while staying in SURVO 84. After finishing the job with the other program we shall be back in our current SURVO 84 session again. Because the commands of the operating system can also be employed in this way, SURVO 84 can be considered an extension of the operating system.

The SURVO 84 system may be compared to any extensive text processing program. However, when using SURVO 84 as a word processor we readily have all other activities available, too.

SURVO 84 is also a tool for making new application programs. It provides several ready-made structures and user-friendly "languages" for such tasks. The SURVO 84 **matrix interpreter** and working modes like **tutorial** and **touch mode** are examples of such an approach.

Basically SURVO 84 is intended for professional users, but it is an easy system even for a beginner, since everything is based on simple text editing. Speaking about "ease" in this context may be misleading. If a system is made easy and friendly just for a beginner, after a short learning period it may turn out to be very frustrating for a user who already knows its characteristics.

A good system should be like a musical instrument that requires a lot from its player before yielding its best. If, for example, the violin were invented in recent days, many people would object to its poor "user interface". However, the violin is far more advanced than mechanical, simple musical instruments, since it gives scope for true skills and even for virtuosity.

If one knows the main ideas and working methods of SURVO 84, there is no need to read manuals and user's guides. The best and always up-to-date source of information is the system's own inquiry and help facility, which is readily available during any SURVO 84 session.

Another way to get acquainted with the system is to watch tutorials recorded during normal SURVO 84 sessions. The users can produce such teaching programs on any topic during the work by turning on **tutorial mode.** This permits saving of all actions selected by the user.

The purpose of this paper is to present some typical working methods of SURVO 84. We feel that it is impossible to transmit ideas of an interactive system by structural and theoretical considerations only. Therefore we try to give many practical examples by presenting displays from SURVO 84 sessions. We know, however, that even in this form, a paper is too rigid a medium to give a true picture of a dynamic system.

A historical note

Many of the ideas and principles appearing in SURVO 84C have been adopted from the earlier versions. The first in line was SURVO 66 originated by the author in 1966 and implemented on Elliott 803. One explanation to the name SURVO is the word "survey", since the first SURVO was primarily planned for analysis of survey data. It can also be derived from the Finnish verb *"survoa"* which means *"compress"*. The SURVO 66 jobs were controlled by a simple command language. The original SURVO 66 was further developed in the University of Tampere and is now known under the name of SURVO/71.

In 1976 the first interactive version SURVO 76 was initiated by the author. It was completed in 1984 by him and his research group. Originally SURVO 76 was made in conversational (menu-based) form. The editorial approach was introduced in 1979. SURVO 76 runs on the Wang 2200 minicomputer.

The work on SURVO 84 started in 1984 on the basis of SURVO 76 by using the interpretative Basic language. This was the the first microcomputer version and could be run on the Wang PC only.

The current SURVO 84C system was originated in 1985. From the user's viewpoint it is much like SURVO 84 and it is also highly compatible with SURVO 76. But the latest version is far more efficient and it allows wider applications, since it is programmed in the C language. SURVO 84C can be run on the MS-DOS microcomputers.

2. SURVO 84 EDITOR

The work area for the SURVO 84 EDITOR is an *edit field* which is entirely located in the central memory of the computer. The edit field has typically 100 lines and 100 columns. Each line is preceded by a control symbol for special notations; this control symbol is always initially '*' on each line.

During the work the user can maintain any number of edit fields. However, only one of them is active at a time (in central memory and partially visible on the screen). The others are in edit files on disks, but they can be scanned in a temporary window (by a SHOW operation) and/or loaded partially to the current edit field.

Usually one is working with various edit fields one after another by saving the current field after editing (SAVE operation) and loading another as a whole (LOAD). Thus it is simple to change the active field when necessary.

In its basic form one edit field corresponds to about two pages of normal text. However, in no application it is necessary to identify one edit field with a page or two in some report. When printing documents consisting of several pages the general PRINT operation of SURVO 84 will automatically take care about proper page division (obeying the wishes given by the user, of course). When defining the printout the user simply tells what are the edit fields and chapters in them which belong to the document. Also pictures made earlier by PLOT schemes may be included and positioned automatically.

For example, all the 20 pages of this paper have been produced by a single PRINT activation.

The edit field (as well as the edit files) normally contains text and tables written by the user, various SURVO 84 operations (commands, work schemes etc.) and their results. Representation of various data structures in the same space formed by the edit field is essential and gives exciting possibilities for combining different activities in a creative manner.

When working with larger data sets the space given in the edit field is not enough for the data itself. Although the dimensions of the edit field may be expanded (up to 600 lines with 100 columns, for example, by the REDIM command), it is not wise to create very large edit fields. For big data sets and tables SURVO 84 supplies special data files.

SURVO 84 also supports some other data representations and permits information from other files to be loaded to the edit field. Even text files created by other systems can be processed in SURVO 84. Furthermore SURVO 84 data and results can be easily moved back to text files.

2.1 Control of the edit field

When starting a new job the upper left corner of an empty edit field is displayed on the screen:

```
 1   1 SURVO 84C EDITOR Sat Dec 06 17:18:01 1986               A: 100 100 0
  1 *_
  2 *
  3 *
  4 *
  5 *
  6 *
  7 *
  8 *
  9 *
 10 *
 11 *
 12 *
 13 *
 14 *
 15 *
 16 *
 17 *
 18 *
 19 *
 20 *
 21 *
 22 *
 23 *
```

On the header line some basic information is given like date and time, the data disk drive designation (A:) and the size of the edit field ('100 100' means 100 lines and 100 columns).

The user can now start writing text as on a standard typewriter. The ENTER key moves the cursor to the next line. A new line is initialized automatically when the visible line becomes full. Correspondingly, when the last visible line has been filled, the visible part of the edit field automatically scrolls upwards giving space for a new line at the bottom.

It is always possible to move the cursor in the field and among the text by using the arrow keys or the PgUp (previous page) and PgDn (next page) keys.

Simple editing takes place by typing over the previous text and by using INSERT and DELETE keys for inserting and deleting texts, respectively. To make room for new empty lines between current text lines, the INSERT key in upper shift has to be pressed. DELETE in upper shift deletes the current line entirely.

In all editing functions the foremost principle is to keep them as simple as possible. The best way to learn these simple actions is to practice them at the computer.

2.2 Operations and commands

Because SURVO 84 includes hundreds of activities that all are invoked from the editor, it would be too heavy to do everything by special keys and key combinations. Therefore each more advanced operation is carried out by writing on any free line some command words that are activated by the ESC key.

Assume that we have written in the edit field:

```
 24  1 SURVO 84C EDITOR Mon Jan 19 17:48:14 1987                     A: 100 100 0
  1 *
  2 *(Conover: Practical nonparametric Statistics, Wiley 1971, p.208)
  3 *Twelve sets of identical twins were given psychological tests to
  4 *determine whether the first-born of the twins tends to be more
  5 *aggressive than the other. The results were as follows, where the
  6 *higher score indicates more aggressiveness.
  7 *
  8 *                 1  2  3  4  5  6  7  8  9 10 11 12
  9 *DATA First:    86 71 77 68 91 72 77 91 70 71 88 87 END
 10 *DATA Second:   88 77 76 64 96 72 65 90 65 80 81 72 END
 11 *
 12 *COMPARE First,Second,14_  / TEST=Pairwise
 13 *
 14 *
 15 *
 16 *
 17 *
 18 *
 19 *
 20 *
 21 *
 22 *
 23 *
```

When writing this text our aim is to make a comparison between two paired samples which are typed on lines 9 and 10. The test statistics will be computed by a COMPARE operation (line 12) referring to samples (First,Second) and giving a line number for the results (14). TEST=Pairwise on the same line is an extra specification which determines the nature of comparison.

The COMPARE operation has now been activated and after completing its task the edit field will be changed into form:

```
24  1 SURVO 84C EDITOR Mon Jan 19 17:50:17 1987                      A: 100 100 0
 1 *
 2 *(Conover: Practical nonparametric Statistics, Wiley 1971, p.208)
 3 *Twelve sets of identical twins were given psychological tests to
 4 *determine whether the first-born of the twins tends to be more
 5 *aggressive than the other. The results were as follows, where the
 6 *higher score indicates more aggressiveness.
 7 *
 8 *                1  2  3  4  5  6  7  8  9 10 11 12
 9 *DATA First:    86 71 77 68 91 72 77 91 70 71 88 87 END
10 *DATA Second:   88 77 76 64 96 72 65 90 65 80 81 72 END
11 *
12 *COMPARE First,Second,14_ / TEST=Pairwise
13 *
14 *Paired comparisons:
15 *Samples: N=12            First           Second          Difference
16 *Mean                     79.08333        77.16667        -1.916667
17 *Standard deviation       8.887768        10.37333         7.153617
18 *Paired t=-0.928 (P=0.1866 one-sided t test df=11)
19 *Wilcoxon signed ranks test=-0.756 (P=0.2247 normal approximation)
20 *Critical levels by simulation:
21 *             Differences Signed rank
22 *Critical level 0.19684     0.23807  N=17100
23 *Standard error 0.00304     0.00326
```

Before yielding the final results (on lines 14-23) the COMPARE operation displays various temporary information on the screen. For example, it tries to estimate the critical level of certain tests by Fisher's randomization principle by simulation. N=17100 (on line 22) gives the number of replicates until the user has interrupted the process.

Many commands and operations refer to edit lines (lines of the edit field) like 14 in the previous COMPARE operation. Instead of line numbers also line labels (of one character like A,B,x,y,+,-) in the control column can be used.

Because the operations (commands) are written into the text like any other information, it is easy to edit them and activate again. Thus the commands do not disappear from the the screen (edit field) after they have been completed. The user can scratch them like any text by the ERASE key, for example.

2.3 Work schemes

In demanding SURVO 84 applications various commands and operations together with various extra **specifications** are written as **work schemes** that to some extent resemble programs.

In a typical work scheme not only the activated operation but also the specifications appearing in the edit field may have particular influence. Each operation has its own specification words. If a specification is not given in the edit field, when the operation (work scheme) is activated by ESC, a default value is automatically entered. By using the specification the user can change the default values.

When planning SURVO 84 operations and program modules much attention has been paid to the selection of specifications and their default values. Good solutions in this respect lessen the burden of the user and make the system more intelligent. They form an essential part of the interface between the user and SURVO 84.

All specifications have the form <specification word>=<values>. For example, when making pictures by a PLOT operation the size of the graph is controlled by the SIZE specification. SIZE=1300,800 indicates that the width of the graph will be 1300 units and the height 800 units. If a laser printer is used, one unit equals 0.1 mm and in this case the dimensions are 13 cm by 8 cm.

Below a typical work scheme with the result is displayed. Note the free setting of specifications (no strict order) and the possibility to alter any detail in the scheme before a new activation.

```
13  1 SURVO 84C EDITOR Sat Dec 06 19:03:02 1986                  A: 100 100 0
*
*
*   Population in Finland (1000)
*
*DATA FINLAND
*Sex      0-14 15-24 25-44 45-64 65-
*Males     506   399   727   458   202
*Females   484   381   693   547   353
*
*HEADER=([HIGH]),Population_of_Finland_in_age_groups
*PLOT FINLAND
*SIZE=1300,800       size of the picture (13*8 cm)
*TYPE=PIE            pie chart
*LEGEND=Age:         text before the legend
*SHADING=0,1,5,8,9P  shadings (colors), P=pull out sector
*XDIV=0,1,0          no vertical margins
*
*
*
*
*
*
*
```

When the PLOT operation on line 11 is activated, the following graph is produced on the laser printer:

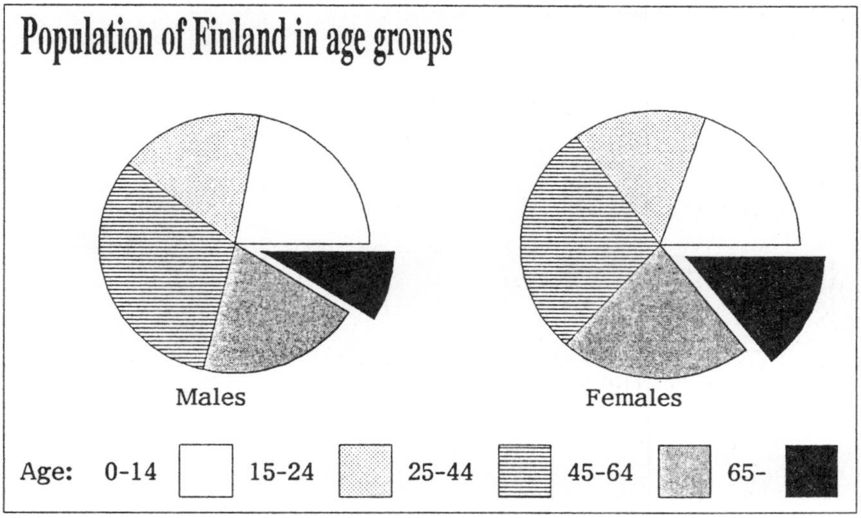

2.4 Subfields

Work schemes which are placed in the same edit field may disturb each other if they are using the same specifications in different ways. To avoid confusions, the schemes may be isolated from each other by typing a border line between them. A border line has the form *.......... (i.e. '*' in the control column followed by at least 10 dots.) If the whole border line (as usual) is filled with dots, the editor displays it as a thicker stripe that clearly reveals the boundary between the work schemes. Then all operations activated between two border lines will adhere to the specifications in this limited area only. The area is called a *subfield.*

In some cases it is desirable to have several work schemes using the same specifications. Those specifications have to be written in the first subfield in the current edit field and this subfield must contain the keyword *GLOBAL* arbitrarily positioned in that subfield. When now some work scheme in the edit field is activated, the specifications are primarily searched for in the local subfield and secondarily in the *GLOBAL* subfield. If neither local nor global specification is found, a default value is maintained.

Below two families of curves are plotted with separate PLOT schemes but using some common specifications given in the *GLOBAL* subfield.

```
12  1 SURVO 84C EDITOR Sun Dec 07 14:03:40 1986                  A: 100 100 0
  1 *
  2 *
  3 *Compound interest on linear and logarithmic scales
  4 *  *GLOBAL* P=5,15,1 SIZE=650,650 PEN=[INDEX]
  5 *           GRID=XY HEADER=
  6 *           XSCALE=0(5)30 XLABEL=Years
  7 *           YLABEL=Rate_of_interest_P=5,6,...,15%
  8 *..........................................................................
  9 *PLOT Y(X)=(1+P/100)^X           / DEVICE=INTER1.CAN
 10 *YSCALE=0(20)100
 11 *..........................................................................
 12 *PLOT Y(X)=(1+P/100)^X           / DEVICE=INTER2.CAN
 13 *YSCALE=*log(y),1,2,5,10,20,50,100
 14 *..........................................................................
 15 *
 16 *PRINT 17,18_
 17 - picture INTER1.CAN,250,100
 18 - picture INTER2.CAN,940,100
 19 *
 20 *
 21 *
 22 *
 23 *
```

The PLOT schemes on lines 9 and 12 save their results in files INTER1.CAN and INTER2.CAN, respectively. The PRINT operation on line 16 yields the following pair of graphs:

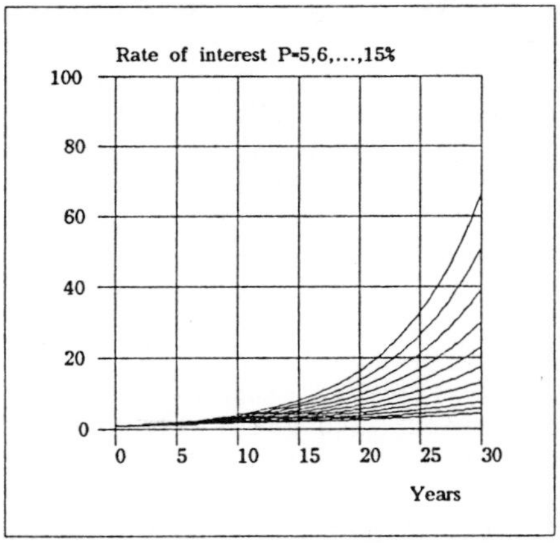

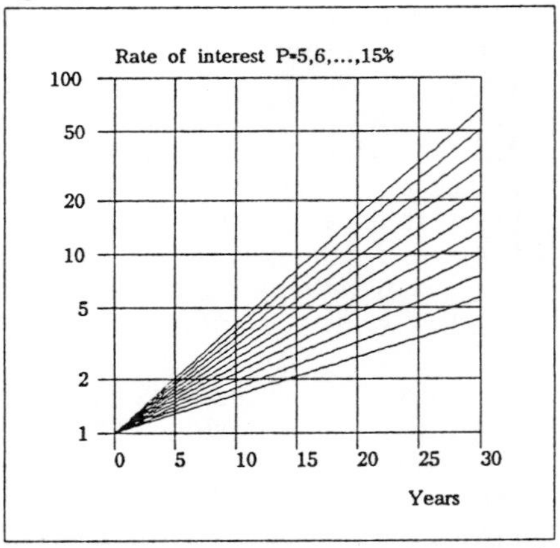

Since the commands, work schemes and other pieces of control information reside in the edit field and are so easy to edit, there is no need to abbreviate keywords as happens in many command languages. By using various copying and editing facilities offered by the editor it is possible to avoid writing of same words several times. It pays to learn how to avoid unnecessary work and how to use old material for new applications. Furthermore, by saving pertinent edit fields the user may create work schemes (or programs) that form a basis for new altered and improved applications.

2.5 Operation sequences

A series of operations and work schemes can be activated by one touch by placing the operations needed on consecutive edit lines. If the first operation is activated by pressing F2:PREFIX and then ESC, then after the execution of the first operation the cursor will automatically be moved to the next line and the operation on that line is activated. The operations are carried out as long as there are feasible operation lines. Usually an empty line is used to interrupt the operation sequence.

It is not necessary that the operations belonging to the operation sequence are written on consecutive lines, since jumps to other lines may be performed by a GOTO command. For example, GOTO 24,30,15 changes the display in the edit field so that edit line 24 will be the first visible one and the cursor will be located on the 15th position of the 30th line. Thus any place in the current edit field can be reached during a sequence of operations.

Likewise jumps to other edit fields may be done by using a LOAD operation. For example, LOAD PART2,1,10 replaces the current edit field by another (PART2) from the data disk and places the cursor there on the 10th line so that the first edit line is the first visible line.

The next display shows a operation sequence that saves a 3x4 matrix A in a matrix file, computes AA' and its inverse matrix and finally displays the result in the edit field.

```
  1  1 SURVO 84C EDITOR Sun Dec 07 14:26:33 1986           A: 100 100 0
 24 *
 25 *Computing the inverse matrix of AA'
 26 *
 27 *MATRIX A ///
 28 *      12.5   4.2  11.0  -8.1
 29 *       0.0  -1.3   5.6   2.4
 30 *       5.2  10.4  -9.3   0.3
 31 *
 32 *MAT SAVE A
 33 *MAT B=MMT(A)           / *B~A*A' S3*3
 34 *MAT B=INV(B)           / *B~INV(A*A') 3*3
 35 *MAT LOAD B,37
 36 *
 37 *MATRIX B
 38 *INV(A*A')
 39 */// 
 40 *  1          0.00347 -0.00662 -0.00200
 41 *  2         -0.00662  0.06308  0.01857
 42 *  3         -0.00200  0.01857  0.00998
 43 *
 44 *
 45 *
 46 *
```

Our operation sequence is on edit lines 32-35 and contains MAT operations only. The situation after the completion of the sequence is displayed. The changes and results due to the operations are indicated in a gray shading. The parameter 37 of the MAT LOAD operation on line 35 refers to the first line of the results.

3. ARITHMETICS AND SPREADSHEET COMPUTING

SURVO 84 provides two different working modes for computing with numbers and tables. Furthermore the statistical operations and general mathematical tools (like the matrix interpreter) give support in more advanced applications. In this section, however, we are mainly considering typical calculations needed in normal work.

Editorial computing permits calculating values for various expressions which have been typed in the edit field. Even more complicated formulae or *computation schemes* (consisting of formulae with instructions) may be entered and activated with given starting values.

In addition to basic arithmetics, various mathematical and statistical functions are readily available. New functions may also be defined by the user either very easily in the edit field or by programming them in the C language.

The functions for editorial computing are also available in the VAR operation (described later) for tasks related to spreadsheet computing. The main applications of VAR are the transformations of variables in statistical data sets.

Touch mode is another unique approach in SURVO 84 for calculations. In this mode various computations are carried out by moving the cursor to touch any number in the edit field and the number is activated by pressing any of the keys +,-,* or / which correspond to standard arithmetical operations. Several numbers can be activated in this way and the resulting numerical expression will appear at each stage on the bottom line of the screen.

To print the current result in the edit field, the cursor is moved to indicate the desired position for output and the key = is pressed.

Touch mode enables very powerful tools for spreadsheet computing, too. Any systematic computing sequence consisting of several steps may be first defined by simply making the sequence in touch mode for the first case. After this definition stage the sequence may be automatically repeated for remaining cases by a single activation. These computation sequences, *touch chains*, can also be saved and used later when needed.

Touch mode provides the same mathematical and statistical library functions as editorial computing.

3.1 Examples on editorial computing

Arithmetic expressions are typed in the edit field according to normal mathematical notation.

For example, to calculate the arithmetic mean of numbers 12, 17 and 25 we enter:

```
 22  1 SURVO 84C EDITOR Sat Jan 03 17:08:54 1987                  A: 100 100 0
  1 *
  2 *
  3 *         (12+17+25)/3=_
  4 *
  5 *
```

Now, when the cursor is blinking immediately after =, we press the activation key ESC. Because there is no command on the current line the SURVO 84 Editor studies the character just before the cursor position. If it is = as in this case, the editor assumes that the user wants to calculate something and calls the editorial computing module which analyzes the current expression, computes its value and writes the result in the edit field. Finally the control is transferred back to the editor and we may continue the work.

In this case we obtain immediately the following display:

```
22  1 SURVO 84C EDITOR Sat Jan 03 17:08:54 1987                A: 100 100 0
  1 *
  2 *
  3 *         (12+17+25)/3=18
  4 *
  5 *
```

When same numbers are used for several computations or when more general expressions are wanted, we may type also

```
31  1 SURVO 84C EDITOR Sat Jan 03 17:45:43 1987                A: 100 100 0
  1 *
  2 *         X=12 Y=17 Z=25
  3 * Arithmetic mean is (X+Y+Z)/3=_
  4 * Geometric mean is (X*Y*Z)^(1/3)=
  5 *
```

After activating both expressions we get the following display:

```
34  1 SURVO 84C EDITOR Sat Jan 03 17:45:43 1987                A: 100 100 0
  1 *
  2 *         X=12 Y=17 Z=25
  3 * Arithmetic mean is (X+Y+Z)/3=18
  4 * Geometric mean is (X*Y*Z)^(1/3)=17.2130062073
  5 *
```

We can also define temporary functions like AM and GM functions in the next display and activate several expressions simultaneously by having .= instead of = at the end of expressions:

```
21  1 SURVO 84C EDITOR Sun Jan 04 18:27:00 1987                A: 100 100 0
  1 *
  2 *
  3 * Arithmetic mean  AM(X,Y,Z):=(X+Y+Z)/3
  4 * Geometric mean   GM(X,Y,Z):=(X*Y*Z)^(1/3)
  5 *
  6 *    AM(12,17,25).=
  7 *    GM(12,17,25).=
  8 *
  9 *    A=11.5 B=14.7 C=16.1
 10 *    AM(A,B,C).=
 11 *    GM(A,B,C).=
 12 *    AM(A+1,B+1,C+1).=
 13 *    GM(A+1,B+1,C+1).=_
 14 *
```

The definitions of function AM and GM appear on lines 3 and 4. On lines 6-13 several expressions using these functions have been written. To evaluate these expressions, it is enough (since all of them are tailed by .=) to activate just one of them and we shall have

```
 21  1 SURVO 84C EDITOR Sun Jan 04 18:27:00 1987              A: 100 100 0
  1 *
  2 *
  3 * Arithmetic mean   AM(X,Y,Z):=(X+Y+Z)/3
  4 * Geometric mean    GM(X,Y,Z):=(X*Y*Z)^(1/3)
  5 *
  6 *   AM(12,17,25).=18
  7 *   GM(12,17,25).=17.2130062073
  8 *
  9 *   A=11.5 B=14.7 C=16.1
 10 *   AM(A,B,C).=14.1
 11 *   GM(A,B,C).=13.9619801763
 12 *   AM(A+1,B+1,C+1).=15.1
 13 *   GM(A+1,B+1,C+1).=14.971612979
 14 *
```

3.2 Computation schemes

It is always up to the user how he/she organizes the computations when working with the SURVO 84 Editor. In many applications all the formulas and operations needed for reaching a specific goal can be expressed as a **computation scheme** which to some extent resembles a computer program. One clear distinction is, however, that in a computation scheme there is no specific order of statements.

Each activation in a SURVO 84 work scheme leads always to a search process where the editor and other programs called for help are looking for the information needed for carrying out the task activated by the user.

In fact all the previous examples of editorial computing have been computation schemes in a modest sense. A real computation scheme, however, usually contains instructions and comments to help the user in applying the scheme.

For example, testing of correlation coefficient by using Fisher's z transformation could be represented as a computation scheme as follows:

```
 33  1 SURVO 84C EDITOR Sun Jan 11 12:22:39 1987                 A: 100 100 0
 47 *  Testing the correlation coefficient
 48 *  The sample correlation coefficient is r and the sample size n.
 49 *  To test the hypothesis that in the population the unknown
 50 *  correlation coefficient rho is r0 against the alternative rho>r0,
 51 *  we form the test statistic
 52 *         U=sqrt(n-3)*(Fisher(r)-Fisher(r0))
 53 *  where
 54 *         Fisher(r):=0.5*log((1+r)/(1-r))
 55 *  is Fisher's transformation of the correlation coefficient.
 56 *
 57 *  If the null hypothesis is true, U is approximately N(0,1).
 58 *  Hence we reject the hypothesis, if P=1-N.F(0,1,U)
 59 *  is less than the risk level (say 0.05).
 60 *
 61 *  Assume now that n=25,  r=0.85 and r0=0.7
 62 *
 63 *  Then U.=1.8238788825   and P.=0.03408519244644
 64 *
 65 *  Thus if P<0.05, we reject the hypothesis that correlation
 66 *  coefficient in the population is r0.=0.7
 67 *..............................................................
 68 *  Instructions: Insert your own values on line 61 and
 69 *                activate P.= on line 63, for example.
```

Above the notation P=1-N.F(0,1,U) on line 58 refers to a library function N.F() which is the standard normal distribution function.

4. DATA ANALYSIS

Tasks related to statistical analysis and statistical computing can be performed in SURVO 84C by various statistical operations. Certainly there is no clear-cut difference between genuine statistical operations and others. For example, the general calculating techniques (editorial computing and touch mode) and the VAR operation for making transformed variables are helpful in many tasks. Similarly SURVO 84C graphics and matrix operations are essential tools in statistical applications.

There are, however, certain structural factors which bind actual statistical procedures together in SURVO 84C.

The primary data values (samples etc.) are always represented as data lists or tables in the edit field or as data files on disk. Data from other sources, like text files, may be easily converted into SURVO 84C representation by special FILE operations.

In some statistical methods, however, the computations include several steps where the output of one step should be studied carefully before continuing with the next step. The common trend in the successive steps of the analysis is to compress the original data to sufficient statistics which often can be represented as matrices with lower dimensions than the raw data.

For example, in standard multivariate analysis based on multinormality of original variables the means, standard deviations and correlations form sufficient statistics. Then matrix files consisting of those statistics can be used as a basis for computations as well. On the other hand it is good to remember that there are often better computational techniques (based, for example, on orthogonalization of data) which do not rely on correlations at all.

In any case a statistical system should provide tools for operating with various intermediate results in matrix form as easily as with primary data. In SURVO 84C we have a large subsystem, the matrix interpreter, for these tasks. Many SURVO 84 operations related to linear models and multivariate analysis are making their matrix computations through the matrix interpreter. An advanced user may as well use the matrix interpreter directly by the MAT commands and chains.

In addition to standard data representation there are other traits which connect various statistical methods in SURVO 84C. Although any statistical operation may have its own special requirements, it is important that the user can expect each operation to work according to the same style as the neighbouring operations at least to some extent.

In sequel we shall present some typical applications. All these examples are based on artificial data sets (files) generated by the VAR operation.

4.1 Correlations

In the next exhibit 200 observations from a first-order autoregressive process are generated and then autocorrelations for lags 1,2,3,4,5 are computed.

The FILE CREATE scheme on lines 2-6 creates a data file AUTO1 for 200 observations of variable X. The VAR operation on line 8 generates the observations. Another VAR scheme on lines 11 and 12 computes the lagged variables and finally CORR AUTO1,15 on line 14 computes the means, standard deviations and correlations. The results are displayed from line 15 onwards. Furthermore they are saved in full precision in matrix files for subsequent analysis.

```
14  1 SURVO 84C EDITOR Sun Feb 22 11:06:33 1987              D:\STAT\ 120  80 0
*
*FILE CREATE AUTO1,24,6,64,7,200
*  200 observations from X(t)=0.8*X(t-1)+eps
*FIELDS:
*1 N 4 X
*END
*
*VAR X=if(ORDER=1)then(eps)else(0.8*X[-1]+eps) TO AUTO1
*       eps=probit(rnd(1))
*..................................................................
*VAR X1,X2,X3,X4,X5 TO AUTO1
*X1=X[-1] X2=X[-2] X3=X[-3] X4=X[-4] X5=X[-5]
*..................................................................
*CORR AUTO1,15
*Means, std.devs and correlations of AUTO1  N=200
*# of missing observations =5
*Variable  Mean        Std.dev.
*X        -0.106722    1.756378
*X1       -0.102504    1.753710
*X2       -0.102143    1.753462
*X3       -0.104000    1.756002
*X4       -0.106121    1.758365
*X5       -0.110815    1.764946
*Correlations:
*              X       X1       X2       X3       X4       X5
* X          1.0000  0.8092  0.6211  0.4683  0.3556  0.2878
* X1         0.8092  1.0000  0.8088  0.6185  0.4665  0.3531
* X2         0.6211  0.8088  1.0000  0.8082  0.6184  0.4664
* X3         0.4683  0.6185  0.8082  1.0000  0.8087  0.6203
* X4         0.3556  0.4665  0.6184  0.8087  1.0000  0.8093
* X5         0.2878  0.3531  0.4664  0.6203  0.8093  1.0000
*
```

4.2 Fitting a univariate distribution

We shall study a mixture of two normal distribution of the form p*N(m1,s1^2)+(1-p)*N(m2,s2^2). We generate 10000 observations with parameters p=0.7, m1=0, s1=1, m2=2, s2=0.5 and try to re-estimate them from starting values INIT=0.5,-1,1.5,3,1.

A file SIMUDATA is first created for 10000 observations of variable X and values from the mixture are then computed by the VAR scheme on lines 8-11.

To estimate the parameters from the sample a frequency distribution of X is formed by a HISTO operation on line 17. The FIT=MIXNORM specification on line 19 implies HISTO to fit the MIXNORM distribution defined on lines 14-16 to SIMUDATA. The results of estimation are displayed on lines 22-35.

```
20  1 SURVO 84C EDITOR Fri Feb 27 15:35:41 1987              D:\STAT\ 120  80 0
 1 *
 2 *FILE CREATE SIMUDATA,4,1,64,7,10000
 3 *  Sample (N=10000) from a mixture of two normal distributions
 4 *FIELDS:
 5 *1 N 4 X
 6 *END
 7 *
 8 *VAR X TO SIMUDATA
 9 *  X=if(rnd(1)<0.7)then(X1)else(X2)
10 *  X1=probit(rnd(1))
11 *  X2=0.5*probit(rnd(1))+2
12 *..............................................................
13 *
14 *DENSITY MIXNORM(p,m1,s1,m2,s2)
15 *y(x)=c*(p/s1*exp(-0.5*((x-m1)/s1)^2)+(1-p)/s2*exp(-0.5*((x-m2)/s2)^2))
16 *      c=0.39894226
17 *HISTO SIMUDATA,X,22
18 *X=-6(0.2)6 XSCALE=-6(1)6 YSCALE=0(100)600
19 *FIT=MIXNORM INIT=0.5,-1,1.5,3,1
20 *SIZE=1300,800
21 *..............................................................
22 *HISTO: Estimated parameters of MIXNORM:
23 *p=0.7003 (0.0125)
24 *m1=0.0301 (0.0296)
25 *s1=1.0077 (0.0193)
26 *m2=2.0095 (0.0180)
27 *s2=0.4906 (0.0137)
28 *logL=16043.335886  # of function evaluations =311
29 *Correlations:
30 *              p      m1      s1      m2      s2
31 * p        1.000  0.845  0.796  0.728 -0.707
32 * m1       0.845  1.000  0.802  0.673 -0.661
33 * s1       0.796  0.802  1.000  0.572 -0.591
34 * m2       0.728  0.673  0.572  1.000 -0.700
35 * s2      -0.707 -0.661 -0.591 -0.700  1.000
```

HISTO also lists the frequency distribution and various related statistics as follows

```
20  1 SURVO 84C EDITOR Fri Feb 27 15:35:41 1987              D:\STAT\ 120  80 0
36 *Frequency distribution of X in SIMUDATA: N=10000
37 *
38 *Class midpoint   f     %    Sum     %     e      e      f     X^2
39 *      <=-3.6     0   0.0      0   0.0    1.1
40 *        -3.5     2   0.0      2   0.0    1.2
41 *        -3.3     1   0.0      3   0.0    2.4
42 *        -3.1     6   0.1      9   0.1    4.5    9.3     9     0.0
43 *        -2.9    12   0.1     21   0.2    8.2    8.2    12     1.7
44 *        -2.7    14   0.1     35   0.4   14.3   14.3    14     0.0
   --------------
71 *         2.7   199   2.0   9829  98.3  199.7  199.7   199     0.0
72 *         2.9    93   0.9   9922  99.2  105.9  105.9    93     1.6
73 *         3.1    44   0.4   9966  99.7   48.4   48.4    44     0.4
74 *         3.3    18   0.2   9984  99.8   19.2   19.2    18     0.1
75 *         3.5    11   0.1   9995 100.0    6.8
76 *         3.7     4   0.0   9999 100.0    2.2
77 *         3.9     1   0.0  10000 100.0    0.7
78 *        > 4.0    0   0.0  10000 100.0    0.4   10.0    16     3.6
79 *Mean=0.623320 Std.dev.=1.267161
80 *Fitted by MIXNORM(0.7003,0.0301,1.0077,2.0095,0.4906) distribution
81 *Chi-square=27.48 df=28 P=0.4921
```

and plots the graph:

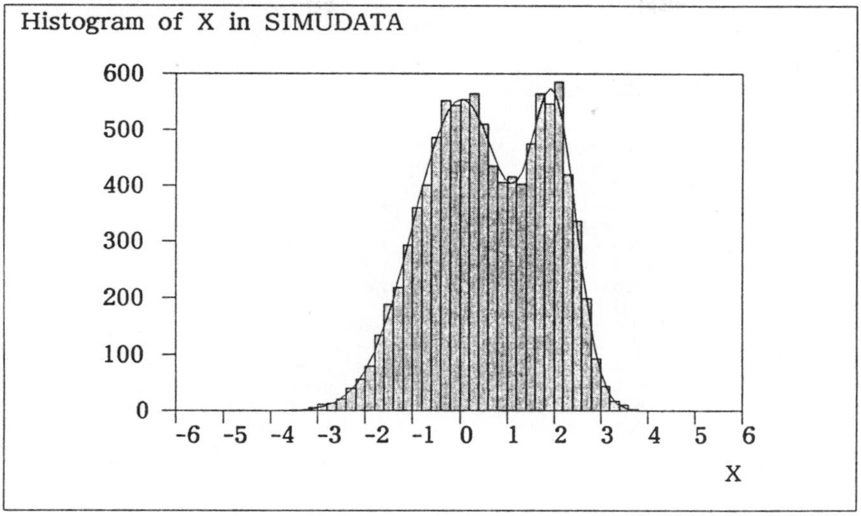

4.3 Nonlinear regression

Our last example deals with a sample of 100 observations from a time series with two sinusoidal components, different amplitudes and phase shifts. Also a noteworthy noise is included. The data set is generated as follows:

```
15  1 SURVO 84C EDITOR Sun Mar 01 13:39:27 1987          C:\S\ 100 100 0
 1 *
 2 *FILE CREATE SOUND,8,2,64,7,100
 3 *
 4 *FIELDS:
 5 *1 N 4 t
 6 *2 N 4 Y
 7 *END
 8 *
 9 *c=2 a1=1 f1=0.1 s1=0 a2=0.7 f2=0.15 s2=2
10 *
11 *VAR t,Y TO SOUND
12 *  t=ORDER-1 Y=c+a1*sin(f1*t+s1)+a2*sin(f2*t+s2)+eps
13 *  eps=0.5*probit(rnd(1))
14 *
```

We shall smoothen this time series by using the ESTIMATE operation which can be applied to various nonlinear estimation problems. It is the user's task to specify the model (see lines 25-26 below). The default method in ESTIMATE is the ordinary least squares technique. ESTIMATE also forms analytically the first and second partial derivatives of the model function and this information is employed for selecting a proper optimization algorithm as well as for computing the gradient and the Hessian matrix.

Starting from crude initial values (on line 28) ESTIMATE gives the following results:

```
 27  1 SURVO 84C EDITOR Sun Mar 01 14:13:08 1987                    C:\S\ 100 100 0
 24
 25 *MODEL INTERVAL
 26 *Y=c+a1*sin(f1*t+s1)+a2*sin(f2*t+s2)
 27 *..........................................................................
 28 *c=2.1 a1=1.1 f1=0.07 s1=1 a2=0.6 f2=0.18 s2=1
 29 *ESTIMATE SOUND,INTERVAL,31
 30 *..........................................................................
 31 *Estimated parameters of model INTERVAL:
 32 *c=1.993927 (0.058301)
 33 *a1=0.922475 (0.116705)
 34 *f1=0.104486 (0.004641)
 35 *s1=-0.300844 (0.241594)
 36 *a2=0.846823 (0.107104)
 37 *f2=0.154264 (0.006873)
 38 *s2=1.540947 (0.333908)
 39 *n=100 rss=27.238543 R^2=0.70012 nf=424
 40 *Correlations:
 41 *                 c      a1      f1      s1      a2      f2      s2
 42 * c           1.000  -0.243   0.054  -0.072  -0.060   0.064  -0.012
 43 * a1         -0.243   1.000   0.519  -0.192   0.509  -0.582   0.342
 44 * f1          0.054   0.519   1.000  -0.772   0.530  -0.390   0.035
 45 * s1         -0.072  -0.192  -0.772   1.000  -0.140  -0.194   0.498
 46 * a2         -0.060   0.509   0.530  -0.140   1.000  -0.589   0.412
 47 * f2          0.064  -0.582  -0.390  -0.194  -0.589   1.000  -0.894
 48 * s2         -0.012   0.342   0.035   0.498   0.412  -0.894   1.000
 49 *
```

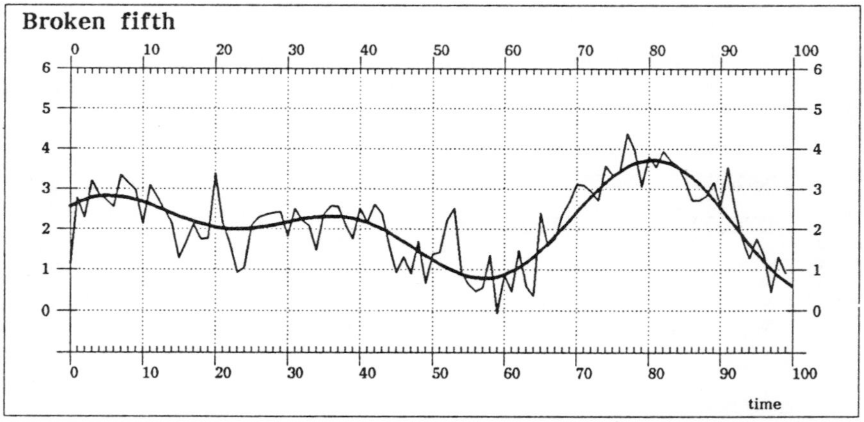

REFERENCES

Mustonen, S. (1980). Interactive analysis in SURVO 76. *Proceedings in Computational Statistics*, ed.by M.M.Barritt and D.Wishart, 253-259. Physica Verlag, Wien.

Mustonen, S. (1981). Statistical computing with a text editor. *Computational Statistics*, ed.by H.Büning and P.Naeve, 327-348. Walter de Gruyter, Berlin.

Mustonen, S. (1982). Statistical computing based on text editing. *Proceedings in Computational Statistics*, ed.by H.Caussinus, P.Ettinger and R.Tomassone, 353-358. Physica Verlag, Wien.

Received 10 March 1987

Department of Statistics
University of Helsinki
Aleksanterinkatu 7
SF-00100 Helsinki
Finland

Proc. Second International Tampere Conference in Statistics
(Tampere, Finland, 1-4 June 1987)
Tarmo Pukkila and Simo Puntanen, *Editors*

pp. 225 - 243

INVITED PAPER

Consistency of invariant tests for the multivariate analysis of variance

T.W. ANDERSON and **Michael D. PERLMAN**
Stanford University, Stanford, CA, U.S.A.
and University of Washington, Seattle, WA, U.S.A.

Key words and phrases: MANOVA, invariant test, consistency, power, monotone acceptance region, noncentral Wishart matrix, characteristic roots.

ABSTRACT

Two notions of consistency of invariant tests for the MANOVA testing problems are examined and compared: sample size consistency (the classical notion) and parameter consistency, which requires that for fixed sample size, the power of the test approaches one for any sequence of alternatives whose distance from the null hypothesis approaches infinity. The Roy, Lawley-Hotelling, and likelihood ratio (≡ Wilks) tests are consistent in both senses, whereas the Bartlett-Nanda-Pillai trace test, although sample size consistent, is not parameter consistent unless the significance level α or the error degrees of freedom n is sufficiently large.

1. INTRODUCTION

Consider the standard multivariate linear regression model (cf. Anderson (1984), Chapter 8):

$$Z = \beta D + \varepsilon , \tag{1.1}$$

where $Z : p \times N$ is the matrix of observations, $\beta : p \times q$ is the matrix of unknown regression coefficients, $D : q \times N$ is the design matrix, and $\varepsilon : p \times N$ is the matrix of unobservable random errors. Assume that $q \leq N$, D is of full rank q, and that

$$\varepsilon \sim \mathrm{N}(0 , \Sigma \otimes I_N) , \tag{1.2}$$

which indicates that the N columns of ε are mutually independent p-variate normal random vectors with zero mean and common unknown covariance matrix $\Sigma: p \times p$, assumed positive definite. We shall study the consistency of invariant tests of the general linear hypothesis

$$H_0: \beta D_0 = 0, \tag{1.3}$$

where $D_0: q \times r$ has full rank r $(1 \le r \le q)$. If β is partitioned as (β_1, β_2) with $\beta_1: q \times r$, an important special case of (1.3) is $H_0: \beta_1 = 0$.

The general multivariate analysis of variance (MANOVA) testing problem is that of testing (1.3) vs. (1.1). It is well known that this testing problem can be reduced by sufficiency and invariance to the following canonical form (cf. Anderson (1984), Section 8.3.3 or Lehmann (1986), Sections 8.1, 8.2): based on the independent observations

$$\begin{aligned} X(p \times r) &\sim \mathrm{N}(\xi, \Sigma \otimes I_r) \\ Y(p \times n) &\sim \mathrm{N}(0, \Sigma \otimes I_n), \end{aligned} \tag{1.4}$$

where $\xi : p \times r$ is a matrix of unknown means, test

$$H_0: \xi = 0 \quad \text{vs.} \quad H: \xi \neq 0 \qquad (\Sigma \text{ unknown}). \tag{1.5}$$

We assume that $n\ (\equiv N - q) \ge p$, so that $\hat{\Sigma} \equiv \frac{1}{n} YY'$ is positive definite with probability one.

The canonical testing problem given by (1.4) and (1.5) is invariant under the group of nonsingular linear transformations

$$(X, Y) \to (BX\Psi_1, BY\Psi_2), \tag{1.6}$$

where $B: p \times p$ is nonsingular and $\Psi_1: r \times r$, $\Psi_2: n \times n$ are orthogonal. Under (1.6), the parameters of the model (1.4) are transformed according to

$$(\xi, \Sigma) \to (B\xi\Psi_1, B\Sigma B'). \tag{1.7}$$

The maximal invariant statistic and maximal invariant parameter may be represented as

$$\begin{aligned} c &\equiv c(X,Y) \equiv (c_1,\ldots,c_t) \\ \lambda &\equiv \lambda(\xi,\Sigma) \equiv (\lambda_1,\ldots,\lambda_t)\,, \end{aligned} \tag{1.8}$$

respectively, where $t = p \wedge r$,

$$\begin{aligned} c_i &= ch_i[XX'(YY')^{-1}] \geq 0 \\ \lambda_i &= ch_i[\xi\xi'\Sigma^{-1}] \geq 0\,, \end{aligned} \tag{1.9}$$

and where, for any real symmetric matrix S, $ch_1(S) \geq ch_2(S) \geq \cdots$ denote its (ordered) characteristic roots (necessarily real). It will be convenient also to use the equivalent representations

$$\begin{aligned} d &\equiv d(c) \equiv (d_1,\ldots,d_t) \\ \delta &\equiv \delta(\lambda) \equiv (\delta_1,\ldots,\delta_t) \end{aligned} \tag{1.10}$$

of the maximal invariant statistic and parameter, respectively, where

$$\begin{aligned} d_i &= \frac{c_i}{c_i+1} = ch_i[XX'(XX'+YY')^{-1}] \\ \delta_i &= \frac{\lambda_i}{\lambda_i+1} = ch_i[\xi\xi'(\xi\xi'+\Sigma)^{-1}]\,. \end{aligned} \tag{1.11}$$

Note that

$$\begin{aligned} c,\lambda \in \mathbf{C}_t &\equiv \{x \equiv (x_1,\ldots,x_t) \mid \infty > x_1 \geq \cdots \geq x_t \geq 0\} \\ d,\delta \in \mathbf{D}_t &\equiv \{x \equiv (x_1,\ldots,x_t) \mid 1 > x_1 \geq \cdots \geq x_t \geq 0\}\,. \end{aligned} \tag{1.12}$$

(More precisely, $c \in \mathbf{C}_t$ and $d \in \mathbf{D}_t$ with probability one.)

The MANOVA problem (1.5) may be expressed in the following equivalent form: test

$$H_0: \lambda = (0,\ldots,0) \quad \text{vs.} \quad H: \lambda \in \mathbf{C}_t\,,\ \lambda \neq (0,\ldots,0)\,. \tag{1.13}$$

We shall be concerned with the consistency of *invariant* tests for (1.5) ≡ (1.13), i.e., tests that depend upon (X,Y) only through c (or, equivalently, through d) and whose power functions therefore depend upon (ξ,Σ) only through $\lambda \equiv (\lambda_1,\ldots,\lambda_t)$, the vector of *noncentrality parameters*. Since c_i estimates λ_i, a "good" invariant test should accept (reject) H_0 for small (large) values of $c_1,\ldots,c_t$ (equivalently, of

$d_1, \ldots, d_t$). In fact, Schwartz (1967b) has shown that every admissible invariant test for (1.5) ≡ (1.13) must have a *monotone acceptance region* A in terms of c or (equivalently) d. That is (in terms of d), if $d \equiv (d_1, \ldots, d_t) \in A \subseteq \mathbf{D}_t$ and $d' \equiv (d_1', \ldots, d_t') \in \mathbf{D}_t$ is such that $d' \leq d$ (i.e., $d_1' \leq d_1, \ldots, d_t' \leq d_t$), then $d' \in A$. Therefore, we shall restrict our attention to the class of *monotone invariant tests*, i.e., those with monotone acceptance regions.

Perlman and Olkin (1980) showed that every monotone invariant test is *unbiased* for testing H_0 vs. H. The criterion of unbiasedness, therefore, does not distinguish among admissible invariant tests. Likewise, neither does the classical notion of consistency, which we shall call *sample size consistency* (SSC). In this paper we introduce the notion of *parameter consistency* and show that it *does* distinguish among admissible invariant tests.

An invariant level α test with acceptance region A is said to be *parameter consistent* (PC) if, for fixed p, r, n, and α, its power

$$P_\lambda\{d \notin A\} \rightarrow 1$$

as one or more noncentrality parameters $\lambda_i \rightarrow \infty$. It will be seen that the well-known Bartlett-Nanda-Pillai trace test, which is both admissible and the locally most powerful invariant test for H_0 vs. H, fails to be PC unless α or n is sufficiently large, whereas the Roy maximum root test, the Lawley-Hotelling trace test, and the likelihood ratio test (≡ Wilks criterion) are both admissible and PC for every α and n (cf. Section 4).

It is important to notice that parameter consistency is defined in terms of the power of a *single* invariant level α acceptance region A at *sequences* of alternatives $\{\lambda\}$ with $\|\lambda\| \rightarrow \infty$, whereas sample size consistency is defined in terms of the limiting power of a *sequence* of invariant level α acceptance regions $\{A^{(n)}\}$ at a *fixed* alternative $\lambda^* \neq (0, \ldots, 0)$. Usually the sequence $\{A^{(n)}\}$ is defined in terms of a single invariant test statistic $f \equiv f(d)$ as follows:

$$A_f^{(n)} = A_f(c_\alpha) \equiv \{d \in \mathbf{D}_t \mid f(d) \leq c_\alpha\}, \tag{1.14}$$

where $c_\alpha \equiv c_\alpha(p, r, n; f)$ satisfies

$$P_{\lambda=0}\{f(d) \le c_\alpha\} = 1 - \alpha. \tag{1.15}$$

If f is *monotone* on $\mathbf{D}_t$ (i.e., nondecreasing in each d_i) then $A_f(c_\alpha)$ is a monotone invariant acceptance region with power function given by $P_\lambda\{f(d) > c_\alpha\}$.

Necessary and sufficient conditions for the parameter consistency of monotone invariant tests are presented in Section 2, while sample size consistency is defined and characterized in Section 3. The relation between parameter consistency and sample size consistency of monotone invariant tests, in particular admissible invariant tests, is examined in Section 4. Few detailed proofs are given, as this paper is primarily expository. The proofs, together with extensions of the results to related multivariate testing problems, will appear in Anderson and Perlman (1988).

2. PARAMETER CONSISTENCY OF MONOTONE INVARIANT TESTS

In a general hypothesis-testing problem, a level α test of H_0 vs. H is said to be parameter consistent if, for fixed sample size, its power approaches one for sequences of alternatives in H whose Kullback-Leibler discrimination distance from H_0 becomes arbitrarily large. For the canonical MANOVA problem (1.5) ≡ (1.13), this definition is equivalent to the following:

DEFINITION 2.1. For fixed p, r, n, and α, an invariant level α test for (1.5) ≡ (1.13) is *parameter consistent* (PC) if its power at $\lambda \equiv (\lambda_1, \ldots, \lambda_t)$ approaches 1 as $\|\lambda\| \to \infty$, where $\|\lambda\| = \Sigma_1^t \lambda_i = \mathrm{tr}\, \xi\xi'\Sigma^{-1}$. For $i = 1, \ldots, t$, the test is *parameter consistent of degree i* (PC(i)) if its power at λ approaches 1 as $\lambda_i \to \infty$. □

Since $\lambda_1 \ge \cdots \ge \lambda_t \ge 0$, obviously PC($i$) $\Longrightarrow$ PC(i+1), and PC $\Longleftrightarrow$ PC(1). It will be seen in Section 4 that parameter consistency is not equivalent to sample size consistency.

In order to study the power of an invariant test at the alternative $\lambda = (\lambda_1, \ldots, \lambda_t)$ we may assume that $(\xi, \Sigma) = (\mu, I_p)$, where $\mu : p \times r$ is any matrix such that $ch_i(\mu\mu') = \lambda_i$, $1 \le i \le t$, and where I_p is the $p \times p$ identity matrix. Under this assumption XX' and YY' have standard Wishart distributions, noncentral and central respectively, and are independent. Define

$$l_i \equiv l_i(X) \equiv ch_i(XX'), \quad 1 \le i \le t .$$

Our characterization in Theorem 2.3 of parameter consistency for monotone invariant tests is based upon the following technical result:

LEMMA 2.2.

(i) $l_i \xrightarrow{p} \infty$ iff $\lambda_i \to \infty$.

(ii) $c_i \xrightarrow{p} \infty$ iff $\lambda_i \to \infty$.

(iii) $d_i \xrightarrow{p} 1$ iff $\lambda_i \to \infty$.

PROOF. The result (i) follows from appropriate stochastic bounds for l_i in terms of λ_i, (ii) follows from (i) by conditioning on Y, while (ii) and (iii) are equivalent by (1.11). See Anderson and Perlman (1988) for details. □

It is most convenient to state our results for acceptance regions A defined in terms of the statistic d. For any subset $A \subseteq \mathbf{D}_t$ (recall (1.12)), we denote the closure of A in $\overline{\mathbf{D}}_t$ by $\overline{A}$, where

$$\overline{\mathbf{D}}_t = \{x \mid 1 \ge x_1 \ge \cdots \ge x_t \ge 0\} . \tag{2.1}$$

If A is monotone then $\overline{A}$ is also monotone and $\overline{A} \setminus A$ has Lebesgue measure 0. Since the distribution of d is absolutely continuous with respect to Lebesgue measure for every $\lambda \in H_0 \cup H$, the power (and hence the consistency) of the invariant test with acceptance region A is the same as that of the test with acceptance region $\overline{A}$.

Let $e_0, e_1, \ldots, e_t$ denote the vertices of $\overline{\mathbf{D}}_t$, i.e.,

$$e_i = (\underbrace{1, \ldots, 1}_{i}, \underbrace{0, \ldots, 0}_{t-i}) , \tag{2.2}$$

and, for $0 \le \eta \le 1$, define $e_i(\eta) \in \overline{\mathbf{D}}_t$ by

$$e_i(\eta) = e_i + \eta(e_t - e_i) = (\underbrace{1,\ldots,1}_{i}, \underbrace{\eta,\ldots,\eta}_{t-i}) . \tag{2.3}$$

If A is a monotone subset of $\mathbf{D}_t$, then $e_i(\eta) \in \overline{A}$ implies $e_{i-1}(\eta) \in \overline{A}$.

THEOREM 2.3. Fix p, r, n, and α, let $A \subseteq \mathbf{D}_t$ be a monotone level α acceptance region for the testing problem (1.5) ≡ (1.13), and fix $i \in \{1, \ldots, t\}$. A necessary and sufficient condition that the invariant test with acceptance region A be PC(i) is that $e_i(\eta) \notin \overline{A}$ for all $\eta > 0$. A sufficient condition is that $e_i \notin \overline{A}$. □

If the upper boundary of A is not too irregular, Theorem 2.3 essentially states that the test with acceptance region A is PC(i) if either $e_i \notin \overline{A}$ (cf. Fig. 2.1a) or e_i lies in the upper boundary of A (Fig. 2.1b), whereas it fails to be PC(i) if $e_i \in A$ but $e_i \notin$ (upper boundary of A) (Fig. 2.1c). Thus, the test is PC(i) if e_i lies either in the *rejection* region or in its lower boundary. These three cases are illustrated in Figures 2.1a,b,c where $i = 1$ and $t = 2$.

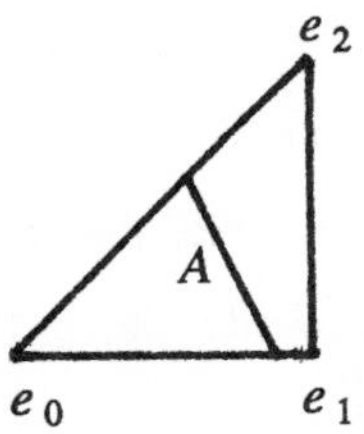

Fig. 2.1a:
test is PC(1)

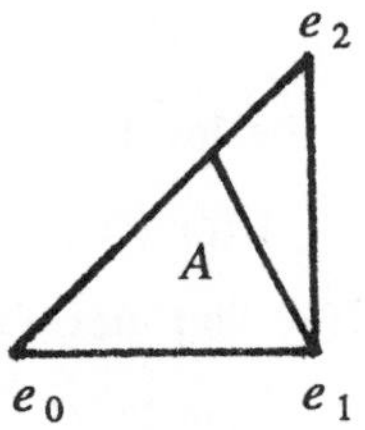

Fig. 2.1b:
test is PC(1)

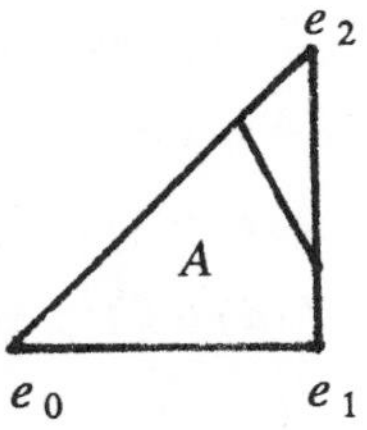

Fig. 2.1c:
test is *not* PC(1)

Note that if $\alpha > 0$, every monotone invariant test must be PC(t).

It is convenient to restate Theorem 2.3 for the case where $A = A_f(c_\alpha)$ as in (1.14) and f is a monotone invariant test statistic defined on $\mathbf{D}_t$. For any extended real-valued function f on $\mathbf{D}_t$ define $\overline{f}$ on $\overline{\mathbf{D}}_t$ as follows:

$$\overline{f}(x) = f(x-) \equiv \lim_{\varepsilon \downarrow 0} f((1-\varepsilon)x) . \tag{2.4}$$

Then $\overline{f}$ is defined (possibly extended real-valued), monotone, and lower semicontinuous on $\bar{\mathbf{D}}_t$, hence $A_{\bar{f}}(c_\alpha) \equiv \overline{A_f(c_\alpha)}$ is a closed and monotone subset of $\bar{D}_t$. Furthermore, $\{x \in \bar{D}_t \mid \bar{f}(x) \neq f(x)\}$ has Lebesgue measure zero, hence so does $A_{\bar{f}}(c_\alpha) \setminus A_f(c_\alpha)$ for every c_α. Thus, the level α tests determined by f and $\bar{f}$ are equivalent and have identical power functions.

COROLLARY 2.4. Fix p, r, n, and α, let f be a monotone test statistic defined on $\mathbf{D}_t$, let $c_\alpha \equiv c_\alpha(p, r, n; f)$ satisfy (1.15), and fix $i \in \{1, \ldots, t\}$.

(i) A necessary and sufficient condition that the invariant level α test based on f be PC(i) is that $\bar{f}(e_i(\eta)) > c_\alpha$ for all $\eta > 0$. A sufficient condition is that $\bar{f}(e_i) > c_\alpha$.

(ii) If $\eta = 0$ is a point of increase of $\bar{f}(e_i(\eta))$ (i.e., $\bar{f}(e_i(\eta)) > \bar{f}(e_i)$ for all $\eta > 0$) then $\bar{f}(e_i) \geq c_\alpha$ is a sufficient condition that the level α test based on f be PC(i). If $\bar{f}(e_i(\eta))$ is continuous at $\eta = 0$ then $\bar{f}(e_i) \geq c_\alpha$ is a necessary condition. Thus, if $\eta = 0$ is both a point of increase and a continuity point of $\bar{f}(e_i(\eta))$, then $\bar{f}(e_i) \geq c_\alpha$ is both necessary and sufficient for the level α test to be PC(i). □

If $f(d)$ depends only on $d_{i+1}, \ldots, d_k$ for some $0 \leq i < k \leq t$, it follows from Corollary 2.4(i) that for $0 < \alpha < 1$, the level α test based on f is PC(k) but not PC(i). It may or may not be PC(j) for $i < j < k$: for example, the level α test based on $f(d) \equiv \prod_{i+1}^{k} d_j$ is PC(k) but not PC$(k-1)$, while those based on $\prod_{i+1}^{k} d_j(1-d_j)^{-1}$ and $\sum_{i+1}^{k} d_j(1-d_j)^{-1}$ are PC$(i+1)$ but not PC(i). Furthermore, parameter consistency may depend on the value of α: the level α test based on $f(d) \equiv \sum_{i+1}^{k} d_j$ is PC(j) but not PC$(j-1)$ for α satisfying $j-i-1 < c_\alpha \leq j-i$, $j = i+1, \ldots, k$. Further examples are considered in Section 4.

3. SAMPLE SIZE CONSISTENCY OF MONOTONE INVARIANT TESTS

For fixed p, r, and α, we shall study the consistency of the sequence of invariant level α tests determined by the acceptance regions $A_f^{(n)} (n \geq p)$ based on a monotone test statistic $f = f(d)$ defined on $\mathbf{D}_t$ (see (1.14)).

DEFINITION 3.1. For the testing problem (1.5) ≡ (1.13), the sequence of invariant level α tests based on f is *sample size consistent* (SSC) at the fixed alternative $\lambda^* \neq (0,\ldots,0)$ if the power

$$P_{\lambda^{(n)}}\{f(d) > c_\alpha(p,r,n;f)\} \to 1$$

as $n \to \infty$ for every sequence of alternatives $\{\lambda^{(n)}\}$ such that $\lambda^{(n)} = (n + o(n))\lambda^*$, i.e., $n^{-1}\lambda^{(n)} \to \lambda^*$. If this holds for every $0 < \alpha < 1$, we say that f is *sample size consistent* at λ^* (SSC *at* λ^*). The test statistic f is called *sample size consistent* (SSC) if it is SSC at every $\lambda^* \neq (0,\ldots,0)$. □

REMARK 3.2. The condition $\lambda^{(n)} = (n + o(n))\lambda^*$ stems from the fact that X in (1.4) is typically of the form $\sqrt{n}\,(1+o(1))\bar{X}$ with $\bar{X}$ a sample mean vector. In the simplest case, for example, $X_1,\ldots,X_N$ are independent univariate observations from $N(\mu, \sigma^2)$ and we wish to test $\mu=0$ vs. $\mu\neq 0$. Here the two-sided t-test rejects $\mu=0$ for large values of $T^2 \equiv (\sqrt{N}\,\bar{X})^2/s^2$, where $s^2 = \Sigma(X_i - \bar{X})^2/N-1$. In this case, $p=q=r=1$, $t=1$, $n=N-1$, and, when $\mu\neq 0$, T^2 has the noncentral F distribution with 1 and n degrees of freedom and noncentrality parameter $\lambda = N\mu^2/\sigma^2$. This is of the form $\lambda = (n+o(n))\lambda^*$ with $\lambda^* = \mu^2/\sigma^2$. □

Theorem 3.5, our main result on the sample size consistency of a monotone invariant test statistic f, is based on the following two elementary lemmas, which summarize the limiting behavior of d and $c_\alpha \equiv c_\alpha(p,r,n;f)$ as $n \to \infty$ with p, r, and α fixed. Their proofs follow directly from the definitions (1.4) and (1.11) of d_i, X, and Y.

LEMMA 3.3. Fix p, r, $\lambda^* \in \mathbf{C}_t$, and $i \in \{1,\ldots,t\}$.

(i) If $\lambda=0$, then $d_i = O_p(1/n)$ as $n \to \infty$.

(ii) If $\lambda = (n+o(n))\lambda^*$, then $d \xrightarrow{p} \delta^*$ as $n \to \infty$, where

$$\begin{aligned} \delta^* &\equiv \delta(\lambda^*) \equiv (\delta_1^*,\ldots,\delta_t^*) \in \mathbf{D}_t\,, \\ \delta_i^* &= \lambda_i^*/(\lambda_i^*+1) \qquad \square. \end{aligned} \tag{3.1}$$

For $0 \le \eta < 1$, define

$$f_0(\eta) = f(e_0(\eta)) \equiv f(\eta, \ldots, \eta). \tag{3.2}$$

To avoid technicalities, we shall assume that

$$f_0 \quad \text{is continuous and strictly increasing,} \tag{3.3}$$

hence the inverse function f_0^{-1} is well-defined, continuous, and strictly increasing. Since

$$f_0(d_t) \le f(d) \le f_0(d_1) \tag{3.4}$$

for $d \equiv (d_1, \ldots, d_t) \in \mathbf{D}_t$, it follows from (1.15) that

$$P_{\lambda=0}\{d_t > f_0^{-1}(c_\alpha)\} \le \alpha \le P_{\lambda=0}\{d_1 > f_0^{-1}(c_\alpha)\}. \tag{3.5}$$

Because $P_{\lambda=0}\{0 < d_t < d_1 < 1\} = 1$, this implies that $0 < f_0^{-1}(c_\alpha) < 1$ for $0 < \alpha < 1$, which, together with Lemma 3.3(i), yields the following result:

LEMMA 3.4. Fix p, r and α $(0 < \alpha < 1)$. Then

$$0 < \liminf_{n \to \infty} n\, f_0^{-1}(c_\alpha) \le \limsup_{n \to \infty} n\, f_0^{-1}(c_\alpha) < \infty. \qquad \square$$

For $\delta \equiv (\delta_1, \ldots, \delta_t) \in \mathbf{D}_t$, define

$$\tilde{f}(\delta) = \lim_{\eta \downarrow 0} \frac{1}{\eta} f_0^{-1}[f(\delta \vee e_0(\eta))] \tag{3.6}$$

provided the limit exists (possibly infinite), where $x \vee y = (x_1 \vee y_1, \ldots, x_t \vee y_t)$. Then by the monotonicity of f,

$$\tilde{f} \quad \text{is monotone on} \quad \mathbf{D}_t, \tag{3.7}$$

$$1 = \tilde{f}(0) \le \tilde{f}(\delta) \le \infty. \tag{3.8}$$

By (3.7),

$$\tilde{f}(\delta-) \le \tilde{f}(\delta) \le \tilde{f}(\delta+), \tag{3.9}$$

where $\tilde{f}(\delta-)$ is defined as in (2.4) and where

$$\bar{f}(\delta+) = \lim_{\varepsilon\downarrow 0} \bar{f}((1+\varepsilon)\delta).$$

By (3.7), $\bar{f}(\delta\pm)$ exists provided that $\bar{f}((1\pm\varepsilon)\delta)$ exists for sufficiently small $\varepsilon > 0$. We say that $\bar{f}$ is *radially continuous* at δ if

$$\bar{f}(\delta-) = \bar{f}(\delta) = \bar{f}(\delta+). \tag{3.10}$$

Radial continuity is a weaker requirement than continuity: for example, if $\bar{f}$ depends on $\delta \equiv (\delta_1, \ldots, \delta_t)$ only through

$$\begin{aligned} \operatorname{rank}(\delta) &\equiv \text{number of nonzero } \delta_i \\ &\equiv \max\{i \mid \delta_i > 0\}, \end{aligned} \tag{3.11}$$

then $\bar{f}$ is radially continuous on $\mathbf{D}_t$ but not necessarily continuous.

The following characterization of the sample size consistency of f in terms of $\bar{f}$ may be proved by applying Lemma 3.3(ii) and Lemma 3.4. It also follows from a slightly stronger result in Anderson and Perlman (1988).

THEOREM 3.5. Fix p, r, and $\lambda^* \in \mathbf{C}_t$, and set $\delta^* = \delta(\lambda^*)$ as in (3.1). Let f be a monotone invariant test statistic defined on $\mathbf{D}_t$ and satisfying (3.3).

(i) If $\bar{f}(\delta^*-) = \infty$, then f is SSC at λ^*.

(ii) If $\bar{f}(\delta^*+) < \infty$, then f is not SSC at λ^*.

(iii) Suppose, in addition, that $\bar{f}$ exists and is radially continuous at δ^*. Then f is SSC at λ^* iff $\bar{f}(\delta^*) = \infty$. □

It is important to note that when Theorem 3.5 is applicable, the sample size consistency of the sequence of level α tests based on f does not depend on the value of α $(0 < \alpha < 1)$.

REMARK 3.6. In addition to the hypotheses of Theorem 3.5, suppose that $f_0(0) = 0$ and that

$$f_0(\eta) \sim a\eta^b \quad \text{as} \quad \eta \downarrow 0 \tag{3.12}$$

for some $a, b > 0$. Then it may be shown that for any $\delta \in \mathbf{D}_t$,

$$\tilde{\tilde{f}}(\delta) \equiv \lim_{\eta \downarrow 0} \frac{f(\delta \vee e_0(\eta))}{f_0(\eta)} = [\tilde{f}(\delta)]^b . \tag{3.13}$$

That is, $\tilde{\tilde{f}}(\delta)$ exists iff $\tilde{f}(\delta)$ exists, in which case (3.13) holds. Then all conclusions of Theorem 3.5 remain valid with $\tilde{f}$ replaced by $\tilde{\tilde{f}}$, which is usually easier to calculate. The condition (3.12) is satisfied, for example, if $f(d_1, \ldots, d_t)$ is monotone on $\mathbf{D}_t$, and admits a power series expansion about $(0, \ldots, 0)$. □

LEMMA 3.7. Suppose that f is monotone on $\mathbf{D}_t$ and satisfies (3.3). Then for $\delta \in \mathbf{D}_t$,

$$f(\delta) > f_0(0) \implies \tilde{f}(\delta) = \infty . \tag{3.14}$$

PROOF. It follows from (3.3) and the monotonicity of f that for every $M > 0$ and sufficiently small $\eta > 0$,

$$\begin{aligned} f(\delta) > f(0) &\implies f(\delta) > f_0(\eta M) \\ &\implies \frac{1}{\eta} f_0^{-1}[f(\delta \vee e_0(\eta))] \geq \frac{1}{\eta} f_0^{-1}[f(\delta)] \geq M . \end{aligned} \tag{3.15}$$

Now let $\eta \downarrow 0$ and $M \uparrow \infty$. □

If $\lambda^* = \lambda(\xi^*, \Sigma^*)$ and $\delta^* = \delta(\lambda^*)$, then from (1.8)-(1.11) and (3.1),

$$\text{rank}(\lambda^*) = \text{rank}(\delta^*) = \text{rank}(\xi^*) . \tag{3.16}$$

COROLLARY 3.8. Suppose that f is monotone on $\mathbf{D}_t$ and satisfies (3.3). Fix $\lambda^* \in \mathbf{C}_t$ and set $\delta^* = \delta(\lambda^*)$.

(i) If $f(\delta^*-) > f_0(0)$, then f is SSC at λ^*.

(ii) If rank $(\delta^*) = t$, then f is SSC at λ^*.

PROOF. (i) For sufficiently small $\varepsilon > 0$,

$$f(\delta^*-) > f_0(0) \implies f((1-\varepsilon)\delta^*) > f_0(0)$$
$$\implies \bar{f}((1-\varepsilon)\delta^*) = \infty$$

by Lemma 3.7, hence $\bar{f}(\delta^*-) = \infty$. By Theorem 3.5(i), f is SSC at λ^*.

(ii) For sufficiently small $\varepsilon > 0$,

$$\text{rank}(\delta^*) = t \iff \text{rank}((1-\varepsilon)\delta^*) = t$$
$$\iff (1-\varepsilon)\delta_t^* > 0$$
$$\implies f((1-\varepsilon)\delta^*) \geq f_0((1-\varepsilon)\delta_t^*) > f_0(0)$$

by (3.4) and (3.3). As in the proof of part (i) it follows that $\bar{f}(\delta^*-) = \infty$, hence that f is SSC at λ^*. □

REMARK 3.9. The converse of Corollary 3.8(i) is not necessarily true; in fact, it is not necessarily true that if $f(\delta^*) = f_0(0)$ then f fails to be SSC at λ^*. For example, if $f(d) = \prod_1^t d_i$ then $f(\delta^*) = f_0(0) = 0$ whenever $\text{rank}(\delta^*) < t$; however, $\bar{f}(\delta^*) = \bar{\bar{f}}(\delta^*) = \infty$ for every $\delta^* \neq (0, \ldots, 0)$, hence f is SSC at every $\lambda^* \neq (0, \ldots, 0)$ by Theorem 3.5(i) or Remark 3.6. □

The following definition is suggested by Corollary 3.8(ii):

DEFINITION 3.10. For $i = 1, \ldots, t$, the invariant test statistic f is said to be *sample size consistent of degree i* (SSC(i)) for the testing problem (1.5) ≡ (1.13) if it is SSC at every λ^* such that $\text{rank}(\lambda^*) \geq i$. □

Clearly, SSC$(i) \implies$ SSC$(i+1)$ and SSC $\iff$ SSC(1). If f is monotone on $\mathbf{D}_t$, satisfies (3.3), and is SSC at every λ^* such that $\text{rank}(\delta^*) = i$, then by Theorem 3.5(ii), $\bar{f}(\delta^*+) = \infty$ for every $\delta^* \in \mathbf{D}_t$ of rank i. By the monotonicity of $\bar{f}$ this implies that $\bar{f}(\delta^*-) = \infty$ for every δ^* of rank i, hence that $\bar{f}(\delta^*-) = \infty$ for every δ^* such that $\text{rank}(\delta^*) \geq i$, and therefore, by Theorem 3.5(i), that f is SSC(i).

The following result is similar to Corollary 3.8. Note, however, that in part (i) only the weaker condition $f(\delta) > f_0(0)$ need be assumed, rather than $f(\delta-) > f_0(0)$.

COROLLARY 3.11. Suppose that f is monotone on $\mathbf{D}_t$ and satisfies (3.3).

(i) If $f(\delta) > f_0(0)$ for every δ such that rank $(\delta) = i$, then f is SSC (i).

(ii) If $f(d) = g(d_1, \ldots, d_i)$ for some g, then f is SSC (i).

(iii) If $f(d) = g(d_{i+1}, \ldots, d_t)$ for some g, then f is *not* SSC (i). In fact, f is not SSC at *any* λ^* such that rank $(\lambda^*) \le i$.

(iv) f is SSC (t).

PROOF. (i) By Lemma 3.7, $\tilde{f}(\delta) = \infty$ for every δ of rank i, hence $\tilde{f}(\delta-) = \infty$ for every δ such that rank $(\delta) \ge i$, so the result follows from Theorem 3.5(i).

(ii) If rank $(\delta) = i$, then $\delta_i > 0$ and $f(\delta) = g(\delta_1, \ldots, \delta_i) \ge g(\delta_i, \ldots, \delta_i) = f_0(\delta_i) > f_0(0)$ by (3.3), so the result follows from part (i).

(iii) If rank $(\delta) \le i$, then $\delta_{i+1} = \cdots = \delta_t = 0$ and $f(\delta \vee e_0(\eta)) = g(\eta, \ldots, \eta) = f_0(\eta)$, hence $\tilde{f}(\delta) = 1$ by (3.6). Thus $\tilde{f}(\delta^* +) = 1$ whenever rank $(\delta^*) \le i$, so f cannot be SSC at the corresponding λ^*.

(iv) This is immediate from Corollary 3.8(ii). □

By Remark 3.9, however, it is quite possible that f is SSC (i) even though $f(\delta) = f_0(0)$ for every δ of rank i. To illustrate this more fully, consider the four test statistics $f(d)$ appearing in the final paragraph of Section 2. Each is monotone on $\mathbf{D}_t$, satisfies (3.3), and, by Corollary 3.11(ii) and (iii), is SSC (k) but not SSC (i). In fact, however, Theorem 3.5 or Remark 3.6 implies that each is SSC $(i+1)$ but not SSC (i), even though the first two statistics ($\prod_{i+1}^k d_j$ and $\prod_{i+1}^k d_j(1-d_j)^{-1}$) satisfy $f(\delta) = f_0(0)$ for every δ of rank $< k$.

4. COMPARISON OF PARAMETER CONSISTENCY AND SAMPLE SIZE CONSISTENCY

When comparing these two properties for a monotone invariant test statistic f defined on $\mathbf{D}_t$, it is important first to examine their differences. Throughout this discussion the dimensions p and r remain fixed, while we write $d=d(n)$, $c_\alpha = c_\alpha(n)$, and $A_f(c_\alpha) = A_f(c_\alpha(n))$ to stress the dependence of these quantities on n (the number of degrees of freedom for estimating Σ) — cf. (1.4), (1.11), (1.14), (1.15).

First (recall Definition 2.1), parameter consistency is defined for the *single* level α acceptance region $A_f(c_\alpha(n))$ with n and α fixed: we say that $A_f(c_\alpha(n))$ is PC(i) (or simply "f is PC(i) for n, α") if

$$\lim_{\lambda_i \to \infty} P_\lambda\{d(n) \notin A_f(c_\alpha(n))\} = 1 . \tag{4.1}$$

It was seen at the end of Section 2 that the PC(i) property may depend non-trivially on the value of α. By Corollary 2.4, this property is determined not only by the values of $\bar{f}(e_i(\eta))$ but also by that of $c_\alpha(n) \equiv c_\alpha(p,r,n;f)$, therefore by the *global* behavior of f on $\mathbf{D}_t$.

On the other hand (recall Definitions 3.1 and 3.10), sample size consistency is defined for the *sequence* of acceptance regions $\{A_f(c_\alpha(n)) \mid n \geq p\}$ for a fixed α: this sequence is said to be SSC(i) if

$$\lim_{n \to \infty} P_{\lambda^{(n)}}\{d(n) \notin A_f(c_\alpha(n))\} = 1 \tag{4.2}$$

for every sequence $\{\lambda^{(n)}\}$ of the form $\lambda^{(n)} = (n+o(n))\lambda^*$ with rank$(\lambda^*) \geq i$. Whenever Theorem 3.5 applies, this property does not depend on the value of α $(0<\alpha<1)$, so we then simply say that f is SSC(i). Again by Theorem 3.5, this property is determined by the values of $\tilde{f}(\delta)$ for every $\delta \in \mathbf{D}_t$ of rank i, hence only by the *local* behavior of f in a neighborhood of the set $\{x \in \mathbf{D}_t \mid x_t = 0\}$. (This is because f is *always* SSC(t) (cf. Corollary 3.11(iv)), while for $1 \leq i \leq t-1$, if rank$(\delta) = i$ then the value of $\tilde{f}(\delta)$ is determined by the values of f in a neighborhood of $\{x \in \mathbf{D}_t \mid x_t = 0\}$.

In view of these differences, it is not surprising that the properties PC(i) and SSC(i) are not equivalent, even for a monotone test statistic f. In general, neither property implies the other, as demonstrated by the following examples. Define (cf. (1.11))

$$
\begin{aligned}
f_1(d) &= d_1 && \equiv ch_{\max}[XX'(XX'+YY')^{-1}] \\
f_2(d) &= \sum_{j=1}^{t} d_j(1-d_j)^{-1} && \equiv \operatorname{tr}[XX'(YY')^{-1}] \\
f_3(d) &= \prod_{j=1}^{t} (1-d_j)^{-1} && \equiv \det(XX'+YY')/\det(YY') \\
f_4(d) &= \prod_{j=1}^{t} d_j(1-d_j)^{-1} && \equiv \det(XX')/\det(YY') \\
f_5(d) &= \sum_{j=1}^{t} d_j && \equiv \operatorname{tr}[XX'(XX'+YY')^{-1}] \\
f_6(d) &= \prod_{j=1}^{t} d_j && \equiv \det(XX')/\det(XX'+YY') \\
f_7(d) &= d_t \prod_{j=1}^{t-1} (1-d_j)^{-1} && \\
f_8(d) &= d_t \prod_{j=1}^{t-1} (1+d_j) && \\
f_9(d) &= d_t && \equiv ch_{\min}[XX'(XX'+YY')^{-1}] .
\end{aligned}
$$

The statistics f_1, f_2, f_3, and f_5 are well-known (cf. Anderson (1984), Chapter 8): f_1 is the Roy maximum root statistic, f_2 is the Lawley-Hotelling trace statistic, f_3 is the likelihood ratio (LR) statistic ($\equiv$ Wilks statistic), and f_5 is the Bartlett-Nanda-Pillai trace statistic. Each of the statistics $f_1 - f_9$ is monotone on $\mathbf{D}_t$ and satisfies (3.3), (3.10), and (3.12) (more precisely, $f_3 - 1$ satisfies (3.12)). The parameter consistency and sample size consistency of the invariant level α tests based on $f_1 - f_9$ are readily determined from the results of Sections 2 and 3:

f_1, f_2, f_3, f_4 are PC(1) for all n, α	*and*	f_1, f_2, f_3, f_4 are SSC(1)
f_5 is PC(i), not PC($i-1$), for $i-1 < c_\alpha(n) \le i,\ i = 1, \ldots, t$	*but*	f_5 is SSC(1)
f_6 is PC(t), not PC($t-1$), for all n, α	*but*	f_6 is SSC(1)
f_7 is PC(1) for all n, α	*but*	f_7 is SSC(t), not SSC($t-1$)
f_8, f_9 are PC(t), not PC($t-1$), for all n, α	*and*	f_8, f_9 are SSC(t), not SSC($t-1$).

Although "f is PC(i) for all n, α" and "f is SSC(i)" are thus seen to be inequivalent, their defining properties (4.1) and (4.2) have an important common feature: since $\lambda_i^{(n)} = (n + o(n))\lambda_i^*$, we see that $\mathrm{rank}(\lambda^*) \ge i$ iff $\lambda_i^{(n)} \to \infty$ as $n \to \infty$. This suggests the following definition:

DEFINITION 4.1. For fixed α, the sequence $\{A_f(c_\alpha(n)) \mid n \ge p\}$ of level α acceptance regions based on f is said to be *eventually parameter consistent of degree i* (*eventually* PC(i)) if there exists $n_0(\alpha)$ such that $A_f(c_\alpha(n))$ is PC(i) for every $n \ge n_0(\alpha)$. If $\{A_f(c_\alpha(n)) \mid n \ge p\}$ is eventually PC(i) for every $0 < \alpha < 1$, then the test statistic f is *eventually* PC(i). □

It is now natural to ask whether, for a monotone test statistic f, the properties "f is eventually PC(i)" and "f is SSC(i)" are equivalent. Again this is not true in general, as shown by the behavior of f_6 and f_7 above, but it is more nearly true: although f_5 is not PC(1) for some values of n, α, it is both eventually PC(1) and SSC(1). It is interesting to note that this is indeed *true for all admissible test statistics* f. More precisely, we introduce the following definition:

DEFINITION 4.2. For fixed α, the sequence $\{A_f(c_\alpha(n)) \mid n \ge p\}$ is said to be *eventually admissible* for the original testing problem (1.5) if there exists $n_1(\alpha)$ such that $A_f(c_\alpha(n))$ is an admissible acceptance region for (1.5) whenever $n \ge n_1(\alpha)$. If $\{A_f(c_\alpha(n)) \mid n \ge p\}$ is eventually admissible for every $0 < \alpha < 1$, then the test statistic f is *eventually admissible* for the testing problem (1.5). □

The proof of the following theorem is based on Schwartz's (1967b) necessary condition for the admissibility of an invariant acceptance region for problem (1.5)—see Anderson and Perlman (1988).

THEOREM 4.3. Let f be a monotone invariant test statistic. If f is eventually admissible for the testing problem (1.5), then f is SSC(1) and eventually PC(1). □

Schwartz's (1967b) sufficient condition for admissibility implies that the level α tests based on f_1, f_2, f_3, f_5 are admissible for every n and α, hence *a fortiori* f_1, f_2, f_3, f_5 are eventually admissible. It has already been noted that f_1, f_2, f_3, f_5 are SSC(1) and eventually PC(1), in agreement with Theorem 4.3. Both f_4 and f_7 are eventually *in*admissible (although admissible for sufficiently small n or α) and both are PC(1), but f_4 is SSC(1) whereas f_7 is not SSC(1). The level α tests based on f_6, f_8, and f_9 are *in*admissible for every n and α and none of f_6, f_8, f_9 is eventually PC(1), but f_6 is SSC(1) whereas f_8, f_9 are not SSC(1).

Thus, neither the requirement that f be eventually PC(1) nor the requirement that f be SSC(1) distinguishes among invariant tests. For *fixed* n and α, however, the requirement that f be PC(1) for n and α is *not* satisfied by every admissible test statistic. In particular, *the Bartlett-Nanda-Pillai statistic f_5 fails to be parameter consistent* unless its critical value satisfies $c_\alpha(p, r, n; f_5) \leq 1$, i.e., unless n or α is sufficiently large. (Values of $c_\alpha(p, r, n; f_5)$ are tabulated in Anderson and Perlman (1988).) Despite the facts that the Bartlett-Nanda-Pillai test is admissible (Schwartz (1967b)), proper Bayes (Kiefer and Schwartz (1965)), locally most powerful invariant and locally minimax (Schwartz (1967a)), and robust (Olson (1974)), its failure to be parameter consistent is a potentially serious drawback. Injudicious or routine use of such a test (for example, in a statistical computer package) could result in failure to detect a sizable departure from the null hypothesis H_0.

ACKNOWLEDGEMENTS

The first author's research was supported in part by U.S. Army Research Office Contract No. DAAG29-85-K-0239. The second author's research was supported in part by U.S. National Science Foundation Grant No. DMS 86-03489.

REFERENCES

Anderson, T. W. (1984). *An Introduction to Multivariate Statistical Analysis*, Second Edition. Wiley, New York.

Anderson, T. W. and Perlman, M. D. (1988). On the consistency of invariant tests for the multivariate analysis of variance and related multivariate problems. In preparation.

Kiefer, J. and Schwartz, R. E. (1965). Admissible Bayes character of T^2, R^2, and other fully invariant tests for classical multivariate normal problems. *Ann. Math. Statist.*, **36**, 747-770.

Lehmann, E. L. (1986). *Testing Statistical Hypotheses*, Second Edition. Wiley, New York.

Olson, C. L. (1974). Comparative robustness of six tests in multivariate analysis of variance. *J. Amer. Statist. Assoc.*, **69**, 894-908.

Perlman, M. D. and Olkin, I. (1980). Unbiasedness of invariant tests for MANOVA and other multivariate problems. *Ann. Statist.*, **8**, 1326-1341.

Schwartz, R. E. (1967a). Locally minimax tests. *Ann. Math. Statist.*, **38**, 340-360.

Schwartz, R. E. (1967b). Admissible tests in multivariate analysis of variance. *Ann. Math. Statist.*, **38**, 698-710.

Received October 1987

Department of Statistics
Stanford University
Stanford, CA 94305
U.S.A.

Department of Statistics, GN-22
University of Washington
Seattle, WA 98195
U.S.A.

Proc. Second International Tampere Conference in Statistics
(Tampere, Finland, 1-4 June 1987)
Tarmo Pukkila and Simo Puntanen, *Editors*

pp. 245 - 260

INVITED PAPER

Dividing the indivisible: using simple symmetry to partition variance explained

John W. Pratt
Harvard University, Boston, MA, U.S.A.

Key words and phrases: Relative importance, regression, explanatory variables, contribution of predictors, influence of factors, variance components.

ABSTRACT

How should one define a measure of the relative importance of the independent variables in explaining the variance of the dependent variable in multiple regression? This paper introduces natural requirements of symmetry and invariance under linear transformation and derives a unique measure from them. The importance this measure assigns to each variable is interpreted as the geometric mean of the variable's direct and total effects on variance, as an average reduction in variance if the variable is held fixed, and in several other ways. Negative importance can occur, but signifies a situation too complex for a single measure, not a defect in the definition. The measure has sometimes been used previously, but apparently with little justification or interpretation.

1. INTRODUCTION

What is the relative importance of education and economic status as determinants of opinion (Williams and Mosteller, 1947)? Of misdeeds or conditions that combine to cause damage (Kruskal, 1986b)? Of home and school in educating children? Of monetary and fiscal policy? Of components of industrial productivity? Theoretical and applied discussion of such questions has a long history. For recent statistical review, including these and other examples and appropriate references, see Kruskal (1986a, 1987).

Questions of this kind can of course be made precise in many different ways, which could properly depend in a particular application on everything from the causal and stochastic structure and the controls available to specific objectives, costs, and benefits. This paper deals only with the multiple regression structure and with the relative importance of the independent variables in explaining the variance of the dependent variable. In this very common situation, it is natural and not unusual to try to measure relative importance without explicitly introducing further specifics. In this spirit, we shall put forth some properties or axioms that one would expect of any general procedure and show that they lead to a unique measure. This measure is old, but has been so lacking in interpretation and justification heretofore that Kruskal (1986a) recently called it "a somewhat strange proposal" which "lends itself to satirical mention" and "Swiftian burlesque" (giving citations), and he conjectured that its genesis "may be a well-known identity for the multiple correlation coefficient ... together with an additive leap of faith."

The measure gives each independent variable an importance proportional to the product of its beta coefficient and its correlation with the dependent variable, or equivalently, the product of the standardized coefficients of its direct and total effects. The measure is also interpretable as the geometric mean of the variances of the direct and total effects (provided the effects have the same sign), and as an average reduction in variance associated with holding the variable fixed (see Section 4). The measure has very convenient additivity properties, but these are not assumed *a priori*. The justification here is based essentially on symmetry and invariance under linear transformation.

The general set-up and the argument for two explanatory variables appear in Section 2. Section 3 discusses the extension to more variables and some properties and interpretations of the proposed measure. Section 4 discusses Kruskal's (1987) averaging approach and Schlaifer's (1985) observation, and shows that applying the former to the latter leads to the proposed measure. Section 5 presents an example and Section 6 contains further summary and remarks.

We conclude the introduction with a sketch of the main argument.

If the situation is completely symmetric in the explanatory variables x_i, that is, if they have equal variances, equal regression coefficients, and equal correlations with one another, then they have equal relative importance. A natural conclusion, further argued in Section 3, is that the relative importance of $X_1 = x_1 + ... + x_m$ and $X_2 = x_{m+1} + ... + x_{m+n}$ is in the ratio m to n. If relative importance is to be invariant under rescaling, the same ratio must apply to any other regression of a dependent variable y on two independent variables X_1 and X_2 with the same standardized regression coefficients β_1 and β_2 and correlation ρ. Some algebra shows that this ratio can be expressed as $\beta_1 B_1/\beta_2 B_2$ where $B_1 = \beta_1 + \rho\beta_2$ and $B_2 = \beta_2 + \rho\beta_1$. This determines relative importance in all 2-variable regressions where $\beta_i B_i > 0$ for $i = 1, 2$. Since B_i is the standardized coefficient in the regression of y on X_i alone, we conclude that each variable has importance proportional to the product of its direct effect β_i and its total effect B_i after the variable is standardized, provided these have the same sign for each variable. This extends to any number of independent variables.

2. MAIN RESULTS FOR TWO EXPLANATORY VARIABLES

2.1 Set-up and notation

We deal throughout with an ordinary linear regression $y = b_0 + b_1 x_1 + ... + b_n x_n + u$ where u has mean 0 and variance σ^2 and is uncorrelated with $x_1, ..., x_n$. We are not concerned with sampling problems, so we treat all parameters as known. We allow relative importance to depend on the regression coefficients, multiple correlations, residual variances, and any other functions of the means, variances, and correlations of the variables, but not on higher moments or other aspects of the marginal or joint distributions. This amounts to the following assumption.

(A1) *Relative importance depends only on the means, variances, and correlations of* $x_1, ..., x_n$ *and* y.

Since relative importance should not depend on the units in which the variables are measured, (A1) almost inevitably entails:

(A2) *Nonconstant linear transformation of any variable leaves relative importance unchanged.*

This allows not only rescaling in the ordinary sense, but also adding a constant and multiplying by −1. For some purposes the latter two might not be innocuous, but in discussions of relative importance of the kind we have in mind, (A2) would be completely acceptable, and indeed is often taken for granted.

Let $\beta_i = |b_i|\sigma_i$ where σ_i is the standard deviation of x_i. Then β_i is the standard deviation of $b_i x_i$, the x_i term in the full regression, which we shall call the direct effect of x_i. If x_i is standardized by linear transformation so as to have variance 1 and a nonnegative regression coefficient, then β_i is this coefficient. If y is also scaled to have variance 1, then β_i is the absolute value of the so-called beta coefficient, but our results and formulas are unaffected by the scaling of y.

Let B_i be the regression coefficient when x_i is standardized as above and y is regressed on x_i alone. Specifically, let the regression of y on the original x_i alone, before standardization, be $c_i + a_i x_i$. Then $B_i = a_i\sigma_i$ if $b_i \geq 0$ while $B_i = -a_i\sigma_i$ if $b_i < 0$. Furthermore, $|B_i|$ is the standard deviation of $a_i x_i$, which we shall call the total effect of x_i, and B_i is positive if the total and direct effects of x_i have the same sign. If y is also scaled to have variance 1, then B_i is the correlation of y with x_i or its negative according as b_i is positive or negative. (Henceforth, when $b_i = 0$, we consider it positive and assume that the standardizing linear transformation of x_i is increasing.)

2.2 Reduction of the parameter space for two explanatory variables

For $n = 2$, let ρ be the correlation of the two explanatory variables after standardization, that is, the correlation of x_1 and x_2 or its negative according as b_1 and b_2 have the same or opposite signs. Then

$$B_1 = \beta_1 + \rho\beta_2, \quad B_2 = \beta_2 + \rho\beta_1, \tag{1}$$

$$Var(b_1x_1 + b_2x_2) = \beta_1^2 + 2\rho\beta_1\beta_2 + \beta_2^2 = \beta_1 B_1 + \beta_2 B_2, \text{ and} \tag{2}$$

$$Var(y) = \beta_1 B_1 + \beta_2 B_2 + \sigma^2. \tag{3}$$

Standardization of x_1 and x_2 leads easily to

PROPOSITION 1. *For two explanatory variables, under assumptions* (A1) *and* (A2), *relative importance depends only on* β_1, β_2, ρ, and σ^2.

Of course other parameters such as B_1 and B_2 could be added or substituted for some of these with no change in meaning. Standardization of y would eliminate one further parameter. If (A2) is assumed for the explanatory variables only, then a location parameter would be needed in

Proposition 1. It would be eliminated by the symmetry assumption and argument below, but such refinements would clutter the exposition more than they are worth.

2.3 The implications of symmetry

Now consider $N = m+n$ explanatory variables x_i, each having mean 0, variance s^2, and the same correlation r with every x_j, $j \neq i$. Suppose that the regression of y on $x_1,\dots, x_N$ is $x_1+\dots+x_N$, with residual variance σ^2. Let

$$X_1 = x_1+\dots+x_m \text{ and } X_2 = x_{m+1}+\dots+x_{m+n}. \tag{4}$$

Then since the regression of y on $x_1,\dots, x_n$ is $X_1 + X_2$, this is also its regression on X_1 and X_2. Assume either directly or as a consequence of other assumptions and arguments given later that

(A3) *In this symmetric situation, the relative importance of* X_1 *and* X_2 *is as* m *to* n.

The standardized coefficients of X_1 and X_2 are easily obtained from their variances $\sigma_1{}^2$ and $\sigma_2{}^2$ as

$$\begin{cases} \beta_1 = \sigma_1 = [m+m(m-1)r]^{\frac{1}{2}}s \text{ and} \\ \beta_2 = \sigma_2 = [n+n(n-1)r]^{\frac{1}{2}}s. \end{cases} \tag{5}$$

Similarly their correlation (before or after standardization) is

$$\rho = mnrs^2/\sigma_1\sigma_2 = mnrs^2/\beta_1\beta_2 \tag{6}$$

and the residual variance is still σ^2. Hence, under assumptions (A1) - (A3), if a two-variable regression has β_1, β_2, and ρ which satisfy these equations for some m, n, s, and r, then the relative importance of the two variables is as m to n. To see what this implies, we need to determine when the equations can be solved for m, n, s, and r; whether m/n is unique; and how to find m/n from β_1, β_2 and ρ. Fortunately all three questions have convenient answers. For the moment we ignore the reguirement that m and n be integers, but we require that m, n, and s be positive and that r satisfy the positive definiteness condition for $m+n$ equally correlated variables, namely, $1 > r > -1/(m+n-1)$.

PROPOSITION 2. *Given* $\beta_1>0$, $\beta_2>0$, *and* ρ, $-1<\rho<1$, *equations* (5) *and* (6) *have solutions* m, n, s, *and* r *with* $m>0$, $n>0$, $s>0$, *and* $1>r>-1/(m+n-1)$ *iff* $\beta_1+\rho\beta_2$ and $\beta_2+\rho\beta_1$ *are positive. In this case*

$$\frac{m}{n}=\frac{\beta_1(\beta_1+\rho\beta_2)}{\beta_2(\beta_2+\rho\beta_1)} . \tag{7}$$

Proof. Equations (5) and (6) imply by straightforward algebra

$$\beta_1(\beta_1+\rho\beta_2)=m[1+(m+n-1)r]s^2 \text{ and} \tag{8}$$

$$\beta_2(\beta_2+\rho\beta_1)=n[1+(m+n-1)r]s^2. \tag{9}$$

The "only if" and the expression (7) for m/n follow immediately. Conversely, suppose β_1, β_2, and ρ are given and $B_1=\beta_1+\rho\beta_2$ and $B_2=\beta_2+\rho\beta_1$ are positive. Then (5) and (6) are equivalent to (6), (8) and (9). For any positive value of $a=[1+(m+n-1)r]s^2$, (8) and (9) are satisfied by

$$m=\beta_1B_1/a \text{ and } n=\beta_2B_2/a. \tag{10}$$

If $\rho=0$, then (6) is satisfied by $r=0$. Otherwise (6) becomes

$$\rho=\frac{B_1B_2r/a}{1+(m+n-1)r} . \tag{11}$$

As r increases from $-1/(m+n-1)$ to 1, the right-hand side of (11) increases from $-\infty$ to

$$\frac{B_1B_2}{a(m+n)}=\frac{B_1B_2}{\beta_1B_1+\beta_2B_2}=\rho+\frac{(1-\rho^2)\beta_1\beta_2}{\beta_1B_1+\beta_2B_2} . \tag{12}$$

which is $>\rho$. Hence a solution r in the required range exists, and $s^2=a/[1+(m+n-1)r]$ is also positive. Q.E.D.

The following corollary shows that the requirement that m and n be integers is not significant.

COROLLARY 1. *In Proposition 2, m and n can be chosen to be integers if the right-hand side of* (7) *is rational.*

Proof. At (10) in the proof of Proposition 2, if $\beta_1 B_1/\beta_2 B_2$ is rational, then one can choose a to make m and n integer. Q.E.D.

Combining the foregoing results gives our main theorem for two explanatory variables. Technically we use the following continuity assumption in the proof. A monotonicity assumption could be used instead.

(A4) *If relative importance is defined for a sequence of two-x regressions, and if the values of β_1, β_2, ρ, and the relative importance all approach limits, then the limiting relative importance is the relative importance for the limiting β_1, β_2, and ρ.*

THEOREM 1. *Under assumptions* (A1) - (A4), *if there are just two explanatory variables x_1 and x_2 and if $\beta_1 + \rho\beta_2$ and $\beta_2 + \rho\beta_1$ are nonnegative, then the importance of x_1 relative to x_2 is as $\beta_1(\beta_1 + \rho\beta_2)$ to $\beta_2(\beta_2 + \rho\beta_1)$.*

Proof. If β_1, β_2, and ρ satisfy the conditions of Corollary 1, the conclusion follows from Proposition 1 and Corollary 1. If β_1, β_2, and ρ do not satisfy the conditions of Corollary 1, they can be approximated arbitrarily closely by values that do, and the conclusion now follows by continuity. Q.E.D.

3. EXTENSION TO n EXPLANATORY VARIABLES, INTERPRETATIONS, AND PROPERTIES

3.1 Simple extension

As before, let the regression of y on $x_1,\ldots, x_n$ be $b_0 + b_1x_1 + \ldots + b_nx_n$ and the regression of y on x_i alone be $c_i + a_ix_i$. We call b_ix_i the direct effect and a_ix_i the total effect of x_i. Recall that the coefficients after standardization are β_i and B_i where $\beta_i = |b_i|\sigma_i$ and $B_i = a_i\sigma_i$ if $b_i \geq 0$ while $B_i = -a_i\sigma_i$ if $b_i < 0$. Also β_i and $|B_i|$ are standard deviations of the direct and total effects, so $|\beta_iB_i|$ is the geometric mean of their variances.

For $n = 2$, the previous section led, by an argument based on symmetry, to assigning importance to x_i in proportion to β_iB_i provided $\beta_iB_i \geq 0$. For all n

$$\begin{aligned}\beta_i B_i &= b_i a_i \sigma^2{}_i \\ &= Cov\,(b_i x_i\,,\ a_i x_i) \\ &= Cov\,(b_i x_i\,,\ \Sigma_1{}^n b_j x_j) \\ &= Cov\,(b_i x_i\,,\ y).\end{aligned} \tag{13}$$

Summing gives the well-known identity for the explained variance

$$\beta_1 B_1 + \ldots + \beta_n B_n = Var(b_1 x_1 + \ldots + b_n x_n). \tag{14}$$

It has occasionally been proposed that

(*) *The portion of the explained variance attributed to* x_i *should be* $\beta_i B_i$.

We see that, for $n = 2$, this rule can be derived from a symmetry argument. The extension to $n > 2$ has several intuitive properties that make it very natural, although some further assumption is needed to derive it formally.

3.2 Properties of $\beta_i B_i$ as a measure of importance

First, we can define the importance of a subset of the explanatory variables, say $(x_1,\ldots, x_m)$, to be the sum of their individual importances. (This is not an assumption. It is merely a conventional definition in itself, gaining force from later use.)

Second, for $X_1 = \Sigma_1{}^m b_i x_i$, the regression of y on X_1 , $x_{m+1},\ldots, x_n$ is the same as its regression on $x_1,\ldots, x_n$, and with the proposed definition, the importance of $(x_1,\ldots, x_m)$ relative to $(x_{m+1},\ldots, x_n)$ is the same as the importance of X_1 relative to $(x_{m+1},\ldots, x_n)$ in the relevant regressions. Also with $X_2 = \Sigma^n{}_{m+1} b_i x_i$, the importance of $x_1,\ldots, x_m$ relative to $x_{m+1},\ldots, x_n$ is the same as that of X_1 relative to X_2, and similarly for a partition of $x_1,\ldots, x_n$ into more than two subsets.

Third, if $x_1,\ldots, x_m$ are uncorrelated with $x_{m+1},\ldots, x_n$, so that the variance explained by $x_1,\ldots, x_n$ equals the variance explained by $x_1,\ldots, x_m$ plus the variance explained by $x_{m+1},\ldots, x_n$, then the importance is assigned correspondingly.

Even a more limited additivity requirement than is embodied in the first two properties above would suffice with the argument of Section 2 to derive the relative importance $\beta_i B_i$ for $n > 2$. We omit such derivations, however, because they are simple and not completely compelling: additivity is more convenient and pleasant than inevitable, and using it to derive relative importance for $n > 2$ from $n = 2$ assumes too much of what we want to prove.

We now state two additional assumptions that come into play when there are more than two explanatory variables. They are satisfied by the rule (*), but we state them as assumptions rather than properties because they are extremely natural assumptions and they strongly suggest and almost imply the extension of the rule (*) from $n = 2$ to $n > 2$. The first is that adding an explanatory variable that is pure noise does not change relative importance.

(A5) *If z is distributed independently of $(x_1,\ldots, x_n, y)$, then $(x_1,\ldots, x_m)$ has the same importance relative to $(x_{m+1},\ldots, x_n)$ as to $(x_{m+1},\ldots, x_n, z)$.*

Even if z is only uncorrelated with $x_1,\ldots, x_n$ and y, the same conclusion is a property of (*) and a consequence of (A1) and (A5). Because independent noise suffices when (A5) is assumed below, however, we state it in this weaker form.

The other extremely natural assumption we use is that the importance of one set of variables relative to another, disjoint set is unaffected by the way the second set is expressed. For example, the importance of a married couple's income relative to their ages is the same as it is relative to the average and difference of their ages.

(A6) *If $(x'_{m+1},\ldots, x_n')$ is a nonsingular linear transformation of $(x_{m+1},\ldots, x_n)$ then the importance of $(x_1,\ldots, x_m)$ relative to $(x'_{m+1},\ldots, x_n')$ when y is regressed on $(x_1,\ldots, x_m, x'_{m+1},\ldots, x_n')$ is the same as its importance relative to $(x_{m+1},\ldots, x_n)$ when y is regressed on $(x_1,\ldots, x_m, x_{m+1},\ldots, x_n)$.*

Assumption (A6) implies (A2) except that adding a constant is included in (A2) but not (A6) since "linear" for one variable becomes "affine" for more. As just mentioned, (*) satisfies (A5) and (A6). A stronger version of (A6) allowing nonlinear transformations would be reasonable, but it will not be needed and is out of keeping with the restriction to two moments (A1).

3.3 Reduction of the parameter space for n explanatory variables

Natural as assumptions (A5) and (A6) are, they have surprisingly strong consequences. One is a result that helps justify assumption (A3).

PROPOSITION 3. *Under assumptions* (A1), (A5) *and* (A6), *in the symmetric situation of assumption* (A3), *the importance of* $(x_1,\ldots, x_m)$ *relative to* $(x_{m+1},\ldots, x_{m+n})$ *is the same as the importance of* $X_1 = \Sigma_1{}^m x_i$ *relative to* $X_2 = \Sigma_1{}^n x_{m+i}$.

Proof. Let $x_i' = x_i - x_{i-1}$. Then $(X_1, x_2',\ldots, x_m')$ is a nonsingular linear transformation of $(x_1,\ldots, x_m)$. Furthermore, $(x_2',\ldots, x_m')$ are uncorrelated with $X_1, X_2, x'_{m+2},\ldots, x'_{m+n}$, and hence also with y. They can be made uncorrelated with each other as well by a further nonsingular transformation. Therefore $(x_1,\ldots, x_m)$ has the same relative importance as $(X_1, x_2',\ldots, x_m')$ by (A6) and hence as X_1 by (A1) and (A5). A similar argument shows that $(x_{m+1},\ldots, x_{m+n})$ has the same relative importance as X_2. Q.E.D.

We would like to generalize Proposition 3 to show that, if the regression of y on $x_1,\ldots, x_n$ is $\Sigma_1{}^n b_i x_i$, then the importance of $(x_1,\ldots, x_m)$ relative to $(x_{m+1},\ldots, x_n)$ is the same as that of $X_1 = \Sigma_1{}^m b_i x_i$ relative to $X_2 = \Sigma^n{}_{m+1} b_i x_i$ when y is regressed on X_1 and X_2. Unfortunately the assumptions made so far apparently don't quite get us there, but we can come close by at least two routes.

We seek to determine the importance of x_1 relative to $x_2,\ldots, x_n$. First, let $x_2',\ldots, x_n'$ be a nonsingular linear transformation of $x_2,\ldots, x_n$ such that x_2' is the regression of x_1 on $x_2,\ldots, x_n$ and $x_3',\ldots, x_n'$ are uncorrelated with x_2'. Then let $x_3'',\ldots, x_n''$ be a nonsingular linear transformation of $x_3',\ldots, x_n'$ such that x_3'' is the regression of y on $x_3',\ldots, x_n'$ and $x_4'',\ldots, x_n''$ are uncorrelated with x_3''. Then $x_4'',\ldots, x_n''$ are uncorrelated with x_1, x_2', x_3'' and y and hence are irrelevant under assumptions (A1), (A5), and (A6). Furthermore, x_3'' is uncorrelated with x_1 and x_2'. Thus we have reduced the problem to determining the relative importance of 3 variables, one of which (x_3'') is uncorrelated with the other two. It is certainly natural now to assign importance to x_3'' and the pair (x_1, x_2') in proportion to the variance explained, and then to assign the importance of the pair to x_1 and x_2' individually as one would in the absence of x_3''. This agrees with and leads to the rule (*).

If we define the transformations as before except that $x_2' = \Sigma_2{}^n b_i x_i$ and x_3'' is the regression of x_1 on $x_3',\ldots, x_n'$, then the problem reduces as before to x_1, x_2', x_3'', but now x_3'' has no direct effect in the regression of y on x_1, x_2', and x_3'', although x_3'' may be correlated with x_1. Theorem 1 gives x_3'' no importance relative to $b_1x_1 + b_2x_2'$. It is natural to assume that its importance relative to (x_1, x_2') is not greater and hence is 0. Then assigning importance to x_1 and x_2' as one would in the absence of x_3''

means that x_1 has the same importance relative to $(x_2,\ldots, x_n)$ as to $x_2' = \Sigma_2{}^n b_i x_i$. This again agrees with and leads to the rule (*).

It appears, then, that to derive the rule (*), only a slight further assumption is needed in addition to (A1), (A2), (A4), (A5) and (A6). Different assumptions may appeal to different people, so no particular one will be advanced here. Note, however, that even though the axioms stated do not quite suffice to derive the rule (*), they do eliminate other prominent proposals – indeed, all others that I am aware of.

4. TWO RECENT PROPOSALS

Though this paper makes no attempt at historical review, two recent proposals merit mention. The first is like the Shapley value in an n-person game. Kruskal (1986a) proposes to define the importance of x_i by averaging the stepwise improvement due to x_i in each of the $n!$ orderings of $x_1,\ldots, x_n$. To measure improvement he uses primarily the fractional reduction in the remaining variance. He mentions other possibilities, however, including actual reduction in variance, which at least avoids giving full credit for reducing a tiny ε to 0: if two uncorrelated x's explain fractions ε and $1-\varepsilon$ of the variance, fractional reduction assigns them relative importance $1+\varepsilon$ and $2-\varepsilon$. Fractional reduction changes if σ^2 changes with no change in explained variances, becoming proportional to actual reduction as $\sigma^2 \to \infty$ $(R^2 \to 0)$. For $n = 2$, this proposal agrees with (*) if a geometric rather than an arithmetic mean is taken, but any version of it differs from (*) in general. With an arithmetic mean, it gives positive importance to any variable having any relevant correlation. This may be desirable, although one may prefer the property of (*) that a variable with no direct effect has no importance even if it can proxy for other variables. The reason the proposal disagrees with (*), despite agreeing in completely symmetric situations, is that it does not satisfy (A3), or more fundamentally, (A6), that is, the relative importance it attaches to x_1 depends on the coordinate system used for $x_2,\ldots, x_n$.

For example, let us start with the regression $y = b_1x_1 + b_2x_2$ and measure improvement by reduction in variance. In the order x_1, x_2, we find that x_1 reduces variance by $B_1{}^2$ and x_2 reduces variance by $(1-\rho^2)\beta_2{}^2$. In the opposite order, x_2 reduces variance by $B_2{}^2$ and x_1 by $(1-\rho^2)\beta_1{}^2$. Thus the importance of x_1 relative to x_2 (in the sense of average improvement) is

$$\frac{B_1^2 + (1-\rho^2)\beta_1^2}{B_2^2 + (1-\rho^2)\beta_2^2} \quad . \tag{15}$$

Now let $x_3,\ldots, x_n$ differ from x_2 by small, independent random errors and regress y on $x_1, x_2,\ldots, x_n$. Then x_1 has $1/n$ chance of being first, otherwise it is approximately as if x_2 were first, and the importance of x_1 relative to $x_2,\ldots, x_n$ is approximately

$$\frac{B_1^2 + (n-1)(1-\rho^2)\beta_1^2}{(n-1)B_2^2 + (1-\rho^2)\beta_2^2} \quad . \tag{16}$$

This ranges from (15) to $(1-\rho^2){\beta_1}^2/{B_2}^2$ as n increases from 2 to ∞. The same formula would apply if a small random error were added to x_2 as well, making it indistinguishable from $x_3,\ldots, x_n$. On the other hand, let x_i be replaced by $x_i' = x_i - x_2$ for $i>2$ (and x_2 by $\Sigma_2^{\,n} x_i/(n-1)$ if desired). Then the importance of x_1 relative to $x_2, x_3',\ldots, x_n'$ is again given by (15) approximately. This is obviously unreasonable: the relative importance of x_1 should not depend on how often x_2 is measured erroneously, or how many copies of it are included, or whether the remaining variables are x_2 and its repetitions or x_2 and the differences from it.

Schlaifer (1985, Section 11.1.5) provides a valuable contrast to the present discussion for both specificity and results. He observes that

> an assertion that $p\%$ of the variance of y is explained by a certain set of predictors would be interpreted by anyone speaking ordinary rather than statistical English as meaning that the variance of y would have been reduced by $p\%$ if those predictors had been held constant. The R^2 of an ordinary regression model does not have this meaning, but given a model in which the coefficients of the factors are true causal effects, there is a well-defined fraction of the variance of y that the factors really explain.
>
> ... For the coefficients of some of the predictors (the factors) to be causal effects, the model must contain other predictors (concomitants) that proxy for excluded determinants of y and thereby prevent the factors from so doing. If however the model does contain the required concomitants, the fraction by which the variance of y would have been reduced if the factors had been held constant is usually neither the first-in nor the last-in contribution of the factors to the R^2 of the model ....

He then shows in effect that holding the factors $x_1,\ldots, x_m$ constant reduces the variance of y by

$$Var(\sum_{1}^{n} b_i x_i) - Var(\sum_{m+1}^{n} b_j x_j) = Var(\sum_{1}^{m} b_i x_i) + 2Cov(\sum_{1}^{m} b_i x_i, \sum_{m+1}^{n} b_j x_j), \tag{17}$$

provided $x_1,..., x_m$ do not affect any other factors or concomitants -- a usual but not inevitable requirement (Pratt and Schlaifer, 1984). He points out that the reduction can be negative because

> The effects of x_1 and x_2 and the correlation between them obviously *may* be such that when x_1 is free to vary its effect tends to offset the effect of x_2; and when this is true the variance of y would obviously increase rather than decrease if x_1 were held constant.

We note that the (last-in) contribution of x_m to (17) is

$$Var(b_m x_m) + 2Cov(b_m x_m, \sum_{m+1}^{n} b_j x_j). \tag{18}$$

This is the reduction in variance due to holding x_m constant after holding $x_1,..., x_{m-1}$ constant. If all variables are factors, one might average (18) over all $n!$ orderings of $x_1,..., x_n$. Since each x_j follows x_m in half of all orderings, we find by (13)

PROPOSITION 4. *If $x_1,..., x_n$ are all factors that do not affect one another, the average over all orderings of the reduction in the variance of y associated with holding x_m constant is*

$$Var(b_m x_m) + Cov(b_m x_m, \sum_{j \neq m} b_j x_j) = \beta_m B_m. \tag{19}$$

Thus averaging over orderings can satisfy (A6), and one man's arithmetic mean is another's geometric mean!

5. EXAMPLE

Yule (1899), in an investigation of the causes of changes in pauperism in England, examines the regression of the percent change in pauperism (y) on the percent changes in out-relief ratio (x_1), proportion of old (x_2), and population (x_3), for each of two intercensal decades and four groups of unions. (See his paper for full definitions and details.) He discusses relative importance mainly in terms of algebraic contributions to average change, and notes the difficulty of combining groups when a variable contributes positively in one group but negatively in another (but not the within-group counterpart). There is much to be said for Yule's approach and his extensive attention to the specific context, and he chooses his variables for this approach and context, but in a post-ANOVA era it is natural to consider also a more general method based on contributions to variance. We do so for one of Yule's eight regressions.

For the group of 236 rural unions in the decade 1871-81, Yule gives the standard deviations σ_i, correlations, and regression coefficients b_i in Table 1, the standard deviation $\sigma_y = 16.17$, and the residual standard deviation 14.12. From this information it is easy to calculate the standardized regression coefficients $\beta_i = |b_i|\sigma_i$ and $B_i = \pm a_i\sigma_i = \pm cor(x_i, y)\sigma_y$ and the contributions $\beta_i B_i$ to variance given in Table 1. B_3 and $\beta_3 B_3$ are negative because b_3 and the correlation of x_3 with y have opposite signs; however x_3 has little relative importance any way you look at it, while x_1 is about 8 times as important as x_2 in terms of $\beta_i B_i$.

The total contribution to variance explained is $58.9 + 7.57 - 0.72 = 65.7$. This agrees less well with the reduction in variance $(16.17)^2 - (14.12)^2 = 62.1$ than one would expect, but I do not know the source of the discrepancy.

TABLE 1

Regression of percent changes in pauperism, rural unions, 1871-81

	Standard deviation σ_i	Correlation with y	x_1	x_2	Regression coefficients b_i	β_i	B_i	Importance $\beta_i B_i$
x_1	25.2	.484			.299	7.53	7.83	58.9
x_2	8.9	.194	.131		.271	2.41	3.14	7.57
x_3	7.3	-.094	-.079	-.575	.064	.47	-1.52	-0.72

6. CONCLUDING SUMMARY AND DISCUSSION

This paper's main effort has been to explore the implications of some axioms or requirements it is natural to impose on measures of relative importance of explanatory variables in regression. Section 2 obtained a unique measure in the 2-variable case from the symmetric n-variable case (as summarized in the introduction). Section 3 reduced the general case by linear transformation to 2 variables plus noise. This gave the measure (*) except for one seemingly minor step.

The measure (*) has many convenient properties. Some of these might well be required of any measure of relative importance. No other measure proposed to date satisfies these minimum requirements, and no other measure *can* satisfy what appears to be slight augmentations of them. Thus it seems hard to avoid adopting (*) except by incorporating specifics of each application, and we need no longer say, like Williams and Mosteller (1947, p. 197, for a binary y) that "our measure is completely arbitrary; many others could have been used."

Possible interpretations of (*) include the following. The importance of variable x_i is the product of the standard deviations of the direct and total effects $b_i x_i$ and $a_i x_i$, or the geometric mean of their variances, with a positive or negative sign according as these effects are in the same or opposite directions. It is the product of the coefficients β_i and B_i of the direct and total effects when x_i is standardized to have variance 1. (The product is the same whether the standardized variable is taken to be x_i/σ_i or $-x_i/\sigma_i$.) It is the product of the beta coefficient of x_i and the correlation of x_i with y if y is standardized. It is the average reduction in the variance of y associated with holding x_i fixed if all independent variables can be viewed as factors, precedence is determined by randomizing their order, and fixing any subset does not effect the remainder, that is, fixing some variables does not effect the conditional distribution of the remaining variables.

The interpretations of (*) and the minimum requirements not met by other measures seem to have been generally overlooked or under-appreciated in the literature. The possibility of negative importance may have been instrumental. But negative importance has a positive side.

The possibility of a negative sign reflects the oversimplification inherent in reducing a multidimensional question to a single number, and the occurrence of a negative sign signifies a situation in which this over-simplification may be especially dangerous. False positives and false negatives can easily occur, however. For example, if β_i is small, the importance of x_i will be relatively small, yet individual terms $cov(b_i x_i, b_j x_j)$ in (13) may be large but happen to largely cancel out. If b_i is small, x_i may nevertheless contribute greatly to the predictive power of some subsets of the other variables. Neither need be signalled by negative importance. On the other hand, the negative signal in Section 5 was unimportant. Perhaps we need next a way to measure how much of an oversimplification it is to represent importance by $\beta_i B_i$, much as standard deviation measures how much of an oversimplification it is to represent a distribution by its mean.

ACKNOWLEDGEMENT

William Kruskal made a great many, extremely helpful comments, for which the author is most grateful. I am also grateful for help from many others, but not in a position to name them all individually.

REFERENCES

Kruskal, William (1986a). Relative importance of determiners. (Chinese). *Journal of the Chinese Statistical Association*, 24 (3) 11042-11060. (English version 1984.)

Kruskal, William (1986b). Terms of reference: singular confusion about multiple causation. *Journal of Legal Studies*, 15, 427-436.

Kruskal, William (1987). Relative importance by averaging over orderings. *The American Statistician*, 41, 6-10.

Pratt, John W. and Schlaifer, Robert (1984). On the nature and discovery of structure. *Journal of the American Statistical Association*, 79, 9-21.

Schlaifer, Robert (1985). *Introduction to Data Analysis*. Harvard University Graduate School of Business Administration, Boston, MA.

Williams, Frederick and Mosteller, Frederick (1947). Education and economic status as determinants of opinion. *Gauging Public Opinion* (Hadley Cantril, *ed.*). 195-208. Princeton University Press.

Yule, G. Udny (1899). An investigation into the causes of changes in pauperism in England, chiefly during the last two intercensal decades. (Part I.) *Journal of the Royal Statistical Society*, 62, 249-286; discussion 287-295.

Received 13 January 1987
Revised 3 June 1987

Graduate School of Business Administration
Harvard University
Soldiers Field
Boston, MA 02163
U.S.A.

Proc. Second International Tampere Conference in Statistics
(Tampere, Finland, 1-4 June 1987)
Tarmo Pukkila and Simo Puntanen, *Editors*

pp. 261 - 274

INVITED PAPER

Majorization orderings for linear regression designs

Friedrich PUKELSHEIM
University of Augsburg, F.R. Germany

Key words and phrases: Group majorization, invariant design problems, information increasing orderings, universal optimality, simultaneous optimality.

ABSTRACT

Classical vector majorization captures the idea of whether the entries of a vector are more nearly equal than those of another one. Much of experimental design theory revolves around the same idea, and is called "balance" here. In the present paper we outline the use of majorization techniques for a general concept of balancedness: Matrix majorization replaces vector majorization, and linear transformation groups which leave the design problem invariant take the place of the permutation group for vector majorization.

1. INTRODUCTION

Majorization has emerged as a powerful tool to describe the notion of *balancedness* in experimental design theory. Its possible usefulness was already alluded to by Kiefer (1974, p. 862). The papers of Giovagnoli & Wynn (1981) and Bondar (1983) study majorization properties of the vector of eigenvalues associated with the information matrices of the designs. Further work of Giovagnoli &

Wynn (1985a,b) suggests a transition from vector majorization to matrix majorization, as elaborated in Giovagnoli, Pukelsheim & Wynn (1987). Applications of this theory to the design problem has been surveyed by Pukelsheim (1987a,b). In the present paper we sketch the essential steps of this recent development.

In Section 2 we first identify the various levels of the design problem:

- designs ξ, i.e. discrete probability distributions on an experimental domain $\mathcal{X} \subset \mathbb{R}^k$,
- moment matrices $M(\xi)$, i.e. $k \times k$ nonnegative definite matrices depending on a design ξ,
- information matrices $C(M)$, i.e. $s \times s$ nonnegative definite matrices representing the information for the s-dimensional parameter system of interest, as a function of moment matrices M, and
- objective functions $\phi(C)$, i.e. real functions of the information matrices C with information-like properties.

Invariance of the design problem under a group $\mathcal{Q}$ of linear transformations has its impact on each of these levels, as discussed in Section 3. We wish to stress that the groups act on moment matrices and on information matrices by congruence, not by similarity.

In Section 4 we turn to the desired orderings for experimental designs. Majorization relative to the group $\mathcal{Q}$ leads to the notion of when one moment matrix is *more balanced* than another one. However, for design applications this has to be built up into a two-stage preordering also involving the Löwner ordering of nonnegative definite matrices. The resulting *information increasing ordering* $\gg$ generalizes Kiefer's (1975) notion of *universal optimality*, and is intimately related to *simultaneous optimality* relative to all objective functions which are invariant.

2. MAXIMIZING INFORMATION

Suppose the *experimental conditions* are given through some k-dimensional vector from a compact *experimental domain* $\mathcal{X} \subset \mathbb{R}^k$. Under experimental conditions $x \in \mathcal{X}$ we may draw a single real-valued observation

$$Y(x) = x'\theta + \sigma e,$$

where we assume that the error e has unit variance, and that the errors are uncorrelated between observations under different experimental conditions as well as between repeated observations under the same experimental conditions. A *design* ξ then is taken to be a discrete probability distribution on the experimental domain $\mathcal{X}$, determining allocation and proportion of the experimental conditions. However, the problem is not really one of dealing with probability distributions ξ themselves, but to study the behaviour of certain matrices associated with designs ξ.

2.1 Moment matrices

The mean vector parameter θ is of dimension k. Accordingly we associate with a design ξ its $k \times k$ *moment matrix*

$$M(\xi) = \int_{\mathcal{X}} xx'd\xi = \sum_{i=1}^{l} \xi(x_i)x_i x_i'.$$

The set of all moment matrices forms a convex compact subset of nonnegative definite matrices, due to the assumed compactness of the experimental domain $\mathcal{X}$. Formally we simply assume to start with a convex compact *feasible set* $\mathcal{M}$ of $k \times k$ nonnegative definite matrices. The primary choice for $\mathcal{M}$ certainly is the set of *all* moment matrices, but other choices are of interest. For instance, $\mathcal{M}$ may be a set of moment matrices obtained from designs with certain restrictions on their marginals, or with restrictions on the support, here $\mathcal{M}$ would be a proper subset of all moment matrices. Bayesian design problems call for a choice of $\mathcal{M}$ to be the set of all moment matrices shifted by a prior information matrix R, see Chaloner (1984, p. 286). Through the introduction of the set $\mathcal{M}$ we are in a position to allow for such cases and others. Optimality will thus always be meant relative to the set $\mathcal{M}$.

2.2 Information matrices

We shall assume that the parameter system of interest $K'\theta$ is given through the $k \times s$ matrix K of rank s. For a nonsingular moment matrix M the $s \times s$ *information matrix for* $K'\theta$ is defined to be

$$C(M) = (K'M^{-1}K)^{-1}.$$

In our earlier work (Pukelsheim 1980, p. 341) we have chosen to define the information matrix for $K'\theta$ to be 0 when $K'\theta$ is not identifiable. The resulting discontinuity is unnecessarily strong, as has transpired in recent work of Gaffke (1985b, p. 73) and Müller-Funk, Pukelsheim & Witting (1985, p. 23). Moreover, the old definition fails to measure identifiability of subsystems of $K'\theta$. A refined definition for singular moment matrices M is

$$C(M) = \lim_{\epsilon \downarrow 0}(K'(M + \epsilon I_k)^{-1}K)^{-1} = \min_L LML',$$

where the minimum is taken over all left inverses L of K, relative to the Löwner matrix ordering. Now identifiability holds if and only if the matrix $C(M)$ in the refined definition is nonsingular, and in this case $C(M)$ admits the simpler representation $(K'M^-K)^{-1}$ where M^- is an arbitrary generalized inverse of M. Thus identifiability is entirely encoded in the rank behaviour of the information matrix, under the refined definition.

2.3 Objective functions

Over the years a considerable amount of work has gone into exploring the frontiers of the class of functionals which may rightly serve as optimality criteria for the experimental design problem. The classical criteria of D-, A-, E-optimality have been embedded into the continuous class of p-means (Kiefer 1974, p. 865). Pukelsheim (1980) admits information functionals which are defined to be concave, positive, and homogeneous; Gaffke (1985a, p. 385; 1985b, p. 69) presents a subgradient theorem covering functionals which are concave and isotonic (under the Löwner ordering).

Such a bewildering variety of optimality criteria does not please every human mind. Yet it serves its purpose. For instance let us discuss uniform optimality of some moment matrix M, i.e. in the Löwner ordering – the 'usual' ordering between symmetric matrices – we have $M \geq A$ for all $A \in \mathcal{M}$. Then optimality is inherited by the huge class of isotonic (i.e. increasing) criterion functions, since evidently $\phi(M) \geq \phi(A)$ whenever ϕ is isotonic. Conversely, if M is ϕ-optimal for every function ϕ in the relatively small class $\phi(C) = z'Cz$, with $z \in \mathbb{R}^s$, then M is uniformly optimal (Pukelsheim 1980, p. 344).

More generally let $\gg$ be a partial ordering for information matrices. Then it is useful to know which class of objective functions is order preserving, and the bigger the class the better. On the other hand it may be helpful to identify a subclass of functionals as small as possible so that simultaneous optimality over the subclass implies optimality under the partial ordering $\gg$. Hence we study wide classes of criteria in order to move away from any single particular criterion in the direction of statistically more reasonable partial orderings $\gg$. Section 4 will illuminate this situation further, but first we must briefly digress into when a design problem is invariant.

3. INVARIANT DESIGN PROBLEMS

The groups which determine our majorization relations originate from the invariance properties of the design problem. Assume that a subgroup $\mathcal{Q}$ of the general linear group $GL(k)$ acts linearly on the experimental conditions x, i.e.

$$x \to Qx, \text{ with } Q \in \mathcal{Q} \subset GL(k).$$

For this to make sense we require that the experimental domain is invariant, i.e. $Q(\mathcal{X}) \subset \mathcal{X}$ for all $Q \in \mathcal{Q}$.

3.1 Moment congruence

The linear group action on the experimental conditions induces a *congruence action* on moment matrices:

$$M(\xi) = \int_{\mathcal{X}} xx' d\xi \longrightarrow \int_{\mathcal{X}} Qxx'Q' d\xi = QM(\xi)Q'$$

For this to make sense we require that the feasible set $\mathcal{M}$ is invariant, i.e. $Q\mathcal{M}Q' \subset \mathcal{M}$ for all $Q \in \mathcal{Q}$. This invariance property is clearly satisfied for instance when $\mathcal{M}$ is the set of all moment matrices because then $QM(\xi)Q' = M(\eta)$, say, with η being the distribution of Qx under ξ.

3.2 Information congruence

The final invariance property focuses on the parameter system of interest $K'\theta$, stipulating

$$Q(\text{range}K) = \text{range}K \qquad \text{for all } Q \in \mathcal{Q}.$$

This simply means that the linear hypothesis $K'\theta = 0$ is invariant (or that the group $\mathcal{Q}$ acts on the parameter space in such a way that the set of parameters of interest remains invariant as does the set of nuisance parameters).

Range invariance of K entails that for every $Q \in \mathcal{Q}$ the $s \times s$ matrix $\check{Q} = K^{+}QK$ is nonsingular, where K^{+} is the Moore-Penrose inverse of K. Furthermore we have

$$QK = K\check{Q}, \qquad \text{and} \qquad C(QMQ') = \check{Q}C(M)\check{Q}'.$$

Since the set $\check{\mathcal{Q}} = \{\check{Q} \in GL(s) \mid Q \in \mathcal{Q}\}$ forms a subgroup of $GL(s)$ the latter property means that the transition C from moment matrices to information matrices is equivariant under the groups $\mathcal{Q}$ and $\check{\mathcal{Q}}$.

3.3 Invariant objective functions

Once on the level of information matrices the reduced group $\check{\mathcal{Q}}$ has been determined it is clear that for an optimality criterion ϕ to be invariant we require

$$\phi(\check{Q}C\check{Q}') = \phi(C) \qquad \text{for all } \check{Q} \in \check{\mathcal{Q}}.$$

As the group $\check{\mathcal{Q}}$ becomes larger the class of invariant criteria will evidently become smaller. For instance for the trivial group $\check{\mathcal{Q}} = \{I_s\}$ all criteria are invariant, for the orthogonal group $\check{\mathcal{Q}} = \mathrm{Orth}(s)$ an invariant criterion $\phi(C)$ must be a function of the ordered eigenvalues of C, and for the group of unimodular linear transformations $\check{\mathcal{Q}} = \mathrm{Unim}(s)$ the only invariant information functional is $(\det C)^{1/s}$.

It ought to be acknowledged that invariance is **not** automatically built into any given design problem. The preceding exposition has dwelt on its mathematical prerequisites. Whether it is meaningful from a statistical point of view must be decided on the ground of the practical poblem in question.

4. INFORMATION INCREASING ORDERINGS

Group majorization underlies the idea that given a point A any other point M in the convex hull of the orbit of A is an average and as such is more balanced. For instance the group $\mathcal{Q}$ acts on the designs themselves through $\xi \to \xi \circ Q^{-1}$. Hence a design ξ is *more balanced* than η when

$$\xi = \sum \alpha_i \, \eta \circ Q_i^{-1},$$

for a finite number of transformations $Q_i \in \mathcal{Q}$, where $\min \alpha_i \geq 0$ and $\sum \alpha_i = 1$. For a finite group $\mathcal{Q} = \{Q_1, \cdots, Q_n\}$ of order n we may choose $\alpha_i = 1/n$ and average over *all* transformations Q_i to obtain a design ξ which is invariant. For an infinite compact group $\mathcal{Q}$ a similar averaging procedure is possible with respect to Haar probability measure, but the resulting invariant measure may no longer be discrete. For instance an equidistant design on the circle averaged over all rotations yields Lebesgue measure on the circle.

Fortunately we rarely work on the level of designs ξ. Rather we let a design ξ inherit its performance properties from the moment matrix $M(\xi)$, and the information matrix $C(M(\xi))$. As an example recall that a design ξ is uniformly optimal for $K'\theta$ when in the Löwner ordering $C(M(\xi)) \geq C(M(\eta))$ for all competing designs η. Thus orderings of moment matrices and of information matrices are of greater interest.

4.1 Balancedness among moment matrices

We shall call a moment matrix B *more balanced than* another moment matrix A when B lies in the convex hull of the orbit of A, i.e.

$$B = \sum \alpha_i Q_i A Q_i'$$

for a finite number of transformation $Q_i \in \mathcal{Q}$, where $\min \alpha_i \geq 0$ and $\sum \alpha_i = 1$. This is the usual concept of group majorization where B is considered "smaller" than A. In the design context B carries more information and therefore we reverse the majorization notation and consider B to be "larger" than A, from the information point of view.

When the group $\mathcal{Q}$ is compact then each moment matrix A has in the convex hull of its orbit a unique invariant and hence most balanced matrix $\overline{A}$, obtained from averaging with respect to Haar probability measure dQ according to

$$\overline{A} = \int_{\mathcal{Q}} QAQ' \, dQ.$$

Due to compactness and convexity the feasible set $\mathcal{M}$ must contain any such matrix $\overline{A}$. Hence the invariant matrix $\overline{A}$ may be obtained as the moment matrix of a design ξ, without necessitating the invariance of ξ! For instance in trigonometric regression (Pukelsheim 1980, p. 360) every uniform distribution on an arbitrary set of equidistant support points leads to the same moment matrix as Lebesgue measure on the circle which is the unique rotation invariant distribution.

An ordering which is always available for comparing information is the Löwner ordering among moment matrices. It is natural to try and combine these two concepts.

The combination appropriate for the design problem produces the *information increasing ordering*, as follows. A moment matrix M is called *at least as informative* as another moment matrix A, denoted by $M \gg A$, when M is larger in the Löwner ordering than some matrix B which is more balanced than A, i.e.

$$M \geq B \in \text{ convex hull of the orbit of } A \text{ under } \mathcal{Q}, \qquad \text{for some } B.$$

The information increasing ordering is *transitive*, in that $M \gg A$ and $A \gg F$ imply $M \gg F$. When the group $\mathcal{Q}$ is compact the ordering is also *antisymmetric 'modulo $\mathcal{Q}$'*, i.e. $M \gg A$ and $A \gg M$ entail $M = QAQ'$ for some $Q \in \mathcal{Q}$ (rather than $M = A$).

Other combinations of the two orderings are feasible and are discussed by Giovagnoli, Pukelsheim & Wynn (1987). That the present combination is appropriate for the design problem becomes apparent as we continue the discussion on the information matrix level.

4.2 Balancedness among information matrices

An information matrix C is called *at least as informative* as another information matrix D, denoted by $C \gg D$, when C is larger in the Löwner ordering than some matrix E which is more balanced than D, i.e.

$$C \geq E \in \text{ convex hull of the orbit of } D \text{ under } \check{\mathcal{Q}}, \qquad \text{for some } E.$$

Notice that we do not insist that the intermediate matrix E lies in $C(\mathcal{M})$: On the level of moment matrices the intermediate matrix B automatically lies in the feasible set $\mathcal{M}$, but due to the lack of convexity of $C(\mathcal{M})$ we have no such knowledge about E.

Consistency of the information increasing ordering from the level of moment matrices to the level of information matrices is shown by the following.

Theorem 1 *(Giovagnoli, Pukelsheim & Wynn 1987, Thm. 4). If M is at least as informative as A then $C(M)$ is at least as informative as $C(A)$.*

Proof. If $M \geq B = \sum \alpha_i Q_i A Q_i'$ then monotonicity, concavity, and equivariance of C yield

$$C(M) \geq C(B) = C(\sum \alpha_i Q_i A Q_i')$$

$$\geq \sum \alpha_i C(Q_i A Q_i') = \sum \alpha_i \check{Q}_i C(A) \check{Q}_i' = E, \text{ say }.$$

Evidently E lies in the convex hull of the orbit of $C(A)$. □

An analysis of the proof shows that when B is more balanced than A it does not generally follow that $C(B)$ is more balanced than $C(A)$, while it does follow that $C(B)$ is more informative than $C(A)$. A similar remark pertains to the "opposite" two-stage ordering: When M is more balanced than some matrix B which is larger in the Löwner ordering than A then $C(M)$ is more informative than $C(A)$ rather than inheriting the "opposite" ordering property.

Using a notion first introduced by Kiefer (1975) we now define an information matrix C to be *universally optimal* when C is at least as informative as D for all competing information matrices D. When the group $\check{\mathcal{Q}}$ is compact then a given information matrix C may be averaged with respect to Haar probability measure $d\check{Q}$ to obtain

$$\overline{C} = \int_{\check{\mathcal{Q}}} \check{Q} C \check{Q}' \, d\check{Q}.$$

However, this matrix $\overline{C}$ need not be a feasible information matrix as the set $C(\mathcal{M})$ may fail to be convex. On the other hand we have the following result.

Theorem 2. *Suppose the group $\check{\mathcal{Q}}$ is compact and the information matrix C is invariant. Then C is universally optimal if and only if C is larger in the Löwner ordering than the invariant matrices $\overline{D}$ obtained from the competing information matrices D.*

Proof. If C is invariant and universally optimal then $C = \overline{C} \geq \overline{E} = \overline{D}$, for every competing information matrix D. Conversely, if $C \geq \overline{D}$ then obviously $C \gg D$, since $\overline{D}$ lies in the convex hull of the orbit of D. □

As an example let $\check{\mathcal{Q}}$ be the permutation group. Then an invariant information matrix is completely symmetric, i.e. it has the form $\alpha\overline{J}+\beta(I-\overline{J})$ where $\overline{J}$ is the $s\times s$ matrix with all entries equal to $1/s$. Comparison in the Löwner ordering thus reduces the task to comparing the two eigenvalues α and β.

4.3 Invariant objective functions

If an information matrix C is at least as informative as D, then $\phi(C) \geq \phi(D)$, for every criterion function ϕ which is isotonic, concave, and invariant. This simply follows from a repetition of the steps used to establish Theorem 1. In many cases the subclass of linear criteria is sufficient to establish universal optimality:

Theorem 3 *(Giovagnoli, Pukelsheim & Wynn 1987, Thm. 2). Suppose the group $\mathcal{Q}$ is compact and the information matrix C is invariant. Then C is universally optimal if and only if C is ϕ-optimal simultaneously for all criteria ϕ which are linear, isotonic, and invariant.*

Proof. The functions $\phi(D) = z'\overline{D}z$ are linear, isotonic, and invariant. Hence $C \geq \overline{D}$, for all competing information matrices D, and we may invoke Theorem 2. □

Whether universal optimality under a noncompact group $\mathcal{Q}$ is equivalent to simultaneous optimality relative to an appropriate subclass of criteria remains an open question.

Acknowledgement. Sincere thanks go to James V. Bondar and the referees for their helpful remarks on the original draft of the paper.

REFERENCES

Bondar, J.V. (1983). Universal optimality of experimental designs: definitions and a criterion. *The Canadian Journal of Statistics,* 11, 325-331.

Chaloner, K. (1984). Optimal Bayesian experimental design for linear models. *The Annals of Statistics,* 12, 283-300.

Gaffke, N. (1985a). Directional derivatives of optimality criteria at singular matrices in convex design theory. *Statistics,* 16, 373-388.

Gaffke, N. (1985b). Singular information matrices, directional derivatives, and subgradients in optimal design theory. In: T. Calinski & W. Klonecki (Eds.) *Linear Statistical Inference, Proceedings, Poznan 1984,* pp. 61-77. Lecture Notes in Statistics Vol. 35. Springer, Berlin Heidelberg New York Tokyo.

Giovagnoli, A. & Wynn, H.P. (1981). Optimum continuous block designs. *Proceedings of the Royal Society London Series A,* 377, 405-416.

Giovagnoli, A. & Wynn, H.P. (1985a). Schur-optimal continuous block designs for treatments with a control. In: L.M. Le Cam & R.A. Olshen (Eds.) *Proceedings of the Berkeley Conference in Honor of Jerzy Neyman and Jack Kiefer,* Vol. 1, pp. 418-433. Wadsworth, Belmont CA.

Giovagnoli, A. & Wynn, H.P. (1985b). G-majorization with applications to matrix orderings. *Linear Algebra and Its Applications,* 67, 111-135.

Giovagnoli, A., Pukelsheim, F. & Wynn, H.P. (1987). Group invariant orderings and experimental designs. *Journal of Statistical Planning and Inference,* forthcoming.

Kiefer, J.C. (1974). General equivalence theory for optimum designs (approximate theory). *The Annals of Statistics*, 2, 849-879.

Kiefer, J.C. (1975). Constructions and optimality of generalized Youden designs. In: J.N. Srivastava (Ed.) *A Survey of Statistical Design and Linear Models*, pp. 333-353. North-Holland, Amsterdam.

Müller-Funk, U., Pukelsheim, F. & Witting, H. (1985). On the duality between locally optimal tests and optimal experimental designs. *Linear Algebra and Its Applications*, 67, 19-34.

Pukelsheim, F. (1980). On linear regression designs which maximize information. *Journal of Statistical Planning and Inference*, 4, 339-364.

Pukelsheim, F. (1987a). Information increasing orderings in experimental design theory. *International Statistical Review*, forthcoming.

Pukelsheim, F. (1987b). Ordering experimental designs. In: *Proceedings of the First World Congress of the Bernoulli Society, Tashkent, September 1986*, forthcoming. VNU Science Press, Utrecht.

Received 12 March 1987
Revised 1 July 1987

Institut für Mathematik
Universität Augsburg
Memminger Strasse 6
D-8900 Augsburg
F.R. Germany

Proc. Second International Tampere Conference in Statistics
(Tampere, Finland, 1-4 June 1987)
Tarmo Pukkila and Simo Puntanen, *Editors*

pp. 275 - 288

INVITED PAPER

On the order determination of time series models

Tarmo PUKKILA and **Arto KALLINEN**
University of Tampere, Finland

Key words and phrases: ARMA models, identification, model adequacy, white noise test, order selection criterion, initial estimation.

ABSTRACT

The identification of ARIMA models can be regarded as the search for a linear filter which transforms the observed time series into white noise. The goal in model building is usually a parsimoniously parametrized model. The selection of an ARMA(p, q) model can be performed by applying an ordinary model selection criterion of the form

$$\delta(p,q) = n\log(\hat{\sigma}^2) + (p+q)g(n), \quad p = 0,1,\ldots,p^*, \quad q = 0,1,\ldots,q^*.$$

One important problem in this approach is that unidentified models have to be estimated.

In this paper a new model selection is introduced in which the model builder will not meet the problem of estimating an unidentified model. The method also produces a model which is adequate and parsimoniously parametrized. Furthermore the method is consistent. The method is based on the white noise test, which is carried out by applying a consistent autoregressive order selection criterion. The method can also be used to select the degree of differencing. It therefore offers an objective way to build ARIMA(p, d, q) models.

1. INTRODUCTION

In this paper we introduce a new method of building autoregressive integrated moving average models which, according to Box and Jenkins (1970), are widely known as ARIMA(p, d, q) models, or commonly also as Box and Jenkins models. These can be expressed compactly in the form

$$\phi(B)(\nabla^d X_t - \mu) = \theta(B) a_t, \qquad (1.1)$$

where B is the backshift operator such that $B^k X_t = X_{t-k}$ for any integer k, and

$$\phi(B) = \sum_{k=1}^{p} \phi_k B^k, \quad \theta(B) = \sum_{k=1}^{q} \theta_k B^k.$$

It will be assumed that the zeroes of the polynomials $\phi(B)$ and $\theta(B)$ are located outside the unit circle. Then the process $\{\nabla^d X_t\}$ is stationary and invertible. In a stationary case $\mu = E\{\nabla^d X_t\}$, the expected value of the time series $\{\nabla^d X_t\}$. Here $\nabla^d = (1 - B)^d$ is the differencing operator which forms the d times differenced series $\{\nabla^d X_t\}$. The model (1.1) is considered to be stationary if $d = 0$ and nonstationary in the case $d > 0$. Finally, $\{a_t\}$ is a white noise process; i.e., a sequence of uncorrelated random variables with mean zero and σ^2 as the variance. Later at the estimation stage we will assume that the random variables a_t are normally distributed.

The paper is organized as follows. In Section 2 we discuss the model building method proposed by Box and Jenkins (1970). Section 3 describes the new model building method to be proposed in this paper. The approach is largely based on autoregressive model selection criteria. In Section 4 we describe the performance of the method in the case of the time series A–F of Box and Jenkins (1970). Finally Section 5 offers some concluding remarks.

2. THE IDENTIFICATION METHOD OF BOX AND JENKINS

Assume that we have observed a time series $X_1, X_2, \ldots, X_n$ generated by the process (1.1). At the identification stage of the model the degree of differencing d as well as p and q, the orders of autoregressive and moving average polynomials $\phi(B)$ and $\theta(B)$, respectively, are selected. Furthermore, the identification of an ARIMA(p, d, q) model also includes the decision whether $\mu = E\{\nabla^d X_t\}$ is zero or not.

In the identification procedure proposed by Box and Jenkins (1970) the first problem is to determine whether an observed time series is stationary or not. In this paper we assume that possible nonstationarity can be removed by differencing the observed time series d times. Box and Jenkins propose the degree of differencing, d, to be selected by looking visually at the graphs of $\nabla^d X_t$ for various values of d. The graph of a time series often suggests that differencing is necessary. Furthermore the behaviour of the estimated autocorrelation function might reveal nonstationarity in the observed time series. This is because for nonstationary time series the estimated autocorrelations approach zero only slowly and often linearly with increasing lag. It is clear that these kinds of "non-stationarity tests" are subjective in their nature.

The next step in the identification of a Box and Jenkins model is the determination of the pair (p, q). In the method of Box and Jenkins (1970) this is done by interpreting the behaviour of the estimated autocorrelation and partial autocorrelation functions of the time series $\nabla^d X_t$. Next, the estimated correlation structures are compared with the theoretical correlation structures of ARMA(p, q) models for various pairs (p, q). The pair (p, q), for which the behaviour of the estimated correlations structure seems to resemble most closely the corresponding theoretical correlation structure, will then be selected. Because of the sampling properties of the estimated autocorrelations and partial autocorrelations, the identification of p and q in this way is very difficult, and especially for short time series the selection of p and q is often more or less only a trial and error process.

As in all model building, in time series analysis, too, one of the leading ideas is to specify a parsimoniously parametrized model. Of course the model should still be adequate. In the method which we propose in this paper, we automatically end up at a model which contains the least possible number of parameters while still being adequate in a definite sense.

The identification of (1.1) can be interpreted as a search of a linear filter which transforms an observed time series $X_1, X_2, \ldots, X_n$ into white noise. Therefore, after the estimation of an ARMA model it is natural to test whether the corresponding residuals behave like a white noise series. Following this idea also in the new method, diagnostic checkings are applied in the residuals of an estimated model. If the null hypothesis of the whiteness of the residuals is not rejected, the estimated model is considered to be adequate.

Besides autocorrelations and partial autocorrelations a number of other tools have been suggested to be used in the identification of autoregressive moving average models. These include inverse autocorrelations and inverse partial autocorrelations (see Chatfield (1979), Cleveland (1972), Bhansali (1980,1983a), Hipel, McLeod and Lennox (1977) and Abraham and Ledolter (1984)). Bequin, Gourieroux and Monfort (1980) proposed the so-called corner method for the identification of an ARMA(p, q) model. For the same purpose Woodward and Gray (1981) defined generalized autocorrelations and partial autocorrelations. Tiao and Tsay (1983a,b) and Tsay and Tiao (1984) defined the so-called extended sample autocorrelations which again can be used to determine the degrees p and q.

Common to all of these methods is that one tries to identify a certain type of pattern in the behaviour of a function. A pattern depends on p and q, and theoretically these functions uniquely determine p and q. However, in practice sampling variation makes the interpretation of the behaviour of various functions very difficult, especially for short time series. It is clear that the use of the above mentioned functions in the identification of ARIMA models also includes subjective elements and often the result depends on the experience of a model-builder.

In practice univariate ARIMA models are often built for forecasting purposes. The resulting models are always autoexplanations, i.e. an observation X_t at the time point t is explained by the behaviour of the past observations of the same time series. If we have to build an ARIMA model for an economic time series, usually, at least at present, there is no economic theory which would help us to determine d, p and q. On the other hand, according to the Wold decomposition (see Wold 1938) every stationary and purely non-deterministic time series can be expressed as a linear process

$$X_t = \mu + \sum_{k=0}^{\infty} \psi_k a_{t-k}, \quad \psi_0 = 1. \tag{2.1}$$

Linear models (2.1) form the basis for the Box and Jenkins models. These are parsimoniously parametrized rational approximations of the linear processes. Therefore it is clear that automatic and objective identification methods of ARIMA models are of importance e.g. in practical forecasting applications. Similarily stationary ARMA models can be used in parametric spectrum estimation.

As was previously mentioned, the selection of p and q is preceded by the selection of d, the degree of differencing. A good review of the problem of testing nonstationarity and its implications is given by Dickey, Bell and Miller (1986), which also contains up-to-date references to the earlier results in the area.

3. A NEW IDENTIFICATION METHOD OF ARIMA MODELS

In our model building approach we estimate ARIMA(p, d, q) models for various combinations of (p, d, q). After estimating a certain model a test of white noise is carried out to determine whether the corresponding residuals are white or not.

The models are estimated in the order $p+d+q=0$, $p+d+q=1$, $p+d+q=2,\ldots$ This means that the first step is to test whether the observed time series is white noise. Later in this paper we will describe the white noise test which will be used for this purpose. The test is based on autoregressive order determination criteria. The test has the property that with increasing sample size its significance level approaches zero. This step corresponds to the case $p+d+q=0$. If the null hypothesis of white noise is not rejected, there is nothing more left to be done but to determine whether the mean of the series is zero or not. If the mean of the series is nonzero, it is estimated by the arithmetic mean of the observed series. Otherwise the model building process is completed.

If, however, the null hypothesis of white noise is rejected, we will take the case $p+d+q=1$ into consideration, i.e. we will take the one-parameter models into consideration. For each one-parameter model we also test whether the mean of the series is nonzero or not. All of the one-parameter models are estimated using the method to be described later in the paper. If the null hypothesis of white noise is not rejected, we have completed the model building process; otherwise we will increase the number of parameters in the model, i.e. we will move to the step $p+d+q=2$, and so on. The model building process will be finished at the stage where the residuals from an estimated model are regarded as white noise. This approach will guarantee the least possible number of parameters such that the resulting model is adequate.

It is clear that a method like this heavily depends on the estimation method of an ARMA model as well as on the white noise test to be used to check the whiteness of the residual series. Before handling these matters two facts should be mentioned:

The first is connected with the degree of differencing. If it is known that a time series is stationary, we can restrict d to be zero. If this decision cannot be made in practice, however, we can often restrict the maximum value of d to be less than 4, say. By restricting d we can restrict the number of models to be estimated. In the model building idea itself there is no need to restrict d.

Secondly, in practice it is often the case that several different ARIMA models with the same number of parameters will be adequate. For this reason we will always estimate all the models with a certain number of parameters and discriminate between adequate models to obtain the unique model. The method for discrimination between adequate models is based on the values of the model selection criteria calculated for adequate models.

3.1 Testing for model adequacy

Invertible ARIMA models (1.1) can always be expressed as an infinite autoregression in the form

$$X_t = \theta_0 + \sum_{k=1}^{\infty} \pi_k X_{t-k} + a_t, \tag{3.1}$$

where θ_0 is a constant and a_t a white noise process. An important special case of (3.1) is the white noise process, i.e. the situation where $\pi_k = 0$ for all $k = 1, 2, \ldots$ On the other hand, we can always approximate an infinite autoregression (3.1) by a finite autoregression. The order of this finite autoregression can be selected by applying order determination criteria. If the selected order of an autoregressive model for an observed time series is zero, the generated process is considered to be a white noise process. This is because an AR(0) process is a white noise process. This test is also the starting point for the proposed model building strategy.

This kind of white noise test for univariate time series was considered by Pukkila and Krishnaiah (1986a). It was generalized to a multivariate white noise test by Pukkila and Krishnaiah (1986b). In the following we will consider univariate white noise tests based on autoregressive order determination criteria.

To carry out a white noise test as described above, autoregressive models of orders $0, 1, 2, \ldots, p^*$ have to be estimated, where p^* is the maximum autoregressive order to be tried. For this purpose the Yule-Walker method will be applied here. Furthermore for our white noise test we have used the BIC criterion, independently derived by Schwarz (1978) and Rissanen (1978), to select the order of an autoregressive model. The BIC criterion is known to be consistent. In a white noise test this means that with increasing n the correct null hypothesis of white noise is rejected with decreasing probability. On the other hand, with increasing n the incorrect null hypothesis is rejected with increasing probability. Using the terminology of the traditional statistical test theory it can be said that with increasing n the significance level of the white noise test approaches zero while the power of the test approaches unity.

To perform the above described white noise test we select the order of an autoregressive model for the time series in question. This is done by minimizing the order selection criterion

$$\delta_1(p) = n \log(\hat{\sigma}_p^2) + p \log n, \quad p = 0, 1, \ldots, p^*, \tag{3.2}$$

where $\hat{\sigma}_p^2$ is the Yule-Walker estimate for the residual variance σ^2, i.e.

$$\hat{\sigma}_p^2 = c(0)(1 - \sum_{k=1}^{p} \hat{\phi}_k r(k)).$$

Here $c(0)$ is the variance of an observed time series and $r(k)$ is the autocorrelation estimate at lag k. Furthermore $\hat{\phi}_k$ are the Yule-Walker estimates of the autoregressive parameters ϕ_k.

If $\delta_1(p) < \delta_1(0) = n \log c(0)$ for some $p = 1, 2, \ldots, p^*$, the null hypothesis of white noise will be rejected. Because

$$\delta_1(p) = n \log c(0) + n \log(1 - \sum_{k=1}^{p} \hat{\phi}_k r(k)) + p \log n$$

we can use

$$\delta(p) = n \log(1 - \sum_{k=1}^{p} \hat{\phi}_k r(k)) + p \log n, \quad p = 1, \ldots, p^*, \tag{3.3}$$

as the test statistics. If $\delta(p) < 0$ for some $p \in \{1, 2, \ldots, p^*\}$, the null hypothesis of white noise will be rejected. Let it be mentioned that $\delta(p)$ is independent of $c(0)$, the variance of the observed time series. This is of course natural, because the whiteness property is by no means dependent on the variance of a time series.

Suppose that H_0, the null hypothesis of white noise, is true. Then under H_0 the autoregressive estimates $\hat{\phi}_k \to r(k)$ in probability, i.e. $p\lim \hat{\phi}_k = r(k)$. Therefore also the asymptotic distribution of $\hat{\phi}_k$ is the same as the asymptotic distribution of $r(k)$. Therefore it can be easily seen that under H_0 the test statistic $\delta(p)$ has the same asymptotic distribution as the statistic

$$\delta_3(p) = -n \sum_{k=1}^{p} r^2(k) + p \log n. \tag{3.4}$$

On the other hand it is known that the autocorrelations $r(k)$ are asymptotically normally distributed with mean zero and with variance $1/n$. Therefore, using (3.4), the distributional properties of the autocorrelations $r(k)$ and results from random walk theory, Pukkila and Krishnaiah (1986a) calculated asymptotic significance levels corresponding to the test statistics (3.3) for various values of n and p^*. It was observed that if $p > p_0$ (say), then the significance level is the same for all $p > p_0$. The result implies that p_0 can be selected large. Numerical results in Pukkila and Krishnaiah (1986a) show that $p_0 = \sqrt{n}$ is a reasonable selection.

3.2 A new estimation method for ARMA models

Any invertible model like (1.1) can be approximated by a finite autoregression

$$X_t = \sum_{k=1}^{L} \beta_k X_{t-k} + a_t. \tag{3.5}$$

The parameters β_k in (3.5) can be estimated efficiently by using e.g. the least squares method. In this study we have, however, applied the Burg algorithm (Burg 1967) for the estimation of autoregressive models.

The residuals $\hat{a}_t$ from an estimated finite autoregression are taken as estimates for the true innovations a_t. These residuals can be expressed in the form $\hat{a}_t = a_t - \epsilon_t$, i.e. $a_t = \hat{a}_t + \epsilon_t$. By defining $Y_t = \nabla^d X_t$ we can express (1.1) in the form

$$Y_t - \mu = \sum_{k=1}^{p} \phi_k (Y_{t-k} - \mu) + \hat{a}_t - \sum_{k=1}^{q} \theta_k \hat{a}_{t-k} + \epsilon_t - \sum_{k=1}^{q} \theta_k \epsilon_{t-k},$$

or

$$Y_t - \hat{a}_t - \mu = \sum_{k=1}^{p} \phi_k (Y_{t-k} - \mu) - \sum_{k=1}^{q} \theta_k \hat{a}_{t-k} + \epsilon_t - \sum_{k=1}^{q} \theta_k \epsilon_{t-k}. \tag{3.6}$$

By defining $Z_t = Y_t - \hat{a}_t$ we can express (3.6) further in the form

$$Z_t - \mu = \sum_{k=1}^{p} \phi_k (Y_{t-k} - \mu) - \sum_{k=1}^{q} \theta_k \hat{a}_{t-k} + \eta_t, \tag{3.7}$$

where

$$\eta_t = \epsilon_t - \sum_{k=1}^{q} \theta_k \epsilon_{t-k}.$$

Thus (3.7) is a regression model with the error term η_t which is a moving average model of order q, shortly MA(q) model. Thus in principle we can estimate the parameters ϕ_k and θ_k by using the generalized least squares(GLS) algorithm. Therefore, using the matrix notations we can estimate the parameters $\gamma^T = (\phi_1, \ldots, \phi_p, \theta_1, \ldots, \theta_q)$, the superscript ^{T} referring to the transpose of a matrix, using the equation

$$\hat{\gamma} = (Y^T \Gamma^{-1} Y)^{-1} Y^T \Gamma^{-1} Z. \tag{3.8}$$

Here Γ is the covariance matrix of the noise term η_t and Y is the design matrix of the model (3.7). Thus Y contains the observations from the explanatory variables $Y_{t-k} - \hat{\mu}$ and $\hat{a}_{t-j}$ where $\hat{\mu}$ is either zero or the average of the differences $\nabla^d X_t$. Similarily Z is the column vector of the dependent variable $Z_t - \hat{\mu} = Y_t - \hat{a}_t - \hat{\mu}$.

In practice the covariance matrix Γ is unknown because it depends on the moving average parameters θ_j. For this reason we cannot apply (3.8) directly. Therefore we begin the estimation procedure by assuming $\Gamma = I$, i.e. we calculate the initial estimates $\theta_{(0)}$ by applying the ordinary least squares(OLS) method. This step is in fact rather close to the estimation method proposed by Hannan and Rissanen (1982). Having obtained the OLS estimates we can calculate the estimator $\Gamma_{(0)}$ for the covariance matrix Γ. This, on the other hand, leads to the GLS estimates

$$\hat{\gamma}_{(1)} = (Y^T \Gamma_{(0)}^{-1} Y)^{-1} Y^T \Gamma_{(0)}^{-1} Z.$$

The GLS estimates for the moving average parameters can be used to obtain the updated covariance matrix. This procedure can be iterated. Thus we have the iterative GLS estimator

$$\hat{\gamma}_{(m+1)} = (Y^T \Gamma_{(m)}^{-1} Y)^{-1} Y^T \Gamma_{(m)}^{-1} Z. \tag{3.9}$$

The last step of (3.9) produces the autoregressive and moving average estimates $\hat{\gamma}$ which are based on the initial autoregressive residuals $\hat{a}_t$ from (3.5). Similarily we have an estimate $\hat{\mu}$. The estimates $\hat{\gamma}$ and $\hat{\mu}$ can be used to obtain the residual series $\hat{a}_t(ARMA)$ from the ARMA model in question. By fitting again a long autoregression of order L_1, which is smaller than L, we can obtain refined residuals $\hat{a}_t$. These can be used then to take the GLS steps (3.9). The idea of selecting $L_1 < L$ is to make the estimates $\hat{\gamma}$ and $\hat{\mu}$ more independent of the order of autoregression to be used to calculate the innovation estimates which are needed in the calculation of the GLS estimates of the parameters.

3.3 Calculation of ARMA residuals

Suppose that we have to calculate the innovations a_t from a stationary and invertible ARMA(p, q) model

$$X_t = \sum_{k=1}^{p} \phi_k X_{t-k} + a_t - \sum_{k=1}^{q} \theta_k a_{t-k}, \tag{3.10}$$

namely,

$$a_t = \sum_{k=1}^{q} \theta_k a_{t-k} + X_t - \sum_{k=1}^{p} \phi_k X_{t-k}. \tag{3.11}$$

From this it is difficult to obtain good estimates for the innovations. This is because the moving average part in an ARMA(p, q) model causes (3.11) to be recursive, and there is an initialization problem to be solved. Depending on the values of the moving average parameters, the initial values used for the unknown a_t's can have a dramatic effect on the estimates $\hat{\gamma}$ and $\hat{\mu}$. For this reason we use a different method to calculate the innovation estimates a_t.

The first step in the calculation of the innovations is to form the time series

$$\eta_t = X_t - \sum_{k=1}^{p} \phi_k X_{t-k}. \tag{3.12}$$

Thus we first eliminate the autoregressive part from the observed time series. Then

$$\eta_t = a_t - \sum_{k=1}^{q} \theta_k a_{t-k}, \tag{3.13}$$

i.e. η_t is a MA(q) process. Let Σ be the covariance matrix of η_t with $\sigma^2 = 1$, and let

$$\Sigma = AA^T \tag{3.14}$$

be the Cholesky decomposition of Σ, where A is a lower triangular matrix. Then the approximation of the innovation series a_t can be obtained from the "innovation transformation"

$$a = A^{-1}\eta, \tag{3.15}$$

where η is the column matrix containing the observations from the time series η_t.

It is interesting to observe that the above method can be, technically speaking, applied to non-stationary and non-invertible models. If we apply (3.13) directly to non-invertible models, we will unavoidably have problems with overflow. This is because in a non-invertible case (3.13) is an explosive autoregression with respect to a_t.

4. NUMERICAL EXAMPLES

To illustrate the modelling procedure illustrated in this paper we built an ARIMA model for each of the time series A–F considered by Box and Jenkins (1970). The nature and the number of the observations n in these series are given in Table 1.

Table 1. The time series A–F considered by Box and Jenkins (1970).

Series	n	Nature of series
A	197	Chemical process concentration readings
B	369	IBM common stock closing prices
C	226	Chemical process temperature readings
D	310	Chemical process viscosity readings
E	100	Wolfer sunspot numbers
F	70	Yields from batch chemical process

In its present form our computer program allows time series of maximum length 200 to be analyzed. The restriction is only technical in its nature, and it can be easily changed so that the length of an observed time series causes no problems. However, because of this restriction, for time series B, C and D we were only able to build ARIMA models which were based on only parts of the observed time series. In Table 2 we report the models which were selected by the proposed method for the time series A–F using at most observations 1–200 in each series. For comparative purposes in the same table we also give the models for the corresponding time series which were tentatively identified by Box and Jenkins (1970) using the estimated autocorrelation and partial autocorrelation functions calculated from all available observations.

Table 2. Identified models for time series A–F using the proposed method and the tentative identifications given by Box and Jenkins (1970).

		Identified model	
Series	n	Proposed method	Box and Jenkins
A	197	$X_t = 7.32 + 0.57 X_{t-1} + \hat{a}_t$	ARIMA(1,0,1) or ARIMA(0,1,1)
B	369	$\nabla X_t = 0.23 \nabla X_{t-1} + \hat{a}_t$	ARIMA(0,1,1)
C	226	$\nabla X_t = 0.81 \nabla X_{t-1} + \hat{a}_t$	ARIMA(1,1,0) or ARIMA(0,2,2)
D	310	$X_t = 1.16 + 0.87 X_{t-1} + \hat{a}_t$	ARIMA(1,0,0) or ARIMA(0,1,1)
E	100	$X_t = 14.30 + 1.40 X_{t-1} - 0.71 X_{t-2} + \hat{a}_t$	ARIMA(2,0,0) or ARIMA(3,0,0)
F	70	$X_t = 72.39 - 0.42 X_{t-1} + \hat{a}_t$	ARIMA(2,0,0)

In Table 3 we give the models which were specified for the time series A–D when different subsections of these time series were analyzed using the proposed method.

Table 3. Identified models for time series A–D using the proposed method using different subsections of the proposed method.

Series	Observations	Identified model
A	1-100	$X_t = 8.19 + 0.52 X_{t-1} + \hat{a}_t$
	101-197	$X_t = 6.14 + 0.64 X_{t-1} + \hat{a}_t$
B	1-100	$\nabla X_t = \hat{a}_t$
	101-200	$\nabla X_t = 0.23 \nabla X_{t-1} + \hat{a}_t$
	201-300	$\nabla X_t = \hat{a}_t$
	301-369	$\nabla X_t = \hat{a}_t$
C	1-100	$\nabla X_t = 0.82 \nabla X_{t-1} + \hat{a}_t$
	101-226	$\nabla X_t = 0.90 \nabla X_{t-1} + \hat{a}_t - 0.27 \hat{a}_{t-1}$
D	1-100	$\nabla X_t = \hat{a}_t$
	101-200	$X_t = 1.78 + 0.81 X_{t-1} + \hat{a}_t$
	201-310	$X_t = 1.69 + 0.82 X_{t-1} + \hat{a}_t$

Table 2 indicates that the proposed method selects a stationary AR(1) model for the time series A using all 197 observations. Similarily, by Table 3 we see that both halves of the time series A lead to the same specification, i.e. to stationary AR(1) models. On the other hand the identification techniques proposed by Box and Jenkins (1970) lead to stationary ARMA(1, 1) model or to the nonstationary model where the first differences follow a MA(1) model.

In the case of the time series B the specifications proposed by the two methods are in fact very close to each other. This can be seen by Table 2. On the other hand Table 3 gives some hints on the fact that the first differences of the series are perhaps not stationary. This is because different sections do not lead to the same specification.

By considering Table 2 and Table 3 it can be seen that for C and D similar conclusions to B can be derived. For all of the time series B–D there are problems in transforming the observed time series into a stationary series. Different subsections of the series clearly indicate the observed time series, even though after differencing they are perhaps not stationary. For these time series varying parameter models might be reasonable alternatives.

For E and F the specified models produced by the two methods are close to each other again. Both of these series are so short that the analysis of subsections were not considered.

5. CONCLUDING REMARKS

In this paper we have introduced a new model building method of ARIMA (p, d, q) models. The idea behind the method is the same as in Box and Jenkins(1970), i.e. to find a parsimoniously parametrized model which transforms an observed time series into white noise. To test the whiteness of the residual series an autoregressive order selection method is used to select the order of autoregression for the residual series. If the selected autoregressive order is zero, the model is considered to be adequate. If a consistent autoregressive order selection criterion such as the log criterion is applied in testing the whiteness of the residual series, the proposed method can be regarded as a consistent order selection criterion in the family of ARIMA(p, d, q) models. The method leads to an adequate model where the sum $p + d + q$ has the smallest possible value, i.e. to a parsimoniously parametrized model.

REFERENCES

Abraham, B. and Ledolter, J. (1984). A note on inverse autocorrelations. *Biometrika*, 71, 609–614.

Beguin, J.M., Gourieroux, C. and Monfort, A. (1980). Identification of a mixed autoregressive-moving average process: The corner method. *Time Series*, Ed. O.D. Anderson. North-Holland, Amsterdam, 423–436.

Bhansali, R.J. (1980). The autoregressive and the window estimate of the inverse correlation function. *Biometrika*, 67, 511–566.

Bhansali, R.J. (1983). The inverse partial correlation function of a time series and its application. *Journal of Multivariate Analysis*, 13, 310–327.

Box, G.E.P. and Jenkins, G.M. (1970). *Time Series Analysis, Forecasting and Control*. Holden-Day, San Francisco.

Burg, J.P. (1967). Maximum entropy spectral analysis. Paper presented at *The 37th Annual International S.E.G. Meeting*. Oklahoma City, October 31, 1967.

Chatfield, C. (1979). Inverse autocorrelations. *J. R. Statist. Soc. Ser. A*, 142, 363–377.

Cleveland, W.S. (1972). The inverse autocorrelations of a time series and their applications. *Technometrics*, 14, 277–298.

de Gooijer,J.G. , Abraham, B., Gould, A., and Robinson, L. (1985). Methods for determining the order of an autoregressive-moving average process: a survey. *International Statistical Review*, 55, 3, 301–329.

Hannan, E.J. and Rissanen, J. (1982). Recursive estimation of mixed autoregressive-moving average order. *Biometrika*, 69, 81–94.

Hipel, K.W., McLeod, A.I. and Lennox, W.C. (1977). Advances in Box-Jenkins modelling 1. Model construction. *Water Resources Res.*, 13, 567–575.

Pukkila, T.M. and Krishnaiah, P.R. (1986a). On the use of autoregressive order determination criteria in univariate white noise tests. Technical Report # 86-15. Center for Multivariate Analysis, University of Pittsburgh, Pittsburgh, U.S.A.

Pukkila, T.M. and Krishnaiah, P.R. (1986b). On the use of autoregressive order determination criteria in multivariate white noise tests. Technical Report # 86-16. Center for Multivariate Analysis, University of Pittsburgh, Pittsburgh, U.S.A.

Rissanen, J. (1978). Modelling by shortest data description. *Automatica*, 14, 465–471.

Schwarz, G. (1978). Estimating the dimension of a model. *Ann. Statist.*, 6, 461–464.

Tiao, G.C. and Tsay, R.S. (1983a). Consistency properties of least squares estimates of autoregressive parameters in ARMA models. *Ann. Statist.*, 11, 856–871.

Tiao, G.C. and Tsay, R.S. (1983b). Multiple time series modelling and extended sample cross-correlations. *J. Business Econ. Statist.*, 1, 43–56.

Tsay, R.S. and Tiao, G.C. (1984). Consistent estimates of autoregressive parameters and extended sample autocorrelation function for stationary and nonstationary ARMA models. *J. Am. Statist. Assoc.*, 79, 84–96.

Wold, H. (1938). *A Study in the Analysis of Stationary Time Series*. Almquist and Wicksell, Uppsala (2nd ed. 1954).

Woodward, W.A. and Gray, H.L. (1978). New ARMA models for Wölfer's sunspot data. *Comm. Statist. B*, 7, 97–115.

Received 15 June 1987

Department of Mathematical Sciences
University of Tampere
P. O. Box 607
SF-33101 Tampere
Finland

Proc. Second International Tampere Conference in Statistics
(Tampere, Finland, 1-4 June 1987)
Tarmo Pukkila and Simo Puntanen, *Editors*

pp. 289 - 302

INVITED PAPER

Complexity and information in contingency tables

Jorma RISSANEN
IBM Almaden Research Center, San Jose, CA, U.S.A.

Key words and phrases: Total maximum likelihood method, least code length, estimation of laws, small sample hypothesis testing.

ABSTRACT

All the statistical constraints in a contingency table, represented by classes of multinomial model distributions, may be found by computing the corresponding stochastic complexities of the data. Formulas for the complexities of 2- and 3-way contingency tables, relative to some of the usual families of multinomial models, are given.

1. INTRODUCTION

In this paper we discuss a new way to study contingency tables, in which the familiar Neyman-Pearson hypothesis testing technique with its asymptotic approximation for the distribution of the test statistic and the arbitrary choice of the confidence level is not needed. Neither do we have any place for individual judgment. The approach is based upon the principle of maximizing a global likelihood of the observed table in each fitted class of models, and selecting that class for which the likelihood is largest. Alternatively, and equivalently, we may compute the *stochastic complexity* of the data, Rissanen (1986), (1987), relative to each class of multinomial models, and we pick that class for which the complexity is smallest as providing the best explanation of the data. An independent interpretation of the stochastic complexity is that it represents the greatest lower bound for the number of binary digits with which the observed data can be encoded, when advantage is taken of the selected models. It is our thesis that the calculation of the complexities, relative to the different classes, gives a full account of all the relationships between the variables in the table that on the whole can be expressed with the fitted models. We illustrate the approach with examples of 2-way and 3-way contingency tables.

Further, a special maximum complexity, resulting when no model is fitted to the data, provides a convenient level of reference, against which the other complexities may be compared. We define the complexity difference from this reference to represent the amount of information about the data that we have been able to learn with the corresponding class.

2. STOCHASTIC COMPLEXITY

Consider first a class M of parametric distributions $\{P(x \mid \theta, M):\theta \in \Omega^k\}$ together with a "prior" $\pi(\theta \mid M)$, where for each positive integer k, Ω^k denotes a subset of the k-dimensional euclidean space with a nonempty interior. The last requirement insures that the k parameters are mathematically independent. Further, we assume a set $\{M\}$ of the model classes, which in the case of the main interest to us in this paper is finite. In this setting, we may ask for a distribution $P(. \mid M)$, somehow determined by the members of the model class M, such that the probability $P(x \mid M)$ of a given string of observations $x = x^n = x_1, \dots, x_n$ is maximum. We consider here only the cases where each *symbol* x_i belongs to a finite set A.

There is a natural candidate for the sought-for distribution, which in fact can be shown to be near optimal for long samples in a certain well defined sense, Rissanen (1986), namely,

$$P(x \mid M) = \int P(x \mid \theta, M) d\pi(\theta \mid M), \tag{2.1}$$

where the integration is over the k-dimensional set of the parameters. Related constructs for various other purposes are also discussed in Aitchison (1975) and Dawid (1984). We call the negative logarithm $I(x \mid M) = -\log P(x \mid M)$ the *stochastic complexity* of the data x, relative to the class M. (All logarithms in this paper are to the base two.) It is a standard fact in coding theory that a binary code can be constructed such that the number of binary digits in the codeword for x is given by the least integer upper bound for $-\log P(x \mid M)$. In Rissanen (1986) and (1987) we showed, in essence, that for long strings of the same length the mean of $I(x \mid M)$, the mean taken relative to any distribution in the class M, except for a null set, gives the shortest mean code length with all ways of doing the coding. Because the distribution of $I(x \mid M)$ for long strings is very peaked about the mean, it follows that $I(x \mid M)$ is close to being the shortest code length for all but rare "atypical" strings for that source. Equivalently, we have been able to assign a near maximum probability to each of the typical strings.

Equation (2.1) might suggest that the idea of stochastic complexity requires a prior and Bayesian arguments. This is not true, for we can define the stochastic complexity without any a priori given distribution $\pi(\theta)$, even when the space of parameters is not compact. Indeed, define first $C(x^t) = \int P(x^t \mid \theta) d\theta$, which for all the usual likelihood functions exists for $t \geq t_0$. For simplicity, take $t_0 = 1$. Next define a distribution $\pi(\theta \mid x^t) = C(x^t)^{-1} P(x^t \mid \theta)$, conditional on the data x^t. Then

$$P(x_{t+1} \mid x^t) = \int P(x_{t+1} \mid x^t, \theta) d\pi(\theta \mid x^t) = C(x^{t+1})/C(x^t)$$

defines a conditional distributions, and by taking their product we get

$$I(x \mid M) = -\log C(x) + \log \int C(x_1) dx_1, \tag{2.2}$$

where we now let x_1 denote a variable rather than the first observed item. For uniform priors the two definitions (2.1) and (2.2) agree. The idea of stochastic complexity differs further from the Bayesian approaches in that the key distribution is $P(x)$, defined for the data, rather than the posterior distribution $\pi(\theta \mid x)$, which is incapable of distinguishing between models with different numbers of parameters; see the discussions following Rissanen (1987).

We are particularly interested in that class for which the stochastic complexity is the smallest or, equivalently, which permits the greatest probability assignment to the observed data. That class we take to provide the best "explanation" of the data in the sense of incorporating the statistical regular features of the studied kind.

Let $|A|$ denote the number of distinct symbols. Then the uniform distribution $P(x_i) = 1/|A|$ on the symbols, extended to the strings by independence, gives a special model defining the singelton class $M_0 = \{P(x)\}$. The associated complexity $I(x \mid M_0) = n \log |A|$ represents a certain worst case complexity, for

$$n \log |A| = \max_x \min_p - \sum_i n_i(x) \log p_i, \tag{2.3}$$

where the strings x run through all strings of length n and $p = \{p_i\}$ through all the symbol distributions. The count $n_i(x)$ denotes the number of times symbol i occures in the string x. We define the difference, Rissanen (1986),

$$U(x \mid M) = I(x \mid M_0) - I(x \mid M) \tag{2.4}$$

to be a measure of the amount of *information* that we have been able to extract or learn from the data with the class M. Hence, in these terms, the class with the smallest stochastic complexity extracts the maximum amount of information from the data. If, in an extreme case, we have been unable to reduce the unmodeled complexity with any of our model classes, we have learned nothing: the proposed model classes do not capture anything worthwhile in the data. This happens, in particular, when the observations define a purely random string, about which there is nothing to learn.

3. 1-WAY TABLES

We begin with a primitive case of contingency tables, namely, the 1-way tables. The resulting formulas are needed repeatedly. A 1-way contingency table T consists of d integers, $n_1, \ldots, n_d$. The interpretation is that a string $x = x_1, \ldots, x_n$ in the symbols $1, \ldots, d$ has been observed, and the symbol i occurs n_i times in the string. Some of these counts may be zero. Pick the positive symbol probabilities $p_i = P(i)$ as parameters, which are extended to probabilities for strings by independence, $P(x \mid p_1, \ldots, p_d) = \Pi\, p_i^{n_i}$. The space of the parameters, is defined by the constraint $\Sigma p_i \leq 1$. The integral needed for $C(x^t)$ can now be carried out, and we get from (2.2) the stochastic complexity of the string as $I(x \mid B) = -\log P(x \mid B)$, where

$$P(x \mid B) = \frac{(d-1)!}{(n+d-1)!} \prod_{i=1}^{d} n_i! = \binom{n}{n_i}^{-1} \binom{n+d-1}{n}^{-1}. \tag{3.1}$$

A simple derivation of this is obtained by partial integration and induction, first for the case $d = 2$, and then by breaking the general case into a series of the binary ones. The probability of the table T is clearly

$$P(T \mid B) = \binom{n+d-1}{n}^{-1}, \tag{3.2}$$

and the corresponding stochastic complexity is

$$I(T \mid B) = I(x \mid B) - \log \binom{n}{n_i} = \log \binom{n+d-1}{n}. \tag{3.3}$$

Hence, in particular, for $d = 2$ we get $I(T \mid B) = \log(n_1 + n_2 + 1)$, reflecting the fact that there are $n + 1$ such tables with a given sum of the counts

$n = n_1 + n_2$. We get an instructive approximation formula for the stochastic complexity using Stirling's approximation:

$$I(x \mid B) \cong nH(\{n_i/n\}) + \frac{d-1}{2} \log n + \sum_{i=1}^{d} \log \frac{n}{n_i}, \tag{3.4}$$

where $H(\{n_i/n\}) = \sum_i \frac{n_i}{n} \log \frac{n}{n_i}$,

In addition to the class B we have the second class of models M_0, corresponding to no free parameters, as defined in the previous section, and the corresponding string complexity is given by $I(x \mid M_0) = n \log d$. The table complexity $I(T \mid M_0)$ is still given by the same difference as in (3.3), or

$$I(T \mid M_0) = I(x \mid M_0) - \log \binom{n}{n_i}.$$

Having defined the two classes for the d-ary strings, we may ask about a particular string to which class it most likely belongs. In other words, we may make the test

$$I(x \mid M_0) \leq I(x \mid B). \tag{3.5}$$

In view of the maximum property (2.3) of $I(x \mid M_0)$, we may define a string as random, relative to the class B, if and only if it satisfies (3.5). This is an instance of an exact definition of randomness, relative to a class of models, which is free from arbitrarily selected thresholds. Evidently, we do not mean the test with this particular model class B to capture our intuitive notion of randomness. For that a far more complex model class is needed. With (3.4) the test becomes (approximately)

$$n(\log d - H(\{n_i/n\})) \leq \frac{d-1}{2} \log n + \sum_{i=1}^{d} \log \frac{n}{n_i}. \tag{3.6}$$

The left hand side is non-negative and it grows as the counts deviate from the equal counts, while the right hand side, representing the "cost" of estimating the parameters, is positive. We therefore accept the randomness hypothesis except when the counts differ sufficiently from the equal counts. Under the randomness hypothesis, represented by the class M_0, the first term is asymptotically χ^2 distributed with $d-1$ degrees of freedom. Hence, we see that our test is like the usual test except that the size of the test automatically decreases to zero as n grows, as it should. For $d = 2$ the size turns out to be a few percent for the sample size ranging from 50 to 1000, which corresponds quite well to the usual subjectively selected value, 5 % . In general, one can show that whenever the null hypothesis is simple the test based upon the stochastic complexity difference is uniformly most powerful or unbiassed such for the automatically defined threshold whenever such test statistics exist, of course.

4. 2-WAY TABLES

A 2-way contingency table $T = \{ n_{ij} \mid i = 1, \ldots, r, j = 1, \ldots, s \}$ is an $r \times s$-array of the occurrence counts of the symbols (i,j), $i = 1, \ldots, r$, $j = 1, \ldots, s$, in a string $x = (i_1, i_2), \ldots, (i_n, j_n)$ of length n. Here, too, we allow some of the counts to be zero. We consider two classes of models, M_1 and M_2, corresponding to the hypotheses "independent" and "free", respectively, together with the third class M_0, consisting of a single model defined by $p_{ij} = 1/rs$.

We begin with the class of "free" models, which is defined by the set $\{ p_{ij} \}$ of the rs cell probabilities $p_{ij} = P((i,j))$, among which there is only the constraint that they are non-negative and add up to unity. Hence, the parameter space is a closed subset of the $rs - 1$-dimensional euclidean space with non-empty interior; the number of free parameters is then $rs - 1$. In the literature these models are called "saturated", and because they have as many parameters as there are free entries in the table, assuming the total count fixed, they are regarded as uninteresting with which nothing useful can be learned about the data. This mistaken conclusion has its roots in the awkward and roundabout way contingency tables are being studied traditionally by fitting the usual likelihood functions. Once the effect of having to estimate the parameters is included, as done in our global likelihood maximization principle, there is nothing pathological in the saturated models, even if we fit them to the counts rather than to the data strings themselves. As we shall see in the next section, the saturated model may well be the best, which simply means that there is an association between the variables, and the data cannot be justifiably imposed any of the various conditional and partial independence constraints. By the formula (3.1) the stochastic complexity in this model class is given by

$$I(x \mid M_2) = \log \binom{n}{n_{ij}} + \log \binom{n + rs - 1}{n}. \qquad (4.1)$$

In the class of independent models the cell probabilities are constrained such that they be the product of the marginal probabilities $p_{ij} = p_{i\cdot}\, p_{\cdot j}$. There are then $r + s - 2$ free parameters, and it is readily shown that

$$I(x \mid M_1) = \log\left[\binom{n}{n_{i\cdot}} \binom{n}{n_{\cdot j}} \right] + \log\left[\binom{n + r - 1}{n} \binom{n + s - 1}{n} \right]. \qquad (4.2)$$

Finally, in the third class the complexity is given by $I(x \mid M_0) = n \log(rs)$, and we then have for the information that class M_i can extract from the table the formula $U(T \mid M_i) = I(T \mid M_0) - I(T \mid M_i)$.

The corresponding stochastic complexities of the table is obtained simply by subtracting the first term in (4.1), the logarithm of the number of strings with the given counts, from the string complexities. Hence,

$$I(T \mid M_k) = I(x \mid M_k) - \log \binom{n}{n_{ij}}, \quad k = 0, 1, 2. \tag{4.3}$$

Of particular interest is the information difference between the free and the independent models, $U = I(T \mid M_1) - I(T \mid M_2)$, which with Stirling's formula may be written approximately as

$$U \cong \sum_{i,j} n_{ij} \log \frac{nn_{ij}}{n_{i.}n_{.j}} - \frac{k}{2} \log n + \frac{1}{2} (\sum_i \log \frac{n_{i.}}{n} + \sum_j \log \frac{n_{.j}}{n} - \sum_{ij} \log \frac{n_{ij}}{n}),$$

where $k = (r - 1)(s - 1)$ denotes the difference between the numbers of free parameters in the two model classes. We may take U/n to represent the amount of association between the two variables, as found in the sample. Ordinarily the dominant term is the first, which has the form of the mutual information between the two random variables, defined by the sample counts, namely

$$H(\{ \frac{n_{i.}}{n} \}) + H(\{ \frac{n_{.j}}{n} \}) - H(\{ \frac{n_{ij}}{n} \}) = \frac{1}{n} \sum_{i,j} n_{ij} \log \frac{nn_{ij}}{n_{i.}n_{.j}}.$$

Alternatively, this term is one half of Kullback's G^2 measure, which was taken to represent the information difference between the two hypotheses in Gokhale and Kullback (1978). Our measure U is seen to differ from this by an amount reflecting the uncertainty about the parameter values which, of course, are not known. We might reason that Kullback's measure is appropriate if we wish to measure the information difference between two 2-variate random variables, corresponding to the two hypotheses, whose distributions are exactly defined by the counts. However, it is less appropriate if the distributions are not known and must be estimated, which is precisely the case which our measure is meant for. As a final remark, many of the traditional measures of association are also related to the mutual information, Kendall and Stuart (1961).

We conclude this section with an example.

Example. (Kendall and Stuart, (1961,p. 552)). The counts 4, 16 (first row) and 1, 21 (second row), defining a two-way table, are obtained from 42 children according to the nature of their teeth and the type of feeding. The objective is to test whether the type of feeding is independent of the nature of teeth. The test statistic $U(x) = I(x \mid M_2) - I(x \mid M_1)$ is given by

$$U(x) = \log \frac{n!n_{11}!n_{12}!n_{21}!n_{22}!}{n_{1.}!n_{2.}!n_{.1}!n_{.2}!} + \log \frac{3!(n+1)}{(n+2)(n+3)} = 0.056.$$

Hence, the independence hypothesis is narrowly rejected. This example is a border line case, for in Kendall and Stuart (1961) the null hypothesis was accepted with one of the two "natural" thresholds and rejected with the other. We also see, Kendall and Stuart (1961, Section 33.24), that this test is uniformly most powerful unbiassed even though neither hypothesis is simple.

5. 3-WAY TABLES

This time the data string is given by $x = (i_1, j_1, k_1), \ldots, (i_n, j_n, k_n)$, where the triplets of indices range over the cells of a $r \times s \times t$- table T. There are now a large number of possible model classes, of which we consider the most usual ones.

For the class of free models we have as above

$$I(x \mid M_5) = \log \binom{n}{n_{ijk}} + \log \binom{n + rst - 1}{n}. \tag{5.1}$$

The next most complex model is the so-called "no (second order) interaction" model, which also is the most difficult to treat, and we discuss it last. There are three permutations of the next class of models, called "zero partial correlation between rows and columns with layers fixed". We give the complexity of one, namely, when

$$p_{ijk} = \frac{p_{\cdot jk}}{p_{\cdot\cdot k}} \times \frac{p_{i\cdot k}}{p_{\cdot\cdot k}} \times p_{\cdot\cdot k}.$$

This states in words that there is independence between the first two variables, the rows and the columns, when the third variable; i.e., the layer, is held fixed. This model class has $t(r + s - 1) - 1$ independent parameters, and the string complexity, as derived in the Appendix, is given by

$$\begin{aligned} I(x \mid M_{31}) = \sum_k \log \left[\binom{n_{\cdot\cdot k}}{n_{i\cdot k}} \binom{n_{\cdot\cdot k}}{n_{\cdot jk}} \right] + \log \binom{n}{n_{\cdot\cdot k}} \\ + \sum_k \log \left[\binom{n_{\cdot\cdot k} + r - 1}{n_{\cdot\cdot k}} \binom{n_{\cdot\cdot k} + s - 1}{n_{\cdot\cdot k}} \right] + \log \binom{n + t - 1}{n}. \end{aligned} \tag{5.2}$$

There are also three permutations of the next class of models, of which we give the complexity of one, namely, when $p_{ijk} = p_{i\cdot\cdot}\, p_{\cdot jk}$. This clearly states that the first variable is independent of the last two, considered to have a bivariate distribution. There are then $st + r - 2$ free parameters, and the complexity is given by

$$I(x \mid M_{21}) = \log\left[\binom{n}{n_{.jk}}\binom{n}{n_{i..}}\binom{n+r-1}{n}\binom{n+st-1}{n}\right]. \quad (5.3)$$

Further, in the class of independent models, where each cell probability is the product of the marginal probabilities, there are $r + s + t - 3$ independent parameters, and the complexity is obtained by repeated use of (3.1) as

$$I(x \mid M_1) = \log\left[\binom{n}{n_{i..}}\binom{n}{n_{.j.}}\binom{n}{n_{..k}}\right] + \log\left[\binom{n+r-1}{n}\binom{n+s-1}{n}\binom{n+t-1}{n}\right]. \quad (5.4)$$

In the simplest model class the complexity is given by $I(x \mid M_0) = n \log(rst)$, which gives the information that class M can extract from table T as $U(T \mid M) = I(T \mid M_0) - I(T \mid M)$, where the table complexities are given in terms of the string complexities by

$$I(T \mid M) = I(x \mid M) - \log\binom{n}{n_{ijk}}. \quad (5.5)$$

We now turn to the remaining class, the "no second order interaction" class, which has $m = (r-1)t + s(t-1) + r(s-1) - 1$ independent parameters, for it is of the form

$$p_{ijk} = r_{ik}s_{jk}t_{ij}$$

where

$$\sum_i r_{ik} = \sum_k s_{jk} = \sum_j t_{ij} = \sum_{uvw} r_{uw}s_{vw}t_{uv} = 1, \quad (5.6)$$

for all k, j, and i, respectively. It is the last equation which causes complications in the evaluation of the integral needed for the stochastic complexity. The role of this equation is to make sure that we arrive at a proper distribution $P(x \mid M_4)$, which we, however, can obtain in a slightly different manner. Indeed, by letting $\theta = \{r_{ik}, s_{jk}, t_{ij}\}$ stand for the parameters, we can evaluate the integral $C(x) = \int P(x \mid \theta) d\theta$ over the space defined by the equations in (5.6), except for the last, and we get

$$C(x) = \int \prod_{ik} r_{ik}^{n_{i.k}} dr_{ik} \int \prod_{jk} s_{jk}^{n_{.jk}} ds_{jk} \int_{ij} t_{ij}^{n_{ij.}} dt_{ij}.$$

Hence from (3.1) we have

$$- \log C(x) = \sum_i \log \binom{n_{i..}}{n_{ij.}} + \sum_j \log \binom{n_{.j.}}{n_{.jk}} + \sum_k \log \binom{n_{..k}}{n_{i.k}}$$
$$+ \log \prod_{ijk} \binom{n_{i..} + s - 1}{n_{i..}} \binom{n_{.j.} + t - 1}{n_{.j.}} \binom{n_{..k} + r - 1}{n_{..k}}. \tag{5.7}$$

$C(x)$ is not a proper distribution but $C(x_{t+1} \mid x^t) = K(x^t)^{-1} C(x^{t+1})/C(x^t)$ is, where

$$K(x^t) = \sum_{x_{t+1}} C(x_{t+1} \mid x^t)$$
$$= \sum_{ijk} \frac{(n_{ij.} + 1)(n_{i.k} + 1)(n_{.jk} + 1)}{(n_{i..} + s)(n_{.j.} + t)(n_{..k} + r)}, \tag{5.8}$$

where all the counts are obtained from the string x^t. Hence, we may approximate the stochastic complexity relative to the class M_4 as

$$I(x^n \mid M_4) = - \log C(x^n) + \sum_{t=0}^{n} \log K(x^t). \tag{5.9}$$

Quite remarkably the term $K(x)$ appears to be very close to unity no matter what x is. We cannot actually prove that it is exactly unity except when all the counts are equal. Therefore, when the second term in (5.9), which depends on the order of the triplets in x, is so ordered that for each t the counts are kept as equal as possible, which is easily done, then this term is very small. This is seen in the examples below, which involve hundreds of evaluations of the coefficient $K(x^t)$.

We conclude this paper by discussing three examples.

Example 1. (Bartlet's data from Bishop, Fienberg, Holland (1975, p.87) on plant growth due to different treatments). All the variables have two values; those of the first are "alive" and "dead"; the values of the second are planting times; and the third variable has two values of cutting length. The eight entries in the table, written in the lexical order $n_{111}, n_{112}, \ldots, n_{222}$, are 156, 107, 84, 31, 84, 133, 156, 209.

Example 2. (Gokhale and Kullback (1978, p. 137)). This is a 4 × 4 × 2 table listing occurrence counts of persons with observed blood pressure, variable 1, serum cholesterol level, variable 2, and coronary heart disease condition, yes or no, variable 3. The entries are in the lexical order 2, 117, 3, 85, 8, 119, 7, 67, 3, 121, 2, 98, 11, 209, 12, 99, 3, 47, 1, 43, 6, 68, 11, 46, 4, 22, 3, 20, 6, 43, 11, 33.

Example 3. (Luoma and Taanonen (1980, p. 9)). This is a 3 × 3 × 2 table of the relationship between drinking and smoking habits among a group of students, the counts collected separately for males and females. The first variable is the frequency of alcohol consumption, the second is the amount of smoking, and the third is sex. The entries are 11, 23, 2, 7, 1, 3, 4, 3, 1, 4, 2, 4, 4, 5, 3, 0, 6, 4.

The following table has the values of the stochastic complexities $I(T|M)$ along with the number of the independent parameters m of nine models for the three examples.

	Example 1		Example 2		Example 3	
model	m	I	m	I	m	I
M5	7	*57.1	31	209.6	17	63.6
M4	5	60.6	21	173.4	12	66.1
M31	5	128.4	13	*167.0	9	70.8
M32	5	91.2	19	193.7	11	66.2
M33	5	58.1	19	195.5	11	63.0
M21	4	157.2	10	181.8	7	70.5
M22	4	124.2	10	185.9	7	66.9
M23	4	87.0	16	212.1	9	*62.4
M1	3	154.1	7	191.4	5	68.3
M0	0	171.7	0	1288.2	0	84.3

Table 1. Three examples

In Example 1 the best model class is the class of the free models, which means that there is an association between all the variables. This is not to be taken as obvious, because it means that imposing any of the suggested constraints with the hope of finding the corresponding "law" simply will not improve the result. Hence, no such constraint is supported by the evidence, and the class of free models is the most "parsimonious" when one takes into account what the model accomplishes. A close second is the class M_{33}, which states that for each outcome of the plant survival the time of planting and the cutting length are independent. In Bishop, Fienberg, Holland (1975) the best model was judged to be M_4, which in truth does not trail the other two by all that much. Incidentally, the second term in (5.9) was -1.22, which suggests strongly that imposing the constraint to force the sum of p_{ijk} to be unity would lead to a longer code length than by performing the normalization after the integral, as done here.

In Example 2 the best model, rather clearly, is M_{31}, which states that the blood pressure and the cholesterol level are independent in each group of people, those with and without heart disease. If we look at the other two models of this kind, M_{32} and M_{33}, we see that there is a clear association between blood pressure and heart disease as well as between cholesterol level and heart disease. All these findings agree well with what one might expect about the association between the three variables. Finally, for the class M_4 the second term in (5.9) was no larger than 1.49.

Finally, in Example 3 the best model class is M_{23}, which states that the drinking and smoking habits do not depend on sex, while there is an association between the frequency of drinking and the amount of smoking. The result agrees with that in Luoma and Taanonen (1980), which was arrived at with a combination of hypothesis testing and "sound judgment". For the model class M_4 the second term in (5.9) was only .02.

In conlusion we may say that our mechanical procedure for studying relationships between the variables in contingency tables appears to be sound, what one might, of course, expect. After all, what better guidance is there than judge model classes by the probability they are able to assign to the table. And the best model corresponds quite well to the intuitive interpretation, "the most likely explanation" of the observed data. By contrast, we know of no other procedure where the intuitively attractive idea of the "most parsimonious" model could even be defined in an objective and meaningful manner let alone computed. After all, the statisticians' tinkering with the "most significant" variables and any personal opinions in their selection mean strictly speaking nothing, and no such technique is acceptable in a scientific investigation. Of course, the investigator may wish to complement the analysis by computing the maximum likelihood parameters, above all in the winning model.

However, this still does not answer the question of how to assess the reliability of the findings. For example, in Example 1 the information difference between the two best models is only one bit, and it is not clear at all whether that represents a "significant" difference or not. Such an assessment, which often affects the decisions we make, can be made as soon as we agree on the subjective yardstick by which to judge the results. Clearly, the issue is to study which part is played by the uncertainty due to sampling. However, since there is no "true" distribution behind the data, such sampling effects must be expressed in terms of the modeled distributions. Such an analysis can be done in the Bayesian decision theoretic framework along the following lines: When each model class is assigned equal probability, which is implicitly taken here for the reason that the same code length to be added to each complexity is dropped, we get a distribution for all the complexities. With Monte Carlo experiments we can get as reliable an idea of this distribution as we please, as well as of any decision function of these that we want to consider. Hence, by checking where the actual table T falls, we can form the desired probabilistic idea of the reliability of our decision. We maintain, however, that it is the test statistics $I(x \mid M)$ and not some others that we ought to study, and that there is no point to replace the Monte Carlo experiments by some asymptotically derived approximations. We also emphasize that all such findings have a meaning only relative to the chosen model classes, and therefore they lack any absolute significance.

Appendix

Derivation of (5.2). For the sake of notational simplicity we pick $r = s = t = 2$. The derivation of the general case is completely similar. From the product formula for p_{ijk} we see that the free parameters are the five binary probabilities

$$q_{1\cdot k} = \frac{p_{1\cdot k}}{p_{\cdot\cdot k}}, \quad q_{\cdot 1k} = \frac{p_{\cdot 1k}}{p_{\cdot\cdot k}}, \quad k = 1,2, \ p_{\cdot\cdot 1},$$

so that for example $q_{2\cdot k} = 1 - q_{1\cdot k}$. Therefore,

$$P(x \mid M_{31}) = \prod_{ijk} (q_{i\cdot k} q_{\cdot jk} p_{\cdot\cdot k})^{n_{ijk}} = \prod_{ijk} q_{i\cdot k}^{n_{i\cdot k}} q_{\cdot jk}^{n_{\cdot jk}} p_{\cdot\cdot k}^{n_{\cdot\cdot k}},$$

which upon writing out in full can be seen to be a product of five binary string probabilities, one for each of the free parameters. Therefore, by applying (3.1) for $d = 2$ on each, (5.2) results.

REFERENCES

Aitchison, J. (1975) "Goodness of prediction fit", Biometrika, 62, 547-554.

Bishop, M.M., Fienberg, S.E., Holland, P.W. (1975). Discrete Multivariate Analysis: Theory and Practice, The MIT Press, Cambridge, Mass., and London, England.

Cox, D.R. and Hinkley, D.V. (1974), Theoretical Statistics. Imperial College, London, 511 pages.

Dawid, A.P. (1984), "Present Position and Potential Developments: Some Personal Views, Statistical Theory, The Prequential Approach", J. Royal Stat. Soc. Series A, Vol. 147, Part 2, 278-292.

Gokhale, D.V. and Kullback, S. (1978), The Information in Contingency Tables, Marcel Decker, New York.

Kendall, M.G. and Stuart, A. (1961), The Advanced Theory of Statistics, Vol. 2., Hafner Publishing Co., New York.

Kullback, S. (1959), Information Theory and Statistics, John Wiley, New York.

Luoma, M., Taanonen, M. (1980), The Analysis of Contingency Tables with Loglinear Models", Research Paper Nr. 69, Vaasa School of Economics, Finland, (in Finnish).

Rissanen, J. (1986), "Stochastic Complexity and Modeling", Ann. of Statistics, Vol 14, No 3, 1080-1100.

Rissanen, J. (1987), "Stochastic Complexity", Royal Statistical Society B, 49, No. 3. (to appear)

Received 16 January 1987

IBM Almaden Research Center
650 Harry Road
San Jose, CA 95120-6099
U.S.A.

Proc. Second International Tampere Conference in Statistics
(Tampere, Finland, 1-4 June 1987)
Tarmo Pukkila and Simo Puntanen, *Editors*

pp. 303 - 316

INVITED PAPER

Generalized linear models with survey data

Alastair J. SCOTT
University of Auckland, New Zealand and University of Southampton, U.K.

Key words and phrases: Cluster sampling, stratification, singular generalized linear models.

ABSTRACT

In this paper we look at generalized linear models for vectors of estimated proportions obtained from a complex sample survey. We look first at iterative procedures for simple designs where there is an explicit expression for the covariance matrix of the estimated proportions in terms of the corresponding vector of population proportions and then at approximate methods for more complex designs. Corrections are developed for the output obtained when the data is run through a standard program ignoring the sampling structure.

1. INTRODUCTION

This paper deals with the analysis of tables of estimated totals or proportions obtained from a sample survey rather than a designed experiment. Let $\hat{\mathbf{p}}$ denote the $k \times 1$ vector of estimated proportions arranged in some order and let $\mathbf{p}$ be the corresponding vector of population proportions. We are interested in fitting models of the form

$$\mathbf{g}(\mathbf{p}) = \mathbf{X}\boldsymbol{\beta}$$

where $\mathbf{g}$ is a $k \times 1$ vector of twice differentiable functions of $\mathbf{p}$, $\mathbf{X}$ is a $k \times r$ matrix of known constants and $\boldsymbol{\beta}$ is a $r \times 1$ vector of unknown parameters.

EXAMPLE 1. The data shown in Table 1 come from a survey of 12 Wisconsin firms carried out by the Department of Rehabilitation Psychology at the University of Wisconsin to study behaviour of employees who have been through a programme to stop smoking.

TABLE 1

Success vs change in level of exercise

Still smoking?	Level of exercise Somewhat less	No change	Somewhat more	Much more
Yes	3	118	73	23
No	5	32	39	22

We are interested in the hypothesis of independence between success in stopping smoking and change in the level of exercise i.e. we want to check the fit of the model

$$\log p_{ij} = a_i + \beta_j \qquad (i = 1, 2;\ j = 1, \ldots, 4).$$

Obviously it could be misleading to use the ordinary chi-squared statistic, pretending that we have a simple random sample of 315 individuals.

EXAMPLE 2. Table 2 is based on interviews with 9918 women in the 1981 Canada Health Survey, a complex multi-stage survey covering about 12,000 households conducted by Statistics Canada. The estimates also have a complex structure involving ratio estimation and post-stratification.

A natural way of analysing such a table is to fit a model involving age and education effects for the logits, $\log p_{ij1}/p_{ij2}$, but again standard methods based on random sampling are inappropriate.

Note that the model in Example 2 is of the form

$$\mathbf{A}\, \mathbf{g}(\mathbf{p}) = \mathbf{X}\boldsymbol{\beta},$$

where $\mathbf{A}$ is a known $u \times k$ matrix with $u < k$. This is equivalent to

$$\mathbf{g}(\mathbf{p}) = \mathbf{C}\boldsymbol{\gamma} + \mathbf{A}^{-}\mathbf{X}\boldsymbol{\beta}$$

TABLE 2

Frequency of breast self-examination by age and education

Age	Post-secondary education	Frequency of breast self-examination	
		Sometimes	Never
15 - 24	Some	.04	.02
	None	.08	.10
25 - 44	Some	.10	.03
	None	.18	.07
45 +	Some	.05	.01
	None	.20	.11

for some **C** of rank $k - u$ with $\mathbf{AC} = \mathbf{0}$ and hence comes under the general framework of this paper. Other applications involve linear functions (particularly for models involving the marginal proportions), power transformations of individual proportions etc. [See, for example, Bonett, Woodward and Bentler (1985), Stirling (1986), Haber (1986).]

We assume throughout this paper that there is a Central Limit Theorem available for the sampling scheme used so that $n^{\frac{1}{2}}(\hat{\mathbf{p}} - \mathbf{p})$ is asymptotically k-variate normal. In most cases $\mathbf{V}(\mathbf{p}) = nCov(\mathbf{p})$ will be singular since the proportions sum to one over the whole table or, more generally, to known constants over strata defined by a partition of the cells of the table. Thus we assume that

$$\mathbf{K}^{\mathrm{T}} \hat{\mathbf{p}} = \mathbf{\Pi}$$

where $\mathbf{\Pi}$ is an $L \times 1$ vector of known constants and **K** is a $k \times L$ matrix of zeros and ones with rank($\mathbf{K}$) $= L$. To avoid problems with constraints on the parameters, we follow the standard convention for log-linear models [see Haberman (1974)] and assume that **K** is included in the column space of **X**. Thus, without further loss of generality, we can set $\mathbf{X} = (\mathbf{K}, \mathbf{X}_1)$ so that

$$\mathbf{g}(\mathbf{p}) = \mathbf{K}\boldsymbol{\beta}_0 + \mathbf{X}_1\boldsymbol{\beta}_1. \qquad (1.1)$$

If $\mathbf{H}(\mathbf{p}) = \partial\mathbf{g}/\partial\mathbf{p}^{\mathrm{T}}$ is non-singular in a neighbourhood of **p** then, for a given value of $\boldsymbol{\beta}_1$, $\boldsymbol{\beta}_0$ is determined exactly by the non-linear constraints

$$\mathbf{K}^T\mathbf{g}^{-1}(\mathbf{K}\boldsymbol{\beta}_0 + \mathbf{X}_1\boldsymbol{\beta}_1) = \boldsymbol{\Pi}, \quad (1.2)$$

but $\boldsymbol{\beta}_1$ is free to vary in an open neighbourhood of the true value.

2. STRATIFIED RANDOM SAMPLING

Consider any survey design for which $\mathbf{V}(\mathbf{p})$ can be expressed as an explicit function of $\mathbf{p}$, the vector of population proportions. In particular, for stratified random sampling with strata formed by a partition of the cells of the table, the vector of observed cell frequencies has a product-multinomial distribution and

$$\mathbf{V}(\mathbf{p}) = \mathbf{D}_p - \mathbf{D}_p\mathbf{K}(\mathbf{K}^T\mathbf{D}_p\mathbf{K})^{-1}\mathbf{K}^T\mathbf{D}_p, \quad (2.1)$$

where $\mathbf{D}_p = \text{diag}(\mathbf{p})$. When $L = 1$ and $\mathbf{K} = \mathbf{1}$ we get the ordinary multinomial covariance matrix, $\mathbf{D}_p - \mathbf{p}\mathbf{p}^T$.

Such models are all special cases of the general class considered by McCullagh (1983) who expresses the generalized least squares equations in the form

$$D(\boldsymbol{\beta}_1)^T\mathbf{V}^-(\boldsymbol{\beta}_1)[\hat{\mathbf{p}} - \mathbf{p}(\boldsymbol{\beta}_1)] = 0$$

where

$$D(\boldsymbol{\beta}_1) = \frac{\partial \mathbf{p}}{\partial \boldsymbol{\beta}_1^T}$$

and $\mathbf{V}^-$ is any generalized inverse of $\mathbf{V}(\mathbf{p})$. (These are the likelihood equations in the case of stratified random sampling.) For the model defined by (1.1) and (1.2),

$$\mathbf{p}(\boldsymbol{\beta}_1) = \mathbf{g}^{-1}(\mathbf{K}\boldsymbol{\beta}_0 + \mathbf{X}_1\boldsymbol{\beta}_1)$$

with $\boldsymbol{\beta}_0$ determined by $\mathbf{K}^T\mathbf{p} = \boldsymbol{\Pi}$. It follows that

$$D(\boldsymbol{\beta}_1) = \mathbf{H}(\mathbf{p})^{-1}(\mathbf{K}\frac{\partial \boldsymbol{\beta}_0}{\partial \boldsymbol{\beta}_1^T} + \mathbf{X}_1).$$

Differentiating the constraint $\mathbf{K}^T\mathbf{p} = \boldsymbol{\Pi}$ leads to

$$\frac{\partial \boldsymbol{\beta}_0}{\partial \boldsymbol{\beta}_1^T} = -(\mathbf{K}^T\mathbf{H}^{-1}\mathbf{K})^{-1}\mathbf{K}^T\mathbf{H}^{-1}\mathbf{X}_1$$

so that

$$D(\boldsymbol{\beta}_1) = \mathbf{H}^{-1}\tilde{\mathbf{X}}_1,$$

where

$$\tilde{\mathbf{X}}_1 = [\mathbf{I} - \mathbf{K}(\mathbf{K}^T \mathbf{H}^{-1} \mathbf{K})^{-1} \mathbf{K}^T \mathbf{H}^{-1}]\mathbf{X}_1, \tag{2.2}$$

Thus the generalized least squares equations become

$$\tilde{\mathbf{X}}_1^T \mathbf{V}_g^- \mathbf{H}(\hat{\mathbf{p}} - \mathbf{p}) = \mathbf{0} \tag{2.3}$$

with

$$\mathbf{K}^T(\hat{\mathbf{p}} - \mathbf{p}) = \mathbf{0} \tag{2.4}$$

where $\mathbf{V}_g^-$ is any generalized inverse of $\mathbf{V}_g = \mathbf{H}(\mathbf{p})\mathbf{V}(\mathbf{p})\mathbf{H}(\mathbf{p})^T$.

Assuming (w.l.o.g.) that $\mathbf{X}_1$ is of full column rank we can obtain $\hat{\boldsymbol{\beta}}_1$, the solution of (2.3), in the standard way by setting

$$\hat{\boldsymbol{\beta}}_{1,i+1} = \hat{\boldsymbol{\beta}}_{1,i} + (\tilde{\mathbf{X}}_{1i}^T \mathbf{V}_{gi}^- \tilde{\mathbf{X}}_{1i})^{-1} \tilde{\mathbf{X}}_{1i}^T \mathbf{V}_{gi}^- \mathbf{H}_i (\hat{\mathbf{p}} - \mathbf{p}_i). \tag{2.5}$$

A convenient starting value can be obtained by inserting

$$\mathbf{g}(\hat{\mathbf{p}}) - \mathbf{X}\boldsymbol{\beta} \approx \mathbf{H}[\hat{\mathbf{p}} - \mathbf{p}(\boldsymbol{\beta})]$$

in (2.3) and setting $\mathbf{p}_0 = \hat{\mathbf{p}}$ to give

$$\hat{\boldsymbol{\beta}}_{10} = (\tilde{\mathbf{X}}_{10}^T \mathbf{V}_{g0}^- \tilde{\mathbf{X}}_{10})^{-1} \mathbf{X}_{10}^T \mathbf{V}_{g0}^- \mathbf{g}(\hat{\mathbf{p}}). \tag{2.6}$$

The asymptotic theory in McCullagh (1983) is not appropriate here since it assumes that $k \to \infty$. However there is a Central Limit Theorem available for the design so that $n^{\frac{1}{2}}(\hat{\mathbf{p}} - \mathbf{p})$ is asymptotically normal, and it is straightforward to show that

$$n^{\frac{1}{2}}(\hat{\boldsymbol{\beta}}_1 - \boldsymbol{\beta}_1) \to N[\mathbf{0}, (\tilde{\mathbf{X}}_1^T \mathbf{V}_g^- \tilde{\mathbf{X}}_1)^{-1}]$$

as $n \to \infty$.

Note that each step of the iteration involves the solution of the (non-linear) equations

$$\mathbf{f}(\boldsymbol{\beta}_0) = \mathbf{K}^T[\mathbf{g}^{-1}(\mathbf{K}\boldsymbol{\beta}_0 + \mathbf{X}_1\boldsymbol{\beta}_{1i}) - \hat{\mathbf{p}}] = \mathbf{0}$$

to obtain $\boldsymbol{\beta}_0$ and hence $\mathbf{p}_i$. Solving these by Newton's method leads to the iterations

$$\boldsymbol{\beta}_{0,j+1} = \boldsymbol{\beta}_{0,j} - \left(\frac{\partial \mathbf{f}}{\partial \boldsymbol{\beta}_0^T}\right)^{-1} \mathbf{f}(\boldsymbol{\beta}_{0,j})$$

$$= \boldsymbol{\beta}_{0,j} + (\mathbf{K}^T \mathbf{H}_j^{-1} \mathbf{K})^{-1} \mathbf{K}^T(\hat{\mathbf{p}} - \mathbf{p}_j). \tag{2.7}$$

The extra iterations can be avoided in a simple way by a careful choice of the generalized inverse. From now on, let $\mathbf{V}_g^-$ denote the particular choice

$$\mathbf{V}_g^- = (\mathbf{V}_g + \mathbf{K}\mathbf{K}^T)^{-1}$$

and consider the iterative scheme defined by

$$\boldsymbol{\beta}_{i+1} = \boldsymbol{\beta}_i + (\mathbf{X}^T\mathbf{V}_{gi}^-\mathbf{X})^{-1}\mathbf{X}^T\mathbf{V}_{gi}^-\mathbf{H}(\hat{\mathbf{p}} - \mathbf{p}_i) \tag{2.8}$$

where $\boldsymbol{\beta} = (\boldsymbol{\beta}_0{}^T, \boldsymbol{\beta}_1{}^T)^T$ and $\mathbf{p}_i = \mathbf{g}^{-1}(\mathbf{X}\boldsymbol{\beta}_i)$. This is the procedure we would obtain if we naively took the GLIM steps for a non-singular model [see Jorgensen (1983)] and simply replaced $\mathbf{V}_g^{-1}$ by $\mathbf{V}_g^-$. After some algebraic manipulation (see Appendix) equations (2.8) can be expressed as

$$\boldsymbol{\beta}_{1,i+1} = \boldsymbol{\beta}_{1,i} + (\tilde{\mathbf{X}}_{1i}{}^T\mathbf{V}_{gi}^-\tilde{\mathbf{X}}_{1i})^{-1}\tilde{\mathbf{X}}_{1i}{}^T\mathbf{V}_{gi}^-\mathbf{H}_i(\hat{\mathbf{p}} - \mathbf{p}_i), \tag{2.9}$$

and

$$\boldsymbol{\beta}_{0,i+1} = \boldsymbol{\beta}_{0,i} + (\mathbf{K}^T\mathbf{H}_i^{-1}\mathbf{K})^{-1}\mathbf{K}^T[(\hat{\mathbf{p}} - \mathbf{p}_i) + \mathbf{H}_i^{-1}(\boldsymbol{\beta}_{1,i} - \boldsymbol{\beta}_{1,i+1})]. \tag{2.10}$$

Assuming the procedure converges, the limiting value obviously satisfies (2.3) and (2.4). Note that this is not necessarily true for an arbitrary choice of generalized inverse in (2.8). The relationship between (2.7) and (2.10) becomes more obvious if we rewrite model (1.1) in the orthogonal form

$$\mathbf{g}(\mathbf{p}) = \mathbf{K}\tilde{\boldsymbol{\beta}}_0 + \tilde{\mathbf{X}}_1\boldsymbol{\beta}_1,$$

where $\mathbf{X}_1$ is given by (2.2) and

$$\tilde{\boldsymbol{\beta}}_0 = \boldsymbol{\beta}_0 + (\mathbf{K}^T\mathbf{H}^{-1}\mathbf{K})^{-1}\mathbf{K}^T\mathbf{H}^{-1}\mathbf{X}_1\boldsymbol{\beta}_1.$$

Then (2.10) is approximately equivalent to

$$\tilde{\boldsymbol{\beta}}_{0,i+1} = \tilde{\boldsymbol{\beta}}_{0,i} + (\mathbf{K}^T\mathbf{H}^{-1}\mathbf{K})^{-1}\mathbf{K}^T(\hat{\mathbf{p}} - \mathbf{p}_i).$$

A convenient starting point for the iteration is given by

$$\hat{\boldsymbol{\beta}}_A = (\mathbf{X}^T\mathbf{V}_{g0}^-\mathbf{X})^{-1}\mathbf{X}^T\mathbf{V}_{g0}^-\mathbf{g}(\hat{\mathbf{p}}). \tag{2.11}$$

The asymptotic efficiency of $\hat{\boldsymbol{\beta}}_A$ is demonstrated in Rao, Scott and Thomas (1986) using the asymptotic normality of $\hat{\mathbf{p}}$ and results for singular linear models in Rao (1971) and Mitra (1973). [See also Rao (1985).]

For stratified random sampling, (2.8) leads to standard procedures in the two most important special cases. For log-linear models, $\mathbf{H} = \mathbf{D}_p^{-1}$ and (2.8) gives algebraically identical results to those obtained by substituting $\mathbf{V}_g^{-} = \mathbf{D}_p^{-1}$, i.e. to those obtained by feeding the data through the GLIM program assuming independent Poisson counts in each cell. [See McCullagh and Nelder (1983).] For linear models, (2.8) gives identical results to those obtained by using the GLIM program assuming independent Poisson counts and using the OFFSET directive to force a constant vector $\mathbf{K}(\mathbf{K}^T\mathbf{K})^{-1}\mathbf{\Pi}$ into the model [see Stirling (1986)]. (Note that $\tilde{\boldsymbol{\beta}}_{0,i}$ and $\tilde{\mathbf{X}}_{1i}$ both remain constant during the iterations in this latter case.)

3. GENERAL SURVEY DESIGNS

Now consider an arbitrary survey design for which there is a Central Limit Theorem ensuring that

$$n^{\frac{1}{2}}(\hat{\mathbf{p}} - \mathbf{p}) \rightarrow N(\mathbf{0}, \mathbf{V}(\mathbf{p})) \qquad (3.1)$$

but with no explicit expression for $\mathbf{V}(\mathbf{p})$ as a function of $\mathbf{p}$ in general. Instead we suppose that a consistent estimator of $\mathbf{V}(\mathbf{p})$, say $\hat{\mathbf{V}}_p$, is available. This means we can still obtain an asymptotically efficient weighted least squares estimator of $\boldsymbol{\beta}_1$ by using (2.6) or (2.11) with $\hat{\mathbf{V}}_p$ replacing $\mathbf{V}(\hat{\mathbf{p}})$ in the expression for $\mathbf{V}_{g0}^{-}$. It follows by standard linearization methods that the resulting estimator is asymptotically normal with large-sample covariance matrix equal to $(n\tilde{\mathbf{X}}_1^T\mathbf{V}_g^{-}\tilde{\mathbf{X}}_1)^{-1}$. We could also obtain an asymptotically equivalent smoothed version by iterating using (2.5) or (2.8) with $\hat{\mathbf{V}}_p$ in place of $\mathbf{V}(\mathbf{p})$ but no investigation has made into the comparative performance of the two estimators in small samples.

An alternative approach is to eliminate the singularity by a suitable linear transformation and then apply standard weighted least squares methods. Specifically, let $\mathbf{y} = \mathbf{C}_K\mathbf{g}(\hat{\mathbf{p}})$, where $\mathbf{C}_K$ is a $(k - L) \times k$ matrix of rank $k - L$ with $\mathbf{C}_K\mathbf{K} = \mathbf{0}$. Using standard delta method arguments,

$$n^{\frac{1}{2}}(\mathbf{y} - \mathbf{C}_K\mathbf{X}_1\boldsymbol{\beta}_1) \rightarrow N(\mathbf{0}, \mathbf{C}_K\mathbf{V}_g\mathbf{C}_K^T)$$

where $\mathbf{C}_K\mathbf{V}_g\mathbf{C}_K^T$ is non-singular. We can then use a standard WLS program to make inferences about $\boldsymbol{\beta}_1$. This is the approach adopted for example, by Grizzle and Williams (1973), Koch *et al.* (1975) and Imrey *et al.* (1982). It can be shown that the resulting estimator (which is independent of the choice of $\mathbf{C}_K$) is algebraically identical to $\hat{\boldsymbol{\beta}}_{10}$ obtained from (2.5) and hence asymptotically efficient. Details are given in the Appendix.

The implementation of either approach depends on a good estimate of $\mathbf{V}_p$ being available. If this is so, $\hat{\mathbf{V}}_p$ will almost always be obtained by assuming that primary sampling units are selected independently within strata and hence will have fewer degrees of freedom than the number of p.s.u.'s. A very common practice, for example, is to chose two p.s.u.'s per stratum and use BRR or the jackknife method to calculate $\hat{\mathbf{V}}_p$ [see Wolter (1985)]. In this case $\hat{\mathbf{V}}_p$ will have rank at most equal to the number of strata. This is not likely to cause problems if there are relatively few cells in the table but k increases rapidly as the dimensions of the table increase. Even for a $5 \times 5 \times 5$ table with $k = 125$, $\hat{\mathbf{V}}_g + \mathbf{K}\mathbf{K}^{\mathrm{T}}$ is likely to be singular in most surveys. In practice, of course, few survey organisations would be willing to provide estimates of a 125×125 covariance matrix no matter how many degrees of freedom were available.

Faced with these problems, most practitioners simply ignore the complexity of the design apart from the stratification and use GLIM or any other standard program based on the assumption of stratified random sampling (i.e. on the product-multinomial distribution). Under (3.1) and the assumptions of the introduction it follows from (2.3) that

$$\hat{\boldsymbol{\beta}}_1 \sim \boldsymbol{\beta}_1 + (\tilde{\mathbf{X}}_1^{\mathrm{T}}\mathbf{V}_{0g}^{-}\tilde{\mathbf{X}}_1)^{-1}\tilde{\mathbf{X}}_1^{\mathrm{T}}\mathbf{V}_{0g}^{-}\mathbf{H}(\hat{\mathbf{p}} - \mathbf{p}),$$

where $\mathbf{V}_{0g}$ is the covariance matrix of $\mathbf{g}(\hat{\mathbf{p}})$ under stratified random sampling. Thus $\hat{\boldsymbol{\beta}}_1$ is a consistent estimator of $\boldsymbol{\beta}_1$ with large sample covariance matrix $\mathbf{V}_1/n$ where

$$\mathbf{V}_1 = \mathbf{F}\mathbf{V}_{01} \tag{3.2}$$

with

$$\mathbf{V}_{01} = (\tilde{\mathbf{X}}_1^{\mathrm{T}}\mathbf{V}_{0g}^{-}\tilde{\mathbf{X}}_1)^{-1} \tag{3.3}$$

and

$$\mathbf{F} = \mathbf{V}_{01}(\tilde{\mathbf{X}}_1^{\mathrm{T}}\mathbf{V}_{0g}^{-}\mathbf{V}_g\mathbf{V}_{0g}^{-}\tilde{\mathbf{X}}_1).$$

The standard programs produce an estimate of $\mathbf{V}_{01}$ so that $\mathbf{F}$ represents the correction factor which needs to be applied to the standard output. Note that, although $\mathbf{F}$ involves $\mathbf{V}_g$, we need only estimate

$$\tilde{\mathbf{X}}_1^{\mathrm{T}}\mathbf{V}_{0g}^{-}\mathbf{V}_g\mathbf{V}_{0g}^{-}\tilde{\mathbf{X}}_1 = Cov(\tilde{\mathbf{X}}_1^{\mathrm{T}}\mathbf{V}_{0g}^{-}\mathbf{H}\hat{\mathbf{p}}).$$

This requires the estimated covariance matrix of $r - L$ linear combinations of the cell proportions which is a straightforward task for moderate values of r with most standard survey designs.

The effect on the standard errors can be substantial. For example, Table 3 shows the estimated coefficients together with their estimated standard errors when a simple model of the form

$$\log p_{ij1}/p_{ij2} = \beta_1 + \beta_2 A_i + \beta_3 E_j,$$

where A_i is equal to the median age for the ith row and E_j is a dummy variable with a value of one indicating some post-secondary education, is fitted to the Canada Health Survey data shown in Table 2 using a standard logistic regression program.

TABLE 3

Estimates for Canada Health Survey Data

i	$\hat{\beta}_i$	Estimated Standard Errors	
		Nominal	True
1	+0.090	.052	.057
2	+0.803	.052	.100
3	+0.115	.024	.031

Note that the standard errors require multipliers ranging from 1.1 to 1.9 to correct them. Thus a common multiplier such as the GLIM correction for overdispersion [see McCullagh and Nelder (1983)], which is equal to 1.2 here, cannot correct all the estimates simultaneously.

We can also look at the behaviour of test statistics produced under the assumption of a product-multinomial distribution. Suppose we are interested in the hypothesis $\mathbf{H}_0 : \mathbf{A}\boldsymbol{\beta}_1 = \mathbf{b}$. Then the standard likelihood-ratio or Pearson chi-squared statistics are both asymptotically equivalent to

$$R = n(\mathbf{A}\hat{\boldsymbol{\beta}}_1 - \mathbf{b})^{\mathrm{T}}(\mathbf{A}\mathbf{V}_{01}\mathbf{A}^{\mathrm{T}})^{-1}(\mathbf{A}\hat{\boldsymbol{\beta}}_1 - \mathbf{b}),$$

where $\mathbf{V}_{01}$ is given by (3.3). From standard results on quadratic forms in asymptotically normal random variables [see Serfling (1980)] it follows that the asymptotic null distribution of the standard test statistics takes the form $\gamma_1 Z_1^2 + \dots + \gamma_v Z_v^2$, where $Z_1, \dots, Z_v$ are independent $N(0, 1)$ random variables and $\gamma_1, \dots, \gamma_v$ are the eigenvalues of

$$\boldsymbol{\Gamma} = (\mathbf{A}\mathbf{V}_{01}\mathbf{A}^{\mathrm{T}})^{-1}(\mathbf{A}\mathbf{V}_1\mathbf{A}^{\mathrm{T}}),$$

where $\mathbf{V}_1$ is given by (3.2). Again computation of an estimate of $\boldsymbol{\Gamma}$ requires the estimation of a covariance matrix of size $r - L$. Once we have obtained estimates of $\gamma_1, \dots, \gamma_v$ we can judge the significance of the observed value of the test statistic using one of the standard approximations to the distribution of linear combinations of chi-squared random variables [see Solomon and Stephens (1977) for example].

Formally, we can treat the standard goodness of fit tests (likelihood ratio or chi-square) as a special case of this by embedding the model (1.1) in a completely saturated model and choosing $\mathbf{A}$ to be a $(k - r) \times k$ matrix of rank $k - r$ with $\mathbf{AX} = \mathbf{0}$. After some simplification [see Rao and Scott (1984) for details in the special case of log-linear models], $\gamma_1, \dots, \gamma_{T-r}$ can be obtained as the eigenvalues of

$$\boldsymbol{\Gamma}^* = (\mathbf{A}\mathbf{V}_{0g}\mathbf{A}^{\mathrm{T}})^{-1}(\mathbf{A}\mathbf{V}_g\mathbf{A}^{\mathrm{T}}). \tag{3.4}$$

A number of important special cases have been treated in the literature. Tests for independence in two-way tables are dealt with in Rao and Scott (1981) [see also Holt, Scott and Ewings (1980)], general log-linear models in Rao and Scott (1984) and logistic regression models in Roberts, Rao and Kumar (1987).

Note that $\boldsymbol{\Gamma}^*$ in (3.4) is a $(k - r) \times (k - r)$ matrix and hence will often be quite large. In such cases the best we can usually hope for is to have estimates of the cell variances (i.e. the diagonal elements of $\mathbf{V}_p$) and perhaps the variances of the marginal proportions. In special cases this can provide useful information on the γ_i's. In particular, for testing independence in a two-way table or for testing the fit of any log-linear model which has explicit estimates of the fitted proportions it is possible to calculate $\bar{\gamma} = (\gamma_1 + \dots + \gamma_{k-r})/(k - r)$ from such information [see Bedrick (1983), Rao and Scott (1984, 1987) and Gross (1985)]. This enables us to make a first-order correction by dividing the test statistics by the estimate of $\bar{\gamma}$ and treating the modified statistic as a χ_{k-r}^2 random variable. Typically the resulting significance level is still higher than the nominal value but much closer to it than we get by ignoring the sampling design [see Holt, Scott and Ewings (1980) for empirical results].

To illustrate the results, consider the data in Example 1. With $k = 8$ and a simple single stage cluster sample, calculation of $\hat{\mathbf{V}}_p$ is straightforward and was carried using PC CARP [see Fuller *et al.* (1986)]. The ordinary χ^2 statistic for testing independence takes the value $\chi^2 =$ 17.72 and the estimated eigenvalues are $\hat{\gamma}_1 = 3.58$, $\hat{\gamma}_2 = 0.83$, $\hat{\gamma}_3 = 0.15$ with $\hat{\bar{\gamma}} = 1.52$. Pretending that the null distribution of χ^2 is $\chi_3{}^2$ gives a P value of approximately .0005 while (ignoring the variability in the estimated γ_i's) using a Sattethwaite approximation yields a P value of .027. Thus we would still judge the result to be significant at the 5 % level but the observed value appears much less extreme than at first sight. Treating $\chi^3/\bar{\gamma} = 11.66$ as a $\chi_3{}^2$ variate leads to a P value of about .009 which is still too low but a considerable improvement on the unmodified value.

4. CONCLUDING REMARKS

In all this work we have taken it for granted that the overall population proportions are the proper target for our inference i.e. that it is appropriate to model the marginal proportions collapsed over clusters and whatever other internal structure the population possesses. As is well-known [see Bishop *et al.* (1975) or Fienberg (1980) for example], this may lead to very different conclusions from those we would get by including the clusters in the analysis so some careful thought is needed before proceeding blindly with the aggregated data. Of course, within-cluster information is often not available because of concerns about confidentiality. Even if it is available, there will often be too many clusters to include each cluster as a separate layer in the table but it might be worthwhile grouping similar clusters together.

The results of the previous section allow us to make valid inferences from the output of a program which ignores the correlation structure, but little is known about the relative efficiency of these procedures. Analogous results for fitting linear models to survey data [Scott and Holt (1983)] suggest that the loss in efficiency should be small but more work needs to be done on this.

Finally, this paper deals just with generalized linear models for tables of fixed size. More general models for continuous variables are considered in Binder (1983) who develops asymptotic theory based on weighted estimating equations.

APPENDIX

We first sketch some basic properties of $\mathbf{V}_g^- = (\mathbf{V}_g + \mathbf{K}\mathbf{K}^T)^{-1}$.

LEMMA 1. $\mathbf{K}^T\mathbf{V}_g^- = (\mathbf{K}^T\mathbf{H}^{-1}\mathbf{K})^{-1}\mathbf{K}^T\mathbf{H}^{-1}$.

This result follows directly from the fact that $\mathbf{K}^T\mathbf{H}^{-1}\mathbf{V}_g = \mathbf{0}$. As a consequence we have:

COROLLARY. (i) $\mathbf{K}^T\mathbf{V}_g^-\mathbf{K} = \mathbf{I}_L$, $\mathbf{K}^T\mathbf{V}_g^-\tilde{\mathbf{X}}_1 = \mathbf{0}$,
(ii) $\mathbf{V}_g\mathbf{V}_g^-\mathbf{K} = \mathbf{0}$, $\mathbf{V}_g\mathbf{V}_g^-\mathbf{X}_1 = \tilde{\mathbf{X}}_1$,
(iii) $\mathbf{V}_g\mathbf{V}_g^-\mathbf{V}_g = \mathbf{V}_g$ (*i.e.* $\mathbf{V}_g^-$ *is indeed a g- inverse of* $\mathbf{V}_g$).

The next result follows from these properties and the standard form for the inverse of a partitioned matrix [Rao (1973a, p.33)].

LEMMA 2.

$$(\mathbf{X}^T\mathbf{V}_g^-\mathbf{X})^{-1} = \begin{pmatrix} \mathbf{I} & \mathbf{0} \\ \mathbf{0} & \mathbf{0} \end{pmatrix} + \begin{pmatrix} \mathbf{F} \\ -\mathbf{I} \end{pmatrix} \mathbf{E}^{-1}(\mathbf{F}^T, -\mathbf{I}), \tag{A1}$$

where $\mathbf{E} = \tilde{\mathbf{X}}_1^T\mathbf{V}_g^-\tilde{\mathbf{X}}_1$ *and* $\mathbf{F} = \mathbf{K}^T\mathbf{V}_g^-\mathbf{X}_1$.

Lemma 2 has two important consequences. The first is that it establishes the equivalence of equations (2.9) and (2.10) as claimed in Section 2. The second is that, since the asymptotic covariance matrix of $n\hat{\boldsymbol{\beta}}$ is equal to the second term on the righthand side of (A1), $Cov(\hat{\boldsymbol{\beta}}_1)$ and $Cov(\hat{\boldsymbol{\beta}}_0, \hat{\boldsymbol{\beta}}_1)$ can be obtained directly from the corresponding blocks of $(\mathbf{X}^T\mathbf{V}_g^-\mathbf{X})^{-1}$.

Next we establish the equivalence between the estimator of (2.6),

$$\hat{\boldsymbol{\beta}}_{10} = (\tilde{\mathbf{X}}_1^T\mathbf{V}_g^-\tilde{\mathbf{X}}_1)^{-1}\tilde{\mathbf{X}}_1^T\mathbf{V}_g^-\mathbf{g}(\hat{\mathbf{p}}),$$

and the weighted least-squares estimator obtained from the Grizzle - Williams approach,

$$\hat{\boldsymbol{\beta}}_{gw} = [\mathbf{X}_1^T\mathbf{C}_K^T(\mathbf{C}_K\mathbf{V}_g\mathbf{C}_K^T)^{-1}\mathbf{C}_K\mathbf{X}_1]^{-1}\mathbf{X}_1^T\mathbf{C}_K^T(\mathbf{C}_K\mathbf{V}_g\mathbf{C}_K^T)^{-1}\mathbf{C}_K\mathbf{g}(\hat{\mathbf{p}}),$$

where $\mathbf{C}_K$ is a $(k-L)\times k$ matrix of rank $k-L$ with $\mathbf{C}_K\mathbf{K} = \mathbf{0}$. Since $\hat{\boldsymbol{\beta}}_{gw}$ is invariant under different choices of $\mathbf{C}_K$ we can make the particular choice with $\mathbf{C}_K^T = (\mathbf{V}_g^-\tilde{\mathbf{X}}_1, \ \mathbf{F})$, where $\mathbf{F}$ is any $k\times(k-r)$ matrix of rank $k-r$ with $\mathbf{F}^T\mathbf{X} = \mathbf{0}$. It then follows from the results above that

$$\mathbf{C}_K\mathbf{X}_1 = \begin{pmatrix} \tilde{\mathbf{X}}_1^{\mathrm{T}}\mathbf{V}_g^{-}\tilde{\mathbf{X}}_1 \\ \mathbf{0} \end{pmatrix}$$

and

$$\mathbf{C}_K\mathbf{V}_g\mathbf{C}_K^{\mathrm{T}} = \begin{pmatrix} \tilde{\mathbf{X}}_1^{\mathrm{T}}\mathbf{V}_g^{-}\tilde{\mathbf{X}}_1 & \mathbf{0} \\ \mathbf{0} & \mathbf{F}^{\mathrm{T}}\mathbf{V}_g\mathbf{F} \end{pmatrix}.$$

Substituting these values into the expression for $\hat{\boldsymbol{\beta}}_{gw}$ leads immediately to $\hat{\boldsymbol{\beta}}_{10}$. The equivalence of the expression for variances and residual sums of squares from the two approaches follows in the same way.

ACKNOWLEDGEMENTS

This paper is partly a review of work carried out jointly with J.N.K. Rao and D.R. Thomas. I am grateful for their help and also the support of D. Holt and T.M.F. Smith at the University of Southampton where the work was completed while on sabbatical leave from the University of Auckland with support from the Science and Engineering Research Council of the U.K. (Grant No. GR/E/39853).

REFERENCES

Bedrick, E.J. (1983). Adjusted chi-squared tests for cross-classified tables of survey data. *Biometrika*, 70, 591-596.

Binder, D.A. (1983). On the variances of asymptotically normal estimators from complex surveys. *Int. Statist. Rev.*, 51, 279-292.

Bishop, Y. M. M., Fienberg, S. E. and Holland, P. W. (1975). *Discrete Multivariate Analysis: Theory and Practice.* MIT Press, Cambridge, Mass.

Bonett, D.G., Woodward, J.A. and Bentler, P.M. (1985). Some extensions of a linear model for categorical variables. *Biometrics*, 41, 745-750.

Fay, R.E. (1985). A jackknifed chi-squared test for complex samples. *J. Amer. Statist. Assoc.*, 80, 148-157.

Fienberg, S.E. (1980). *The Analysis of Cross-Classified Categorical Data*, 2nd ed. MIT Press, Cambridge, Mass.

Fuller, W.A., Kennedy, W., Schell, D., Sullivan, G. and Park, H.J. (1986). *PC CARP*, Iowa State University, Ames, Iowa.

Grizzle, J.E. and Williams, O.D. (1972). Log-linear models and tests of independence for contingency tables. *Biometrics*, 28, 137-156.

Gross, W.F. (1984). A note on chi-squared tests with survey data. *J. Roy. Statist. Soc. B*, 46, 207-272.

Haber, M. (1985). Maximum likelihood methods for linear and log-linear models in categorical data. *Comput. Statist. Data Anal.*, 3, 1-10.

Haberman, S.J. (1974). *The Analysis of Frequency Data.* University of Chicago Press, Chicago.

Holt, D., Scott, A.J. and Ewings, P.D. (1980). Chi-squared tests with survey data. *J. Roy. Statist. Soc. A*, 143, 302-320.

Imrey, P. B., Koch, G. G. and Stokes, M. E. (1982). Categorical data analysis: some reflections on the log linear model and logistic regression. Part II : data analysis. *Int. Statist. Rev.*, 50, 35-64 (in collaboration with J.N. Darroch, D.H. Freeman, Jr. and H.D. Tolley).

Jorgensen, B. (1983). Maximum likelihood estimation and large-sample inference for generalized linear and non-linear regression models. *Biometrika*, 70, 19-28.

Koch, G.G., Freeman, D.H.Jr. and Freeman, J.L. (1975). Strategies in the multivariate analysis of data from complex surveys. *Int. Statist. Rev.*, 43, 59-78.

McCullagh, P. (1983). Quasi-likelihood functions. *Ann. Statist.*, 11, 59-67.

McCullagh, P. and Nelder, J. A. (1983). *Generalized Linear Models.* Chapman and Hall, London.

Mitra, S.K. (1973). Unified least squares approach to linear estimation in a general Gauss-Markov model. *SIAM J. Appl. Math.*, 25, 671-680.

Rao, C.R. (1971). Unified theory of linear estimation. *Sankhyā*, 33, 371-394.

Rao, C.R. (1973). *Linear Statistical Inference and Its Applications.* Second Edition. Wiley, New York.

Rao, C.R. (1985). A unified approach to inference from linear models. *Proc. First International Tampere Seminar on Linear Statistical Models and their Applications.* Dept. of Mathematical Sciences, University of Tampere, Finland, 9-36.

Rao, J.N.K. and Scott, A.J. (1981). The analysis of categorical data from complex sample surveys: chi-squared tests for goodness of fit and independence of two-way tables. *J. Amer. Statist. Assoc.*, 76, 221-230.

Rao, J.N.K. and Scott, A.J. (1984). On chi-squared tests for multi-way contingency tables with cell proportions estimated from survey data. *Ann. Statist.*, 12, 46-60.

Rao, J.N.K. and Scott, A.J. (1987). On simple adjustments to chi-squared tests with sample survey data. *Ann. Statist.*, 15, 385-397.

Rao, J.N.K., Scott, A.J. and Thomas, D.R. (1986). Generalized linear models with conditional Poisson distributions. *Proc. Amer. Statist. Assoc.* (to appear).

Roberts, G. A., Rao, J. N. K. and Kumar, S. (1987). Logistic regression analysis of sample survey data. *Biometrika*, 74, 1-12.

Scott, A. J. and Holt, D. (1982). The effect of two-stage sampling on ordinary least-squares methods. *J. Amer. Statist. Assoc.*, 72, 848-855.

Serfling, R. J. (1980). *Approximation Theorems in Mathematical Statistics.* Wiley, New York.

Solomon, H. and Stephens, M.A. (1977). Distribution of a weighted sum of chi-squared variables. *J. Amer. Statist. Assoc.*, 72, 881-885.

Stirling, W.D. (1986). Testing linear hypotheses in contingency tables with zero cell counts. *Comp. Statist. Data Anal.*, 4, 1-13.

Wolter, K. M. (1985). *Introduction to Variance Estimation.* Springer- Verlag.

Received 21 April 1987

Department of Mathematics and Statistics
University of Auckland
Private Bag
Auckland
New Zealand

Proc. Second International Tampere Conference in Statistics
(Tampere, Finland, 1-4 June 1987)
Tarmo Pukkila and Simo Puntanen, *Editors*

pp. 317 - 342

INVITED PAPER

Robust optimum invariant unbiased tests for variance components

Rita DAS and **Bimal K. SINHA**
Northern Illinois University, De Kalb, IL, U.S.A.
and University of Maryland Baltimore County, Catonsville, MD, U.S.A.

Key words and phrases: Uniformly most powerful invariant unbiased test, locally best invariant unbiased test, maximal invariant, analysis of variance, robustness, balanced, unbalanced, Wijsman's representation theorem, fixed effects model, random effects model, mixed effects model.

ABSTRACT

In one-way random effects unbalaced model the locally best invariant unbiased test for the equality of the treatment effects is derived. Surprisingly, this is different from the widely used familiar F-test. In the balanced case, however, the two tests coincide and represent the uniformly most powerful invariant unbiased test. For two-way random effects and mixed effects balanced models, the uniformly most powerful invariant unbiased test for the equality of the treatment effects is derived both with and without interaction, and shown to be equivalent to the usual F-tests under fixed effects models. The optimum invariant tests derived here are shown not to depend on the assumption of normality. Different aspects of null, nonnull and optimality robustness of these tests [Kariya and Sinha, Annals of Statistics, 1985] are studied.

A generalization to the multivariate case is also considered in this paper. We consider the canonical form MANOVA setup with $X : n \times p = (X_1', X_2', X_3')' = (M_1', M_2', 0)' + E, X_i : n_i \times p, i = 1,2,3, M_i : n_i \times p, i = 1,2, n_1 + n_2 + n_3 = n, n_3 \geq p$, where E is a random error matrix with location 0 and unknown scale matrix $\Sigma > 0$ (p.d.). Assume, unlike in the usual sense, that M_1 is random with location 0 and scale matrix $\sigma_1^2\Sigma$, M_2 is either fixed or random with location 0 and a different scale matrix $\sigma_2^2\Sigma, \sigma_1^2, \sigma_2^2$ being unknown. For testing $H_0 : \sigma_1^2 = 0$ versus $H_1 : \sigma_1^2 > 0$ under a left orthogonally invariant distribution of X, it is shown that when either $n_2 = 0$ or M_2 fixed if $n_2 > 0$, the trace test of Pillai (1955) is UMPIU if $min(n_1, p) = 1$ and LBIU if $min(n_1, p) > 1$. The test is null, nonnull and optimality robust. However, such a result does not hold if $n_2 > 0$ and M_2 random.

1. INTRODUCTION

In the analysis of variance (fixed, random or mixed effects models) a common method is to decompose the total amount of variation in the entire data into different meaningful components of variation corresponding to distinct sources like treatments, blocks, treatment-block interaction, error etc. and use appropriate F-tests for different hypotheses. Some excellent references on this account are the books by Kempthorne (1952), Federer (1955), Cochran and Cox (1957), Chakravarty (1962), Das and Giri (1979), Montgomery (1984), Raghavarao (1971), Scheffé (1959) etc. While such F-tests beyond being merely valid also represent optimum invariant unbiased tests under the assumption of a fixed effects model, vide Lehmann (1959) under normality and independence of the errors, and Kariya (1981), Kariya and Sinha (1985) under more general error distributions, the properties of these widely used F-tests under random and mixed effects models are far from clear. In the literature the use of these F-tests in the two latter cases is justified only as valid tests and no mention is made of their optimum properties, if any (vide Imhof (1962), Khuri (1984), Mostafa (1967), Scheffé (1956), Arnold (1981)). See, however, the two references mentioned below.

It is the object of this paper to derive optimum invariant tests for the equality of the treatment effects in the one-way and two-way random effects, and two way mixed effects models and verify if these tests coincide with the standard F-tests. Our findings reveal that, surprisingly, in one-way unbalanced random effects model, the locally best invariant unbiased test is different from the usual F-test, the latter being asymptotically (as the alternative $\to \infty$) best invariant (vide Spjøtvoll (1967)). The new test statistic we have derived coincides with the F-statistic only in the balanced case where it provides a uniformly most powerful invariant unbiased test. In the two-way random effects balanced model, we have established the appropriate F-tests as being uniformly most powerful invariant unbiased. This was attempted by Herbach (1959). In the two-way mixed effects model, the appropriate F-tests have been established as locally best invariant unbiased. It is shown that, unlike the two-way unbalanced fixed effects model, the two-way unbalanced random or mixed effects model presents a considerable difficulty towards derivation of an optimum invariant unbiased test even with the so called orthogonality condition, $n_{ij} \alpha n_{i\cdot} n_{\cdot j}$, for all i and j. This is attributed to the incompleteness of the family of sufficient statistics under the null hypothesis of no treatment effects.

Another problem considered in this paper is concerned with a multivariate generalization to the random MANOVA model. This is described below.

The usual MANOVA model in the canonical form consists of an $n \times p$ random data matrix X decomposed as $X = (X_1', X_2', X_3')'$ with

$X_i : n_i \times p,\ i=1,2,3, n_1+n_2+n_3=n,\ n_3 \geq p,$ following the structure

$$\begin{bmatrix} X_1 \\ X_2 \\ X_3 \end{bmatrix} = \begin{bmatrix} M_1 \\ M_2 \\ 0 \end{bmatrix} + E. \tag{1.1}$$

Here $M_i : n_i \times p$ is the mean of X_i, $i=1,2$, and $E : n \times p$ is the random error matrix. Under the distributional assumption $E \sim N(0, I_n \otimes \Sigma)$ for some unknown p.d. $p \times p$ matrix Σ, many tests of the MANOVA hypothesis $H_0 : M_1 = 0$ versus $H_1 : M_1 \neq 0$ are well known, e.g. the likelihood ratio test, Roy's maximum root test, Lawley-Hotelling's trace test and Pillai's trace test. All these tests ignore X_2 and are functions of $X_1(X_3'X_3)^{-1}X_1'$ (vide Anderson (1984)). Moreover, the trace test of Pillai (1955) is known to be LBI in general (Schwartz (1967)) and UMPI if $\min(p,n_1)=1$ (Lehmann (1959)). On the other hand, if M_1 and M_2 are assumed to be independent normal with zero mean and dispersion $\sigma_1^2\Sigma$ and $\sigma_2^2\Sigma$ respectively, Roy and Gnanadesikan (1959) considered the problem of testing $H_{01} : \sigma_1^2 = 0$ versus $H_{11} : \sigma_1^2 > 0$ and proposed the maximum root test, $\lambda_{\max}(X_1(X_3'X_3)^{-1}X_1')$. See also Roy and Cobb (1960) for some related results. However, so far no optimum test is known.

The second objective of this paper is to derive an optimum invariant test for testing H_{01} versus H_{11} under the model

$$X \sim f(x|\sigma_1^2, M_2, \Sigma) = |\Sigma|^{-n/2}(1+\sigma_1^2)^{-n_1/2} \tag{1.2}$$
$$q(\Sigma^{-1}(X_1'X_1/(1+\sigma_1^2) + (X_2-M_2)'(X_2-M_2) + X_3'X_3))$$

for some $q \in Q$. Here Q is the class of functions from the set of $p \times p$ matrices into $[0,\infty)$ such that $q \in Q$ satisfies

$$\int_{R^{np}} q(X'X)\,dX = 1, \tag{1.3}$$
$$\int_{Gl(p)} \int_{R^{n_2p}} q(AA' + F'F)\,|AA'|^{(n_1+n_3-p)/2}\,dF\,dA < \infty$$

and

$$q(BV) = q(VB) \text{ for } all\ V \in L(p) \text{ and } B \in Gl(p) \tag{1.4}$$

where F is a matrix of order $n_2 \times p$ with elements in R^{n_2p}, dF is Lebesgue over R^{n_2p}, $L(p)$ is the set of $p \times p$ nonnegative definite matrices and $Gl(p)$ is the group of $p \times p$ nonsingular matrices. The model (1.2) corresponds to (1.1) with only M_1 as random. This can be thought of as a mixed MANOVA model for $n_2 > 0$ and a random MANOVA model for $n_2 = 0$. Of course, unlike in previous papers, the normality of X has been replaced by a very general left orthogonally invariant distribution. We show that whatever be $q \in Q$, the trace test of Pillai (1955) is UMPIU if $\min(n_1,p) = 1$ and LBIU otherwise. In particular, for $n_2 = 0$ which makes the model (1.2) comparable to Roy and Gnanadesikan's (1959), the trace test is superior to the invariant maximum root test. Under normality of X, it is mentioned in Lehmann (1959, page 344) that when $n_2 = 0$ and $n_1 = n_3 = 1$, there exists a UMPI test under the group G_T of all $p \times p$ nonsingular lower triangular matrices with positive diagonal elements. This test is based on X_{11}^2/X_{31}^2 and, therefore, not very appealing due to its asymmetry. Here X_{11} and X_{31} are the first components in the vectors $\underset{\sim 1}{X}$ and $\underset{\sim 3}{X}$ respectively. It seems to us that for the above problem the group $Gl(p)$ rather than G_T is the right group to use. For some discussion on propoerties of $q \in Q$, we refer to Kariya (1981).

The optimum invariant trace test is shown to be null, nonnull and optimality robust (vide Kariya and Sinha (1985)). It is interesting to compare our results with those of Kariya (1981) and Kariya and Sinha (1985) who proved similar results under the fixed effects MANOVA model (i.e. M_1, M_2 fixed matrices), Kariya (1981) requiring $q \in Q$ to be covex for the UMPI property to hold when $\min(n_2,p) = 1$, while Kariya and Sinha (1985) restricting q to belong to the class of elliptically symmetric distributions and satisfying some other conditions for the LBI property to hold when $\min(n_1,p) > 1$. However here we do not impose any condition on q other than the integrability condition (1.3) and the condition (1.4). Moreover, our proof of the LBIU property of the trace test for $min(n_1,p) > 1$ is extremely simple due to the nature of the model (1.2). We refer to Schwartz (1967) and Kariya and Sinha (1985) for the LBI property of the trace test under fixed effects MANOVA model for normal q and elliptically symmetric q respectively.

If $n_2 > 0$ and M_2 random with mean zero and scale matrix $\sigma_2^2\Sigma$ so that the distribution of X follows

$$X \sim f(x \mid \sigma_1^2,\sigma_2^2,\Sigma) = |\Sigma|^{-n/2}(1+\sigma_1^2)^{-n_1/2}(1+\sigma_2^2)^{-n_2/2} \quad (1.5)$$
$$q(\Sigma^{-1}(X_1'X_1/(1+\sigma_1^2) + X_2'X_2/(1+\sigma_2^2) + X_3'X_3))$$

a difficulty in the derivation of an optimum invariant test is pointed out.

The paper is organized as follows. In Section 2, one-way unbalanced random effects model is discussed and the locally best invariant unbiased test is derived. That the test coincides with the usual F-test and becomes UMPIU in the balanced case is pointed out (also shown in Herbach (1959)). Section 3 deals with two-way balanced random effects model while that under mixed effects model is presented in Section 4. The standard UMPIU F-tests are shown to be again UMPIU in the first case and LBIU in the second case.

Section 5 deals with a multivariate generalization to the random MANOVA model. Throughout Sections 2 through 4 the assumption of normality with mean vector μ and dispersion matrix Σ of the underlying random effects and error components has been replaced by the more general assumption of an elliptically symmetric distribution with μ as the location vector and Σ as the scale matrix. It turns out that the optimum property of the F-tests is preserved for a fairly general class of such distributions. Different aspects of robustness of these tests, namely, null, nonnull and optimality (vide Kariya and Sinha, 1985), are studied.

As a technical tool, Wijsman's representation theorem (1967) is used.

2. ONE WAY UNBALANCED RANDOM EFFECTS MODEL

The statistical model in this setup involves $n = n_1 + \cdots + n_k$ observations $\{x_{ij}\}$ following the structure

$$x_{ij} = \mu + \tau_i + e_{ij},\ j = 1,\ldots,n_i,\ i = 1,\ldots,k \tag{2.1}$$

where μ represents a fixed unknown general effect, $\tau_1, \ldots, \tau_k$ represent k random treatment effects with mean 0 and variance σ_τ^2, and e_{ij}'s denote the random error components with mean 0 and variance σ^2. Denote by $\underset{\sim}{x} = (x_{11}, x_{12}, \ldots, x_{1n_1}, x_{21}, \ldots, x_{2n_2}, \ldots, x_{k1}, x_{k2}, \ldots, x_{kn_k})$ the $n \times 1$ observation vector, its mean by $E(\underset{\sim}{x}) = \mu \underset{\sim}{1}_n$ and its variance-covariance matrix by $\Sigma : n \times n$ where,

$$\Sigma = block\ diagonal\ (\Lambda_1, \ldots, \Lambda_k),\ \Lambda_i : n_i \times n_i = \sigma^2 I_{n_i} + \sigma_\tau^2 E_{n_i}, \qquad i = 1,\ldots k, \tag{2.2}$$

where $\underset{\sim}{1}_n{}' : (1,\ldots 1) : 1 \times n$, I_{n_i} denotes the identity matrix of order n_i and $E_{n_i} = \underset{\sim}{1}_{n_i} \underset{\sim}{1}_{n_i}{}'$. We assume that $\underset{\sim}{x}$ has an elliptically symmetric distribution with density (with respect to the Lebesgue measure on R^n)

$$f(\underset{\sim}{x} | \mu, \sigma_\tau^2, \sigma^2) = |\Sigma|^{-1/2} \phi((\underset{\sim}{x} - \mu \underset{\sim}{1}_n)' \Sigma^{-1} (\underset{\sim}{x} - \mu \underset{\sim}{1}_n)) \tag{2.3}$$

for some $\phi : [0,\infty) \to [0,\infty)$ satisfying $\int_{R^n} \phi(\underset{\sim}{x}'\underset{\sim}{x}) d\underset{\sim}{x} = 1$.

To reduce the model to a canonical form, let $D_i : n_i \times n_i$ be an orthogonal matrix with its first row $(n_i^{-1/2}, \ldots, n_i^{-1/2})$, $i = 1,\ldots k$, and $D : n \times n =$ block-diagonal $(D_1, \ldots, D_k)$. Define

$$\underset{\sim}{z} = (z_{11}, z_{12}, \ldots z_{1n_1}, z_{21}, \ldots z_{2n_2}, \ldots, z_{k1}, \ldots, z_{kn_k})' = D\underset{\sim}{x}, \Delta = \sigma_\tau^2/\sigma^2. \tag{2.4}$$

Then $\underset{\sim}{z}$ has a elliptically symmetric distribution with density

$$f(\underset{\sim}{z}|\mu,\sigma^2,\Delta) = \sigma^{-n}(\prod_{i=1}^{k}(1 + \Delta n_i)^{-1/2})\phi[\{\sum_{i=1}^{k}(z_{i1} - \mu\sqrt{n_i})^2/ \tag{2.5}$$

$$(n_i\Delta + 1) + \sum_{i=1}^{k}\sum_{j=2}^{n_i} z_{ij}^2\}/\sigma^2]$$

This is the canonical form of the model and the hypothesis of interest is $H_0 : \Delta = 0$ vs $H_1 : \Delta > 0$. To derive an optimum invariant test, we note that the problem remains invariant under the group G of transformations acting on $\underset{\sim}{z}$ as

$$g\underset{\sim}{z} = (c,a)\underset{\sim}{z} = c\underset{\sim}{z} + a(\sqrt{n_1},0,\ldots 0,\sqrt{n_2},0\ldots 0,\ldots,\sqrt{n_k},0,\ldots 0) \tag{2.6}$$

$$c > 0, -\infty < a < \infty$$

As a left invariant measure ν on G, we take $d\nu(c,a) = dcda/c^2$. Applying Wijsman's representation theorem (1967), the ratio of the nonnull to null distributions of a maximal invariant statistic $T(\underset{\sim}{z})$ under G is obtained as

$$R_\Delta = dP_\Delta^T/dP_0^T(t(\underset{\sim}{z})) = \frac{\int_G f(g\cdot\underset{\sim}{z}|\mu,\sigma^2,\Delta)c^n d\nu(c,a)}{\int_G f(g\cdot\underset{\sim}{z}|\mu,\sigma^2,\Delta = 0)c^n d\nu(c,a)}. \tag{2.7}$$

The quantity R_Δ is simplified in the following lemma.

Lemma 2.1: The ratio R_Δ in (2.7) is evaluated as

$$R_\Delta = \left\{\prod_{i=1}^{k}(1 + \Delta n_i)^{-1/2}\right\}\left\{\sum_{i=1}^{k} n_i(1 + \Delta n_i)^{-1}\right\}^{-1/2} n^{1/2}\left\{\frac{Q(\underset{\sim}{z},\Delta)}{Q(\underset{\sim}{z},0)}\right\}^{-\frac{n-1}{2}} \tag{2.8}$$

where

$$Q(\underset{\sim}{z},\Delta) = \sum_{i=1}^{k} z_{i1}^2 (n_i\Delta + 1)^{-1} + \sum_{i=1}^{k}\sum_{j=2}^{n_i} z_{ij}^2 - \left(\sum_{i=1}^{k} z_{i1}\sqrt{n_i}(1 + \Delta n_i)^{-1}\right)^2 \cdot \left(\sum_{i=1}^{k} n_i(1 + \Delta n_i)^{-1}\right)^{-1} \tag{2.9}$$

and

$$Q(\underset{\sim}{z},0) = Q(\underset{\sim}{z},\Delta)\,|_{\Delta=0} = \sum_{i=1}^{k}\sum_{j=1}^{n_i} z_{ij}^2 - \left(\sum_{i=1}^{k} z_{i1}\sqrt{n_i}\right)^2/n.$$

Proof. Without loss of generality, due to the invariance of the problem, we take $\mu = 0$ and $\sigma = 1$ in (2.7). The argument of ϕ in (2.5), after the substitution $\underset{\sim}{z} \to g\cdot\underset{\sim}{z}$ displayed in (2.6) and completion of square, reduces to

$$\left\{\sum_{i=1}^{k} n_i(1 + \Delta n_i)^{-1}\right\}\left\{a + c\left(\sum_{i=1}^{k} z_{i1}\sqrt{n_i}(1 + \Delta n_i)^{-1}\right)\cdot \left(\sum_{i=1}^{k} n_i(1 + \Delta n_i)^{-1}\right)^{-1}\right\}^2 + c^2 Q(\underset{\sim}{z},\Delta). \tag{2.10}$$

The numerator of R_Δ in (2.7) is then evaluated as

$$\int_G f(g\cdot\underset{\sim}{z}\,|\mu = 0,\ \sigma = 1,\Delta)\, c^{n-2}\, da\,dc \tag{2.11}$$

$$= \left\{\prod_{i=1}^{k}(1 + \Delta n_i)^{-1/2}\right\}\left\{\sum_{i=1}^{k} n_i(1 + \Delta n_i)^{-1}\right\}^{-1/2}\left\{Q(\underset{\sim}{z},\Delta)\right\}^{-(n-1)/2}\cdot$$

$$\int_0^\infty \tilde{\phi}(c^2)\, c^{n-2} dc \qquad \textit{using a result of Dawid (1977), for some } \tilde{\phi}.$$

Since the denominator of (2.7) corresponds to its numerator with $\Delta = 0$, the lemma follows.

Remark 2.1: From (2.8) it is clear that the ratio R_Δ of the nonnull to null p.d.f. of $T(\underset{\sim}{z})$ is independent of ϕ. This proves the nonnull robustness of any null robust invariant test. In particular, the locally best invariant test derived below is null and hence nonnull robust.

In the balanced case when $n_1 = \cdots = n_k$ it is easy to verify that R_Δ is increasing in

$$\left\{\sum_{i=1}^{k} z_{i1}^2 - \left(\sum_{i=1}^{k} z_{i1}\sqrt{n_i}\right)^2/n\right\}\Big/\left(\sum_{i=1}^{k}\sum_{j=2}^{n_i} z_{ij}^2\right) = \psi(\underset{\sim}{z})\ (say),$$

for every $\Delta > 0$ and that $\psi(\underset{\sim}{z})$ satisfies

$$\psi((\underset{\sim}{z} - \mu \underset{\sim n}{l})c) = \psi(\underset{\sim}{z}) \text{ for } all\ \mu \in R,\ c > 0$$

$\underset{\sim n}{l} = (\sqrt{n_1},0...0,...\sqrt{n_k},0,...0)'$. The latter condition is equivalent to $\chi(\underset{\sim}{x}) = \chi((\underset{\sim}{x} - \mu \underset{\sim n}{l})c)$ for all $\mu \in R$, $c > 0$ where $\chi(\underset{\sim}{x}) = \psi(\underset{\sim}{z})$, and in turn implies the null robustness of the test based on $\psi(\underset{\sim}{z})$ (vide corollary 2.1, Kariya (1981)). This proves the following result.

<u>Theorem 2.1:</u> Under the model (2.5) with $n_1 = \cdots = n_k$, the UMPI test rejects $H_0 : \Delta = 0$ vs $H_1 : \Delta > 0$ for large values of

$$\left\{\sum_{i=1}^{k} z_{i1}^2 - \left(\sum_{i=1}^{k} z_{i1}\sqrt{n_i}\right)^2 / n\right\} / \left(\sum_{i=1}^{k}\sum_{j=2}^{n_i} z_{ij}^2\right).$$

The test is null, nonnull, and optimality robust.

Expressed in terms of the original observations $\underset{\sim}{x}, \chi(\underset{\sim}{x})$ is nothing but a function of the usual F-statistic define by

$$F = (n-k)\sum_{i=1}^{k}\sum_{j=1}^{n_i}(\bar{x}_{i\cdot} - \bar{x}_{\cdot\cdot})^2 / \left(\sum_{i=1}^{k}\sum_{j=1}^{n_i}(x_{ij} - \bar{x}_{i\cdot})^2(k-1)\right) \text{ where}$$

$$\bar{x}_{i\cdot} = \sum_{j=1}^{n_i} x_{ij}/n_i \text{ and } \bar{x}_{\cdot\cdot} = \sum_{i=1}^{k} n_i\bar{x}_{i\cdot}/n. \quad Of\ course,\ here\ n_i = J\forall i$$

$$\left[\chi(\underset{\sim}{x}) = \frac{J(n-J)F}{(n-k) + (n-J)F}\right]$$

In the unbalanced case it turns out that the preceding observation does not hold in the sense that the ratio R_Δ is not increasing in $\psi(\underset{\sim}{z})$ or any other statistic for every $\Delta > 0$. This demonstrates that an UMPI test does not exist. To derive a LBI test, we expand R_Δ around $\Delta = 0$. Towards this end, a straightforward computation yields

$$Q(\underset{\sim}{z},\Delta) = Q(\underset{\sim}{z},0) + \Delta\{-\sum_{i=1}^{k} z_{i1}^2 n_i - \left(\sum_{i=1}^{k} n_i^2\right)\left(\sum_{i=1}^{k} z_{i1}\sqrt{n_i}\right)^2 + 2\left(\sum_{i=1}^{k} n_i\right)\left(\sum_{i=1}^{k} z_{i1}\sqrt{n_i}\right)\left(\sum_{i=1}^{k} z_{i1}n_i^{3/2}\right)\} + \delta(\underset{\sim}{z},\Delta) \tag{2.12}$$

where the remainder term $\delta(\underset{\sim}{z},\Delta)$ is given by

$$\delta(\underset{\sim}{z},\Delta) = \sum_{i=1}^{k}(1 + \Delta n_i)^{-1}\{z_{i1} - \sqrt{n_i}\left(\sum_{i=1}^{k} z_{i1}\sqrt{n_i}(1 + \Delta n_i)^{-1}\right) / \sum_{i=1}^{k} n_i(1 + \Delta n_i)^{-1})\}^2 - \sum_{i=1}^{k}(1 - \Delta n_i)\left(z_{i1} - \sqrt{n_i}\left(\sum_{i=1}^{k} z_{i1}\sqrt{n_i}\right)/n\right)^2 \tag{2.13}$$

Using $(1+\Delta n_i)^{-1} = 1 - \Delta n_i + \{\Delta^2 n_i^2/(1+\Delta n_i)\}$, it can be shown after simplification that,

$$\delta(\underset{\sim}{z},\Delta)/Q(\underset{\sim}{z},0) = o(\Delta), \text{ uniformly in } \underset{\sim}{z} \tag{2.14}$$

and hence from (2.12) we get,

$$Q(\underset{\sim}{z},\Delta)/Q(\underset{\sim}{z},0) = 1 - \Delta\psi(\underset{\sim}{z}) + o(\Delta) \tag{2.15}$$

where

$$\psi(\underset{\sim}{z}) = \{\sum_{i=1}^{k} n_i(z_{i1} - \sqrt{n_i}(\sum_{i=1}^{k} z_{i1}\sqrt{n_i})/n)^2\}/ \tag{2.16}$$
$$\{\sum_{i=1}^{k}(z_{i1} - \sqrt{n_i}(\sum_{i=1}^{k} z_{i1}\sqrt{n_i})/n)^2 + \sum_{i=1}^{k}\sum_{j=2}^{n_i} z_{ij}^2\}$$

Finally, since $\psi(\underset{\sim}{z})$ is uniformly bounded in $\underset{\sim}{z}$, the ratio R_Δ in (2.8) is evaluated as

$$R_\Delta = 1 + \Delta\{\psi(\underset{\sim}{z})(n-1)/2 + c\} + r(\underset{\sim}{z},\Delta) \tag{2.17}$$

where c is a constant depending on $n_1,\ldots n_k$ and $r(\underset{\sim}{z},\Delta)$ is a term satisfying $\sup_{\underset{\sim}{z}} r(\underset{\sim}{z},\Delta) = o(\Delta)$.

Consider now an invariant level α test $\phi(t(\underset{\sim}{z}))$. Then its local power can be evaluated as

$$\int \phi(t(\underset{\sim}{z}))dP_\Delta^T(t(\underset{\sim}{z})) = \alpha + \Delta\int \{\psi(\underset{\sim}{z})(n-1)/2 + c\}dP_0^T(t(\underset{\sim}{z})) + o(\Delta) \tag{2.18}$$

By applying the Neyman-Pearson Lemma, and proceeding as in Theorem 2.1, we have therefore proved the following main result.

Theorem 2.2: The locally best invariant test of $H_0 : \Delta = 0$ vs $H_1 : \Delta > 0$ under the model (2.5) rejects H_0 for large values of $\psi(\underset{\sim}{z})$. The test is null, nonnull and optimality robust.

Remark 2.2: Expressed in terms of the original observations $\underset{\sim}{x}$, the test statistic $\psi(\underset{\sim}{z})$ in (2.16) assumes the form

$$\psi(\underset{\sim}{z}) = \{\sum_{i=1}^{k} n_i^2(\bar{x}_{i.} - \bar{x}_{..})^2\} / \left\{ \sum_{i=1}^{k} \sum_{j=1}^{n_i} (x_{ij} - \bar{x}_{i.})^2 + \sum_{i=1}^{k} n_i(\bar{x}_{i.} - \bar{x}_{..})^2 \right\} \tag{2.19}$$

which is clearly different from the traditional F-statistic. Under the normality assumption, Spjøtvoll (1967) proved that the F-test is asymptotically (as $\Delta \to \infty$) best invariant. We may note that, under H_o, the test statistic $\psi(\underset{\sim}{z})$ can be expressed as a ratio of linear combinations of independent chisquare variates and, unlike F, is not readily tabulated. Our simulation study (not reported here) shows that $\psi(\underset{\sim}{z})$ has a definite edge over the F-statistic for local alternatives.

Remark 2.3. It is interesting to note that the tests derived in Arvesen and Layard (1975) are different from the one given above. The reader may also look at Roebruck (1982) for some related results.

3. TWO-WAY BALANCED RANDOM EFFECTS MODEL

We shall treat here two cases: with and without interaction. The first one is explained in detail while the second is briefly mentioned.

3.1 With interaction. The statistical model here involves $n = bvr$ observations $\{x_{ijk}\}$ following the structure

$$x_{ijk} = \mu + \beta_i + \tau_j + \gamma_{ij} + e_{ijk}, \; i = 1,\ldots b, \; j = 1,\ldots v, \; k = 1,\ldots r \tag{3.1}$$

where μ represents a fixed unknown general effect, β_i's represent random block effects each with mean 0 and variance σ_β^2, τ_j's random treatment effects each with mean 0 and variance σ_τ^2, γ_{ij}'s random interaction effects each with mean 0 and variance σ_γ^2, and e_{ijk}'s random errors each with mean 0 and variance σ^2. If we denote by $\underset{\sim}{x} = (x_{111}, x_{211}, \cdots x_{b11}, \ldots x_{bvr})'$ the $n \times 1$ observation vector, then its mean $E(\underset{\sim}{x}) = \mu \underset{\sim n}{1}$ and its variance-covariance matrix Σ is given by

$\Sigma = [R/B]$, $R = [A_1/B_1]$, $B = [A_2/B_2]$,where $[C/D]$ is a matrix with diagonal blocks as C and off diagonal blocks as D (3.2)

$A_1 = (\sigma_\beta^2 + \sigma_r^2 + \sigma_\gamma^2 + \sigma^2/\sigma_r^2)$, $B_1 = (\sigma_\beta^2/0) = B_2$

$$A_2 = (\sigma_\beta^2 + \sigma_r^2 + \sigma_\gamma^2/\sigma_r^2) \text{ and } A = (a+b/a) \implies A = \begin{bmatrix} a+b & a & a \\ a & a+b & a \\ \cdots & \cdots & \cdots \\ \cdots & \cdots & \cdots \\ a & a & a+b \end{bmatrix}$$

We assume that $\underset{\sim}{x}$ has an elliptically symmetric distribution with density (with respect to Lebesgue measure on R^n)

$$f(\underset{\sim}{x}|\mu,\sigma_\beta^2,\sigma_r^2,\sigma_\gamma^2,\sigma^2) = |\Sigma|^{-1/2}\,\phi((\underset{\sim}{x} - \mu \underset{\sim n}{1})'\,\Sigma^{-1}(\underset{\sim}{x} - \mu \underset{\sim n}{1})) \qquad (3.3)$$

for some $\phi : [0,\infty) \to [0,\infty)$ satisfying $\int_{R^n} \phi(\underset{\sim}{u}'\underset{\sim}{u})\,d\underset{\sim}{u} = 1$.

To reduce the problem to a canonical form (vide Herbach (1959)), note that there exists an orthogonal matrix $\Gamma : n \times n$ with its first row as $(n^{-1/2}, \ldots, n^{-1/2})$ such that $\Gamma\Sigma\Gamma'$ reduces to a diagonal matrix with diagonal elements given by $\lambda_1 = \sigma^2 + r\sigma_\gamma^2 + br\sigma_r^2 + vr\sigma_\beta^2$ with multiplicity 1, $\lambda_2 = \sigma^2 + r\sigma_\gamma^2 + vr\sigma_\beta^2$ with multiplicity $(b-1)$, $\lambda_3 = \sigma^2 + r\sigma_\gamma^2 + br\sigma_r^2$ with multiplicity $(v-1)$, $\lambda_4 = \sigma^2 + r\sigma_\gamma^2$ with multiplicity $(b-1)(v-1)$, and $\lambda_5 = \sigma^2$ with multiplicity $bv(r-1)$. Denote $\underset{\sim}{z} = \Gamma\underset{\sim}{x} = (z_{111}, z_{211}, \cdots z_{b11}, \cdots z_{bvr})'$. Then $\underset{\sim}{z}$ has the elliptically symmetric distribution with density

$$f(\underset{\sim}{z}|\mu,\lambda_1,\lambda_2,\cdots\lambda_5) = \lambda_1^{-1/2}\lambda_2^{-(b-1)2}\lambda_3^{-(v-1)/2}\lambda_4^{-(b-1)(v-1)/2}\lambda_5^{-bv(r-1)/2}. \qquad (3.4)$$

$$\phi\left(\frac{(z_{111} - \mu\sqrt{n})^2}{\lambda_1} + \frac{s_2}{\lambda_2} + \frac{s_3}{\lambda_3} + \frac{s_4}{\lambda_4} + \frac{s_5}{\lambda_5}\right)$$

where

$$s_2 = \sum_{i=2}^{b} z_{i11}^2, \quad s_3 = \sum_{j=2}^{v} z_{1j1}^2, \quad s_4 = \sum_{j=2}^{v}\sum_{i=2}^{b} z_{ij1}^2 \qquad (3.5)$$

$$s_5 = \sum_{i=2}^{b}\sum_{j=2}^{v}\sum_{k=2}^{r} z_{ijk}^2.$$

This is the canonical form of the model and the hypotheses of interest are $H_{o\gamma} : \lambda_4 = \lambda_5$, $H_{o\beta} : \lambda_2 = \lambda_4$ and $H_{or} : \lambda_3 = \lambda_4$. The quantities s_2, s_3, s_4 and s_5 represent respectively the block sum of squares, the treatment sum of squares, the block-treatment interaction sum of squares and the error or residual sum of squares.These are expressed in terms of the original observations $\{x_{ijk}\}$ as,

$$s_2 = rv \sum_i (\bar{x}_{i..} - \bar{x}_{...})^2, \quad s_3 = br \sum_j (\bar{x}_{.j.} - \bar{x}_{...})^2, \tag{3.6}$$
$$s_4 = r \sum_i \sum_j (\bar{x}_{ij.} - \bar{x}_{i..} - \bar{x}_{.j.} + \bar{x}_{...})^2, \quad s_5 = \sum_i \sum_j \sum_k (x_{ijk} - \bar{x}_{ij.})^2$$

where

$$\bar{x}_{i..} = \sum_{j=1}^{v} \sum_{k=1}^{r} x_{ijk}/rv, \quad \bar{x}_{.j.} = \sum_{i=1}^{b} \sum_{k=1}^{r} x_{ijk}/br$$
$$\bar{x}_{ij.} = \sum_{k=1}^{r} x_{ijk}/r, \quad \bar{x}_{...} = \sum_{i=1}^{b} \sum_{j=1}^{v} \sum_{k=1}^{r} x_{ijk}/n.$$

We derive below an optimum invariant unbiased test of $H_{o\tau} : \lambda_3 = \lambda_4$ under the model (3.4). The other hypotheses $H_{o\beta}$ and $H_{o\gamma}$ can be similarly dealt with and are mentioned in Theorems 3.2(a) and 3.2(b). Towards this end, note that the testing problem $H_{o\tau}$ remains invariant under the group G of transformations acting on $\underset{\sim}{z}$ as

$$g \cdot \underset{\sim}{z} = (c, a)\underset{\sim}{z} + a(1,0,...,0)', \; c > 0, \; a \in R. \tag{3.7}$$

We take $d\nu(c,a) = dcda/c^2$ as a left invariant measure on G. Denoting by $T(\underset{\sim}{z})$ a maximal invariant statistic induced by the group G and by Δ a maximal invariant parameter, and applying Wijsman's representation theorem (1967), the ratio $R_{\Delta\Delta_o}(t(\underset{\sim}{z}))$ of the nonnull to null pdf's of $T(\underset{\sim}{z})$ is obtained as

$$R_{\Delta\Delta_o}(t(\underset{\sim}{z})) = dP^T_\Delta / dP^T_{\Delta_o}(t(\underset{\sim}{z})) = \frac{\int_G f(g \cdot \underset{\sim}{z} | \mu, \lambda_1, \lambda_2, \lambda_3, \lambda_4, \lambda_5) c^n d\nu(c,a)}{\int_G f(g \cdot \underset{\sim}{z} | \mu, \lambda_1, \lambda_2, \lambda_3 = \lambda_4, \lambda_5) c^n d\nu(c,a)} \tag{3.8}$$

The following lemma provides a simple expression of the ratio $R_{\Delta\Delta_o}(t(\underset{\sim}{z}))$.

<u>Lemma 3.1:</u> The ration $R_{\Delta\Delta_o}(t(\underset{\sim}{z}))$ in (3.8) is evaluated as

$$R_{\Delta\Delta_o}(t(\underset{\sim}{z})) = \theta^{-(v-1)/2} \{(1 + u + \frac{v}{\psi_2} + \frac{w}{\psi_5})/(1 + \frac{u}{\theta} + \frac{v}{\psi_2} + \frac{w}{\psi_5})\}^{\frac{n-1}{2}} \tag{3.9}$$

where

$$\theta = \lambda_3/\lambda_4, \; \psi_2 = \lambda_2/\lambda_4, \; \phi_5 = \lambda_5/\lambda_4, \; u = s_3/s_4, \; v = s_2/s_4, \; w = s_5/s_4. \tag{3.10}$$

<u>Proof:</u> Proceeding exactly similarly as in the proof of Lemma 2.1 we obtain the result.

Remark 3.1: Since the ratio $dP^T_\Delta/dP^T_{\Delta_o}(t(\underset{\sim}{z}))$ is independent of ϕ, it follows that any null robust invariant test is also nonnull robust. In particular, the optimum invariant test derived below is null and hence nonnull robust.

Remark 3.2: It is clear from the group G that $t(\underset{\sim}{z}) = (u,v,w)$ and $\Delta = (\theta,\psi_2,\psi_5)$ with $\Delta_o = (\theta = 1,\psi_2,\psi_5)$. Thus ψ_2 and ψ_5 are the two nuisance parameters in the null distribution of $t(\underset{\sim}{z})$.

A direct calculation shows that, whatever be ϕ in (3.4), the null distribution of $t(\underset{\sim}{z})$ has the form (vide Kariya (1981) and straightforward normal theory)

$$dP^T_{\Delta_o}(t(\underset{\sim}{z}))/dudvdw = c\,(1 + u + \frac{v}{\psi_2} + \frac{w}{\psi_5})^{-(n-1)/2} \cdot \frac{u^{\frac{\nu_2-2}{2}}\, v^{\frac{\nu_3-2}{2}}\, w^{\frac{\nu_5-2}{2}}}{\psi_2^{\nu_3/2}\,\psi_5^{\nu_5/2}} \tag{3.13}$$

where

$$c = \frac{\Gamma\left(\frac{n-1}{2}\right)}{\Gamma\left(\frac{b-1}{2}\right)\Gamma\left(\frac{v-1}{2}\right)\Gamma\left(\frac{(b-1)(v-1)}{2}\right)\Gamma\left(\frac{bv(r-1)}{2}\right)}, \qquad \begin{array}{l} \nu_2 = v-1 \\ \nu_3 = b-1 \\ \nu_5 = bv(r-1) \\ 0 < u,\ v,\ w < \infty \end{array}$$

and consequently $(t_1(\underset{\sim}{z}) = v/(1+u), t_2(\underset{\sim}{z}) = w/(1+u))$ are sufficient for (ψ_2,ψ_5) under H_o.

It is not difficult to verify that the family of joint distributions of $(t_1(\underset{\sim}{z}),t_2(\underset{\sim}{z}))$ is complete. This suggests that any level α invariant unbiased test $\psi(t(\underset{\sim}{z}))$ of $H_o : \theta = 1$ has the Neyman Structure (vide Lehmann (1959)). Using (3.8) and (3.9), the power of such a $\psi(t(\underset{\sim}{z}))$ is expressed as

$$\int \psi(t(\underset{\sim}{z}))\,dP^t_\Delta(t(\underset{\sim}{z})) = \int \psi(t(\underset{\sim}{z}))R_{\Delta\Delta_o}(t(\underset{\sim}{z}))\cdot dP^T_{\Delta_o}(t(\underset{\sim}{z})) \tag{3.14}$$

$$= \theta^{-(n-1)/2} c \int \psi(t(\underset{\sim}{z}))(1 + \frac{u}{\theta} + \frac{v}{\psi_2} + \frac{w}{\psi_5})^{-(n-1)/2} \cdot u^{(\nu_2-2)/2} v^{(\nu_3-2)/2} w^{(\nu_5-2)/2} \frac{1}{\psi_2^{\nu_3/2}\,\psi_5^{\nu_5/2}}\, dudvdw$$

Comparing (3.14) and (3.13), it follows that the conditional power of $\psi(t(\underset{\sim}{z}))$, conditional on $(t_1(\underset{\sim}{z}),t_2(\underset{\sim}{z}))$, is a maximum for all $\theta > 1$ if $\psi(t(\underset{\sim}{z}))$ is chosen as the indicator function of the set $\{t(\underset{\sim}{z}) : u > u_o(t_1(\underset{\sim}{z}), t_2(\underset{\sim}{z}))\}$ where $u_o(t_1(\underset{\sim}{z}), t_2(\underset{\sim}{z}))$ satisfies $\alpha = E_{H_o}\{\psi(t(\underset{\sim}{z})) | t_1(\underset{\sim}{z}), t_2(\underset{\sim}{z})\}$ a.e. However, from (3.13), it is evident that the null distribution of u is independent of ψ_2, ψ_5, and hence, by Basu's theorem (1955), u_o is a constant. In fact, under H_o, $(b-1)u$ has a centeral $F_{(v-1),(b-1)(v-1)}$ distribution. Thus we have proved the following main result.

Theorem 3.1: Under the model (3.4), the UMPIU test of $H_{o\tau} : \lambda_3 = \lambda_4$ vs $H_{1\tau} : \lambda_3 > \lambda_4$ rejects H_o for large values of u. The test is null, nonnull, and optimality robust.

Proceeding analogously, we can prove the following.

Theorem 3.2(a): Under the model (3.4), the UMPIU test of $H_{o\beta} : \lambda_2 = \lambda_4$ vs $H_{1\beta} : \lambda_2 > \lambda_4$ rejects H_o for large values of v. The test is null,nonnull and optimality robust.

Theorem 3.2(b): Under the model (3.4), the UMPIU test of $H_{o\gamma} : \lambda_4 = \lambda_5$ vs $H_{1\gamma} : \lambda_4 > \lambda_5$ rejects H_o for small value of w. The test is null, nonnull and optimality robust.

3.2 Without interaction: This is a simplified form of the model with interaction when $r = 1$ and the interaction effects $\gamma_{ij}'s$ are zero, leading to the model

$$x_{ij} = \mu + \beta_i + \tau_j + e_{ij},\ i = 1,\dots b,\ j = 1,\dots v \tag{3.15}$$

where μ, $\beta's$, $\tau's$ and $e's$ have interpretations as in section 3.1. Writing $n = bv$ and $\underset{\sim}{x} = (x_{11}, x_{21}, \dots x_{bv})'$, we assume that $\underset{\sim}{x}$ has an elliptically symmetric distribution with density (with respect to the Lebesgue measure on R^n)

$$f(\underset{\sim}{x} | \mu, \sigma_\beta^2, \sigma_\tau^2, \sigma^2) = |\Sigma|^{-1/2}\phi((\underset{\sim}{x} - \mu \underset{\sim n}{1})'\Sigma^{-1}(\underset{\sim}{x} - \mu \underset{\sim n}{1})) \tag{3.16}$$

for some $\phi : [0,\infty) \to [0,\infty)$ satisfying $\int_{R^n} \phi(\underset{\sim}{u}'\underset{\sim}{u})d\underset{\sim}{u} = 1$, where

$$\begin{aligned} \Sigma &= [R/B],\ R = (\sigma_\beta^2 + \sigma_\tau^2 + \sigma^2/\sigma_\tau^2) \\ B &= [\sigma_\beta^2/0]. \end{aligned} \tag{3.17}$$

The hypothesis of interest is $H_{o\tau} : \sigma_\tau^2 = 0$. As before, a canonical form of the model is obtained by defining $\underset{\sim}{z} = \Gamma \underset{\sim}{x}$ and noting that $\underset{\sim}{z}$ has the elliptically symmetric density :

$$f(\underset{\sim}{z}|\mu,\lambda_1,\lambda_2,\lambda_3,\lambda_4) = \lambda_2^{-1/2}\lambda_2^{-\frac{b-1}{2}}\lambda_3^{-\frac{v-1}{2}}\lambda_4^{-\frac{(b-1)(v-1)}{2}} \cdot \phi\left(\frac{(z_{11}-\mu\sqrt{n})^2}{\lambda_1}+\frac{s_2}{\lambda_2}+\frac{s_3}{\lambda_3}+\frac{s_4}{\lambda_4}\right) \tag{3.18}$$

where

$$s_2 = \sum_{i=2}^{b} z_{i1}^2,\ s_3 = \sum_{j=2}^{v} z_{1j}^2,\ s_4 = \sum_{j=2}^{v}\sum_{i=2}^{b} z_{ij}^2 \tag{3.19}$$

and

$$\lambda_1 = \sigma^2 + b\sigma_\tau^2 + v\sigma_\beta^2,\ \lambda_2 = \sigma^2 + v\sigma_\beta^2,\ \lambda_3 = \sigma^2 + b\sigma_\tau^2,\ \lambda_4 = \sigma^2$$

Stated in terms of $\lambda's$, the problem is to test $H_o : \lambda_3 = \lambda_4$ vs $H_1 : \lambda_3 > \lambda_4$. The group G and its action on $\underset{\sim}{z}$ here is the same as in (3.7). The following result, stated without proof, is similar to lemma 3.1.

Lemma 3.2: The Ratio $R_{\Delta\Delta_o}(t(\underset{\sim}{z})) = dP_\Delta^T/dP_{\Delta_o}^T(t(\underset{\sim}{z}))$ of the nonnull to null pdf's of a maximal invariant $T(\underset{\sim}{z})$ under G is expressed as

$$R_{\Delta\Delta_o}(t(\underset{\sim}{z})) = \theta^{-(v-1)/2}\{(1+u+v/\psi)/(1+\frac{u}{\theta}+\frac{v}{\psi})\}^{(n-1)/2} \tag{3.20}$$

where

$$\theta = \lambda_3/\lambda_4,\ \psi = \lambda_2/\lambda_4,\ u = s_3/s_4,\ v = s_2/s_4. \tag{3.21}$$

Remark 3.3: As in section 3.1, the ratio $dP_\Delta^T/dP_{\Delta_o}^T(t(\underset{\sim}{z}))$ is independent of ϕ. This suggests that any null robust invariant test is also nonnull robust.

Remark 3.4: From (3.18) and the group G, it is clear that $t(\underset{\sim}{z}) = (u,v)$ and $\Delta = (\theta,\psi)$ with $\Delta_o = (\theta = 1,\psi)$. Thus, ψ is a nuisance parameter in the null distribution of $t(\underset{\sim}{z})$.

Proceeding as in the previous section, it can be easily shown that, whatever be ϕ in (3.16), $t_1(\underset{\sim}{z}) = v/(1+u)$ is a complete sufficient statistic for ψ under $H_{o\tau}$ and that the null distribution of u is independent of ψ. This follows from the following null distribution of $t(\underset{\sim}{z})$ whose proof is straightforward.

$$dP^T_{\Delta_o}(t(\underset{\sim}{z}))/dudv = c(1 + u + \frac{v}{\psi})^{-(n-1)/2} u^{(\nu_3-2)/2} v^{(\nu_2-2)/2}, \qquad (3.22)$$

$$0 < u,\ v < \infty$$

where $c = \dfrac{\Gamma\frac{n-1}{2}}{\Gamma\frac{b-1}{2}\Gamma\frac{v-1}{2}\Gamma\frac{(b-1)(v-1)}{2}} \cdot \dfrac{1}{\psi^{(b-1)/2}}$, $\nu_2 = b-1$ $\nu_3 = v-1$.

Using (3.20) & (3.22) and proceeding as in section 3.1, we prove the following result.

Theorem 3.3: Under the model (3.18), the UMPIU test of $H_{or}: \lambda_3 = \lambda_4$ vs $H_{1r}: \lambda_3 > \lambda_4$ rejects H_o for large values of u. The test is null, nonnull and optimality robust.

Remark 3.5: Unlike the balanced designs, the two-way unbalanced random effects models pose a problem which is pointed out below. Consider, for simplicity, the simplest unbalanced design with $b = 2$, $v = 2$, and replications $n_{11} = n_{12} = 2$, $n_{21} = n_{22} = 4$ where n_{ij} is the number of observations in the ith block corresponding to the jth treatment. Note that this is the "orthogonality" condition in an unbalanced design (vide Charkraborty (1962)). The model without interaction, again for simplicity, can be written as

$$x_{ijk} = \mu + \beta_i + \tau_j + e_{ijk},\ k = 1,\dots,n_{ij},\ i = 1,2,\ j = 1,2 \qquad (3.23)$$

where the rotations μ,β,τ,e have their usual significance as in section 3.1. Write $\underset{\sim}{x} = (x_{111}, x_{112}, x_{121}, x_{122}, x_{211}, x_{212}, x_{213}, x_{214}, x_{221}, x_{222}, x_{223}, x_{224})'$ so that $E(\underset{\sim}{x}) = \mu \underset{\sim 12}{1}$ and the variance-covariance matrix of $\underset{\sim}{x} \equiv \Sigma$ where

$$\Sigma = \begin{bmatrix} D_2 & E_2 & E_\tau & 0 \\ E_2' & D_2 & 0 & E_\tau \\ E_\tau' & 0' & D_4 & E_4 \\ 0 & E_\tau' & E_4' & D_4 \end{bmatrix} \qquad (3.24)$$

with $D_k = (\sigma^2_\beta + \sigma^2_\tau + \sigma^2 \not{/} \sigma^2_\beta + \sigma^2_\tau)$ of order $k \times k$,

$E_k = (\sigma_\beta^2)\cdot$ *the matrix of order* $k \times k$ *with all entries as* 1

$$E_\tau = \sigma_\tau^2 \begin{bmatrix} 1 & 1 & 1 & 1 \\ 1 & 1 & 1 & 1 \end{bmatrix},\quad 0 = \begin{bmatrix} 0 & 0 & 0 & 0 \\ 0 & 0 & 0 & 0 \end{bmatrix}$$

To reduce the problem to a canonical form let $P = \begin{bmatrix} P_1 & & & 0 \\ & P_2 & & \\ & & P_3 & \\ 0 & & & P_4 \end{bmatrix}$ be a block orthogonal matrix with first rows of P_1, P_2, P_3, P_4 as $(2^{-1/2}, 2^{-1/2}), (2^{-1/2}, 2^{-1/2})$, $(4^{-1/2}, 4^{-1/2}, 4^{-1/2}, 4^{-1/2})$ and $(4^{-1/2}, 4^{-1/2}, 4^{-1/2}, 4^{-1/2})$ respectively and define $\underset{\sim}{z} = P \cdot \underset{\sim}{x}$. Then $E(\underset{\sim}{z}) = (\sqrt{2}\mu, 0, \sqrt{2}\mu, 0, \sqrt{4}\mu, 0, 0, 0, \sqrt{4}\mu, 0, 0, 0)'$ and under the null hypothesis $H_{o\tau} : \sigma_\tau^2 = 0$ of no treatment effects, the variance-covariance matrix of $\underset{\sim}{z}$ is Σ^* where

$$\Sigma^* = \begin{bmatrix} A_1 & B_1 & 0 & 0 \\ B_1' & A_1 & 0 & 0 \\ 0' & 0' & A_2 & B_2 \\ 0' & 0' & B_2' & A_2 \end{bmatrix}$$

where $A_1 = \begin{bmatrix} \sigma^2 + 2\sigma_\beta^2 & 0 \\ 0 & \sigma^2 \end{bmatrix}$, $B_1 = \begin{bmatrix} \sigma_\beta^2 & 0 \\ 0 & 0 \end{bmatrix}$, $0 =$ matrix of proper order with all entries as 0 ,

$$A_2 = \begin{bmatrix} \sigma^2 + 4\sigma_\beta^2 & 0 & 0 & 0 \\ 0 & \sigma^2 & 0 & 0 \\ 0 & 0 & \sigma^2 & 0 \\ 0 & 0 & 0 & \sigma^2 \end{bmatrix},\quad B_2 = \begin{bmatrix} \sigma_\beta^2 & 0 & 0 & 0 \\ 0 & 0 & 0 & 0 \\ 0 & 0 & 0 & 0 \\ 0 & 0 & 0 & 0 \end{bmatrix}.$$

Under the null hypothesis the p.d.f. of $\underset{\sim}{z}$ assuming normality can be written as

$$f(\underset{\sim}{z} \mid \mu, \sigma_\beta^2, \sigma^2) = c \exp\{-1/2(\underset{\sim}{z} - \underset{\sim}{\xi})'\Sigma^{*-1}(\underset{\sim}{z} - \underset{\sim}{\xi})\} \tag{3.26}$$

where c is the normalizing constant and $\underset{\sim}{\xi} = E(\underset{\sim}{z})$.

Now,

$$(\underset{\sim}{z} - \underset{\sim}{\xi})'\Sigma^{*-1}(\underset{\sim}{z} - \underset{\sim}{\xi}) = a_1\{(z_1 - \sqrt{2}\mu)^2 + (z_3 - \sqrt{2}\mu)^2\} +$$
$$a_2\{(z_5 - \sqrt{4}\mu)^2 + (z_9 - \sqrt{4}\mu)^2\} + 2b_1(z_1 - \sqrt{2}\mu)(z_3 - \sqrt{2}\mu) +$$
$$2b_2(z_5 - \sqrt{4}\mu)(z_9 - \sqrt{4}\mu) + (z_2^2 + z_4^2 + z_6^2 + z_7^2 + z_8^2 + z_{10}^2 + z_{11}^2 + z_{12}^2)/\sigma^2$$

where $\underset{\sim}{z} = (z_1, z_2, \ldots, z_{12})'$, $a_1 = \dfrac{-\gamma}{\sigma_\beta^4 - \gamma^2}$, $b_1 = \dfrac{\sigma_\beta^2}{\sigma_\beta^4 - \gamma^2}$,

$$a_2 = -\frac{\delta}{\sigma_\beta^4 - \delta^2},\ b_2 = \frac{\sigma_\beta^2}{\sigma_\beta^4 - \sigma^2},\ \gamma = \sigma^2 + 2\sigma_\beta^2 \text{ and } \delta = \sigma^2 + 4\sigma_\beta^2.$$

From the above expression it is clear that although there are three nuisance parameters μ, σ_β^2 and σ^2 under $H_{o\tau}$, there exist five sufficient statistics for them namely,

$$(z_2^2 + z_4^2 + z_6^2 + z_7^2 + z_8^2 + z_{10}^2 + z_{11}^2 + z_{12}^2),\ (z_1^2 + z_3^2),\ (z_5^2 + z_9^2),\ z_1 z_3, z_5 z_9.$$

It is easy to show that $E(z_1 - z_3)^2 = 2(\sigma_\beta^2 + \sigma^2)$, $E(z_5 - z_9)^2 = 2(3\sigma_\beta^2 + \sigma^2)$. Therefore completeness is violated because $E\left\{T_1 - 6(T_2 - 2T_4) + 2(T_3 - 2T_5)\right\} = 0$ where

$$T_1 = z_2^2 + z_4^2 + z_6^2 + z_7^2 + z_8^2 + z_{10}^2 + z_{11}^2 + z_{12}^2,$$
$$T_2 = z_1^2 + z_3^2,\ T_3 = z_5^2 + z_9^2,\ T_4 = z_1 z_3 \text{ and } T_5 = z_5 z_9.$$

This means there exist similar tests without having Neyman-structure. It is possible to construct such similar tests along the lines of Wijsman (1958), Sinha et al. (1983) but these tests are randomized and subject to criticism.

4. TWO-WAY BALANCED MIXED EFFECTS MODEL

The statistical model without interaction in this setup involves $n = bv$ observations $\{x_{ij}\}$ with the following structure

$$x_{ij} = \mu + \beta_i + \tau_j + \epsilon_{ij},\ i = 1, \ldots, b,\ j = 1, \ldots v. \tag{4.1}$$

where μ is a constant general effect, τ's are fixed treatment effects with $\sum_{j=1}^{v} \tau_j = 0$, β's are random block effects with mean 0 and variance σ_β^2, and ϵ's are random error components with mean 0 and variance σ^2. Denote by $\underset{\sim}{x} = (x_{11}, x_{21}, \ldots, x_{b1}, \ldots, x_{bv})'$ the $n \times 1$ observation vector, with its mean vector as $E(\underset{\sim}{x}) = \underset{\sim}{\xi} = (\mu + \tau_1, \ldots \mu + \tau_v)' \otimes \underset{\sim}{1}_b$ and variance-covariance matrix Σ given by

$$\Sigma = R \otimes I_v,\quad R = (\sigma_\beta^2 + \sigma^2/\sigma_\beta^2) : b \times b \text{ where } A = (a + b/b) \text{ is} \tag{4.2}$$

defined as in section 3 and $\underset{\sim}{1}_k$ = *a k-dimensional vector with all entries as* 1.

We assume as in previous section that $\underset{\sim}{x}$ has an elliptically symmetric distribution with density (with respect to Lebesgue measure on R^n)

$$f(\underset{\sim}{x}|\mu,\sigma_\beta^2,\sigma^2)=|\Sigma|^{-1/2}\phi((\underset{\sim}{x}-\underset{\sim}{\xi})'\Sigma^{-1}(\underset{\sim}{x}-\underset{\sim}{\xi})) \tag{4.3}$$

for some $\phi : [0,\infty)\rightarrow[0,\infty)$ satisfying $\int_{R^n}\phi(\underset{\sim}{u}'\underset{\sim}{u})\,d\underset{\sim}{u}=1$. It is easy to verify that

$$(\underset{\sim}{x}-\underset{\sim}{\xi})'\Sigma^{-1}(\underset{\sim}{x}-\underset{\sim}{\xi})=\frac{1}{\sigma^2}\sum_{i=1}^{b}\sum_{j=1}^{v}(x_{ij}-\mu-\tau_j)^2-\frac{v^2\sigma_\beta^2}{\sigma^2(\sigma^2+v\sigma_\beta^2)}\sum_{i=1}^{b}(\bar{x}_{i.}-\mu-\bar{\tau})^2 \tag{4.4}$$

The hypothesis of interest is $H_o : \tau_j=0,\ \forall j=1,...v$ against the alternative H_1: not all τ_j's are zeros. To derive an optimum invariant unbiased test for this problem, note that it remains invariant under the following group G of transformations:

(i) permutation of treatments within each block
(ii) addition of a constant 'a' to each observation
(iii) multiply each observation by a nonzero constant 'c'.

An element $g \in G$ can be denoted by $g \cong(\alpha,a,c)$ with its action on x_{ij} by

$$x_{ij}\rightarrow gx_{ij}=cx_{i\alpha_j}+a,\ c\neq0,\ a \in R \text{ and } \underset{\sim}{\alpha}=(\alpha_1,...\alpha_v) \textit{ denotes a} \tag{4.5}$$
$$\textit{permutation of } (1,.......v).$$

We take $d\nu(\underset{\sim}{\alpha},c,a)=d\underset{\sim}{\alpha}\cdot dc\cdot da/c^2$ as a left invariant measure on G, where $d\underset{\sim}{\alpha}$ denotes the discrete uniform probability measure with mass $\frac{1}{v!}$ at each of the $v!$ permutations of $(1,...v)$. Denoting by $T(\underset{\sim}{x})$ a maximal invariant statistic induced by the group G and applying Wijsman's representation theorem (1967), the ratio $R_{\tau,\tau_0}(t(\underset{\sim}{x}))$ of the nonnull to null p.d.f.'s of $T(\underset{\sim}{x})$ is obtained as:

$$R_{\tau,\tau_o}(t(\underset{\sim}{x}))=dP_\tau^T/dP_{\tau_o}^T(t(\underset{\sim}{x})) \tag{4.6}$$
$$=\frac{\int_G f(g\cdot\underset{\sim}{x}|\underset{\sim}{\tau})\,|c|^n\,d\nu(\underset{\sim}{\alpha},c,a)}{\int_G f(g\cdot\underset{\sim}{x}|0)\,|c|^n\,d\nu(\underset{\sim}{\alpha},c,a)}$$

The following lemma provides a simple expression of the ratio $R_{\tau,\tau_o}(t(\underset{\sim}{x}))$.

<u>Lemma 4.1:</u> The ratio $R_{\tau,\tau_o}(t(\underset{\sim}{x}))$ in (4.6) is evaluated as

$$R_{\tau,\tau_o}(t(\underset{\sim}{x}))=\frac{\Sigma_\alpha\int_{-\infty}^{\infty}\tilde{\phi}(c^{*2}-\frac{2c^*}{\sqrt{Q}}\sum_{i=1}^{b}\sum_{j=1}^{v}x_{ij}\tau_{\alpha_j}+\frac{b}{\sigma^2}\sum_{j=1}^{v}\tau_j^2)\,|c^*|^{n-2}dc^*}{v!\int_{-\infty}^{\infty}\tilde{\phi}(c^{*2})\,|c^*|^{n-2}dc^*} \tag{4.7}$$

where Q is defined as

$$Q = \frac{1}{\sigma^2}\sum_{i=1}^{b}\sum_{j=1}^{v}(x_{ij}-\bar{x}_{..})^2 + \frac{v^2\sigma_\beta^2}{\sigma^2(\sigma^2+v\sigma_\beta^2)}\sum_{i=1}^{b}(\bar{x}_{i.}-\bar{x}_{..})^2$$

Proof: The proof goes along the similar lines of the proof of Lemma 2.1.

Remark 4.1: Since $dP_\tau^T/dP_o^T(t(\underset{\sim}{x}))$ is dependent on $\tilde{\phi}$ we cannot have nonnull robustness even though null robust tests exist.

We now proceed to evaluate the expression in (4.7). It turns out that there is no UMPIU test for this problem. To derive an LBIU test we need to expand $\tilde{\phi}$ around $\underset{\sim}{\tau} = \underset{\sim}{0}$ using Taylor series expansion. We assume that $\tilde{\phi}$ admits continuous derivatives up to third order and

$$\int_{-\infty}^{\infty}\tilde{\phi}'''(c^{*2}+\delta)\,|c^*|^{n-2}dc^* < \infty \text{ for some } \delta > 0\,. \tag{4.8}$$

This leads to after simplification,

$$R_{\tau,\tau_o}(t(\underset{\sim}{x})) = 1 + k_1\sum_{j=1}^{v}\tau_j^2 + k_2\sum_{j=1}^{v}\tau_j^2\,\frac{\sum_{j=1}^{v}(\bar{x}_{.j}-\bar{x}_{..})^2}{Q} + k_3\Big(\sum_{j=1}^{v}\tau_j^2\Big)^2 + r(\underset{\sim}{x},\underset{\sim}{\tau}) \tag{4.9}$$

where k_1, k_2, k_3 are constants and the remainder term $r(\underset{\sim}{x},\underset{\sim}{\tau}) \equiv \dfrac{\Sigma_\alpha\int_{-\infty}^{\infty}\delta(\underset{\sim}{x},\underset{\sim}{\tau})\,|c^*|^{n-2}dc^*}{v!\int_{-\infty}^{\infty}\tilde{\phi}(c^{*2})\,|c^*|^{n-2}dc^*}$ is given by

$$r(\underset{\sim}{x},\underset{\sim}{\tau}) = \frac{\frac{1}{3!}\Sigma_\alpha\int_{-\infty}^{\infty}\Big(-\frac{2c^*}{\sqrt{Q}}\sum_{i=1}^{b}\sum_{j=1}^{v}x_{ij}\tau_{\alpha_j}+\frac{b}{\sigma^2}\sum_{j=1}^{v}\tau_j^2\Big)^3\tilde{\phi}'''(c^{*2}+\eta)\,|c^*|^{n-2}dc^*}{v!\int_{-\infty}^{\infty}\tilde{\phi}(c^{*2})\,|c^*|^{n-2}dc^*} \tag{4.10}$$

where η lies between 0 and $-\frac{2c^*}{\sqrt{Q}}\sum_{i=1}^{b}\sum_{j=1}^{v}x_{ij}\tau_{\alpha_j} + \frac{b}{\sigma^2}\sum_{j=1}^{v}\tau_j^2$. Using the fact that $\left|\dfrac{\sum_{i=1}^{b}\sum_{j=1}^{v}x_{ij}\tau_{\alpha_j}}{\sqrt{Q}}\right| \le \sigma\sqrt{b}\sqrt{\sum_{j=1}^{v}\tau_j^2}$ uniformly in $\underset{\sim}{x}$, it follows that

$$\sup_{\underset{\sim}{x}}\ r(\underset{\sim}{x},\underset{\sim}{\tau}) = o\Big(\sum_{j=1}\tau_j^2\Big). \tag{4.11}$$

Therefore, the local power of a level α invariant unbiased test $\psi(t(\underset{\sim}{x}))$ can be written as

$$\int \psi(t(\underset{\sim}{x}))\,dP_\tau^T(t(\underset{\sim}{x})) = \alpha + k_1\alpha \sum_{j=1}^{v} \tau_j^2 \tag{4.12}$$

$$+ \frac{k_2\sigma^2}{v\sigma_\beta^2} \sum_{j=1}^{v} \tau_j^2 \int \psi(t(\underset{\sim}{x})) \frac{\sum_{j=1}^{v}(\bar{x}_{.j}-\bar{x}_{..})^2}{\frac{s_2+s_3}{\lambda_2-\lambda_3}+\frac{s_2}{\lambda_2}} dP_o^T(t(\underset{\sim}{x}))$$

$$+ o\left(\sum_{j=1}^{v} \tau_j^2\right)$$

where $\lambda_2 = \sigma^2 + v\sigma_\beta^2$, $\lambda_3 = \sigma^2$, $s_2 = v\sum_{i=1}^{b}(\bar{x}_{i.} - \bar{x}_{..})^2$, $s_3 = \sum_{i=1}^{b}\sum_{j=1}^{v}(x_{ij} - \bar{x}_{i.})^2$, and $dP_o^T(t(\underset{\sim}{x}))$ denotes the null distribution of the maximal invariant $t(\underset{\sim}{x})$. To describe the nature of $\psi(t(\underset{\sim}{x}))$, note that, under H_o, $(\bar{x}_{..}, s_2, s_3)$ are jointly sufficient for $(\mu, \lambda_2, \lambda_3)$. Moreover, due to null robustness of $t(\underset{\sim}{x})$ (vide Kariya (1981)), working with the normal distribution of $\underset{\sim}{x}$, it is easy to show that the joint distribution of $(\bar{x}_{..}, s_2, s_3)$ is complete. Thus every unbiased $\psi(t(\underset{\sim}{x}))$ has Neyman-structure i.e., has level α conditionally given $\bar{x}_{..}$, s_2 and s_3. Therefore the conditional and hence the unconditional local power of such a $\psi(t(\underset{\sim}{x}))$ is a maximum when $\psi(t(\underset{\sim}{x}))$ is chosen as

$$\psi(t(\underset{\sim}{x})) = \begin{cases} 1 & \text{if } \sum_{j=1}^{v}(\bar{x}_{.j}-\bar{x}_{..})^2 > k(s_2,s_3) \\ 0 & \textit{otherwise} \end{cases} \tag{4.13}$$

where $k(s_2,s_3)$ is such that $\alpha = E_{H_o}\{\psi(t(\underset{\sim}{x}))\,|s_2,s_3\}$ a.e.. (4.13) can be equivalently expressed as

$$\psi(t(\underset{\sim}{x})) = \begin{cases} 1 \text{ if } u = \dfrac{b\sum_{j=1}^{v}(\bar{x}_{.j}-\bar{x}_{..})^2}{(s_3 - b\sum_{j=1}^{v}(\bar{x}_{.j}-\bar{x}_{..})^2)/(b-1)} > k_1(s_2,s_3) \\ 0 \quad \textit{otherwise} \end{cases} \tag{4.14}$$

where $k_1(s_2,s_3)$ satisfies $\alpha = E_{H_o}\{\psi(t(\underset{\sim}{x}))\,|s_2,s_3\}$ a.e. However, it is easy to see that the null distribution of $u = b(b-1)\sum_{j=1}^{v}(\bar{x}_{.j}-\bar{x}_{..})^2/\{\sum_{i=1}^{b}\sum_{j=1}^{v}(x_{ij} - \bar{x}_{i.} - \bar{x}_{.j}+\bar{x}_{..})^2\}$ is independent of λ_2, λ_3, and hence by Basu's theorem (vide Lehmann (1959)) $k_1(s_2,s_3)$ is an absolute constant. In fact, under H_o, u has a central F-distribution with $(v-1), (b-1)(v-1))$ d.f. We have, therefore, proved the following result.

Theorem 4.1: Under the model (4.1) and assumption (4.8), the LBIU test of $H_{or} : \tau_j = 0, \forall j = 1,\ldots,v$ vs H_{1r}: not all τ_j's are zero rejects H_o for large values of u. The test is null and optimality robust.

Remark 4.2. It has been shown in Seifert (1979) that the above test is UMPIU for H_{or} under normality of X when a more general group is considered.

5. RANDOM MANOVA MODEL

Consider the model (1.2) and the problem of testing $H_0 : \sigma_1^2 = 0$ versus $H_1 : \sigma_1^2 > 0$ where $M_2 \in R^{n_2 p}$ and $\Sigma > 0$ are unknown. It is easy to see that the problem is left invariant under the group $G = Gl(p) \times R^{n_2 p}$ acting on X and $(M_2, \Sigma, \sigma_1^2)$ as

$$gX = (X_1 A', X_2 A' + F, X_3 A') \tag{5.1}$$

and

$$g(M_2, \Sigma, \sigma_1^2) = (M_2 A' + F, A\Sigma A', \sigma_1^2)$$

where $g = (A,F) \in G$, $A \in Gl(p)$, $F \in R^{n_2 \times p}$. Here $Gl(p)$ is the group of $p \times p$ nonsingular matrices and $R^{n_2 \times p}$ is the (additive) group of matrices of order $n_2 \times p$. As a left invariant measure ν on G, we take $d\nu(A,F) = dFdA / |AA'|^{p/2}$ where dA and dF are Lebesgue measures on $R^{n_2 p}$ and R^{p^2} respectively. Let $T(X)$ be a maximal invariant under G and denote its distribution under H_1 by $dP^T_{\sigma_1}$ and under H_0 by dP^T_0. Then applying Wijsman's representation theorem (1967), the ratio $dP^T_{\sigma_1}/dP^T_0(t(x)) = R_{\sigma_1}(t(x))$ is given by

$$R_{\sigma_1}(t(x)) = \frac{\int_G f(gx | \sigma_1^2, M_2, \Sigma) \, |AA'|^{(n_1+n_3)/2} d\nu(A,F)}{\int_G f(gx | 0, M_2, \Sigma) \, |AA'|^{(n_2+n_3)/2} d\nu(A,F)} . \tag{5.2}$$

The quantity $R_{\sigma_1}(t(x))$ is simplified in the following lemma.

Lemma 5.1. The ratio $R_{\sigma_1}(t(x))$ in (2.2) is evaluated as

$$R_\eta(t(x)) = (1-\eta)^{n_1/2} |I_p - \eta X_1' X_1 (X_1' X_1 + X_3' X_3)^{-1}|^{-(n_1+n_3)/2} \tag{5.3}$$

where $\eta = \sigma_1^2/(1 + \sigma_1^2)$.

Proof. The numerator $N_{\sigma_1}(t(x))$ of (5.2) is given by

$$N_{\sigma_1}(t(x)) = |\Sigma|^{-n/2}(1+\sigma_1^2)^{-n_1/2}. \tag{5.4}$$

$$\int_{Gl(p)}\int_{R^{n_2p}} q(\Sigma^{-1}A(X_1'X_1/(1+\sigma_1^2) + X_3'X_3)A'$$

$$+ \Sigma^{-1}(X_2A' + F - M_2)'(X_2A' + F - M_2))\,|AA'|^{(n_1+n_3-p)/2} dFdA$$

$$(using\,(1.4))$$

$$= |\Sigma|^{-n/2}(1+\sigma_1^2)^{-n_1/2}.$$

$$\int_{Gl(p)}\int_{R^{n_2p}} q(\Sigma^{-1/2}A(X_1'X_1/(1+\sigma_1^2) + X_3'X_3)A'\Sigma^{-1/2}$$

$$+ \Sigma^{-1/2}(X_2A' + F - M_2)'(X_2A' + F - M_2)\Sigma^{-1/2})\,|AA'|^{(n_1+n_3-p)/2} dFdA$$

$$= |\Sigma|^{-(n_1+n_3)/2}(1+\sigma_1^2)^{-n_1/2}.$$

$$\int_{Gl(p)} \tilde{q}(\Sigma^{-1/2}A(X_1'X_1/(1+\sigma_1^2) + X_3'X_3)A'\Sigma^{-1/2})\,|AA'|^{(n_1+n_3-p)/2} dA$$

$$=(1+\sigma_1^2)^{-n_1/2}\,|X_1'X_1/(1+\sigma_1^2) + X_3'X_3|^{-(n_1+n_3)/2}\int_{Gl(p)} \tilde{q}(AA')\,|AA'|^{(n_1+n_3-p)/2} dA$$

where $\tilde{q}(V) = \int_{R^{n_2p}} q(V + F'F)\,dF$.

Since the denominator of (5.2) corresponds to $N_{\sigma_1}(t(x))$ with $\sigma_1 = 0$, the result follows upon simplification.

Remark 5.1. Since the ratio $dP_{\sigma_1}^T/dP_0^T(t(x))$ is independent of q, it follows that any null robust test is also nonnull robust (vide Kariya and Sinha (1985)). In particular the optimum invariant test derived below is null and hence nonnull robust.

If $p = 1$ or $n_1 = 1$, the ratio $R_\eta(t(x))$ is evidently monotone increasing in $tr\, X_1(X_1'X_1 + X_3'X_3)^{-1}X_1'$, the Pillai's trace statistic, which is the familiar F-statistic for $p = 1$ and Hotelling's T^2-statistic for $n_1 = 1$. Its null robustness for arbitrary p and n_1, under the model (1.2), follows from Kariya (1981). This proves the following result.

Theorem 5.1. When $\min(n_1, p) = 1$, for testing $H_0 : \sigma_1^2 = 0$ versus $H_1 : \sigma_1^2 > 0$ under the model (1.2), the test which rejects H_0 for large values of $tr\, X_1(X_1'X_1 + X_3'X_3)^{-1}X_1'$ is UMPI, whatever be $q \in Q$. The test is null, nonnull and optimality robust.

If $\min(n_1, p) > 1$, no UMPI test exists. But a Taylor series expansion of $R_\eta(t(x))$ with respect to $\eta = 0$, coupled with the observation that

$$\sup_{X_1, X_3} \|X_1'X_1(X_1'X_1 + X_3'X_3)^{-1}\| < 1$$

where $\|\cdot\|$ denotes the Euclidean norm, yields

$$R_\eta(t(x)) = 1 + \eta\{K + tr\ X_1(X_1'X_1 + X_3'X_3)^{-1}X_1'\} + o(\eta) \tag{5.5}$$

where K is a constant. For an invariant test $\psi(t)$ of size α, its local power is then evaluated as

$$\int \psi(t(x))\,dF_\eta^T(t(x)) = \tag{5.6}$$
$$\alpha + \eta \int \{K + tr\ X_1(X_1'X_1 + X_3'X_3)^{-1}X_1'\}dP_0^T(t(x)) + o(\eta).$$

An application of the Neyman-Pearson lemma gives the following result.

Theorem 5.2. When $\min(n_1,p) > 1$, for testing $H_0 : \sigma_1^2 = 0$ versus $H_1 : \sigma_1^2 > 0$ under the model (1.2), the test which rejects H_0 for large values of Pillai's trace statistic $tr\ X_1(X_1'X_1 + X_3'X_3)^{-1}X_1'$ is LBI, whatever be $q \in Q$. The test is null, nonnull and optimality robust.

Remark 5.2. Under the above setup when $\min(n_1,p) > 1$, if $A \in G_T$ is used in (5.1) where G_T is the group of $p \times p$ nonsingular lower triangular matrices with positive diagonal elements, it is not difficult to show that the corresponding ratio $r_\eta(t(x))$ takes the form

$$r_\eta(t(x)) = 1 + \eta\{K_1 + K_2\delta(x)\} + o_x(\eta) \tag{5.7}$$

where K_1, K_2 (> 0) are constants, $o_x(\eta)$ is $o(\eta)$ uniformly in x and

$$\delta(x) = \sum_{i=1}^{p-1} tr\ X_{1[i]}(X_{1[i]}'X_{1[i]} + X_{3[i]}'X_{3[i]})^{-1}X_{1[i]}'$$
$$+ \left(\frac{n_1+n_3-p+1}{2}\right) tr\ X_1(X_1'X_1 + X_3'X_3)^{-1}X_1' \tag{5.8}$$

where $X_{j[i]}$ is the $n \times i$ submatrix of the $n \times p$ matrix X_j consisting of the first i columns of X_j, $j = 1,3$. It, therefore, follows that the test which rejects H_0 for large values of $\delta(x)$ is LBI under this group. However, as noted in the Introduction, this test suffers from a serious drawback due to its asymmetry in the use of the p columns of X_1 and X_3.

Going back to the other model (1.5), we note that under the null hypothesis $H_0 : \sigma_1^2 = 0$, $T_1 = X_1'X_1 + X_3'X_3$ and $T_2 = X_2'X_2$ are sufficient for the nuisance parameters Σ and σ_2^2. However, their joint distribution is not complete which can be seen as follows. Assume for simplicity $q(u) = 1/2\ exp(-\ tr\ u/2)$. Write $T_1 = ((t_{ij}^{(1)}))$, $T_2 = ((t_{ij}^{(2)}))$. Then $E(t_{11}^{(1)}t_{22}^{(2)} - t_{22}^{(1)}t_{11}^{(2)}) = 0$ but "$t_{11}^{(1)}t_{22}^{(2)} - t_{22}^{(1)}t_{11}^{(2)} = 0$, a.e." does not hold. This lack of completeness leads to the obvious difficulty of constructing an optimum invariant test.

ACKNOWLEDGEMENT

Our sincere thanks are due to the three referees for their constructive criticism and suggestions.

REFERENCES

Anderson ,T.W. (1984). An introduction to Multivariate Statistical Analysis,Wiley, New York.

Arnold ,S.F. (1981). The theory of linear models and multivariate Analysis,Wiley, New York.

Arvesen ,J.N. and W.J. Layard (1975). Asymptotically Robust Tests in Unbalanced Variance Component Models,Ann.Statist,3,1122-1134.

Chakraborty ,M.C. (1962). Mathematics of Design and Analysis of Experiments,Asia Publishing House, New York.

Cochran ,W.G. and Cox G. (1957). Experimental Designs,Second Ed., Wiley, New York.

Das ,M.N. and Giri ,N.C. (1979). Design and Analysis of Experiments,Wiley, New York.

Dawid ,A.P. (1977). Spherical matrix distributions and a multi- variate model,JRSS.B 39 ,254-261.

Federer ,W.T. (1955). Experimental Design,Theory and Application, Macmillan and Company.

Herbach ,L.H. (1959). Properties of model II-type analysis of variance tests,A: optimum nature of the F-test for model II in the balanced case. Ann.Math.Statist. 30,939-959.

Imhof ,J.P. (1962). Testing the hypothesis of no fixed main effects in Scheffe's mixed model. Ann.Math.Statist. 33,1085-1095.

Kariya ,T. (1981). Robustness of multivariate tests,Ann.Statist. 9,1267-1275.

Kariya ,T. and Sinha ,B.K. (1985). Nonnull and optimality robust- ness of some tests, Ann.Statist.,13,1182-1197.

Kempthorne ,O. (1952). The Design and Analysis of Experiments, Wiley, New York.

Khuri ,A.I. (1984). Interval estimation of fixed effects and of functions of variance components in balanced mixed models, Sankhyā B 46,10-28.

Lehmann ,E.L. (1959). Testing Statistical Hypothesis, Wiley, New York.

Montgomery ,D.C. (1984). Design and Analysis of Experiments , Wiley ,New York.

Mostafa ,M.G. (1967). Note on testing hypotheses in an unbalanced random effects model ,Biometrika 54,659-662.

Raghavarao ,D. (1971). Constructions and Combinatorial Problems in Design of Experiments , Wiley,New York.

Scheffé ,H. (1956). A 'mixed model' for analysis of variance, Ann.Math.Statist,27, 23-36.

----------- (1959). The Analysis of Variance ,Wiley,New York.

Searle ,S.R. (1971). Linear Models ,Wiley,New York.

Seely ,J. and El-Bassiouni ,Y. (1983). "Applying Wald's Variance Component Test" , Ann.Statist,11,197-201.

Seifert ,B. (1979). "Optimal Testing for Fixed Effects in General Balanced Mixed Classification Models",Math.Oper.Statist.,Ser. Statistics,10,237-255.

----------- (1985). "Estimation and Test of Variance Components Using the MINQUE-Method",Statistics,16,621-635.

Spjøtvoll ,E. (1967). Optimum invariant tests in unbalanced variance components models,Ann.Math.Statist.,38,422-428.

Wijsman ,R.A. (1967). Cross-sections of orbits and their application to densities of maximal invariants. Fifth Berkley Symp. Math. Statist. Prob. I,389-400.

Pillai ,K.C.S. (1955). Some new test critera in multivariate analysis ,Ann.Math.Statist.,26,117-121.

Roebruck ,P. (1982). Canonical Forms and Tests of Hypotheses, Statistica Neerlandica,36,63-74.

Roy ,S.N. and Cobb ,W. (1960). Mixed model variance analysis with normal error and possibly non-normal other random effects: Part I: The univariate case, Part II: The multivariate case. Ann. Math. Statist.,31,939-968.

Roy ,S.N. and Gnanadesikan ,R. (1959). Some contributions to ANOVA in one or more dimensions: I,II. Ann. Math. Statist. ,30, 304-330.

Schwartz ,R.E. (1967). Locally minimax tests, Ann.Math.Statist., 38,340-360.

Received 31 March 1987
Revised 28 October 1987

Department of Mathematical Sciences
Northern Illinois University
De Kalb, IL 60115
U.S.A.

Department of Mathematics
University of Maryland Baltimore County
Catonsville, MD 21228
U.S.A.

Proc. Second International Tampere Conference in Statistics
(Tampere, Finland, 1-4 June 1987)
Tarmo Pukkila and Simo Puntanen, *Editors*

pp. 343 - 360

INVITED PAPER

Generalized variance component models

T. P. SPEED

CSIRO Division of Mathematics and Statistics, Canberra, Australia

Key words and phrases: Analysis of variance, estimation of means, best linear prediction, factorial dispersion model, restricted maximum likelihood estimation.

ABSTRACT

The case for using models in the analysis of variance more general than the familiar normal random effects models is reviewed. Factorial dispersion models are introduced and their use in the following problems illustrated: the estimation of a population mean; the study of variability; the estimation of fixed parameters; and the prediction of random variables. The discussion applies only to balanced data, and some remarks are made concerning the methods applicable to and the need for more research on unbalanced data.

1. INTRODUCTION

This paper discusses some aspects of the analysis of variance, where we use the term in the narrow sense, assuming that there are at least two variances in the problem that are of direct interest, cf. Cox (1960, p.210). In recent years discussions of this topic have almost invariably been based upon linear models with independent, normally distributed random effects, but it is important to remember that this was not always so. From the earliest days of the subject, in the writings of Fisher, Yates, Neyman, Pitman, Welch and Daniels, right through to the late 1950s, in work of Kempthorne (and his students), Cornfield, Tukey, Nelder and others, alternative approaches to and models for the analysis of variance have been used. A concise survey of much of this research can be found in Plackett (1960). Scheffé (1959) covers a good deal of the same ground. Key papers by the authors mentioned are listed in the references.

The title 'generalized variance component models' of this paper refers to the dispersion models for multiply-indexed arrays of random variables discussed recently by the author and his coworkers, which endeavour to incorporate the essential features of the models proposed by the abovementioned writers within a unified framework. In what follows we attempt to show that many of the practical problems which make use of the analysis of variance can be satisfactorily addressed using these more general models, rather than using the common linear models with independent normally distributed effects, and we try to show that the advantages of these more general models can be obtained in practice with little extra difficulty.

We begin by reviewing some problem areas in which variance component models commonly arise, and then briefly outline some of the reasons why more general models have been found desirable or necessary. After an examination of just what a component of variance should be, we examine in turn four of these problem areas, exemplifying our approach in each case. A discussion of what we term 'the problem of unbalanced arrays' then follows, and the paper concludes with some final remarks.

2. USES OF ANOVA

This section repeats some of the discussion from Cox (1960) and Thompson (1980).

2.1 Estimation of a population mean

One of the earliest uses of the analysis of variance arises when we wish to estimate the mean of a population stratified or cross-classified in some way, see e.g. Yates and Zacopanay (1935), Youden and Mehlich (1937), Cochran (1939). Typically we have a preliminary sample which is analysed in order to determine an optimum allocation scheme for the main sample; at other times we may simply be interested in the relative efficiencies of different sampling schemes. In these problems the variance parameters are not of intrinsic interest, but are a tool for calculating the variance of the mean of an arbitrary sample.

2.2 Study of variability

In other problems the variability of the observations is of direct interest, important examples come from genetics, industrial statistics and agriculture, see e.g. Falconer (1977), Tippett (1931) and Daniels (1938), and Patterson and Silvey (1980). In these contexts, the variance parameters are of intrinsic interest and so as Cox (1960) points out, "we have to arrange, if possible, that the components are not only clearly defined, but also that they measure physically relevant things ... , [and that they] have the right practical interpretation."

2.3 Efficient estimation of fixed parameters

In incomplete block experiments, information on contrasts amongst a set of treatment means will be available in both the within-block and between-block strata, and in order to combine these two sources of information efficiently, the relative magnitude of the within and between block variability is required, cf. Yates (1940). The same issue arises quite generally, see Nelder (1965b, 1968) and Houtman and Speed (1982).

Another interesting application described in Thompson (1980) involves cereal variety trials carried out at centres across Great Britain, see Patterson and Silvey (1980) for fuller details. Not all varieties are grown in all years or at all sites. There is interest in efficiently estimating not only variety comparisons, as above, but also in estimating the year, centre, year by centre, variety by year and variety by centre components of variability.

2.4 Prediction of random variables

It is now common across the world to have breeding programs which involve the testing and subsequent selection of young dairy bulls (sires) based upon the milk yield of their daughters. A suitable basis for this sire evaluation is the predicted yield of future daughters of the bulls and this can be computed once the herd-year-season effects of the daughters' yields have been removed. As Thompson (1980) points out, "There is also interest in estimating the bull and residual variance components so that the effects of different selection strategies might be investigated and optimal schemes implemented." The use of prediction in the context of selection goes back to Smith (1936).

3. WHY GENERALIZED MODELS?

The reasons why many writers have chosen models which are different from — and usually more general than — the now-common normal linear random effects models are many and varied, and they are discussed in detail in the works cited. A common feature noted by Plackett (1960) is the feeling that

> "the reference set for statistical inference should be clearly specified, and that the model should more accurately reflect the experimental situation rather than supply conditions under which the classical tests remained valid."

Cornfield and Tukey (1956) elaborate this view slightly, writing (p.908):

> "The question of what assumptions to make seems, at first glance, to be a purely empirical question, one that should be referred to the subject-matter knowledge of the experimenter, who is the expert on such matters. Sometimes this is helpful and sometimes not. But closer study shows that the choice of assumptions depends on more than empirical questions about the behaviour of experimental material. It depends on the nature of the sampling and randomization involved in obtaining the data (as has been recognized by many statisticians, and recently emphasized by Kempthorne and by Wilk). Moreover, it often depends on the purpose of the analysis, as expressed by the situations or populations to which one wishes to make *statistical* inference."

Wilk and Kempthorne (1955, p.1145) felt that the following items have tended to get obscured in the identification of anova with linear models:

> "The normality assumptions are always false in an actual situation.
> The independence assumptions often bear no relationship to the physical situation.
> Many writers lean on randomization as a justification for the assumptions, but very few have examined in detail just what randomization accomplishes and how it does so.
> Because the models employed are often not explicitly related to the experimental situation, there has been some difficulty in deciding just what the analysis of variance measures".

One can hardly argue that blanket assumptions of normality are reasonable, although we should be more concerned with evidence that deviations from normality can significantly affect our conclusions. The assumption that interaction terms (when they are included at all!) are distributed independently of the corresponding main effects seems to run counter to intuition, which suggests that they depend on them. And the complete absence of any role or place for randomization in the usual linear models with random effects is a point which continues to cause concern to many even today.

A general and flexible class of dispersion models for multiply-indexed arrays was described in Speed (1985) following the work of Nelder (1965a), see also Speed and Bailey (1987) and Speed (1987). These *factorial dispersion models* seem to us to provide a general framework incorporating the generalized models of most other authors, and within which their concerns can be addressed. In this paper we will only discuss such models for the simplest data arrays, and refer the reader to the papers cited for general theory along the same lines. As stated in the Introduction above, our aim is to show that it is possible to study the topics mentioned in §2 (and others) in a unified way and at a level of generality far greater than that associated with the usual linear models with random effects.

Authors not yet mentioned who have had similar aims to those of the present writer, but whose detailed development differs from what follows, include H. Smith (1955), G. Zyskind (1962) and J.A. Nelder (1977). Unfortunately our allotted space prevents us giving a detailed comparison between the work of these (and other) writers and our approach, so that we can do little more than recommend that the reader consult these stimulating and valuable contributions to our subject.

4. WHAT IS A COMPONENT OF VARIANCE?

Let us suppose that we have an array $y = (y_{ij})$ of random variables where the index $j = 1, \ldots, n$ *is nested within* the index $i = 1, \ldots, m$; such arrays have been called *balanced one-way layouts* by Scheffé (1959). For example, i may label sires (bulls) and j daughters (cows), with y_{ij} being a measure of milk production. Is it *necessary* to suppose that

$$y_{ij} = \mu + \alpha_i + \beta_{ij} \tag{4.1}$$

where μ is constant, the (α_i) are normally and independently distributed with mean zero and variance σ_α^2, the (β_{ij}) are normally and independently distributed with mean zero and variance σ_β^2 independently of the (α_i), *before* we can speak of the *within* and *between* class *components of variation*? Of course normality is readily recognized as being inessential here, and we could replace 'independent' by 'uncorrelated' without loss for a second-order analysis, but don't we need *some* linear model such as (4.1) to give us 'variance component parameters' such as σ_α^2 and σ_β^2?

Generalizing (4.1), let us suppose that our array $y = (y_{ij})$ satisfies the second-order assumptions:

$$\mathbf{E}y_{ij} = \mu \quad \text{(constant)}$$
$$\mathbf{cov}\,(y_{ij}, y_{i'j'}) = \begin{cases} \sigma^2 & \text{if } i = i',\ j = j'\ ; \\ \gamma_{\{1\}} & \text{if } i = i',\ j \neq j'\ ; \\ \gamma_{\emptyset} & \text{if } i \neq i'\ . \end{cases} \tag{4.2}$$

In other words, we are supposing that the dispersion matrix $\Gamma = \mathbf{D}y$ is factorial, see Speed and Bailey (1987) for further details, including an explanation of the use of subsets of the set of factors as subscripts. Let us further suppose that each sire could have a large (potentially infinite) number of daughters; this is possible through artificial insemination. We may or may not wish to suppose that the m bulls we are considering come from a large (potentially infinite) herd of bulls.

From these assumptions and no more we can proceed as follows. The equation

$$y_{ij} = y_{\cdot -} + (y_{i-} - y_{\cdot -}) + (y_{ij} - y_{i-}) \tag{4.3}$$

decomposes each element of our observed array of random variables into three uncorrelated parts; here $y_{i-} = m.s.\lim_n n^{-1}\sum_{j=1}^n y_{ij}$, and $y_{\cdot -} = m^{-1}\sum_{i=1}^m y_{i-}$. If we wish to let $m \to \infty$, then (4.3) and the stated properties will remain valid when the '$\cdot$' (average over $i = 1, m$) is replaced by '$-$' (m.s. limit or average over $i = 1, \infty$). Furthermore, the terms $y_{ij} - y_{i-}$ are all uncorrelated, and the terms $y_{i-} - y_{\cdot -}$ are also uncorrelated if $m = \infty$. The first term in (4.3) has expectation μ and the other two have expectation zero. Simple formulae exist for the variances of these three terms; in particular $\mathrm{var}\,(y_{i-} - y_{\cdot -}) = \frac{m-1}{m}(\gamma_{\{1\}} - \gamma_{\emptyset})$ and $\mathrm{var}\,(y_{ij} - y_{i-}) = \sigma^2 - \gamma_{\{1\}}$. We note in passing that it is easy to pass from the parameters $(\sigma^2, \gamma_{\{1\}}, \gamma_{\emptyset})$ which we use here to those defined by other writers in the field; see Speed and Bailey (1987) for details.

A final point about (4.3) is worth making. With only the assumptions already made, it can be proved that the usual within-group mean square

$$W = \frac{1}{m(n-1)}\sum_{i=1}^{m}\sum_{j=1}^{n}(y_{ij} - y_{i\cdot})^2 \tag{4.4}$$

is an unbiased estimate of $\sigma^2 - \gamma_{\{1\}}$; with the further assumption of what we term generalized exchangeability we can prove that it converges almost surely. Definitions and details can be found in Speed (1986b). In general its limit will be a random variable whose expectation is $\sigma^2 - \gamma_{\{1\}}$; additional conditions can be imposed — one such is assumption (4.1) — to ensure that this limit is a.s. constant. Similarly, the usual estimate of the between-group component of variance

$$\frac{1}{n}\left[\frac{n}{m-1}\sum_{i=1}^{m}(y_{i\cdot} - y_{\cdot\cdot})^2 - \frac{1}{m(n-1)}\sum_{i=1}^{m}\sum_{j=1}^{n}(y_{ij} - y_{i\cdot})^2\right] \tag{4.5}$$

is an unbiased estimate of $\gamma_{\{1\}} - \gamma_{\emptyset}$. If m can become infinite and the abovementioned condition of generalized exchangeability holds, then (4.5) also converges a.s. as $m, n \to \infty$, also to a random variable, with expectation $\gamma_{\{1\}} - \gamma_{\emptyset}$. Again conditions can be described which imply that this limit is a constant.

What can we conclude from the foregoing? Simply this: that it is *not necessary* to assume a model of the form (4.1) in order to have a well-defined notion of within- and between-class components of variation. The more general (4.2) — which includes randomization models, random sampling models, and symmetry or generalized exchangeability models such as are common in Bayesian statistics, as well as (4.1) — determines parameters which can play the role of σ_α^2 and σ_β^2 in (4.1). Exactly which parameters are used will depend upon the context, for example, on the extent to which the observed array is regarded as extendable (in i or j), on the meaning attached to the parameters, and on the nature of the statistical inferences envisaged. By 'extendable' we mean that our observed array may reasonably be viewed as a subarray of a larger, finite or possibly infinite, array of random variables; the finiteness or infiniteness of this larger array may depend on one or more of the indices involved.

The parameters $f_{\{1,2\}} = \sigma^2 - \gamma_{\{1\}}$, $f_{\{1\}} = \gamma_{\{1\}} - \gamma_{\emptyset}$ and $f_{\emptyset} = \gamma_{\emptyset}$ are a convenient set under certain circumstances, although these are not necessarily non-negative; in our example of sires and dams, when n is potentially extendable to ∞, the non-negative parameters $\phi_W = \sigma^2 - \gamma_{\{1\}}$ and $\phi_B = \frac{m-1}{m}(\gamma_{\{1\}} - \gamma_{\emptyset})$ are obviously of interest, and they are completed by $\phi_{\emptyset} = m^{-1}[\gamma_{\{1\}} + (m-1)\gamma_{\emptyset}]$. It is easy to check that when (4.1) holds and both m and n are potentially infinite, $f_{\{1,2\}} = \phi_W = \sigma_\beta^2$, $f_{\{1\}} = \phi_B = \sigma_\alpha^2$ and $f_{\emptyset} = \phi_{\emptyset} = 0$. Other examples will be given in later sections.

5. ESTIMATION OF A POPULATION MEAN

Probably the earliest discussion of the use of anova in the estimation of the mean of a stratified or cross-classified population is that found in Tippett (1931, Chap.X). We will briefly review these well-known ideas in a way which shows clearly that no models such as (4.1) are required in order to reach the familiar conclusions.

We suppose given a finite array $y = (y_{ij})$ whose second-order properties are as in (4.2), and the object of our study is to gain insight into the estimation of μ. It is not hard to show that, using the f-parameters of §4:

$$\mathbf{E}\{(y_{\cdot\cdot} - \mu)^2\} = \frac{1}{mn} f_{\{1,2\}} + \frac{1}{m} f_{\{1\}} + f_{\emptyset} \,. \tag{5.1}$$

The fact that these parameters are not necessarily non-negative will be seen to have no impact on our conclusions, although we would typically suppose that $f_{\emptyset} (= \gamma_{\emptyset}) = 0$, permitting the consistent estimation of μ to be envisaged. Under this assumption, an unbiased estimator of $\mathrm{var}\,(y_{\cdot\cdot})$ is $\frac{1}{mn} B$ where $B = \frac{n}{m-1} \sum_i (y_{i\cdot} - y_{\cdot\cdot})^2$ is the between-class mean-square in the familiar anova table, and as already observed in §4, unbiased estimates of $f_{\{1,2\}}$ and $f_{\{1\}}$ are given by (4.4) and (4.5) respectively. With estimates of $f_{\{1,2\}}$ and $f_{\{1\}}$ based upon an initial sample, optimal allocation of effort for any subsequent sampling can be readily determined. (The presence of $f_{\emptyset}$ in (5.1) causes no problems; it is not estimable and it is not affected by any sampling strategy.)

For a somewhat different example, let us turn to the paper Yates and Zacopanay (1935). Here y_{ij} denotes a measure of yield whilst $i = 1, \ldots, m$ indexes agricultural field plots and $j = 1, \ldots, h$ indexes sampling units within plots. Their object was to compare the efficiency of estimating the plot means $y_{i-} = \frac{1}{h} \sum_{j=1}^{h} y_{ij}$ using means $y_{i\cdot} = \frac{1}{n} \sum_1^n y_{ij}$ based upon samples of size $n < h$. In order to make this comparison Yates and Zacopanay introduced a linear model generalizing (4.1) by including a term allowing for possible competition — negative correlation — between samples within the same plot. Their model is included within (4.2) and we will now show that their conclusions continue to hold at this level of generality. A straightforward calculation shows that if $\gamma_{\emptyset} = 0$,

$$\frac{\mathrm{var}\,(y_{i-})}{\mathrm{var}\,(y_{i\cdot})} = \frac{f_{\{1\}} + \frac{1}{h} f_{\{1,2\}}}{f_{\{1\}} + \frac{1}{n} f_{\{1,2\}}} = 1 - L$$

where

$$L = \left(1 - \frac{n}{h}\right) \frac{\mathbf{E}\{W\}}{\mathbf{E}\{B\}} \,. \tag{5.2}$$

Not surprisingly, the authors went on to estimate L by replacing the expectations in (5.2) with their observed counterparts.

Before leaving this example, let us note that neither our more general model (4.2) nor Yates and Zacopanay's special model permits the separation of the effect of competition from other sources of variation.

However, if $B - W$ is clearly and significantly negative, the existence of competition is demonstrated.

6. STUDY OF VARIABILITY

We turn now to an interesting account given by Daniels (1938) of an investigation of variability arising in the process of *carding* wool prior to spinning. The reader is referred to the reference cited for fuller details of the process and a diagrammatic representation of the lay-out of the card. Data are presented from two tests on a particular type of card in which threads of what is known as slubbing were wound onto four bobbins, 25 ends being wound side by side on each bobbin. The data $y = (y_{ijk})$, where $i = 1,2$ labels tests, $j = 1,\ldots,4$ labels bobbins and $k = 1,\ldots,25$ labels the position or trend along the bobbin, represent weights (in gms.) of 95 yard lengths of slubbing run off under carefully controlled conditions from the corresponding positions on all the 100 ends. Interest focussed on variation between individual ends of slubbing coming from the card.

The analysis offered in Daniels (1938) was amplified in the later paper Daniels (1939), which distinguishes between factors which introduce variation in a *random* way and those which give rise to *systematic* variation. The following second-order model for $y = (y_{ijk})$ generalizes that of Daniels slightly, whilst giving essentially the same analysis of variance table and estimated components of variation:

$$\mathbf{E}y_{ijk} = \mu_{jk}$$

$$\mathbf{cov}\,(y_{ijk}, y_{i'j'k'}) = \begin{cases} \sigma^2 & \text{if } i = i',\, j = j',\, k = k' \;; \\ \gamma_{\{1,2\}} & \text{if } i = i',\, j = j',\, k \neq k' \;; \\ \gamma_{\{1,3\}} & \text{if } i = i',\, j \neq j',\, k = k' \;; \\ \gamma_{\{1\}} & \text{if } i = i',\, j \neq j',\, k \neq k' \;; \\ \gamma_{\emptyset} & \text{if } i \neq i' \,. \end{cases} \qquad (6.1)$$

It follows from the theory outlined in Speed and Bailey (1987) (following Nelder (1965a)) that there are 5 common eigenspaces or strata for the class of dispersion matrices $\Gamma = \mathbf{D}y$ described in (6.1) and these are labelled and named as follows: $\emptyset$ (grand mean), $\{1\}$, $\{1,2\}$, $\{1,3\}$, and $\{1,2,3\}$, where 1, 2 and 3 abbreviate Tests, Bobbins and Trend respectively. The systematic effects of Bobbins, Trend and Trend$\times$Bobbins

ANOVA TABLE FOR DANIELS' DATA

Stratum	Term	D.f.	Sum of Squares	Mean Square
$\emptyset$		1	$abc\,y_{\cdot\cdot\cdot}^2$	
$\{1\}$		$a-1$	$bc\sum_i(y_{i\cdot\cdot}-y_{\cdot\cdot\cdot})^2$	MS_1
$\{1,2\}$	Bobbins	$b-1$	$ac\sum_j(y_{\cdot j\cdot}-y_{\cdot\cdot\cdot})^2$	MS_2
	Residual	$(a-1)(b-1)$	By difference	$MS_{1\times 2}$
	Subtotal	$a(b-1)$	$c\sum_{i,j}(y_{ij\cdot}-y_{i\cdot\cdot})^2$	
$\{1,3\}$	Trend	$c-1$	$ab\sum_k(y_{\cdot\cdot k}-y_{\cdot\cdot\cdot})^2$	MS_3
	Residual	$(a-1)(c-1)$	By difference	$MS_{1\times 3}$
	Subtotal	$a(c-1)$	$b\sum_{i,k}(y_{i\cdot k}-y_{i\cdot\cdot})^2$	
$\{1,2,3\}$	Bobbins ×Trend	$(b-1)(c-1)$	$a\sum_{j,k}(y_{\cdot jk}-y_{\cdot j\cdot}-y_{\cdot\cdot k}+y_{\cdot\cdot\cdot})^2$	$MS_{2\times 3}$
	Residual	$(a-1)(b-1)(c-1)$	By difference	$MS_{1\times 2\times}$
	Subtotal	$a(b-1)(c-1)$	$\sum_{i,j,k}(y_{ijk}-y_{ij\cdot}-y_{i\cdot k}+y_{i\cdot\cdot})^2$	
	Total	abc	$\sum_{i,j,k}y_{ijk}^2$	

TABLE 1

Mean Square	Expectation
MS_1	$f_{\{1,2,3\}}+cf_{\{1,2\}}+bf_{\{1,3\}}+bcf_{\{1\}}$
MS_2	$f_{\{1,2,3\}}+cf_{\{1,2\}}+\frac{abc}{b-1}\sigma_b^2$
$MS_{1\times 2}$	$f_{\{1,2,3\}}+cf_{\{1,2\}}$
MS_3	$f_{\{1,2,3\}}+bf_{\{1,3\}}+\frac{abc}{c-1}\sigma_c^2$
$MS_{1\times 3}$	$f_{\{1,2,3\}}+bf_{\{1,3\}}$
$MS_{2\times 3}$	$f_{\{1,2,3\}}+\frac{abc}{(b-1)(c-1)}\sigma_{bc}^2$
$MS_{1\times 2\times 3}$	$f_{\{1,2,3\}}$

TABLE 2

are separated out in the strata labelled $\{1,2\}$, $\{1,3\}$ and $\{1,2,3\}$ respectively, see the anova of Table 1 where we have supposed that $i = 1,\ldots,a$; $j = 1,\ldots,b$ and $k = 1,\ldots,c$ to make the structure a little clearer.

In Table 2 we have used the f-parameters, which are derived from the γ-parameters according to the rules: $f_{\{1\}} = \gamma_{\{1\}} - \gamma_{\emptyset}$, $f_{\{1,2\}} = \gamma_{\{1,2\}} - \gamma_{\{1\}}$, $f_{\{1,3\}} = \gamma_{\{1,3\}} - \gamma_{\{1\}}$ and $f_{\{1,2,3\}} = \sigma^2 - \gamma_{\{1,2\}} - \gamma_{\{1,3\}} + \gamma_{\{1\}}$, whilst σ_b^2, σ_c^2 and σ_{bc}^2 are (following Daniels) defined by $\sigma_b^2 = \frac{1}{b}\sum_j(\mu_{j\cdot} - \mu_{\cdot\cdot})^2$, $\sigma_c^2 = \frac{1}{c}\sum_k(\mu_{\cdot k} - \mu_{\cdot\cdot})^2$ and $\sigma_{bc}^2 = \frac{1}{bc}\sum_j\sum_k(\mu_{jk} - \mu_{j\cdot} - \mu_{\cdot k} + \mu_{\cdot\cdot})^2$. As Daniels points out, it is *only* with this definition that $\frac{1}{bc}\sum_j\sum_k(\mu_{jk} - \mu_{\cdot\cdot})^2 = \sigma_b^2 + \sigma_c^2 + \sigma_{bc}^2$.

The various parameters are readily estimated by equating observed with expected mean squares, a procedure which generally works with balanced and orthogonal arrays, and which corresponds with the REML method under the assumption of normality, see §9 below. Of course exhibits such as Tables 1 and 2 have been given in many places, with very similar if not identical interpretations, the first being Daniels (1939)! Our purpose in presenting them was to demonstrate that it is not necessary to assume a standard random effects model in order to define and estimate components of variation.

7. EFFICIENT ESTIMATION OF FIXED PARAMETERS

To illustrate the fact that models such as (4.1) are not necessary for the analysis of designed experiments, let us consider an incomplete block experiment with yields y_{ij}, the index $i = 1, m$ labelling blocks and $j = 1, n$ labelling plots within blocks. The restriction to equal-sized blocks is important and will be remarked upon later. A general model for $y = (y_{ij})$ which permits the efficient estimation of treatment means is the following: $\mathbf{E}y_{ij} = \tau_{[i,j]}$ where $[i,j]$ denotes the label of the treatment applied to plot j in block i, and $\mathbf{D}y = \Gamma$ as in (4.2). In such situations we generally make no assumptions concerning the extendability of y. For simplicity we will further assume that the incomplete block design is balanced, with t treatments replicated $r = mn/t$ times, and each pair occurring λ times in the same block.

In order to explain the *within* and *between* block analysis, and go on to the full *combined* or weighted least squares analysis of the array y, it is important to note that $\Gamma = \mathbf{D}y$ may be reparametrized and rewritten as

$$\Gamma = \xi_{\emptyset}G + \xi_{\{1\}}(B - G) + \xi_{\{1,2\}}(I - B) \tag{7.1}$$

where $G = \frac{1}{m}J_m \otimes \frac{1}{n}J_n$ is the grand mean averaging matrix, $B = I_m \otimes \frac{1}{n}J_n$ is the block averaging matrix and $I = I_m \otimes I_n$ is the identity matrix appropriate to our data array $y = (y_{ij})$. Here I_m and J_m are the $m \times m$ identity and matrix of 1s, respectively. The ξ-parameters are expressed in terms of the f-parameters of §4 by $\xi_{\{1,2\}} = f_{\{1,2\}}$, $\xi_{\{1\}} = f_{\{1,2\}} + nf_{\{1\}}$ and $\xi_\emptyset = f_{\{1,2\}} + nf_{\{1\}} + mnf_\emptyset$, and these turn out to be eigenvalues of Γ of multiplicity $m(n-1)$, $m-1$ and 1, respectively, with the projectors onto the corresponding eigenspaces being $I - B$ (intra-block), $B - G$ (inter-block) and G (grand mean). We refer to Nelder (1965a,b), Houtman and Speed (1982) and Speed and Bailey (1987) for further details and generalizations of these ideas.

Introducing the projector T which acts on arrays $y = (y_{ij})$ by averaging within levels of the treatment factor: $(Ty)_{ij} = \frac{1}{r}\sum_{[i',j']=k} y_{i'j'}$ if $[i,j] = k$, we can write explicit expressions for the within- and between-block estimates $\hat{\tau}_W$ and $\hat{\tau}_B$ of τ as follows:

$$\hat{\tau}_W = e^{-1}T(I-B)y\,, \qquad \hat{\tau}_B = (1-e)^{-1}T(B-G)y \tag{7.2}$$

where $e = \lambda t/rm$ is the so-called *efficiency factor* of the balanced incomplete block design. The weighted least squares estimate $\hat{\tau}$ of τ — assuming that Γ is *known* — then turns out to be

$$\hat{\tau} = w\hat{\tau}_W + (1-w)\hat{\tau}_B \tag{7.3}$$

where the *weight* $w = e\xi^{-1}_{\{1,2\}}/[e\xi^{-1}_{\{1,2\}} + (1-e)\xi^{-1}_{\{1\}}]$. In practice $\xi_{\{1\}}$ and $\xi_{\{1,2\}}$ will not be known and a natural way to estimate them is by equating the sums of squares of the within and between stratum residuals to their expectations, i.e. by solving

$$\begin{aligned} \left|(B-G)(y-\hat{\tau})\right|^2 &= \xi_{\{1\}}d'_{\{1\}} \\ \left|(I-B)(y-\hat{\tau})\right|^2 &= \xi_{\{1,2\}}d'_{\{1,2\}} \end{aligned} \tag{7.4}$$

for $\xi_{\{1\}}$ and $\xi_{\{1,2\}}$ where $d'_{\{1\}}$ and $d'_{\{1,2\}}$ are certain *effective* degrees of freedom, see Nelder (1968) for details. It should not be forgotten that the fitted values $\hat{\tau}$ on the left-hand-side are also functions of $\xi_{\{1\}}$ and $\xi_{\{1,2\}}$ through (7.3) and so an iterative solution of (7.4) is required. Under normality the method just described is equivalent to the REML method of §9 below.

The foregoing discussion indicates how the weighted least squares estimation of treatment means in a designed experiment can be carried out in an entirely straightforward way with a dispersion model more general than that implicit in a linear model with independent random effects. Unfortunately our ability to test hypotheses concerning these treatment means at the same level of generality is far less clear. The famous early work of Welch (1937) and Pitman (1938) gives insight into the testing of the null hypothesis of *no* treatment differences in randomized complete block and Latin square experiments under a randomization model for the observed data, and in the former case this has been extended to give an Edgeworth expansion for the distribution of the F-ratio for treatments, see Davis and Speed (1987). But in general little is known about the performance of the standard F-tests when models like (4.1) are not assumed; indeed if information on treatment contrasts is available in more than one stratum, but the ratio of the stratum variances is unknown, this is an extremely difficult problem even assuming (4.1) and normality!

8. PREDICTION OF RANDOM VARIABLES

Our final illustration of the way in which linear models such as (4.1) can be replaced by a more general model within which the same questions can be addressed concerns the prediction of unobserved or unobservable random variables. Let us suppose that we have an array of random variables $(y_{IJ} : I = 1, \ldots, M;\ J = 1, \ldots, N)$ — the *population* — whose second-order characteristics are as in (4.2) above. Here we adopt the idea of Cornfield and Tukey (1956) of labelling the population with I, J and the sample i, j, and we hope no confusion will occur with the matrices I and J of the previous section. The numbers M and N could be finite or arbitrarily large (infinite), in which case the population averages defined below will be regarded as mean-square limits. Further, let us suppose that we observe a finite sub-array $y = (y_{ij} : i = 1, \ldots, m; j = 1, \ldots, n)$ — the *sample* — of the larger array, and that on the basis of this sample we wish to *predict* certain population characteristics such as $y_{I-} = \frac{1}{N}\sum_1^N y_{IJ}$, $y_{--} = \frac{1}{MN}\sum_1^M \sum_1^N y_{IJ}$, or $y_{I-} - y_{--}$. We refer to Speed and Bailey (1987) for a discussion of the way in which this framework includes the appropriate notions of random *sampling* from structured populations, incorporating notions such as 'tied interactions' discussed by Cornfield and Tukey (1956).

Our results, which will be stated here without proof, take a slightly different form depending on whether we assume that $\mu = 0$ is known, or that μ is an unknown parameter estimated by $y_{\cdot\cdot} = \frac{1}{mn}\sum_1^m \sum_1^n y_{ij}$. One can get the results for the latter case by putting $f_\emptyset = \infty$ in formulae obtained under the former assumption. The best (= minimum variance) linear predictor (BLP) of $y_{h-} - y_{--}$ for $h \in \{1, \ldots, m\}$ based upon the sample $y = (y_{ij})$, under the assumption $\mu = 0$, is

$$\frac{Nf_{\{1\}} + f_{\{1,2\}}}{nf_{\{1\}} + f_{\{1,2\}}} \cdot \frac{n}{N}(y_{h\cdot} - y_{\cdot\cdot}) \; + \; \frac{Nf_{\{1\}} + f_{\{1,2\}}}{mnf_\emptyset + nf_{\{1\}} + f_{\{1,2\}}} \cdot \frac{n}{N}\left(1 - \frac{m}{M}\right) y_{\cdot\cdot} \tag{8.1}$$

which converges, as $M, N \to \infty$, to

$$\frac{nf_{\{1\}}}{nf_{\{1\}} + f_{\{1,2\}}}(y_{h\cdot} - y_{\cdot\cdot}) \; + \; \frac{nf_{\{1\}}}{mnf_\emptyset + nf_{\{1\}} + f_{\{1,2\}}} y_{\cdot\cdot} \quad .$$

On the other hand, if μ is unknown and estimated from the sample by $y_{\cdot\cdot}$, or if $m = M$, and $\gamma_\emptyset = 0$ (8.1) becomes

$$\frac{Nf_{\{1\}} + f_{\{1,2\}}}{nf_{\{1\}} + f_{\{1,2\}}} \cdot \frac{n}{N}(y_{h\cdot} - y_{\cdot\cdot})$$

which converges, as $N \to \infty$, to

$$\frac{nf_{\{1\}}}{nf_{\{1\}} + f_{\{1,2\}}}(y_{h\cdot} - y_{\cdot\cdot}) \quad . \tag{8.2}$$

In this case these expressions are more properly termed best linear *unbiased* predictors (BLUPs) of the population characteristic $y_{h-} - y_{--}$, and (8.2) is a familiar expression from the literature on animal breeding, see e.g. Thompson (1980) although use of this formula goes back more than fifty years. Our derivation permits the underlying model to be based on a randomization distribution, or random sampling, or on a Bayesian generalized exchangeability assumption. An alternative form of (8.2) is obtained by using the intra-class correlation coefficient $\rho = f_{\{1\}}/(f_{\{1\}} + f_{\{1,2\}})$, and then the BLUP becomes

$$\frac{n\rho}{1 + (n-1)\rho}(y_{h\cdot} - y_{\cdot\cdot}) \quad .$$

Similarly, if we suppose that $\mu = 0$, then the BLP of y_{--} is $y_{..}$, and that of $y_{hk} - y_{h-}$, where $h \in \{1, \ldots, m\}$ and $k \in \{1, \ldots, n\}$, is

$$y_{hk} - y_{h\cdot} + \frac{f_{\{1,2\}}}{f_{\{1,2\}} + nf_{\{1\}}}(1 - \frac{n}{N})(y_{h\cdot} - y_{..}) + \frac{f_{\{1,2\}}}{f_{\{1,2\}} + nf_{\{1\}} + mnf_{\emptyset}} y_{..} \, .$$

This expression simplifies appropriately if $m = M$ or $n = N$, or if $M, N \to \infty$, and gives the corresponding BLUP upon putting $f_{\emptyset} = \infty$.

9. THE PROBLEM OF UNBALANCED ARRAYS

If we include the assumption that our arrays are jointly normally distributed, and that they are infinitely extendable, then linear models such as (4.1) are equivalent to dispersion models such as (4.2), see Speed (1986a). And with joint normality we can estimate fixed parameters and components of variance and predict unobservable or unobserved random variables, regardless of whether the array y is "balanced" or not, see Patterson and Thompson (1971) and Thompson (1980) for details and further references concerning what has come to be known as the REML method: restricted or reduced maximum likelihood. This method has been implemented in programs which have become widely used in the United Kingdom, Australia and New Zealand.

In particular we can efficiently estimate treatment means in block designs with unequal block sizes, and we can predict sire effects when the number of daughters differs across sires and estimate genetic parameters with essentially arbitrary pedigrees, all in the presence of fixed effects. We no longer have any need to feel apologetic, as Scheffé did in 1959, see p.vii of his book, about the shortage of material "on the unbalanced cases of the random effects models and mixed models," at least if we are prepared to assume the joint normality of our arrays.

Where does this leave us if we are not willing to assume joint normality? Can we estimate means, variances, covariances and related parameters, and predict unobserved random variables, with *unbalanced* arrays under the more general models which have been the subject of this paper? Unfortunately, here we seem to be in the same situation as Scheffé was in 1959, for there are currently very few practically useful results along these lines. And although space prevents us making a critical review, none seem to be at the stage where they can be implemented for widespread or routine use, see Rao and Kleffe (1980) for a survey of some relevant theory.

If the insights, understanding and generalizations contributed by the scientists of the 1930s, 1940s and 1950s whose work has been discussed in this paper are to become (or remain) relevant to the analysis of variance in the 1980s and beyond, then their models and methods must be adapted to the unbalanced arrays which are so commonly met in practice. Algorithms need to be developed, asymptotic and other theory derived, and the advantages resulting from building upon this fine early work understood and publicized. Only then will we be able to go beyond the normal linear model in practice to a truly generalized variance component analysis.

REFERENCES

Cochran, W. G. (1939). The use of the analysis of variance in enumeration by sampling. *J. Amer. Statist. Assoc.*, 34, 492–510.

Cornfield, J. and Tukey, J. W. (1956). Average values of mean squares in factorials. *Ann. Math. Statist.*, 27, 907–949.

Cox, D. R. (1960). Contribution to the discussion of Plackett (1960), pp.209–210.

Daniels, H. E. (1938). Some problems of statistical interest in wool research. *Supp. J. R. Statist. Soc.*, 5, 89–128.

Daniels, H. E. (1939). The estimation of components of variance. *Supp. J. R. Statist. Soc.*, 6, 186–197.

Davis, A. W. and Speed, T. P. (1987). An Edgeworth expansion for the distribution of the F-ratio under a randomization model for the randomized block design. *Proceedings of the Fourth Purdue Symposium on Statistical Decision Theory and Related Topics* (to appear).

Falconer, D. R. (1977). *Introduction to Quantitative Genetics.* Longman, London.

Fisher, R. A. (1925). *Statistical Methods for Research Workers.* Oliver & Boyd, Edinburgh.

Houtman, A. M. and Speed, T. P. (1982). Balance in designed experiments with orthogonal block structure. *Ann. Statist.*, 11, 1069–1085.

Kempthorne, O. (1952). *The Design and Analysis of Experiments.* John Wiley & Sons, Inc., New York.

Nelder, J. A. (1965a). The analysis of randomized experiments with orthogonal block structure. I. Block structure and the null analysis of variance. *Proc. R. Soc. (London) Ser. A*, 273, 147–162.

Nelder, J. A. (1965b). The analysis of randomized experiments with orthogonal block structure. II. Treatment structure. *Proc. R. Soc. (London) Ser. A*, 273, 163–178.

Nelder, J. A. (1968). The combination of information in generally balanced designs. *J. R. Statist. Soc. B*, 30, 303–311.

Nelder, J. A. (1977). A reformulation of linear models (with discussion). *J. R. Statist. Soc. A*, 140, 1–47,

Neyman, J., with the cooperation of Iwaszkiewicz, K., and Kolodziejczyk, St. (1935). Statistical problems in agricultural experimentation (with discussion). *Supp. J. R. Statist. Soc.*, 2, 107–180.

Patterson, H. D. and Silvey, V. (1980). Statutory and recommended list trials of crop varieties in the United Kingdom. *J. R. Statist. Soc. A*, 143, 219–252.

Patterson, H. D. and Thompson, R. (1971). Recovery of inter-block information when block sizes are unequal. *Biometrika*, 58, 545–554.

Pitman, E. J. G. (1938). Significance tests which may be applied to samples from any population. III. The analysis of variance test. *Biometrika*, 29, 322–335.

Plackett, R. L. (1960). Models in the analysis of variance (with discussion). *J. R. Statist. Soc.*, 22, 195–217.

Rao, C. Radhakrishna and Kleffe, J. (1980). Estimation of variance components. *Handbook of Statistics*, 1, 1–40. P.R. Krishnaiah, ed. North Holland Publishing Company, Amsterdam.

Scheffé, H. (1956). Alternative models for the analysis of variance. *Ann. Math. Statist.*, 27, 251–271.

Scheffé, H. (1959). *The Analysis of Variance.* John Wiley & Sons, Inc., New York.

Smith, H. F. (1936). A discriminant function for plant selection. *Ann. Eugen. (London)*, 7, 240–260.

Smith, H. F. (1955). Variance components, finite populations, and experimental inference. *Institute of Statistics, North Carolina State College. Mimeo Series No. 135.*

Speed, T. P. (1985). Dispersion models for factorial experiments. *Bull. Internat. Statist. Inst.*, 51, 24.1, 16pp.

Speed, T. P. (1986a). Anova models with random effects: an approach via symmetry. In *Essays in Time Series and Allied Processes: Papers in honour of E.J. Hannan.* Eds. J. Gani & M.B. Priestley, pp.355–368. Applied Probability Trust, Sheffield.

Speed, T. P. (1986b) Cumulants and partition lattices IV: A.s. convergence of generalized k-statistics. *J. Austral. Math. Soc. (Series A)*, 41, 79–94.

Speed, T. P. (1987). What is an analysis of variance? *Ann. Statist.*, 15, 885-910.

Speed, T. P. and Bailey, R. A. (1987). Factorial dispersion models. *Internat. Statist. Rev.*, 55. (to appear).

Thompson, R. (1980). Maximum likelihood estimation of variance components. *Math. Operationsforsch. Statist., Ser. Statistics*, 11, 545–561.

Tippett, L. H. C. (1931). *The Methods of Statistics.* Williams & Norgate, Ltd., London.

Welch, B. L. (1937). On the z test in randomized blocks and Latin squares. *Biometrika*, 29, 21–52.

Wilk, M. B. and Kempthorne, O. (1955). Fixed, mixed, and random models. *J. Amer. Statist. Assoc.*, 50, 1144–1167.

Yates, F. (1940). The recovery of inter-block information in balanced incomplete block designs. *Ann. Eugen. (London)*, 10, 317–325.

Yates, F. and Zacopanay, I. (1935). The estimation of the efficiency of sampling, with special reference to sampling for yield in cereal experiments. *J. Agric. Sci.*, 25, 545–577.

Youden, W. J. and Mehlich, A. (1937). Selection of efficient methods for soil sampling. *Contrib. Boyce Thompson Inst.*, 9, 59–70.

Zyskind, G. (1962). On structure, relation, Σ and expectation of mean squares. *Sankhyā, Series A*, 24, 115–148.

Received 14 April 1987
Revised 19 June 1987

CSIRO
Division of Mathematics and Statistics
G.P.O. Box 1965
Banks Street, Yarralumla
Canberra, ACT 2601
Australia

Current address:
Department of Statistics
University of California
Berkeley, CA 94720
U.S.A.

Proc. Second International Tampere Conference in Statistics
(Tampere, Finland, 1-4 June 1987)
Tarmo Pukkila and Simo Puntanen, *Editors*

pp. 361 - 383

INVITED PAPER

Partially ordered idempotent matrices

Robert E. HARTWIG and George P. H. STYAN
North Carolina State University, Raleigh, NC, USA
and McGill University, Montreal, QC, Canada

Key words and phrases: Matrix partial orderings, Löwner order, Drazin's "star" order, rank subtractivity, generalized inverses, orthogonal projectors.

ABSTRACT

In this paper we characterize those idempotent matrices that "lie below" or that "lie above" a given idempotent matrix relative to a particular matrix partial ordering. We consider three matrix partial orderings: (1) $\mathbf{A} \leq_{\mathsf{L}} \mathbf{B}$ whenever $\mathbf{B} - \mathbf{A}$ is Hermitian nonnegative definite, (2) $\mathbf{A} \leq_{*} \mathbf{B}$ whenever $\mathbf{A}^*\mathbf{A} = \mathbf{A}^*\mathbf{B}$ and $\mathbf{A}\mathbf{A}^* = \mathbf{B}\mathbf{A}^*$, and (3) $\mathbf{A} \leq_{\mathsf{rs}} \mathbf{B}$ whenever $\text{rank}(\mathbf{B} - \mathbf{A}) = \text{rank}(\mathbf{B}) - \text{rank}(\mathbf{A})$. In particular we show that for two complex idempotent matrices $\mathbf{E}$ and $\mathbf{F}$, neither necessarily Hermitian: $\mathbf{E} \leq_{\mathsf{L}} \mathbf{F} \Rightarrow \mathbf{E} \leq_{*} \mathbf{F} \Rightarrow \mathbf{E} \leq_{\mathsf{rs}} \mathbf{F}$, while $\mathbf{E} \leq_{\mathsf{rs}} \mathbf{F}$ and $\mathbf{F} - \mathbf{E} = (\mathbf{F} - \mathbf{E})^*$ together imply $\mathbf{E} \leq_{\mathsf{L}} \mathbf{F}$. When both $\mathbf{G}$ and $\mathbf{H}$ are Hermitian and if $\mathbf{H}$ is also nonnegative definite, we find that $\mathbf{G} \leq_{*} \mathbf{H} \Rightarrow \mathbf{G} \leq_{\mathsf{rs}} \mathbf{H} \Rightarrow \mathbf{G} \leq_{\mathsf{L}} \mathbf{H}$. It then follows that our three matrix partial orderings coincide when both $\mathbf{G}$ and $\mathbf{H}$ are orthogonal projectors (Hermitian and idempotent).

1. INTRODUCTION

For complex matrices there exist three dominant matrix partial orderings, namely, the Löwner or nonnegative definite order, Drazin's "star" order, and rank subtractivity. The Löwner order may be defined by

$$\mathbf{A} \leq_{\mathsf{L}} \mathbf{B} \text{ whenever } \mathbf{B} - \mathbf{A} \text{ is Hermitian nonnegative definite,} \tag{1.1}$$

i.e., whenever there exists a matrix **C**, say, so that $\mathbf{B} - \mathbf{A} = \mathbf{CC}^*$, where $\mathbf{C}^*$ denotes the conjugate transpose of **C**. This ordering (1.1), due to Löwner (1934, p. 177), has usually been applied (particularly in statistics) when both **A** and **B** are Hermitian; this is, however, not necessary and will not, in general, be assumed in this paper.

Drazin's "star" ordering is defined by

$$\mathbf{A} \leq_* \mathbf{B} \text{ whenever } \mathbf{A}^*\mathbf{A} = \mathbf{A}^*\mathbf{B} \text{ and } \mathbf{AA}^* = \mathbf{BA}^*. \tag{1.2}$$

This partial ordering (1.2) is due to Drazin (1978). Matrices **A** and **B** satisfying (1.2) were, however, also considered earlier by Hestenes (1961, Lemma 3.4), who then defined **A** to be a "section" of **B**; the matrices **A** and $\mathbf{B} - \mathbf{A}$ are then also said to be "star-orthogonal".

Rank subtractivity:

$$\mathbf{A} \leq_{\mathsf{rs}} \mathbf{B} \text{ whenever } \operatorname{rank}(\mathbf{B} - \mathbf{A}) = \operatorname{rank}(\mathbf{B}) - \operatorname{rank}(\mathbf{A}) \tag{1.3}$$

was shown by Hartwig (1980) to be a matrix partial ordering; he then called (1.3) the "plus" order since it is equivalent to

$$\mathbf{A}^+\mathbf{A} = \mathbf{A}^+\mathbf{B} \text{ and } \mathbf{AA}^+ = \mathbf{BA}^+ \tag{1.4}$$

for *some* reflexive generalized inverse $\mathbf{A}^+$ of **A**. It was later observed (Baksalary 1986) that (1.4) is equivalent to

$$\mathbf{A}^-\mathbf{A} = \mathbf{A}^-\mathbf{B} \text{ and } \mathbf{AA}^= = \mathbf{BA}^=, \tag{1.5}$$

for *some* (possibly distinct) "inner" generalized inverses $\mathbf{A}^-$ and $\mathbf{A}^=$ (satisfying $\mathbf{AA}^-\mathbf{A} = \mathbf{A} = \mathbf{AA}^=\mathbf{A}$). The ordering (1.3) has, therefore, more recently been referred to as the "minus" order (Baksalary 1986, Baksalary, Pukelsheim and Styan 1987, Carlson 1987, Hartwig and Styan 1986, Mitra 1986). In this paper we will refer to (1.3) as the rank-subtractive ordering (or just as rank subtractivity).

If we replace $\mathbf{A}^+$ with the Moore-Penrose inverse $\mathbf{A}^\dagger$ in (1.4) then

$$\mathbf{A}^\dagger\mathbf{A} = \mathbf{A}^\dagger\mathbf{B} \text{ and } \mathbf{AA}^\dagger = \mathbf{BA}^\dagger, \tag{1.6}$$

which is equivalent to Drazin's "star" ordering (1.2), cf. Drazin (1978) and Hartwig (1978).

It was shown by Hartwig and Styan (1986) that

$$\mathbf{A} \leq_* \mathbf{B} \Leftrightarrow \mathbf{A} \leq_{rs} \mathbf{B} \text{ and } \mathbf{B}^\dagger - \mathbf{A}^\dagger = (\mathbf{B} - \mathbf{A})^\dagger \tag{1.7}$$

and so Drazin's "star" ordering (1.2) always implies rank subtractivity (1.3), cf. also Lemma 2.2 and (2.4) below.

For the orderings (1.2) and (1.3) the matrices **A** and **B** may be rectangular but for the Löwner ordering (1.1) the matrices **A** and **B** must be square (but as observed earlier not necessarily Hermitian).

Our purpose in this paper is to examine:

(a) properties of idempotent matrices under these three partial orders,
(b) the role played by idempotent matrices in linking up the three partial orders, and
(c) how one partial order can be used to investigate another partial order via idempotent matrices.

In statistics the Löwner order (1.1) has often been used to compare covariance matrices and mean-squared error matrices of competing vector estimators. We may say that the estimator $\hat{\boldsymbol{\beta}}$ is "better" or at least "no worse" than the estimator $\tilde{\boldsymbol{\beta}}$ of the parameter vector $\boldsymbol{\beta}$ whenever

$$\hat{\boldsymbol{\Sigma}} = \operatorname{cov} \hat{\boldsymbol{\beta}} \leq_{L} \operatorname{cov} \tilde{\boldsymbol{\beta}} = \tilde{\boldsymbol{\Sigma}} \tag{1.8}$$

for this is equivalent to

$$\mathbf{a}'\hat{\boldsymbol{\Sigma}}\mathbf{a} = \operatorname{var} \mathbf{a}'\hat{\boldsymbol{\beta}} \leq \operatorname{var} \mathbf{a}'\tilde{\boldsymbol{\beta}} = \mathbf{a}'\tilde{\boldsymbol{\Sigma}}\mathbf{a} \tag{1.9}$$

for every conformable vector **a**. In (1.9) the prime denotes transpose and everything is real, and the $\leq$ compares real scalars and therefore means the usual "less than or equal to". The well–known Gauss-Markov theorem for the optimality of the ordinary least-squares estimator $\hat{\boldsymbol{\beta}}$ of $\boldsymbol{\beta}$ in the general linear model

$$E\mathbf{y} = \mathbf{X}\boldsymbol{\beta}, \quad \operatorname{cov} \mathbf{y} = \sigma^2\mathbf{I} \tag{1.10}$$

states that (1.8) holds for all linear unbiased estimators $\tilde{\boldsymbol{\beta}}$ of $\boldsymbol{\beta}$. In (1.10) the design matrix **X** is assumed to have full column rank (so that $\boldsymbol{\beta}$ is estimable).

A matrix generalization, using the rank-subtractive order (1.3), of Cochran's theorem (Cochran 1934) for the chi-squaredness of quadratic forms in normal random variables is as follows:

Suppose that

$$\mathbf{E}' = \mathbf{E} \leq_{\mathsf{rs}} \mathbf{F} = \mathbf{F}', \tag{1.12}$$

where $\mathbf{E}$ and $\mathbf{F}$ are real symmetric matrices; if

$$\mathbf{x}'\mathbf{F}\mathbf{x} \sim \chi^2, \tag{1.13}$$

where the random vector $\mathbf{x}$ follows the multivariate normal distribution $\mathsf{N}(\mathbf{0}, \mathbf{I})$, then

$$\mathbf{x}'\mathbf{E}\mathbf{x} \sim \chi^2. \tag{1.14}$$

The condition (1.13) is equivalent to $\mathbf{F} = \mathbf{F}^2$, and so

$$\mathbf{E} \leq_{\mathsf{rs}} \mathbf{F} = \mathbf{F}^2 \Rightarrow \mathbf{E} = \mathbf{E}^2 \tag{1.15}$$

when $\mathbf{E}$ and $\mathbf{F}$ are both real and symmetric. Chipman and Rao (1964, p. 4) showed that (1.15) also holds when neither $\mathbf{E}$ nor $\mathbf{F}$ is necessarily symmetric. It follows that (1.15) holds for complex matrices $\mathbf{E}$ and $\mathbf{F}$ with neither necessarily Hermitian. For related results see the recent papers by Baksalary and Hauke (1984) and the survey by Anderson and Styan (1982).

On the other hand, if

$$\mathbf{E} \leq_{\mathsf{L}} \mathbf{F}, \tag{1.16}$$

and if (1.13) and (1.14) also hold with $\operatorname{rank}(\mathbf{F}) > \operatorname{rank}(\mathbf{E})$, then

$$\mathbf{x}'\mathbf{F}\mathbf{x} - \mathbf{x}'\mathbf{E}\mathbf{x} \sim \chi^2. \tag{1.17}$$

This result is due to Ogasawara and Takahashi (1951) and Hogg and Craig (1958). See also Baksalary and Kala (1980). A matrix generalization, due to Mäkeläinen and Styan (1976), is that

$$\mathbf{E}^2 = \mathbf{E} \leq_{\mathsf{L}} \mathbf{F} = \mathbf{F}^2 \;\Rightarrow\; (\mathbf{F} - \mathbf{E})^2 = \mathbf{F} - \mathbf{E}, \tag{1.18}$$

where the complex matrices $\mathbf{E}$ and $\mathbf{F}$ are not necessarily Hermitian. In fact Mäkeläinen and Styan (1976) showed that (1.18) also holds when the Löwner order $\mathbf{E} \leq_{\mathsf{L}} \mathbf{F}$ is replaced by

$$\operatorname{rank}[(\mathbf{F} - \mathbf{E})^2] = \operatorname{rank}(\mathbf{F} - \mathbf{E}) \quad \text{and all} \quad \operatorname{ch}(\mathbf{F} - \mathbf{E}) \geq 0\,, \tag{1.19}$$

cf. Lemma 2.4 below. In (1.19) ch denotes characteristic root or eigenvalue.

It is well known, cf. e.g., Clifford and Preston (1961), that projectors and idempotents play a dominant role in any study of semigroups with partial orders. One of the main reasons for the algebraic success of projectors is that for these types of elements all three partial orders coincide, given a suitable involution. In addition all three partial orders then agree with the partial order on principal right ideals, i.e., ranges, in a semigroup S. This is defined by $aS^1 \leq bS^1 \Leftrightarrow aS^1 \subseteq bS^1$. Although, in general, the star order has a much stronger "horizontal" character than the two other partial orders, for idempotents the three orders are much more closely related than one might initially expect. Indeed, as we shall see, if **E** and **F** are idempotent matrices, then (Theorem 2.3 below)

$$\mathbf{E} \leq_{\mathsf{L}} \mathbf{F} \Rightarrow \mathbf{E} \leq_{*} \mathbf{F} \Rightarrow \mathbf{E} \leq_{\mathsf{rs}} \mathbf{F} \tag{1.20}$$

and when **E** and **F** are also Hermitian, and so are orthogonal projectors, then all three partial orders coincide. In this paper we shall concentrate exclusively on complex matrices, however, even though many of the results that follow will hold in more general settings, such as *-semigroups with proper involution (Drazin 1978).

We will address three different types of problems in this paper:

(a) Given an idempotent matrix **F**, we will try (in Section 3) to characterize all matrices **A** that "lie below" **F**, i.e.,

$$\mathbf{A} \leq_{?} \mathbf{F}. \tag{1.21}$$

Here in (1.21), as well as in (1.22) and (1.23) below, $\leq_?$ denotes any one of the three partial orders $\leq_{\mathsf{L}}$, $\leq_*$ and $\leq_{\mathsf{rs}}$, defined respectively by (1.1), (1.2) and (1.3).

(b) Given an idempotent matrix **E**, we will try (in Section 4) to characterize all matrices **B** that "lie above" **E**, i.e.,

$$\mathbf{E} \leq_{?} \mathbf{B}. \tag{1.22}$$

(c) Given *two* idempotent matrices, **E** and **F**, such that

$$\mathbf{E} \leq_{?} \mathbf{F}, \tag{1.23}$$

we will see (Section 5) how this forces **E** and **F** to be related. We will also consider what happens when **E** and/or **F** in (1.21), (1.22) and/or (1.23) are orthogonal projectors (Hermitian idempotent matrices).

2. PRELIMINARY RESULTS

We begin with some characterizations of the three matrix partial orderings $\leq_{\mathsf{L}}$, $\leq_*$ and $\leq_{\mathsf{rs}}$, defined respectively by (1.1), (1.2) and (1.3).

LEMMA 2.1 *Let* **A** *and* **B** *be* $n \times n$ *complex matrices, neither necessarily Hermitian. Then the following are equivalent:*

(a) $\mathbf{A} \leq_{\mathsf{L}} \mathbf{B}$, *i.e.*, $\mathbf{B} - \mathbf{A} = \mathbf{CC}^*$ *for some* **C**,

(b) $\mathbf{B} - \mathbf{A} = (\mathbf{B} - \mathbf{A})^*$ *and all* $\mathrm{ch}(\mathbf{B} - \mathbf{A}) \geq 0$,

(c) $\mathbf{k}^*(\mathbf{B} - \mathbf{A})\mathbf{k} \geq 0$ *for all* $n \times 1$ *complex vectors* **k**.

It follows from (a) and (c) that

$$\mathbf{A} \leq_{\mathsf{L}} \mathbf{B} \Leftrightarrow \mathbf{K}^*\mathbf{AK} \leq_{\mathsf{L}} \mathbf{K}^*\mathbf{BK} \tag{2.1}$$

for every complex matrix **K** with n rows. Moreover, when **A** and **B** are both Hermitian, then

$$\mathbf{0} \leq_{\mathsf{L}} \mathbf{A} \leq_{\mathsf{L}} \mathbf{B} \Rightarrow \mathrm{rank}(\mathbf{A}, \mathbf{B}) = \mathrm{rank}(\mathbf{B}), \tag{2.2}$$

cf. Baksalary and Hauke (1984), and when **A** and **B** are Hermitian then

$$\mathbf{A} \leq_{\mathsf{L}} \mathbf{B} \Rightarrow \mathrm{ch}_i(\mathbf{A}) \leq \mathrm{ch}_i(\mathbf{B}); \quad i = 1, ..., n, \tag{2.3}$$

where ch_i denotes the ith largest (real) eigenvalue. Equality holds on the right-hand side of (2.3) for all $i = 1, ..., n$ if and only if $\mathbf{A} = \mathbf{B}$. It would be of interest to find a "nice" condition that **A** and **B** should satisfy in addition to the conditions on the right-hand sides of (2.2) and (2.3) in order that $\mathbf{A} \leq_{\mathsf{L}} \mathbf{B}$. For further properties of the Löwner order that $\mathbf{A} \leq_{\mathsf{L}} \mathbf{B}$ see the survey paper by Gaffke and Krafft (1982) and the book by Horn and Johnson (1985, pp. 469-476).

LEMMA 2.2. *Let* **A** *and* **B** *be* $m \times n$ *complex matrices with ranks* a *and* b, *respectively, with* $b > a \geq 1$. *Then the following are equivalent*:

(a) $\mathbf{A} \leq_* \mathbf{B}$, *i.e.*, $\mathbf{A}^*\mathbf{A} = \mathbf{A}^*\mathbf{B}$ *and* $\mathbf{A}\mathbf{A}^* = \mathbf{B}\mathbf{A}^*$,

(b) *There exist unitary matrices* $\mathbf{U}$ *and* $\mathbf{V}$, *respectively* $m \times m$ *and* $n \times n$, *so that*

$$\mathbf{A} = \mathbf{U}\begin{pmatrix} \mathbf{D}_a & \mathbf{0} \\ \mathbf{0} & \mathbf{0} \end{pmatrix}\mathbf{V}^*$$

and

$$\mathbf{B} = \mathbf{U}\begin{pmatrix} \mathbf{D}_a & \mathbf{0} & \mathbf{0} \\ \mathbf{0} & \mathbf{D} & \mathbf{0} \\ \mathbf{0} & \mathbf{0} & \mathbf{0} \end{pmatrix}\mathbf{V}^*,$$

where $\mathbf{D}_a$ *and* $\mathbf{D}$ *are positive definite diagonal matrices, respectively* $a \times a$ *and* $(b-a)\times(b-a)$,

(c) $\mathbf{A} \leq_{\mathrm{rs}} \mathbf{B}$ *and* $(\mathbf{B}-\mathbf{A})^\dagger = \mathbf{B}^\dagger - \mathbf{A}^\dagger$.

Moreover, if $\mathbf{A} \leq_* \mathbf{B}$ *and if* $\operatorname{rank}(\mathbf{A}) = \operatorname{rank}(\mathbf{B})$ *or if* $\mathbf{A}$ *has full rank then* $\mathbf{A} = \mathbf{B}$.

From (a) $\Leftrightarrow$ (b) in Lemma 2.2 we see that $\mathbf{A} \leq_* \mathbf{B}$ if and only if $\mathbf{A}$ and $\mathbf{B}$ have simultaneous singular value decompositions where every nonzero singular value of $\mathbf{A}$ is also a singular value of $\mathbf{B}$; if $\mathbf{A} \leq_* \mathbf{B}$ (or just $\mathbf{A} \leq_{\mathrm{rs}} \mathbf{B}$) and $a = b$ then $\mathbf{A} = \mathbf{B}$. That (a) $\Leftrightarrow$ (c) was established by Hartwig and Styan (1986), who also gave several other extra conditions that must be added in order that rank subtractivity $\mathbf{A} \leq_{\mathrm{rs}} \mathbf{B}$ become equivalent to the stronger star order $\mathbf{A} \leq_* \mathbf{B}$. See also Baksalary (1986).

LEMMA 2.3. *Let* $\mathbf{A}$ *and* $\mathbf{B}$ *be* $m \times n$ *complex matrices. Then the following are equivalent*:

(a) $\mathbf{A} \leq_{\mathrm{rs}} \mathbf{B}$, *i.e.*, $\operatorname{rank}(\mathbf{B}-\mathbf{A}) = \operatorname{rank}(\mathbf{B}) - \operatorname{rank}(\mathbf{A})$,

(b) $\mathbf{A} = \mathbf{A}\mathbf{B}^-\mathbf{A}$ *and* $\operatorname{rank}(\mathbf{A}, \mathbf{B}) = \operatorname{rank}(\mathbf{B}) = \operatorname{rank}\begin{pmatrix} \mathbf{A} \\ \mathbf{B} \end{pmatrix}$,

(c) $\mathbf{A} = \mathbf{A}\mathbf{B}^-\mathbf{A} = \mathbf{A}\mathbf{B}^-\mathbf{B} = \mathbf{B}\mathbf{B}^-\mathbf{A}$,

(d) $\operatorname{rank}(\mathbf{B}-\mathbf{A}) = \operatorname{rank}[(\mathbf{I}-\mathbf{A}\mathbf{A}^-)\mathbf{B}] = \operatorname{rank}[\mathbf{B}(\mathbf{I}-\mathbf{A}^-\mathbf{A})]$,

(e) $\mathbf{B}-\mathbf{A} \leq_{\mathrm{rs}} \mathbf{B}$,

where $\mathbf{A}^-$ *and* $\mathbf{B}^-$ *are any choices of "inner" generalized inverses. If* (b), (c) *or* (d) *holds for any particular "inner" generalized inverses* $\mathbf{A}^-$ *and* $\mathbf{B}^-$ *then all conditions hold for every "inner" generalized inverse. Moreover, if* $\mathbf{A} \leq_{rs} \mathbf{B}$ *and if* $\text{rank}(\mathbf{A}) = \text{rank}(\mathbf{B})$ *or if* $\mathbf{A}$ *has full rank then* $\mathbf{A} = \mathbf{B}$.

Lemma 2.3 was established by Marsaglia and Styan (1974, Theorem 17); see also Hartwig and Styan (1986, Lemma 1) and Cline and Funderlic (1979).

It follows at once from Lemma 2.2 that

$$\mathbf{A} \leq_* \mathbf{B} \;\Rightarrow\; \mathbf{A} \leq_{rs} \mathbf{B}. \tag{2.4}$$

We may, however, use rank subtractivity to characterize the star order as follows:

PROPOSITION 2.1. *Let* $\mathbf{A}$ *and* $\mathbf{B}$ *be* $m \times n$ *complex matrices. Then*

$$\mathbf{A} \leq_* \mathbf{B} \quad\Leftrightarrow\quad \mathbf{A}^*\mathbf{A} \leq_{rs} \mathbf{A}^*\mathbf{B} \text{ and } \mathbf{A}\mathbf{A}^* \leq_{rs} \mathbf{B}\mathbf{A}^*. \tag{2.5}$$

We recall the definition (1.2) of $\mathbf{A} \leq_* \mathbf{B}$, which is the right-hand side of (2.5) with $\leq_{rs}$ replaced by $=$. To prove Proposition 2.1 it, therefore, suffices to note that

$$\mathbf{A}^*(\mathbf{B}-\mathbf{A}) \leq_{rs} \mathbf{A}^*\mathbf{B} \quad\Leftrightarrow\quad \mathbf{A}^*\mathbf{A} \leq_{rs} \mathbf{A}^*\mathbf{B} \quad\Rightarrow\quad \mathbf{A}^*\mathbf{A} = \mathbf{A}^*\mathbf{B}, \tag{2.6}$$

since the left-hand side of (2.6) implies that

$$\text{rank}(\mathbf{A}^*\mathbf{A}) \leq \text{rank}(\mathbf{A}^*\mathbf{B}) \leq \text{rank}(\mathbf{A}^*) = \text{rank}(\mathbf{A}^*\mathbf{A}). \tag{2.7}$$

When both $\mathbf{A}$ and $\mathbf{B}$ are square we enquire as to where the Löwner order sits *vis-à-vis* (2.4). A partial answer is provided by:

THEOREM 2.1. *Let* $\mathbf{G}$ *and* $\mathbf{H}$ *be* $n \times n$ *complex Hermitian matrices and let* $\mathbf{H}$ *be nonnegative definite. Then*

$$\mathbf{G} \leq_* \mathbf{H} \;\Rightarrow\; \mathbf{G} \leq_{rs} \mathbf{H} \;\Rightarrow\; \mathbf{G} \leq_L \mathbf{H}. \tag{2.8}$$

Proof. We need only prove the second implication in (2.8). Clearly $\mathbf{G} \leq_{rs} \mathbf{H} \Leftrightarrow \mathbf{H} - \mathbf{G} \leq_{rs} \mathbf{H}$, and so from Lemma 2.3: (a) $\Rightarrow$ (b) we see that $\mathbf{H} - \mathbf{G} = (\mathbf{H} - \mathbf{G})\mathbf{H}^-(\mathbf{H} - \mathbf{G}) \geq_L \mathbf{0}$, which follows since we may choose $\mathbf{H}^-$ Hermitian nonnegative definite. □

The second implication in (2.8) generalizes the statement (17) of Baksalary, Kala and Klaczyński (1983), who assume that both $\mathbf{G}$ and $\mathbf{H}$ are Hermitian nonnegative definite.

From Theorem 2.1 we see that when the matrices being ordered are Hermitian and the "larger" one is also nonnegative definite then the Löwner order is the weakest. In Theorem 2.2 below we will see that when the matrices are idempotent but not necessarily Hermitian then the Löwner order is the strongest. To establish Theorems 2.2 and 2.3 we need three interim results: the first follows at once from Lemma 2.4, which is due to Mäkeläinen and Styan (1976, Theorem 2).

LEMMA 2.4. *Let* $\mathbf{E}$ *and* $\mathbf{F}$ *be* $n\times n$ *complex idempotent matrices, and let* $\mathbf{F}-\mathbf{E}$ *have nonnegative (real) eigenvalues with*

$$\operatorname{rank}[(\mathbf{F}-\mathbf{E})^2] = \operatorname{rank}(\mathbf{F}-\mathbf{E}). \tag{2.9}$$

Then $\mathbf{F}-\mathbf{E}$ *is idempotent.*

We will also need the following Lemma 2.5, which is a matrix version of an extension due to James (1952) of Cochran's theorem, cf. (1.15) above and Anderson and Styan (1982, p. 3); see also Theorem 5.1 below.

LEMMA 2.5. *Let* $\mathbf{E}$ *and* $\mathbf{F}$ *be* $n\times n$ *complex idempotent matrices, neither necessarily Hermitian. Then the following are equivalent:*

$$\left.\begin{array}{ll}\text{(a)} & \mathbf{F}-\mathbf{E} \text{ is idempotent,}\\ \text{(b)} & \mathbf{EF}+\mathbf{FE}=2\mathbf{E},\\ \text{(c)} & \mathbf{E}=\mathbf{EF}=\mathbf{FE}.\end{array}\right\} \tag{2.10}$$

Proof. Clearly $\mathbf{F}-\mathbf{E}=(\mathbf{F}-\mathbf{E})^2 \Leftrightarrow \mathbf{EF}+\mathbf{FE}=2\mathbf{E} \Leftarrow$ (b). To go the other way we premultiply $\mathbf{EF}+\mathbf{FE}=2\mathbf{E}$ by $\mathbf{E}$ to yield $\mathbf{EF}+\mathbf{EFE}=2\mathbf{E}$; postmultiplication by $\mathbf{E}$ yields $\mathbf{EFE}+\mathbf{FE}=2\mathbf{E}$ and so $\mathbf{EF}=\mathbf{FE}=\mathbf{E}$. □

We will also use certain rank cancellation rules (Marsaglia and Styan 1974, p. 271) given in

LEMMA 2.6. *Suppose that the complex matrices* $\mathbf{A}$ *and* $\mathbf{X}$, *respectively* $m\times n$ *and* $n\times k$ *satisfy*

$$\operatorname{rank}(\mathbf{AX}) = \operatorname{rank}(\mathbf{A}). \tag{2.11}$$

Then for every $p \times m$ *complex matrix* $\mathbf{P}$,

$$\operatorname{rank}(\mathbf{PAX}) = \operatorname{rank}(\mathbf{PA}), \tag{2.12}$$

and for any $p \times m$ *complex matrices* $\mathbf{P}$ *and* $\mathbf{Q}$,

$$\mathbf{PAX} = \mathbf{QAX} \;\Rightarrow\; \mathbf{PA} = \mathbf{QA}. \tag{2.13}$$

When $\mathbf{X} = \mathbf{A}^*$ then (2.11) holds and we call (2.12) and (2.13) "star-cancellation".

We are now ready to establish

THEOREM 2.2. *Let* $\mathbf{E}$ *and* $\mathbf{F}$ *be* $n \times n$ *complex idempotent matrices, neither necessarily Hermitian. Then*

$$\mathbf{E} \leq_{\mathsf{L}} \mathbf{F} \;\Rightarrow\; \mathbf{E} \leq_{*} \mathbf{F} \;\Rightarrow\; \mathbf{E} \leq_{\mathsf{rs}} \mathbf{F}. \tag{2.14}$$

Proof. We need only prove the first implication. Since $\mathbf{F} - \mathbf{E}$ is Hermitian nonnegative definite it has nonnegative eigenvalues and (2.9) holds. Thus Lemma 2.4 applies and so $\mathbf{F} - \mathbf{E}$ is idempotent. From Lemma 2.5 we see that (2.10) holds and so

$$\mathbf{E}(\mathbf{F}-\mathbf{E})^* = \mathbf{E}(\mathbf{F}-\mathbf{E}) = \mathbf{0} = (\mathbf{F}-\mathbf{E})\mathbf{E} = (\mathbf{F}-\mathbf{E})^*\mathbf{E} \tag{2.15}$$

since $\mathbf{F} - \mathbf{E}$ is Hermitian. Thus $\mathbf{E} \leq_* \mathbf{F}$ follows using the basic definition (1.2). □

Combining Theorems 2.1 and 2.2 yields the first part of Theorem 2.3; see also Theorem 5.8 at the end of this paper.

THEOREM 2.3. *Let* $\mathbf{E}$ *and* $\mathbf{F}$ *be* $n \times n$ *complex Hermitian idempotent matrices (orthogonal projectors). Then*

$$\mathbf{E} \leq_{\mathsf{L}} \mathbf{F} \;\Leftrightarrow\; \mathbf{E} \leq_{*} \mathbf{F} \;\Leftrightarrow\; \mathbf{E} \leq_{\mathsf{rs}} \mathbf{F}. \tag{2.16}$$

And any one of the orderings in (2.16) *holds if and only if*

$$\mathbf{E} = \mathbf{EFE}, \tag{2.17}$$

i.e., if and only if $\mathbf{F}$ *is an "inner" generalized inverse of* $\mathbf{E}$.

Proof. In view of (1.2) it suffices to note that

$$\mathbf{E} \leq_* \mathbf{F} \Leftrightarrow \mathbf{E} = \mathbf{EF} \Leftrightarrow (\mathbf{E}-\mathbf{EF})(\mathbf{E}-\mathbf{EF})^* = \mathbf{0} \Leftrightarrow \mathbf{E} = \mathbf{EFE}. \quad (2.18)$$

□

We now examine whether the (Hermitian) idempotent property is passed down (Section 3) or up (Section 4) under each of the three partial orders $\leq_L$, $\leq_*$ and $\leq_{rs}$, defined respectively by (1.1), (1.2) and (1.3). At the same time we characterize all matrices that lie above or below an (Hermitian) idempotent matrix under one of these three partial orders.

3. PROPERTIES PASSED DOWN

In this section we consider the first type (a) of problem stated at the end of Section 1. Given an idempotent matrix **F**, how does one characterize the matrices **A** that lie *below* **F**? We will, therefore, try to characterize all **A** such that $\mathbf{A} \leq_? \mathbf{F}$, where $\leq_?$ denotes any one of the three partial orders $\leq_L$, $\leq_*$ and $\leq_{rs}$, defined respectively by (1.1), (1.2) and (1.3). We begin with

THEOREM 3.1. *Let* **F** *be an* $n \times n$ *complex idempotent matrix, not necessarily Hermitian. Let* **A** *be an* $n \times n$ *complex matrix. Then*

$$\mathbf{A} \leq_* \mathbf{F} \Rightarrow \mathbf{A}^2 = \mathbf{A} \quad (3.1)$$

$$\mathbf{A} \leq_{rs} \mathbf{F} \Rightarrow \mathbf{A}^2 = \mathbf{A} \quad (3.2)$$

$$\mathbf{A} \leq_L \mathbf{F} \nRightarrow \mathbf{A}^2 = \mathbf{A}. \quad (3.3)$$

Proof. From (2.4) we know that $\mathbf{A} \leq_* \mathbf{F} \Rightarrow \mathbf{A} \leq_{rs} \mathbf{F}$, which in turn implies $\mathbf{A}^2 = \mathbf{A}$ from Cochran's theorem cf.(1.15). This establishes (3.1) and (3.2). To see that the Löwner order $\mathbf{A} \leq_L \mathbf{F}$ is not sufficient to ensure that **A** be idempotent, consider

$$\mathbf{A} = \begin{pmatrix} 0 & 1 \\ 0 & 0 \end{pmatrix} \leq_L \begin{pmatrix} 1 & 1 \\ 0 & 0 \end{pmatrix} = \mathbf{F} = \mathbf{F}^2. \quad (3.4)$$

In (3.4) $\mathbf{A}^2 = \mathbf{0}$ and so $\mathbf{A}^2 \neq \mathbf{A}$ though here $\mathbf{F}-\mathbf{A}$ is both idempotent and Hermitian (an orthogonal projector). □

We note that (3.1) is due to Drazin (1978), cf. also Hartwig and Spindelböck (1983), while (3.2) may be traced back to Chipman and Rao (1964).

We may strengthen (3.2) by noting that from Lemma 2.3: (a) $\Leftrightarrow$ (c) with $\mathbf{B}^- = \mathbf{F}^- = \mathbf{I}$

$$\mathbf{A} \le_{\mathrm{rs}} \mathbf{F} \Leftrightarrow \mathbf{A}^2 = \mathbf{A} = \mathbf{AF} = \mathbf{FA}, \tag{3.5}$$

where, as in Theorem 3.1, $\mathbf{F}$ is an $n \times n$ complex idempotent matrix, not necessarily Hermitian.

THEOREM 3.2. *Let* $\mathbf{F}$ *be an* $n \times n$ *complex Hermitian idempotent matrix (orthogonal projector). Then*

$$\mathbf{A} \le_* \mathbf{F} \quad \Rightarrow \mathbf{A}^2 = \mathbf{A} = \mathbf{A}^* \tag{3.6}$$

$$\mathbf{A} \le_{\mathrm{rs}} \mathbf{F} \begin{cases} \Rightarrow & \mathbf{A}^2 = \mathbf{A} \\ \not\Rightarrow & \mathbf{A} = \mathbf{A}^* \end{cases} \tag{3.7}$$

$$\mathbf{A} \le_{\mathrm{L}} \mathbf{F} \begin{cases} \not\Rightarrow & \mathbf{A}^2 = \mathbf{A} \\ \Rightarrow & \mathbf{A} = \mathbf{A}^*. \end{cases} \tag{3.8}$$

Proof. The first parts of (3.6), (3.7) and (3.8) follow at once from Theorem 3.1. To see that $\mathbf{A} \le_* \mathbf{F} \ \Rightarrow \ \mathbf{A} = \mathbf{A}^*$ we note that in view of (1.2), (2.4) and (3.5)

$$\mathbf{A}^*\mathbf{A} = \mathbf{A}^*\mathbf{F} = \mathbf{FA} = \mathbf{A}. \tag{3.9}$$

To see that $\mathbf{A} \le_{\mathrm{rs}} \mathbf{F} \not\Rightarrow \mathbf{A} = \mathbf{A}^*$ we consider

$$\mathbf{A} = \begin{pmatrix} 1 & 1 \\ 0 & 0 \end{pmatrix} \le_{\mathrm{rs}} \begin{pmatrix} 1 & 0 \\ 0 & 1 \end{pmatrix} = \mathbf{F}. \tag{3.10}$$

That $\mathbf{A} \le_{\mathrm{L}} \mathbf{F} \Rightarrow \mathbf{A} = \mathbf{A}^*$ follows at once from both $\mathbf{F}$ and $\mathbf{F} - \mathbf{A}$ being Hermitian. □

We note that (3.6) is due to Drazin (1978), and that the second part of (3.7) and an example like (3.10) have been given recently by Baksalary and Hauke (1987).

When the Hermitian idempotent matrix $\mathbf{F}$ in Theorem 3.2 is equal to the identity matrix $\mathbf{I}$, then we see (Theorem 3.3) that *all* matrices that lie below in the rank-subtractive order, cf. (3.7), *must* be idempotent, while (Theorem 3.4) *all* matrices that lie below in the star order, cf. (3.6), *must* be both idempotent and Hermitian (orthogonal projectors). Little can be said about matrices $\mathbf{A}$ that lie below $\mathbf{I}$ in the Löwner order — though such $\mathbf{A}$ must be Hermitian, cf. (3.8), and their eigenvalues cannot exceed 1, cf. (2.3).

THEOREM 3.3. *Let* $\mathbf{A}$ *be an* $n \times n$ *complex matrix, not necessarily Hermitian. Then the following five conditions are equivalent:*

(a) $\mathbf{A} \leq_{\mathrm{rs}} \mathbf{I}$,

(b) $\mathbf{A} = \mathbf{A}^2$,

(c) $\mathbf{A} \leq_{*} \mathbf{F}$ *for some* $\mathbf{F} = \mathbf{F}^2$,

(d) $\mathbf{A} \leq_{\mathrm{rs}} \mathbf{F}$ *for some* $\mathbf{F} = \mathbf{F}^2$,

(e) $\mathbf{A} \leq_{\mathrm{L}} \mathbf{F}$ *and* $\mathbf{AF} = \mathbf{FA}$ *for some* $\mathbf{F} = \mathbf{F}^2$, *and all* $\mathrm{ch}(\mathbf{A}) = 0$ *or* 1.

It is easy to see that the first four conditions (a)-(d) in Theorem 3.3 are equivalent and imply (e). To show that (e) $\Rightarrow$ (b) we may write:

$$\mathbf{F} = \begin{pmatrix} \mathbf{I} & \mathbf{G} \\ \mathbf{0} & \mathbf{0} \end{pmatrix} \quad \text{and} \quad \mathbf{A} = \begin{pmatrix} \mathbf{A}_{11} & \mathbf{A}_{12} \\ \mathbf{A}_{21} & \mathbf{A}_{22} \end{pmatrix}. \tag{3.11}$$

Then (e) $\Rightarrow \mathbf{A}_{11} = \mathbf{A}_{11}{}^* = \mathbf{A}_{11}{}^2$, $\mathbf{A}_{12} = \mathbf{A}_{11}\mathbf{G}$, $\mathbf{A}_{21} = \mathbf{0}$, and $\mathbf{A}_{22} = \mathbf{0}$; hence $\mathbf{A} = \mathbf{A}^2$.

THEOREM 3.4. *Let* $\mathbf{A}$ *be an* $n \times n$ *complex matrix, not necessarily Hermitian. Then the following six conditions are equivalent:*

(a) $\mathbf{A} \leq_{*} \mathbf{I}$,

(b) $\mathbf{A} = \mathbf{A}^2 = \mathbf{A}^*$,

(c) $\mathbf{A} \leq_{\mathrm{rs}} \mathbf{A}^*\mathbf{A}$,

(d) $\mathbf{A} \leq_* \mathbf{F}$ *for some orthogonal projector* $\mathbf{F}$,

(e) $\mathbf{A} \leq_{\mathsf{L}} \mathbf{F}$ *for some orthogonal projector* $\mathbf{F}$, *and all* $\mathrm{ch}(\mathbf{A}) = 0$ *or* 1,

(f) $\mathbf{0} \leq_{\mathsf{L}} \mathbf{A} \leq_{\mathsf{L}} \mathbf{F}$ *and* $(\mathbf{F} - \mathbf{A})^2 = \mathbf{F} - \mathbf{A}$ *for some orthogonal projector* $\mathbf{F}$.

4. PROPERTIES PASSED UP

We now consider the second type (b) of problem stated at the end of Section 1: Given an idempotent matrix $\mathbf{E}$ how does one characterize the matrices $\mathbf{B}$ that lie *above* $\mathbf{E}$? We will, therefore, try to characterize those $\mathbf{B}$ such that $\mathbf{E} \leq_? \mathbf{B}$, where $\leq_?$ denotes any one of the three partial orders $\leq_{\mathsf{L}}$, $\leq_*$ and $\leq_{\mathsf{rs}}$, defined respectively by (1.1), (1.2) and (1.3).

Unfortunately, very little can be said to characterize such matrices $\mathbf{B}$. We do, however, have the following

THEOREM 4.1. *Let* $\mathbf{E}$ *and* $\mathbf{B}$ *be* $n \times n$ *complex matrices and let* $\mathbf{E}$ *be idempotent, not necessarily Hermitian. Then*

$$\mathbf{E} \leq_{\mathsf{rs}} \mathbf{B} \Leftrightarrow \mathrm{rank}(\mathbf{B} - \mathbf{E}) = \mathrm{rank}(\mathbf{B} - \mathbf{BE}) = \mathrm{rank}(\mathbf{B} - \mathbf{EB}). \tag{4.1}$$

If, in addition, $\mathbf{E}$ *is also Hermitian (and thus an orthogonal projector) then*

$$\mathbf{E} \leq_* \mathbf{B} \Leftrightarrow \mathbf{E} = \mathbf{BE} = \mathbf{EB} \tag{4.2}$$

and

$$\mathbf{E} \leq_* \mathbf{B} \text{ and } \mathbf{I} - \mathbf{E} \leq_* \mathbf{B} \Leftrightarrow \mathbf{B} = \mathbf{I}. \tag{4.3}$$

Proof. The characterization (4.1) follows at once from Lemma 2.3: (a) $\Leftrightarrow$ (d), while (1.2) $\Rightarrow$ (4.2) $\Rightarrow$ (4.3). □

If we substitute $\mathbf{E} = \mathbf{I}$ in (4.1) and (4.2) we obtain

$$\mathbf{I} \leq_{\mathsf{rs}} \mathbf{B} \Leftrightarrow \mathbf{I} \leq_* \mathbf{B} \Leftrightarrow \mathbf{B} = \mathbf{I}, \tag{4.4}$$

cf. (4.3), and so the only matrix that lies "above" the identity matrix in either the rank-subtractive or star ordering is the identity matrix itself. Indeed if $\mathbf{A} \leq_{\mathsf{rs}} \mathbf{B}$ or if $\mathbf{A} \leq_* \mathbf{B}$ for some $m \times n$ complex matrices $\mathbf{A}$ and $\mathbf{B}$, and if $\mathbf{A}$ has full rank then $\mathbf{B} = \mathbf{A}$, cf. the last sentence in each of Lemmas 2.2 and 2.3. On the other hand, when (4.2) holds and $\mathrm{rank}(\mathbf{E}) = n - 1$ then $\mathbf{B} = \mathbf{B}^*$.

5. TWO IDEMPOTENT MATRICES

We end our paper by considering the third (and last) type (c) of problem stated at the end of Section 1: Given two idempotent matrices **E** and **F** such that

$$\mathbf{E} \leq_{?} \mathbf{F}, \tag{5.1}$$

where $\leq_{?}$ denotes any one of the three partial orders $\leq_{\mathsf{L}}$, $\leq_{*}$ and $\leq_{\mathsf{rs}}$, defined respectively by (1.1), (1.2) and (1.3), how does (5.1) force these two matrices **E** and **F** to be related? We begin with the situation where neither of the idempotent matrices **E** nor **F** is necessarily Hermitian.

THEOREM 5.1. *Let* **E** *and* **F** *be* $n\times n$ *complex idempotent matrices, neither necessarily Hermitian. Then the following nine conditions are equivalent:*

(a) $\mathbf{E} \leq_{\mathsf{rs}} \mathbf{F}$, *i.e.*, $\operatorname{rank}(\mathbf{F}-\mathbf{E}) = \operatorname{rank}(\mathbf{F}) - \operatorname{rank}(\mathbf{E})$,

(b) $\mathbf{F}-\mathbf{E} = (\mathbf{F}-\mathbf{E})^2$,

(c) $\mathbf{E} \leq_{*} \mathbf{E} + (\mathbf{I}-\mathbf{F})^*$,

(d) $\mathbf{I}-\mathbf{F} \leq_{\mathsf{rs}} \mathbf{I}-\mathbf{E}$,

(e) $\mathbf{E} = \mathbf{EF} = \mathbf{FE}$,

(f) $\mathbf{EF} + \mathbf{FE} = 2\mathbf{E}$,

(g) $\operatorname{rank}(\mathbf{F}-\mathbf{E}) = \operatorname{rank}(\mathbf{F}-\mathbf{EF}) = \operatorname{rank}(\mathbf{F}-\mathbf{FE})$,

(h) $\operatorname{rank}(\mathbf{E}, \mathbf{F}) = \operatorname{rank}(\mathbf{F}) = \operatorname{rank}\begin{pmatrix}\mathbf{E}\\ \mathbf{F}\end{pmatrix}$,

(i) $\operatorname{rank}[(\mathbf{F}-\mathbf{E})^2] = \operatorname{rank}(\mathbf{F}-\mathbf{E})$ *and all* $\operatorname{ch}(\mathbf{F}-\mathbf{E}) \geq 0$.

And then

$$\operatorname{rank}(\mathbf{E}) = \operatorname{rank}(\mathbf{EF}^*) = \operatorname{rank}(\mathbf{E}^*\mathbf{F}) \tag{5.2}$$

and

$$\operatorname{rank}(\mathbf{F}-\mathbf{E}) = \operatorname{rank}[(\mathbf{F}-\mathbf{E})\mathbf{F}^*] = \operatorname{rank}[(\mathbf{F}-\mathbf{E})^*\mathbf{F}]. \tag{5.3}$$

Proof. That (a) $\Leftrightarrow$ (e) $\Leftrightarrow$ (g) $\Leftrightarrow$ (h) follows directly from Lemma 2.3: (a) $\Leftrightarrow$ (c) $\Leftrightarrow$ (d) $\Leftrightarrow$ (b), choosing $\mathbf{A}^- = \mathbf{B}^- = \mathbf{I} = \mathbf{E}^- = \mathbf{F}^-$. From (a) $\Leftrightarrow$ (e) we see that (d) $\Leftrightarrow$ $\mathbf{I}-\mathbf{F} = (\mathbf{I}-\mathbf{F})(\mathbf{I}-\mathbf{E}) = (\mathbf{I}-\mathbf{E})(\mathbf{I}-\mathbf{F})$, which reduces to (e). For (i) see (1.19). To see that (a) $\Rightarrow$ (5.2) we note that since (a) $\Rightarrow$ (e) we have $\mathbf{E} = \mathbf{EF}$ and so $\operatorname{rank}(\mathbf{EF}^*) = \operatorname{rank}(\mathbf{EFF}^*) = \operatorname{rank}(\mathbf{EF}) =$

rank(**E**), using star-cancellation, cf. (2.12). That (a) implies the rest of (5.2) and all of (5.3) follows similarly. □

THEOREM 5.2. *Let* **E** *and* **F** *be* $n\times n$ *complex idempotent matrices, neither necessarily Hermitian. Then the following nine conditions are equivalent*:

(a) $\mathbf{E} \leq_{*} \mathbf{F}$, *i.e.*, $\mathbf{E}^*(\mathbf{F}-\mathbf{E}) = \mathbf{0} = (\mathbf{F}-\mathbf{E})\mathbf{E}^*$,

(b) $\mathbf{EE}^* \leq_{*} \mathbf{FF}^*$ *and* $\mathbf{E}^*\mathbf{E} \leq_{*} \mathbf{F}^*\mathbf{F}$,

(c) $\mathbf{EE}^* \leq_{\mathsf{rs}} \mathbf{FF}^*$ *and* $\mathbf{E}^*\mathbf{E} \leq_{\mathsf{rs}} \mathbf{F}^*\mathbf{F}$,

(d) $\mathbf{EE}^* \leq_{\mathsf{L}} \mathbf{FF}^*$ *and* $\mathbf{E}^*\mathbf{E} \leq_{\mathsf{L}} \mathbf{F}^*\mathbf{F}$,

(e) $\mathbf{FF}^* - \mathbf{EE}^* = (\mathbf{F}-\mathbf{E})(\mathbf{F}-\mathbf{E})^*$ *and* $\mathbf{F}^*\mathbf{F} - \mathbf{E}^*\mathbf{E} = (\mathbf{F}-\mathbf{E})^*(\mathbf{F}-\mathbf{E})$,

(f) $\mathbf{F}-\mathbf{E} \leq_{\mathsf{rs}} \mathbf{I}-\mathbf{E}^*$,

(g) $\mathbf{EE}^*\mathbf{E} = \mathbf{EF}^*\mathbf{F} = \mathbf{FF}^*\mathbf{E}$,

(h) $\mathbf{EE}^*\mathbf{E} = \mathbf{EF}^*\mathbf{E}$ *and* $(\mathbf{EE}^*)^2 = \mathbf{EF}^*\mathbf{FE}^*$ *and* $(\mathbf{E}^*\mathbf{E})^2 = \mathbf{E}^*\mathbf{FF}^*\mathbf{E}$,

(i) $\mathbf{E}^* + \mathbf{F} - \mathbf{E} \leq_{\mathsf{rs}} \mathbf{I}$ *and* $\mathbf{F}-\mathbf{E} \leq_{\mathsf{rs}} \mathbf{I}$.

Proof. We note that (a) ⇒ (b) for any (not necessarily idempotent) matrices **E** and **F**, cf. Drazin (1977, Prop. 7.2, page 55) and Baksalary and Hauke (1987). Clearly (a) ⇒ (e) ⇒ (d). That (b) ⇒ (c) ⇒ (d) follows at once from Theorem 2.1. From (2.2) we see that

$$\text{(d)} \Rightarrow \operatorname{rank}(\mathbf{EE}^*, \mathbf{FF}^*) = \operatorname{rank}(\mathbf{FF}^*) = \operatorname{rank}(\mathbf{F}) = \operatorname{rank}(\mathbf{E}, \mathbf{F}); \tag{5.4}$$

similarly $\operatorname{rank}\begin{pmatrix}\mathbf{E}\\ \mathbf{F}\end{pmatrix} = \operatorname{rank}(\mathbf{F})$, and so from Theorem 5.1: (h) ⇒ (e) we obtain $\mathbf{E} = \mathbf{EF} = \mathbf{FE}$. Hence $\mathbf{E}(\mathbf{FF}^* - \mathbf{EE}^*)\mathbf{E}^* = \mathbf{0}$, and since $\mathbf{FF}^* - \mathbf{EE}^*$ is nonnegative definite from (d) we obtain $\mathbf{E}(\mathbf{FF}^* - \mathbf{EE}^*) = \mathbf{0} \Leftrightarrow \mathbf{EF}^* = \mathbf{EE}^*$. Similarly $\mathbf{E}^*\mathbf{F} = \mathbf{E}^*\mathbf{E}$ and so (d) ⇒ (a). From Theorem 5.1: (a) ⇔ (e) we find that (f) ⇔ $\mathbf{F}-\mathbf{E} = (\mathbf{F}-\mathbf{E})(\mathbf{I}-\mathbf{E}^*) = (\mathbf{I}-\mathbf{E}^*)(\mathbf{F}-\mathbf{E})$ ⇔ $(\mathbf{F}-\mathbf{E})\mathbf{E}^* = \mathbf{0} = \mathbf{E}^*(\mathbf{F}-\mathbf{E})$ and so (f) ⇔ (a). Clearly (a) ⇒ (g) ⇒ last two equalities in (h). Moreover (g) ⇒ $C(\mathbf{E})\subset C(\mathbf{F})$, where C denotes column space; hence $\mathbf{E} = \mathbf{FE}$. Similarly $\mathbf{E} = \mathbf{EF}$. Postmultiplying $\mathbf{EE}^*\mathbf{E} = \mathbf{EF}^*\mathbf{F}$ by **E** then yields $\mathbf{EE}^*\mathbf{E} = \mathbf{EF}^*\mathbf{E}$. Finally (h) ⇒ $\mathbf{E}^*(\mathbf{E}-\mathbf{F})(\mathbf{E}-\mathbf{F})^*\mathbf{E} = \mathbf{0} = \mathbf{E}(\mathbf{E}-\mathbf{F})^*(\mathbf{E}-\mathbf{F})\mathbf{E}^*$ ⇒ (a). □

As we noted already at the beginning of this paper, cf. (1.7) and (2.4), the star partial ordering $\mathbf{E} \leq_* \mathbf{F}$ is stronger than the rank-subtractive ordering $\mathbf{E} \leq_{rs} \mathbf{F}$. In Theorem 5.3 we assemble some conditions that when coupled with $\mathbf{E} \leq_{rs} \mathbf{F}$ turn it into $\mathbf{E} \leq_* \mathbf{F}$.

THEOREM 5.3. *Let* $\mathbf{E}$ *and* $\mathbf{F}$ *be* $n \times n$ *complex idempotent matrices, neither necessarily Hermitian. Then*

$$\mathbf{E} \leq_* \mathbf{F}, \quad i.e., \quad \mathbf{E}^*(\mathbf{F}-\mathbf{E}) = \mathbf{0} = (\mathbf{F}-\mathbf{E})\mathbf{E}^*, \tag{5.5}$$

if and only if

$$\mathbf{E} \leq_{rs} \mathbf{F}, \quad i.e., \quad \operatorname{rank}(\mathbf{F}-\mathbf{E}) = \operatorname{rank}(\mathbf{F}) - \operatorname{rank}(\mathbf{E}) \tag{5.6}$$

and any one of the following five conditions hold:

(a) $\mathbf{EF}^* \leq_* \mathbf{FF}^*$ *and* $\mathbf{F}^*\mathbf{E} \leq_* \mathbf{F}^*\mathbf{F}$,

(b) $\mathbf{EF}^* \leq_L \mathbf{FF}^*$ *and* $\mathbf{F}^*\mathbf{E} \leq_L \mathbf{F}^*\mathbf{F}$,

(c) $\mathbf{E}^*\mathbf{F} \leq_{rs} \mathbf{E}^*\mathbf{E}$ *and* $\mathbf{FE}^* \leq_{rs} \mathbf{EE}^*$,

(d) $\mathbf{EE}^\dagger\mathbf{F} = \mathbf{FE}^\dagger\mathbf{E}$,

(e) $\mathbf{EE}^\dagger\mathbf{F}\cdot\mathbf{FE}^\dagger\mathbf{E} = \mathbf{FE}^\dagger\mathbf{E}\cdot\mathbf{EE}^\dagger\mathbf{F}$.

Proof. From (1.2) we see that (5.5) $\Rightarrow$ (a), and that $\mathbf{EF}^* \leq_* \mathbf{FF}^* \Rightarrow \mathbf{FE}^*(\mathbf{F}-\mathbf{E})\mathbf{F}^* = \mathbf{0}$, and so when $\mathbf{E} \leq_{rs} \mathbf{F}$ we have $\mathbf{E}^*(\mathbf{F}-\mathbf{E}) = \mathbf{0}$ using Lemma 2.6 and (5.3). Similarly $\mathbf{E} \leq_{rs} \mathbf{F}$ and $\mathbf{F}^*\mathbf{E} \leq_* \mathbf{F}^*\mathbf{F} \Rightarrow (\mathbf{F}-\mathbf{E})\mathbf{E}^* = \mathbf{0}$, and so (a) $\Rightarrow$ (5.5). Clearly (a) $\Rightarrow$ (b) using Theorem 2.1 since here $\mathbf{EF}^* \leq_* \mathbf{EE}^*$ and $\mathbf{F}^*\mathbf{E} \leq_* \mathbf{E}^*\mathbf{E}$ are both Hermitian. From Theorem 5.1 (e) we see that (5.5) $\Rightarrow \mathbf{E}[(\mathbf{F}-\mathbf{E})\mathbf{F}^*]\mathbf{E}^* = \mathbf{0}$; when (b) holds, therefore, $(\mathbf{F}-\mathbf{E})\mathbf{F}^*\mathbf{E}^* = \mathbf{0} = (\mathbf{F}-\mathbf{E})\mathbf{E}^*$. Similarly $\mathbf{E}^*(\mathbf{F}-\mathbf{E}) = \mathbf{0}$, and so (b) $\Rightarrow$ (5.5). Clearly (1.6) shows that (5.5) $\Rightarrow$ (d), and obviously (d) $\Rightarrow$ (e), which reduces to $\mathbf{EE}^\dagger\mathbf{FE}^\dagger\mathbf{E} = \mathbf{FE}^\dagger\mathbf{F}$. Pre- and postmultiplying this by $\mathbf{E}$ reduces it to $\mathbf{FE}^\dagger\mathbf{F} = \mathbf{E}$, using Theorem 5.1. Hence $\mathbf{EFE}^\dagger\mathbf{F} = \mathbf{E}^2 = \mathbf{E} = \mathbf{EE}^\dagger\mathbf{F} \Rightarrow \mathbf{E}^\dagger\mathbf{E} = \mathbf{E}^\dagger\mathbf{F}$. Similarly $\mathbf{EE}^\dagger = \mathbf{FE}^\dagger$ and (e) $\Rightarrow$ (5.5), using (1.6). □

Several additional conditions may be given so that when coupled with (5.6) turn it into (5.5), cf. Hartwig and Styan (1986, Theorem 2).

THEOREM 5.4. *Let* $\mathbf{E}$ *and* $\mathbf{F}$ *be* $n \times n$ *complex idempotent matrices, neither nesessarily Hermitian. Then the following nine conditions are equivalent:*

(a) $\mathbf{E} \leq_{\mathsf{L}} \mathbf{F}$, *i.e.*, $\mathbf{F}-\mathbf{E}$ *is Hermitian nonnegative definite,*

(b) $\mathbf{F}-\mathbf{E} = (\mathbf{F}-\mathbf{E})^2 = (\mathbf{F}-\mathbf{E})^*$,

(c) $\mathbf{I}-\mathbf{F} \leq_{\mathsf{L}} \mathbf{I}-\mathbf{E}$,

(d) $\mathbf{E} \leq_{\mathsf{rs}} \mathbf{F}$ *and* $\mathbf{F}-\mathbf{E} = (\mathbf{F}-\mathbf{E})^*$,

(e) $\mathbf{E} \leq_{\mathsf{rs}} \mathbf{F}$ *and* $\mathbf{F}-\mathbf{E} \leq_{\mathsf{L}} \mathbf{I}$,

(f) $\mathbf{E} \leq_{*} \mathbf{F}$ *and* $\mathbf{I}-\mathbf{F} \leq_{*} \mathbf{I}-\mathbf{E}$,

(g) $\mathbf{F}-\mathbf{E} \leq_{*} \mathbf{I}$,

(h) $\mathbf{F}-\mathbf{E} \leq_{\mathsf{rs}} \mathbf{I}-\mathbf{E}^*$ *and* $\mathbf{F}-\mathbf{E} \leq_{\mathsf{rs}} \mathbf{F}^*$,

(i) $\mathbf{E}^* + \mathbf{F}-\mathbf{E} \leq_{\mathsf{rs}} \mathbf{I}$ *and* $\mathbf{F}-\mathbf{E} \leq_{\mathsf{rs}} \mathbf{F}^*$.

And then the parallel sum of $\mathbf{E}$ *and* $\mathbf{F}$

$$\mathbf{E}(\mathbf{F}-\mathbf{E})^{\dagger}\mathbf{F} = \mathbf{0}. \tag{5.7}$$

Proof. From Lemma 2.4 we see that (a) $\Rightarrow$ (b), while clearly (b) $\Rightarrow$ (a) $\Leftrightarrow$ (c). To see that (d) $\Rightarrow$ (b) $\Rightarrow$ (e) we use Theorem 5.1: (a) $\Leftrightarrow$ (b) and $\mathbf{I}-(\mathbf{F}-\mathbf{E})$ Hermitian idempotent, while (e) $\Rightarrow$ (d) is immediate. That (f) $\Rightarrow$ (g) $\Rightarrow$ (b) $\Rightarrow$ (f) follows directly from (1.2). To see that (f) $\Leftrightarrow$ (h) we use Theorem 5.2: (a) $\Leftrightarrow$ (f) twice. To show that (i) $\Leftrightarrow$ (f) we use Theorem 5.2: (i) $\Leftrightarrow$ (f) *sic*. □

So far our results have covered the situation when neither of the idempotent matrices $\mathbf{E}$ nor $\mathbf{F}$ is necessarily Hermitian. When $\mathbf{E} \leq_{\mathsf{rs}} \mathbf{F}$ and either $\mathbf{E}$ or $\mathbf{F}$ is also Hermitian then no new characterizations are obtained over and beyond those given in Theorem 5.1. When $\mathbf{E} \leq_{\mathsf{L}} \mathbf{F}$ and either $\mathbf{E}$ or $\mathbf{F}$ is also Hermitian then both $\mathbf{E}$ and $\mathbf{F}$ are Hermitian and we defer our consideration of this situation till Theorem 5.7 — when $\mathbf{E}$ and $\mathbf{F}$ are both Hermitian and idempotent then our three partial orders coincide, cf. Theorem 2.3.

We turn now to $\mathbf{E} \leq_{*} \mathbf{F}$ with both $\mathbf{E}$ and $\mathbf{F}$ idempotent and either $\mathbf{E}$ or $\mathbf{F}$ Hermitian; we consider first (Theorem 5.5) what happens when $\mathbf{E}$ is

Hermitian (but **F** is not necessarily so), and then (Theorem 5.6) the situation when **F** is Hermitian (but **E** is not necessarily so).

We need the following:

LEMMA 5.1. *Let* **E** *and* **F** *be* $n \times n$ *complex idempotent matrices and let* **E** *also be Hermitian, but* **F** *not necessarily so. Then the following four conditions are equivalent*:

(a) $\mathbf{E} \leq_{\mathsf{L}} \mathbf{F}\mathbf{F}^*$,

(b) $\mathbf{E} = \mathbf{F}\mathbf{E}$,

(c) $\mathbf{F}\mathbf{F}^* = \mathbf{E} + \mathbf{F}(\mathbf{I}-\mathbf{E})\mathbf{F}^*$,

(d) $\mathbf{F}\mathbf{F}^* = \mathbf{E} + (\mathbf{F}-\mathbf{E})(\mathbf{F}-\mathbf{E})^*$.

Proof. From (2.2) we see that (a) ⇒ (b) and it is easy to show that (b) ⇒ (c) ⇒ (a) and that (b) ⇒ (d) ⇒ (a). □

THEOREM 5.5. *Let* **E** *and* **F** *be* $n \times n$ *complex idempotent matrices and let* **E** *also be Hermitian, but* **F** *not necessarily so. Then*

$$\mathbf{E} \leq_* \mathbf{F} \Leftrightarrow \mathbf{E} \leq_{\mathsf{rs}} \mathbf{F} \Leftrightarrow \mathbf{E} \leq_* \mathbf{F}\mathbf{F}^* \Leftrightarrow \mathbf{E} \leq_{\mathsf{rs}} \mathbf{F}\mathbf{F}^* \tag{5.8}$$

which imply $\mathbf{E} \leq_{\mathsf{L}} \mathbf{F}\mathbf{F}^*$. *Dual results hold with* **F** *replaced by* $\mathbf{F}^*$.

Proof. It suffices to prove that $\mathbf{E} \leq_{\mathsf{rs}} \mathbf{F}\mathbf{F}^* \Rightarrow \mathbf{E} \leq_{\mathsf{rs}} \mathbf{F}$. To do this we use (2.8) and Lemma 5.1(d) to observe that $\mathbf{E} \leq_{\mathsf{rs}} \mathbf{F}\mathbf{F}^* \Rightarrow \mathbf{F}\mathbf{F}^* - \mathbf{E} = (\mathbf{F}-\mathbf{E})(\mathbf{F}-\mathbf{E})^*$; taking ranks of both sides completes the proof. □

In Theorem 5.5 we saw that

$$\mathbf{E} \leq_* \mathbf{F} \Rightarrow \mathbf{E} \leq_{\mathsf{L}} \mathbf{F}\mathbf{F}^*. \tag{5.9}$$

In Theorem 5.6 below we present three equivalent conditions which when coupled with $\mathbf{E} \leq_{\mathsf{L}} \mathbf{F}\mathbf{F}^*$ make it equivalent to $\mathbf{E} \leq_* \mathbf{F}$.

THEOREM 5.6. *Let* **E** *and* **F** *be* $n \times n$ *complex idempotent matrices and let* **E** *also be Hermitian, but* **F** *not necessarily so. Then* $\mathbf{E} \leq_* \mathbf{F}$ *is equivalent to any one of the four statements* (a) – (d) *in Lemma 5.1 coupled with any one of the following three conditions:*

(i) $\mathbf{F}^{\dagger} - \mathbf{E} = (\mathbf{F} - \mathbf{E})^{-}$, *i.e.*, $(\mathbf{F} - \mathbf{E})(\mathbf{F}^{\dagger} - \mathbf{E})(\mathbf{F} - \mathbf{E}) = \mathbf{F} - \mathbf{E}$,

(ii) $\mathbf{E}\mathbf{F}^{\dagger}\mathbf{F} = \mathbf{F}^{\dagger}\mathbf{F}\mathbf{E}$,

(iii) $\mathbf{E}\mathbf{F} = (\mathbf{E}\mathbf{F})^{*}$.

Proof. When $\mathbf{E} \leq_{*} \mathbf{F}$ holds then Theorem 5.5 shows that the four conditions in Lemma 5.1 hold; in addition (i) follows from (1.7). Furthermore when (i) and condition (b) in Lemma 5.1 hold, i.e., $\mathbf{E} = \mathbf{F}\mathbf{E}$, then $\mathbf{E} + \mathbf{E}\mathbf{F}^{\dagger}\mathbf{E} = \mathbf{E}\mathbf{F}^{\dagger}\mathbf{F} + \mathbf{F}\mathbf{F}^{\dagger}\mathbf{E}$; since $\mathbf{F}\mathbf{F}^{\dagger}\mathbf{E} = \mathbf{E}$ it follows that $\mathbf{E}\mathbf{F}^{\dagger}\mathbf{E} = \mathbf{E}\mathbf{F}^{\dagger}\mathbf{F} = \mathbf{E}\mathbf{F}^{\dagger}\mathbf{F}\mathbf{E}$, which is Hermitian and so (ii) holds. When (ii) and condition (b) in Lemma 5.1 hold, then $\mathbf{F}^{\dagger}\mathbf{E} = \mathbf{F}^{\dagger}\mathbf{F}\mathbf{E} = \mathbf{E}\mathbf{F}^{\dagger}\mathbf{F} = \mathbf{E}\mathbf{F}^{\dagger}\mathbf{E} = \mathbf{E}\mathbf{F}^{\dagger}\mathbf{F}\mathbf{E}$, which is Hermitian; thus $\mathbf{E}\mathbf{F} = \mathbf{E}\mathbf{F}^{*}\mathbf{F} = \mathbf{E}\mathbf{F}^{\dagger}\mathbf{F}\mathbf{F}^{*}\mathbf{F} = \mathbf{F}^{\dagger}\mathbf{E}\mathbf{F}^{*}\mathbf{F} = \mathbf{F}^{\dagger}\mathbf{E}\mathbf{F} = \mathbf{E}\mathbf{F}^{\dagger}\mathbf{F} = \mathbf{E}\mathbf{F}^{\dagger}\mathbf{F}\mathbf{E}$, which is Hermitian and so (iii) holds. When (iii) and condition (b) in Lemma 5.1 hold then $(\mathbf{E}\mathbf{F} - \mathbf{E})^{*}(\mathbf{E}\mathbf{F} - \mathbf{E}) = (\mathbf{E}\mathbf{F} - \mathbf{E})^{2} = \mathbf{0}$, and hence $\mathbf{E}\mathbf{F} = \mathbf{E} = \mathbf{F}\mathbf{E}$ and so by (4.2) it follows that $\mathbf{E} \leq_{*} \mathbf{F}$. □

THEOREM 5.7. *Let* $\mathbf{E}$ *and* $\mathbf{F}$ *be* $n \times n$ *complex idempotent matrices and let* $\mathbf{F}$ *also be Hermitian, but* $\mathbf{E}$ *not necessarily so. Then the following seven conditions are equivalent:*

(a) $\mathbf{E} \leq_{*} \mathbf{F}$,

(b) $\mathbf{E}\mathbf{E}^{*} \leq_{*} \mathbf{F}$,

(c) $\mathbf{E}\mathbf{E}^{*} \leq_{\mathrm{rs}} \mathbf{F}$,

(d) $\mathbf{E}\mathbf{E}^{*} \leq_{\mathrm{L}} \mathbf{F}$,

(e) $\mathbf{E}\mathbf{E}^{*} \leq_{\mathrm{rs}} \mathbf{E}\mathbf{F}$,

(f) $\mathbf{E}\mathbf{E}^{*} = \mathbf{E}\mathbf{F}$,

(g) $\mathbf{E}\mathbf{E}^{*}\mathbf{E} = \mathbf{E}\mathbf{F}$.

And then $\mathbf{E} = \mathbf{E}^{*}$. *Dual results holds with* $\mathbf{E}$ *replaced by* $\mathbf{E}^{*}$.

Proof. We only prove that (g) $\Rightarrow$ (a). When $\mathbf{E}\mathbf{E}^{*}\mathbf{E} = \mathbf{E}\mathbf{F}$ then $\mathbf{E} = \mathbf{E}\mathbf{F}$ and $\mathbf{E}\mathbf{E}^{*}\mathbf{E} = \mathbf{E}$; hence $(\mathbf{E} - \mathbf{E}^{*})^{3} = \mathbf{0}$ and thus $\mathbf{E} = \mathbf{E}^{*}$. □

In addition to the six conditions (b) – (g) in Theorem 5.7 that are each equivalent to $\mathbf{E} \leq_{*} \mathbf{F}$, when both $\mathbf{E}$ and $\mathbf{F}$ are idempotent and $\mathbf{F}$ is Hermitian, we also note that

$$\mathbf{E} \leq_* \mathbf{F} \;\Leftrightarrow\; \mathbf{E} = \mathbf{EF} \;\; \textit{and} \;\; \mathbf{F} - \mathbf{E}^\dagger = (\mathbf{F} - \mathbf{E})^-, \tag{5.10}$$

cf. Theorem 5.6 (i). It is straightforward to show that establishing (5.10) reduces to proving that when $\mathbf{E}$ is idempotent and

$$\mathbf{E} + \mathbf{E}^\dagger = \mathbf{E}\mathbf{E}^\dagger + \mathbf{E}^\dagger\mathbf{E} \tag{5.11}$$

then $\mathbf{E}$ is Hermitian. We may do this by taking traces throughout (5.11) to yield

$$\mathrm{tr}\mathbf{E}^\dagger = \mathrm{rank}(\mathbf{E}). \tag{5.12}$$

We note that in general for an idempotent matrix $\mathbf{E}$

$$\mathrm{tr}\mathbf{E}^\dagger \leq \mathrm{rank}\,(\mathbf{E}) \tag{5.13}$$

with equality if and only if $\mathbf{E} = \mathbf{E}^*$ and/or $\mathbf{E} = \mathbf{E}^\dagger$.

We end this paper by considering the situation when the matrices $\mathbf{E}$ and $\mathbf{F}$ are both both idempotent and Hermitian (orthogonal projectors), cf. Theorem 2.3.

THEOREM 5.8. *Let* $\mathbf{E}$ *and* $\mathbf{F}$ *be* $n \times n$ *complex idempotent Hermitian matrices (orthogonal projectors). Then the following nine conditions are equivalent:*

(a) $\mathbf{E} \leq_{\mathsf{L}} \mathbf{F}$,

(b) $\mathbf{E} \leq_* \mathbf{F}$,

(c) $\mathbf{E} \leq_{\mathsf{rs}} \mathbf{F}$,

(d) $\mathbf{F} = \mathbf{E}^-$, *i.e.*, $\mathbf{E} = \mathbf{EFE}$,

(e) $\mathbf{F} - \mathbf{E} \leq_* \mathbf{I}$,

(f) $\mathbf{E} \leq_{\mathsf{rs}} \mathbf{EF}$,

(g) $\mathbf{I} - \mathbf{F} \leq_{\mathsf{L}} \mathbf{I} - \mathbf{E}$,

(h) $\mathbf{I} - \mathbf{F} \leq_* \mathbf{I} - \mathbf{E}$,

(i) $\mathbf{I} - \mathbf{F} \leq_{\mathsf{rs}} \mathbf{I} - \mathbf{E}$.

We note that the condition (a) in Theorem 5.8 is equivalent to the idempotency of $\mathbf{F} - \mathbf{E}$ and so the equivalence of (e) and (f) is closely related to Theorem 5.1.3 in Rao and Mitra (1971, p. 108); see also Baksalary (1987).

ACKNOWLEDGEMENTS

This research was begun when the authors met at the Auburn Matrix Theory Conference, held at Auburn, Alabama, in March 1980. We are extremely grateful to Jerzy K. Baksalary for many very helpful comments and suggestions. Thanks also go to Friedrich Pukelsheim for useful discussions. This research was supported in part by the Academy of Finland, the Natural Sciences and Engineering Research Council of Canada and by the Fonds pour la Formation de Chercheurs et l'Aide à la Recherche du Gouvernement du Québec.

REFERENCES

Anderson, T.W. and Styan, George P.H. (1982). Cochran's theorem, rank additivity and tripotent matrices. *Statistics and Probability: Essays in Honor of C. R. Rao* (G. Kallianpur, P.R. Krishnaiah and J.K. Ghosh, *eds.*), North-Holland, Amsterdam, 1-23.

Baksalary, Jerzy K. (1986). A relationship between the star and minus orderings. *Linear Algebra and Its Applications*, 82, 163-167.

Baksalary, Jerzy K. (1987). Algebraic characterizations and statistical implications of the commutativity of orthogonal projectors. *Proc. Second International Tampere Conference in Statistics* (Tarmo Pukkila and Simo Puntanen, *eds.*), Dept. of Mathematical Sciences, University of Tampere, Finland, 113-142.

Baksalary, Jerzy K. and Hauke, Jan (1984). Inheriting independence and chi-squaredness under certain matrix orderings. *Statistics & Probability Letters*, 2, 35-38.

Baksalary, Jerzy K. and Hauke, Jan (1987). Partial orderings of matrices referring to singular values or eigenvalues. *Linear Algebra and Its Applications*, 96, 17-26.

Baksalary, Jerzy K. and Kala, Radoslav (1980). On the difference between two second degree polynomials, each following a chi-square distribution. *Sankhyā Ser. A*, 42, 123-127.

Baksalary, Jerzy K., Kala, Radoslav and Klaczyński, K. (1983). The matrix inequality $\mathbf{M} \geq \mathbf{B}^*\mathbf{M}\mathbf{B}$. *Linear Algebra and Its Applications*, 54, 77-86.

Baksalary, Jerzy K., Pukelsheim, Friedrich and Styan, George P. H. (1987). Some properties of matrix partial orderings. *Linear Algebra and Its Applications*, to appear.

Carlson, David (1987). Generalized inverse invariance, partial orders, and rank-minimization problems for matrices. *Current Trends in Matrix Theory* (F. Uhlig and R. Grone, *eds.*), Elsevier Science Publishing, New York.

Chipman, John S. and Rao, M.M. (1964). Projections, generalized inverses, and quadratic forms. *Journal of Mathematical Analysis and Applications*, 9, 1-11.

Clifford, A.H. and Preston, G.B. (1961). *The Algebraic Theory of Semigroups, Volume I.* Mathematical Surveys, No. 7, American Mathematical Society, Providence, R.I.

Cline, R.E. and Funderlic, R.E. (1979). The rank of a difference of matrices and associated generalized inverses. *Linear Algebra and its Applications*, 24, 185-215.

Cochran, William G. (1934). The distribution of quadratic forms in a normal system, with applications to the analysis of covariance. *Proceedings of the Cambridge Philosophical Society*, 30, 178-191.

Drazin, Michael P. (1977). Natural structures on rings and semigroups with involution. Unpublished manuscript.

Drazin, Michael P. (1978). Natural structures on semigroups with involution. *Bulletin of the American Mathematical Society*, 84, 139-141.

Gaffke, N. and Krafft, O. (1982). Matrix inequalities in the Löwner ordering. *Modern Applied Mathematics – Optimization and Operations Research* (B. Korte, *ed.*), 596-622.

Hartwig, Robert E. (1978). A note on the partial ordering of positive semi-definite matrices. *Linear and Multilinear Algebra*, 6, 223-226.

Hartwig, Robert E. (1980). How to partially order regular elements. *Mathematica Japonica*, 25, 1-13.

Hartwig, Robert E. and Spindelböck, Klaus (1983). Some closed form formulae for the intersection of two special matrices under the star order. *Linear and Multilinear Algebra*, 13, 323-331.

Hartwig, Robert E. and Styan, George P.H. (1986). On some characterizations of the "star" partial ordering for matrices and rank subtractivity. *Linear Algebra and Its Applications*, 82, 145-161.

Hestenes, Magnus R. (1961). Relative Hermitian matrices. *Pacific Journal of Mathematics*, 11, 225-245.

Hogg, Robert V. and Craig, Allen T. (1958). On the decomposition of certain χ^2 variables. *Annals of Mathematical Statistics*, 29, 608-610.

Horn, Roger A., and Johnson, Charles R. (1985). *Matrix Analysis*. Cambridge University Press.

James, G.S. (1952). Notes on a theorem of Cochran. *Proceedings of the Cambridge Philosophical Society*, 48, 443-446.

Löwner, Karl (1934). Über monotone Matrixfunktionen. *Mathematische Zeitschrift*, 38, 177-216.

Mäkeläinen, Timo and Styan, George P. H. (1976). A decomposition of an idempotent matrix where nonnegativity implies idempotence and none of the matrices need be symmetric. *Sankhyā, Ser. A*, 38, 400-403.

Marsaglia, George and Styan, George P.H. (1974). Equalities and inequalities for ranks of matrices. *Linear and Multilinear Algebra*, 2, 269-292.

Marshall, Albert W. and Olkin, Ingram (1979). *Inequalities: Theory of Majorization and Its Applications*. Academic Press, New York.

Mitra, Sujit Kumar (1986). The minus partial order and the shorted matrix. *Linear Algebra and Its Applications*, 83, 1-27.

Ogasawara, Tōzirō and Takahashi, Masayuki (1951). Independence of quadratic quantities in a normal system. *Journal of Science of the Hiroshima University, Ser. A*, 15, 1-9.

Rao, C. Radhakrishna and Mitra, Sujit Kumar (1971). *Generalized Inverse of Matrices and Its Applications*. Wiley, New York.

Received 23 August 1986
Revised 15 October 1987

Department of Mathematics
North Carolina State University
P. O. Box 8205
Raleigh, NC 27695-8205
U.S.A.

Department of Mathematics and Statistics
McGill University, Burnside Hall
805 ouest, rue Sherbrooke Street West
Montréal, Québec
Canada H3A 2K6

Proc. Second International Tampere Conference in Statistics
(Tampere, Finland, 1-4 June 1987)
Tarmo Pukkila and Simo Puntanen, *Editors*

pp. 385 - 392

Fortran8x as a language for statistical computation

Alan J.B. ANDERSON
The University of Aberdeen, Scotland

Key words and phrases: Preprocessing, flexible data analysis, software design.

ABSTRACT

Although languages for computational statistics can be developed *ab initio*, this paper argues that it is more effective to extend one of the well-known high-level languages. Now is an opportune time to reconsider the appropriateness of Fortran as best choice for this in view of the imminent appearance of the next version of the language – Fortran8x. In particular, the new dialect is considered in relation to the use of preprocessing as a means of language extension.

1. INTRODUCTION

Two somewhat opposing schools of thought have considered the problem of providing software for statistical analysis. The first has spawned large packages, usually directive-driven; at the other extreme, pleas for the extension of Fortran to include statistical features have been made by Anderson (1980, 1984). The arguments for and against each of these approaches can be briefly rehearsed as follows.

A purpose-built statistical package such as SPSS concentrates on essentials and is therefore easy to use. It is, moreover, eminently suitable for conversational working because its operation is basically task-by-task. Both these strengths are enhanced by the fact that few operands need be identifiably declared since tasks are more or less self-contained. However, the price to be paid for this apparent ease of use is inefficiency. On the one hand, the fact that the software must be compiled once and for all implies a rigid space allocation that is wasteful for the majority of jobs yet occasionally prohibitively inadequate. On the other hand, some types of analysis – particularly the larger surveys – can suffer time penalties by the overheads of interpretive working. But, above all, the standard packages impose such limitations on flexibility that the more skilful user is frequently frustrated in any attempt to achieve the most appropriate analysis of the data under consideration. (Although the population of "users" is not polarised in a one-dimensional space labelled "novice-skilled", these terms provide a conceptually useful shorthand.)

With a high-level language, on the other hand, the core demanded by a job is exactly what the analysis requires, the compiled code is dedicated entirely to the task in hand and, most importantly, the user is relatively unconstrained as to the nature of possible analyses. Superficially, it might appear that this flexibility is achieved at the expense of ease of use since statistical manipulations have to be grafted on to the high-level language in ways that are not always elegant. At the simplest level, a subroutine library might be provided (Healy, 1981); such was the case with the Rothamsted General Survey Program (Yates, 1973, 1975). A much more appealing mechanism from the cosmetic point of view derives from the interposition of a preprocessor to handle language extensions (Anderson, 1984). It is the purpose of this paper to examine in the statistical context the facilities added to the latest version of Fortran – Fortran8x – the release of which is imminent (ANSI, 1987).

Since the flavour of Fortran is unpalatable to some, it may be worthwhile to begin with a review of potential competitors as host. We discard APL since, although it has some appeal as a language for experimentation with novel statistical ideas, it has little to commend it for large-scale data analysis. Indeed, we assume the choice to be limited to those languages familiar to a large number of possible users. The other obvious candidates are therefore Pascal and ANSI Basic. We compare all three first in terms of the kernel of features required in data analysis which the three languages already provide and then in relation to specific advantages and disadvantages.

2. CHOICE OF HOST LANGUAGE

An important feature of statistical analysis is the volume of data and the variety of formats in which data can be presented. Of the three contenders, Pascal is the weakest since even the syntax of its "read" (and "write") procedures is non-standard. However, the Pascal concept of a record could be useful particularly for hierarchical surveys. Fortran77 and ANSI Basic are fairly dissimilar in what they provide for input formatting, with Basic accepting only free-format data whereas the

Fortran emphasis is on fixed. It must be accepted that, for any high-level language, an input protocol must be evolved from scratch for statistical work. Some statistical software includes database facilities but the main database systems are hosted in Fortran or Cobol, supporting the suggestion that analysis software be similarly hosted (Anderson, 1982).

The syntax of expressions (and hence of assignment) is reasonably consistent across all high-level languages and statistical software. Anderson (1984) has argued that entities representing statistical variates must be added to the basic lists of operand types supported by the three languages under review and noted difficulties in the evaluation of expressions involving variates due to the missing data problem.

Program control statements include facilities for looping, (conditional) branching, procedure or macro calls etc. The first of these is of considerable importance since statistical analysis is particularly dependent on the provision of good repetition commands. Fortran, Pascal and Basic are all adequate in this respect though lacking the sophistication of Genstat which allows parallel cycling such as

```
'FOR' dummy1 = list1 [; dummy2 = list2; ...]
:
:
'REPEAT'
```

where the list(s) can be of scalar constants or identifiers of any kind. The form of loop command underlines the importance of list specification in statistical analysis and the attendant semantic difficulties of identifier "order" in syntactical shorthand such as in BMDP's "tabbing" facility or in the "VARZ, VARA TO VARD" and hence "VARZ TO VARC" forms of SPSSX.

Pascal has good program control, some potentially useful features (types, sets, records) and is widely available on micros. Against this, block structure and nested programming are neither useful nor helpful to novices and the language is excessively declarative. The language is not well standardised and portability seems likely to deteriorate. Furthermore, little is available in procedure libraries.

Basic is, like Pascal, deficient in procedure libraries. Moreover, the ANSI dialect is not (yet) widely available and other dialects are inferior. However, the language is very widely known (at least in some version), is reasonably modular, supports MAT operations and (in the ANSI dialect at anyrate) possesses good program control features.

Of the three languages, Fortran is certainly the most modular and, because of this, many useful subroutine libraries (NAG, GINO, IMSL, database systems, etc.) have been created. One result is the ease with which the latest statistical techniques can be explored and novel analyses made. (Of course, it is to some extent feasible nowadays to link Fortran libraries to programs in other languages.) The arrival of Fortran77 eliminated many of the blemishes inherent in a pioneering design; it is the contention of this paper that Fortran8x demonstrates very much greater advances and the provision of a development path for future versions ensures continuing active commitment to the language.

The above comments ignore the possibility of eliminating deficiencies in any of the languages by some preprocessing mechanism. Whichever might be chosen as host or indeed, if a hybrid or completely new language were to be preferred, Fortran comes out top as target, being uniquely able to access widely-used subroutine libraries. We now go on to examine the facilities offered in Fortran8x, considering in particular how these will simplify any preprocessing that might still be felt necessary.

3. FORTRAN8X

The new facilities in Fortran8x are far too numerous to list here. If we mention only the provision of extra control constructs (e.g. SELECT CASE), the rationalisation of source code format and the considerable enhancement of input/output, the reader will appreciate the variety of improvements. However, the main additional features can be summarised under four headings.

3.1 Language evolution

The problem of a long-standing language like Fortran has been the stultifying requirement of upwards compatibility for each new version. A scheme has now been evolved that allows obsolescent or deprecated features to be flagged as intended for removal at the subsequent revision. This should greatly assist the continuing development of Fortran and gives it a secure future.

3.2 Numerical computations

Many portability problems have now been considerably lessened by introducing control over numeric precision. In particular, a number of inquiry functions are now available to enable the programmer to ascertain properties of the machine on which the program is running.

3.3 Array operations

Fortran8x incorporates operations for processing whole arrays and array sections, MAT(7:10,:) representing the matrix consisting of rows 7 to 10 of MAT. Arrays can have their space requirement allocated and de-allocated dynamically and parts of arrays can be dynamically aliassed to separate identifiers e.g.

```
ALLOCATE (table(1:nrows+1,1:ncols+1))
IDENTIFY (body(I,J) = table(I,J), I=1:nrows,J=1:ncols)
IDENTIFY (row_margin(I) = table(I,ncols+1), I=1:nrows)
```

All these features have been designed to take advantage of parallel processing computer architecture when such is available; this will be of enormous benefit in simulation studies. Functions, intrinsic or user-defined can be array-valued; for example, if X is a vector, then SUM(X) is the sum of its elements; if T is a two-dimensional array, then SUM(T,DIM=2) is the vector of row totals. Thus we might have

```
row_margin=SUM(body,DIM=2)
```

The corrected sum of squares for rows would similarly be obtained as

```
SUM((SUM(body,DIM=2)/ncols-SUM(body)/(nrows*ncols))**2)
```

A "shift" function provides a mean of producing lagged variables when the whole data matrix can be held in core. Or, for example, the vector of K-period moving averages derived from a data vector, OBS, could be calculated by the somewhat revolting looking code

```
SUM(EOSHIFT(SPREAD(OBS,1,K),1,[0:1-K:-1])(K:,:),2)/K
```

If "total_rent" and "total_sq_metres" are tables (i.e. arrays) summarising information on business properties categorised in some way, we might have

```
WHERE(total_sq_metres > 0.0)
    rent_per_sq_metre = total_rent/total_sq_metres
ELSEWHERE
    rent_per_sq_metre = -1.0
END WHERE
```

This computes a table of average rents per square metre except for any categories in which no properties are found where the cells are set to -1.0.

Vector constants such as	[6,2[11:9:-1],4:7]
i.e.	[6 11 10 9 11 10 9 4 5 6 7]

go a considerable way towards satisfying the statistical need for lists. Unfortunately, a proposal to permit arrays to be indexed by vectors has been dropped; it *would* have been useful to be able to define a subset of variates in the data matrix by, for example, data__matrix([1:7,10,16])

or to reorder the rows of a table by this means.

3.4 Modules and derived types

These two features are considered together because of their interrelationship but the concept of modules is one of greater generality and usefulness than is indicated here. Anderson (1984) has stressed the need in statistical analysis for entities of type "variate" capable of correctly handling missing values. When the data matrix can be held in core, variates can be considered as complete vectors; otherwise a variate refers only to the current unit of a matrix that is being scanned row-by-row. Fortran8x provides a ready means for the introduction of new operand types, defining operators for those types and incorporating such definitions in library modules. This enhancement alone will be very valuable.

Fig. 1 shows a module that defines what is meant by scalar type "variate" for a single (current) unit of data and how the binary subtraction operator is to be applied to two variates; obviously, the full set of unary and binary operators would be similarly incorporated. The example should make the mechanism reasonably clear:

- the derived type "variate" is defined to consist of two basic components, the second being a logical flag to indicate whether or not the first – the value of the variate – is known.
- the function MINUS with two variate arguments defines the operation of subtraction.
- the suffix "OPERATOR(–)" means that if the compiler finds the binary operator "–" between two variates in an expression, it is as though the MINUS function were invoked with those variates as arguments.
- the whole will be linked to the compiler by the directive

  USE VAR

  after which there might be, for instance, the declaration

  TYPE (VARIATE) age__of__father, age__of__mother, age__difference

  An assignment such as

  age__difference = age__of__father – age__of__mother

  will then be correctly executed, the variate result being unknown if either of the right-hand-side variates is unknown.

FIGURE 1

Module to define type "variate" and variate subtraction.

```
MODULE VAR                                              !module is called "VAR"
   TYPE VARIATE
      REAL VALUE                                        !value of variate if known
      LOGICAL MISSING                                   !missing value indicator flag
   END TYPE
   TYPE (VARIATE) FUNCTION MINUS (V1,V2) OPERATOR(-)
      TYPE (VARIATE) V1,V2
      IF(V1%MISSING.OR.V2%MISSING) THEN
         MINUS%MISSING = .TRUE.                         !flag result unknown
      ELSE
         MINUS%VALUE = V1%VALUE - V2%VALUE              !value of difference
         MINUS%MISSING = .FALSE.                        !flag result known
      END IF
      RETURN
   END FUNCTION
END MODULE
```

It will be obvious how type variate can be defined to accomodate further attributes such as range of permitted values or, for categorical variates, names of levels. The reader may also care to consider the trivial amendments necessary to operate on whole columns of the data matrix rather than just the current unit.

4. CONCLUSIONS

Clearly, module libraries could be built up making available a repertoire of such facilities. Statisticians could readily define their own personal versions of statistical constructs (though diversity of this kind with no recognised standard might be another recipe for chaos). However, although these enhancements to Fortran (similar to features in ADA) will be of great use in *ad hoc* computing, it cannot be said that the result amounts to a "statistical language". This is principally because the main constituents of such a language must be procedures and the Fortran8x syntax remains a little less than ideal in this respect. The reader may well be repelled by the verbosity of the array-handling features even if impressed by their power. Likewise, although keywords are now permitted for arguments in procedure calls, the CALL syntax remains and, as a result, task requests are still clumsy by comparison with the various statistical packages.

Therefore it seems that, even with the advent of Fortran8x, the need for preprocessing will remain in order to achieve a syntax sufficiently smoothed for the naive user yet maintaining the flexibility necessary for serious analyses. In this way, it should be possible to stimulate increased collaboration between scientist and professional statistician by providing a software environment within which analysis could be shared using a single system equally friendly to each.

The design and construction of such software will benefit enormously from the arrival of Fortran8x in ways that the above examples only hint at. The key features of the new dialect that will be of greatest use in preprocessing into Fortran (whether from extended Fortran or from an independent language) are the array-handling facilities, the powerful module constructs and, most importantly, the availability of derived types linked with their own operands. Unhappily, an event-handling capability that was to have been included has now been dropped so that run-time diagnostics will remain a problem.

REFERENCES

Anderson, A.J.B. (1980). The use of preprocessors in statistical software. *COMPSTAT 1980*. Physica-Verlag, Vienna, 490-496.

Anderson, A.J.B. (1982). Software to link database interrogation and statistical analysis. *COMPSTAT 1982*. Physica-Verlag, Vienna, 139-144.

Anderson, A.J.B.(1984). STATFOR: An extended Fortran for statistical analysis. *Proceedings of the III International Symposiun on Data Analysis and Informatics*. North Holland, Amsterdam, 491-501.

ANSI (1987). Fortran. Draft S8, version 103. Committee X3J3 of American National Standards Institute.

Healy, M.J.R. (1981). A Subroutine Library for Statisticians. *BIAS*, 8, 12-25.

Yates, F. (1973). The analysis of surveys on computers – features of the Rothamsted General Survey Program. *Applied Statistics*, 22, 161-171.

Yates, F. (1975). The design of computer programs for survey analysis – a contrast between the Rothamsted General Survey Program (RGSP) and SPSS. *Biometrics*, 31, 573-584.

Received 31 December 1986
Revised 1 July 1987

Department of Statistics
University of Aberdeen
King's College
Aberdeen AB9 2UB
Scotland

Proc. Second International Tampere Conference in Statistics
(Tampere, Finland, 1-4 June 1987)
Tarmo Pukkila and Simo Puntanen, *Editors*

pp. 393 - 401

On regression analysis under the intrinsic inference model for stratified dependent normal random variables

Bruno BALDESSARI

Dipartimento di Statistica, Probabilità e Statistica Applicata,
Università degli Studi "La Sapienza", Rome, Italy

Key words and phrases: Mixtures, quadratic forms, Fisher F-distributions.

ABSTRACT

In a Regression Analysis problem with one response and p predictors, a stratified random sample is drawn without replacement from a finite set of stratified dependent normal random variables. We determine necessary and sufficient conditions on the covariance matrix of the normal random variables, under which the numerator and the denominator of the F-statistic used to test the hypothesis of zero regression coefficients are independent and have non-central chi-square distributions.

The same conditions are sufficient for the F-statistics that are used to test equality to zero of any set of linear forms of the regression coefficients to have Fisher F-distributions.

1. INTRODUCTION

The purpose of this paper is to determine a general form of the covariance matrix of a finite set of dependent normal random variables that permits a Regression Analysis under the "intrinsic inference model for stratified dependent normal random variables". The model assumes that:

(i) the population of interest in a given statistical study is finite and may be represented by the random N vector $\mathbf{Z} = (Z_1,..., Z_N)'$ of the dependent (population) response random variables $Z_1,..., Z_n$;

(ii) $\mathbf{Z} \sim N(\boldsymbol{\mu}, \boldsymbol{\Sigma})$, i.e., $\mathbf{Z}$ is normally distributed with unknown mean vector $\boldsymbol{\mu}$ and covariance matrix $\boldsymbol{\Sigma}$, where $\boldsymbol{\Sigma}$ is positive definite, and $\mathbf{Z}$ is stratified in k sub-vectors $\mathbf{Z}_{(1)},..., \mathbf{Z}_{(k)}$ of dimensions $N_1,..., N_k$, respectively, with $N_1 + ... + N_k = N$;

(iii) a sample $\mathbf{Y} = (Y_1,..., Y_n)'$, i.e., a random n vector of dependent (sample) response random variables $Y_1,..., Y_n$ is selected from $Z_1,..., Z_N$ without replacement and with selection probabilities $p(\mathbf{Y})$ which assign fixed sample sizes $n_1,..., n_k$, say, to each stratum $\mathbf{Z}_{(1)},..., \mathbf{Z}_{(k)}$;

(iv) the random variables $Y_1,..., Y_n$ are realized, in the observed values $y_1,..., y_n$, independently of the selection of the random variables $Y_1,..., Y_n$.

Justifications for the assumption of dependence of the random variables $Z_1,..., Z_N$ can be found in Baldessari (1985), Baldessari and Weber (1985), and, Weber and Baldessari (1986).

To study the Regression Analysis procedures under the above intrinsic inference model for stratifield dependent normal random variables, we assume that there exists a "population regression model"

$$\mathbf{Z} = \mathbf{W}\boldsymbol{\beta} + \boldsymbol{\varepsilon}, \tag{1.1}$$

where: $\boldsymbol{\varepsilon} \sim N(\mathbf{0}, \boldsymbol{\Sigma})$; $\boldsymbol{\beta} = (\beta_1,..., \beta_{p+1})'$ is a $p + 1$ vector of unknown constants; and $\mathbf{W} = (\mathbf{1}_N, \mathbf{W}_1)$, with $\boldsymbol{\mu} = \mathbf{W}\boldsymbol{\beta}$, is a $N \times (p + 1)$ matrix with elements one in the first column ($\mathbf{1}_N$, say) and elements $x_{jh} - \bar{x}_h$ (with $\bar{x}_h = N^{-1}\Sigma x_{jh}$, $h = 1,..., p$) in the remaining columns. We assume, also, that the predictors are under control of the investigator (so that $\mathbf{W}$ is known) and that $|\mathbf{W}\mathbf{W}'| > 0$. Consequently, $\mathbf{Z} \sim N(\mathbf{W}\boldsymbol{\beta}, \boldsymbol{\Sigma})$. Moreover, we assume that the units of the same stratum have identical values of the predictors and so that the same variables are used both as predictors and stratification variables.

In section 2, we prove that, under the intrinsic inference model for stratified dependent normal random variables, (1.1) implies that there exists a "sample regression model"

$$\mathbf{Y} = \mathbf{X}\boldsymbol{\beta} + \mathbf{e}, \tag{1.2}$$

where: $\mathbf{X} = (\mathbf{1}_n, \mathbf{X}_1)$ is an $n \times (p+1)$ sub-matrix of $\mathbf{W}$ which is common to all samples and, for $n > p + 1$, $\mathbf{X}$ is such that $|\mathbf{XX}'| > 0$; and, $\mathbf{e}$ is distributed as a mixture of the n component normal marginal distributions of $(\varepsilon_1,..., \varepsilon_N)'$. Consequently, $\mathbf{Y}$ is distributed as a mixture of the n component marginal distributions of $(Z_1,..., Z_N)$.

We, then, consider the F-statistic, given in equation (3.1), for testing H_0: $\beta_2 = ... = \beta_{p+1} = 0$ in the models (1.1) and (1.2). Let $\boldsymbol{\Sigma}$ be defined by

$$\boldsymbol{\Sigma} = a\mathbf{I}_N + \boldsymbol{\Gamma} + \boldsymbol{\Gamma}', \tag{1.3}$$

where $a > 0$, $\boldsymbol{\Gamma} = \mathbf{1}_N(\gamma_1,..., \gamma_N) = \mathbf{1}_N\boldsymbol{\gamma}'$, say, with $\gamma_1,..., \gamma_N$ arbitrary.

In section 3, we prove that, under the intrinsic inference model for stratified dependent normal random variables, with simple random sampling in each stratum, (1.3) is a necessary and sufficient condition for:

(i) the numerator and the denominator of the F-statistic (3.1) are independent;

(ii) the denominator has the distribution of a central chi-square random variable times a; and,

(iii) the numerator has the distribution of a non-central chi-square random variable times a.

It follows that, if equation (1.3) in true, the F-statistic for testing H_0: $\beta_2 = ... = \beta_{p+1} = 0$ has the exact Fisher F-distribution.

We consider, also, the F-statistics for testing the hypotheses H_0: $\mathbf{A}\boldsymbol{\beta} = \mathbf{0}$, where $\mathbf{A}$ is any $q \times (p+1)$ matrix of rank q with elements equal to zero in the first column. These statistics are given in equation (3.10). In section 3, we prove that (1.3) is sufficient for the F-statistics to have the exact Fisher F-distributions. Consequently, under the intrinsic inference model for stratified dependent normal random variables (with simple random sampling from each stratum where the predictor variables are constant) if the dependence of $Z_1,..., Z_N$ is of the type (1.3), any Regression Analysis can be performed on the $Y_1,..., Y_n$.

2. REGRESSION ANALYSIS UNDER THE INTRINSIC INFERENCE MODEL FOR STRATIFIED DEPENDENT NORMAL RANDOM VARIABLES

We prove that, under the intrinsic inference model for stratified normal random variables, the sample regression model (1.2) is true.

Let $\mathbf{S}$, ($\mathbf{S} = \mathbf{S}[j_1,...,j_n]$), be the $n \times N$ matrix with elements equal to one in the cases $(1, j_1),..., (n, j_n)$, where $\{j_1,..., j_n\} \subset \{1,..., N\}$, and elements equal to zero otherwise. Denote by $\mathbf{Y}_s$ the conditional random vector $(\mathbf{Y}|Y_1 = Z_{j1},\ldots, Y_n = Z_{jn})$, and we have $\mathbf{Y}_s = \mathbf{SZ}$. Also, let $\mathcal{S}$ be the set of the $\mathbf{S}$ matrices such that $p(\mathbf{S}) = p(\mathbf{Y}_s) > 0$. Model (1.2) exists in the sense that, for any given stratified design and for every $\mathbf{S} \in \mathcal{S}$, $\mathbf{Y}_s$ satisfies the "sample regression models"

$$\mathbf{Y}_s = \mathbf{X}\boldsymbol{\beta} + \mathbf{e}_s, \tag{2.1}$$

where $\mathbf{X} = \mathbf{SW}$ and $\mathbf{e}_s = \mathbf{S}\boldsymbol{\varepsilon}$. This follows from (1.1), by $\mathbf{Y}_s = \mathbf{SZ} = \mathbf{SW}\boldsymbol{\beta} + \mathbf{S}\boldsymbol{\varepsilon}$. Since the predictors are constant and the sample sizes are fixed in each population stratum, we have exactly n_1 equal predictors from the first stratum, ..., n_k equal predictors from the k-th stratum. Therefore, for every $\mathbf{S} \in \mathcal{S}$, the $n\times p$ matrix $\mathbf{X}= \mathbf{SW}$ and the distribution of $\mathbf{e}_s$ are constant with respect to $\mathbf{S}$. Consequently, for every $\mathbf{S} \in \mathcal{S}$, the model $\mathbf{Y}_s = \mathbf{X}\boldsymbol{\beta} + \mathbf{e}_s$ is constant and the model (1.2) is true.

Also, note that, since $n > p + 1$ the $n\times(p + 1)$ matrix $\mathbf{X} = (\mathbf{1}_n, \mathbf{X}_1)$ is of full rank. In fact, for every $\mathbf{S} \in \mathcal{S}$, $\mathbf{X}'\mathbf{X} = \mathbf{W}'\mathbf{S}'\mathbf{SW}$ is the matrix composed of the elements $t_{j_u j_v}$, $u, v = 1,\ldots, n$ of $\mathbf{T} = \mathbf{W}'\mathbf{W}$. Thus, since $n > p + 1$, $\mathbf{W}$ is of full rank, $\mathbf{T}$ is positive definite, $\mathbf{X}'\mathbf{X}$ is positive definite and, so, $\mathbf{X}$ is of full rank.

Finally, we consider the sample distribution of $\mathbf{Y}$. By simple conditioning,

$$\mathbf{e} \sim \Sigma_s p(\mathbf{S})N(\mathbf{0}, \mathbf{S}\boldsymbol{\Sigma}\mathbf{S}'), \tag{2.2}$$

that is, $\mathbf{e}$ is distributed as the mixture, over the matrices $\mathbf{S} \in \mathcal{S}$, with weights $p(\mathbf{S})$, of the stated normal random variables. Consequently,

$$\mathbf{Y} \sim \Sigma_s p(\mathbf{S})N(\mathbf{X}\boldsymbol{\beta}, \mathbf{S}\boldsymbol{\Sigma}\mathbf{S}'), \tag{2.3}$$

and we study the distributions of the F-statistics of Regression Analysis in the next section.

3. FISHER F-DISTRIBUTIONS OF THE F-STATISTICS OF REGRESSION ANALYSIS UNDER THE INTRINSIC INFERENCE MODEL FOR STRATIFIED DEPENDENT NORMAL RANDOM VARIABLES

Suppose that, having selected n observations under the intrinsic inference model for stratified dependent normal random varables, the classical F-statistic for testing H_0: $\beta_2 = \ldots = \beta_{p+1} = 0$ is used. This statistic is (see, for example, Tranquilli and Baldessari, 1987)

$$F = p^{-1}(n-p-1)[\mathbf{Y}'(\mathbf{P}-n^{-1}\mathbf{U})\mathbf{Y}][\mathbf{Y}'(\mathbf{I}_n-\mathbf{P})\mathbf{Y}]^{-1}, \tag{3.1}$$

where $\mathbf{P} = \mathbf{X}(\mathbf{X}'\mathbf{X})^{-1}\mathbf{X}'$, and $\mathbf{U} = \mathbf{1}_n\mathbf{1}_n'$. To establish general sufficient conditions on $\mathbf{\Sigma}$ under which (3.1) has the Fisher F-distribution, a first step is the following theorem.

THEOREM 1. *Under the intrinsic inference model for stratified dependent normal random variables, the following statements are equivalent:*

(a) (i) $\mathbf{Y}'(\mathbf{I}_n-\mathbf{P})\mathbf{Y}$ *is independent of* $\mathbf{Y}'(\mathbf{P}-n^{-1}\mathbf{U})\mathbf{Y}$,

(ii) $\mathbf{Y}'(\mathbf{I}_n-\mathbf{P})\mathbf{Y} \sim a\chi^2(n-p-1),\ a>0,$ (3.2)

(iii) $\mathbf{Y}'(\mathbf{P}-n^{-1}\mathbf{U})\mathbf{Y} \sim \delta\chi^2(p, \delta^{-1}\boldsymbol{\theta}'(\mathbf{P}-n^{-1}\mathbf{U})\boldsymbol{\theta}),\ \delta>0,$
where $\boldsymbol{\theta} = \mathbf{S}\boldsymbol{\mu} = \mathbf{X}\boldsymbol{\beta}$;

(b) $\mathbf{S\Sigma S}' = a\mathbf{I}_n + \mathbf{\Gamma}_s + \mathbf{\Gamma}_s' + (\delta-a)\mathbf{P},\ \forall\, \mathbf{S} \in \mathcal{S},$ (3.3)
where $\mathbf{\Gamma}_s = \mathbf{1}_n(\gamma_{1s},\ldots,\gamma_{ns}) = \mathbf{1}_n\boldsymbol{\gamma}_s'$, *say, and the parameters* γ_{is}, $i = 1,\ldots,n$, *are arbitrary.*

Also, if $a = \delta$ *and if* $p(\mathbf{Y})$ *assigns positive selection probabilities to every pair* $(X_j, X_t), j \neq t = 1,\ldots, N$, *then* (a) *and* (b) (*with* $a = \delta$) *are equivalent to equation* (1.3).

Proof. (b) is sufficient for (a). In fact, by Theorem 2 of Tranquilli and Baldessari (1987), equations (3.3) imply that, for every $\mathbf{S} \in \mathcal{S}$,

$$\mathbf{Y}_s'(\mathbf{I}_n-\mathbf{P})\mathbf{Y}_s \text{ is independent of } \mathbf{Y}_s'(\mathbf{P}-n^{-1}\mathbf{U})\mathbf{Y}_s,$$

$$\mathbf{Y}_s'(\mathbf{I}_n-\mathbf{P})\mathbf{Y}_s \sim a\chi^2(n-p-1),\ a>0, \tag{3.4}$$

$$\mathbf{Y}_s'(\mathbf{P}-n^{-1}\mathbf{U})\mathbf{Y}_s \sim \delta\chi^2(p, \delta^{-1}\boldsymbol{\theta}'(\mathbf{P}-n^{-1}\mathbf{U})\boldsymbol{\theta}),\ \delta>0,$$

where the non-centrality parameter of $\chi^2(n-p-1)$ is zero since $\boldsymbol{\theta}'(\mathbf{I}_n-\mathbf{P})\boldsymbol{\theta} = \boldsymbol{\beta}'\mathbf{X}'(\mathbf{I}_n-\mathbf{X}(\mathbf{X}'\mathbf{X})^{-1}\mathbf{X}')\mathbf{X}\boldsymbol{\beta} = 0$.

Let $D[\cdot]$ be the distribution of the random variable appearing in the argument. By simple conditioning and by statements (3.4),

$$[\mathbf{Y}'(\mathbf{I}_n-\mathbf{P})\mathbf{Y}, \mathbf{Y}_s'(\mathbf{P}-n^{-1}\mathbf{U})\mathbf{Y}]$$
$$\sim \mathbf{\Sigma}_s p(\mathbf{S})D[\mathbf{Y}_s'(\mathbf{I}_n-\mathbf{P})\mathbf{Y}_s, \mathbf{Y}_s'(\mathbf{P}-n^{-1}\mathbf{U})\mathbf{Y}_s]$$

$$\sim \Sigma_S p(\mathbf{S}) D[\mathbf{Y}_S'(\mathbf{I}_n - \mathbf{P})\mathbf{Y}_S] D[\mathbf{Y}_S'(\mathbf{P} - n^{-1}\mathbf{U})\mathbf{Y}_S] \tag{3.5}$$
$$\sim \Sigma_S p(\mathbf{S}) D[a\chi^2(n-p-1)] D[\delta\chi^2(p, \delta^{-1}\boldsymbol{\theta}'(\mathbf{P}-n^{-1}\mathbf{U})\boldsymbol{\theta})]$$
$$\sim D[\mathbf{Y}'(\mathbf{I}_n-\mathbf{P})\mathbf{Y}]\mathbf{Y}' D[\mathbf{Y}'(\mathbf{P}-n^{-1}\mathbf{U})\mathbf{Y}],$$

which proves that (b) is sufficient for (a).

(a) is sufficient (b). In fact, by Theorems 1 and 2 of Baldessari (1986), statements (3.2) imply statements (3.4), which, by Theorem 2 of Tranquilli and Baldessari (1987), imply equations (3.3).

Also, if $a = \delta$ and if $p(\mathbf{Y})$ assigns positive selection probability to every pair (X_j, X_t), $j \neq t = 1,\ldots, N$, (b) is sufficient for equation (1.3). In fact, suppose that (b) is true while there exists an $N \times N$ matrix $\mathbf{M}$ such that

$$\boldsymbol{\Sigma} = a\mathbf{I}_N + \boldsymbol{\Gamma} + \boldsymbol{\Gamma}' + \mathbf{M}. \tag{3.6}$$

Left and right multiplication by $\mathbf{S}$ and $\mathbf{S}'$ of (3.6) shows that, for every $\mathbf{S} \in \mathcal{S}$, $\mathbf{S\Gamma S}' = \mathbf{1}_n(\gamma_{1S},\ldots, \gamma_{nS}) = \boldsymbol{\Gamma}_S$, and so $\mathbf{SMS}' = \mathbf{0}$. Therefore, since every pair (X_j, X_t), $j \neq t = 1,\ldots, N$, has a positive probability of being selected, $\mathbf{M} = \mathbf{0}$, i.e., (b) implies equation (1.3).

On the other hand, if equation (1.3) is true, left and right multiplication by $\mathbf{S}$ and $\mathbf{S}'$ implies that (b) is true.

Since the selection probabilities $p(\mathbf{Y})$ that correspond to simple random sampling in each population stratum assign positive selection probabilities to every pair (X_j, X_t), $j \neq t = 1,\ldots, N$, the following corollary holds.

COROLLARY 1. *Under the intrinsic model for stratified dependent normal random varibles, with simple random sampling in each stratum, equation* (1.3) *is necessary and sufficient for* (a) *and, therefore, we have*

$$F \sim F(n-p-1, p, a^{-1}\boldsymbol{\theta}'(\mathbf{P} - n^{-1}\mathbf{U})\boldsymbol{\theta}), \tag{3.7}$$

where $F(.,.,.)$ denotes the non-central Fisher F-distribution.

We now prove that if equation (1.3) is true, the F-statistic for testing H_0: $\mathbf{A}\boldsymbol{\beta} = \mathbf{0}$ (in which the $q \times (p + 1)$ matrix $\mathbf{A}$ has rank q, $q < p + 1$, and has the first columns composed of zero elements) has exact Fisher F-distribution.

Denote by $\mathbf{F}_A$ the F-statistic for testing H_0: $\mathbf{A}\boldsymbol{\beta} = \mathbf{0}$ and, Seber (1977),

$$\mathbf{F}_A = q^{-1}(n-p-1)\mathbf{Y}'\mathbf{X}(\mathbf{X}'\mathbf{X})^{-1}\mathbf{A}'[\mathbf{A}(\mathbf{X}'\mathbf{X})^{-1}\mathbf{A}']^{-1}\mathbf{A}(\mathbf{X}'\mathbf{X})^{-1}\mathbf{Y}'\mathbf{X}[\mathbf{Y}'(\mathbf{I}_n - \mathbf{P})\mathbf{Y}]^{-1}$$

$$=q^{-1}(n-p-1)\mathbf{Y}'\mathbf{H}_{\mathrm{A}}\mathbf{Y}[\mathbf{Y}'(\mathbf{I}_n-\mathbf{P})\mathbf{Y}]^{-1}, \text{ say.} \tag{3.8}$$

THEOREM 2. *Under the intrinsic inference model for stratified dependent normal random variables, with simple random sampling in each stratum, if equations (1.3) is true, then*

$$\text{(i) } \mathbf{Y}'(\mathbf{I}_n-\mathbf{P})\mathbf{Y} \textit{ is independent of } \mathbf{Y}'\mathbf{H}_{\mathrm{A}}\mathbf{Y},$$

$$\text{(ii) } \mathbf{Y}'(\mathbf{I}_n-\mathbf{P})\mathbf{Y} \sim a\chi^2(n-p-1),\ a>0, \tag{3.9}$$

$$\text{(iii) } \mathbf{Y}'\mathbf{H}_{\mathrm{A}}\mathbf{Y} \sim a\chi^2(p, a^{-1}\boldsymbol{\theta}'\mathbf{H}_{\mathrm{A}}\boldsymbol{\theta}),\ a>0.$$

Proof. We begin by proving that, if equation (1.3) is true, statements (3.9) are true for every $\mathbf{Y}_{\mathrm{S}}$ such that $\mathbf{S} \in \mathsf{S}$.

In fact, equation (1.3) implies

$$\mathbf{S}[a\mathbf{I}_N + \boldsymbol{\Gamma} + \boldsymbol{\Gamma}']\mathbf{S}' = a\mathbf{I}_n + \boldsymbol{\Gamma}_{\mathrm{S}} + \boldsymbol{\Gamma}_{\mathrm{S}}',\ \forall\, \mathbf{S} \in \mathsf{S}. \tag{3.10}$$

Since $\mathbf{S}\boldsymbol{\mu} = \boldsymbol{\theta}$ for every $\mathbf{S} \in \mathsf{S}$, equations (3.10) imply, by Theorem 3 of Tranquilli and Baldessari (1987), that, for every $\mathbf{Y}_{\mathrm{S}}$ such that $\mathbf{S} \in \mathsf{S}$,

$$\mathbf{Y}_{\mathrm{S}}'(\mathbf{I}_n-\mathbf{P})\mathbf{Y}_{\mathrm{S}} \sim a\chi^2(n-p-1),\ \forall\, \mathbf{S} \in \mathsf{S}. \tag{3.11}$$

Moreover, equation (1.3) implies

$$\mathbf{H}_{\mathrm{A}}[a\mathbf{I}_N + \boldsymbol{\Gamma}_{\mathrm{S}} + \boldsymbol{\Gamma}_{\mathrm{S}}']\mathbf{H}_{\mathrm{A}} = a\mathbf{H}_{\mathrm{A}},\ \forall\, \mathbf{S} \in \mathsf{S}, \tag{3.12}$$

which is a consequence of the idempotency of $\mathbf{H}_{\mathrm{A}}$ (which implies $\mathbf{H}_{\mathrm{A}}a\mathbf{I}_n\mathbf{H}_{\mathrm{A}} = a\mathbf{H}_{\mathrm{A}}$) and of the check that

$$\mathbf{H}_{\mathrm{A}}(\boldsymbol{\Gamma}_{\mathrm{S}} + \boldsymbol{\Gamma}_{\mathrm{S}}')\mathbf{H}_{\mathrm{A}} = \mathbf{0}_{q,q},\ \forall\, \mathbf{S} \in \mathsf{S}, \tag{3.13}$$

where $\mathbf{0}_{q,q}$ is the $q \times q$ matrix of zeros.

Equation (3.13) is checked by the self-defining decompositions

$$\mathbf{A} = \begin{pmatrix} 0 & , \mathbf{a}' \\ \mathbf{0}_{q-1}, & \mathbf{A}_1 \end{pmatrix}, (\mathbf{X}'\mathbf{X})^{-1} = \begin{pmatrix} n^{-1}, & \mathbf{0}_p' \\ \mathbf{0}_p & , (\mathbf{X}_1'\mathbf{X}_1)^{-1} \end{pmatrix}, \tag{3.14}$$

which imply

$$\mathbf{A}(\mathbf{X}'\mathbf{X})^{-1}\mathbf{X}'(\mathbf{1}_n\mathbf{Y}_\mathbf{S}' + \mathbf{Y}_\mathbf{S}\mathbf{1}_n')\mathbf{X}(\mathbf{X}'\mathbf{X})^{-1}\mathbf{A}'$$

$$= \begin{pmatrix} 0 & , \mathbf{a}'(\mathbf{X}_1'\mathbf{X}_1)^{-1} \\ \mathbf{0}_p, & \mathbf{A}_1(\mathbf{X}_1'\mathbf{X}_1)^{-1} \end{pmatrix} \begin{pmatrix} n\mathbf{Y}_\mathbf{S}' + \mathbf{1}_n'\mathbf{Y}_\mathbf{S}\mathbf{1}_n \\ \mathbf{X}_1'\mathbf{Y}_\mathbf{S}\mathbf{1}_n' \end{pmatrix} (\mathbf{1}_n, \mathbf{X}_1)(\mathbf{X}'\mathbf{X})^{-1}\mathbf{A}'$$

$$= n \begin{pmatrix} \mathbf{a}'(\mathbf{X}_1'\mathbf{X}_1)^{-1}\mathbf{X}'\mathbf{Y}_\mathbf{S}, & \mathbf{0}_p' \\ \mathbf{A}_1(\mathbf{X}_1'\mathbf{X}_1)^{-1}\mathbf{X}_1\mathbf{Y}_\mathbf{S}, & \mathbf{0}_{q-1,p} \end{pmatrix} \begin{pmatrix} 0 & , & \mathbf{0}_{q-1} \\ (\mathbf{X}_1'\mathbf{X}_1)^{-1}\mathbf{a}, & & (\mathbf{X}_1'\mathbf{X}_1)^{-1}\mathbf{A}_1 \end{pmatrix}$$

$$= \mathbf{0}_{q,q}, \quad \forall\, \mathbf{S} \in \mathcal{S}. \tag{3.15}$$

Equations (3.12) imply, by Theorem 3 of Tranquilli and Baldessari (1987), that, for every $\mathbf{Y}_\mathbf{S}$ such that $\mathbf{S} \in \mathcal{S}$,

$$\mathbf{Y}_\mathbf{S}'\mathbf{H}_\mathbf{A}\mathbf{Y}_\mathbf{S} \sim a\chi^2(n-p-1, a^{-1}\,\boldsymbol{\theta}'\mathbf{H}_\mathbf{A}\boldsymbol{\theta}), \quad \forall\, \mathbf{S} \in \mathcal{S}. \tag{3.16}$$

Also, equation (1.3) implies

$$\mathbf{H}_\mathbf{A}(a\mathbf{I}_N + \boldsymbol{\Gamma}_\mathbf{S} + \boldsymbol{\Gamma}_\mathbf{S}')(\mathbf{I}_n - \mathbf{P}) = \mathbf{0}_{q,n}, \quad \forall\, \mathbf{S} \in \mathcal{S}. \tag{3.17}$$

In fact, $\mathbf{X}' = \mathbf{X}'\mathbf{P}$ implies $\mathbf{H}_\mathbf{A}(\mathbf{I}_n - \mathbf{P}) = \mathbf{0}$. On the other hand,

$$\mathbf{A}(\mathbf{X}'\mathbf{X})^{-1}\mathbf{X}'\mathbf{1}_n\mathbf{Y}_\mathbf{S}'(\mathbf{I}_n - \mathbf{P})$$

$$= \begin{pmatrix} 0 & , & \mathbf{a}' \\ \mathbf{0}_{q-1}, & & \mathbf{A}_1 \end{pmatrix} \begin{pmatrix} \mathbf{1}_n' \\ \mathbf{X}_1' \end{pmatrix} \mathbf{1}_n\mathbf{Y}_\mathbf{S}'(\mathbf{I}_n - \mathbf{P})$$

$$= \begin{pmatrix} 0 & , & \mathbf{a}' \\ \mathbf{0}_{q-1}, & & \mathbf{A}_1 \end{pmatrix} \begin{pmatrix} n \\ \mathbf{0}_p \end{pmatrix} \mathbf{Y}_\mathbf{S}'(\mathbf{I}_n - \mathbf{P}) = \mathbf{0}_{q,n}. \tag{3.18}$$

Equations (3.17) imply, by Theorem 2 of Tranquilli and Baldessari (1987), that, for every $\mathbf{Y}_\mathbf{S}$ such that $\mathbf{S} \in \mathcal{S}$,

$$\mathbf{Y}_\mathbf{S}'(\mathbf{I}_n - \mathbf{P})\mathbf{Y}_\mathbf{S} \text{ is independent of } \mathbf{Y}_\mathbf{S}'\mathbf{H}_\mathbf{A}\mathbf{Y}_\mathbf{S}, \forall\, \mathbf{S} \in \mathcal{S}. \tag{3.19}$$

We now prove that equation (1.3) implies statements (3.9). In fact, by classical results, statements (3.11) and (3.16) are equivalent, respectively, to

$$(\mathbf{I}_n - \mathbf{P})\mathbf{S}(a\mathbf{I}_N + \mathbf{\Gamma} + \mathbf{\Gamma}')\mathbf{S}'(\mathbf{I}_n - \mathbf{P}) = a(\mathbf{I}_n - \mathbf{P}), \quad \forall\, \mathbf{S} \in \mathcal{S}, \tag{3.20}$$

and,

$$\mathbf{H}_{\mathbf{A}}\mathbf{S}(a\mathbf{I}_N + \mathbf{\Gamma} + \mathbf{\Gamma}')\mathbf{S}'\mathbf{H}_{\mathbf{A}} = a\mathbf{H}_{\mathbf{A}}, \quad \forall\, \mathbf{S} \in \mathcal{S}. \tag{3.21}$$

These equations, by Theorem 1 of Baldessari (1986), are equivalent to statements (ii) and (iii) of (3.9).

Moreover, by classical results, statements (3.19) are equivalent to

$$\mathbf{H}_{\mathbf{A}}\mathbf{S}(a\mathbf{I}_N + \mathbf{\Gamma} + \mathbf{\Gamma}')\mathbf{S}'(\mathbf{I}_n - \mathbf{P}) = \mathbf{0}, \quad \forall\, \mathbf{S} \in \mathcal{S}, \tag{3.22}$$

which, since statements (ii) and (iii) of (3.9) are true, by Theorem 2 of Baldessari (1986), prove statement (i) of (3.9).

Theorem 2 implies that, under the intrinsic inference model for stratified dependent normal random variables, with simple random sampling in each stratum, if (1.3) is true then

$$\mathbf{F}_{\mathbf{A}} \sim F(n-p-1, q, a^{-1}\,\boldsymbol{\theta}'\mathbf{H}_{\mathbf{A}}\boldsymbol{\theta}) \tag{3.23}$$

and, therefore, that the statistics $\mathbf{F}_{\mathbf{A}}$ may be used for testing H_0: $\mathbf{A}\boldsymbol{\beta} = \mathbf{0}$.

REFERENCES

Baldessari, B. (1985). Some aspects of intrinsic inference. *Metron*, XLIII, 1-2, 41-53.

Baldessari, B. (1986). Intrinsic interence. 11. On quadratic forms and analysis of variance under the intrinsic inference model for stratified normal populations. *University of Arizona, Department of Statistics. Technical Report*, 86-8.

Baldessari, B. and Weber, J. (1985). Classical, Bayesian and intrinsic inference. *Metron*, XLIII, 3-4, 95-115.

Seber, G.A.F. (1977). *Linear Regression Analysis*. Wiley, New York.

Tranquilli, G.B. and Baldessari, B. (1987). Regression analysis with dependent observations: conditions for invariance of the distribution of an *F*-statistic. *Statistica* (submitted for publication).

Weber, J. and Baldessari B. (1986). Statistical models, intrinsic dependence and intrinsic inference. *Metron*, XLIV, 1-4, 83-100.

Received 29 December 1986
Revised 28 July 1987

Dipartimento di Statistica,
Probabilità e Statistica Applicata
Università degli Studi "La Sapienza"
Rome
Italy

Proc. Second International Tampere Conference in Statistics
(Tampere, Finland, 1-4 June 1987)
Tarmo Pukkila and Simo Puntanen, *Editors*

pp. 403 - 410

Control in the duration model framework

Kurt BRÄNNÄS
University of Umeå, Sweden

Key words and phrases: Expectation, quantile, loss function.

ABSTRACT

Optimal control methods play an important role in policy related econometrics. Here, we are concerned with duration models and suggest approaches to the control of the expected duration and of quantiles. In special cases goal values can be exactly achieved. Generally, loss functions have to be introduced and trade-offs between target satisfaction for different variables have to be accepted. A numerical example illustrates some of the issues raised.

1. INTRODUCTION

We consider the problem of assigning values to a vector of control variables for the control of an expected duration or of quantiles. Problems of this sort arise inter alia in setting saleries to minimize turnover of labour (maximize e.g. the expected employment duration) and in promotion expenditure and break-even relations (e.g. minimum expected duration to a required sales volume) for a new product (e.g. Kamien and Schwartz 1972). An analogous problem is faced in a biometrical multidose response modelling process.

The problem can also be approached as a prediction exercise (Brännäs 1986). The difference between the two approaches is particularly important when several control variables are available, and is analogous to the difference between optimal control and direct model simulation in macro-econometrics.

Here, the model is taken as known, but both known and estimated parameter cases are dealt with. We consider a log-linear duration model

$$y = \log t = \mathbf{x\beta} + \mathbf{z\delta} + \sigma\omega, \tag{1}$$

where t is a continous duration variable, $\mathbf{x}$ is a $(1\times k)$ vector of fixed, time-invariant, exogenous variables, and $\mathbf{z}$ is the $(1\times g)$ control variable vector. The parameters are $\boldsymbol{\beta}$, $\boldsymbol{\delta}$ and σ, while ω is a random disturbance term with expectation μ and variance η. If $\tau = \exp(\omega)$ has hazard function $(hf)\lambda_0(\tau)$, the hf of τ is

$$\lambda(t) = \lambda_0(t^{\alpha}e^{-\alpha\mathbf{x\beta}-\alpha\mathbf{z\delta}})\alpha t^{\alpha-1}e^{-\alpha\mathbf{x\beta}-\alpha\mathbf{z\delta}} \tag{2}$$

where $\alpha = \sigma^{-1}$. The corresponding density function for t is

$$f(t) = \lambda(t)\exp(-\Lambda(t)), \tag{3}$$

where $\Lambda(t)$ is the integrated hf. The expected duration is given by

$$E(t) = \int_0^{\infty} \exp(-\Lambda(t))dt, \tag{4}$$

and the quantile, t_p, is obtained from the relationship

$$1-F(t_p) = 1-p = \exp(-\Lambda(t_p)), \tag{5}$$

where $1 - p$ is the probability of surviving t_p for given parameters and values on $\mathbf{x}$ and $\mathbf{z}$, and $F(\cdot)$ is the distribution function.

In Section 2 the known parameter case is considered, while Section 3 deals with the case of estimated parameters. Section 4 contains an illustration of some of the technical issues raised in a wage control context.

2. CONTROL WITH KNOWN PARAMETERS

More structure is needed for the basic $hf\,\lambda_0(\tau)$ or density function $f(t)$ in order to achieve a desired expected duration, the target value, using (4).

In the Weibull case, we have $\lambda_0(\tau) = 1$ and obtain a target value m^* by adjusting the control variable vector $\mathbf{z}$ in the expression for the expectation $m = E(t)$, $m^* = \exp(\mathbf{x\beta} + \mathbf{z\delta})\Gamma(1 + \sigma)$. In the one control variable case we have $z = \delta^{-1}(\log m^* - \mathbf{x\beta} - \log\Gamma(1 + \sigma))$(cf. Table 1). The exponential case follows as a special case ($\sigma = 1$).

With two or more control variables there is no unique solution. It is unique only up to a linear combination of control variables. Uniqueness can be obtained only by bringing in extraneous restrictions. In this respect, the conventional optimal control approach of introducing a quadratic loss function seems reasonable (e.g. Chow 1975). Other types of loss functions can, of course, also be used.

Let $\mathbf{K}$ be a symmetric positive semi-definite weight matrix, $\mathbf{a} = (\mathrm{a}_1, \mathbf{a}_2')'$ a $1+g$ vector of target values, and $\mathbf{d} - \mathbf{a} = (\log m - a_1, \mathbf{z} - \mathbf{a}_2')'$. Then, a quadratic loss function is

$$\mathbf{W} = (\mathbf{d} - \mathbf{a})'\mathbf{K}(\mathbf{d} - \mathbf{a}). \tag{6}$$

In the Weibull case

$$\mathbf{d} - \mathbf{a} = \begin{pmatrix} \mathbf{x\beta} + \log\Gamma(1 + \sigma) - \mathrm{a}_1 \\ -\mathbf{a}_2 \end{pmatrix} + \begin{pmatrix} \boldsymbol{\delta}' \\ \mathbf{I} \end{pmatrix}\mathbf{z}' = \mathbf{A} + \mathbf{Bz}'$$

so that $\mathbf{W} = \mathbf{A'KA} + \mathbf{zB'KBz'} + 2\mathbf{A'KBz'}$. From this we obtain the optimal control solution

$$\mathbf{z} = -(\mathbf{B'KB})^{-1}\mathbf{B'KA} \tag{7}$$

with loss $\mathbf{W}^* = \mathbf{A'KA} - \mathbf{A'KB}(\mathbf{B'KB})^{-1}\mathbf{B'KA}$.

In the case of a scalar z variable, $\mathbf{K}_{11} = 1$ and all other $\mathbf{K}_{ij} = 0$, we obtain the result given above with $\mathrm{a}_1 = \log m^*$. This corresponds to minimizing ($\log m/m^*$, $\mathbf{z} - \mathbf{a}_2'$) with the above weights. The alternative loss based on ($m - \mathrm{a}_1$, $\mathbf{z} - \mathbf{a}_2'$) leads to nonlinearities that are eliminated by the present approach. Holbrook (1974) suggests a simple numerical solution.

Note that $y - \mathrm{a}_1 = (y - \log m) + (\log m - \mathrm{a}_1) = (\sigma\omega - \log\Gamma(1 + \sigma)) + (\log m - \mathrm{a}_1)$ implies the same optimal control solution, but with a higher expected loss. An analogous solution is obtained if we utilize $y - \mathrm{a}_1 = (y - E(y)) + (E(y) - \mathrm{a}_1)$, the only change being that $\log\Gamma(1 + \sigma)$ is replaced by $\sigma\mu$ in the $\mathbf{A}$ matrix.

Several distributions have related expectations (cf. Table 1). For these, (7) is applicable with appropriate substitutions for the $\log\Gamma(1 + \sigma)$ in the **A** matrix.

In an analogous manner, we want to get close to a target value, t_{p*}, of a quantile, t_p, with probability $1 - p$ using the control vector **z**. For a number of distributions (cf. Table 1) the general structure of the relationship to be used is $t_p = c \cdot \exp(\mathbf{x}\boldsymbol{\beta} + \mathbf{z}\boldsymbol{\delta})$, where c varies with distribution. Let $\mathbf{d} - \mathbf{a} = (\log t_p - a_1, \mathbf{z} - \mathbf{a}_2')$ and let the loss function be given by (6). The optimal control solution is then given by (7), with $\mathbf{A}' = (\mathbf{x}\boldsymbol{\beta} + \log c - a_1, -\mathbf{a}_2')$.

The approach can be extended to a control problem involving more that one quantile, simply by reinterpreting $\log t_p$ as a vector $(\log t_{p1},\ldots, \log t_{pm})'$ and by expanding a_1 accordingly. The single control variable results of Table 1 are obtained with $\mathbf{K}_{11} = 1$, $\mathbf{K}_{ij} = 0$ $(i, j \neq 1)$ and $a_1 = \log t_{p*}$. The control of two quantiles using two control variables is not feasible without a loss function. This restates Tinbergen's classic proposition that a target is attainable only if there are as many linearly independent targets as there are control variables (e.g. Preston 1974). The latter ones are in this setting linearly dependent. If the control vector **z** is time dependent, $\mathbf{z}(t)$, this problem disappears.

3. CONTROL WITH UNKNOWN PARAMETERS

In this section we consider the control of an expected duration and quantiles when parameters are estimated. The loss functions are now always random. We adopt the conventional approach of minimizing the expected loss and utilize $E(\mathbf{W}) = E_{\hat{\theta}}[E(\mathbf{W}|\hat{\boldsymbol{\theta}})]$, where $\hat{\boldsymbol{\theta}}$ is the parameter estimator. Employing this in the Weibull case leading to (7), yields

$$E(\mathbf{W}) = E_{\hat{\theta}}(\hat{\mathbf{A}}'\mathbf{K}\hat{\mathbf{A}}) + \mathbf{z}E_{\hat{\theta}}(\hat{\mathbf{B}}'\mathbf{K}\hat{\mathbf{B}})\mathbf{z}' + 2E_{\hat{\theta}}(\hat{\mathbf{A}}'\mathbf{K}\hat{\mathbf{B}})\mathbf{z}'$$

so that the optimal control solution is

$$\mathbf{z} = -[E_{\hat{\theta}}(\hat{\mathbf{B}}'\mathbf{K}\hat{\mathbf{B}})]^{-1}E_{\hat{\theta}}(\hat{\mathbf{A}}'\mathbf{K}\hat{\mathbf{B}}) \tag{8}$$

where $\hat{}$ signifies a matrix function of $\hat{\boldsymbol{\theta}}$.

The control solutions, both in terms of expected duration and of quantiles, have a related structure (cf. Table 1). In evaluating the expectations in (8) e.g. the asymptotic properties of estimators can be used. The expectations needed are $E(\hat{\boldsymbol{\delta}}\mathbf{K}_{11}\hat{\boldsymbol{\delta}}')$, $E(\hat{\boldsymbol{\delta}}\mathbf{K}_{11}\mathbf{x}\hat{\boldsymbol{\beta}})$, $E(\hat{\boldsymbol{\delta}}\mathbf{K}_{11}\hat{c})$, $E(\hat{\boldsymbol{\beta}})$, $E(\hat{\boldsymbol{\delta}})$, and $E(\hat{c})$, where $\mathbf{K}_{11}$ is the first element of matrix $\mathbf{K}$, and $\hat{c}$ varies with distribution (e.g. $\log\Gamma(1 + \hat{\sigma})$ in the expected Weibull case).

For a single target variable we have $E(\hat{\boldsymbol{\beta}}\mathbf{K}_{11}\hat{\boldsymbol{\beta}}') = \mathbf{K}_{11}[\boldsymbol{\Sigma}_{\beta} + E(\hat{\boldsymbol{\beta}})E(\hat{\boldsymbol{\beta}}')]$ and $E(\hat{\boldsymbol{\beta}}\mathbf{K}_{11}\mathbf{x}\hat{\boldsymbol{\delta}}) = \mathbf{K}_{11}[\boldsymbol{\Sigma}_{\beta\delta} + E(\hat{\boldsymbol{\beta}})E(\hat{\boldsymbol{\delta}}')]\mathbf{x}'$, where $\boldsymbol{\Sigma}_{\beta} = \text{cov}(\hat{\boldsymbol{\beta}})$ and $\boldsymbol{\Sigma}_{\beta\delta} = \text{cov}(\hat{\boldsymbol{\beta}}, \hat{\boldsymbol{\delta}})$. If the estimator $\hat{\boldsymbol{\theta}}' = (\hat{\boldsymbol{\beta}}', \hat{\boldsymbol{\delta}}', \hat{\sigma})$ is consistent we have asymptotically

$$E_{\hat{\theta}}(\hat{\mathbf{B}}'\mathbf{K}\hat{\mathbf{B}}) = \mathbf{K}_{11}[\boldsymbol{\Sigma}_{\delta} + \boldsymbol{\delta}\boldsymbol{\delta}'] + 2\mathbf{K}_{21}\boldsymbol{\delta}' + \mathbf{K}_{22}, \tag{9}$$

$$E_{\hat{\theta}}(\hat{\mathbf{B}}'\mathbf{K}\hat{\mathbf{A}}) = \mathbf{K}_{11}[\boldsymbol{\Sigma}_{\delta\beta} + \boldsymbol{\delta}\boldsymbol{\beta}']\mathbf{x}' + \mathbf{K}_{11}E(\hat{\boldsymbol{\delta}}\hat{c}) + \mathbf{K}_{21}\mathbf{x}\boldsymbol{\beta} + \mathbf{K}_{21}c - (\mathbf{K}_{21} + \boldsymbol{\delta}\mathbf{K}_{11})\mathbf{a}_1 - (\boldsymbol{\delta}\mathbf{K}_{12} + \mathbf{K}_{22})\mathbf{a}_2, \tag{10}$$

where $E(\hat{\boldsymbol{\delta}}\hat{c})$ has to be obtained for each special case. When $c = \log\Gamma(1+\sigma)$, we obtain $E(\hat{\boldsymbol{\delta}}\hat{c}) \approx \boldsymbol{\delta}\log\Gamma(1+\sigma) + \boldsymbol{\Sigma}_{\hat{\delta}\sigma}\Gamma'(1+\sigma)/\Gamma(1+\sigma)$ by a Taylor expansion of $\hat{c}$ around σ, and $\boldsymbol{\Sigma}_{\delta\sigma} = \text{cov}(\hat{\boldsymbol{\delta}}, \hat{\sigma})$. Both (9) and (10) are larger than the components of (7). Whether the solution (8) is more conservative, as anticipiated, cannot be implied from the general case.

For practical use of (8), estimates are substituted for the parameters of (9)-(10). When several quantiles are controlled (9)-(10) are not valid, as they are based on a scalar $\mathbf{K}_{11}$. An extension is , however, straightforward.

4. AN ILLUSTRATION

To illustrate with an example, we consider a small set of observations on blue collar employees. The spells of employment for 60 employees are observed. No individual is followed for more than 8 quarters (720 days), giving 10 censored observations. Maximum likelihood estimates and variable definitions are given in Table 2. A Weibull model specification received support in Brännäs (1986).

Control solutions using one control variable are shown graphically in Figure 1. If a male younger than 25 years is to remain employed for more than 400 days with probability .9 he should be offered an internal relative wage of 1.57 when the internal-external wage is 1.1. If the employing firm pays more than the external labour market the internal relative wage offer can be lowered.

In Figure 2, the trade-offs between two control variables are shown for m^* arbitrarily chosen as 200 and 400 days. In the same graph we present a few optimal control solutions using different loss functions, both for known and estimated parameters. In controlling the expected duration under known parameter assumptions the solutions are close to the trade-off lines with the used weights. With more weight given to the control variables these would come closer to the target. For estimated parameters the control solution is more conservative, particularly in the internal-external wage ratio. Moreover, the weighting scheme has to be substantially altered to obtain realistic solutions. Also given are the outcomes of controlling one quantile.

TABLE 1

Expressions for a single control variable z for the control of an expected duration and of a single quantile

Distribution of t	Expected Duration	Quantile
Weibull	$\delta^{-1}\{\log m - \mathbf{x\beta} - \log\Gamma(1+\sigma)\}$	$\delta^{-1}\{\log t_p - \mathbf{x\beta} - \sigma\log[-\log(1-p)]\}$
Gamma	$\delta^{-1}\{\log m - \mathbf{x\beta} - \log[\Gamma(m_1+1)/\Gamma(m_1)]\}$	$\delta^{-1}\{\log t_p - \mathbf{x\beta} + \log\Gamma_{m_1}(p)\}$
Lognormal	$\delta^{-1}\{\log m - \mathbf{x\beta} + \sigma^2\eta/2\}$	$\delta^{-1}\{\log t_p - \mathbf{x\beta} - \sigma\phi^{-1}(p)\}$
Generalized Gamma	$\delta^{-1}\{\log m - \mathbf{x\beta} - \log[\Gamma(m_1+\sigma)/\Gamma(m_1)]\}$	$\delta^{-1}\{\log t_p - \mathbf{x\beta} - \sigma\log\Gamma_{m_1}(p)\}$
Log-logistic	$\delta^{-1}\{\log m - \mathbf{x\beta} - \log[\sigma\pi/\sin\sigma\pi]\}$	$\delta^{-1}\{\log t_p - \mathbf{x\beta} + \sigma\log[p/(1-p)]\}$
Generalized F	$\delta^{-1}\{\log m - \mathbf{x\beta} - \log[m_2/(m_2-2)]\}$ ($\sigma = 1$)	—

TABLE 2

Parameter estimates (s.e. in parenthesis)

Age[a] <25	Age[a] 25-40	Sex[b]	Internal[c] Relative Wage	Internal[d] External Wage Ratio	Constant	σ^2
1.88	2.48	–.43	2.72	10.13	–9.42	.42
(.34)	(.37)	(.18)	(.61)	(3.36)	(3.78)	(.09)

[a] Dummy variables, 1 in the class and 0 otherwise

[b] Dummy variable, 1 for males and 0 for females

[c] Individual wage relatively the mean wage in the same category in the firm

[d] Mean wage ratio for the category in the firm and the local labour market

FIGURE 1

Expected duration and quantile control using the internal relative wage for males under 25 years

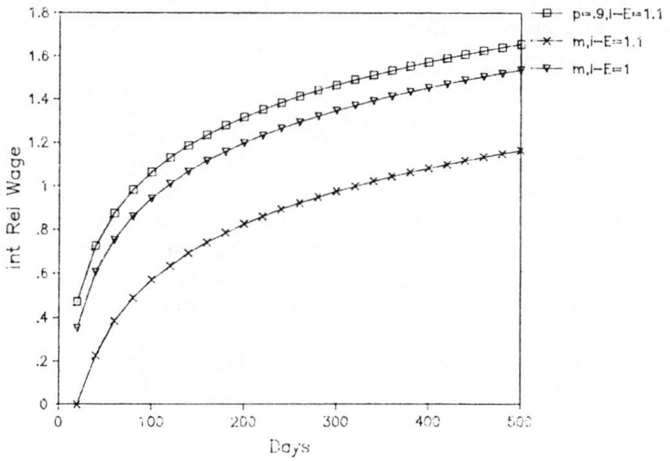

FIGURE 2

Trade-offs between control variables and optimal solutions for males under 25 years (Known parameters with weights a:($\mathbf{K}_{11}$ = 1, $\mathbf{K}_{22}$ = 1, $\mathbf{K}_{33}$ = 1); b:(1, 1, 2); c:(1, 1, 4), estimated parameters d:(.05, 2, 2); e:(.05, 1, 1), and quantile $1-p$ = .9 in f:(1, 1, 1))

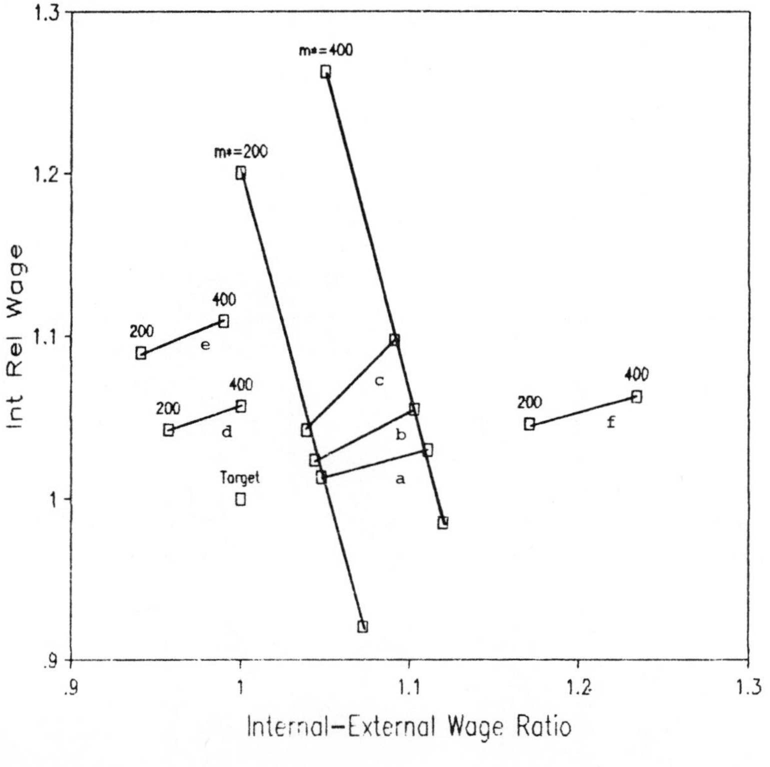

ACKNOWLEDGEMENT

Research in part supported by the Swedish Council for Research in the Humanities and Social Sciences.

REFERENCES

Brännäs, K. (1986). Prediction in a duration model. *Journal of Forecasting*, 5, 97-103.

Chow, G.C. (1975). *Analysis and Control of Dynamic Economic Systems*. Wiley, New York.

Holbrook, R.S. (1974). A practical method for controlling a large non-linear stochastic system. *Annals of Economic and Social Measurement*, 3, 155-175.

Kamien, M.I. and Schwartz, N.L. (1972). Timing of innovations under rivalry. *Econometrica*, 40, 43-60.

Preston, A.J. (1974). Dynamic generalization of Tinbergen's theory of policy. *Review of Economic Studies*, 41, 65-74.

Received 29 December 1986
Revised 27 August 1987

Department of Statistics
University of Umeå
S-901 87 Umeå
Sweden

Proc. Second International Tampere Conference in Statistics
(Tampere, Finland, 1-4 June 1987)
Tarmo Pukkila and Simo Puntanen, *Editors*

pp. 411 - 416

The asymptotic variances of the necessary and sufficient conditions for the infinite divisibility of discrete distributions through multivariate analysis

Edward J. Danial and **S.K. Katti**

Kearney State College, NE, U.S.A. and University of Missouri at Columbia, MO, U.S.A.

Key words and phrases: Necessary and sufficient conditions, Taylor's formula in 'i' dimensions, partial differentiation, corn-borers infestation.

ABSTRACT

Necessary and sufficient conditions for a discrete random variable X to be infinitely divisible are given by:

$$\pi_i = i p_i^* - \sum_{j=1}^{i-1} p_j^* \pi_{i-j} \geq 0$$

for $i = 1, 2, \ldots$, where $p_i^* = p_i/p_0$ and $p_i = P(X = i)$. Multivariate analysis is used to derive the asymptotic variance of $\hat{\pi}_i$, the estimator of π_i. This is done with the aide of Taylor's formula in 'i' dimension and partial differentiation. The knowledge of these variances is useful for constructing a test for a class of infinitely divisible distributions. An area of application is larvae infestation in farms and similar infestation problems.

1. INTRODUCTION

Let $\{p_i\}$, $i = 0, 1, 2,\ldots$, $p_0 \neq 0$, $p_1 \neq 0$ represent the probability distribution of a random variable (r.v.) X with support contained in $\{0, 1, 2,\ldots\}$. Katti (1967) shows that a necessary and sufficient set of conditions for X to be infinitely divisible (i.d.) is that

$$\pi_i = i p_i^* - \sum_{j=1}^{i-1} \pi_{i-j}\, p_j^* \geq 0, \qquad i = 1,2,\ldots, \tag{1}$$

where $p_i^* = p_i/p_0$. These conditions constitute a convenient operational scheme for checking the infinite divisibility of discrete distributions via analysis and the computer. Consequently, they could serve as a basis for constructing a test statistic for the purpose of inferring the infinite divisibility of a certain empirical distribution. The estimate of π_i is

$$\hat{\pi}_i = i\hat{p}_i^* - \sum_{j=1}^{i-1} \hat{\pi}_{i-j}\, \hat{p}_j^*, \tag{2}$$

where $\hat{p}_i^*$ is the estimate of p_i^*.

The knowledge of the variance of $\hat{\pi}_i$ is important for constructing the test statistics mentioned above. The aim of this paper is to derive this variance asymptotically. Applications are discussed.

2. THE ASYMPTOTIC VARIANCES OF $\hat{\pi}_i$

Let $x_1,\ldots, x_n$ be a random sample from an integer valued distribution with f_i standing for frequency of count i. It follows that $\hat{p}_i$ appearing in the statistics $\hat{\pi}_i$ is given

$$\hat{p}_i = f_i / n\,; \qquad i = 0, 1,\ldots$$

This makes $\hat{\pi}_i$ asymptotically consistent statistic for π_i. Equivalently,

$$\lim_{n\to\infty} \hat{\pi}_i(n) \overset{\mathrm{P}}{=} \pi_i = E(\hat{\pi}_i). \tag{3}$$

Two lemmas are needed in order to derive the asymptotic variance of $\hat{\pi}_i$.

LEMMA 1. (i) $Cov(\hat{p}_j, \hat{p}_k) = -p_j p_k / n, \quad k, j = 1, 2, \ldots, i$ *and* $j \neq k$.

(ii) $V(\hat{p}_j) = p_j(1 - p_j)/n, \quad j = 1, 2, \ldots, i.$

Proof. The proof follows directly from the fact that the joint probability distribution of f_j and f_k is multinomial (trinomial). □

LEMMA 2.

(i) $$\frac{\partial \pi_i}{\partial p_j} = 0; \quad i < j$$

(ii) $$\frac{\partial \pi_i}{\partial p_j} = \frac{i}{p_0}; \quad i = j$$

(iii) $$\frac{\partial \pi_i}{\partial p_j} = \frac{\partial \boldsymbol{\pi}_{i-1}}{\partial p_j} \cdot {}_0\mathbf{p}_{i-j}^{*} - \frac{\pi_{i-j}}{p_0}; \quad i > j, i = 2, 3, \ldots,$$ and $j = 1, 2, \ldots, i-1$;

where

$$\boldsymbol{\pi}_i = (\pi_1, \pi_2, \ldots, \pi_i)$$

and

$${}_0\mathbf{p}_{i-j}^{*} = (0, 0, \ldots, 0, p_{i-j}^{*}, p_{i-j+1}^{*}, \ldots, p_1^{*})'.$$

Proof. We can rewrite recursion (1) as

$$\pi_i = i p_i^{*} - (\pi_{i-1} p_1^{*} + \pi_{i-2} p_2^{*} + \ldots + \pi_1 p_{i-1}^{*}).$$

Hence, part (i) and (ii) are clear.

Suppose now $i > 1$. Then, by computing $\frac{\partial \pi_2}{\partial p_1}, \frac{\partial \pi_3}{\partial p_1}, \ldots$ and so on, we can conclude by induction on 'i', that

$$\frac{\partial \pi_i}{\partial p_1} = -\frac{\partial \pi_1}{\partial p_1} p_{i-1}^{*} - \frac{\partial \pi_2}{\partial p_1} p_{i-2}^{*} - \ldots - \frac{\partial \pi_{i-1}}{\partial p_1} p_1^{*} - \frac{\pi_{i-1}}{p_0}$$

$$= -\frac{\partial \boldsymbol{\pi}_{i-1}}{\partial p_1} (p_{i-1}^{*}, p_{i-2}^{*}, \ldots, p_1^{*})' - \frac{\pi_{i-1}}{p_0}.$$

Similarly, by computing $\frac{\partial \pi_3}{\partial p_2}, \frac{\partial \pi_4}{\partial p_2},\ldots$ and so on, we can show by induction on $'i'$, that

$$\frac{\partial \pi_i}{\partial p_2} = \frac{\partial \boldsymbol{\pi}_{i-1}}{\partial p_2} (0, p_{i-2}^*, p_{i-3}^*, \ldots, p_1^*)' - \frac{\pi_{i-2}}{p_0}$$

Similarly,

$$\frac{\partial \pi_i}{\partial p_3} = \frac{\partial \boldsymbol{\pi}_{i-1}}{\partial p_3} (0, 0, p_{i-3}^*, p_{i-4}^*, \ldots, p_1^*)' - \frac{\pi_{i-3}}{p_0}.$$

Consequently, part (iii) follows by induction. □

THEOREM 1. *For large sample size n, the asymptotic variance of $\hat{\pi}_i$ has the following form:*

$$V(\hat{\pi}_i) = \left(\frac{\partial h}{\partial p_1} \frac{\partial h}{\partial p_2} \cdots \frac{\partial h}{\partial p_i}\right) \boldsymbol{\Sigma} \left(\frac{\partial h}{\partial p_1} \frac{\partial h}{\partial p_2} \cdots \frac{\partial h}{\partial p_i}\right)',$$

where $\frac{\partial h}{\partial p_j} = \left.\frac{\partial h(\mathbf{p}_i)}{\partial p_j}\right|_{\mathbf{p}_i = \hat{\mathbf{p}}_i}$ $;j = 1, 2, \ldots, i,$

$$\mathbf{p}_i = (p_1, p_2, \ldots, p_i)', \ \hat{\mathbf{p}}_i = (\hat{p}_1, \hat{p}_2, \ldots, \hat{p}_i)', \ h(\mathbf{p}_i) = \pi_i = h(p_1, p_2, \ldots, p_i),$$

and $\boldsymbol{\Sigma} = Cov(\hat{\mathbf{p}}_i) = E\{(\hat{\mathbf{p}}_i - \mathbf{p}_i)(\hat{\mathbf{p}}_i - \mathbf{p}_i)'\}$ *is the variance-covariance matrix of the vector* $\hat{\mathbf{p}}_i$.

Proof. Let n be large. Without loss of generality, we will assume p_0 is known from side consideration and hence ignore its variability. However, $(\hat{p}_i - p_i)$ are infinitisimal. Thus, we can neglect the quadratic terms of Taylor's formula in $'i'$ dimensions. This implies

$$\begin{aligned}\hat{\pi}_i = \hat{\pi}_i(n) &= h(\hat{p}_1 \hat{p}_2 \cdots \hat{p}_i) \\ &= h(\hat{p}_1 - p_1 + p_1, \hat{p}_2 - p_2 + p_2, \ldots, \hat{p}_1 - p_1 + p_i)\end{aligned}$$

$$= h(p_1 p_2 \dots p_i) + (\hat{p}_1 - p_1)\frac{\partial h}{\partial p_1} + (\hat{p}_2 - p_2)\frac{\partial h}{\partial p_2} + \dots + (\hat{p}_i - p_i)\frac{\partial h}{\partial p_i}.$$

Hence,

$$\hat{\pi}_i(n) - \pi_i = (\hat{p}_1 - p_1)\frac{\partial h}{\partial p_1} + (\hat{p}_2 - p_2)\frac{\partial h}{\partial p_2} + \dots + (\hat{p}_i - p_i)\frac{\partial h}{\partial p_i}.$$

Now,

$$\begin{aligned}
V(\hat{\pi}_i) &= E\{\hat{\pi}_i - E(\hat{\pi}_i)\}^2 \\
&= E(\hat{\pi}_i - \pi_i)^2 \quad ; \text{ by (3)} \\
&= \mathrm{E}\{(\hat{p}_1 - p_1)\frac{\partial h}{\partial p_1} + (\hat{p}_2 - p_2)\frac{\partial h}{\partial p_2} + \dots + (\hat{p}_i - p_i)\frac{\partial h}{\partial p_i}\}^2 \\
&= V(\hat{p}_1)\left(\frac{\partial h}{\partial p_1}\right)^2 + V(\hat{p}_2)\left(\frac{\partial h}{\partial p_2}\right)^2 + \dots + V(\hat{p}_i)\left(\frac{\partial h}{\partial p_i}\right)^2 \\
&\quad + 2\,\mathrm{Cov}(\hat{p}_1 ; \hat{p}_2)\frac{\partial h}{\partial p_1}\frac{\partial h}{\partial p_2} \\
&\quad + \dots + 2\,\mathrm{Cov}(\hat{p}_{i-1} ; \hat{p}_i)\frac{\partial h}{\partial p_{i-1}}\frac{\partial h}{\partial p_i} \\
&= \left(\frac{\partial h}{\partial p_1}\frac{\partial h}{\partial p_2}\dots\frac{\partial h}{\partial p_i}\right) \boldsymbol{\Sigma} \left(\frac{\partial h}{\partial p_1}\frac{\partial h}{\partial p_2}\dots\frac{\partial h}{\partial p_i}\right)'
\end{aligned}$$

where

$$\boldsymbol{\Sigma} = (\sigma_{jk}) \quad ; \ j, k = 1, 2, \dots, i$$

such that

$$\sigma_{jk} = \mathrm{Cov}(\hat{p}_j, \hat{p}_k) = E\{(\hat{p}_j - p_j)(\hat{p}_k - p_k)\} \ ; \ j, k = 1, 2, \dots, i; \ j \neq k$$

and

$$\sigma_{jj} = V(\hat{p}_j) = E\{(\hat{p}_j - p_j)^2\} \ ; \ j = 1, \dots, i.$$

3. CONCLUDING REMARKS ON APPLICATIONS

The intention is to use infinite divisibility in order to test if corn borer infestation from plant to plant is "statistically independent". This is the physical meaning of infinite divisibility. Independent infestation requires general spraying of insecticide, while local spraying would be enough in case of dependence. Thus, testing whether the distribution of the corn borers on corn fields is i.d. or not could help farmers in controlling the spread of larvae (corn borers). To perform such testing, the problem must be put in mathematical terms.

The mathematical meaning of infinite divisibility is represented by recursion (1). If X represents the number of borers on the whole farm, then the interest is to test if its distribution is i.d. Consequently, the null hypothesis of our interest is

$$H_0 : \text{ the class of distributions is i.d.},$$

where the $\hat{\pi}_i$ of recursion (2) is the main tool for testing it.

Katti (1977) presents an infinite divisibility test based on $\hat{\pi}_2$. Danial and Katti (1986) introduce another test based on the first ten $\hat{\pi}_i$'s; without including their variances. The knowledge of these variances could produce a better result. This will be done in future work.

Finally, it is worth to note that this infinite divisibility concept could also be easily applied in social sciences, technology and medicine. In medicine for instance, it could test for contageous diseases. Only time will tell what are the best applications of it.

REFERENCES

Katti, S. (1967). Infinite divisibility of discrete distributions. *Annals of Mathematical Statistics*, 38, 1306-1308.

Katti, S. (1977). Infinite divisibility of discrete distributions III. *Colloquia Mathematica Societatis Ianos Bolyai*, Debrecen, Hungary, 165-171.

Danial, E. and Katti, S. (1986). Controlling larvae infestation through safe and conservative consumption of insecticide. Submitted to the *American Journal of Agricultural Economics*.

Received 12 December 1986
Revised 21 August 1987

Department of Mathematics and Statistics
Kearney State College
Kearney, NE 68849
U.S.A.

Department of Statistics
University of Missouri at Columbia
Columbia, MO 65211
U.S.A.

Proc. Second International Tampere Conference in Statistics
(Tampere, Finland, 1-4 June 1987)
Tarmo Pukkila and Simo Puntanen, *Editors*

pp. 417 - 428

Computing the maximum likelihood estimates of the hyper-Poisson parameters interactively through APL computer language

Edward J. Danial and S.K. Katti
Kearney State College, NE, U.S.A. and University of Missouri at Columbia, MO, U.S.A.

Key words and phrases: Newton-Raphson iterations, Monte Carlo simulation, initial vector, proper convergence.

ABSTRACT

The purpose of this paper is to introduce computational techniques for overcoming the numerous hurdles encountered in finding the maximum likelihood estimates for the hyper-Poisson parameters. This is accomplished by various matrix manipulations with the aide of the interactive power and simplicity of the APL computer language.

1. INTRODUCTION

The probability of the hyper-Poisson has the following form:

$$p_i = P(X = i) = f(x; \lambda, \theta) = \frac{\Gamma(\lambda)\theta^i}{{}_1F_1(1; \lambda, \theta)\Gamma(\lambda + i)} \qquad (1)$$

where $i = 1, 2, 3, \ldots, \lambda > 0, \theta > 0$, and

$$_1F_1(1;\lambda,\theta) = 1 + \frac{\theta}{\lambda} + \frac{\theta^2}{\lambda(\lambda+1)} + \frac{\theta^3}{\lambda(\lambda + 1(\lambda+2)} + \ldots$$

$$= 1 + \sum_{i=1}^{\infty} \frac{\theta^i}{\lambda(\lambda+1)\ldots(\lambda+i-1)} \qquad (2)$$

The above r.v. is a two-parameter generalization of the Poisson distribution and forms a sub-class of the three-parameter family of confluent hypergeometric series distribution. This r.v. is used to provide satisfactory fits to some observed distributions for which the Poisson does not. Crow and Bardwell (1963) considers examples for its application in this area. Danial and Katti (1986) uses the hyper-Poisson in order to construct a critical region for testing if a certain empirical distribution is infinitely divisible. Such testing is important for controlling larvae infestation in farms and similar infestation problems. Hence, the knowledge of its maximum likelihood estimates (MLE's) is valuable.

To find the MLE's for the hyper-Poisson parameters is not a straightforward process. First, we are dealing with two parameters. Secondly, the probability distribution function given in (1) is not ordinary. The aim of this paper is to find these estimates interactively through the APL conputer language. This is done in three stages. First, the likelihood equations are put into the simplest form possible. Secondly, Newton-Raphson's iterative matrix form is employed to solve this system. Finally, Monte Carlo simulation is used to find the MLE's via Newton-Raphson. The choice of APL is based on its high interactive power and convenience for matrix manipulation.

2. MATHEMATICAL DEVELOPMENT

2.1 Simplifying the likelihood system

Crow and Bardwell (1963) presented the likelihood function for the distribution in (1) as

$$L = ns\bar{x}\ln\theta - ns\ln{}_1F_1 + ns\ln\Gamma(\lambda) - \sum_{i=1}^{ns} \ln\Gamma(\lambda + x_i), \qquad (3)$$

and hence, the likelihood equations are

$$\frac{\theta}{{}_1F_1}\,\frac{\partial_1F_1}{\partial\theta} = \bar{x}, \tag{4}$$

$$\frac{\theta}{{}_1F_1}\,\frac{\partial_1F_1}{\partial\theta} - \psi(\lambda) + \frac{1}{ns}\sum_{i=1}^{ns}\psi(\lambda + x_i) = 0, \tag{5}$$

where $\psi(\lambda) = \Gamma'(\lambda)/\Gamma(\lambda)$ is the digamma function.

Our objective is to solve equations (4) and (5) simultaneously for θ and λ. For that, we need to simplify some of the expressions in both of these equations. Well,

$${}_1F_1 = 1 + \sum_{i=1}^{\infty}\frac{\theta^i}{(\lambda)_i}, \tag{2}$$

where

$$(\lambda)_i = \lambda(\lambda + 1)\ldots(\lambda + i - 1) = \frac{\Gamma(\lambda + i)}{\Gamma(\lambda)}, \tag{6}$$

such that $i = 1, 2, \ldots$ Now,

$$\frac{\partial_1F_1}{\partial\theta} = \sum_{i=1}^{\infty}\frac{i\theta^{i-1}}{(\lambda)_i}. \tag{7}$$

Secondly,

$$\frac{\partial_1F_1}{\partial\lambda} = \sum_{i=1}^{\infty}\theta^i\cdot\left\{\frac{1}{(\lambda)_i}\right\}'. \tag{8}$$

where

$$\left\{\frac{1}{(\lambda)_i}\right\}' = \frac{\partial\left\{\frac{1}{(\lambda)_i}\right\}}{\partial\lambda} = \frac{-\{(\lambda)_i\}'}{\{(\lambda)_i\}^2}$$

$$= \frac{-\{(\lambda)_i\}'}{(\lambda)_i}\cdot\frac{1}{(\lambda)_i} = -\{\ln(\lambda)_i\}'\frac{1}{(\lambda)_i}$$

$$= -[\ln\{\lambda(\lambda + 1)\ldots(\lambda + i - 1)\}]'\cdot\frac{1}{(\lambda)_i}$$

$$= -\{\ln\lambda + \ln(\lambda+1)\ldots\ln(\lambda+i-1)\}'\cdot\frac{1}{(\lambda)_i}$$

$$= -\left(\frac{1}{\lambda}+\frac{1}{\lambda+1}+\ldots+\frac{1}{\lambda+i-1}\right)\frac{1}{(\lambda)_i}$$

$$= -\left(\sum_{j=1}^{i}\frac{1}{\lambda+j-1}\right)\frac{1}{(\lambda)_i}$$

$$= -\frac{S_i}{(\lambda)_i}, \tag{9}$$

where

$$S_i = \begin{cases}\sum_{j=1}^{i}\frac{1}{\lambda+j-1}; & i = 1, 2, \ldots \\ 0 & ; i = 0.\end{cases} \tag{10}$$

Substituting equation (9) in (8), we deduce that

$$\frac{\partial_1F_1}{\partial\lambda} = -\sum_{i=1}^{\infty}\frac{\theta^i}{(\lambda)_i}S_i. \tag{11}$$

Finally,

$$\frac{1}{ns}\sum_{i=1}^{ns}\psi(\lambda+x_i)-\psi(\lambda) = \frac{1}{ns}\sum_{i=1}^{ns}\psi(\lambda+x_i)-\frac{ns\psi(\lambda)}{ns}$$

$$= \frac{1}{ns}\sum_{i=1}^{ns}\psi(\lambda+x_i)-\frac{1}{ns}\sum_{i=1}^{ns}\psi(\lambda)$$

$$= \frac{1}{ns}\left[\sum_{i=1}^{ns}\{\psi(\lambda+x_i)-\psi(\lambda)\}\right]$$

$$= \frac{1}{ns}\left[\sum_{i=1}^{ns}\left\{\frac{\Gamma'(\lambda+x_i)}{\Gamma(\lambda+x_i)}-\frac{\Gamma'(\lambda)}{\Gamma(\lambda)}\right\}\right]$$

$$= \frac{1}{ns}\left[\sum_{i=1}^{ns} \{\ln\Gamma(\lambda + x_i) - \ln\Gamma(\lambda)\}'\right]$$

$$= \frac{1}{ns}\left[\sum_{i=1}^{ns} \left\{\ln\frac{\Gamma'(\lambda + x_i)}{\Gamma(\lambda)}\right\}'\right]$$

$$= \frac{1}{ns}\left[\sum_{i=1}^{ns} \{\ln(\lambda)_{x_i}\}'\right]; \text{ by (6)}$$

$$= \frac{1}{ns}\sum_{i=1}^{ns} S_{x_i}; \tag{12}$$

where the last step is deduced by equation (9).

Hence, substituting equations (7), (11) and (1) in our two likelihood equations (4) and (5), we deduce the most convenient equations, for solving for λ and θ, in the following

$$f(\lambda, \theta) = \left(\frac{1}{ns}\sum_{i=1}^{ns} x_i\right)_1F_1 - \sum_{i=1}^{\infty} \frac{\theta^i i}{(\lambda)_i} = 0 \tag{13}$$

$$g(\lambda, \theta) = \left(\frac{1}{ns}\sum_{i=1}^{ns} S_{x_i}\right)_1F_1 - \sum_{i=1}^{\infty} \frac{\theta^i S_i}{(\lambda)_i} = 0. \tag{14}$$

Observe the analogies between equations (13) and (14): $x_i \leftrightarrow S_{x_i}$, $i \leftrightarrow S_i$ and remaining terms are identical in both equations. This phenomenon itself is very valuable in solving this system with the aid of the computer.

2.2 Preparing the likelihood system for APL

The intention is to solve the simplified hyper-Poisson system given in equations (13) and (14) through APL. This is done by putting the system in matrix form. This is a system of two non-linear equations. There are various methods for solving such system. Newton-Raphson's method is the most convenient for us. From Taylor's formula in two dimensions and by neglecting the quadratic terms, we obtain

$$f(\lambda,\theta) = f(\lambda_0,\theta_0) + (\lambda-\lambda_0)f_\lambda(\lambda_0,\theta_0) + (\theta-\theta_0)f_\theta(\lambda_0,\theta_0) = 0,$$
$$g(\lambda,\theta) = g(\lambda_0,\theta_0) + (\lambda-\lambda_0)g_\lambda(\lambda_0,\theta_0) + (\theta-\theta_0)g_\theta(\lambda_0,\theta_0) = 0.$$

In matrix form, the last system is equivalent to

$$\begin{bmatrix} f_\lambda & f_\theta \\ g_\lambda & g_\theta \end{bmatrix} \begin{bmatrix} \lambda-\lambda_0 \\ \theta-\theta_0 \end{bmatrix} = \begin{bmatrix} -f \\ -g \end{bmatrix},$$

or

$$\begin{bmatrix} \lambda \\ \theta \end{bmatrix} = \begin{bmatrix} \lambda_0 \\ \theta_0 \end{bmatrix} + \begin{bmatrix} f_\lambda & f_\theta \\ g_\lambda & g_\theta \end{bmatrix}^{-1} \begin{bmatrix} -f \\ -g \end{bmatrix}.$$

Equivalently

$$\mathbf{x} = \mathbf{x}_0 + A^{-1}\mathbf{b}, \tag{15}$$

where $\mathbf{x}_0 = (\lambda_0,\theta_0)'$ is the initial vector estimate of the vector $\mathbf{x} = (\lambda, \theta)'$, A^{-1} is the inverse of the matrix

$$A = \begin{bmatrix} f_\lambda & f_\theta \\ g_\lambda & g_\theta \end{bmatrix},$$

and the vector $\mathbf{b} = (-f, -g)'$; such that f_λ, f_θ, g_λ and g_θ are the partial derivatives of $f(\lambda, \theta)$ and $g(\lambda, \theta)$, with respect to λ and θ respectively, at the point (λ_0, θ_0).

Hence, to solve the iterative equation given in (15), we must find nice expressions for the partial derivatives. For convenience and clarity, we denote

$$Q_{x_i} = \sum_{j=1}^{\mathbf{x}_i} \frac{1}{(\lambda + j - 1)^2}, \tag{16}$$

$$S_1 = \sum_{j=1}^{\infty} \frac{\theta^i}{(\lambda)_i} S_i, \tag{17}$$

$$S_2 = \sum_{j=1}^{\infty} \frac{i\theta^i}{(\lambda)_i}, \tag{18}$$

$$S_3 = \sum_{j=1}^{\infty} \frac{\theta^i}{(\lambda)_i} \{S_i^2 + Q_{x_i}\}, \tag{19}$$

$$S_4 = \sum_{j=1}^{\infty} \frac{i\theta^{i-1}}{(\lambda)_i} S_i, \tag{20}$$

and

$$S_5 = \sum_{j=1}^{\infty} \frac{i^2\theta^{i-1}}{(\lambda)_i}, \tag{21}$$

then, after some manipulations, we deduce the following:

$$f = \bar{\mathbf{x}}_1 F_1 - S_2, \tag{22}$$

$$g = (\frac{1}{ns}\sum_{i=1}^{ns} S_{x_i})_1 F_1 - S_1, \tag{23}$$

$$f_\lambda = S_1 \bar{\mathbf{x}} + \theta S_4, \tag{24}$$

$$f_\theta = \frac{S_2}{\theta} \bar{\mathbf{x}} - S_5, \tag{25}$$

$$g_\lambda = -(\frac{1}{ns}\sum_{i=1}^{ns} S_{x_i}) S_1 - (\frac{1}{ns}\sum_{i=1}^{ns} Q_{x_i})_1 F_1 + S_3, \tag{26}$$

and

$$g_\theta = (\frac{1}{ns}\sum_{i=1}^{ns} S_{x_i}) S_2 - S_4. \tag{27}$$

With these tools in our hands, we can proceed in solving for $\hat{\lambda}$ and $\hat{\theta}$, the maximum likelihood estimates for the hyper-Poisson parameters, by employing Newton-Raphson's iterative matrix form given in (15). This matrix form is very convenient to use in APL.

An equivalent form of equation (15) can be written as in the following,

$$\mathbf{x}_{i+1} = \mathbf{x}_i + A_i^{-1} \mathbf{b}_i, \qquad i = 1, 2, 3, \ldots \tag{28}$$

where

$$\mathbf{x}_i = (\lambda_i, \theta_i)', \quad \mathbf{b}_i = (-f(\lambda_i, \theta_i), \ -g(\lambda_i, \theta_i))'$$

and

$$A_i = \begin{bmatrix} f_\lambda(\lambda_i, \theta_i) & f_\theta(\lambda_i, \theta_i) \\ g_\lambda(\lambda_i, \theta_i) & g_\theta(\lambda_i, \theta_i) \end{bmatrix}.$$

Newton's method in two dimensions converges provided that $\mathbf{x}_0$ lies sufficiently close to the solution $\mathbf{x}_{m1} = (\hat{\lambda}, \hat{\theta})'$.

3. NUMERICAL RESULTS THROUGH APL

Two major steps are needed. First, generating Monte Carlo random samples from hyper-Poisson. Secondly, iterating equation (28) until proper convergence occurs.

The simulated sample is obtained in three stages. First, the APL secondary function (subroutine) MONFREQ generates the frequencies f_0, f_1,... . This is done by calling it inside the secondary function HPOIMOND. The resultant of HPOIMOND is the distribution of the Monte Carlo run $\hat{p}_i$, $i = 1, 2,...$, from hyper-Poisson with parameters λ_t and θ_t. With this empirical distribution, a random sample $x_1, x_2,..., x_{ns}$ is generated via the secondary functions MONSAMP.

The above random sample is used to evaluate the various quantities needed for iterating equation (28). The crucial question is to decide on the initial estimates for $\hat{\lambda}$ and $\hat{\theta}$. In order to achieve proper convergence, these estimates must be close enough to the likelihood system solutions. In other words, if these estimates are λ_0 and θ_0, then the distance

$$\| \mathbf{x}_0 - \mathbf{x}_{m1} \|, \tag{29}$$

must be sufficiently small. There are two methods for deciding on such a vector $\mathbf{x}_0$ that leads to the convergence of the $\mathbf{x}_i$'s in equation (28). The first requires choosing $\mathbf{x}_0 = \mathbf{x}_t = (\lambda_t, \theta_t)'$, where $\mathbf{x}_t$ is the true vector. If convergence does not occur, then it is needed to look for $\mathbf{x}_0 \neq \mathbf{x}_t$. This is done in the second method.

3.1 Iteration from the true vector

The simulated random samples comes from hyper-Poisson with parameters vector $\mathbf{x}_t$. The MLE's vector $\mathbf{x}_{m1}$ is the best and hence should be close enough to the true vector $\mathbf{x}_t$. Thus, choosing $\mathbf{x}_0 = \mathbf{x}_t$ should be a successful guess in order to achieve proper convergence.

Random samples, each of size $ns = 100$, are simulated. For each sample, the series given by equations (2), (10) and (16) through (21) are computed by the secondary functions F11, SXI1, SXI2, S1, S2, S3, S4, and S5, respectively. In the infinite series, an accuracy of eight correct decimals has been achieved. All these calculations are performed in the secondary function ITIRAT. ITIRAT, in turn, itirates the matrix equation (28).

Table I shows the convergence rate when different samples are simulated from the same $\mathbf{x}_t = (1.5, 2.0)'$. The convergence rate is in truth good, the fourth and fifth iterations are very close. The fifth iteration, $\mathbf{x}_5 = (\lambda_5, \theta_5)'$ is the MLE for $\mathbf{x}_t$. The second digit after the decimal point of every iteration shown has been rounded off.

However, for some samples convergence never occurs; even more than ten iterations are not enough. In these instances, the initial vector is far away from the MLE vector. This phenomena is overcomed in the following section.

3.2 Iteration from the screened vector

The objective is to find a vector $\mathbf{x}_0$ that is close enough to $\mathbf{x}_{m1}$. The value of the likelihood function at such vector $\mathbf{x}_0$ is higher than its values at vectors far from $\mathbf{x}_{m1}$. Based on this concept, the initial vector is screened out. This is done by computing the likelihood function $L(\lambda, \theta)$ of equation (3) for various vectors $\mathbf{x} = (\lambda, \theta)'$. The results are tabulated in a square matrix of a certain order, where its ij-th element $l_{ij} = L(\lambda_i, \theta_j)$. If the maximum that occurs in the center is $L(\lambda_{max}, \theta_{max})$, then $\mathbf{x}_0 = (\lambda_{max}, \theta_{max})'$ is decided. On the other hand, if $L(\lambda_{max}, \theta_{max})$ is observed at one of the likelihood matrix edges, then an alternative matrix for different vectors is computed. The secondary function SCREE tabulates the likelihood matrix, while MAXMAT locates $(\lambda_{max}, \theta_{max})$.

Table II screens out one such vector. This is a situation where the choice $\mathbf{x}_0 = \mathbf{x}_t = (1.5, 2.0)'$ leads to divergence. However, screening out $\mathbf{x}_0 = (0.50, 1.50)'$ leads to convergence, only after 16 iterations, to $\mathbf{x}_{m1} = (0.94, 1.86)'$.

TABLE I

CONVERGENCE RATE OF THE HYPER-POISSON LIKELIHOOD SYSTEM FOR DIFFERENT MONTE CARLO SAMPLES WITH $X_0 = X_t$

Vector # i	The Initial Value $\lambda_0 = \lambda_t$	The 4th Iteration λ_4	The 5th Iteration λ_5	The Initial Value $\theta_0 = \theta_t$	The 4th Iteration θ_4	The 5th Iteration θ_5
1	1.50	1.48	1.47	2.00	1.97	1.96
2	1.50	1.64	1.66	2.00	2.06	2.07
3	1.50	1.77	1.76	2.00	2.09	2.08
4	1.50	1.61	1.60	2.00	2.01	2.01
5	1.50	-0.09	-0.08	2.00	-5.75	-6.59

TABLE II

THE LIKELIHOOD FUNCTION MATRIX FOR SCREENING OUT AN INITIAL VECTOR $(\lambda_0, \theta_0)' = X_0 \neq X_t = (1.50, 2.00)'$
$L(\lambda, \theta)$

λ \ θ	0.50	1.00	1.20	$\theta_0 \downarrow$ 1.50	2.00	2.20
0.15	-10650.92	-10621.29	-10621.20	-10626.93	-10646.00	-10655.81
0.20	-10268.55	-10234.02	-10232.81	-10237.25	-10254.74	-10264.05
$\lambda_0 \rightarrow$ 0.50	-9910.54	-9852.30	-9845.01	-9842.16	-9850.51	-9865.84
1.00	-12118.98	-12036.59	-12021.77	-12009.15	-12004.20	-12006.00
1.50	-16046.69	-15950.69	-15930.93	-15911.40	-15896.10	-15894.13
2.00	-21125.86	-21021.56	-20998.54	-20974.19	-20951.18	-20946.25

4. THE APL FUNCTION NEWTRAPH

All the secondary functions used in section (3) are called in the APL primary function (main program) NEWTRAPH. NEWTRAPH coordinates the logic of execution. First, the vector $\mathbf{x}_t$ is read and a sample is simulated from it. Secondly, few iterations are performed with $\mathbf{x}_0 = \mathbf{x}_t$. In case of convergence, the result is accepted and step (1) is repeated, otherwise additional iterations are performed. Thirdly, if after 5 to 10 iterations convergence is not apparent, a new $\mathbf{x}_0 \neq \mathbf{x}_t$ is screened out and step (2) is repeated. Finally, the accepted result is printed.

The primary function NEWTRAPH is presented in the following. Lines [25] and [36] demonstrate all possible decisions while executing NEWTRAPH. The interactive power of APL has been crucial for making these decisions. (The list of all APL functions is available from the first author.)

```
      ∇NEWTRAPH[⎕]∇
    ∇ NEWTRAPH
[1]     A←(2,2)ρ0
[2]     B←XO←XN←(2,1)ρ0
[3]    READ:'READ VECTOR XT(TRUE VALUES OF LAMDA'
[4]     'AND THETA, RESPECTIVELY)'
[5]     XIN←XI←XT←⎕
[6]     'READ NS(SIMULATED SAMPLE SIZE FROM'
[7]     'HYPER-POISSON)'
[8]     NS←⎕
[9]     P←XT HPOIMOND 10
[10]    SAMP←P MONSAMP NS
[11]    XBR←+/SAMP÷NS
[12]   ITIRATE:'READ NITS(NUMBER OF ITIRATIONS)'
[13]    NITS←⎕
[14]    XOXN←ITIRAT NITS
[15]    'TRUE VALUES OF LAMDA AND THETA:'
[16]    XT
[17]    'INITIAL VALUES OF LAMDA AND THETA:'
[18]    XIN
[19]    'VALUES OF LAMDA AND THETA AFTER THE'
[20]    '(NITS-1)TH ITIRATION:'
[21]    XOXN[1],XOXN[3]
[22]    'VALUES OF LAMDA AND THETA AFTER THE'
[23]    'LAST ITIRATION:'
[24]    XOXN[2],XOXN[4]
[25]    '1 TO READ, 2 TO ITIRATE, 3 TO SCREEN AND'
[26]    '4 TO RESULT'
[27]    TEST1←⎕
[28]    →(TEST1=⍳4)/READ,ITIRATE,SCREEN,RESULT
[29]   SCREEN:'READ LMV(VECTOR OF LAMDA VALUES)'
[30]    LMV←⎕
[31]    'READ THV(VECTOR OF THETA VALUES)'
[32]    THV←⎕
[33]    LKHM←LMV SCREE THV
[34]    'THE MAXIMUM LIKELIHOOD MATRIX IS:'
[35]    LKHM
[36]    '0 TO SCREEN, OTHERWISE GET THE NEW'
[37]    '(SCREENED) INITIAL VALUES.'
[38]    TEST2←⎕
[39]    →(TEST2=0)/SCREEN
[40]    MAXIJ←(ρLMV)MAXMAT LKHM
[41]    XI←LMV[MAXIJ[1]],THV[MAXIJ[2]]
[42]    'NEW INITIAL VALUES(THE SCREENED) OF'
[43]    'LAMDA AND THETA:'
[44]    XI
[45]    XIN←XI
[46]    →ITIRATE
[47]   RESULT:'MAXIMUM LIKELIHOOD ESTIMATES OF'
[48]    'LAMDA AND THETA:'
[49]    XOXN[2],XOXN[4]
    ∇
```

REFERENCES

Abramowitz, M. and Stegun, I. (1972). *Handbook of Mathematical Functions.* National Bureau of Standards.

Crow, E. and Bardwell, G. (1963). Estimation of the parameters of the hyper-Poisson distributions. *Proceedings of the International Symposium*, McGill University, Montreal, Canada, August 15-20.

Danial, E. and Katti, S. (1986). Controlling larvae infestation through safe and conservative consumption of insecticide. Submitted to the *American Journal of Agricultural Economics.*

Hastings, C. (1955). *Approximations for Digital Computers.* Princeton University Press.

Received 12 December 1986

Department of Mathematics and Statistics
Kearney State College
Kearney, NE 68849
U.S.A.

Department of Statistics
University of Missouri at Columbia
Columbia, MO 65211
U.S.A.

Proc. Second International Tampere Conference in Statistics
(Tampere, Finland, 1-4 June 1987)
Tarmo Pukkila and Simo Puntanen, *Editors*

pp. 429 - 438

Factor structure in a multivariate ARCH model of exchange rate fluctuations

Francis X. Diebold and **Marc Nerlove**

Board of Governors of the Federal Reserve System, Washington, DC, U.S.A.
and University of Pennsylvania, Philadelphia, PA, U.S.A.

Key words and phrases: GARCH, heteroskedasticity, latent variable, volatility.

ABSTRACT

A multivariate ARCH model with factor structure is developed and illustrated. The underlying latent variable leads to a rich, yet parsimonious, variance-covariance structure and facilitates tractable parameter estimation. A two-stage estimation and testing framework is proposed and illustrated.

1. INTRODUCTION

In this paper we report on a multivarite time-series model which we have found useful for modeling asset prices in general, and exchange rates in particular. The key feature is that innovations are characterized by multivariate autoregressive conditional heteroskedasticity. A detailed application to floating exchange rate movements, as well as an estimation strategy based on the Kalman filter, is contained in Diebold and Nerlove (1986). In this paper we briefly summarize the approach, focusing on a simple two-stage estimation procedure and numerical methods which aid its convergence.

The existing literature on ARCH effects in asset markets is growing rapidly; see in particular Engle and Bollerslev (1986) and the many references therein. Existing univariate results point to the need for a multivariate specification, but successful multivariate ARCH modeling has to date proved elusive due to the large number of parameters which must be estimated. [See Kraft and Engle (1982)]. Our approach involves a multivariate latent-variable model in which the common factor may display ARCH. The rich conditional variance-covariance structure of the observed variables arises from their joint dependence on a common factor: this leads to *commonality* in temporal volatility movements across variables, which is in fact observed. The model is developed in Section 2, and concluding remarks are contained in Section 3.

2. THE MULTIVARIATE MODEL

While univariate ARCH models often offer good statistical descriptions of asset price movements, they are not satisfying relative to a full multivariate model. They do, however, provide a useful guide to multivariate specification.

The move to a multivariate framework is important for a number of reasons. First, nonzero covariances among exchange rate innovations require simultaneous multivariate estimation if full efficiency in parameter estimation is to be achieved. This is particularly relevant in the case of vector asset prices: "news" effects may lead to substantial contemporaneous correlation.

Second, the conditional covariances may not be constant. Specifically, they may display temporal persistence, exactly like the conditional variances. If this is found to be the case, examination of the time paths of conditional covariances may provide useful information.

Third, further insight into the nature of interaction may be gained via a multivariate parameterization motivated by latent variable considerations. The role of "news" in asset price determination suggests not only correlation among shocks, but also commonality in the conditional variance movements of the shocks (regardless of correlation). The common factor approach thus has strong substantive motivation and leads to an elegantly parsimonious (and testable) variance-covariance structure.

Consider first the multivariate ARCH model of Kraft and Engle (1982), in which the temporal evolution of an entire variance-covariance matrix is modelled. If Ω_{t-1} is an information set containing:

$$\{\varepsilon_{i,\,t-1},\dots,\varepsilon_{i,\,t-p}\},\ i = 1,\dots,N,$$

then the N-variate ARCH(p) system is given by:

$$\boldsymbol{\varepsilon}_t/\Omega_{t-1} \sim N(\mathbf{0},\mathbf{H}_t(\Omega_{t-1})),$$

where $\boldsymbol{\varepsilon}_t = (\varepsilon_{1t},\dots,\varepsilon_{N,\,t})'$. $\mathbf{H}_t$ is an $(N\times N)$ symmetric positive definite variance-covariance matrix, where:

$$H_{ij,\,t} = a_{ij,\,0} + \boldsymbol{\varepsilon}'_{t-1}C_{1,\,ij}\boldsymbol{\varepsilon}_{t-1} + \dots + \boldsymbol{\varepsilon}'_{t-p}C_{p,\,ij}\boldsymbol{\varepsilon}_{t-p}.$$

Each element of $\mathbf{H}_t$ is therefore a sum of quadratic forms in $\boldsymbol{\varepsilon}_{t-1},\dots,\boldsymbol{\varepsilon}_{t-p}$, and depends on the p-th order past histories of all innovation squares and cross products. The full variance-covariance matrix may then be written as:

$$\mathbf{H}_t = \mathbf{H}_0 + (\mathbf{I}\otimes\boldsymbol{\varepsilon}_{t-1})'\mathbf{C}_1(\mathbf{I}\otimes\boldsymbol{\varepsilon}_{t-1}) + \dots + (\mathbf{I}\otimes\boldsymbol{\varepsilon}_{t-p})'\mathbf{C}_p(\mathbf{I}\otimes\boldsymbol{\varepsilon}_{t-p}),$$

where $\mathbf{C}_k$ is an $(N^2\times N^2)$ matrix with $(N\times N)$ blocks $\mathbf{C}_{k,\,ij}$. It will be convenient for our purposes to vectorize the lower triangle of $\mathbf{H}_t$ and write:

$$\mathbf{h}_t = \mathbf{a}_0 + \mathbf{A}_1\boldsymbol{\eta}_{t-1} + \dots + \mathbf{A}_p\boldsymbol{\eta}_{t-p},$$

where:

$$\mathbf{h}_t = \mathrm{vec}(LT(H_t)),$$

$$\dim\mathbf{h}_t = \dim\mathbf{a}_0 = \dim\boldsymbol{\eta}_{t-i} = \frac{N^2+N}{2},\quad i = 1,\dots,p,$$

$$\dim\mathbf{A}_i = \left(\frac{N^2+N}{2}\ \mathrm{x}\ \frac{N^2+N}{2}\right),\quad i = 1,\dots,p,$$

and "vec", "dim", and "LT" are the vectorization, dimension and lower triangle operators, respectively.

Consistent, asymptotically efficient and normal parameter estimates are obtained by maximizing the log likelihood, which is a direct multivariate analog of the univariate case. Immediately:

$$\ln L = \sum_{t=1}^{T} \ln \ell_t$$

where $\ln \ell_t = -(N/2) \ln 2\pi + \frac{1}{2} \ln |\mathbf{H}_t^{-1}| - \frac{1}{2}\boldsymbol{\varepsilon}_t{}'\mathbf{H}_t^{-1}\boldsymbol{\varepsilon}_t$.

The model developed thus far has $K = \dfrac{N^2+N}{2} + p(\dfrac{N^2+N}{2})^2$ parameters to be estimated. Even for systems of moderate dimension, estimation is infeasible. Our task, then, is to impose various restrictions designed to reduce the number of free parameters, while simultaneously not imposing too much prior information on the data.

The system as written above allows each conditional variance and covariance to depend on p lags of the squared innovations and innovation cross products of *every* variable in the system. A more manageable and intuitively reasonable parameterization is obtained by allowing each conditional variance to depend only on *own* lagged squared innovations and each conditional covariance to depend only on own lagged innovation cross products. This corresponds to diagonal **A** matrices. Formally, then:

$$\mathbf{A}_i = \text{diag}\,(a_{i,1},\ldots, a_{i,D}), \quad i = 1,\ldots, p,$$

where $D \equiv \dfrac{N^2+N}{2}$. This reduces the number of parameters by two orders of magnitude, since we now have only $K = \dfrac{N^2+N}{2} + p(\dfrac{N^2+N}{2})$ parameters to estimate.

Fortunately, however, we may use a linearly decreasing ARCH lag weight structure to provide further parametric economy. We retain $\mathbf{A}_i$ = diag, $i = 1,\ldots, p$, but we now require that the $\mathbf{A}_i$ be *scalar* matrices given by:

$$\mathbf{A}_i = \text{scal}\,(p - i + 1).$$

Thus,

$$\mathbf{A}_i = \text{diag}\,(p-i+1,\ldots, p-i+1), \quad i = 1,\ldots, p,$$

and we write:

$$\mathbf{h}_t = \mathbf{a}_0 + \mathbf{M}(\mathbf{A}_1\boldsymbol{\eta}_{t-1} + \ldots + \mathbf{A}_p\boldsymbol{\eta}_{t-p}),$$

where M is an $(\frac{N^2 + N}{2} \times \frac{N^2 + N}{2})$ diagonal matrix given by:

$$\mathbf{M} = \mathrm{diag}\,(m_1, \ldots, m_D).$$

By the well-known properties of scalar matrices, we can rewrite the system as:

$$\mathbf{h}_t = \mathbf{a}_0 + \mathbf{M} \sum_{i=1}^{p} (p - i + 1)\boldsymbol{\eta}_{t-i}.$$

This leads to $K = (N^2 + N)$ parameters to be estimated. This is still a very large number of parameters; in addition, such a specification cannot account for the fact that substantial commonality exists often in volatility movements.

A factor analytic approach, as suggested by Diebold (1986) and Diebold and Nerlove (1986), enables us to address both problems simultaneously. For concreteness, consider the 7-variate exchange rate system studied by Diebold and Nerlove (1986):

$$\begin{bmatrix} \Delta \ln S_{CDt} \\ \Delta \ln S_{FFt} \\ \Delta \ln S_{DMt} \\ \Delta \ln S_{LIRt} \\ \Delta \ln S_{YENt} \\ \Delta \ln S_{SFt} \\ \Delta \ln S_{BPt} \end{bmatrix} = \begin{bmatrix} \lambda_{CD} \\ \lambda_{FF} \\ \lambda_{DM} \\ \lambda_{LIR} \\ \lambda_{YEN} \\ \lambda_{SF} \\ \lambda_{BP} \end{bmatrix} F_t + \begin{bmatrix} \varepsilon_{CDt} \\ \varepsilon_{FFt} \\ \varepsilon_{DMt} \\ \varepsilon_{LIRt} \\ \varepsilon_{YENt} \\ \varepsilon_{SFt} \\ \varepsilon_{BPt} \end{bmatrix},$$

where:

$$E\,F_t = E\,\varepsilon_{it} = 0, \text{ for all } i \text{ and } t,$$

$$E\,F_t F_{t'} = 0, \mathrm{t} \neq \mathrm{t}', E\,F_t \varepsilon_{it'} = 0, \text{ for all } i, t, t',$$

$$F_t / \Omega_{t-1} \sim N(0, \sigma_t^2),$$

$$\sigma_t^2 = a_0 + \theta \sum_{i=1}^{p} (p - i + 1) F_{t-i}^2,$$

$$E\,\varepsilon_{it}\varepsilon_{jt'} = \begin{cases} \gamma_i & \text{if } i = j, t = t' \\ 0 & \text{otherwise.} \end{cases}$$

In obvious matrix notation, we write the model as:

$$\Delta \ln S_t = \boldsymbol{\lambda} \mathbf{F}_t + \boldsymbol{\varepsilon}_t.$$

It follows that:

$$\boldsymbol{\Sigma}_t \equiv cov(\Delta \ln S_t / \Omega_{t-1}) = \sigma_t^2 \boldsymbol{\lambda\lambda}' + \boldsymbol{\Gamma},$$

where $\boldsymbol{\Gamma} = cov(\boldsymbol{\varepsilon}_t)$. Writing this out,

$$\boldsymbol{\Sigma}_t = \sigma_t^2 \begin{bmatrix} \lambda^2_{CD} & \lambda_{CD}\lambda_{FF} \cdots & \lambda_{CD}\lambda_{BP} \\ \lambda_{FF}\lambda_{CD} & \lambda_{FF} \cdots\cdots & \lambda_{FF}\lambda_{BP} \\ \cdot & & \cdot \\ \cdot & & \cdot \\ \cdot & & \cdot \\ \lambda_{BP}\lambda_{CD} \cdots & \cdots\cdots & \lambda^2_{BP} \end{bmatrix} + \begin{bmatrix} \gamma_{CD} & & \\ & \gamma_{FF} & 0 \\ & & \cdot \\ & & \cdot \\ 0 & & \cdot \\ & & \gamma_{BP} \end{bmatrix}.$$

Thus, the j-th time-t conditional variance is given by:

$$H_{jjt} = (\lambda_j^2 \alpha_0 + \gamma_j) + \lambda_j^2 \theta \sum_{i=1}^{p} (p - i + 1) F^2_{t-i}.$$

and the jk-th time-t conditional covariance is:

$$H_{jkt} = (\lambda_j \lambda_k \alpha_0) + \lambda_j \lambda_k \theta \sum_{i=1}^{p} (p - i + 1) F^2_{t-i}.$$

The intuitive motivation of such a model is strong. The "common factor" F represents general influences which tend to affect *all* exchange rates. The impact of the common factor on exchange rate i is reflected in the value λ_i. The "unique factors", represented by the ε_j's, reflect uncorrelated country-specific shocks.

The conditional variance-covariance structure (Σ_t) of the observed variables arises from their joint dependence on the common factor F. All conditional variance and covariance functions depend on the *common* innovations of F. The parameters of those functions are different, however, depending on the λ's and γ's. In this paper we highlight a two-stage estimation and testing procedure.

First, we abstract from ARCH effects so that $\sigma_t^2 = \sigma^2$. Without loss of generality, we normalize the unconditional variance $\sigma^2 = 1$ to set the scale of the common factor. Then:

$$\boldsymbol{\Sigma} = \boldsymbol{\lambda\lambda}' + \boldsymbol{\Gamma}.$$

There are more independent covariance equations than parameters to be estimated, which is necessary, but not sufficient, for identification. The Jöreskog (1979) sufficient conditions for idenfication are also satisfied, however, since $\sigma^2 = 1$ and $\boldsymbol{\Gamma}$ is diagonal. The identification ensures that meaningful parameter estimates may be obtained by any of a number of methods to obtain estimates of the λ's and γ's, as well as an estimated time series of the unobserved common factor values $\{F_t\}$, expressed as a linear combination of observables.

Second, we test for and model ARCH effects in F so that σ^2 is replaced by:

$$\sigma_t^2 = a_0 + \theta \sum_{i=1}^{p} (p - i + 1) F_{t-i}^2 .$$

Note that the two-stage nature of the estimator renders it, in general, suboptimal relative to a fully simultaneous estimator. The simplicity and extreme convenience of the approach, however, makes it worthy of serious consideration. At the very least, it should be capable of generating good "startup" estimates for more sophisticated estimation techniques.

We have found it helpful to scale all data prior to estimation to help avoid the formation of non-positive definite point conditional covariance matrices ($\mathbf{H}_t$) while iterating. The importance of this cannot be overstated; if at any time a non-p.s.d. $\mathbf{H}_t$ is constructed, it will cause algorithmic failure, because the determinant of its inverse enters the objective function in log form. In other words, it is not enough to constrain the maximization algorithm to *converge* to p.s.d. matrices $\mathbf{H}_t$; rather, all $\mathbf{H}_t$ must stay p.s.d. at all times while iteration is in process.

To see why rescaling the data (by a factor of 1000, say) helps avoid singularity, consider a diagonal system (i.e., no covariances). Then $\det \mathbf{H}_t = \sigma^2_{t,1} \cdots \sigma^2_{t,N}$. Because high-frequency percent changes are typically quite small, being roughly $0(10^{-3})$, $\sigma^2_{t,i}$ is roughly $0(10^{-6})$, $i = 1,\ldots, N$ and $\det \mathbf{H}_t = 0(10^{-6N})$. The fact that the *true* det $\mathbf{H}_t$ is so close to the non-p.s.d. region makes it very easy for numerical optimizational algorithms to wander there; rescaling by 1000 makes typical data values

0(10), so $\sigma^2_{t,i}$ is roughly 0(100) and det $\mathbf{H}_t = 0(10^{2N})$, which is far from the non-p.s.d. region.

It is of interest to note that rescaling will change only the intercept parameters in the ARCH variance and covariance equations. Consider, for example, the bivariate system:

$$\begin{bmatrix} Y_{1t} \\ \\ Y_{2t} \end{bmatrix} = \begin{bmatrix} \sum_{i=1}^{p} \rho_{1i} Y_{i,t-i} \\ \\ \sum_{i=1}^{p} \rho_{2i} Y_{2,t-i} \end{bmatrix} + \begin{bmatrix} \varepsilon_{1t} \\ \\ \varepsilon_{2t} \end{bmatrix},$$

$$(\varepsilon_{1t}, \varepsilon_{2t})' / \Omega_{t-1} \sim N(\mathbf{0}, \mathbf{H}_t),$$

$$\mathbf{H}_t = \begin{bmatrix} a_{01} + L_{11}(\varepsilon^2_{1,t-i}) & \cdot \\ \gamma_0 + L_{21}(\varepsilon_{1,t-i}\varepsilon_{2,t-i}) & a_{02} + L_{22}(\varepsilon^2_{2,t-i}) \end{bmatrix},$$

where the L_{ij} are functions linear in the stated arguments.

Now let $(X_{1t}, X_{2t})' = K(y_{1t}, y_{2t})$. Immediately,

$$\begin{bmatrix} X_{1t} \\ \\ X_{2t} \end{bmatrix} = \begin{bmatrix} \sum_{i=1}^{p} \rho_{1i} X_{i,t-i} \\ \\ \sum_{i=1}^{p} \rho_{2i} X_{2,t-i} \end{bmatrix} + \begin{bmatrix} V_{1t} \\ \\ V_{2t} \end{bmatrix}$$

where $(V_{1t}, V_{2t})' = K(\varepsilon_{1t}, \varepsilon_{2t})'$. Thus,

$$(V_{1t}, V_{2t})' / \Omega_{t-1} \sim N(\mathbf{0}, K^2 \mathbf{H}_t).$$

But

$$K^2 \mathbf{H}_t = \begin{bmatrix} K^2 a_{01} + L_{11}(K^2 \varepsilon^2_{1,t-i}) & \cdot \\ K^2 \gamma_0 + L_{21}(K^2 \varepsilon_{1,t-i}\varepsilon_{2,t-i}) & K^2 a_{02} + L_{22}(K^2 \varepsilon^2_{2,t-i}) \end{bmatrix}$$

by linearity of the functions L_{ij}. We can rewrite this as:

$$K^2\mathbf{H}_t = \begin{bmatrix} K^2a_{01} + L_{11}(V^2_{1,t-i}) & \cdot \\ K^2\gamma_0 + L_{21}(V_{1,t-i}V_{2,t-i}) & K^2a_{02} + L_{22}(V^2_{2,t-i}) \end{bmatrix}.$$

Thus, the K-scaled variables $(X_{1t}, X_{2t})'$ follow a multivariate ARCH process identical to that of the original variables $(y_{1t}, y_{2t})'$, except that the intercept of each ARCH variance and covariance function is now multiplied by K^2.

3. CONCLUDING REMARKS

In this paper we have concentrated on theoretical issues related to multivariate ARCH modeling; in particular, we have argued for a factor-analytic approach. In other work, much of which is contained in Diebold and Nerlove (1986), we estimate the model for a wide variety of exchange rates, and use the results to explain the leptokurtic nature of unconditional exchange rate distributions and their convergence to normality under temporal aggregation, measure exchange rate volatility and associated time-varying risk premia, and provide time-varying prediction intervals.

ACKNOWLEDGEMENTS

The research on which this paper is based was supported by a grant from the National Science Foundation to the University of Pennsylvania. Our work draws on Diebold's Ph.D. dissertation, "The Time-Series Structure of Exchange Rate Fluctuations," University of Pennsylvania, 1986. The views expressed herein are those of the authors and do not necessarily reflect those of the Federal Reserve System or its staff. We are indebted to T. Bollerslev, R.F. Engle, N. Ericsson, W. Gaab, H. Garbers, L.R. Klein, H. Koenig, J.H. McCullogh, P. Pauly, A. Rose, P. Spindt, J.H. Stock, P.A.V.B. Swamy, P. Tinsley, M. Watson, and participants in the 1985 and 1986 NBER/NSF Time-Series Analysis Conferences for helpful comments.

REFERENCES

Diebold, F.X. (1986). Modeling persistence in conditional variances: a comment. *Econometrics Reviews*, 5, 51-56.

Diebold, F.X. and Nerlove, M. (1986). The dynamics of exchange rate volatility: a multivariate latent-factor ARCH model. Board of Governors of the Fedral Reserve System, Special Studies Working Paper #205.

Engle, R.F. and Bollerslev, T. (1986). Measuring the persistence in conditional variances. *Econometrics Reviews*, 5, 1-50.

Jöreskog, K.G. (1979). Author's addendum to: A general approach to confirmatory maximum likelihood factor analysis. In K.G. Jöreskog, and D. Sörbom (eds.) *Advances in Factor Analysis and Structural Equation Models.* Cambridge, MA: Abt Books.

Kraft, D. and Engle, R.F. (1982). ARCH in multiple time-series models. USCD Discussion Paper #82-23.

Received 27 January 1987
Revised 15 May 1987

Federal Reserve Board
Special Studies Section, Stop 180
Division of Research & Statistics
20th & Constitution Ave., N.W.
Washington, DC 20551
U.S.A.

Department of Economics
University of Pennsylvania
3718 Locust Walk
Philadelphia, PA 19104-6297
U.S.A.

Proc. Second International Tampere Conference in Statistics
(Tampere, Finland, 1-4 June 1987)
Tarmo Pukkila and Simo Puntanen, *Editors*

pp. 439 - 444

Simultaneous confidence intervals when the regressor matrix is unbalanced

R.W. FAREBROTHER
University of Manchester, U.K.

Key words and phrases: Farebrother's linear unbiased estimator, statistical tests.

ABSTRACT

In this paper we will use Farebrother's (1976) method to correct the balance of an unbalanced regressor matrix. This technique generates a linear unbiased estimator of the parameters which may be used in place of the ordinary least squares estimator as the basis of certain implicit or explicit statistical tests. In particular we will develop a set of simultaneous confidence intervals based on this estimator.

1. INTRODUCTION

Consider the model

$$\mathbf{y} = \mathbf{X}\boldsymbol{\beta} + \boldsymbol{\varepsilon}, \tag{1.1}$$

where $\mathbf{y}$ is an $n \times 1$ matrix of observations on the dependent variable, $\mathbf{X}$ is an $n \times k$ matrix of observations on the k regressors, $\boldsymbol{\beta}$ is a $k \times 1$ matrix of parameters, and $\boldsymbol{\varepsilon}$ is an $n \times 1$ matrix of disturbances which are independently and identically normally distributed with zero means and common variance σ^2

$$\boldsymbol{\varepsilon} \sim N(\mathbf{0}, \sigma^2 \mathbf{I}_n). \tag{1.2}$$

We shall also suppose that the $n \times k$ matrix $\mathbf{X}$ has full column rank k so that the least squares estimator of $\boldsymbol{\beta}$ is given by

$$\hat{\boldsymbol{\beta}} = (\mathbf{X}'\mathbf{X})^{-1}\mathbf{X}'\mathbf{y} \tag{1.3}$$

and the corresponding least squares residuals by

$$\hat{\boldsymbol{\varepsilon}} = \mathbf{y} - \mathbf{X}\hat{\boldsymbol{\beta}}. \tag{1.4}$$

Now $\hat{\boldsymbol{\beta}}$ is the best (minimum variance) linear unbiased estimator of $\boldsymbol{\beta}$ and $\hat{\boldsymbol{\varepsilon}}$ is the best (minimum mean squared error) linear unbiased approximator of $\boldsymbol{\varepsilon}$ but

$$\hat{\boldsymbol{\varepsilon}} = \mathbf{My} = \mathbf{M}\boldsymbol{\varepsilon}, \tag{1.5}$$

where

$$\mathbf{M} = \mathbf{I}_n - \mathbf{X}(\mathbf{X}'\mathbf{X})^{-1}\mathbf{X}' \tag{1.6}$$

so that

$$\hat{\boldsymbol{\varepsilon}} \sim N(\mathbf{0}, \sigma^2\mathbf{M}) \tag{1.7}$$

and $\hat{\boldsymbol{\varepsilon}}$ has a variance matrix, $\sigma^2\mathbf{M}$, whose elements are functions of the elements of the matrix of regressors $\mathbf{X}$. This feature is inconvenient for testing purposes and there is a considerable literature based on the use of linear unbiased approximators of $\boldsymbol{\varepsilon}$ with scalar variance matrices (LUS approximators)

$$\boldsymbol{\eta} = \mathbf{Z}'\mathbf{y} = \mathbf{Z}'\boldsymbol{\varepsilon}, \tag{1.8}$$

where $\mathbf{Z}$ is an $n\times(n-k)$ matrix of rank $n-k$ satisfying $\mathbf{Z}'\mathbf{X} = \mathbf{0}$ and $\mathbf{Z}'\mathbf{Z} = \mathbf{I}_{n-k}$ so that $\mathbf{M} = \mathbf{ZZ}'$ and

$$\boldsymbol{\eta} \sim N(\mathbf{0}, \sigma^2\mathbf{I}_{n-k}). \tag{1.9}$$

See Farebrother (1985) for a recent survey.

It is less well known that $\hat{\boldsymbol{\beta}}$ is also an inconvenient foundation for certain explicit or implicit tests. In this paper we will show that there are alternative linear unbiased estimators of $\boldsymbol{\beta}$ which form a more convenient basis for such tests and, in particular, serve as the basis for a set of simultaneous confidence intervals.

2. STANDARD RESULTS

Let $\mathbf{R}$ be a $q\times k$ matrix of rank $p \le q$ and suppose that we are interested in obtaining a set of simultaneous confidence intervals for the q linear combinations of the parameters $\mathbf{R}_{1\cdot}\boldsymbol{\beta}$, $\mathbf{R}_{2\cdot}\boldsymbol{\beta}$,..., $\mathbf{R}_{q\cdot}\boldsymbol{\beta}$ where $\mathbf{R}_{i\cdot}$ represents the ith row of $\mathbf{R}$. Without loss of generality we may suppose that the rows of $\mathbf{R}$ have been normalised so that $\mathbf{R}_{i\cdot}\mathbf{VR}'_{i\cdot} = 1$ where $\mathbf{V} = (\mathbf{X}'\mathbf{X})^{-1}$. Thus we have

$$\mathbf{R}\hat{\boldsymbol{\beta}} \sim N(\mathbf{R}\boldsymbol{\beta}, \sigma^2\mathbf{P}), \tag{2.1}$$

where $\mathbf{P} = \mathbf{RVR}'$ is a $q \times q$ correlation matrix.

Let

$$\hat{\sigma}^2 = \hat{\boldsymbol{\varepsilon}}' \hat{\boldsymbol{\varepsilon}} / (n-k) \tag{2.2}$$

be the usual least squares estimator of σ^2 then

$$(\mathbf{R}_{i\cdot}\hat{\boldsymbol{\beta}} - \mathbf{R}_{i\cdot}\boldsymbol{\beta}) / \hat{\sigma} \tag{2.3}$$

has a univariate t distribution with $n-k$ degrees of freedom and

$$Pr[\,|\mathbf{R}_{i\cdot}\hat{\boldsymbol{\beta}} - \mathbf{R}_{i\cdot}\boldsymbol{\beta}| < \hat{\sigma}\, t(n-k, 1-\gamma)\,] = 1 - 2\gamma \tag{2.4}$$

represents a 1-2γ level confidence interval for $\mathbf{R}_{i\cdot}\boldsymbol{\beta}$ where $t(n\text{-}k, 1\text{-}\gamma)$ is the upper γ level critical value of a t distribution with n-k degrees of freedom.

Using the Bonferroni inequality to combine q intervals of the form (2.4) we have

$$Pr[\,|\mathbf{R}_{i\cdot}\hat{\boldsymbol{\beta}} - \mathbf{R}_{i\cdot}\boldsymbol{\beta}| < \hat{\sigma}\, t(n-k, 1-\gamma) \text{ for } i=1,2,\ldots, q\,] \geq 1 - 2q\gamma \tag{2.5}$$

or

$$Pr[\,\mathbf{R}_{i\cdot}\boldsymbol{\beta} \in \mathbf{R}_{i\cdot}\hat{\boldsymbol{\beta}} \pm \hat{\sigma}\, t(n-k, 1-\alpha/2q) \text{ for } i=1,2,\ldots, q] \geq 1 - \alpha \tag{2.6}$$

where $\alpha = 2q\gamma$.

Now equation (2.5) may be written in the form

$$Pr[\max|\,\mathbf{R}_{i\cdot}\hat{\boldsymbol{\beta}} - \mathbf{R}_{i\cdot}\boldsymbol{\beta}\,| / \hat{\sigma} < t(n-k,\ 1-\alpha/2q)] \geq 1 - \alpha \tag{2.7}$$

and $(\mathbf{R}\hat{\boldsymbol{\beta}} - \mathbf{R}\boldsymbol{\beta})/\hat{\sigma}$ has a multivariate t distribution so that an exact confidence region of the form (2.5) namely

$$Pr[\,\max|\,\mathbf{R}_{i\cdot}\hat{\boldsymbol{\beta}} - \mathbf{R}_{i\cdot}\boldsymbol{\beta}\,| / \hat{\sigma} < m(q, n-k, \mathbf{P}, 1-\alpha)] = 1 - \alpha \tag{2.8}$$

or

$$Pr[\mathbf{R}_{i\cdot}\boldsymbol{\beta} \in \mathbf{R}_{i\cdot}\hat{\boldsymbol{\beta}} \pm \hat{\sigma} m(q, n-k, \mathbf{P}, 1-\alpha) \quad \text{for } i=1,2,\ldots, q] = 1 - \alpha \tag{2.9}$$

may be obtained if the user is willing to evaluate the upper α level critical value of the studentised maximum modulus distribution for the particular correlation matrix $\mathbf{P}$ he happens to have chosen.

In practice most users would prefer to approximate the unknown exact value $m(q, n-k, \mathbf{P}, 1-\alpha)$ by the value $m(q, n-k, \mathbf{I}, 1-\alpha)$ tabulated by Bechhofer and Dunnett (1982) and use Šidák's (1967) result to obtain the confidence region

$$Pr[\mathbf{R}_{i\cdot}\boldsymbol{\beta} \in \mathbf{R}_{i\cdot}\hat{\boldsymbol{\beta}} \pm \hat{\sigma} m(q, n-k, \mathbf{I}, 1-\alpha) \text{ for } i=1,2,\ldots, q] \geq 1 - \alpha. \tag{2.10}$$

Further, we note that equation (2.7) may be written as

$$Pr\left[\max_i \frac{(\mathbf{R}_{i\cdot}\hat{\boldsymbol\beta} - \mathbf{R}_{i\cdot}\boldsymbol\beta)^2}{\hat\sigma^2\mathbf{R}_{i\cdot}\mathbf{V}\mathbf{R}'_{i\cdot}} > t^2(n-k, 1-\alpha/2q)\right] \geq 1-\alpha \qquad (2.11)$$

so that a confidence region of the form

$$\Pr[\mathbf{R}_{i\cdot}\boldsymbol\beta \in \mathbf{R}_{i\cdot}\hat{\boldsymbol\beta} \pm [p\hat\sigma^2 F(p, n-k, 1-\alpha)]^{1/2} \quad \text{for } i=1,2,\ldots,q] \geq 1-\alpha \qquad (2.12)$$

can be deduced from the exact confidence region

$$Pr\left[\max_c \frac{(\mathbf{c}'\mathbf{R}\hat{\boldsymbol\beta} - \mathbf{c}'\mathbf{R}\boldsymbol\beta)^2}{\hat\sigma^2\mathbf{c}'\mathbf{R}\mathbf{V}\mathbf{R}'\mathbf{c}} < pF(p, n-k, 1-\alpha)\right] = 1-\alpha \qquad (2.13)$$

or

$$Pr[(\mathbf{R}_0\hat{\boldsymbol\beta}-\mathbf{R}_0\boldsymbol\beta)'(\mathbf{R}_0\mathbf{V}\mathbf{R}'_0)^{-1}(\mathbf{R}_0\hat{\boldsymbol\beta}-\mathbf{R}_0\boldsymbol\beta)/\hat\sigma^2 < pF(p, n-k, 1-\alpha)] = 1-\alpha \qquad (2.14)$$

where the $q \times k$ matrix $\mathbf{R}$ has rank p and $\mathbf{R}_0$ is a $p \times k$ submatrix of $\mathbf{R}$ of rank p.

Finally we note that Richmond (1982) has shown that equation (2.8) implies that

$$Pr[\,|\mathbf{c}'\mathbf{R}\hat{\boldsymbol\beta} - \mathbf{c}'\mathbf{R}\boldsymbol\beta| < \hat\sigma\phi(c)\, m(q, n-k, \mathbf{P}, 1-\alpha) \text{ for all } c\,] = 1-\alpha, \qquad (2.15)$$

where

$$\phi(c) = \min \sum_{i=1}^{q} |d_i| \quad \text{for } \mathbf{c}'\mathbf{R} = \mathbf{d}'\mathbf{R}. \qquad (2.16)$$

3. NONSTANDARD RESULTS

The major difficulty with the standard results on simultaneous confidence intervals outlined in section 2 is that the widths of the confidence intervals are determined by the values of $(\mathbf{R}_{i\cdot}\mathbf{V}\mathbf{R}'_{i\cdot})^{1/2}$ and are identifical if $\mathbf{RVR}'$ is a correlation matrix, as we shall assume. In this section we shall address the problem of obtaining simultaneous confidence intervals for $\mathbf{R}_{1\cdot}\boldsymbol\beta$, $\mathbf{R}_{2\cdot}\boldsymbol\beta,\ldots,\mathbf{R}_{q\cdot}\boldsymbol\beta$ whose widths are proportional to given values $w_1, w_2,\ldots,w_q$.

Suppose that $t(n-k, 1-\gamma_i) = hw_i$ for $i=1,2,\ldots, q$ then (2.5) may be replaced by

$$Pr[\,|\mathbf{R}_{i\cdot}\hat{\boldsymbol\beta} - \mathbf{R}_{i\cdot}\boldsymbol\beta| < \hat\sigma hw_i \text{ for } i=1,2,\ldots,q] \geq 1-2\sum_{i=1}^{q}\gamma_i \qquad (3.1)$$

and we have a set of simultaneous confidence intervals of the desired form provided $2\Sigma\gamma_i \leq \alpha$.

For our generalisations of equations (2.9), (2.10) and (2.12) we shall require an estimator of $\mathbf{R\beta}$ of the form

$$\tilde{\boldsymbol{\delta}} = \mathbf{R}\hat{\boldsymbol{\beta}} + \mathbf{S}\boldsymbol{\eta}_1 \tag{3.2}$$

where $\mathbf{S}$ is a $q \times f$ matrix of rank f and $\boldsymbol{\eta}_1$ is an $f \times 1$ matrix containing the first f elements of the $(n-k) \times 1$ matrix $\boldsymbol{\eta}$ defined in equation (1.8). Now $\tilde{\boldsymbol{\delta}}$ is a linear unbiased estimator of $\mathbf{R\beta}$ with variance matrix

$$var(\tilde{\boldsymbol{\delta}}) = \sigma^2\mathbf{RVR}' + \sigma^2\mathbf{SS}' \tag{3.3}$$

so that we may choose the elements of $\mathbf{S}$ in such a way that $\tilde{\boldsymbol{\delta}}$ has variance matrix $\sigma^2\,\mathbf{Q}$ provided only that $\mathbf{Q} - \mathbf{RVR}'$ is positive semidefinite and of rank $f \leq n-k$. In particular we may choose a matrix $\mathbf{Q}$ with diagonal elements proportional to $w_1^{\,2}, w_2^{\,2}, \ldots, w_q^{\,2}$.

Let $\mathbf{D}$ be a $q \times q$ diagonal matrix with ith diagonal element $d_{ii} = q_{ii}^{\,1/2}$, let $\mathbf{P}_0 = \mathbf{D}^{-1}\mathbf{Q}\mathbf{D}^{-1}$ and let

$$\tilde{\sigma}^2 = \boldsymbol{\eta}_2{}'\boldsymbol{\eta}_2/(n-k-f). \tag{3.4}$$

Then $(\tilde{\sigma}\mathbf{D})^{-1}(\tilde{\boldsymbol{\delta}} - \mathbf{R\beta})$ has a multivariate t distribution and we have

$$Pr[\mathbf{R}_{i}.\boldsymbol{\beta} \in \tilde{\delta}_i \pm d_{ii}\tilde{\sigma}m(q, n-k-f, \mathbf{P}_0, 1-\alpha) \text{ for } i=1,2,\ldots,q] = 1-\alpha \tag{3.5}$$

$$Pr[\mathbf{R}_{i}.\boldsymbol{\beta} \in \tilde{\delta}_i \pm d_{ii}\tilde{\sigma}m(q, n-k-f, \mathbf{I}, 1-\alpha) \text{ for } i=1,2,\ldots,q] \geq 1-\alpha \tag{3.6}$$

and

$$Pr[\mathbf{R}_{i}.\boldsymbol{\beta} \in \tilde{\delta}_i \pm d_{ii}[p_0\tilde{\sigma}^2F(p_0, n-k-f, 1-\alpha)]^{1/2} \text{ for } i=1,2,\ldots,q] \geq 1-\alpha \tag{3.7}$$

corresponding to equations (2.9), (2.10) and (2.12) respectively where p_0 denotes the rank of $\mathbf{Q}$. For choices of $w_1, w_2, \ldots, w_q$ remote from equality the confidence regions given by equations (3.5), (3.6) or (3.7) may be considerably narrower than the regions with the same dimensions deduced from the standard results of section 2.

4. THE ESTIMATOR $\tilde{\boldsymbol{\beta}}$

The nature of the estimator $\tilde{\boldsymbol{\delta}}$ is perhaps controversial. We shall therefore now examine this estimator in greater detail when $\mathbf{R} = \mathbf{I}_k$. That is we shall examine the nature of the estimator

$$\tilde{\boldsymbol{\beta}} = \hat{\boldsymbol{\beta}} + \mathbf{S}\boldsymbol{\eta}_1 \tag{4.1}$$

Let $\mathbf{Z} = [\mathbf{Z}_1 : \mathbf{Z}_2]$ be partitioned by its first f columns and the remaining $n-k-f$ columns and let

$$\mathbf{A} = \mathbf{X}(\mathbf{X}'\mathbf{X})^{-1} + \mathbf{Z}_1\mathbf{S}' \tag{4.2}$$

then $\tilde{\boldsymbol{\beta}} = \mathbf{A}'\mathbf{y}$ and $\mathbf{A}'\mathbf{X} = \mathbf{I}$ so that $\tilde{\boldsymbol{\beta}}$ is a linear unbiased estimator of $\boldsymbol{\beta}$.

Let $\mathbf{B} = \mathbf{A}(\mathbf{A}'\mathbf{A})^{-1/2}$ and let $\mathbf{C}$ be an $n \times f$ matrix satisfying $\mathbf{C}'\mathbf{B}=\mathbf{0}$, $\mathbf{C}'\mathbf{Z}_2=\mathbf{0}$ and $\mathbf{C}'\mathbf{C}=\mathbf{I}$ then $[\mathbf{B} : \mathbf{C} : \mathbf{Z}_2]$ is an $n \times n$ orthogonal matrix and we may premultiply model (1.1) by its transpose to obtain

$$\begin{pmatrix} \mathbf{B}'\mathbf{y} \\ \mathbf{C}'\mathbf{y} \\ \mathbf{Z}_2'\mathbf{y} \end{pmatrix} = \begin{pmatrix} \mathbf{B}'\mathbf{X} \\ \mathbf{C}'\mathbf{X} \\ \mathbf{0} \end{pmatrix} \boldsymbol{\beta} + \begin{pmatrix} \mathbf{B}'\boldsymbol{\varepsilon} \\ \mathbf{C}'\boldsymbol{\varepsilon} \\ \mathbf{Z}_2'\boldsymbol{\varepsilon} \end{pmatrix} \tag{4.3}$$

Deleting the second block of observations we have

$$\tilde{\boldsymbol{\beta}} = (\mathbf{B}'\mathbf{X})^{-1}\mathbf{B}'\mathbf{y} \tag{4.4}$$

from the first block and

$$\tilde{\sigma}^2 = (\mathbf{Z}_2'\mathbf{y})'\mathbf{Z}_2'\mathbf{y} / (n-k-f) \tag{4.5}$$

from the third.

Thus the simultaneous confidence intervals developed in section 3 are based on an inefficient use of the information contained in model (1.1).

The technique developed in this paper may also be used to simplify multiple comparisons tests and tests of $\mathbf{R}\boldsymbol{\beta} = \boldsymbol{\gamma}$ against $\mathbf{R}\boldsymbol{\beta} \geq \boldsymbol{\gamma}$.

REFERENCES

Bechhofer, R. E. and Dunnett, C.W. (1982). Multiple comparisons for orthogonal contrasts: Examples and tables. *Technometrics*, 24, 213-222.

Farebrother, R.W. (1976). The minimum proportional variance unbiased linear estimator of β and simultaneous confidence intervals. *Journal of the American Statistical Association*, 71, 761-762.

Farebrother, R. W. (1978). Simultaneous confidence intervals with an application to the Parkin, Sumner and Ward model of wage inflation. *The Manchester School*, 46, 360-363.

Farebrother, R. W. (1985). Linear unbiased approximators of the disturbances in the standard linear model. *Linear Algebra and Its Applications*, 67, 259-274.

Richmond, J. (1982). A general method for constructing simultaneous confidence intervals. *Journal of the American Statistical Association*, 77, 455-460.

Šidák, Z. (1967). Rectangular confidence regions for the means of multivariate normal distributions. *Journal of the American Statistical Association*, 62, 626-633.

Received 16 January 1987

Department of Econometrics & Social Statistics
University of Manchester
Manchester M13 9PL
U.K.

Proc. Second International Tampere Conference in Statistics
(Tampere, Finland, 1-4 June 1987)
Tarmo Pukkila and Simo Puntanen, *Editors*

pp. 445 - 456

Some Fourier procedures for time domain analysis of parametric and semi-parametric time series models

Andrey FEUERVERGER
University of Toronto, Toronto, Canada

Key words and phrases: Empirical characteristic function, nonlinear time series, testing independence, asymptotic efficiency.

ABSTRACT

We study an extension of the empirical characteristic function for stationary time series. A general theorem on the asymptotic efficiency of related inference procedures is given for the time series context. Fourier methods are proposed for testing independence in the time series context. Estimation procedures in time series models based on independence testing are then developed that are adaptive in the sense that the distribution of the error terms need not be specified. A bound relating to the asymptotic efficiency of such procedures is given which depends on a measure of the intrinsic nonlinearity of the time series model.

1. INTRODUCTION AND SUMMARY

The object of this paper is to explore some aspects of the idea that analysis of stationary time series may be carried out effectively by means of certain characteristic function (cf) quantities and their estimates. Two principal themes underly this effort, namely testing for independence by means of the empirical cf (ecf) -- particularly its extension to stationary processes -- and notions of the identification and modelling of linear and nonlinear stationary time series. One of the motivations for this work is to search for new practical, adaptive procedures for the identification and analysis of linear and nonlinear stationary processes without restrictive assumptions on the distribution of the error terms appearing in the models, as their distribution, in practice, is generally unknown. A second motivation is to introduce what may be a useful diagnostic procedure; this is based on examining the degree of *independence* attained by the error terms of the fitted model using certain cf based test procedures which are quite natural in the general context.

If $X_j = (X_j^1, \ldots, X_j^p)'$, $j = 1,2,\ldots,n$ is an iid sample from a p-variate distribution whose cf is $c(t) = E\, e^{it'X_1}$ where $t = (t_1, \ldots, t_p)'$, then the ecf is defined as

$$c_n(t) = \frac{1}{n}\sum_{j=1}^{n} e^{it'X_j} . \tag{1.1}$$

The properties of this estimator are now fairly well understood; a detailed review is provided, for example, in Feuerverger (1987a). If however $\{X_j\}$ is a univariate strictly stationary process then certain analogous quantities may be defined as

$$c^p(t) = E\, e^{it'Y_1^p} \quad \text{and} \quad c_n^p(t) = \frac{1}{n}\sum_{j=1}^{n} e^{it'Y_j^p} \tag{1.2}$$

where $Y_j^p = (X_j, X_{j-1}, \ldots, X_{j-p})'$. Feuerverger and McDunnough (1981) have termed these the poly-cf (pcf) and the empirical pcf (epcf) to distinguish from the iid context, for while $c^p(t)$ is a cf in the ordinary sense, $c_n^p(t)$ is unlike the ecf owing to the statistical dependence amongst its terms. Some further investigation into the properties of the epcf is undertaken here in section 3. The testing for process independence -- which has a natural characterization by means of the pcf -- is of special interest here and a treatment of testing for independence in the time series context using the epcf is given in section 3. Preparatory to this, in section 2 we briefly discuss some ecf based tests for independence in the multivariate iid context.

While not all of these are addressed here, our motivating interest concerns the identification, fitting, and diagnosis of univariate, discrete time, strictly stationary series $\{X_j\}_{j=-\infty}^{\infty}$ which satisfy certain mixing and invertability conditions and lend themselves to physically well-behaved prediction procedures provided that we can observe a sufficiently long segment of a realization. Typical processes of this type will admit, for example, infinite Volterra functional expansions of the form

$$X_j = \sum_{k=1}^{\infty} a_k X_{j-k} + \sum_{k=1}^{\infty}\sum_{l=1}^{\infty} a_{kl} X_{j-k} X_{j-l} + \sum_{k=1}^{\infty}\sum_{l=1}^{\infty}\sum_{m=1}^{\infty} a_{klm} X_{j-k} X_{j-l} X_{j-m} + \cdots + \xi_j \tag{1.3}$$

where ξ_j is an iid sequence and independent of the process past. The restrictions associated with a representation such as (1.3) are more or less essential in order that a process be generally amenable to effective statistical analysis. In section 4 procedures are presented which depend upon prediction of a process given its past, where estimation of the required prediction is based upon measuring (and minimizing) -- by means of the epcf functions -- the dependency in the fitted noise. We also report on some investigations concerning asymptotic efficiency of such procedures in the adaptive context where the error distribution is not specified. It turns out that a measure of the degree of nonlinearity of the model plays an important role and that asymptotically arbitrarily high efficiency is generally possible only when the model is *intrinsically linear*. However, even in nonlinear cases the measure of efficiency is typically high, and in any case a useful diagnostic tool emerges.

To conserve space, a number of proofs are omitted from the text below. The omitted proofs may be found in Feuerverger (1987a,b,c). Additional details are given in an unpublished technical report available from the author.

2. AN ECF TEST FOR INDEPENDENCE

The use of the ecf in testing for independence is considered in Csorgo (1985) and Feuerverger (1987a). In this section we briefly describe such a test for the iid case. This is adapted to the time series context in section 3 and applied in section 4.

Following the notation of (1.1) and introducing the obvious marginal quantities $c^l(t^l)$ and $c_n^l(t^l)$ for $l = 1,2,\ldots,p$, the null hypothesis H_0 of independence may be expressed as H_0: $c(t) = \prod_{l=1}^{p} c^l(t^l)$, all $t \in R^p$. Now the central quantity in our tests for independence is $\Gamma(t) = c(t) - \prod_{l=1}^{p} c^l(t^l)$ and its asymptotically normal empirical counterpart $\Gamma_n(t) = c_n(t) - \prod_{l=1}^{p} c_n^l(t^l)$. Of course $\Gamma_n(t) \rightarrow \Gamma(t)$ a.s. for all $t \in R^p$ and the independence hypothesis may be expressed as H_0: $\Gamma(t) = 0$, $t \in R^p$. It is also readily shown that $E\,\Gamma_n(t) = \Gamma(t) + O(n^{-1})$ uniformly in t. The finite sample and asymptotic first and second moment structure for $\Gamma_n(t)$ are given, for example, in theorems 4.1 and 4.2 of Feuerverger (1987a). In particular, under H_0, $\mathrm{cov}\,(\Gamma_n(s),\Gamma_n(t))$ depends only upon the quantities $c^l(s^l-t^l)$, $c^l(s^l)$, $c^l(t^l)$ and is readily computed even for s, t ranging over some appropriate grid. We remark that the process $\Gamma_n(t)$ will inherit certain weak convergence characteristics from the ecf, but it suffices here that the finite-dimensional marginals of $\Gamma_n(t)$ are, by virtue of the central limit theorem, asymptotically multivariate normal. We may now describe our ecf based tests for independence in the iid case as follows: First select k fixed p-vectors t_j, $j = 1, \ldots, k$ such that the 2k points $-t_k, \ldots, -t_1, t_1, \ldots, t_k$ are all distinct, and form the 2k-dimensional vector $\hat{\xi} = (\Gamma_n(-t_k), \ldots, \Gamma_n(-t_1), \Gamma_n(t_1), \ldots, \Gamma_n(t_k))'$. Now the complex covariance structure Σ_ξ for $\hat{\xi}$ under H_0 is known (indeed, we may use its asymptotic form) and further may be estimated as $\hat{\Sigma}_\xi$ by replacing all cf's by their respective ecf's. Our test will then consist of rejecting H_0 for large values of the test statistic $\hat{\xi}^* \hat{\Sigma}_\xi^{-1} \hat{\xi}$ (here $*$ is conjugate transpose) which has, under H_0, asymptotically a chi-squared distribution with 2k degrees of freedom. In accordance with the usual multivariate theory, this test is asymptotically the uniformly most powerful invariant test based upon the statistics $\Gamma_n(\pm t_j)$. We remark here that this Γ_n-based statement can be referred to the class of tests based on the $c_n^p(\pm t_j)$ and their marginals (see Feuerverger, 1987a). Finally we remark that while the grid $\{t_j\}$ is regarded here and throughout as fixed and finite, in practice we may wish to use only a moderate number of data-dependent gridpoints. Heuristic arguments, as well as numerical experience as in Feuerverger and McDunnough (1981), and Feuerverger (1987a) suggest no practical difficulties in this.

3. THE EPCF AND THE TIME SERIES CONTEXT

In this section $\{X_j\}_{j=-\infty}^{\infty}$ is a univariate, strictly stationary, discrete time series of which a finite segment has been observed. Let $Y_j^p = (X_j, X_{j-1}, \cdots, X_{j-p})'$ -- a vector of length (p+1). When the underlying value of p is understood, we may write Y_j instead of Y_j^p; $\{Y_j\}$ is a stationary vector process. The cf of a stretch of length (p+1) of our time series is called the p-th order poly-cf (pcf):

$$c^p(t) = Ee^{it'Y_1^p} \tag{3.1}$$

where $t = (t^0, t^1, \cdots, t^p)'$; and the corresponding empirical quantity,

$$c_n^p(t) = \frac{1}{n} \sum_{j=1}^{n} e^{it'Y_j^p} \tag{3.2}$$

is the p-th order empirical pcf (epcf). Note that p may equal zero. We remark here again that while the pcf is an ordinary cf, the epcf is unlike the ecf owing to the statistical dependence amongst its terms.

For our purposes we invariably require that $\{X_j\}$ be ergodic (so that a single realization suffices to determine the underlying probability distribution) and further that $\{X_j\}$ satisfy certain mixing conditions as well. The following statistically intuitive condition will be more than sufficient for our needs: Define the maximal correlation $\rho(m)$ to be the supremum of the correlations between square integrable random variables U and V measurable with respect to the σ-fields $\sigma\{\cdots, X_{-1}, X_o\}$ and $\sigma\{X_m, X_{m+1}, \cdots\}$ respectively. (Thus $\rho(m)$ is the maximum correlation possible between functions of the X_j's separated in time by a distance m.) The mixing condition is

$$\sum_{m=1}^{\infty} \rho(m) < \infty . \tag{3.3}$$

We remark that (3.3) implies ergodicity.

We state first a basic consistency result.

Theorem 3.1 If $\{X_j\}$ is a stationary, ergodic time series then

(i) $c_n^p(t) \to c^p(t)$ a.s. for all $t \in R^{p+1}$

(ii) $\sup_{-T \le t^o, \ldots, t^p \le T} |c_n^p(t) - c^p(t)| \to 0$ a.s. for fixed $0 \le T < \infty$.

The proof of this result is given in Feuerverger (1987c).

Define now the *epcf process*

$$W_n^p(t) = \sqrt{n}\left(c_n^p(t) - c^p(t)\right) . \tag{3.4}$$

Since $E\, c_n^p(t) = c^p(t)$ we have $E\, W_n^p(t) = 0$. Further

$$\operatorname{cov}\left(W_n^p(s), W_n^p(t)\right) = n \cdot \operatorname{cov}\left(c_n^p(s), c_n^p(t)\right)$$

$$= \frac{1}{n} \sum_{j=1}^{n} \sum_{l=1}^{n} \operatorname{cov}\left(e^{is'Y_j^p}, e^{it'Y_l^p}\right) \tag{3.5}$$

$$= \sum_{r=n+1}^{n-1} \frac{n-|r|}{n} \operatorname{cov}\left(e^{is'Y_o^p}, e^{it'Y_r^p}\right) \tag{3.6}$$

since by stationarity the terms in (3.5) depend only on $r = l-j$, and so

$$\lim_{n\to\infty} \operatorname{cov}\left(W_n^p(s), W_n^p(t)\right) = \lim n\cdot\operatorname{cov}\left(c_n^p(s), c_n^p(t)\right)$$

$$= \sum_{r=-\infty}^{\infty} \operatorname{cov}\left(e^{is'Y_o^p}, e^{it'Y_r^p}\right) \tag{3.7}$$

since (3.7) converges absolutely due to (3.3). Now in the case where the process $\{X_j\}$ is an iid sequence, only the terms $-p \le r \le p$ will contribute in (3.7) in which case (3.7) becomes

$$\prod_{l=0}^{p} c(s^l - t^l) - \prod_{l=0}^{p} c(s^l)\overline{c(t^l)}$$

$$+ \sum_{r=1}^{p}\left\{\prod_{l=0}^{r-1} c(s^l)\cdot\prod_{l=r}^{p} c(s^l - t^{l-r})\cdot\prod_{l=p-r+1}^{p}\overline{c(t^l)} - \prod_{l=0}^{p} c(s^l)\overline{c(t^l)}\right\}$$

$$+ \sum_{r=1}^{p}\left\{\prod_{l=0}^{r-1}\overline{c(t^l)}\cdot\prod_{l=0}^{p-r} c(s^l - t^{l+r})\cdot\prod_{l=p-r+1}^{p} c(s^l) - \prod_{l=0}^{p} c(s^l)\overline{c(t^l)}\right\}. \tag{3.8}$$

In this expression we have taken $s = (s^0, ..., s^p)'$, $t = (t^0, ..., t^p)'$ and have written $c(\cdot)$ throughout in place of the 0-th order pcf function $c^o(\cdot)$. An examination of this expression shows that its computation is in fact straightforward, and further that even if s and t range over some appropriate grid in R^{p+1}, the tensor of covariance values defined by (3.8) is readily computed, a fact which plays a helpful role in the procedures proposed below.

The following distributional result is also required:

Theorem 3.2 Let $W^p(t)$, $t \in R^{p+1}$ be a zero-mean complex valued Gaussian process with covariance structure as given by (3.7) and assume $\{X_j\}$ satisfies (3.3). Then $W_n^p(t_1), ..., W_n^p(t_k)$ converges in distribution to $W^p(t_1), ..., W^p(t_k)$ for all k and $t_1, t_2, ..., t_k$ in R^{p+1}.

A proof of this result is given in Feuerverger (1987c). We remark that the necessary and sufficient conditions for weak convergence of the epcf process is a difficult question of considerable probabilistic interest, but falls outside the scope of our work.

Turning now to questions of inference, we next give a result which states that the epcf lends itself, under general conditions, to asymptotically efficient inference in stationary processes. In this context, the distribution of the process $\{X_j\}$ depends on a real vector parameter θ having unknown true value θ_o. Being rather lengthy, our proof will be given elsewhere (see Feuerverger 1987c). The 'general conditions' referred to below are of the type generally required for maximum likelihood estimation to be asymptotically efficient in stationary processes; to save space these are not repeated here.

Theorem 3.3 Let $\{X_j\}$ be a stationary process having distribution belonging to the class defined by the pcf functions $c_\theta^p(t)$. Under general conditions, the procedure based on fitting $c_\theta^p(t)$ to the epcf $c_n^p(t)$ at a grid of points $t_j \in R^{p+1}$, $j=1,2,...,k$ by non-linear least squares, weighted in accordance with a consistent estimator of the covariances as given by (3.7) results in an estimator which is asymptotically normal and which can be made to have arbitrarily high asymptotic efficiency provided that p is sufficiently large and the grid $\{t_j\}$ is sufficiently fine and extended.

While theorem 3.3 is of theoretical interest, implementation of the implied procedure requires the pcf functions and their covariances (3.7). This can limit the feasability of the procedure. On the other hand, the covariance structure in the iid case (3.8) is seen, on examination, to be readily computed and consistently estimated, even for s, t ranging over an appropriate grid in R^{p+1}. This suggests studying time series estimation procedures based on testing for independence amongst the fitted residuals in time series models by means of the epcf functions and their properties.

Turning therefore to aspects of testing for process independence, define here

$$\Gamma^p(t) = c^p(t) - \prod_{l=1}^{p+1} c^o(t^l), \quad t \in R^{p+1} \tag{3.9}$$

where $c^p(\cdot)$ and $c^o(\cdot)$ are the p-th and 0-th order pcf's given by (3.1). Define also the corresponding empirical quantity

$$\Gamma_n^p(t) = c_n^p(t) - \prod_{l=1}^{p+1} c_n^o(t^l). \tag{3.10}$$

In view of theorem 3.1 we know that $\Gamma_n^p(t) \to \Gamma^p(t)$ a.s. and indeed uniformly on compact sets. Our underlying null hypothesis H_o here is that $\{X_j\}$ is an iid sequence. This may be expressed as H_o: $\Gamma^p(t) \equiv 0$, all p, $t \in R^{p+1}$.

Consider now the evaluation of $E\,\Gamma_n^p(t)$. Although it is clear that we may obtain an exact expansion here, in the general case this will involve the pcf functions of all orders up to n and so will be rather impractical. We have, however, the following:

Theorem 3.4 If the time series $\{X_j\}$ is stationary and satisfies the mixing condition (3.3) then

$$E\,\Gamma_n^p(t) = \Gamma^p(t) + 0(n^{-1}) \tag{3.11}$$

where the order term is uniform in t. The null hypothesis H_o of process independence is equivalent to the assertion that $E\,\Gamma_n^p(t) = 0(n^{-1})$, all p.

Proof. To determine $E\,\Gamma_n^p(t)$ we first observe that $E\,c_n^p(t) = c^p(t)$ and then proceed to examine

$$E \prod_{l=0}^{p} c_n^o(t^l) = E \prod_{l=0}^{p} \left(\frac{1}{n} \sum_{j_l=1}^{n} e^{it^l X_{j_l}} \right) . \tag{3.12}$$

There are n^{p+1} terms that arise upon expansion.

Consider now, for some fixed k, a subset $X_{i_1}, X_{i_2}, \ldots, X_{i_k}$ of k distinct terms from $\{X_j\}$ and let $M = \inf_{\substack{1 \le a,b \le k \\ a \ne b}} | i_a - i_b |$ be the minimum time separation between these k observations. Then from (3.3) we have, after some simple analysis, that

$$\left| c^{X_{i_1}, \ldots, X_{i_k}} (t_1, \ldots, t_k) - \prod_{j=1}^{k} c^o(t_j) \right| < \epsilon_M \tag{3.13}$$

uniformly in t, where $\epsilon_M \to 0$ as $M \to \infty$ (indeed $\sum \epsilon_M$ converges).

Now for a fixed M, of the n^{p+1} terms that arise on expanding (3.12), we have, to an adequate approximation, that about $n^{p+1} - n(n - 2M)(n - 4M) \ldots (n - 2pM)$ of the terms will involve j_l subscripts that admit a time separation less that M. But due to the n^{-p-1} factor their entire contribution is of order $0(n^{-1})$ uniformly. The remaining terms are subject to the condition (3.13). Hence clearly $E \prod_{l=0}^{p} c_n^o(t^l)$ can be made arbitratily close to

$\prod_{l=0}^{p} c^{o}(t^{l})$ by first selecting M large enough and then letting $n \to \infty$. The last assertion of the theorem hold because $\Gamma_n^p(t) \to \Gamma^p(t)$ a.s.

Since under H_o mixing is automatic, then under H_o we have by this theorem that $E\,\Gamma_n^{(p)}(t) = O(n^{-1})$. The exact result is easily given in this case. To state it we define for $q = 1,2,\ldots,p+1$

$$\nu_q = \sum \prod_{a=1}^{q} c^{(o)}(t^{i_{a1}} + \cdots + t^{i_{ap_a}}) \tag{3.14}$$

where the sum is over all partitions of $(0,1,\ldots,p)$ into q sets $(i_{11}, i_{12}, \ldots, i_{1p_1})$, $(i_{21}, i_{22}, \ldots, i_{1p_2})$, …, $(i_{q1}, i_{q2}, \ldots, i_{qp_q})$. Note that $\nu_{p+1} = \prod_{l=0}^{p} c^{(o)}(t^{l})$ and $\nu_1 = c^{(o)}(t^0 + t^1 + \cdots + t^p)$.

Theorem 3.5 Under H_o, and for arbitrary $p \geq 0$, we have

$$E\,\Gamma_n^{(p)}(t^0,\ldots,t^p) = \frac{n^{p+1} - n(n-1)(n-2)\cdots(n-p)}{n^{p+1}}\cdot\nu_{p+1} \tag{3.15}$$

$$- \sum_{s=1}^{p} \frac{n(n-1)(n-2)\cdots(n-p+s)}{n^{p+1}}\cdot\nu_{p+1-s}\,.$$

The proof is straightforward and omitted.

The covariance structure of $\Gamma_n^p(t)$ in general is rather complicated. We have however the following useful result:

Theorem 3.6 Under the null hypothesis H_o of process independence the asymptotic covariance structure or $\Gamma_n^p(t)$ is given by

$$\lim_{n\to\infty} n\cdot\mathrm{cov}(\Gamma_n^p(s),\Gamma_n^p(t)) = A - B \tag{3.16}$$

where A is the expression given in (3.8) and B is the expression following:

$$\sum_{l=0}^{p}\sum_{m=0}^{p}\left[c^0(s^l - t^m)\cdot\prod_{l'\neq l} c^0(s^{l'})\cdot\prod_{m'\neq m}\overline{c^0(t^{m'})} - \prod_{k=0}^{p} c^0(s^k)\overline{c^0(t^k)}\right] \tag{3.17}$$

Proof. Expand $\mathrm{cov}(\Gamma_n^p(s), \Gamma_n^p(t))$ to obtain four terms. The term $\mathrm{cov}(c_n^p(s), c_n^p(t))$ leads of course to expression A. The remaining three terms can each be evaluated for example by differential expansion followed by application of expression A. In each of the three cases B is the resulting expression giving $A - B - B + B = A - B$ for the final result.

Our proposal for a test of mutual independence of the terms in the stationary process $\{X_j\}$ now follows along the lines of the procedure outlined in section 2. We first select a value of p considered sufficiently large in relation to the mixing characteristics of the process and then select k points $t_1, t_2, \ldots, t_k \in R^{p+1}$ such that the 2k points $\pm t_1, \ldots, \pm t_k$ are distinct. Then defining here $\hat{\xi} = (\Gamma_n^p(-t_k), \ldots, \Gamma_n^p(+t_k))'$ and $\hat{\Sigma}_\xi$ as the empirical version of the complex covariance as specified in theorem 3.6 we reject the null hypothesis H_0 of independence for large values of $\hat{\xi}^* \Sigma_\xi^{-1} \xi$ based on its (asymptotic) chi-squared distribution with 2k degrees of freedom under H_0. The remarks made in section 2 concerning asymptotic properties of the test procedure (see also Feuerverger, 1987a), as well as the remarks concerning grid selection carry over unchanged to the time series context here.

4. APPLICATION TO LINEAR AND NONLINEAR TIME SERIES ANALYSIS

One motivation for this work was to initiate exploration of the applicability of certain new Fourier procedures to problems of identification, fitting, diagnosis and prediction in linear and nonlinear time series. This general subject area, especially in the nonlinear case, remains full of unresolved difficulties; some general discussion of these may be found in Priestley (1981, chapter 11) and also in the unpublished technical report referred to previously. To sidestep such difficulties here we take as our starting point an ergodic, suitably mixing, stationary, univariate series $\{X_j\}$ assumed to have been generated by a model of the form

$$X_j = h_\beta(X_{j-1}, X_{j-2}, \ldots) + \xi_j \tag{4.1}$$

where the ξ_j are iid and independent of the process past, and where h_β is an appropriate function that depends smoothly on a collection of parameters β whose unknown true value is β_0. The form of h_β must be such that its values can be suitably approximated given only finite segments of the past.

Consider now estimation of the parameters β in a model of the form (4.1). It should be noted that a procedure such as estimating β as the value which minimizes $\sum[X_j - h_\beta(X_{j-1}, \cdots)]^2$ will not be efficient except in the case that the ξ_j are Gaussian in which case we would essentially be maximizing the likelihood. (Such a procedure would however generally yield a $\sqrt{n}$-consistent estimator.) Asymptotic efficiency would be attained by minimizing $\sum \rho(X_j - h_\beta(X_{j-1}, \cdots))$ where $\rho(x) = -\log f(x)$ and $f(x)$ is the density of the ξ_j. However we are especially interested in cases where $f(x)$ is unknown.

We describe here a procedure for estimating β_0 based on the epcf functions. For arbitrary β define the 'error terms' $\xi_j(\beta) = X_j - h_\beta(X_{j-1},...)$. Select p and a grid $t_j \in R^p$, $j=1,2,...k$ and compute the $\Gamma_n^{(p)}(t)$ as defined in section 3 at $t = -t_k,...-t_1,t_1,...,t_k$ based on the $\xi_n(\beta)$ series. We shall denote by $\Gamma_n^{(p)}(\beta)$ the 2k-dimensional column vector of these $\Gamma_n^{(p)}(t)$ values, and by $\hat{\Sigma}_\beta$ the empirical version of the complex covariance matrix of $\Gamma_n^{(p)}(\beta)$ given in theorem 3.6. The proposed estimator for β_o is essentially a minimum chi-squared type estimator obtained by minimizing the expression $\Gamma_n^{(p)}(\beta)^* \hat{\Sigma}_\beta^{-1} \Gamma_n^{(p)}(\beta)$ except that β in the covariance matrix is held fixed at a $\sqrt{n}$ -consistent estimate. (One way to do this would be to carry out the minimization twice -- the first time being to obtain the needed $\sqrt{n}$ -consistent estimate using any reasonable Σ .) The minimization of this statistic is an essentially standard numerical problem and some effort can be saved by exploiting the special form of Σ . We remark in passing that the epcf also provides an interesting diagnostic tool for assessing the fit of a time series model and may also be useful in model identification even though our discussion here focusses mainly on estimation.

We need finally to obtain some idea of the asymptotic efficiency of the independence-based procedure described above for the estimation of β_o in the model (4.1). It should be noted that the optimality statement of theorem 3.3 does *not* automatically carry over to the procedure proposed above. An investigation of the efficiency of independence-testing approaches in general is carried out in Feuerverger (1987b) where it is shown that the efficiency of independence-based procedures for estimation in the model (4.1) is bounded above by

$$e = \frac{\mathrm{VAR}_{\beta_o}\left(\dfrac{\partial h(X_{j-1}, \cdots)}{\partial \beta_o}\right)}{E_{\beta_o}\left(\dfrac{\partial h_\beta(X_{j-1}, \cdots)}{\partial \beta_o}\right)^2} \tag{4.2}$$

where we are assuming for simplicity here that β is a univariate parameter. The expression (4.2) appears to measure the *intrinsic* nonlinearity of the model (4.1). Specifically, when $h_\beta(X_j,...)$ is a linear function in the past X's we obtain $e = 1$ regardless of the distribution of the error terms and that bound can be shown to be attainable in that case. For models having essentially nonlinear character (4.2) may yield values $e < 1$, but not necessarily poor efficiencies.

The numerical procedure implied here may appear involved, however a careful examination of its components reveals that it is entirely manageable, and the procedure has been successfully implemented in FORTRAN and tested on a number of models using the University of Toronto IBM 360/170. Report of a simulation study involving the estimation of time series parameters as detailed here will be made elsewhere.

Acknowledgements. This research was supported by a grant from the National Sciences and Engineering Research Council of Canada. Portions of this work were carried out at Tel Aviv University and the Hebrew University, Jerusalem whose generous hospitality and facilities provided are gratefully acknowledged. It is a pleasure to acknowledge stimulating conversations with L. Brown, P. Feigin, and G. Schwarz.

REFERENCES

Csorgo, S. (1985). Testing independence by the empirical characteristic function. J. Multivar. Anal. Vol. 16, pp. 290-299.

Feuerverger, A. (1987a). On some ecf procedures for testing independence. In *Time series and Econometric Modelling* (I.B. MacNeill and G.J. Umphrey, eds.), 189-206. Reidel, New York.

Feuerverger, A. (1987b). A bound for estimation in nonlinear time series models by independence testing methods. Letters in Probab. and Statist. To appear.

Feuerverger, A. (1987c). An efficiency result for the ecf in stationary time series models. Submitted for publication.

Feuerverger, A. (1984). On some new procedures for the analysis of stationary time series. Hebrew University, Jerusalem, June 1984. Unpublished technical report.

Feuerverger, A. and McDunnough, P. (1981). On some Fourier methods for inference. J. Amer. Statist. Assoc., Vol. 76, No. 374, pp. 379-387.

Priestley, M.B. (1981). *Spectral Analysis and Time Series.* Vol. 1. Academic Press, London.

Received 10 February 1987
Revised 29 May 1987

Department of Statistics
University of Toronto
Toronto, Ontario
Canada M5S 1A1

Proc. Second International Tampere Conference in Statistics
(Tampere, Finland, 1-4 June 1987)
Tarmo Pukkila and Simo Puntanen, *Editors*

pp. 457 - 466

Estimation of a restricted variance ratio

Alan E. GELFAND
The University of Connecticut, Storrs, CT, U.S.A.

Key words and phrases: Truncated parameter space, admissibility, Bayesian viewpoint.

ABSTRACT

The estimation of a variance ratio $\theta = \tau_2/\tau_1$ is studied under restrictions $\theta \geq \theta_0$ or $\theta \leq \theta_0$. We assume first that we observe $U_i \sim \tau_i \chi_{n_i}^2$, independent. A Bayesian viewpoint is taken. We then assume additional information on τ_i is available in the form of an independent observation from a noncentral χ^2 distribution. A natural application arises when the U_i are sums of squares in a variance components model.

1. INTRODUCTION

In this paper we consider the estimation of a variance ratio over a truncated parameter space. Suppose that $U_i \sim \tau_i \chi_{n_i}^2$, $i = 1, 2$, independent, $n_1 \geq 5$, $n_2 \geq 3$. Let $\theta = \tau_2/\tau_1$. Under the scale invariant quadratic loss function

$$L(\theta, a) = \theta^{-2}(\theta - a)^2 \tag{1.1}$$

the estimator

$$\delta_0 = \frac{n_1 - 4}{n_2 + 2} \frac{U_2}{U_1} \tag{1.2}$$

is best invariant in the class based upon U_2/U_1. In fact, using the approach of Brown and Fox (1974), it is straightforward to establish that δ_0 is admissible under (1.1). [See Gelfand and Dey (1987) for details.] Suppose we restrict θ, $\theta \leq \theta_0$ or $\theta \geq \theta_0$ Such restrictions arise naturally when, for example, the U_i are sums of squares in a variance components model. Then δ_0 is no longer admissible. A usual approach to dominating δ_0 is to restrict δ_0 in the same fashion that θ is restricted. Such estimators are no longer smooth, hence inadmissible. From a Bayesian viewpoint, such an approach is essentially taking the posterior mode resulting from a prior over the restricted space. An alternative is to take the posterior mean. Such estimators will be admissible, but exhibit strange behavior. In Section 2 we look into all of these issues drawing upon ideas of Hill (1965) and recent work of Loh (1986).

A broader setting presumes that we have additional information about the τ_i in the form of $V_i \sim \tau_i \chi^2_{m_i,\lambda_i}$ or $V_i \sim (\tau_i + \phi_i)\chi_{m_i}^2$, $i = 1, 2$, V_i independent of U_i and of each other. Now δ_0 is no longer admissible for θ under (1.1). An example is that of $X_{ij} \sim N(\mu_i, \tau_i)$, $i = 1, 2$, $j = 1,\ldots, n_i$, with $U_i = \Sigma(X_{ij} - \bar{X}_i)^2$. Then (1.2) is not admissible for θ. In fact, $(\bar{X}_1, \bar{X}_2, U_1, U_2)$ is a version of the complete, sufficient statistic and the $\bar{X}_i$ (hence $\bar{X}_i^2$) contain information about τ_i. [Gelfand and Dey (1987) discuss this example at length.] In this broader setting, we again seek to estimate a restricted θ. This problem is the issue of Section 3, drawing upon ideas dating to Stein (1964), Brown (1968) and Klotz, Milton and Zacks (1969).

The seminal paper by Katz (1961) on admissibility for estimators of restricted parameters is inapplicable here since the distribution of $W = U_2/U_1$ does not belong to the exponential family.

2. THE BAYESIAN APPROACH

In the spirit of variance components models, let $\tau_1 = a\eta_1 + b\eta_2$, $\tau_2 = c\eta_1 + d\eta_2$ where $\eta_1 \geq 0$, $a, b, c, d \geq 0$ and $r = ad - bc \neq 0$. We necessarily have $\theta_1 \leq \theta \leq \theta_2$ where $\theta_1 = \min(a^{-1}c, b^{-1}d)$, $\theta_2 = \max(a^{-1}c, b^{-1}d)$. If a or $b = 0$ (c or $d = 0$), we obtain a one-sided restriction below (above).

EXAMPLE. In the balanced one-way ANOVA, i.e., $Y_{ij} = \mu + a_i + \varepsilon_{ij}$, $i = 1,\ldots, I$, $j = 1,\ldots, J$, with $a_i \sim N(0, \sigma_a^2)$, $\varepsilon_{ij} \sim N(0, \sigma_e^2)$ all independent, we have $U_1 = \Sigma\Sigma(Y_{ij} - \bar{Y}_{i.})^2$, $U_2 = J\Sigma(\bar{Y}_{i.} - \bar{Y}_{..})^2$, $n_1 = I(J-1)$, $n_2 = I-1$, $\eta_1 = \sigma_e^2$, $\eta_2 = \sigma_a^2$, $a = 1$, $b = 0$, $c = 1$, $d = J$ and $\theta \geq 1$.

We now develop the relevant distribution theory. We assume a prior over η_1, η_2 of the form $\pi(\eta_1)\cdot\pi(\eta_2)$ with η_i^{-1} having a gamma distribution, i.e.,

$$\pi(\eta_1) \propto \eta_i^{-(k_i+1)} e^{-\lambda_i/\eta_i}, \quad i = 1, 2. \tag{2.1}$$

Hill (1965, p.811) argues for the plausibility of the independence assumptions in the context of variance components. The resulting prior for τ_1, τ_2 on the domain $\theta_1\tau_1 \leq \tau_2 \leq \theta_2\tau_2$ is

$$\pi(\tau_1, \tau_2) \propto r^{-(k_1+k_2+2)} (d\tau_1 - b\tau_2)^{-(k_1+1)} (a\tau_2 - c\tau_1)^{-(k_2+1)}$$

$$\cdot \exp\left\{ -r \frac{\lambda_1(a\tau_2 - c\tau_1) + \lambda_2(d\tau_1 - b\tau_2)}{(d\tau_1 - b\tau_2)(a\tau_2 - c\tau_1)} \right\}. \tag{2.2}$$

The noninformative prior on η_i arises as the limiting case $k_i = 0$, $\lambda_i = 0$ in (2.1) and induces $[(d\tau_1 - b\tau_2)\cdot(a\tau_2 - c\tau_1)]^{-1}$ as the prior for $\tau_1\cdot\tau_2$. This differs from the noninformative prior of Box and Tiao (1973, p. 253), $(\tau_1\cdot\tau_2)^{-1}$ which cannot arise from independent η_i unless the transformation from η_i to τ_i is trivial and has been criticized in the variance components case for its dependence in the η space upon the sample size J.

From (2.2)

$$\pi(\theta) \propto \frac{(a\theta - c)^{k_1-1} (d - b\theta)^{k_2-1} r^{k_1+k_2-2}}{[\lambda_1(a\theta - c) + \lambda_2(d - b\theta)]^{k_1+k_2}}, \quad \theta_1 \leq \theta \leq \theta_2, \tag{2.3}$$

i.e., θ follows a generalized Beta distribution. Two cases of (2.3) which we study in greater detail are:

$$\bar{\pi}(\theta) \propto \frac{(\theta - \theta_0)^{k_1-1}}{[1 + \lambda(\theta - \theta_0)]^{k_2-2}}, \quad \theta \geq \theta_0, \tag{2.4}$$

i.e., $\lambda(\theta - \theta_0)$ follows a nonstandardized F distribution or limit thereof and

$$\underline{\pi}(\theta) \propto \theta^{k_1 - 1}(\theta_0 - \theta)^{k_2 - 1}, 0 \le \theta \le \theta_0, \tag{2.5}$$

i.e., θ/θ_0 follows a Beta distribution or limit thereof.

Under (2.1) the posterior distribution of θ given the data is

$$\xi(\theta|U_1, U_2) \propto$$

$$\frac{(d-b\theta)^{\frac{n_1+n_2}{2}+k_2-1}(a\theta-c)^{\frac{n_1+n_2}{2}+k_1-1} r^{k_1+k_2-2}}{\theta^{\frac{n_2}{2}}\left[\frac{(U_1\theta+U_2)(d-b\theta)(a\theta-c)}{2}+\lambda_1(a\theta-c)+\lambda_2(d-b\theta\right]^{\frac{n_1+n_2}{2}+k_1+k_2}}. \tag{2.6}$$

Since (2.6) is analytically intractable, we consider instead the simpler posterior of $\theta|W = U_1/U_2$. The distribution of $W|\theta$ is a nonstandardized F, i.e.,

$$f(w|\theta) = \frac{\Gamma\left(\frac{n_1+n_2}{2}\right)}{\Gamma\left(\frac{n_1}{2}\right)\Gamma\left(\frac{n_2}{2}\right)} \frac{\theta^{\frac{n_1}{2}} w^{\frac{n_1}{2}-1}}{(1+\theta w)^{\frac{n_1+n_2}{2}}} \tag{2.7}$$

whence under (2.1)

$$\xi(\theta|w) \propto \frac{(a\theta-c)^{k_1-1}(d-b\theta)^{k_2-1}\theta^{\frac{n_1}{2}} r^{k_1+k_2-2}}{[\lambda_1(a\theta-c)+\lambda_2(d-b\theta)]^{k_1+k_2}(1+\theta w)^{\frac{n_1+n_2}{2}}}. \tag{2.8}$$

In the special case (2.4) with $\lambda = \theta_0^{-1}$, we obtain

$$\bar{\xi}(\theta|w) \propto \frac{(\theta-\theta_0)^{k_1-1}\,\theta^{\frac{n_1}{2}-(k_1+k_2)}}{(1+\theta w)^{\frac{n_1+n_2}{2}}}, \quad \theta \geq \theta_0. \tag{2.9}$$

In the special case (2.5), we obtain

$$\underline{\xi}(\theta|w) \propto \frac{\theta^{\frac{n_1}{2}+k_1-1}\,(\theta_0-\theta)^{k_2-1}}{(1+\theta w)^{\frac{n_1+n_2}{2}}}, \quad 0 \leq \theta \leq \theta_0. \tag{2.10}$$

The case $k_1 = 1$, $k_2 = 0$ in (2.9) or $k_1 = 0$, $k_2 = 1$ in (2.10) produces the posterior associated with the Box-Tiao prior.

Turning to estimation of θ, we first consider the posterior mode in (2.9) and (2.10). If $k_1 < 1$ in (2.9) or $k_2 < 1$ in (2.10), the mode occurs at θ_0. If $k_1 > 1$ in (2.9) or $k_2 > 1$ in (2.10), the posterior need not be unimodal. However, if in (2.9), $k_1 = 1$ and $k_2 < (n_1 - 2)/2$, we obtain a unique mode at

$$\bar{d}_{k_2,\theta_0} = \max\left(\theta_0, W^{-1}\frac{n_1-2(k_2+1)}{n_2+2(k_2+1)}\right). \tag{2.11}$$

If in (2.10), $k_2 = 1$ and $k_1 < (n_2 + 2)/2$, we obtain a unique mode at

$$\underline{d}_{k_1,\theta_0} = \min\left(\theta_0, W^{-1}\frac{n_1+2(k_1-1)}{n_2-2(k_1-1)}\right). \tag{2.12}$$

Note that $k_1 = 1$ in (2.4), $k_2 = 1$ in (2.5) asserts prior information concentrated near θ_0.

As remarked earlier (2.11) and (2.12) are not admissible. In particular for $k_2 < (n_1 + n_2 - 2)^{-1}(n_2 + 2)$, $\max(\theta_0, \delta_0)$, δ_0 as in (1.2), dominates (2.11) using the fact that δ_0 is best invariant in the unrestricted problem and Lemma 1 of the appendix. However, $\min(\theta_0, \delta_0)$ does not dominate (2.12) by using the same argument as in Loh (1986, p. 700).

Consider now formal Bayes rules extending the loss (1.1) to

$$L(\theta, a) = \theta^c(\theta - a)^2 \tag{2.13}$$

where c is arbitrary.

We recall that if $0 < a < b$

$$\int_0^{\theta_0} \frac{\theta^{a-1}}{(1+\theta w)^b} d\theta = w^{-1} \frac{\Gamma(a)\Gamma(b-a)}{\Gamma(b)} I_{\eta_0}(a, b-a) \tag{2.14}$$

where $I_{\eta_0}(a, b-a)$ is the incomplete Beta function evaluated at $\eta_0 = (1+\theta_0 w)^{-1}\theta_0 w$.

Consider first the case $\theta \geq \theta_0$. Under (2.9) with loss (2.13), let $m_j = n_1/2 + c + 1 - (k_1 + k_2) + j$ and $\ell_j = (n_1 + n_2)/2 - m_j$. If k_1 is a positive integer and $m_0 > 0$ using (2.14), we can obtain the unique Bayes rule as

$\bar{\delta}_{k_1, k_2, c, \theta_0}$

$$= W^{-1} \frac{\sum_{j=0}^{k_1-1} \binom{k_1-1}{j} \theta_0^{-j} W_j^{-m_j} \Gamma(m_j+1)\Gamma(\ell_j-1) I_{1-\eta_0}(\ell_j-1, m_j+1)}{\sum_{j=0}^{k_1-1} \binom{k_1-1}{j} \theta_0^{-j} W_j^{-m_j} \Gamma(m_j)\Gamma(\ell_j) I_{1-\eta_0}(\ell_j, m_j)} \tag{2.15}$$

Similarly, if $0 \leq \theta \leq \theta_0$, under (2.10) with loss (2.13), let $m_j = n_1/2 + c + k_1 + j$ and $\ell_j = n_1 + n_2/2 - m_j$. If k_2 is a positive integer using (2.14), we can obtain the unique Bayes rule as

$\underline{\delta}_{k_1, k_2, c, \theta_0}$

$$= W^{-1} \frac{\sum_{j=0}^{k_2-1} \binom{k_2-1}{j} \theta_0^{-j} w_j^{-m_j} \Gamma(m_j+1)\Gamma(\ell_j-1) I_{\eta_0}(m_j+1, \ell_j-1)}{\sum_{j=0}^{k_2-1} \binom{k_2-1}{j} \theta_0^{-j} w_j^{-m_j} \Gamma(m_j)\Gamma(\ell_j) I_{\eta_0}(m_j, \ell_j)}. \tag{2.16}$$

REMARK 1. Expression (2.15) depends upon c and k_2 through $c - k_2 \equiv \gamma_2$ so we can denote it by $\bar{\delta}_{k_1, \delta_2, \theta_0}$; expression (2.16) depends upon c and k_1 through $c + k_1 \equiv \gamma_1$ so we can denote it by $\underline{\delta}_{\gamma_1, k_2, \theta_0}$.

REMARK 2. Computation of (2.15) and (2.16) requires calculation of the incomplete Beta function only for, say, the denominator by using the well-known relationship (see, e.g., Abramovitz and Stegun 1965)

$$I_x(c+1, d-1) = I_x(c, d) - \frac{\Gamma(c+d)}{\Gamma(c+1)\Gamma(d)} x^c (1-x)^{d-1}.$$

REMARK 3. Using Lemma 2 of the appendix, we may show that

$$\lim_{w \to \infty} \bar{\delta}_{k_1, \gamma_2, \theta_0} = \theta_0 \frac{\sum_{j=0}^{k_1-1} \binom{k_1-1}{j} (\ell_j - 1)^{-1}}{\sum_{j=0}^{k_1-1} \binom{k_1-1}{j} \ell_j^{-1}} > \theta_0,$$

$$\lim_{w \to 0} \underline{\delta}_{\gamma_1, k_2, \theta_0} = \theta_0 \frac{\sum_{j=0}^{k_2-1} \binom{k_2-1}{j} \theta_0^j (m_j + 1)^{-1}}{\sum_{j=0}^{k_2-1} \binom{k_2-1}{j} \theta_0^j (m_j)^{-1}} < \theta_0.$$

In this sense the behavior of $\bar{\delta}$, $\underline{\delta}$ differs strikingly from that of $\bar{d}$, $\underline{d}$ as W approaches the extremes of its domain. The latter have positive probability of equaling θ_0. Loh's Theorem 3.1 shows a special case of this.

REMARK 4. Paralleling (2.11) and (2.12) when $k_1 = 1$, we obtain

$$\bar{\delta}_{1, \gamma_2, \theta_0} = W^{-1} \left(\frac{n_1 + 2\gamma_2}{n_2 - 2(\gamma_2 + 1)} \right) \cdot \frac{I_{1-\eta_0}\left(\frac{n_1}{2} + \gamma_2 + 1, \frac{n_2}{2} - \gamma_2 - 1 \right)}{I_{1-\eta_0}\left(\frac{n_1}{2} + \gamma_2, \frac{n_2}{2} - \gamma_2 \right)}$$

and when $k_2 = 1$ in (2.16) we obtain

$$\underline{\delta}_{\gamma_1, 1, \theta_0} = W^{-1} \frac{n_1 + 2\gamma_1}{n_2 - 2(\gamma_1 + 1)} \frac{I_{\eta_0}\left(\frac{n_1}{2} + \gamma_1 + 1, \frac{n_2}{2} - \gamma_1 - 1\right)}{I_{\eta_0}\left(\frac{n_1}{2} + \gamma_1, \frac{n_2}{2} - \gamma_1\right)}.$$

The estimators (2.15) and (2.16) are admissible under (1.1) within the class of rules based upon W. However, admissibility in the larger class based upon U_1, U_2 (equivalently W, U_1) is a more difficult question. The approach of Brown and Fox (1974) mentioned after (1.2) is not applicable to a restricted parameter space.

3. ADDITIONAL INFORMATION ABOUT τ_i

Recalling the notation of Section 1, suppose $V_i \sim \tau_i \chi^2_{m_i, \lambda_i}$. We have the following results whose proof is essentially contained in Stein (1964) or Strawderman (1974).

RESULT 1. In estimating τ_2 under squared error loss, $\min\{(n_2 + 2)^{-1}U_2, (m_2 + n_2 + 2)^{-1}(U_2 + V_2)\}$ dominates $(n_2 + 2)^{-1}U_2$.

RESULT 2. In estimating τ_1^{-1} under squared error loss, $\max\{(n_1 - 4)U_1^{-1}, (m_1 + n_1 - 4)(U_1 + V_1)^{-1}\}$ dominates $(n_1 - 4)U_1^{-1}$.

If instead $V_i \sim (\tau_i + \phi_i)\chi_{m_i}^2$, Results 1 and 2 still hold. That is, we may think of V_i arising from $V_i | Z_i = z_i \sim \tau_i \chi^2_{m_i, z_i}$ where $Z_i \sim (\phi_i/2\tau_i)\chi_{m_i}^2$. Since Results 1 and 2 holds regardless of Z_i, they hold unconditionally. Klotz, Milton and Zacks (1969, p. 1392) allude to this idea in a special case.

Taking these ideas further to the estimation of an unrestricted variance ratio using essentially the proof of Theorem 3.1 in Gelfand and Dey (1987), we can show

RESULT 3. In estimating θ under squared error loss.

$$\underline{\Delta} = \min\left(\delta_0, \frac{(n_1 - 4)}{U_1} \cdot \frac{(U_2 + V_2)}{m_2 + n_2 + 2}\right) \leq \delta_0 \text{ dominates } \delta_0$$

$$\bar{\Delta} = \max\left(\delta_0, \frac{U_2}{n_2 + 2} \cdot \frac{m_1 + n_1 - 4}{U_1 + V_1}\right) \geq \delta_0 \text{ dominates } \delta_0.$$

By Lemma 1 of the appendix, we have immediately that with squared error loss under the restriction $\theta \geq \theta_0$,

$$\max(\underline{\Delta}, \theta_0) \text{ dominates } \max(\delta_0, \theta_0) \tag{3.1}$$

and under the restriction $0 \leq \theta \leq \theta_0$,

$$\min(\bar{\Delta}, \theta_0) \text{ dominates } \min(\delta_0, \theta_0). \tag{3.2}$$

In Gelfand and Dey other estimators (e.g., using ideas of Brown 1968) which dominate δ_0 in the unrestricted problem are given. These estimators may be used to obtain results similar to (3.1) and (3.2). We omit the details.

APPENDIX

LEMMA 1. *Under squared error loss*

(a) *If T dominates U on $\theta_0 \leq \theta$ and $T \leq U$, then $max(T, \theta_0)$ dominates $max(U, \theta_0)$ on $\theta_0 \leq \theta$.*

(b) *If T dominates U on $0 \leq \theta \leq \theta_0$ and $T \geq U$, then $min(T, \theta_0)$ dominates $min(U, \theta_0)$ on $0 \leq \theta \leq \theta_0$.*

Proof. The proof is essentially that of a lemma in Klotz, Milton and Zacks (1969, p. 1394).

LEMMA 2. *In the notation of* (2.14),

$$\lim_{w \to 0} w^{-a} I_{\eta_0}(a, b-a) = \frac{\Gamma(b)}{\Gamma(a+1)\Gamma(b-a)} \theta_0^{a} \tag{A.1}$$

or equivalently

$$\lim_{w \to 0} w^{a} I_{1-\eta_0}(a, b-a) = \frac{\Gamma(b)}{\Gamma(a+1)\Gamma(b-a)} \theta_0^{-a}. \tag{A.2}$$

Proof. Since the limit as $w \to 0$ of the left-hand side in (2.14) is θ_0^a/a, we obtain (A.1). But (A.2) follows by replacing w with w^{-1} and θ_0 with θ_0^{-1} in (A.1).

REFERENCES

Abramowitz, M. and Stegun, I. (1965). *Handbook of Mathematical Functions 1.* Dover, New York.

Box, G.E.P. and Tiao, G. (1973). *Bayesian Inference in Statistical Analysis.* Addison Wesley, Reading, Mass.

Brown, L.D. (1968). Inadmissibility of the usual estimators of scale parameters in problem with unknown location and scale parameters. *Annals of Mathematical Statistics*, 39, 29-48.

Brown, L.D. and Fox, M. (1974). Admissibility in statistical problems involving a location or scale parameter. *Annals of Statistics*, 2, 807-814.

Gelfand, A.E. and Dey, D.K. (1987). On the estimation of a variance ratio. To appear in *Journal of Statistical Planning and Inference.*

Hill, B.M. (1965). Inference about variance components in the one-way model. *Journal of the American Statistical Association*, 60, 806-825.

Katz, M. (1961). Admissible and minimax estimators of parameters in truncated spaces. *Annals of Mathematical Statistics*, 32, 136-142.

Klotz, J., Milton, R. and Zacks, S. (1969). Mean square efficiency of estimators of variance components. *Journal of the American Statistical Association*, 64, 1383-1402.

Loh, W. (1986). Improved estimators for ratios of variance components. *Journal of the American Statistical Association*, 81, 699-702.

Stein, C. (1964). Inadmissibility of the usual estimator for the variance of a normal distribution with unknown mean. *Annals of the Institute of Statistical Mathematics*, 42, 385-388.

Strawderman, W.E. (1974). Minimax estimation of powers of the variance of normal population. *Annals of Statistics*, 2, 190-198.

Received 2 December 1986
Revised 21 April 1987

Department of Statistics
University of Connecticut
U-120, 196 Auditorium Road
Storrs, CT 06268
U.S.A.

Proc. Second International Tampere Conference in Statistics
(Tampere, Finland, 1-4 June 1987)
Tarmo Pukkila and Simo Puntanen, *Editors*

pp. 467 - 478

Higher order moments of bilinear time series processes with symmetrically distributed errors

Jan G. de GOOIJER and Ruud M. J. HEUTS
University of Amsterdam, Holland and Tilburg University, Holland

Key words and phrases: Bilinear time series, diagonal, superdiagonal and subdiagonal models, kurtosis, skewness.

ABSTRACT

For the stationary superdiagonal, diagonal and subdiagonal bilinear time series model with symmetrically distributed errors we derive formulas for the standardized third and fourth order central moments. These results contain useful information to distinguish a bilinear process from a white noise process if the errors are Gaussian distributed. To some extent these moments can also be used to discriminate between different types of bilinear time series models.

1. INTRODUCTION

Linear time series models such as the well-known class of ARIMA models provide a remarkable effective and versatile range of possibilities to adequately approximate many time series observed in practice. However, despite reported evidence of the advantages of these models, it is increasingly recognized that there are time series in economics and operation research which are unlikely to be well represented by any linear model; see, e.g., Maravall (1983) and Hinich and Patterson (1985). A number of authors have studied various tractable classes of non-linear models. Among these models are the so-called bilinear models discussed by Granger and Andersen (1978) and Subba Rao and Gabr (1984).

The successful application of a particular class of non-linear models hinges heavily on the determination of the most appropriate model within its class with respect to some prespecified loss function. For bilinear models the use of Akaike's information criterion has been suggested to determine the structure of the model. However, it is well-known that for linear ARMA models this criterion asymptotically overestimates the "true" orders with a non-zero probability. This feature of AIC may also extend to bilinear models.

Another way of identifying the structure of a bilinear model has been considered by Granger and Andersen (1978) and Maravall (1983). They use the sample autocorrelations of the unsquared and squared time series as identification tools. By relating these statistics to the known behaviour of their corresponding theoretical quantities inferences can be made about the most appropriate bilinear model. A crucial element in this approach is that knowledge should be available on the theoretical behaviour of the autocorrelations for different bilinear time series models. However these results have only been obtained for a limited number of models. Moreover, one may wonder whether autocorrelations of unsquared and squared time series provide enough information about the system under study. In particular, these statistics do not completely determine the structure of the process when the series are generated by a bilinear model with non-Gaussian errors. In such a situation it is necessary to analyse the higher order moments of the process.

In this paper we study the theoretical properties of the standardized third and fourth central moments of three bilinear time series models having symmetrically distributed errors. If $\{Y_t\}$ denotes a time series with $E(Y_t)=\mu$, these moments are, respectively, given by

$$\beta_1(i,j) = [c(i,j) - \mu\{E(Y_{t-i}Y_{t-j}) + E(Y_tY_{t-j}) + E(Y_tY_{t-i})]\} + 2\mu^3] / \{var(Y_t)\}^{3/2}$$

and

$$\beta_2(0,i,j) = [c(0,i,j) - \mu\{c(0,i) + c(0,j) + 2c(i,j)\} + \mu^2\{E(Y_t^2) + E(Y_{t-i}Y_{t-j}) + 2E(Y_tY_{t-j}) + 2E(Y_tY_{t-i})\} - 3\mu^4] / \{var(Y_t)\}^2$$

where $c(i,j) = E(Y_tY_{t-i}Y_{t-j})$ and $c(u,i,j) = E(Y_tY_{t-u}Y_{t-i}Y_{t-j})$. It will be shown that these moments can be used to distinguish a bilinear process from a white noise process. Moreover, they provide useful information about the most appropriate bilinear time series model.

The paper is organized as follows: In Section 2 we summarize (without proofs) the main result on the higher order moments of three bilinear time series processes assuming that the errors of these processes are symmetrically distributed around mean zero. Estimators of these moments will be considered in Section 3 together with approximate expressions for their variances. Also some simulations are performed in this section to study the sampling properties of the standardized third and fourth central moments. The results can be used to detect bilinear relations in empirical time series data.

2. HIGHER ORDER MOMENTS

Throughout this section it will be assumed that the errors $\{A_t\}$ we consider are distributed symmetrically around mean zero. For ease of reference, we define $E(A_t^i)=\mu_i$ if i is even and $\mu_i=0$ if i is odd. Also, for ease of notation, we will use the definition $\lambda=\beta^2\mu_2$.

LEMMA 2.1. *If* $\{Y_t\}$ *is generated by a superdiagonal bilinear time series model of the form* $Y_t = \beta Y_{t-\ell}A_{t-k} + A_t$ $(0<k<\ell)$ *with* $\{A_t\}$ *distributed symmetrically around mean zero then* $c(i,j) = \beta\mu_2^2/(1-\lambda)$ *for* $(i,j) = (k,\ell)$ *or* (ℓ,k) *and* $c(i,j) = 0$ *otherwise;* see Guegan (1984, Table 1) and Kumar (1986, Lemma 2.1).

LEMMA 2.2. *If* $\{Y_t\}$ *is generated by a superdiagonal bilinear time series model of the form* $Y_t = \beta Y_{t-\ell}A_{t-k} + A_t$ $(0<k<\ell)$ *with* $\{A_t\}$ *distributed symmetrically around mean zero then* $cov(Y_t^2, Y^2_{t-j}) = \lambda^m[\mu_4 - \mu_2^2 - 2\lambda(\mu_4 - 3\mu_2^2)(1-\lambda)]/(1-\lambda)^2(1-\beta^4\mu_4)$ *for* $j=m\ell$ $(m=0, 1, \ldots)$; $cov(Y_t^2, Y^2_{t-j}) = \lambda^{m+1}(\mu_4 - \mu_2^2)/(1-\lambda)^2(1+\lambda)$ *for* $j=m\ell\pm k$ *and* $\ell\neq 2k$ $(m=0, 1, \ldots)$; $cov(Y_t^2, Y^2_{t-j}) = \lambda^{m+1}(\mu_4 - \mu_2^2)/(1-\lambda)^2$ *for* $j=m\ell+k$ *and* $\ell=2k$ $(m=0, 1, \ldots)$ *and* $cov(Y_t^2, Y^2_{t-j}) = 0$ *otherwise.*

Note that for zero-mean Gaussian distributed $\{A_t\}$ and $m=0$ the second part of Lemma 2.2 is identical to the first part of Lemma 4 of Li (1984), while for $m=1$ the second part of Lemma 2.2 conforms to Lemma 3 of Li (1984). Also note that for zero-mean Gaussian distributed $\{A_t\}$ and $m=0$ the third part of Lemma 2.2 is similar to the second part of Lemma 4 of Li (1984). Furthermore, it is easy to see that the results of Lemma 2.2 are in agreement with the results given by Guegan (1984, Table 2) for the superdiagonal model $Y_t = \beta Y_{t-2}A_{t-1} + A_t$ and zero-mean Gaussian distributed errors.

THEOREM 2.1. *If $\{Y_t\}$ is generated by a superdiagonal bilinear time series model of the form $Y_t = \beta Y_{t-\ell}A_{t-k} + A_t$ $(0<k<\ell)$ with $\{A_t\}$ distributed symmetrically around mean zero then the standardized third and fourth central moments of $\{Y_t\}$, respectively, are given by*

$$\beta_1(i,j) = c(i,j)/\{E(Y_t^2)\}^{3/2} \text{ and } \beta_2(u,i,j) = c(u,i,j)/\{E(Y_t^2)\}^2$$

where $E(Y_t^2) = \mu_2/(1-\lambda)$, $c(i,j)$ and $c(0,i,j)$ follow from, respectively, Lemma 2.1 and Lemma 2.2, and where for $i\neq j$ $c(u,i,j) = 2\lambda^{m+2}\mu_2^2/(1-\lambda)$ for $u=0, i=m\ell+k, j=(m+2)\ell$ and $\ell=2k$ $(m=-1,0,1,\ldots)$; and $c(u,i,j)= \lambda\mu_2^2/(1-\lambda)$ for $u=k, i=\ell+k$ and $j=2\ell$; and where $c(u,i,j) = 0$ otherwise.

Clearly, the coefficient of skewness $\beta_1(0,0) = 0$ for every $\beta\neq 0$. Hence, it will be difficult to distinguish this superdiagonal model from a pure white noise process on the basis of this coefficient. The coefficient of kurtosis $\beta_2(0,0,0) = (\mu_4 - \lambda\mu_4 + 6\lambda\mu_2^2)(1-\lambda)/\{\mu_2^2(1-\beta^4\mu_4\}$ is more useful for this purpose, provided $\beta^4\mu_4<1$. For zero-mean Gaussian distributed errors this reduces to $\beta_2(0,0,0) = 3(1-\lambda^2)/(1-3\lambda^2)$. This last result is identical to relation (5.14) given by Granger and Andersen (1978). In this case $\beta_2(0,0,0)>3$, for every $\beta\neq 0$, which indicates that this process has a higher degree of peakedness and thick-tailness than a pure Gaussian distributed white noise.

Let us now assume that $\{Y_t\}$ is generated by a diagonal bilinear time series model of the form $Y_t = \beta Y_{t-k}A_{t-k} + A_t$ $(k>0)$. For this model the series $\{Y_t\}$ is a function of $A_t, A_{t-k}, A_{t-2k}, \ldots$, so there are actually k independent series within $\{Y_t\}$. Therefore, unless $j=mk$, where m is an integer, $cov(Y_t, Y_{t-j}) = 0$ for $\{A_t\}$ distributed symmetrically around mean zero. Hence, $cov(Y_t^2, Y_{t-j}^2) = 0$, unless $j=mk$. This implies that Lemma 1 of Li (1984) is true even for errors distributed symmetrically around mean zero. Furthermore, under this assumption for $\{A_t\}$, $E[(Y_t-\mu)(Y_{t-i}-\mu)(Y_{t-j}-\mu)]$ and $E[(Y_t-\mu)(Y_{t-u}-\mu)(Y_{t-i}-\mu)(Y_{t-j}-\mu)]$ are both identically zero, unless $u=m_1k$, $i=m_2k$ and $j=m_3k$, where m_1, m_2 and m_3 are integers. Thus the problem of deriving higher order moments of this diagonal bilinear time series process reduces to obtaining moments of series generated by the diagonal model $Y_t = \beta Y_{t-1}A_{t-1} + A_t$.

LEMMA 2.3. *If $\{Y_t\}$ is generated by a diagonal bilinear time series model of the form $Y_t = \beta Y_{t-1}A_{t-1} + A_t$ with $\{A_t\}$ distributed symmetrically around* *mean zero and* $\lambda<1$ *then* $c(i,j) = 3\beta\mu_2^{\,2}+\beta^3\mu_6+3\beta^5\mu_4^{\,2}/(1-\lambda)$ *for* $i=j=0$; $c(i,j) = \beta\mu_2^{\,2} + 3\beta\lambda\mu_4 + \beta^3\lambda\mu_6 + 3\beta^5\lambda\mu_4^{\,2}/(1-\lambda)$ *for* $i=0, j=1$; $c(i,j) = \beta\mu_2^{\,2} + \lambda^{j-2}(\beta\lambda\mu_4+\beta^3\lambda^2\mu_6+3\beta\lambda^2\mu_4) + \beta\lambda\mu_4\{1+\lambda^{j-2}(3\beta^4\lambda\mu_4-1)\}/(1-\lambda)$ *for* $i=0, j>1$; $c(i,j) = \beta\mu_4(1+2\lambda)/(1-\lambda)$ *for* $i=j=1$; $c(i,j) = 4\beta\lambda\mu_2^{\,2}$ *for* $i=1, j=2$; $c(i,j) = 2\lambda\mu_2^{\,2}$ *for* $i=1, j>2$; *and* $c(i,j) = \beta\mu_2 E(Y_{t-i}Y_{t-j})$ *for* $i>1, j>1$ *where* $E(Y_tY_{t-i}) = 2\lambda\mu_2$ *for* $i=1$ *and* $E(Y_tY_{t-i}) = E(Y_tY_{t+i}) = \lambda\mu_2$ *for* $i>1$.

LEMMA 2.4. *If* $\{Y_t\}$ *is generated by a diagonal bilinear time series model of the form* $Y_t = \beta Y_{t-1}A_{t-1}+A_t$ *with* $\{A_t\}$ *distributed symmetrically around mean zero and* $\beta^4\mu_4<1$ *then*

$$c(0,i,i) = \begin{cases} [\beta^6(-\mu_2\mu_8-5\mu_2\mu_4^{\,2}+6\mu_4\mu_6) + \beta^4(\mu_8-\mu_4^{\,2}) + 5\lambda\mu_4+\mu_4]/(1-\lambda)(1-\beta^4\mu_4) & \text{for } i=0; \\[2ex] [\beta^8(7\mu_2\mu_4\mu_6-\mu_2^{\,2}\mu_8-6\mu_4^{\,3}) + \beta^6(\mu_2\mu_8-\mu_4\mu_6+\mu_2^{\,3}\mu_4-\mu_2\mu_4^{\,2}) + \\ \quad \beta^4(6\mu_4^{\,2}-\mu_2\mu_6-\mu_2^{\,2}\mu_4) + \beta^2(\mu_6+\mu_2\mu_4-\mu_2^{\,3}) + \mu_2^{\,2}]/(1-\lambda)(1-\beta^4\mu_4) & \text{for } i=1; \\[2ex] \lambda c(0,i-1,i-1) + (1-\lambda)E(Y_t^{\,2})^2 & \text{for } i>1; \end{cases}$$

where $E(Y_t^{\,2}) = (\mu_2-\lambda\mu_2+\beta^2\mu_4)/(1-\lambda)$.

From Lemma 2.4 it is easy to see that for zero-mean Gaussian distributed $\{A_t\}$ we get

$$var(Y_t^{\,2}) = 2\mu_2^{\,2}(1+4\lambda+40\lambda^2+18\lambda^3-54\lambda^4)/(1-\lambda)^2(1-3\lambda^2) \quad (2.1)$$

and

$$cov(Y_t^{\,2}, Y^2_{t-1}) = 6\lambda\mu_2^{\,2}(2+3\lambda+5\lambda^2+\lambda^3-8\lambda^4)/(1-\lambda)^2(1-3\lambda^2) \quad (2.2)$$

provided $\lambda^2<1/3$. Substituting (2.1) and (2.2) into $\rho(1) = cov(Y_t^{\,2}, Y^2_{t-1})/var(Y_t^{\,2})$ leads to relation (6.35) of Granger and Andersen (1978). Furthermore, from the third part of Lemma 2.4, we have $\rho(i) = \lambda\rho(i-1)$ $i>1$, which agrees with relation (6.34) of Granger and Andersen (1978, p. 55). Note, however, that their statement saying that this result can be obtained " ... without any assumptions being made about the distribution of $\{A_t\}$ ... " is not correct. From Lemma 2.2 and 2.4 the following theorem can be straightforwardly obtained.

THEOREM 2.2. *If* $\{Y_t\}$ *is generated by a diagonal bilinear time series model of the form* $Y_t = \beta Y_{t-1}A_{t-1} + A_t$ *with* $\{A_t\}$ *distributed symmetrically around mean zero then* $\beta_1(i,j)$ *and* $\beta_2(0,i,i)$, *respectively, are given by*

$$\beta_1(i,j) = \begin{cases} \beta^3[2\mu_2^{\,3} + \mu_6 + 3\mu_4(\beta^2\mu_4 - \mu_2)/(1-\lambda)]/\{var(Y_t)\}^{3/2} & \text{for } i=j=0; \\ [c(i,1) - \beta\mu_2^{\,2} - 2\beta\lambda\mu_2^{\,2} - \beta\lambda\mu_4/(1-\lambda)]/\{var(Y_t)\}^{3/2} & \text{for } i=0, j=1 \\ & \text{for } i=1, j=1; \\ [c(0,j) - \beta\mu_2^{\,2} - \beta\lambda\mu_4/(1-\lambda)]/\{var(Y_t)\}^{3/2} & \text{for } i=0, j>1; \\ \beta\lambda\mu_2^{\,2}/\{var(Y_t)\}^{3/2} & \text{for } i=1, j=2; \\ 0 & \text{for } i\geq 1, j>2, \end{cases}$$

and

$$\beta_2(0,i,i) = \begin{cases} [c(0,0,0) - 4\beta\mu_2 c(0,0) + 3\lambda\mu_2\{2\mu_2 - \lambda\mu_2 + 2\beta^2\mu_4/(1-\lambda)\}]/\{var(Y_t)\}^2 \\ \qquad \text{for } i=0; \\ [c(0,1,1) - 2\lambda\{c(0,1) + c(1,1)\} + \lambda\mu_2\{2\mu_2 + 5\lambda\mu_2 + 2\beta^2\mu_4/(1-\lambda)\}]/\{var(Y_t)\}^2 \\ \qquad \text{for } i=1; \\ [c(0,i,i) - 2\lambda\{c(0,i) + c(i,i)\} + \lambda\mu_2\{2\mu_2 + \lambda\mu_2 + 2\beta^2\mu_4/(1-\lambda)\}]/\{var(Y_t)^2\}^2 \\ \qquad \text{for } i>1, \end{cases}$$

where $var(Y_t) = (\mu_2^{\,2} - 2\lambda\mu_2 + \beta^2\mu_4 + \lambda^2\mu_2)/(1-\lambda)$.

Using Theorem 2.2 $\beta_1(0,0)$ and $\beta_2(0,0,0)$ for zero-mean Gaussian distributed $\{A_t\}$ are, respectively, given by $[2\beta^3\mu_2^{\,3}(4+5\lambda^2)/(1-\lambda^2)]/\{\mu_2^{\,2}(1+\beta^2+\beta^4\mu_2)/(1-\lambda^2)\}^{3/2}$ provided $\lambda^2 < 1$, and $[\mu_2^{\,2}(3+9\lambda^2+57\lambda^4+93\lambda^6+117\lambda^8+135\lambda^{10})/(1-\lambda^2)(1-3\lambda^4)]/\{\mu_2^{\,2}(1+\beta^2+\beta^4\mu_2)(1-\lambda^2)\}^2$ provided $3\lambda^4 < 1$. Clearly, the expression between squared brackets of $\beta_1(0,0)$ is not in agreement with the third central moment given by Granger and Andersen (1978, p. 52). Also the expression between squared brackets of $\beta_2(0,0,0)$ differs a factor 3 with the fourth central moment (6.26) given by these authors. This last result may explain the high values of the coefficient of kurtosis given in column 5 of Table 1 of Granger and Andersen (1978).

LEMMA 2.5. *If* $\{Y_t\}$ *is generated by a subdiagonal bilinear time series model of the form* $Y_t = \beta Y_{t-k}A_{t-k-1} + A_t$ $(k>0)$ *with* $\{A_t\}$ *distributed symmetrically around mean zero then* $c(i,j) = \beta\lambda\mu_4 + \beta\mu_2^2$ *for* $k=1$ *and* $i=2, j=1$ *or* $i=1, j=2$; $c(i,j) = \beta\mu_2^2$ *for* $k=2$ *and* $i=2, j=3$ *or* $i=3$, $j=2$; $c(i,j) = \lambda\beta\mu_2^2 + \lambda\mu_2^2/(1-\lambda)$ *for* $k=3, 4, \ldots$ *and* $i=k, j=k+1$ *or* $i=k+1, j=k$; *and* $c(i,j) = 0$ *otherwise.*

LEMMA 2.6. *If* $\{Y_t\}$ *is generated by a subdiagonal bilinear time series model of the form* $Y_t = \beta Y_{t-k}A_{t-k-1} + A_t$ $(k>0)$ *with* $\{A_t\}$ *distributed symmetrically around mean zero and* $\beta^4\mu_4 < 1$ *then*

$$c(0,i,i) =$$

$$\begin{cases} \beta^8 E(Y_t^4 A_t^4 A_{t-1}^4) + 6\lambda^2\mu_4(1+\beta^4\mu_4)/(1-\lambda) + \\ 6\lambda\mu_2(\mu_2+\beta^4\mu_6) + \mu_4(1+\beta^4\mu_4) & \text{for } i=0, k=1; \\[2ex] \beta^8\lambda E(Y_t^4 A_t^4 A_{t-1}^4) + \lambda\mu_4\{1+6(\lambda+\beta^8\mu_4^2)\}/(1-\lambda) + \\ \beta^6\mu_4\mu_6(6\lambda+1) + \mu_2^2(\lambda+1) & \text{for } i=1, k=1; \\[2ex] \beta^8\lambda^2 E(Y_t^4 A_t^4 A_{t-1}^4) + \{\lambda^2\mu_4(1+\lambda) + 2\beta^4\lambda\mu_4^2(1+3\beta^4\lambda\mu_4)\}/(1-\lambda) + \\ \beta^6\lambda\mu_4\mu_6(6\lambda+1) + \mu_2^2(\lambda^2+\lambda+1) + \lambda(\beta^2\mu_6+\mu_4) & \text{for } i=2, k=1; \\[2ex] \beta^6\mu_4^2\{6\lambda^4/(1-\lambda) + \beta^{10}\mu_4^2\}/(1-\beta^4\mu_4) + 6\lambda\mu_2^2(1+\beta^4\mu_4+\beta^8\mu_4^2)/(1-\lambda) + \\ \mu_4 + \beta^4\mu_4^2(1+\beta^4\mu_4) & \text{for } i=0, k\geq 2; \\[2ex] \lambda c(0, i-1, i-1) + \mu_2^2\{1+\lambda+\beta^4\mu_4/(1-\lambda)\} + \beta^4(\mu_4-\mu_2^2)E(Y^2_{t-i}A^2_{t-3}) \\ & \text{for } i\geq 3, k=1; \\[2ex] \lambda c(0, i-k, i-k) + \mu_2^2/(1-\lambda) & \text{for } i-1>k\geq 2, \end{cases}$$

where $E(Y_t^4 A_t^4 A_{t-1}^4) = \{6\beta^2\lambda\mu_4{}^2\mu_6/(1-\lambda)+6\lambda\mu_6{}^2+\mu_4\mu_8\}/(1-\beta^4\mu_4)$ *for* $k=1$, *and* $E(Y^2_{t-3}A^2_{t-3}) = \mu_4+\lambda\mu_2{}^2+\lambda^2\mu_4/(1-\lambda)$ *and* $E(Y^2_{t-i}A^2_{t-3}) = \mu_2{}^2\{1+\lambda+\beta^4\mu_4/(1-\lambda)\}$ *for* $i\geq 4$.

THEOREM 2.3. *If* $\{Y_t\}$ *is generated by a subdiagonal bilinear time series model of the form* $Y_t = \beta Y_{t-k}A_{t-k-1} + A_t$ $(k>0)$ *with* $\{A_t\}$ *distributed symmetrically around mean zero then the standardized third and fourth central moment, respectively, are given by*

$$\beta_1(i,j) = c(i,j)/\{E(Y_t^2)\}^{3/2} \quad \textit{and} \quad \beta_2(0,i,i) = c(0,i,i)/\{E(Y_t^2)\}^2$$

where $c(i,j)$ *and* $c(0,i,i)$ *follow from, respectively, Lemma 2.5 and 2.6, and where* $E(Y_t^2) = \mu_2+\lambda\mu_2+\beta^2\lambda\mu_4/(1-\lambda)$ *for* $k=1$ *and* $E(Y_t^2) = \mu_2/(1-\lambda)$ *for* $k\geq 2$.

It is clear from Theorem 2.3 and Lemma 2.5 that the coefficient of skewness is equal to zero for every $\beta\neq 0$ and $\{A_t\}$ distributed symmetrically around mean zero. Also, it is obvious from the last part of Lemma 2.6 that $cov(Y_t^2, Y^2_{t-i}) = \lambda cov(Y_t^2, Y^2_{t-i+k})$, for $i-1>k\geq 2$ and symmetrically distributed errors. This generalizes relation (7.22) of Granger and Andersen (1978) for Gaussian distributed $\{A_t\}$.

3. SOME SIMULATION RESULTS

Let $\{Y_t\!: t=1, \ldots, n\}$ be a set of realizations of $\{Y_t\}$ then the natural estimators of $\beta_1(i,j)$ and $\beta_2(0,i,i)$, respectively, are given by

$$b_1(i,j) = \frac{1}{n}\sum_{t=1}^{n-\max(i,j)} (Y_t-\overline{Y})(Y_{t+i}-\overline{Y})(Y_{t+j}-\overline{Y})/\{\frac{1}{n}\sum_{t=1}^{n}(Y_t-\overline{Y})^2\}^{3/2}$$

and

$$b_2(0,i,i) = \frac{1}{n}\sum_{t=1}^{n-i}(Y_t-\overline{Y})^2(Y_{t+i}-\overline{Y})^2/\{\frac{1}{n}\sum_{t=1}^{n}(Y_t-\overline{Y})^2\}^2$$

where $\overline{Y} = n^{-1}(Y_1+Y_2+\ldots+Y_n)$. If $\{Y_t\}$ follows a zero-mean Gaussian white noise process then it follows from Kendall and Stuart (1969, p. 297-298 and p. 305-306) that, if $i=j=0$,

$$var[b_1(i,j)] = 6(n-2)/(n+1)(n+3)$$

and

$$var[b_2(0, i, i)] = 24n(n-2)(n-3)/(n+1)^2(n+3)(n+5).$$

If, however, $i \neq 0, j \neq 0, i \neq j$, then it can be easily deduced that

$$var[b_1(i,j)] \simeq 1/n \quad \text{and} \quad var[b_2(0, i, i)] \simeq 9/n.$$

Finally, if $0 = i \neq j$ or $0 \neq i = j$ or $0 = j \neq i$, we have $var[b_1(i,j)] \simeq 3/n$. On the basis of a simulation experiment it was noticed that the above approximations can be considered as satisfactory when $n \geq 200$.

To study the behaviour of $b_1(i, j)$ and $b_2(0, i, i)$ for bilinear models with standard normally distributed errors a large scale simulation experiment was carried out. Table 1 and 2, respectively, contain results of $\bar{b}_1(i, j)$ and $\bar{b}_2(0, i, i)$, based on 100 replications of length $n = 200$ for the bilinear model $Y_t = .5Y_{t-\ell}A_{t-k} + A_t$ with $(k, \ell) = (1, 2), (1, 1)$ and $(2, 1)$.

TABLE 1. Mean standardized third central sample moment $\bar{b}_1(i,j)$ from 100 length 200 simulations of the bilinear model $Y_t = .5Y_{t-\ell}A_{t-k} + A_t$ with $(k, \ell) = (1, 2), (1, 1)$ and $(2, 1)$.

		Lag i				
(k, ℓ)	Lag j	0	1	2	3	4
(1, 2)	0	–.061				
	1	–.026	–.026			
	2	–.012	.394*	–.035		
	3	–.093	.022	.001	.034	
	4	–.007	.065	–.009	.051	–.009
(1, 1)	0	.531*				
	1	.695*	.362*			
	2	.120	.121	.068		
	3	.034	–.021	.013	.021	
	4	.018	.001	–.018	–.064	–.020
(2, 1)	0	–.073				
	1	–.054	.057			
	2	.006	.546*	–.128		
	3	.007	–.050	–.008	–.049	
	4	–.089	.034	.123	–.007	.105

Note: * denotes statistically significant at the 5% level.

TABLE 2. Mean standardized fourth central sample moment $\bar{b}_2(0, i, i)$ from 100 length 200 simulations of the bilinear model $Y_t = .5Y_{t-\ell}A_{t-k} + A_t$ with $(k, \ell) = (1, 2), (1, 1)$ and $(2, 1)$.

(k, ℓ)	Lag i				
	0	1	2	3	4
(1, 2)	3.246	1.336	1.429	1.053	.999
(1, 1)	4.030	2.195	1.254	.996	.952
(2, 1)	3.499	1.708	1.495	1.111	.907

The simulated results exhibit a pattern similar to that of $\beta_1(i, j)$ and $\beta_2(0, i, i)$. From Theorem 2.1 and 2.3 it follows directly that the only non-zero values of $\beta_1(i, j)$ are at lag $(i, j) = (j, i) = (1, 2)$ for the super- and subdiagonal model. For standard normally distributed $\{A_t\}$ and $\beta = .5$ they are, respectively, given by $\beta_1(1, 2) = .433$ for the superdiagonal model and $\beta_1(1, 2) = .408$ for the subdiagonal model. The non-zero values of $\beta_1(i, j)$ for the diagonal model follow from Theorem 2.2. Among the non-zeroes are $\beta_1(0, 0) = .756$, $\beta_1(0, 1) = \beta_1(1, 0) = .486$, $\beta_1(1, 1) = .756$, $\beta_1(0, 2) = \beta_1(2, 0) = .148$ and $\beta_1(1, 2) = \beta_1(2, 1) = .054$. The values of the kurtosis for the super- and subdiagonal model are, respectively, 3.462 and 5.205.

Clearly, the estimates of the skewness and kurtosis are much smaller than might be expected from the theoretical results for the three simulated models. It illustrates a characteristic of these series noticed earlier by, for example, Granger and Andersen (1978) and Maravall (1983). They found that stationary bilinear time series usually behave much like a linear series for a long period of time. However, on occasion, a different regime seems to operate on the series resulting in increased activity for short periods. This increased activity will show up in the estimated values of $\beta_1(0, 0)$ and $\beta_2(0, 0, 0)$ only if the series is extremely long. In our case a sample size of length 200 is too short to adequately capture the full bilinear structure of the series. Indeed, when we reestimated $\beta_1(0, 0)$ and $\beta_2(0, 0, 0)$ for series of length 800, using 30 replications, we obtained much better results. For instance, the estimates of the kurtosis for the super- and subdiagonal model with $\beta = .5$ are then, respectively, given by 3.565 and 5.488. We may conclude from this that it will be difficult to accurately identify a particular model on the basis of the statistics $b_1(i, j)$ and $b_2(0, i, i)$ unless the series is much longer than is usually the case in economic time series analysis. In particular, it will be hard to distinguish superdiagonal models from subdiagonal models using these statistics for moderately large sample sizes.

This last point was also confirmed by a blind discrimination experiment between 60 simulated bilinear time series. Each series of length 200 was generated by one of the three previously discussed bilinear models with $\beta = .5$. The choice of a model was determined by an unseen sequence of numbers 1, 2 and 3 which we randomly obtained at the start of the experiment. The mean and variance of the generated $\{A_t\}$ was set, respectively, to zero and one throughout the simulations. Using the characteristic pattern of $\rho(i)$, $\beta_1(i, j)$ and $\beta_2(0, i, i)$ we tried to discriminate between the three bilinear models. Of the 60 simulated series, we correctly identified 43 series which represents about 72% success. Of the 18 series generated by the diagonal model 16 were correctly identified, whilst the other 2 were incorrectly thought to be generated by the superdiagonal model. This successful identification was based on the typical pattern of $\beta_1(i, j)$ and the fact that $E(Y_t) \neq 0$ for diagonal models. By comparing the pattern of the sample statistics $b_1(i, j)$ and $b_2(0, i, i)$ with their theoretical counter parts we obtained 7 out of 19 correct identifications of the subdiagonal model. For the remaining 12 series, 10 were incorrectly identified as being generated by the superdiagonal model. Finally, for the superdiagonal model 20 correct decisions were achieved and 3 were incorrectly identified as coming from the subdiagonal model.

Although the scope of this simulation experiment is rather limited the results seem to suggest that correct discrimination between series of length 200 generated by a super- and subdiagonal model, using $b_1(i, j)$ and $b_2(0, i, i)$, is doubtful. There is hardly any marked difference in the behaviour of these statistics for these two bilinear models which makes them not very useful for model discrimination. Perhaps by estimating both a super- and subdiagonal model one could more easily tell which model generated the data. On the other hand, the results of the simulation experiment also indicated that a quick and relative accurate method for distinguishing a diagonal model from a non-diagonal model can be based on the set of non-zero values of the statistic $b_1(i, j)$. We hope to report elsewhere on the practical usefulness of the results obtained in this paper.

REFERENCES

Granger, C.W.J. and Andersen, A.P. (1978). *An Introduction to Bilinear Time Series Models.* Vandenhoeck & Ruprecht, Göttingen.

Guegan, D. (1984). Test de modèles non linéaires. *Alternative Approaches to Time Series Analysis* (J.P. Florens, M. Mouchart, J.P. Raoult and L. Simar, *eds.*). Facultés Universitaires Saint-Louis, Bruxelles, 45-77.

Hinich, M.J. and Patterson, D.M. (1985). Evidence of nonlinearity in daily stock returns. *Journal of Business & Economic Statistics*, 3, 69-77.

Kendall, M.G. and Stuart, A. (1969). *The Advanced Theory of Statistics.* Vol. 1, Charles Griffin & Company Ltd., London.

Kumar, K. (1986). On the identification of some bilinear time series models. *Journal of Time Series Analysis,* 7, 117-122.

Li, W.K. (1984). On the autocorrelation structure and identification of some bilinear time series. *Journal of Time Series Analysis,* 5, 173-181.

Maravall, A. (1983). An application of nonlinear time series forecasting. *Journal of Business & Economic Statistics*, 1, 66-74.

Subba Rao, T. and Gabr, M.M. (1984). *An Introduction to Bispectral Analysis and Bilinear Time Series Models.* Springer-Verlag, New York.

Received 25 February 1987
Revised 17 September 1987

Department of Economic Statistics
University of Amsterdam
Jodenbreestraat 23
1011 NH Amsterdam
Holland

Department of Econometrics
Tilburg University
P.O. Box 90153
5000 LE Tilburg
Holland

Proc. Second International Tampere Conference in Statistics
(Tampere, Finland, 1-4 June 1987)
Tarmo Pukkila and Simo Puntanen, *Editors*

pp. 479 - 489

On the computation of distributions of serial correlation coefficients through B-splines

Z.G. IGNATOV and V.K. KAISHEV
Institute of Mathematics, Bulgarian Academy of Sciences, Sofia, Bulgaria

Key words and phrases: Dirichlet distribution, divided difference, normalized B-spline, incomplete Beta function, significance points.

ABSTRACT

Formulae for the distribution of linear combinations of random variables having Dirichlet distribution with parameters, all integer except one or two are given. These formulae, related to B-splines are applied to compute significance points of certain serial correlation coefficients.

1. INTRODUCTION

Serial correlation coefficients (SCC) have been introduced to measure the relationship between lag j successive members of a series of observations in time or space. Problems related to serial correlation have been widely considered under the assumption of normality of the underlying population. Among the large number of references we will mention those of von Neumann (1941), Hart and von Neumann (1942), R.L. Anderson (1942), Rubin (1945), Hsu (1946), T.W. Anderson (1948). Important applications of SCC to hypothesis testing in analysis of time series are studied and thoroughly discussed by T.W. Anderson (1971) (see Chapters 6, 10).

Several versions of SCC have been proposed by different authors. Each of these can be represented in a general form as a ratio of two quadratic forms, i.e.,

$$\tau = y'Ay/y'y\,, \tag{1.1}$$

where $y' = (y_1, y_2, \ldots, y_N)$ is a vector of independent normal variables with certain mean and constant variance and A is a real symmetric matrix. The exact distribution of SCC given by (1.1) has been derived only when the latent roots of A all have even multiplicities or when at most one root is of odd multiplicity (see T.W. Anderson 1971).

It is known (see e.g. Margolin 1977) that the distribution of every SCC which admits representation (1.1) coincides with the distribution of a linear combination of a set of random variables having joint Dirichlet distribution with parameters correspondingly equal to 1/2 times the multiplicities of the latent roots of the matrix A. The latent roots appear as real coefficients in such a linear combination.

In the present paper we give computationally convenient formulae for the distribution function of a linear combination of Dirichlet distributed random variables with all save one or two parameters integer. These formulae, involving polynomial B-splines are applied to compute significance points of the distribution of certain SCC given in Section 4. Different expressions for the distribution function of a linear combination of Dirichlet random variables with only one parameter noninteger were obtained by Bloch and Watson (1967) and Margolin (1977). However, they characterize them as "too cumbersome for practical use".

Imhof (1961) suggested to compute the distribution of quadratic forms in normal variables by means of computing the Fourier inversion integral. This method can be applied to compute the distribution of serial correlation coefficients as indicated by Ramasubban (1972). Practical computation of the probabilities of SCC is also possible using a method due to Pan Jie-Jian (1968). However, approaches taken by Imhof, Pan Jie-Jian, Ramasubban and perhaps others lay basically upon approximation of the desired distribution function. This causes an "error of approximation" apart from three other sources of inaccuracy, namely that of "rounding off errors", "error of truncation" and "error of numerical integration".

The formulae presented here are exact, recurrent and their implementation requires numerical integration only, hence they are free of "errors of approximation and truncation".

2. PRELIMINARIES

Since we encounter divided differences and B-splines recall some of their properties which will be needed further.

Let $\{t_i\}$, $-\infty < i < \infty$ be a bi-infinite sequence of points on the real line, i.e., $t_i \in \mathbb{R}^1$.

The n-th order divided difference of a sufficiently smooth function $\phi(u)$ over the points $t_i, \ldots, t_{i+n}$ is defined as

$$[t_i, \ldots, t_{i+n}]_u \phi(u) = \frac{[t_{i+1}, \ldots, t_n]_u \phi(u) - [t_i, \ldots, t_{i+n-1}]_u \phi(u)}{t_{i+n} - t_i} \tag{2.1}$$

provided $t_i \neq t_{i+n}$. If $t_i = t_{i+1} = \ldots = t_{i+n}$ then

$$[t_i, \ldots, t_{i+n}]_u \phi(u) = D^n \phi(t_i)/n!$$

where $D^n\phi(t_i)$ is the n-th derivative of $\phi(u)$ at $u = t_i$, $n \geq 0$, $[D^0\phi(t_i) := \phi(t_i)]$.

Using the notion of a divided difference Curry and Schoenberg (1966) defined the polynomial B-spline $M(t; t_i, \ldots, t_{i+n})$ of degree $n-1$ with knots $t_i, \ldots, t_{i+n} \in \mathbb{R}^1$ as

$$M(t; t_i, \ldots, t_{i+n}) := [t_i, \ldots, t_{i+n}]_u \phi(u), \tag{2.2}$$

where $\phi(u) = n(u-t)_+^{\,n-1}$, $(z)_+ = max\{0, z\}$.

If some of the knots $t_i, \ldots, t_{i+n}$ coincide the B-spline will be denoted by $M(t; \underbrace{t_i, \ldots, t_i}_{g_i}, \ldots, \underbrace{t_{i+\ell}, \ldots, t_{i+\ell}}_{g_{i+\ell}})$ where $g_i, \ldots, g_{i+\ell}$ are called multiplicities of the knots $t_i, \ldots, t_{i+\ell}$.

Commonly used for computations is the normalized B-spline

$$N_{i,n}(t) = (t_{i+n} - t_i) M(t; t_i, \ldots, t_{i+n}) / n, \tag{2.3}$$

where for the sake of brevity the dependence of $N_{i,n}(t)$ on the knots $t_i, \ldots, t_{i+n}$ is not explicitly indicated.

For the j-th derivative ($j \geq 0$) of $N_{i,n}(t)$ the well known de Boor-Cox efficient, numerically stable recurrence relation holds (see e.g. de Boor 1976)

$$\frac{n-j-1}{n-1}N_{i,n}^{(j)}(t)=\frac{t-t_i}{t_{i+n-1}-t_i}N_{i,n-1}^{(j)}(t)+\frac{t_{i+n}-t}{t_{i+n}-t_{i+1}}N_{i+1,n-1}^{(j)}(t). \quad (2.4)$$

The integral of a B-spline can also be recurrently computed using the formula (c.f. de Boor 1976)

$$\int_{-\infty}^{t} M(x;t_i,\dots,t_{i+n})dx=\sum_{j=i}^{i+\tau} N_{j,n+1}(t), \qquad t\le t_{i+\tau+1}. \quad (2.5)$$

For a detailed survey on properties of B-splines we refer to de Boor (1978) and Schumaker (1981).

3. B-SPLINES AND DIRICHLET RANDOM VARIABLES

In this section we give formulae convenient for the computation of the distribution of linear combinations of Dirichlet distributed random variables with all integer parameters except one or two.

Let $(\theta_0,\dots,\theta_m)$ be a random vector having Dirichlet distribution with parameters ρ, $g_1,\dots,g_m$, $\rho>0$, g_i -positive integer, $i=1,\dots,m$, i.e., $(\theta_0,\dots,\theta_m)\in D(\rho,g_1,\dots,g_m)$.

Consider the linear combination

$$S=\lambda_0\theta_0+\dots+\lambda_m\theta_m. \quad (3.1)$$

As shown by the authors (see Ignatov and Kaishev 1986) if $\lambda_0<\lambda_i$, $i=1,\dots,m$ then

$$P(S>x)=\begin{cases}[\underbrace{\lambda_0,\dots,\lambda_0}_{\hat\rho},\underbrace{\lambda_1,\dots,\lambda_1}_{g_1},\dots,\underbrace{\lambda_m,\dots,\lambda_m}_{g_m}]_u\phi(u), & \lambda_0\le x\\ 1, & \lambda_0>x\end{cases} \quad (3.2)$$

where $\phi(u)=(u-\lambda_0)^{-\bar\rho}(u-x)_+^{\rho+g_1+\dots+g_m-1}$, $(\phi(\lambda_0):=0)$, $\bar\rho=\rho-\hat\rho$, $\hat\rho$ is the integer part of ρ, and the density function of S

$$f_S(x)=\begin{cases}[\underbrace{\lambda_0,\dots,\lambda_0}_{\hat\rho},\underbrace{\lambda_1,\dots,\lambda_1}_{g_1},\dots,\underbrace{\lambda_m,\dots,\lambda_m}_{g_m}]_u H(u), & x\in B\\ 0, & x\notin B\end{cases} \quad (3.3)$$

where $H(u) = (\ell + \bar{\rho} - 1)(u - \lambda_0)^{-\bar{\rho}}(u - x)_+^{\ell+\bar{\rho}-2}$, $\ell = \hat{\rho} + g_1 + \ldots + g_m$, B is the convex hull of the set $\{\lambda_0,\ldots, \lambda_m\}$.

In what follows we will assume $\lambda_0 < \lambda_1 <\ldots< \lambda_m$ and let us introduce the set of knots

$$\ldots\leq t_0 \leq t_1 \leq t_2 \leq\ldots\leq t_{\ell-1} < t_\ell = t_{\ell+1} =\ldots= t_{\ell+\hat{\rho}-1} < t_{\ell+\hat{\rho}} = t_{\ell+\hat{\rho}+1}$$
$$=\ldots= t_{\ell+\hat{\rho}+g_1-1} <\ldots< t_{\ell+\hat{\rho}+g_1+\ldots+g_{m-1}} = t_{\ell+\hat{\rho}+g_1+\ldots+g_{m-1}+1}$$
$$= t_{2\ell-1} < t_{2\ell} \leq t_{2\ell+1} \leq\ldots\leq t_{3\ell-2} \leq t_{3\ell-1} \leq\ldots$$

where $t_\ell = \lambda_0$, $t_{\ell+\hat{\rho}} = \lambda_1$, $t_{\ell+\hat{\rho}+g_1} = \lambda_2,\ldots, t_{\ell+\hat{\rho}+g_1+\ldots+g_{m-1}} = \lambda_m$, the rest of the knots, necessary to introduce the basis of B-splines, $N_{i,\ell}(t)$, $i = 1,\ldots, 2\ell - 2$, (c.f. 2.3), being arbitrary. As known, (see C. de Boor 1978), the choice of these extra knots does not affect computations involving this B-spline basis.

We shall note that if $\bar{\rho} = 0$ then $f_S(x)$ from (3.3) coincides with the classical B-spline given by (2.2), hence using (2.5) we obtain

$$P(S > x) = 1 - \int_{-\infty}^{x} f_S(t)dt = 1 - \int_{-\infty}^{x} M(t; t_\ell,\ldots, t_{2\ell-1})dt$$

$$= 1 - \sum_{j=\ell}^{\ell+\tau} N_{j,\ell+1}(x), \qquad x \leq t_{\ell+\tau+1}. \tag{3.4}$$

So, combining (3.4) with (2.4) for $j = 0$ we get efficient means of computing $P(S > x)$.

To compute $P(S > x)$ in the general case (i.e. $\bar{\rho} > 0$) one can use (3.2). However, when ℓ is large direct computation of the power $(u - x)_+^{\rho+g_1+\ldots+g_m-1}$ from (3.2) may lead to loss of accuracy. To avoid this we will rewrite $\phi(u)$ as

$$\phi(u) = \xi(u)(u - x)_+^{\ell-1}, \tag{3.5}$$

where $\xi(u) = [(u - x)_+/(u - \lambda_0)]^{\bar{\rho}}$, $(\xi(\lambda_0):=0)$ and express $(u - x)_+^{\ell-1}$ via normalized B-splines. The latter can be easily and accurately computed using the recurrence relation (2.4).

LEMMA 1. *The probability* $P(S > x)$ *for* S *from* (3.1) *is given as*

$$P(S > x) = \sum_{i=\ell}^{2\ell-1} [t_\ell,\ldots, t_i]_u \xi(u) \sum_{k=1}^{2\ell-2} \psi_{k,\ell}(x)[t_i,\ldots, t_{2\ell-1}]_u N_{k,\ell}(u), \tag{3.6}$$

where $\psi_{k,\ell} = \prod_{\tau=1}^{\ell-1} (t_{k+\tau} - x)_+$.

Proof. We can apply Leibniz's rule for the divided difference of a product to (3.5) to obtain

$$P(S > x) = \sum_{i=\ell}^{2\ell-1} [t_\ell,\ldots, t_i]_u \xi(u)[t_i,\ldots, t_{2\ell-1}]_u (u - x)_+^{\ell-1}. \tag{3.7}$$

Now using Marsden's identity (see de Boor 1972)

$$(u - x)_+^{\ell-1} = \sum_{k=1}^{2\ell-2} \psi_{K,\ell}(x) N_{k,\ell}(u) \tag{3.8}$$

we substitute $(u - x)_+^{\ell-1}$ from (3.8) in (3.7) and get the assertion of Lemma 1. □

Now let us treat the case when the random vector

$$(\theta_0,\ldots, \theta_m) \in D(\rho_0, \rho_1, g_2,\ldots, g_m),$$

where ρ_0, ρ_1 are positive real, $g_2,\ldots, g_m$ -positive integer. For the linear combination S from (3.1) when $\lambda_0 < \lambda_i < \lambda_1$ $i = 2,\ldots, m$ we have

LEMMA 2. *The probability* $P(S > x)$

$$= \frac{1}{B(\bar\rho_0, \bar\rho_1 + \ell)} \int_0^d y^{\bar\rho_0 - 1} (1 - y)^{\bar\rho_0 + \ell - 1} dy$$

$$- [\underbrace{\lambda_0,\ldots, \lambda_0}_{\hat\rho_0}, \underbrace{\lambda_1,\ldots, \lambda_1}_{\hat\rho_1}, \underbrace{\lambda_2,\ldots, \lambda_2}_{g_2},\ldots, \underbrace{\lambda_m,\ldots, \lambda_m}_{g_m}]_u L(u), \tag{3.9}$$

where $B(\bar\rho_0, \bar\rho_1 + \ell) = \Gamma(\bar\rho_0)\Gamma(\bar\rho_1 + \ell)/\Gamma(\ell + \bar\rho_0 + \bar\rho_1)$, $\Gamma(\cdot)$ *is the Gamma function,* $\ell = \hat\rho_0 + \hat\rho_1 + g_2 + \ldots + g_m$,

$$L(u) = \frac{1}{B(\bar\rho_0, \bar\rho_1 + \ell)} \int_c^d y^{\bar\rho_0 - 1} \left(\frac{u - x + (\lambda_0 - u)y}{u - \lambda_1} \right)^{\bar\rho_1} (u - x + (\lambda_0 - u)y)^{\ell - 1} dy,$$

$$c = \left(\frac{u - x}{u - \lambda_0} \right)_+ , (c = 0 \text{ for } u = \lambda_0), d = \frac{\lambda_1 - x}{\lambda_1 - \lambda_0}.$$

For a proof of this result we refer to a forthcoming paper due to the authors (see Ignatov and Kaishev 1987). There, a formula for $P(S > x)$ for arbitrary parameters of the Dirichlet distribution is also derived.

Formula (3.9) represents a difference between the distribution function of the beta distribution and a divided difference term. At present, several routines for the computation of the incomplete beta function are readily available (see e.g. Kennedy and Gentle 1980, p. 106). Whereas the divided difference term can be computed using the recurrence relation (2.1) and expressing the integral function as a Gauss quadrature.

4. THE DISTRIBUTION OF SCC

We shall now consider particular serial correlation coefficients which admit representation (1.1).

Let $y_1, \ldots, y_N$ be variables representing a random sample of N successive observations from a population whose distribution is normal $N(\mu, \sigma^2)$.

The lag j circular serial correlation coefficient in the case of unknown population mean is defined as

$$\tau_j = \sum_{i=1}^{N} (y_i - \bar{y})(y_{i+j} - \bar{y}) / \sum_{i=1}^{N} (y_i - \bar{y})^2, \tag{4.1}$$

where $y_{N+i} = y_i$, $\bar{y} = (y_1 + \ldots + y_N)/N$ and with zero mean as

$$\tau_j = \sum_{i=1}^{N} y_i y_{i+j} / \sum_{i=1}^{N} y_i^2, \quad y_{N+i} = y_i, \quad j < N. \tag{4.2}$$

To test for serial correlation the following two noncircular test statistics have been introduced

$$\tau_j = \sum_{i=1}^{N-j} (\Delta^j y_i)^2 / \sum_{i=1}^{N} y_i^2, \qquad j < N, \tag{4.3}$$

where Δ^j the successive difference operator of order j is given as

$$\Delta^j y_i = \sum_{\tau=0}^{j} (-1)^{j-\tau} \binom{j}{\tau} y_{i+\tau},$$

and

$$\tau_j = \sum_{i=j+1}^{N} y_i y_{i-j} / \sum_{i=1}^{N} y_i^2. \tag{4.4}$$

The distribution of τ_j in (4.1) has been given by R.L. Anderson (1942) while SCC in (4.2), (4.3) and (4.4) are studied by T.W. Anderson (1971) and others.

It can be shown that SCC in (4.1), (4.2), (4.3) and (4.4) admit the canonical representation

$$\tau_j = \sum_i \lambda_{ji} z_i^2 / \sum_i z_i^2, \qquad j = 1, 2, \ldots,$$

where $z_1, z_2, \ldots$ are independent standard normal random variables, λ_{ji}, $i = 1, 2, \ldots$ are the latent roots of the matrix A_j. Hence τ_j coincides in distribution with

$$S_j = \sum_{i=0}^{p} \lambda_{ji} \theta_i, \qquad j = 1, 2, \ldots,$$

where $(\theta_0, \ldots, \theta_p) \in D(\frac{1}{2}, \ldots, \frac{1}{2})$, p depends on the corresponding serial correlation coefficient.

We can use formulae (3.4), (3.6) and (3.9) to compute significance points of SCC from (4.1), (4.2), (4.3) and (4.4) having no more than two roots of odd multiplicity. We shall consider in more details SCC from (4.1).

The latent roots of the matrix A_j in representation (1.1) of τ_j from (4.1) are

$$\lambda_{ji} = cos(2\pi ji/N), \quad i = 1, \ldots, N-1, \quad j < N.$$

When N is odd (i.e. $N = 2w + 1$) the roots are equal in pairs and $P(\tau_j > x)$ can be computed according to formula (3.4) where $\ell = w$, $t_\ell = \lambda'_{j1}$, $t_{\ell+1} = \lambda'_{j2}, \ldots, t_{2\ell-1} = \lambda'_{jw}$ and $\{\lambda'_{j1} \le \lambda'_{j2} \le \ldots \le \lambda'_{jw}\}$ is the set $\{\lambda_{j1}, \lambda_{j2}, \ldots, \lambda_{jw}\}$ arranged in nondecreasing order.

When N is even (i.e. $N = 2w$) the roots are equal in pairs with the esception of the root $\lambda_{jw} = (-1)^j$. If j is odd then $P(\tau_j > x)$ can be computed following formula (3.6) with $\ell = w - 1$, $t_\ell = \lambda'_{j2}$, $t_{\ell+1} = \lambda'_{j3}, \ldots, t_{2\ell-1} = \lambda'_{jw}$, $\lambda_0 = (-1)^j = -1$ and $-1 = \lambda'_{j1} \le \ldots \le \lambda'_{jw}$ are the quantities $\lambda_{j1}, \lambda_{j2}, \ldots, \lambda_{jw}$ arranged in nondecreasing order.

In Table 1 we present significance points of τ_j from (4.1) for $j = 1$ computed by the authors (first two columns) and by R.L. Anderson (1942). The accuracy of the computations is 6 correct digits while for Anderson's data it is 2 digits.

TABLE 1

Significance points of the circular serial correlation coefficient-positive tail

			R.L. ANDERSON'S DATA	
N	1%	5%	1%	5%
5	.2978367	.2531153	.297	.253
6	.4468671	.3446384	.447	.345
7	.5099245	.3695500	.510	.370
8	.5307256	.3713381	.531	.371
9	.5321655	.3661171	.533	.366
10	.5252934	.3596500	.525	.360
11	.5153293	.3537825	.515	.353
12	.5047043	.3477849	.505	.348
13	.4944493	.3412710	.495	.341
14	.4845674	.3346155	.485	.335
15	.4749282	.3280912	.475	.328
20	.4319585	.2991205	.432	.299
25	.3978567	.2759038	.398	.276
30	.3704718	.2571762	.370	.257
35	.3479875	.2417606	.347 *	.242 *
40	.3291410	.2288178	.329 *	.229 *
45	.3130633	.2177634	.314	.218
50	.2991437	.2081839	.301 *	.208 *
55	.2869419	.1997803	.289 *	.199 *
60	.2761327	.1923311	.278 *	.191 *
65	.2664705	.1856685	.268 *	.184 *
70	.2577659	.1796632	.259 *	.178 *
75	.2498704	.1742137	.250	.173
80	.2426659	.1692390		
85	.2360571	.1646738		
90	.2299660	.1604647		
95	.2243281	.1565675		
100	.2190899	.1529453		

* These values were obtained by R.L. Anderson by graphical interpolation.

In Table 2 we summarize the cases in which the distribution of SCC τ_j from (4.2), (4.3) and (4.4) could be computed using formulae (3.4), (3.6) and (3.9). For the latent roots in Table 2 we refer to T.W. Anderson (1971).

TABLE 2

Formulae for computing $P(\tau_j > x)$

SCC – τ_j defined in	Sample size N	Lag j	Latent roots λ_{ji}, $j < N$	Formula defined in
(4.1)	odd	arbitrary	$cos(2\pi ji/N)$, $i = 1,..., N-1$	(3.4)
"	even	odd	"	(3.6)
(4.2)	"	even	$1, (-1)^j, cos(2\pi ji/N), cos(2\pi ji/N),$ $i = 1,..., (N-2)/2$	(3.4)
"	"	odd	"	(3.9)
(4.3)	"	multiple of 4	$cos\ \pi ji/N$ $i = 0,..., N-1$	(3.4)
"	"	even non-multiple of 4	"	(3.9)
(4.4)	"	even	$cos\ \pi ji/(N+1)$ $i = 1,..., N$	(3.4)
"	odd	even non-multiple of 4	"	(3.6)

Finally, let us note that FORTRAN routines for the computation of significance points of all SCC from Table 2 are available from the authors.

ACKNOWLEDGEMENT

We wish to thank the referees for valuable comments and suggestions and for pointing to us the works of Imhof (1961), Pan Jie-Jian (1968) and Ramasubban (1972). It is worth noting, that a thorough comparison of their approach and its computer program realization with the method proposed here is interesting to be accomplished in the future.

REFERENCES

Anderson, R.L. (1942). Distribution of serial correlation coefficient. *Annals of Mathematical Statistics*, 13, 1-13.

Anderson, T.W. (1948). On the theory of testing serial correlation. *Skandinavisk Aktuarietidskrift*, 31, 88-116.

Anderson, T.W. (1971). *The Statistical Analysis of Time Series*. Wiley, New York.

Bloch, D.A. and Watson, G.S. (1967). A Bayesian study of the multinomial distribution. *Annals of Mathematical Statistics* 38, 1423-35.

de Boor, C. (1972). On calculating with B-splines. *Journal of Approximation Theory*, 6, 50-62.

de Boor, C. (1976). Splines as linear combinations of B-splines. *Approximation Theory II* (G.G. Lorentz, C.K. Chui and L.L. Schumaker, *eds.*). Academic Press, New York, 1-47.

de Boor, C. (1978). *A Practical Guide to Splines*. Springer Verlag, New York.

Curry, H.B. and Schoenberg, I.J. (1966). On Polya frequency functions. IV, the fundamental spline functions and their limits. *Journal d'Analyse Mathematique*, 17, 71-107.

Hart, B.J. and von Neumann, J. (1942). Tabulation of the probabilities of the ratio of the mean square successive difference to the variance. *Annals of Mathematical Statistics*, 13, 207-214.

Hsu, P.L. (1946). On the asymptotic distribution of certain statistics used in testing independence between successive observations from a normal population. *Annals of Mathematical Statistics*, 13, 14-33.

Ignatov, Z.G. and Kaishev, V.K. (1986). Multivariate B-splines, analysis of contingency tables and serial correlation. *Proceedings of the Sixth Pannonian Symposium on Mathematical Statistics*. D. Reidel Publ. Co. Dordrecht, Holland, in press.

Ignatov, Z.G. and Kaishev, V.K. (1987). Linear combinations of Dirichlet distributed random variables and B-splines, to appear.

Imhof, P.J. (1961). Computing the distribution of quadratic forms in normal variables. *Biometrika*, 48, 419-426.

Kennedy, Jr. W.J. and Gentle, J.E. (1980). *Statistical Computing*. Marcel Dekker, New York.

Margolin, B.H. (1977). The distribution of internally studentized statistics via Laplace transform inversion. *Biometrika*, 64, 3, 573-82.

von Neumann, J. (1941). Distribution of the ratio of the mean square successive difference to the variance. *Annals of Mathematical Statistics*, 12, 367-95.

Pan Jie-Jian (1968). Distribution of the noncircular correlation coefficients. *Am. Math. Soc. and Inst. Math. Statist. Selected Translations in Probability and Statistics*, 7, 281-291.

Ramasubban, T.A. (1972). An approximate distribution of a noncircular serial correlation coefficient. *Biometrika*, 59, 1, 79-84.

Rubin, H. (1945). On the distribution of the serial correlation coefficient. *Annals of Mathematical Statistics*, 16, 211-15.

Received 25 February 1987
Revised 12 August 1987

Institute of Mathematics
Bulgarian Academy of Sciences
P.O. Box 373
1090 Sofia
Bulgaria

Proc. Second International Tampere Conference in Statistics
(Tampere, Finland, 1-4 June 1987)
Tarmo Pukkila and Simo Puntanen, *Editors*

pp. 491 - 498

Elliptically symmetric distributions and their application to classification and regression

Krzysztof JAJUGA

Economic Cybernetics Institute, Academy of Economics, Wroclaw, Poland

Key words and phrases: Bayes classification rule, mixture of distributions, spherical coordinates.

ABSTRACT

The definition and examples of elliptically symmetric distributions are given. Simple test of ellipsoidal symmetry is then proposed. Finally, the problems of the use of elliptically symmetric distributions in classification and regression are discussed.

1. INTRODUCTION – DEFINITION AND SOME EXAMPLES

The family of elliptically symmetric distributions has recently been frequently considered in statistical papers. It seems that elliptically symmetric distributions may be of use in real applications.

The paper gives a brief discussion of some problems referring to this family of distributions. We consider such distributions for which the second moment matrix exists. The definition of elliptically symmetric (elliptical) distributions is a generalization of the definition of spherically symmetric distributions.

DEFINITION. *An m-dimensional random vector* $\mathbf{X} = [X_1, X_2, \ldots, X_m]^T$ *has elliptical distribution with mean vector* $\boldsymbol{\mu}$ *and covariance matrix* $\boldsymbol{\Sigma}$, *if a random vector*

$$\mathbf{Y} = \boldsymbol{\mu} + \boldsymbol{\Sigma}^{\frac{1}{2}}\mathbf{A}\boldsymbol{\Sigma}^{-\frac{1}{2}}(\mathbf{X} - \boldsymbol{\mu})$$

has the same distribution as $\mathbf{X}$ *for each orthogonal matrix* $\mathbf{A}$.

It is easy to prove the following theorem.

THEOREM. *The isodensity contours of m-variate elliptical distributions are the m-dimensional hyperellipsoids of the form*

$$(\mathbf{x} - \boldsymbol{\mu})^T \boldsymbol{\Sigma}^{-1} (\mathbf{x} - \boldsymbol{\mu}) = c,$$

where c is positive constant.

Cooper (1963) suggested that the density of elliptically symmetric distribution can be presented by the formula:

$$f(\mathbf{x}) = c\,|\mathbf{B}|^{\frac{1}{2}}\,g(r),$$

where $r = [(\mathbf{x} - \boldsymbol{\mu})^T \mathbf{B}\,(\mathbf{x} - \boldsymbol{\mu})]^{\frac{1}{2}}$, $\boldsymbol{\mu}$ is the mean vector, and g is the non-negative valued function defined on the interval $[0, \infty]$ and such that:

$$\int_0^\infty r^M g(r)\,dr < \infty, \qquad M \leq m+1.$$

In addition, it can be proved that the normalizing constant c may be given as:

$$c = \tfrac{1}{2}\pi^{-\frac{1}{2}m}\,\Gamma(\tfrac{1}{2}m)\,[\int_0^\infty r^{m-1} g(r)\,dr]^{-1}.$$

Furthermore, Haralick (1977) proved that the positive definite matrix $\mathbf{B}$ is related to covariance matrix $\boldsymbol{\Sigma}$ in the following way:

$$\mathbf{B} = \boldsymbol{\Sigma}^{-1}\,[\int_0^\infty r^{m+1} g(r)\,dr]\,/\,[m\int_0^\infty r^{m-1} g(r)\,dr]^{-1}.$$

By means of this presentation it is easy to obtain the densities of particular forms of elliptically symmetric distributions. They are given in the following examples.

Example 1. Let $g(r) = e^{-\frac{1}{2}r^2}$. Then

$$c = (2\pi)^{-\frac{1}{2}m}, \qquad \mathbf{B} = \mathbf{\Sigma}^{-1},$$

and

$$f(\mathbf{x}) = (2\pi)^{-\frac{1}{2}m} |\mathbf{\Sigma}|^{-\frac{1}{2}} \exp[-\tfrac{1}{2}(\mathbf{x} - \boldsymbol{\mu})^{\mathrm{T}} \mathbf{\Sigma}^{-1} (\mathbf{x} - \boldsymbol{\mu})].$$

This is the m-variate normal distribution with mean vector $\boldsymbol{\mu}$ and covariance matrix $\mathbf{\Sigma}$.

Example 2. Let $g(r) = e^{-r}$. Then

$$c = \Gamma(\tfrac{1}{2}m)\,[2\pi^{\frac{1}{2}m}\Gamma(m)]^{-1}, \qquad \mathbf{B} = (m+1)\mathbf{\Sigma}^{-1},$$

and

$$f(\mathbf{x}) = \Gamma(\tfrac{1}{2}m)\,[2\pi^{\frac{1}{2}m}\Gamma(m)]^{-1}\,(m+1)^{\frac{1}{2}m}\,|\mathbf{\Sigma}|^{-\frac{1}{2}}\exp\{-(m+1)^{\frac{1}{2}}[(\mathbf{x}-\boldsymbol{\mu})^{\mathrm{T}}\,\mathbf{\Sigma}^{-1}\,(\mathbf{x}-\boldsymbol{\mu})]^{\frac{1}{2}}\}.$$

This is the m-variate Laplace distribution with mean vector $\boldsymbol{\mu}$ and covariance matrix $\mathbf{\Sigma}$.

Example 3. Let

$$g(r) = (1+r^2)^{-k}, \qquad k > \tfrac{1}{2}m+1.$$

Then

$$c = \Gamma(k)\,[\pi^{\frac{1}{2}m}\Gamma(k - \tfrac{1}{2}m)]^{-1}, \qquad \mathbf{B} = (2k - m - 2)^{-1}\,\mathbf{\Sigma}^{-1},$$

and

$$f(\mathbf{x}) = \Gamma(k)[\pi^{\frac{1}{2}m}\Gamma(k-\tfrac{1}{2}m)]^{-1}(2k-m-2)^{-\frac{1}{2}m}|\mathbf{\Sigma}|^{-\frac{1}{2}}[1+(2k-m-2)^{-1}(\mathbf{x}-\boldsymbol{\mu})^{\mathrm{T}}\mathbf{\Sigma}^{-1}(\mathbf{x}-\boldsymbol{\mu})]^{-k}.$$

This is the m-variate Pearson type VII distribution with mean vector $\boldsymbol{\mu}$ and covariance matrix $\mathbf{\Sigma}$.

Example 4. Let

$$g(r) = \begin{cases} (1-r^2)^k & r^2 \le 1, \quad k > 0, \\ 0 & \text{otherwise.} \end{cases}$$

Then

$$c = \Gamma(k+\tfrac{1}{2}m+1)[\pi^{\frac{1}{2}m}\Gamma(k+1)]^{-1}, \qquad \mathbf{B} = (2k+m+2)^{-1}\,\mathbf{\Sigma}^{-1}.$$

Thus, for $\mathbf{x}$ satisfying the inequality

$$(\mathbf{x} - \boldsymbol{\mu})^{\mathrm{T}} \boldsymbol{\Sigma}^{-1} (\mathbf{x} - \boldsymbol{\mu}) \leq 2k + m + 2$$

the density is given as:

$$f(\mathbf{x}) = \Gamma(k + \tfrac{1}{2}m + 1)[\pi^{\frac{1}{2}m}\Gamma(k+1)]^{-1} (2k + m + 2)^{-\frac{1}{2}m} |\boldsymbol{\Sigma}|^{-\frac{1}{2}} \cdot$$
$$\cdot [1 - (2k + m + 2)^{-1} (\mathbf{x} - \boldsymbol{\mu})^{\mathrm{T}} \boldsymbol{\Sigma}^{-1} (\mathbf{x} - \boldsymbol{\mu})]^{k}.$$

Otherwise, the density is equal to zero.

This is the m-variate Pearson type II distribution with mean vector $\boldsymbol{\mu}$ and covariance matrix $\boldsymbol{\Sigma}$.

2. SIMPLE TEST FOR ELLIPTICAL SYMMETRY

The hypothesis that the set of m-dimensional observations:

$$\mathbf{x}_i = [x_{i1}, x_{i2}, \ldots, x_{im}]^{\mathrm{T}} \qquad i = 1, \ldots, n$$

is drawn from the population with elliptically symmetric distribution is tested.

Testing for elliptical (ellipsoidal) symmetry is performed by the following procedure.

1. The estimates of mean vector $\mathbf{m}$ and covariance matrix $\mathbf{S}$ are obtained according to the formulas:

$$m_j = n^{-1} \sum_{i=1}^{n} x_{ij}, \qquad j = 1, \ldots, m,$$

$$s_{jk} = n^{-1} \sum_{i=1}^{n} (x_{ij} - m_j)(x_{ik} - m_k), \qquad j, k = 1, \ldots, m.$$

2. The observations are transformed according to the formula:

$$\mathbf{z}_i = \mathbf{S}^{-\frac{1}{2}}(\mathbf{x}_i - \mathbf{m}), \qquad i = 1, \ldots, n.$$

3. The observations are transformed from Cartesian coordinates to spherical coordinates:

$$[z_{i1}, z_{i2}, \ldots, z_{im}]^{\mathrm{T}} \rightarrow [r_i, \phi_i, \alpha_{i1}, \alpha_{i2}, \ldots, \alpha_{i,m-2}]^{\mathrm{T}},$$

where:

$$z_{i1} = r_i \cos \phi_i \prod_{j=1}^{m-2} \cos \alpha_{ij} ,$$

$$z_{i2} = r_i \sin \phi_i \prod_{j=1}^{m-2} \cos \alpha_{ij} ,$$

$$z_{i,k+2} = r_i \sin \alpha_{ik} \prod_{j=k+1}^{m-2} \cos \alpha_{ij} , \qquad k = 1,..., m\text{-}3,$$

$$z_{im} = r_i \sin \alpha_{i,m-2} .$$

4. The range of variation for each angular coordinate is divided into ℓ equal intervals. So in the $(m\text{-}1)$ - dimensional space of angular coordinates $m_0 = \ell^{m-1}$ regions are obtained.

5. Finally, two hypotheses are verified:

H_{01}: the observations $[\phi_i, \alpha_{i1}, \alpha_{i2},..., \alpha_{i,m-2}]^T$ are drawn from the population with the uniform distribution on a hypersphere of a unit radius.

H_{02}: the samples containing the observations r_i corresponding to particular regions come from the same population.

The hypothesis H_{01} is tested by means of χ^2-goodness-of-fit test and the hypothesis H_{02} by Kruskal-Wallis nonparametric test. The hypothesis of elliptical symmetry is rejected if at least one of the two hypotheses, H_{01} or H_{02}, is rejected.

The proposed procedure is simple. In the first two steps the observations are standardized (paying regard to covariance of variables). Under null hypothesis the transformed set of data may be regarded as drawn from the population, its distribution is spherically symmetric with zero mean vector. In the third step standardized data are transformed to spherical coordinates. It is well known that the density of spherically symmetric distribution with zero mean vector depends only on radius coordinate. Thus under null hypothesis the variables corresponding to angular coordinates are uniformly distributed. This implies the subhypothesis H_{01}. On the other hand, under null hypothesis the distribution of a random variable, corresponding to radius coordinate, does not depend on angular coordinates. This implies the subhypothesis H_{02}.

In statistical papers other tests for directional data are proposed. A survey is given by Mardia (1972). However, these tests are suitable only for the bivariate and trivariate cases.

The proposed test suffers from the fact, that it is large sample test. The requirements for the χ^2 goodness-of-fit test cause the serious limitations of the use of proposed procedure if the dimension of space exceeds three. In such a case, other goodness-of-fit tests should be used, for example Cramér-von Mises test.

Since the proposed test is a nonparametric one and the tested hypothesis is very general, not much is to be said as far as the power of the test is concerned. It is likely to get some results for particular cases of elliptically symmetric distributions by Monte Carlo studies.

3. BAYES CLASSIFICATION RULE FOR SOME ELLIPTICAL DISTRIBUTIONS

The common method of classification of observations is Bayes classification rule. Suppose that the observation $\mathbf{x}$ comes from one of K subpopulations, distributed with density functions: $f_1(\mathbf{x}), f_2(\mathbf{x}),..., f_K(\mathbf{x})$. A priori probabilities that the observation belongs to the particular subpopulations are: P_1, P_2,..., P_K. For the 0-1 loss function Bayes classification rule minimizes the average probability of error. This rule can be stated in the following way: the observation $\mathbf{x}$ is assigned to the jth class (subpopulation) if

$$g_j(\mathbf{x}) = \max_{\ell} g_\ell(\mathbf{x}),$$

where $g_\ell(\mathbf{x})$ is the discriminant function for the ℓth class, given as:

$$g_\ell(\mathbf{x}) = P_\ell f_\ell(\mathbf{x})$$

or as any monotonic transformation of this function.

For particular forms of elliptically symmetric distributions the following discriminant functions are obtained (after simple calculations):

– for normal distribution:

$$g_\ell(\mathbf{x}) = -(\mathbf{x}-\boldsymbol{\mu}_\ell)^{\mathrm{T}} \boldsymbol{\Sigma}_\ell^{-1} (\mathbf{x}-\boldsymbol{\mu}_\ell) - \log|\boldsymbol{\Sigma}_\ell| + 2\log P_\ell$$

– for Laplace distribution:

$$g_\ell(\mathbf{x}) = -(m+1)^{\frac{1}{2}}[(\mathbf{x}-\boldsymbol{\mu}_\ell)^T \boldsymbol{\Sigma}_\ell^{-1} (\mathbf{x}-\boldsymbol{\mu}_\ell)]^{\frac{1}{2}} - \tfrac{1}{2}\log|\boldsymbol{\Sigma}_\ell| + \log P_\ell$$

– for Pearson type VII distribution:

$$g_\ell(\mathbf{x}) = -(2k - m - 2)^{-1} P_\ell^{-1/k} |\boldsymbol{\Sigma}_\ell|^{1/(2k)} (\mathbf{x}-\boldsymbol{\mu}_\ell)^T \boldsymbol{\Sigma}_\ell^{-1} (\mathbf{x}-\boldsymbol{\mu}_\ell) - P_\ell^{-1/k} |\boldsymbol{\Sigma}_\ell|^{1/(2k)}$$

– for Pearson type II distribution:

$$g_\ell(\mathbf{x}) = -(2k + m + 2)^{-1} P_\ell^{1/k} |\boldsymbol{\Sigma}_\ell|^{-1/(2k)} (\mathbf{x}-\boldsymbol{\mu}_\ell)^T \boldsymbol{\Sigma}_\ell^{-1} (\mathbf{x}-\boldsymbol{\mu}_\ell) + P_\ell^{1/k} |\boldsymbol{\Sigma}_\ell|^{-1/(2k)}.$$

In all discriminant functions, $\boldsymbol{\mu}_\ell$ is the mean vector for the ℓth class and $\boldsymbol{\Sigma}_\ell$ is the covariance matrix for the ℓth class.

4. ELLIPTICALLY SYMMETRIC DISTRIBUTIONS IN LINEAR REGRESSION

Elliptically symmetric distributions are of real usefulness in linear regression analysis. The first argument in support of this claim comes from the theorem given by Kelker (1970), who proved that for elliptically symmetric distributions the conditional mean is a linear function of regressors. The second argument comes from the fact that the set of observations drawn from the elliptically distributed population composes the area of a hyperellipsoidal shape in the space of variables. For such an area a hyperplane (which is a linear function) is a good approximation.

Thus, if the hypothesis of ellipticity of a set of observations is not rejected, it is reasonable to assume a linear regression function for this set of observations.

Otherwise, we can assume that the observations are a random sample drawn from the population, its distribution being a mixture of K elliptically symmetric distributions. To determine the maximum-likelihood estimates of the parameters of this mixture, iterative algorithms may be applied. The one proposed by Bezdek and Dunn (1975) gives the estimates as well as a posteriori probabilities $\hat{P}(j \mid \mathbf{x}_i)$ that the observation $\mathbf{x}_i$ belongs to the jth class. Then a classification of observations may be obtained according to the following rule:

The observation $\mathbf{x}_i$ *is assigned to the* jth *class if*

$$\hat{P}(j \mid \mathbf{x}_i) = \max_\ell \hat{P}(\ell \mid \mathbf{x}_i).$$

If the set of observations is not drawn from the population with elliptical distribution, the number of subpopulations (classes) $K = 2$ is assumed. Then the maximum-likelihood estimation and classification of observations are obtained.

Finally, for each class the mentioned test for ellipticity of observations is performed. If both classes of observations may be considered as drawn from a population with elliptical distribution, then the separate regressions for each class are determined. Otherwise, the procedure is continued for $K = 3, 4, \ldots,$ until the hypotheses of ellipticity for all classes of observations are not rejected.

The assuming of a mixture of elliptically symmetric distributions does not seem strong. This class of distributions is rather extensive. Therefore, this assumption - because of its generality - may be often valid in real applications, since the mixture of elliptically symmetric distributions may produce very "unpleasant" sets of data.

REFERENCES

Bezdek, J.C. and Dunn, J.C. (1975). Optimal fuzzy partitions, a heuristic for estimating the parameters in a mixture of normal distributions. *IEEE Transactions on Computers*, 24, 835–838.

Cooper, P.W. (1963). Multivariate extension of univariate distributions.*IEEE Transactions on Electronic Computers*, 12, 572–573.

Haralick, R.M. (1977). Pattern discrimination using ellipsoidally symmetric multivariate density functions. *Pattern Recognition*, 9, 89–94.

Kelker, D. (1970). Distribution theory of spherical distributions and a location-scale parameter generalization, *Sankhyā, Ser. A*, 32, 419-430.

Mardia, K.V. (1972). *Statistics of Directional Data*. Academic Press, New York.

Received 8 January 1987
Revised 20 April 1987

Instytut Cybernetyki Ekonomicznej
Akademia Ekonomiczna
ul. Komandorska 118/120
53-345 Wroclaw
Poland

Proc. Second International Tampere Conference in Statistics
(Tampere, Finland, 1-4 June 1987)
Tarmo Pukkila and Simo Puntanen, *Editors*

pp. 499 - 510

Asymptotic distributions of functions of sample means and covariances

T. KOLLO

Tartu State University, Estonia, USSR

Key words and phrases: Asymptotic normal distribution, matrix derivative, Edgeworth expansion.

ABSTRACT

Asymptotic normal distributions of different statistics, when these statistics are functions of sample means or covariances have been derived by many authors. In this paper we derive asymptotic distributions of statistics which depend on both sample means and covariances, using matrix derivative. We investigate the distribution of the function $\overline{X}'S^{-1}\overline{X}$, where $\overline{X}$ is the sample mean vector and S is the sample covariance matrix, for a general population as well as for normally and elliptically distributed populations.

1. PRELIMINARIES AND NOTATIONS

Let $X = (X_1, \ldots, X_n)$ be a random sample of size n from a p-dimensional population with population mean μ and covariance matrix Σ, and their sample estimators, the p-vector

$$\overline{X} = \frac{1}{n} \sum_{i=1}^{n} X_i \tag{1}$$

and the $p \times p$-matrix

$$S = \frac{1}{n-1} \sum_{i=1}^{n} (X_i - \overline{X})(X_i - \overline{X})' \tag{2}$$

respectively. Let us suppose the existence of the first four moments M_k of the p-vector X_i. The third and fourth central moments of X_i are given by the following equalities:

$$M_3 = E[(X_i - \mu)\otimes(X_i - \mu)'\otimes(X_i - \mu)] \tag{3}$$

and

$$M_4 = E[(X_i - \mu)\otimes(X_i - \mu)'\otimes(X_i - \mu)\otimes(X_i - \mu)']. \tag{4}$$

Following MacRae (1974) we denote the permuted identity matrix (or commutation matrix) by $I_{p,q}$. If A is a p-vector, we use the notation of Traat (1986) for the k-th Kroneckerian power of A

$$A^k = A \otimes \ldots \otimes A.$$

We need a notion of a matrix derivative dY/dX, which is defined by Neudecker (1969) for an $r\times s$ -matrix Y and a $p\times q$ -matrix X in the following form:

$$\frac{dY}{dX} = \frac{d}{d\,vecX} \otimes (vecY)'.$$

It means that the i-th row of the $pq\times rs$-matrix dY/dX consists of the partial derivatives of the coordinates of $vecY$ with respect to the i-th coordinate of $vecX$. The main properties of the matrix derivative are (see Kollo and Kinkar (1984), for example)

(i) $$\frac{dX}{dX} = I_{pq},$$

(ii) $$\frac{dX'}{dX} = I_{p,q},$$

(iii) $$\frac{dZ}{dX} = \frac{dY}{dX}\frac{dZ}{dY},$$ where $Z = Z(Y)$ and $Y = Y(X)$,

(iv) $$\frac{d(AYB)}{dX} = \frac{dY}{dX}(B\otimes A'),$$ where A and B are constant matrices and $Y = Y(X)$,

(v)

$$\frac{dW}{dX} = \frac{dY}{dX}\frac{dW}{dY}\bigg|_{Z\,constant} + \frac{dZ}{dX}\frac{dW}{dZ}\bigg|_{Y\,constant},$$

where $W = W(Y, Z)$, $Y = Y(X)$, $Z = Z(X)$,

(vi)

$$\frac{d(ZY)}{dX} = \frac{dZ}{dX}(Y \otimes I_m) + \frac{dY}{dX}(I_s \otimes Z'),$$

where $Z = Z(X)$ is an $m \times r$-matrix and $Y = Y(X)$ is an $r \times s$-matrix,

(vii)

$$\frac{dX^{-1}}{dX} = -X^{-1} \otimes (X')^{-1},$$

and

(viii)

$$\frac{d(Y \otimes Z)}{dX} = \left(\frac{dY}{dX} \otimes (vecZ)' + \frac{dZ}{dX} \otimes (vecY)' I_{mn,rs} \right)(I_n \otimes I_{m,s} \otimes I_r),$$

where $Y = Y(X)$ is an $m \times n$-matrix and $Z = Z(X)$ is an $r \times s$-matrix.

The main properties of the Kronecker product and the *vec*-operator can be found in Graham (1981), for example. Here we present only two useful equalities,

$$(A \otimes B)(C \otimes D) = (AC) \otimes (BD), \tag{5}$$

and

$$vec(ABC) = (C' \otimes A)\, vecB. \tag{6}$$

The convergence in distribution is denoted by $\rightarrow^D$.

2. THE ASYMPTOTIC DISTRIBUTION OF $g(\bar{X}, S)$

It is a well-known fact that the sequences $\{n^{\frac{1}{2}}(\bar{X}-\mu)\}$ and $\{n^{\frac{1}{2}}\,vec(S-\Sigma)\}$ converge in distribution to the normal distributions, if $n \to \infty$,

$$n^{\frac{1}{2}}(\bar{X}-\mu) \to^{D} N(0,\Sigma),$$

and

$$n^{\frac{1}{2}}\,vec\,(S-\Sigma) \to^{D} N(0,\Pi),$$

respectively, where

$$\Pi = M_4 - vec\,\Sigma\,(vec\,\Sigma)'. \tag{7}$$

Similar convergences hold for any sufficiently smooth functions $g(\bar{X})$ and $h(S)$. For example, due to Theorem 4.2.5 [Anderson (1958)] $n^{\frac{1}{2}}[g(\bar{X})-g(\mu)]$ and $n^{\frac{1}{2}}[h(S) - h(\Sigma)]$ converge to certain normally distributed random vectors if there exist all the first and second order partial derivatives of these functions $g(\bar{X})$ and $h(S)$ in the neighbourhood of the points $\bar{X} = \mu$ and $S = \Sigma$ respectively. But there are many cases when the observed function depends not only on the sample mean vector or the sample covariance matrix, but on both of them. As an example we here refer to the quadratic form $\bar{X}'S^{-1}\bar{X}$, widely used in multivariate analysis.

Parring (1979) proved that if $M_4 < \infty$, with $n \to \infty$

$$n^{\frac{1}{2}}\begin{pmatrix} \bar{X}-\mu \\ vec\,(S-\Sigma) \end{pmatrix} \to^{D} N(0,\psi), \tag{8}$$

where

$$\psi = \begin{pmatrix} \Sigma & M_3' \\ M_3 & \Pi \end{pmatrix}, \tag{9}$$

and M_3 and Π are defined by (3) and (7) respectively. Let Z and Z_0 be the following $p(p+1)$-vectors:

$$Z = \begin{pmatrix} \bar{X} \\ vec\,S \end{pmatrix} \text{ and } Z_0 = \begin{pmatrix} \mu \\ vec\,\Sigma \end{pmatrix}.$$

From the Anderson´s (1958) Theorem 4.2.5 we obtain immediately the following result.

THEOREM 1. *Let* $X = (X_1,..., X_n)$ *be a random sample of size* n *with the moments* $EX_i = \mu$, $DX_i = \Sigma$ *and* $M_4 < \infty$ *and assume that the function* $g(x)$ *from* $\mathbb{R}^{p(p+1)}$ *to* $\mathbb{R}^q$ *has second order partial derivatives in the neighbourhood of* Z_0. *Then, if* $n \to \infty$, *we have*

$$n^{\frac{1}{2}}[g(Z)-g(Z_0)] \to^D N(0,\Omega),$$

where

$$\Omega = \left(\frac{dg(Z)}{dZ}\bigg|_{Z=Z_0} \right)' \Psi \frac{dg(Z)}{dZ}\bigg|_{Z=Z_0},$$

$\dfrac{dg(Z)}{d(Z)}$ *is the matrix derivative and* ψ *is defined by* (9).

It is not possible to go further with the general approach, because the form of the matrix Ω depends on the concrete function $g(x)$. More precisely the distribution of $g(Z)$ can be approximated by different kinds of asymptotic expansions, for example by the Edgeworth expansion. In the matrix form the Edgeworth expansion for a multidimensional statistic is presented by Traat (1986). For the density function $p(x)$ of $n^{\frac{1}{2}}[g(Z) - g(Z_0)]$ we have the Edgeworth expansion in the following form:

$$p(x) = f(x) - n^{-\frac{1}{2}}[\Gamma'_1 d_1 + \frac{1}{6}(vec\,\Gamma_3)' vec\, d_3] + o(n^{-\frac{1}{2}}).$$

Here $f(x)$ is the density of $N(0, \Omega)$

$$f(x) = (2\pi)^{-\frac{q}{2}}|\Omega|^{-\frac{1}{2}} e^{-\frac{1}{2}x'\Omega^{-1}x}, \quad d_k = \frac{d^k f(x)}{dx^k}$$

and the Γ_i are the matrix expressions depending on the population cumulants but not on n in the expressions of the cumulants κ_i of the statistic $n^{\frac{1}{2}}[g(Z) - g(Z_0)]$

$$\kappa_1 = n^{-\frac{1}{2}}\Gamma_1 + o(n^{-1}),$$

$$\kappa_2 = \Omega + n^{-1}\Gamma_2 + o(n^{-1}),$$

and

$$\kappa_3 = n^{-\frac{1}{2}}\Gamma_3 + o(n^{-1}).$$

The matrices d_k are determined by the multivariate Hermite polynomials $H_k(x)$, defined by the equality

$$\frac{d^k f(x)}{dx^k} = (-1)^k H_k(x) f(x).$$

For $k = 1, 2, 3$, we have

$$H_1(x) = \Omega^{-1}x,$$

$$H_2(x) = \Omega^{-1}xx'\Omega^{-1} - \Omega^{-1},$$

and

$$H_3(x) = \Omega^{-1}x[(x'\Omega^{-1}) \otimes (x'\Omega^{-1})] - \Omega^{-1}x\,(vec\Omega^{-1})' - \Omega^{-1}[(x'\Omega^{-1}) \otimes I_q] - \Omega^{-1}[I_q \otimes (x'\Omega^{-1})].$$

For the distribution function $F(x)$ of $n^{\frac{1}{2}}[g(Z) - g(Z_0)]$ we have the formal expansion

$$F(x) = \int_{-\infty}^{x_1} \cdots \int_{-\infty}^{x_q} [f(x) - n^{-\frac{1}{2}}(\Gamma_1{}'d_1 + \frac{1}{6}(vec\,\Gamma_3)'vec\,d_3)]dx_1 \ldots dx_q + o(n^{-\frac{1}{2}}),$$

where $x = (x_1, \ldots, x_q)'$.

The formal Edgeworth expansions approximate the real distribution of $n^{\frac{1}{2}}[g(Z) - g(Z_0)]$ with given accuracy, if the function $g(x)$ and the population distribution satisfy certain conditions [see Bhattacharya and Ghosh (1978)].

3. ASYMPTOTIC DISTRIBUTION OF $\bar{X}'S^{-1}\bar{X}$

Let us investigate the asymptotic behaviour of Hotelling's T^2 statistic $g(Z) = \bar{X}'S^{-1}\bar{X}$, widely used in testing problems and discriminatory analysis. From Theorem 1 we obtain the convergence

$$n^{\frac{1}{2}}(\overline{X}'S^{-1}\overline{X} - \mu'\Sigma^{-1}\mu) \rightarrow^{D} N(0, \Xi),$$

where

$$\Xi = \left(\frac{d(\overline{X}'S^{-1}\overline{X})}{dZ} \bigg|_{Z=Z_0} \right)' \Psi \frac{d(\overline{X}'S^{-1}\overline{X})}{dZ} \bigg|_{Z=Z_0}.$$

Using the properties (v) – (vii) of the matrix derivative

$$\frac{d(\overline{X}'S^{-1}\overline{X})}{dZ}=\frac{d\overline{X}'}{dZ}(S^{-1}\overline{X}\otimes 1)+\frac{d(S^{-1}\overline{X})}{dZ}(1\otimes\overline{X})$$

$$=\frac{d\overline{X}'}{dZ}(S^{-1}\overline{X}\otimes 1)+\frac{dS^{-1}}{dZ}(\overline{X}\otimes I_p)(1\otimes\overline{X})+\frac{d\overline{X}}{dZ}(1\otimes S^{-1})(1\otimes\overline{X})$$

$$=2\begin{pmatrix} I_p \\ 0 \end{pmatrix}S^{-1}\overline{X}+\begin{pmatrix} 0 \\ -S^{-1}\otimes S^{-1} \end{pmatrix}(\overline{X}\otimes\overline{X})=\begin{pmatrix} 2S^{-1}\overline{X} \\ -(S^{-1}\overline{X})\otimes(S^{-1}\overline{X}) \end{pmatrix}.$$

Then

$$\Xi=\begin{pmatrix} 2\Sigma^{-1}\mu \\ -(\Sigma^{-1}\mu)\otimes(\Sigma^{-1}\mu) \end{pmatrix}'\begin{pmatrix} \Sigma & M_3 \\ M_3 & \Pi \end{pmatrix}\begin{pmatrix} 2\Sigma^{-1}\mu \\ -(\Sigma^{-1}\mu)\otimes(\Sigma^{-1}\mu) \end{pmatrix}. \tag{10}$$

The matrix Ξ can be simplified for some special cases. If the X_i are symmetric random vectors, we have $M_3 = 0$ and hence

$$\Xi = 4\mu'\Sigma^{-1}\mu+(\mu'\Sigma^{-1})\otimes(\mu'\Sigma^{-1})\Pi(\Sigma^{-1}\mu)\otimes(\Sigma^{-1}\mu),$$

which is simplified, by (5) and (6), as

$$\Xi = 4\mu'\Sigma^{-1}\mu+(\mu'\Sigma^{-1})^4 vec\Pi. \tag{11}$$

So we have obtained the following result.

THEOREM 2. *Let $X = (X_1,..., X_n)$ be a random sample of size n, with the moments $EX_i = \mu \neq 0$, $DX_i = \Sigma$ and $M_4 < \infty$. For the function $\bar{X}'S^{-1}\bar{X}$, we have, if $n \to \infty$*

$$n^{\frac{1}{2}}(\overline{X}'S^{-1}\overline{X}-\mu'\Sigma^{-1}\mu) \to^D N(0,\Xi), \tag{12}$$

where Ξ is defined by (10). *If the X_i are symmetrically distributed random vectors, the asymptotic covariance matrix Ξ is given by* (11).

From Theorem 2 it is easy to derive asymptotic normal distributions for $\bar{X}'S^{-1}\bar{X}$ in particular cases.

COROLLARY 2.1. *Let $X = (X_1,\dots, X_n)$ be a random sample of size n from a p-dimensional normal population; $X_i \sim N(\mu, \Sigma)$. If $n \to \infty$ and $\mu \neq 0$, we have*

$$n^{\frac{1}{2}}(\overline{X}'S^{-1}\overline{X} - \mu'\Sigma^{-1}\mu) \to^{D} N(0, \phi),$$

where

$$\phi = 2(\mu'\Sigma^{-1}\mu)(2+\mu'\Sigma^{-1}\mu).$$

Proof. For the normal distribution $M_3 = 0$ and

$$\Pi = (I_{p^2} + I_{p,p})(\Sigma \otimes \Sigma).$$

Then

$$\phi = \begin{pmatrix} 2\Sigma^{-1}\mu \\ -(\Sigma^{-1}\mu)\otimes(\Sigma^{-1}\mu) \end{pmatrix}' \begin{pmatrix} \Sigma & 0 \\ 0 & (I_{p^2}+I_{p,p})(\Sigma\otimes\Sigma) \end{pmatrix} \begin{pmatrix} 2\Sigma^{-1}\mu \\ -(\Sigma^{-1}\mu)\otimes(\Sigma^{-1}\mu) \end{pmatrix}$$

$$= 4\mu'\Sigma^{-1}\mu + 2(\mu'\Sigma^{-1}\mu)\otimes(\mu'\Sigma^{-1}\mu) = 2\mu'\Sigma^{-1}\mu(2+\mu'\Sigma^{-1}\mu). \qquad \square$$

In the next corollary we describe the asymptotic behaviour of $\bar{X}'S^{-1}\bar{X}$ for an elliptically distributed population. A random p-vector X is said to have an elliptical distribution with parameters μ and V, if the density function $f(x)$ is of the form

$$f(x) = C_p |V|^{-\frac{1}{2}} h[(x-\mu)'V^{-1}(x-\mu)]$$

for some function h, where V is positive definite and C_p is the normalizing constant. For more details of the elliptical distribution see, for example, Muirhead (1982). We need only the fact that all the marginal distributions of an elliptical distribution have zero skewness and the same kurtosis.

COROLLARY 2.2. *Let $X = (X_1,\dots, X_n)$ be a random sample of size n from an elliptically distributed population with $EX_i = \mu$ and $DX_i = \Sigma$. If $n \to \infty$ and $\mu \neq 0$, we have*

$$n^{\frac{1}{2}}(\overline{X}S^{-1}\overline{X} - \mu'\Sigma^{-1}\mu) \to^{D} N(0, \Lambda),$$

where

$$\Lambda = 4\mu'\Sigma^{-1}\mu + 2(1+K)(\mu'\Sigma^{-1}\mu)^2 + K(vec\Sigma)'vec(\Sigma^{-1}\mu\mu'\Sigma^{-1}\mu\mu'\Sigma^{-1})$$

and K is defined by the equality

$$3K = E(X_{ij} - \mu_j)^4/\sigma_{jj}^{\;2} - 3. \tag{13}$$

Proof. The following convergence has been shown [Tyler (1981)]: if $n \to \infty$, we have

$$n^{\frac{1}{2}}\, vec\,(S - \Sigma) \to^D N(0, \Delta),$$

where

$$\Delta = (1+K)(I_{p^2} + I_{p,p})(\Sigma \otimes \Sigma) + K\, vec\Sigma(vec\Sigma)',$$

and K is determined by (13). Then

$$\Lambda = \begin{pmatrix} 2\Sigma^{-1}\mu \\ -(\Sigma^{-1}\mu)\otimes(\Sigma^{-1}\mu) \end{pmatrix}' \begin{pmatrix} \Sigma & 0 \\ 0 & \Delta \end{pmatrix} \begin{pmatrix} 2\Sigma^{-1}\mu \\ -(\Sigma^{-1}\mu)\otimes(\Sigma^{-1}\mu) \end{pmatrix}$$

$$= 4\mu'\Sigma^{-1}\mu + (\mu'\Sigma^{-1})\otimes(\mu'\Sigma^{-1})[(1+K)(I_{p^2} + I_{p,p})(\Sigma\otimes\Sigma)$$

$$+ K vec\Sigma(vec\Sigma)'](\Sigma^{-1}\mu)\otimes(\Sigma^{-1}\mu)$$

$$= 4\mu'\Sigma^{-1}\mu + 2(1+K)(\mu'\Sigma^{-1}\mu)^2 + K(vec\Sigma)vec(\Sigma^{-1}\mu\mu'\Sigma^{-1}\mu\mu'\Sigma^{-1}).$$

4. THE CASE $\mu = 0$

Unfortunately the convergence (12) is not valid, when $\mu = 0$. Let us find the Taylor expansion for the function $\bar{X}'S^{-1}\bar{X}$.

The Taylor expansion for a function $f\colon \mathbb{R}^p \to \mathbb{R}^q$ at the point x_0 has the following form:

$$f(x) = f(x_0) + \left(\frac{df(x)}{dx}\bigg|_{x=x_0} \right)'(x - x_0)$$

$$+ \tfrac{1}{2}[I_q \otimes (x - x_0)'] \frac{d^2 f(x)}{dx^2}\bigg|_{x=x_0} (x - x_0) + \ldots$$

For the function $\bar{X}'S^{-1}\bar{X}$ we have the expansion

$$\overline{X}'S^{-1}\overline{X} = \mu'\Sigma^{-1}\mu + \left(\frac{d(\overline{X}'S^{-1}\overline{X})}{dZ}\right)'\Bigg|_{Z=Z_0}(Z-Z_0)$$

$$+\tfrac{1}{2}(Z-Z_0)'\frac{d^2(\overline{X}'S^{-1}\overline{X})}{dZ^2}\Bigg|_{Z=Z_0}(Z-Z_0)+\ldots$$

The first derivative has the following form:

$$\left(\frac{d(\overline{X}'S^{-1}\overline{X})}{dZ}\right)\Bigg|_{Z=Z_0} = \begin{pmatrix} 2S^{-1}\overline{X} \\ -(S^{-1}\overline{X})\otimes(S^{-1}\overline{X}) \end{pmatrix}\Bigg|_{Z=Z_0}$$

$$= \begin{pmatrix} 2\Sigma^{-1}\mu \\ -(\Sigma^{-1}\mu)\otimes(\Sigma^{-1}\mu) \end{pmatrix}.$$

For the second derivative we have

$$\frac{d^2(\overline{X}'S^{-1}\overline{X})}{dZ^2}\Bigg|_{Z=Z_0}$$

$$= 2\frac{d(S^{-1}\overline{X})}{dZ}\Bigg|_{Z=Z_0}(I_p|0) - \left[\frac{d(S^{-1}\overline{X})}{dZ}\otimes(\overline{X}'S^{-1})\right]\Bigg|_{Z=Z_0}\cdot[(I_{p^2}+I_{p,p})|0]$$

$$= 2\begin{pmatrix} S^{-1} \\ -(S^{-1}\overline{X})\otimes S^{-1} \end{pmatrix}\Bigg|_{Z=Z_0}\cdot(I_p|0)$$

$$-\left[\begin{pmatrix} S^{-1} \\ -(S^{-1}\overline{X})\otimes S^{-1} \end{pmatrix}\otimes\overline{X}'S^{-1}\right]\Bigg|_{Z=Z_0}\cdot[(I_{p^2}+I_{p,p})|0]$$

$$= 2\begin{pmatrix} \Sigma^{-1} & 0 \\ -(\Sigma^{-1}\mu)\otimes\Sigma^{-1} & 0 \end{pmatrix} - \begin{pmatrix} \Sigma^{-1}\otimes(\mu\Sigma^{-1}) \\ -(\Sigma^{-1}\mu)\otimes\Sigma^{-1}\otimes(\mu'\Sigma^{-1}) \end{pmatrix}[(I_{p^2}+I_{p,p})|0].$$

When $\mu = 0$,

$$\frac{d(\overline{X}'S^{-1}\overline{X})}{dZ}\Bigg|_{Z=Z_0} = 0$$

and

$$\left.\frac{d^2(\bar{X}'S^{-1}\bar{X})}{dZ^2}\right|_{Z=Z_0} = 2\begin{pmatrix} \Sigma^{-1} & 0 \\ 0 & 0 \end{pmatrix}.$$

The first nonzero term in the Taylor expansion is equal to $\bar{X}'\Sigma^{-1}\bar{X}$. It means that $\bar{X}'S^{-1}\bar{X}$ has the same limiting distribution as $\bar{X}'\Sigma^{-1}\bar{X}$ does. By Moore (1978) we have

$$n\bar{X}'\Sigma^{-1}\bar{X} \to^D \chi_p^2,$$

if $n \to \infty$.

So, when $\mu = 0$, we obtain for the function $\bar{X}'S^{-1}\bar{X}$ the following result.

THEOREM 3. *Let $X = (X_1,..., X_n)$ be a random sample of size n with $EX_i = 0$ and $DX_i = \Sigma$. If $n \to \infty$, we have*

$$n\bar{X}'S^{-1}\bar{X} \to^D \chi_p^2.$$

ACKNOWLEDGEMENTS

The author is grateful to the referees, who carefully read the manuscript and gave valuable suggestions, both from mathematical and editorial point of view. As remarked by them, it is possible to generalize this method of deriving asymptotic distributions of $g(\bar{X}, S)$ to the more general situation, where the function g depends on m multivariate statistics, $m \geq 2$.

REFERENCES

Anderson, T.W. (1958). *An Introduction to Multivariate Statistical Analysis*. Wiley, New York.

Bhattacharya, R.N., Ghosh, J.K. (1978). On the validity of the formal Edgeworth expansion. *Ann. Statist.*, 6, 434–451.

Graham, A. (1981). *Kronecker Products and Matrix Calculus: With Applications*. Ellis Horwood Limited, Chichester.

Kollo, T. and Kinkar, T. (1984). The matrix derivative with application to block-matrices. Contributions of the Computer Centre of Tartu State University, 51, 96-107 (in Russian).

MacRae, E. (1974). Matrix derivatives with an application to an adaptive linear decision problem. *Ann. Statist.*, 2, 337–346.

Moore, D.S. (1978). Chi-square tests. *Studies in Statistics* (Hogg, R.V., *ed.*), MAA Studies in Mathematics, 19, 66–106.

Muirhead, R.J. (1982). *Aspects of Multivariate Statistical Theory*. Wiley, New York.

Neudecker, H. (1969). Some theorems on matrix differentiation with special reference to Kronecker matrix products. *J. Amer. Statist. Assoc.*, 64, 953-963.

Parring, A.-M. (1979). Calculation of asymptotic characteristics for the sample functions. *Acta et Commentationes Universitatis Tartuensis*, 492, 86–90 (in Russian).

Traat, I. (1986). Matrix calculus for multivariate distributions. *Acta et Commentationes Universitatis Tartuensis*, 733, 64–84.

Tyler, D.E. (1981). Asymptotic inference for eigenvectors. *Ann. Statist.*, 9,725–736.

Received 8 January 1987
Revised 14 August 1987

Tartu State University
J. Liivi Street 2
202400 Tartu
Estonian SSR
USSR

Proc. Second International Tampere Conference in Statistics
(Tampere, Finland, 1-4 June 1987)
Tarmo Pukkila and Simo Puntanen, *Editors*

pp. 511 - 519

Nonparametric density estimation: L_∞-approach

V.D. KONAKOV

CEMI, Academy of Sciences of the U.S.S.R., Moscow, U.S.S.R.

Key words and phrases: Family of functions, normalized deviation, smoothness, analytical densities.

ABSTRACT

In this paper we study the optimal rates of convergence for nonparametric density estimates when the closeness is measured in L_∞-norm. For different families of functions among which there is the density function we find the optimal orders of such rates of convergence.

L_∞-APPROACH TO DENSITY ESTIMATION

Let X_1, X_2,... be a sequence of independent identically distributed random variables with a common density $f(x)$. We shall consider the estimates

$$f_n(x) = h^{-1} \int k\left(\frac{x-y}{h} \right) dF_n(y), \tag{1}$$

where $F_n(y)$ is an empirical distribution function and $k(x)$ is a "kernel" function with the properties $\operatorname{supp} k = (-\tau, \tau)$, $k^{(i)}(x) \in L_2(-\infty, \infty)$, $i = 1, 2, 3$, $\int k^2(x)\, dx = \int (k^{(1)}(x))^2\, dx = 1$. Suppose that the density is known to belong to a family Σ of functions on [0, 1] (some examples will be given later) and consider the sequence of random processes

$$\xi_n(x) = (nh)^{\frac{1}{2}} f^{-\frac{1}{2}}(x)(f_n(x) - f(x)), \quad x \in [0, 1].$$

Let $G_n^{(2)}(y)$ be a distribution function of a random variable

$$l_n \sup_{[0,1]} |\xi_n(x)| - l_n^2$$

and $G_n^{(1)}(y)$ be a distribution function of a random variable

$$l_n \max_{[0, h^{-1}]} |Y_n(t)| - l_n^2,$$

where $l_n = (2\ln(1/2\pi h))^{\frac{1}{2}}$, $Y_n(t) = \mu_n(t) + \int k(t-s)\, dW(s)$, $W(s)$ is a two-sided Wiener process on $(-\infty, \infty)$, $\mu_n(t) = \lambda_n(ht)$, $t \in [0, h^{-1}]$, $\lambda_n(x) = (nh/f(x))^{\frac{1}{2}} \int k(u)(f(x+hu) - f(x))\, du$. Suppose that the smoothing parameter h is known to satisfy either

(i) $A_1 n^{-\delta_1} \le h \le A_2 n^{-\delta_2}$ for some $0 < A_1, A_2 < \infty$ and $0<\delta_2<\delta_1<1$ (2)

or

(ii) $h \asymp 1/\ln n$, i.e., h is weakly equivalent to $1/\ln n$.

Applying the estimates from Konakov and Piterbarg (1979, p. 167) [inequalities (1.11) with appropriate $\varepsilon(l_n, h)$] we obtain in case (i)

$$G_n^{(1)}(y - C_1 n^{-\alpha}) - C_2 e^{-C_3 n^{\beta}} \le G_n^{(2)}(y) \le G_n^{(1)}(y + C_1 n^{-\alpha}) + C_2 e^{-C_3 n^{\beta}} \tag{3}$$

for some positive α, β, C_1, C_2, C_3, and in case (ii)

$$G_n^{(1)}\left(y - C_4 \frac{\ln\ln n}{(\ln n)^{\frac{1}{2}}}\right) - C_5(\ln n)^{-\gamma} \le G_n^{(2)}(y) \le G_n^{(1)}\left(y + C_4 \frac{\ln\ln n}{(\ln n)^{\frac{1}{2}}}\right) + C_5(\ln n)^{-\gamma} \tag{4}$$

for some C_4, C_5 and γ. We now introduce necessary notation. In (x, y)-plane we consider a curve defined by the equation $r(\phi) = u_n - \mu_n(\phi)$, $0 \le \phi \le h^{-1}$, in polar coordinates. The straight line perpendicular to the radius-vector $\vec{r}(\phi)$ at the endpoint divides the plane in two half-planes. Let $S_{\mu,\phi}$ be the half-plane containing the origin. Denote $I_k = [(k-1)t_0, kt_0]$, $k = 1, 2, \ldots, N_1$, $N_1 = [h^{-1}/t_0]$, $0 < t_0 < \pi/4$, and define

$$p(k+1) = P\{(X,Y) \in \bigcap_{I_{k+1}} S_{\mu,\phi}\} + P\{(X,Y) \in \bigcap_{I_{k+1}} S_{-\mu,\phi}\},$$
$$p(k,k+1) = P\{(X,Y) \in \bigcap_{I_k \cup I_{k+1}} S_{\mu,\phi}\} + P\{(X,Y) \in \bigcap_{I_k \cup I_{k+1}} S_{-\mu,\phi}\}, \tag{5}$$

where (X, Y) is standard two-dimensional normal vector. Theorems formulated below follow from (3), (4) and the asymptotic behavior of $G_n^{(1)}$ [see Konakov (1986)].

THEOREM 1. *Let* $\lambda_n(x) \in C^2[0, 1]$, *h satisfy* (2). *Then it is possible to indicate* $\varepsilon > 0$ *depending only on* $k(u)$ *and* $\delta > 0$, $C < \infty$ *such that* $l_n^{-1} \| \mu_n \|_2 < \varepsilon$ *implies*

$$G_n^{(2)}(y) = \exp\left\{ - \sum_{k=1}^{N_1 - 1} (p(k+1) - p(k,k+1))\right\} (1 + L(n,y)), \tag{6}$$

where $|L(n, y)| \le Ch^\delta \exp(-(1-2\varepsilon)y)$, $p(k+1)$ *and* $p(k, k+1)$ *are defined in* (5), $\|\mu_n\|_2 = \max_{[0,h^{-1}]} \{|\mu_n(t)|, |\mu_n^{(1)}(t)|, |\mu_n^{(2)}(t)|\}$. *If the level* $u_n = l_n + y l_n^{-1}$ *is sufficiently high, namely* $u_n = (N+1)l_n$, *then*

$$G_n^{(2)}(Nl_n^2) = \exp\left\{ -2e^{-(\frac{N^2}{2}+N)l_n^2} \int e^{-\frac{v^2}{2}} ch((N+1)l_n v) d\eta_n(v)\right\}(1 + L(n)), \tag{7}$$

where $|L(n)| \le Ch^{2N(1-2\varepsilon)+\delta}$, $\eta_n(v) = \mu\{s \in [0, 1], \lambda_n(s) \le v\}$, μ *is Lebesgue measure.*

In Theorem 1 it is supposed that trends are twice continuously differentiable. Of course it would be interesting to obtain an analogous result for the case of differentiable or even nonsmooth trends. We next state a result of such a type for one special sequence of trends $\mu_n(t)$; such sequences arise in statistical applications to nonparametric density estimation.

Let us consider a functional class $\mathbf{\Sigma}_0$:

$$\mathbf{\Sigma}_0 = \mathbf{\Sigma}(C, r, \omega) \cap \mathbf{D}_L \cap \mathbf{D}_l, \quad r = 0, 1, 2, \dots, \tag{8}$$

where

$$\mathbf{\Sigma}(C, r, \omega) = \{x(t), t \in [0, 1]: \omega(\delta, x^{(r)}) = \sup_{|x' - x''| \le \delta} |x^{(r)}(t') - x^{(r)}(t'')| \le C\omega(\delta)\},$$

ω is a given concave modulus of continuity and C is a constant such that $\omega(\delta)\geq\delta$, $\mathbf{D}_L=\{x(t),\ t\in[0,1],\ \max_{[0,1]}|x^{(r\wedge 2)}(t)|\leq L\}$, $\mathbf{D}_l=\{x(t),\ x(t)\geq l,\ t\in[0,1]\}$. Suppose $f\in\mathbf{\Sigma}_0$. Let $\Phi_h(x;f)=(f(x)-f_{[h]}(x))/(f(x))^{\frac{1}{2}}$ be a normalized deviation of $f(x)$ from its generalized Steklov's function $f_{[\delta]}(x)=\int k(u)f(x+\delta u)du$, $k(u)$ satisfies the same conditions as the kernel in (1). Define $\mu_n(t;f)=\lambda_h\Phi_h(ht;f)$, $t\in[0,h^{-1}]$, and let λ_h be a coefficient which depends on the smoothness of f. The constant ε_1 will be chosen later. Denote

$$\delta_h=\begin{cases}\omega(h^2;f)/\omega(h)+h/\omega(h),\ r=0,\\ h/\omega(h),\ r=1.\end{cases}\tag{9}$$

THEOREM 2. *Let* $f\in\mathbf{\Sigma}_0$, $r=0$ *or* 1, $u_n=l_n+yl_n^{-1}$, $\delta_h=o(h^\gamma)$, $\gamma>0$ *as* $h\to 0$. *Then for any n there exists a trigonometric polynomial* $\Pi(n,x)$ *whose period* $\tau>0$ *depends only on the modulus of continuity* $\omega(\cdot)$ *and whose degree n is less than* $[r/2\pi h]$ *such that*

$$\sup_{[0,1]}|f(x)-\Pi(n,x)|\leq C\omega(h;f^{(r)})\,h^r$$

and there exist $c<\infty$, $\beta>0$ *and* $\varepsilon>0$ *such that* $l_n^{-1}\,\|\,\mu_n(t;\Pi)\,\|_2<\varepsilon$ *implies*

$$P(\max_{[0,h^{-1}]}|\mu_n(t;f)+\int k(t-s)dW(s)|\leq u_n)=$$

$$\exp\left\{-\frac{1}{2\pi h}\int\left(e^{-\frac{(u_n-x)^2}{2}}+e^{-\frac{(u_n+x)^2}{2}}\right)dv_n(x)\right\}(1+L(n,u_n)),\tag{10}$$

where

$$v_n(x)=h\lambda\{s\in[0,h^{-1}]:\mu_n(s;\Pi)\leq x\},\quad |L(n,u_n)|\leq ch^{\beta-(1-2\varepsilon)y}.$$

Suppose the density function f belongs to a family $\mathbf{\Sigma}$. The accuracy of f_n^* as an estimate of f from observations $X_1,X_2,\ldots,X_n$ will be characterized by $\sup_{[0,1]}|(f_n^*(x)-f(x))/(f(x))^{\frac{1}{2}}|$. We are interested in the following question: at which maximal rate $\{a_n\}$ can vanish while satisfying the condition

$$\lim_{n\to\infty}\ \inf_{f_n^*}\ \sup_{f\in\Sigma}P_f\left\{\sup_{[0,1]}\frac{|f_n^*(x)-f(x)|}{(f(x))^{\frac{1}{2}}}>a_n\right\}=0.\tag{11}$$

The answer to this question will be given below for a number of different families Σ. In what follows we shall use the consequence from one general result of Ibragimov and Khas'minskii (1980).

THEOREM 3. *Let Σ be the set of densities $f(x)$, $x \in \mathbb{R}$, such that for any δ there exist $f_{0\delta}$ and $f_{i\delta} = f_i$, $i = 1, 2,..., \rho(\delta)$, of Σ with the properties*

1) $\| f_{i\delta} - f_{j\delta} \|_\infty \geq \delta, \quad i,j = 1, 2,..., \rho(\delta)$

2) $\| \frac{f_{i\delta}}{f_{0\delta}} - 1 \|_\infty \leq \frac{1}{2}$.

Denote

$$\delta(n, \Sigma) = \sup \left\{ \delta: \sup_{i=1,2,...,\rho(\delta)} \frac{\|(f_{i\delta} - f_{0\delta})/(f_{0\delta})^{\frac{1}{2}}\|_2^2}{ln\, \rho(\delta)} \leq 1/2n \right. . \tag{12}$$

Then

$$\inf_{f_n^*} \sup_{f \in \Sigma} P_f(\sup |f - f^*| > \tfrac{1}{2}\delta(n, \Sigma)) \geq \tfrac{1}{2}. \tag{13}$$

We consider now some concrete examples of classes Σ.

(a) FUNCTIONS OF FINITE SMOOTHNESS

Let us consider the class Σ_0. To obtain the lower bounds on the sequence $\{a_n\}$ we shall construct the sequence of density functions $f_{i\delta}$ in the Ibragimov and Khas'minskii theorem cited above. Denote

$$g_0(x) = \begin{cases} (C/2)\, \omega(x), & 0 \leq x \leq \varepsilon, \\ (C/2)\, \omega(2\varepsilon - x), & \varepsilon < x \leq 2\varepsilon, \quad g_0(-x) = -g_0(x), \\ 0, & x > 2\varepsilon, \end{cases}$$

and define the sequence $g_k(x)$ by recurrence relations

$$g_k(x) = g_k^*(x - 2^k\varepsilon) - g_k^*(x + 2^k\varepsilon),\; g_k^*(x) = \int_{-2^k\varepsilon}^{x} g_{k-1}(y)dy,\; k = 1,..., r.$$

It is easy to see that $g_r(x)$ is finite, supp $g_r = (-2^{r+1}\varepsilon, 2^{r+1}\varepsilon)$, odd function and $g_r(x) \in \Sigma(C/2, r, \omega)$. It easily follows from the definition of $g_r(x)$ that there exist C_r and C_r' such that

$$\sup |g_r(x)| = C_r \varepsilon^{r-1} \int_0^{\varepsilon} \omega(y)dy, \quad \int g_r^2(x)dx \le C_r{}' \varepsilon^{2r-1} \left(\int_0^{\varepsilon} \omega(y)dy \right)^2. \tag{14}$$

Consider the function $\gamma(x) = x^{r+\frac{1}{2}}\omega(x)$, denote $\varepsilon = \varepsilon_m = \gamma^{-1}(1/m)$ and define the family of functions

$$f_a(x) = f_0(x) + \sum_{j=1}^{n_m} a(j)\phi_{mj}(x), \tag{15}$$

where $f_0 \in \Sigma(C/2, r, \omega) \cap \mathbf{D}_{L/2} \cap \mathbf{D}_{2l}$, $\phi_{mj}(x) = g_r(x - (2j-1)2^{r+1}\varepsilon_m)$, $j = 1, 2, \ldots, n_m = [1/2^{r+2}\varepsilon_m]$, $a(j)$ takes on values 0 or 1. Let $\mathbf{D}_0$ be a set of functions $a(j)$ which are equal to 0 everywhere except one point at which $a(j) = 1$. We shall consider the densities f_a, $a \in \mathbf{D}_0$ numbered arbitrarily as $f_{i\delta}$, $f_{i\delta} \in \Sigma_0$. From the concavity of $\omega(\delta)$, the definition of $\gamma(x)$, and (14) it follows that

$$\|f_a - f_{a'}\|_\infty = C_r \varepsilon_m^{r-1} \int_0^{\varepsilon_m} \omega(y)dy \ge \frac{C_r}{2} \varepsilon_m^r \omega(\varepsilon_m) = \frac{C_r{}'}{2} m^{-1} (\gamma^{-1}\left(\frac{1}{m} \right))^{-\frac{1}{2}} = \delta_m,$$

$$\|(f_{i\delta} - f_{0\delta})/(f_{0\delta})^{\frac{1}{2}}\|_2^2 \le \frac{C_r{}'}{2l} m^{-2}.$$

In accordance with (12) we choose $m(n)$ from the condition

$$m(n) = \max \left\{ m: \frac{-C_r{}'}{2lm^2 \ln(2^{r+2}\gamma^{-1}(1/m))} < \frac{1}{2n} \right\}.$$

Hence $m(n) \asymp \left(\frac{n}{\ln n} \right)^{\frac{1}{2}}$, $\delta_{m(n)} \asymp \left[\frac{\ln n}{n} / \gamma^{-1}\left(\frac{\ln n}{n} \right)^{\frac{1}{2}} \right]^{\frac{1}{2}}$. Taking into account (13) we obtain that the sequence $\{a_n\}$ in (11) cannot tend to zero faster than $\left[\frac{\ln n}{n} / \gamma^{-1}\left(\frac{\ln n}{n} \right)^{\frac{1}{2}} \right]^{\frac{1}{2}}$.

We now show that this order is attained for kernel density estimates (1) with h and k chosen in an appropriate way, moreover, (11) tends to 0 at power rate. The kernel k will be chosen in such a way that

$$\operatorname{supp} k(u) = (-1, 1), \ \exists\ k^{(i)}(u), \ i = 1, 2, 3, \ k^{(3)}(u) \in L_2(-\infty, \infty),$$

$k \perp u, u^2, \ldots, u^r$. From the definition of Σ_0 we have $\sup_{[0,\, h^{-1}]} |\mu_n(t)| = \sup_{[0,1]} |\lambda_n(x)| \le C_r \gamma(h) n^{\frac{1}{2}}$. Suppose that $r \ge 2$. After differentiation and simple calculations we obtain that for $f \in \Sigma_0$, $\| \mu_n \|_2 \leq Cn^{\frac{1}{2}}\gamma(h)$, $C < \infty$. Choose h^* from the condition $Cn^{\frac{1}{2}}\gamma(h) = \varepsilon(2\ln(1/2\pi h))^{\frac{1}{2}}$, ε is the same as in Theorem 1. The definition of $\gamma(h)$ implies $h^* \asymp \gamma^{-1}((\ln n)/n)^{\frac{1}{2}}$ and the case (i) in (2) takes place. Theorem 1 implies the inequality

$$G_n^{(2)}(Nl_n^2) \geq \exp\{-2\exp(-(N+\frac{N^2}{2})l_n^2)\int \exp(\varepsilon(N+1)^2 l_n^2)\, d\eta_n(v)\}(1+L(n))$$

$$\geq (1-(2\pi h)^N)(1+L(n)). \tag{16}$$

It is easy to see that

$$a_n^* = \frac{u_n^*}{\sqrt{nh^*}} = \frac{(N+1)\sqrt{2\ln(1/2\pi h^*)}}{\sqrt{nh^*}} \asymp \left[\frac{\ln n}{n\gamma^{-1}(\sqrt{\ln n/n})}\right]^{\frac{1}{2}} \tag{17}$$

and it follows from (16) that for kernel estimate constructed with h^* and $k(u)$ pointed above

$$\lim_{n\to\infty} \sup_{f\in\Sigma_0} P_f\left\{\max_{[0,1]} \frac{|f_n^*(x)-f(x)|}{\sqrt{f(x)}} > a_n^*\right\} = 0. \tag{18}$$

On multiplying a_n^* by a constant, we attain the rate of convergence in (18) faster than any given power of $1/n$. Consider now the case $r=0, 1$, a_n^* has the same order of decreasing as in (17). Thus for the class Σ_0

$$a_n \asymp \left[\frac{\ln n}{n\gamma^{-1}(\sqrt{\ln n/n})}\right]^{\frac{1}{2}}, \; \gamma(x) = x^{r+\frac{1}{2}}\omega(x).$$

Let us consider some simple examples.

EXAMPLE 1. $r=0$, $\omega(x) = 1/\ln_j(1/x)$ where $\ln_j x = \underbrace{\ln\ln\ldots\ln}_{j\text{ times}} x$.

In this case

$$h^* \asymp \frac{\ln n(\ln_j n)^2}{n}, \quad a_n^* \asymp \frac{1}{\ln_j n}.$$

This example shows that very slow rates may occur (those being optimal as we have already seen). Of course, this reflects the fact that L_∞-metric is bad for such families of "almost discountinuous" densities.

EXAMPLE 2. $r = k$, $\omega(x) = x^\alpha$, $0 < \alpha \le 1$. In this case

$$h^* \asymp \left(\frac{\ln n}{n}\right)^{1/(2\beta+1)}, \quad a_n^* \asymp \left(\frac{\ln n}{n}\right)^{\beta/(2\beta+1)}, \beta = k + \alpha .$$

EXAMPLE 3. Let $\Sigma_D = D \cap \{x(t), t \in [0,1] \sup_{[0,1]} |x(t)| \le M\}$, where D is a Skorokhod's space. It follows from Rejtö and Révész (1973) and from theorem of Ibragimov and Khas'minskii formulated above, that

$$h^* \asymp \frac{\ln n}{n}, \quad a_n^* \asymp 1.$$

(b) CLASSES OF ANALYTICAL DENSITIES

Let $A(L, \delta)$ be a set of functions $F(w)$, $w = t + iu$, which are
1) real on the real axis ($u = 0$),
2) regular in a neighbourhood of the band $-\delta < u < \delta$ and $\|\mathrm{Re}F(\cdot \pm i\delta)\|_2 < L$.
Let $\Sigma(L, \delta)$ be a class of real-valued functions of real argument t, each function representing values of some function of the class A on the real axis. Suppose that $f \in \Sigma_1 = \Sigma(L, \delta) \cap \mathbf{D}_l$ for some fixed L, δ, l. From Ibragimov and Khas'minskii (1980, p. 81), it follows that for the class Σ_1 the sequence $\{a_n\}$ in (11) cannot tend to zero faster than $(\ln n \ \ln \ln n/n)^{\frac{1}{2}}$. To attain this rate we choose the kernel of Feier type as $k(u)$, namely, suppose $k(u) = V(u) = (\cos u - \cos 2u)/\pi u^2$. Well known properties of $V(u)$ (see Ibragimov and Khas'minskii (1980, p. 63)) imply

$$\max_{[0, h^{-1}]} |\mu_n(t)| = \max_{[0,1]} \sqrt{nh} \frac{|\int V_h(x-y)f(y)dy - f(x)|}{\sqrt{f(x)}} \le 4\sqrt{nh/l}\, E_{1/h}[f]$$

where $V_h(u) = (1/h)V(u/h)$ and $E_{1/h}[f]$ the value of the best approximation of the functions f by entire functions of the exponentional type $1/h$ in L_∞-metric. For the class Σ_1 (see Akhiezer (1947, p. 231)) $E_{1/h}[f] \le (8/\pi)\exp(-\delta/h)$. From 2) and Akhiezer (1947, p. 230-234) it also follows that $\max(E_{1/h}[f'], E_{1/h}[f'']) \le (8/\pi)\, e^{-\delta/h}$ and hence $\| \mu_n \|_2 \le C(nh)^{\frac{1}{2}}\exp(-\delta/h)$.

Inequality $C(nh)^{\frac{1}{2}}\exp(-\delta/h)\leq \varepsilon[2\ln(1/2\pi h)]^{\frac{1}{2}}$ will be satisfied if $h = h^* = 2\delta/\ln n$ (this corresponds to the case (ii) in (2)). Using Theorem 1 we obtain

$$\lim_{n\to\infty} \sup_{f\in\Sigma_1} P_f\{\max_{[0,1]} \frac{|f_n^*(x)-f(x)|}{(f(x))^{\frac{1}{2}}} > a_n^*\} = 0\,, \tag{19}$$

where

$$a_n^* = \frac{(N+1)\sqrt{2\ln(1/2\pi h^*)}}{(nh^*)^{\frac{1}{2}}} \qquad \left[\frac{\ln n\ \ln\ln n}{n}\right]^{\frac{1}{2}} .$$

By multiplying a_n^* by a corresponding constant, we can attain the rate of convergence in (19) faster than any given degree of $1/\ln n$. Thus for Σ_1 the optimal order is equal to $[\ln n \ln\ln n/n]^{\frac{1}{2}}$.

REFERENCES

Akhiezer, N.I. (1947). Lectures on approximation theory. Gostekhizdat, Moscow.

Bickel, P. and Rosenblatt, M. (1973). On some global measures of deviations of density function estimates. *Ann. Statist.*, 1, 1071-1095.

Billingsley, P. (1968). *Convergence of Probability Measures.* Wiley, New York.

Ibragimov, I.A. and Khas'minskii, R.Z. (1980). On estimate of the density function (in Russian). Zap. Naucn. Sem. Leningrad. Otdel. Mat. Inst. Steklov (LOMI), 98, 61-85.

Konakov, V.D. and Piterbarg, V.I. (1981). Rate of convergence of maximal deviations for Gaussian processes and empirical density functions (in Russian). *Teor. Verojatnost. i Primenen*, 26, no.4, 702-719.

Konakov, V.D. (1986). Extrema of some nonstationary Gaussian processes and optimal asymptotic confidence regions for density function. The Fifth Japan-USSR symposium on probability theory. Abstr. of comm., Kyoto, 25-26.

Rejto, L. and Révész, P. (1973). Density estimation and pattern classification. *Problems of Control and Inform. Theory*, 2, 67-80.

Received 21 April 1987

Central Economics-Mathematical Institute
Academy of Sciences of USSR
ul. Krasikova 32
Moscow 117418
USSR

Proc. Second International Tampere Conference in Statistics
(Tampere, Finland, 1-4 June 1987)
Tarmo Pukkila and Simo Puntanen, *Editors*

pp. 521 - 530

On subjectivity in statistics

Pekka J. KORHONEN and **Subhash C. NARULA**
Helsinki School of Economics, Helsinki, Finland and
Virginia Commonwealth University, Richmond, VA, U.S.A.

Key words and phrases: Multiple criteria, interactive.

ABSTRACT

Subjectivity is introduced in a problem whenever a user has to make a choice. Usually, subjectivity has not been dealt with explicity in statistics, even though it occurs implicitly in many statistical settings, e.g., in tests of hypothesis, confidence interval estimation, selection of variables in multiple regression, selection of a rotation method in factor analysis, etc. However, subjectivity plays a central role in many statistical analyses. As such instead of trying to eliminate the subjective elements, we should try to take full advantages of these. It may give us a deeper understanding of some of the problems than we currently have with the traditional approaches.

In this paper, we discuss the role of subjectivity in several statistical settings and recommend techniques for incorporating it into the analysis process. We show that the multiple criteria problem formulation and the interactive approaches can be especially useful in many situations.

1. INTRODUCTION

Statistical analysis starts with model building. Its beauty lies in the fact that the properties of the procedures can be investigated and studied even before a single observation is taken. The estimates are "optimal" and the conclusions reliable if the underlying assumptions of the model are fully satisfied; otherwise the results may be far from optimal. In the words of Hogg (1974, pp. 909-910):

> "While model building is certainly desirable, we must remember that these models are not sacred and we are obliged to question their reliability.
>
> As a matter of fact, we know in practice that most models will seldom fit exactly the real situations. Thus, for the sake of applications, it seems ridiculous to try to get the last ounce of mathematical efficiency out of some assumed situation. A more realistic approach would be to seek statistical procedures good for a broad class of underlying models, but which are not necessarily best for any one of them."

When we apply any statistical theory or a procedure to a practical problem, we have to decide what assumptions may be valid. More often than one would like to believe or admit, it is a very subjective decision since for a given data set all the assumptions underlying a statistical procedure are rarely, if ever, satisfied. Moreover, it is impossible to prove by statistical analysis whether the perceptions of the user are valid or not. The user generally looks at the computed statistics as "hard cold facts" or objective facts, when in reality these are based on a number of subjective choices right from the start.

We would like to point out that the idea of subjectivity in statistics is not new. A number of statistical procedures explicitly take it into consideration, e.g., Bayesian analysis and adaptive robust procedures. In Bayesian analysis, the inference is based on the data and the apriori knowledge of the investigator in the form of prior distribution. For example, Dickey (1973) has shown how to use a visual representation to describe the functional dependence of the inference on the prior distribution. In adaptive robust procedures the final estimator depends on the results of a preliminary analysis. However, we feel that the role of subjectivity is not very obvious in many statistical settings. For example, even in such objective procedures as confidence interval estimation and hypothesis testing, subjectivity is introduced in the procedures because an investigator has to choose the confidence coefficient and the level of significance, respectively.

Our objectives in this paper are: (i) to discuss explicitly the element of subjectivity inherent in some statistical procedures, and (ii) suggest "interactive" approaches for some statistical problems to deal with it explicitly in ways that are useful to an investigator. It is not our intention here to develop either new statistical procedures or Bayesian analysis for such problems. We also do not review the literature on this topic.

The rest of the paper is organized as follows: In Section 2, we consider subjectivity in confidence interval estimation and hypothesis testing problems. In Section 3 we discuss subjectivity in regression analysis from the theoretical and data analytical point of view. In Section 4, we will consider principal component analysis as an example of a data analytic method, in which introducing subjectivity in the analysis improves its usability. We conclude the paper with a few remarks in Section 5.

2. ESTIMATION AND TESTING PROBLEMS

Let the elements of an $n \times 1$ vector $\mathbf{x}$ represent a random sample of size n from a probability distribution with probability function $f(x; \theta)$, $\theta \in \Omega$, where θ denotes a parameter and Ω, the parameter space. Note that for the sake of clarity and without any loss of generality, we assume that θ is a scalar. Let $\mathbf{x}$, an $n \times 1$ vector, denote the observed value of x.

2.1 Interval estimation

The interval $[\hat{\theta}_L, \hat{\theta}_U]$, where $\hat{\theta}_L$ and $\hat{\theta}_U$ are functions of $\mathbf{x}$, is called a *confidence interval* for θ if $P\{\hat{\theta}_L \leq \theta \leq \hat{\theta}_U\} = 1 - a$. The quantity $1 - a$ is called the *confidence coefficient*. Let $L = \hat{\theta}_U - \hat{\theta}_L$ denote the length of the interval and where for a given a: $\hat{\theta}_L = \hat{\theta}_L(a)$ and $\hat{\theta}_U = \hat{\theta}_U(a)$. Clearly, we would like to maximize the confidence coefficient $1 - a$; and minimize the expected length of the interval, $E(L)$. Even for a given sample size n, the problem is difficult to solve. The problem is usually solved as follows:

Minimize $E(L)$

Subject to

$$a \leq a_0$$

for a preselected value of a equal to $a_0 \in (0, 1)$. For a given value of a_0, the procedure does provide an interval with minimum $E(L)$.

All this gives the procedure the appearance of objectivity. However, there are no explicit and clear guidelines for the selection of a_0. Traditionally, the values .01, .05, and .10 have been used for a_0. In this respect, the procedure is subjective and even arbitrary.

Since we want to minimize $E(L)$ and a, simultaneously, the problem is a bicriteria optimization problem. One way to resolve the problem is to determine $E(L)$ for all acceptable values of a and then plot a graph of $E(L)$ vs. a. From this plot, the decision maker can see how the confidence interval $E(L)$ depends on a. The decision maker may also decide which pair $(a, E(L))$ is desirable to him/her before the data is taken, and thus maintain the objectivity of the procedure, if it is desirable. The plot gives him/her a better method to evaluate the reliability of an estimate calculated from the data.

The problem becomes even more complicated, if the "positioning" of the interval is not made in advance. Then we are faced with three criteria problem: max $\hat{\theta}_L$, min $\hat{\theta}_U$, and min a, which may lead to complex considerations.

2.2 Hypothesis testing

Let Ω' denote the subspace of Ω specified by the null hypothesis H_0. Note $\Omega' \neq \emptyset$ and $\Omega' \subset \Omega$. Based on $\mathbf{x}$, we need to decide whether to reject H_0 or not. To do so, we divide the sample space into two regions, i.e., the critical region K and its complement. The null hypothesis H_0 is rejected whenever $\mathbf{x} \in K$; otherwise, it is not rejected.

Based on the sample, it is possible to arrive at the correct decision; commit a Type I error, i.e., reject H_0 when it is true; or commit a Type II error, i.e., not reject H_0 when it is false. The probabilities of the two errors are:

$$P_1(K) = P\{\mathbf{x} \in K; \theta \in \Omega'\} \text{ and}$$
$$P_2(K) = P\{\mathbf{x} \notin K; \theta \in \Omega - \Omega'\}.$$

It is desirable to minimize $P_1(K)$ and $P_2(K)$ simultaneously. Although no values of $P_1(K)$ and $P_2(K)$ may be acceptable, the problem is resolved by selecting a level of significance $a_0 \in (0, 1)$, which provides an upper bound on $P_1(K)$ and solving the following problem:

$$\text{Minimize } P_2(K) \tag{1}$$

Subject to

$$P_1(K) \leq a_0$$

for $\theta \in \Omega - \Omega'$. If the critical region K is such that $P_1(K) \leq a_0$ and for every other critical region A with $P_1(A) \leq a_0$, and if $P_2(K) \leq P_2(A)$ for all $\theta \in \Omega - \Omega'$, with strict inequality for at least one θ, then K is the *best critical region* of size a_0 and the test the *best test* of size a_0. For a specified value a_0, the procedure gives the best critical region or the best test of size a_0, if it exists. It may be pointed out that using this procedure one can find a complete class of nondominated tests of size a. This makes the procedure appear objective.

At present there are no explicit and clear guidelines available for selecting a value of a_0. The values .01, .05, and .10 are commonly used for

a_0. Since the choice of a_0 is arbitrary, the procedure is subjective. To fully exploit the subjectivity of the procedure without taking anything away from it, it seems more reasonable to determine the best test for each desirable value of $a \in (0, 1)$ and then provide the decision maker with a graph of $P_1(K)$ vs. $P_2(K)$. The graph will clearly show the trade off between the probabilities of the two types of errors and help a decision maker select a pair that seems most reasonable or acceptable.

For $\Omega' = \{\theta_0\}$ and $\Omega = \{\theta_0, \theta_1\}$, Narula (1976) proposed that for a given value of n, (1) be solved for various values of $a \in (0, 1)$. Plot a graph of $P_1(K)$ vs. $P_2(K)$ for $a \in (0, 1)$ and let the decision maker choose the acceptable pair $(P_1(K), P_2(K))$.

The decision maker can use the above graph in a different subjective way. After seeing the sample, he can evaluate, whether he is willing to make an accept/reject decision or not. The graph as such is the way to argue for his choice, not necessary any "cold" number. By representing this graph to the readers a statistician gives a possibility to evaluate the reasonability of his decision.

3. REGRESSION ANALYSIS

Let $\mathbf{y}$, an $n \times 1$ vector, denote the values of the response variable corresponding to $\mathbf{X}$, an $n \times k$ matrix of the values of the regressor (or predictor) variables. Then

$$\mathbf{y} = \mathbf{X}\boldsymbol{\beta} + \boldsymbol{\varepsilon} \tag{2}$$

represents the multiple linear regression model, where $\boldsymbol{\beta}$ is a $k \times 1$ vector of the unknown parameters and $\boldsymbol{\varepsilon}$ is an $n \times 1$ vector of unobservable random errors.

3.1 Estimating the regression coefficients

Our objective is to estimate $\boldsymbol{\beta}$ such that the predicted value of $\mathbf{y}$ is close to the observed value of $\mathbf{y}$. This "closeness" can be stated in terms of the L_p-norm and the problem of estimating $\boldsymbol{\beta}$ as

$$\text{Minimize} \quad \| e_1 \|^p + \ldots + \| e_n \|^p$$

for $p \geq 1$, where $e_i = y_i - \mathbf{X}_i\mathbf{b}$, y_i is the ith element of the vector $\mathbf{y}$, $\mathbf{b}$ is an estimator of $\boldsymbol{\beta}$, and $\mathbf{X}_i$ is the ith row of matrix $\mathbf{X}$. For $p=1$, we minimize the sum of absolute errors MSAE; for $p=2$, we minimize the sum of the squared errors MSSE (popularly known as the least squares); and for $p=\infty$, we minimize the maximum absolute error MMAE. It has been shown that the MSAE, the MSSE, and the MMAE criteria are optimum for the long-tailed, the medium-tailed, and the short-tailed error distributions, respectively. In practice, the underlying assumptions for any value of p are rarely, if ever, fully satisfied.

The least squares regression has dominated the statistical literature for a long time. It is optimal and results in the maximum likelihood estimators of the unknown parameters of the model if the errors are independent and follow a normal distribution with mean zero and a common (though unknown) variance σ^2. However, it is far from optimal in many non-Gaussian situations, especially when the errors follow distribution with longer tail than normal. Outliers are much harder to spot in the regression than in the simple location case (Huber 1973). Futher, outliers occuring with extreme values of the regressor variables can be especially disruptive. Its use is also not recommended if errors follow distribution with nonfinite variance. It may not be very satisfactory measure of loss. Loss denotes the seriousness of the non zero prediction error to the investigator, where the prediction error is the difference between the predicted and the observed value of the response variable.

In an excellent paper, Beckman and Cook (1983) discuss the diagnostic techniques and the robust regression procedures that have been developed to overcome the drawbacks of the least squares regression. The diagnostic techniques identify the influential observations, whereas the robust regression procedures accommodate the possibility of outliers. In fact, at present there are several procedures available to estimate $\boldsymbol{\beta}$ in (2). For adaptive robust regression procedures that take subjectivity into consideration (see, e.g., Hogg 1974). However, the least squares regression still continues to be the most often used procedure. It may appear to be an objective procedure if we do not know alternative procedures. However, in most cases our willingness to be objective does not give the "best" result. On the other hand, if we have chosen it from among a set of alternative procedures, the choice is in most instances purely subjective.

The problem really is a multiobjective optimization problem. Narula and Wellington (1979) proposed that an experimenter may simultaneously minimize two or more of the desired criteria. As an example, they proposed that he/she may want to minimize the sum of absolute errors and the maximum absolute error. The problem is then a bicriteria optimization problem and may be solved by $\boldsymbol{\varepsilon}$-constrained approach. Arthanari and Dodge (1981) proposed the minimization of the weighted sum of the sum of the squared errors and the sum of absolute errors.

In a specific application it may be reasonable to weigh over-prediction and under-prediction of the response variable differently. In such a case, the problem can be stated as

$$\text{Minimize } \mu\mathbf{1}'\mathbf{e}^+ + (1-\mu)\mathbf{1}'\mathbf{e}^- \qquad (3)$$

Subject to

$$\begin{aligned} &\mathbf{Xb} + \mathbf{e}^+ - \mathbf{e}^- = \mathbf{y}, \\ &\mathbf{e}^+, \mathbf{e}^- \geq 0, \\ &\mathbf{b} \text{ unrestricted in sign,} \end{aligned}$$

where $\mathbf{e} = \mathbf{e}^+ - \mathbf{e}^-$ and $\mu \in [0, 1]$. Note the problem formulation assumes absolute error loss function. For a given value of μ, the goal programming formulation (3) gives the μth regression quantile (Koenker and Basset 1978). However, if it is desired to weigh each prediction error differently, we obtain

$$\text{Minimize } \sum_i w_i^+ e_i^+ + \sum_i w_i^- e_i^-$$

Subject to

$$\begin{aligned} &\mathbf{Xb} + \mathbf{e}^+ - \mathbf{e}^- = \mathbf{y}, \\ &\mathbf{e}^+, \mathbf{e}^- \geq 0, \\ &\mathbf{b} \text{ unrestricted in sign,} \end{aligned}$$

where w_i^+ and w_i^- are nonnegative constants representing the relative weights to be assigned to positive and negative deviations for each observation (Charnes and Cooper 1977). However, the problem of choosing the weights is again subjective. There are no objective criteria to choose the weights. Technically, the above problem formulation is easy to solve using the program VIG (Korhonen 1987).

3.2 Selection of variables

Irrespective of the procedure used to estimate $\boldsymbol{\beta}$, it may be desirable for practical, economic and statistical reasons, to include fewer than k variables, say m variables, in the model. In terms of the statistical properties of the estimators, a model with fewer than k variables will have smaller predictive mean square error, i.e., variance + square of the bias for the predicted value of the response (Narula and Ramberg 1971 and Walls and Weeks 1969). Thus the problem is to decide the value of m and then the "best" set of m variables to be included in the model. At present, there are a number of procedures available to select a model with fewer then k variables, viz., forward selection procedure, backward elimination procedure; stepwise procedures and implicit enumeration procedures. For these procedures, the formal tests with F-values to enter & the F-value to leave the model; R^2, $R^2(adj.)$ [$R^2(adj.)$ = corrected for the number of variables in the model], the C_m-statistic, etc., give these procedures the appearance of objectivity in selecting a model with fewer variables. However, the choice of F-values to enter and to leave are arbitrary as are the choice and final selection of the value of R^2 and C_m-statistic. It is important to recognize these facts explicitly in model building and also take into consideration items such as the cost of data collection and maintenance of the final model. Clearly, the problem is a highly subjective multicriteria problem.

The selection of the final set is often based on both statistical analysis and the perception of the user, and thus it is really subjective. Therefore, it is very dangerous to "hide" behind the seemingly objective statistical tests.

4. PRINCIPAL COMPONENT ANALYSIS

The principal components are obtained as a result of an orthogonal transformation of the original set of variables into a new set of variables which are uncorrelated with each other and whose variances decrease from the first to the last component. The first k components explain more of the total variation of the original variables than any other set of k components.

The main objectives of the principal component analysis are (i) to reduce the dimensionality of the data; and (ii) to identify new meaningful underlying variables. When the first few components account for most of the variation in the original data, it is often a good idea to use these components in subsequent analyses while losing as little information as possible in reduction. This may also help overcome some computational problems. (See, e.g., Mardia, Kent and Bibby 1979.)

If the new components, the linear combinations of the original variables, are intuitively meaningful, they may help the user better understand the correlation structure of the original variables and thus attach meaningful "labels" to the principal components. However, in practice, the interpretation of the principal components is seldom easy.

Therefore they are rotated to find new set of components which can be easily interpreted. Although not always, the rotation is usually orthogonal. In some cases the rotation may improve the meaningfulness of the components. The choice of the method for rotation is purely subjective.

Recently, Korhonen (1984) introduced subjectivity explicitly in the analysis and proposed the concept of subjective principal components to determine the most "preferred" and orthogonal linear combinations of the original variables. In this analysis, the investigator may evaluate the correlation between certain (efficient) linear combinations and the original variables and maximize the absolute values of the desired correlations.

Let $\mathbf{X}$, an $n \times p$ matrix, represent n observations on p variables. For the sake of convenience, we assume that the original variables have been standardized, i.e., $\mathbf{X}'\mathbf{X} = \mathbf{R}$, where $\mathbf{R}$ denotes the correlation matrix. Let $\mathbf{Y}$, an $n \times q$ matrix, $q \leq p$, represent a q-dimensional linear transformation $\mathbf{Y} = \mathbf{XB}$, where $\mathbf{B}$, a $p \times q$ matrix, is the weight matrix. We further assume that the columns of $\mathbf{Y}$ are uncorrelated, i.e., $\mathbf{Y}'\mathbf{Y}$ is diagonal. The sum of squares due to $\mathbf{y}_m$, the mth principal components is $\Sigma\, r^2(\mathbf{x}_i, \mathbf{y}_m)$, where $r(\mathbf{x}_i, \mathbf{y}_m)$ denotes the correlation between the original variable $\mathbf{x}_i$ and $\mathbf{y}_m$. The variable $\mathbf{y}_m$, $m = 1, 2, \ldots, q$, is called the mth subjective principal component if the correlation between $\mathbf{y}_m$ and the p original variables are the most preferred (either maximal or minimal) and $\mathbf{y}_m$ is uncorrelated with the first $m - 1$ subjective principal components. Korhonen (1984) formulates the problem as a multiple criteria optimization problem and suggests a visual interactive procedure to solve it.

5. CONCLUDING REMARKS

In this paper we have discussed the element of subjectivity inherent in statistical procedures for problems of confidence interval estimation, hypothesis testing, regression analysis and principal component analysis. In each case we have shown how the procedure can be modified such that it explicitly takes subjectivity into consideration instead of naive objectivity, but without losing the objectivity of the procedure, if it is necessary. We think this approach helps the investigators make choices with full information. As earlier, the properties of the procedures can still be investigated before taking a single observation.

The role of subjectivity in statistics does not have to be passive as is often the case in classical and Bayesian analyses. However, by making the user an integral part of an interactive process, the statistical analysis takes subjectivity explicitly into consideration. It also gives a user a better understanding of the analysis and more flexibility, which may lead to innovative findings on the basis of data. This should make the statistical procedures more useful to the users.

It is important that the use of subjectivity is transparent in the analysis. Other investigators should be able to reproduce the analysis in order to evaluate the choices.

REFERENCES

Arthanari, T.S. and Dodge, Y. (1981). *Mathematical Programming in Statistics.* Wiley, New York.

Beckman, R.J. and Cook, R.D. (1983). Outliers. *Technometrics*, 25, 119-149.

Charnes, A. and Cooper, W.W. (1977). Goal programming and multiple objective optimization. Part 1. *European Journal of Operational Research,* 1, 39-54.

Dickey, J.M. (1973). Scientific reporting and personal probabilities: Student's hypothesis. *Journal of the Royal Statistical Society*, B35, 285-303.

Hogg, R.V. (1974). Adaptive robust procedures: A partial review and some suggestions for future applications and theory. *Journal of American Statistical Association*, 69, 309-323.

Huber, P.J. (1973). Robust regression: Asymptotics, conjectures and Monte Carlo. *The Annals of Statistics*, 1, 799-821.

Koenker, R. and Bassett, G. (1978). Regression quantiles. *Econometrica*, 46, 33-50.

Korhonen, P.J. (1984). Subjective principal component analysis. *Computational Statistics and Data Analysis*, 2, 243-255.

Korhonen, P. (1987). VIG – A Visual Interactive Support System for Multiple Criteria Decision Making. *Belgian Journal of Operations Research, Statistics and Computer Science*, 27, 3-15.

Mardia, K.V., Kent, J.T. and Bibby, J.M. (1979). *Multivariate Analysis*, Acadedmc Press.

Narula, S.C. (1976). Sample size precision function. *Journal of Quality Technology*, 8, 49-52.

Narula, S.C. and Ramberg, J.S. (1972). Letter to the editor. *The American Statistician*, 26, 42.

Narula, S.C. and Wellington, J.F. (1979). Linear regression using multiple-criteria. (G. Fandel and T. Gal, *eds.*), *Multiple criteria decision making: Theory and Application*, Springer-Verlag, 266-277.

Walls, R.E. and Weeks, D.L. (1969). A note on the variance of a predicted response in regression. *The American Statistician*, 23, 24-26.

Received 17 February 1987
Revised 1 June 1987

Helsinki School of Economics
Runeberginkatu 14-16
SF-00100 Helsinki
Finland

School of Business
Virginia Commonwealth University
Box 4000
Richmond, VA 23284-0001
U.S.A.

Proc. Second International Tampere Conference in Statistics
(Tampere, Finland, 1-4 June 1987)
Tarmo Pukkila and Simo Puntanen, *Editors*

pp. 531 - 540

Optimal designs in linear models and Hadamard matrices

Christos KOUKOUVINOS and **Stratis KOUNIAS**
Aristotle University of Thessaloniki, Greece

Key words and phrases: Construction, Baumert-Hall arrays, *T*-matrices.

ABSTRACT

In two-level factorial experiments and in other linear models the coefficients of the unknown parameters can take one out of two values. When the number of observations is a multiple of 4, the *D*-optimal design is an Hadamard matrix. A new method is given for constructing Hadamard matrices. An algorithm is described, two tables with the corresponding results are given, and a table with *T*-matrices of all odd order $t \leq 21$ is presented.

1. INTRODUCTION

In certain linear models the coefficients of the unknown parameters can take one out of two values, which can be taken as $+1$ and -1. The problem we face is how to design the experiment so that the estimators of the unknown parameters are "efficient" in some sense.

To be more precise, suppose we have the linear model

$$y_i = x_{i1}b_1 + \ldots + x_{ik}b_k + e_{ik}, \quad i = 1, \ldots, n,$$

where the errors e_i are uncorrelated with zero mean and common variance σ^2.

The $n \times k$ array $\mathbf{X} = (x_{ij})$, $i = 1,\dots, n$, $j = 1,\dots, k$, is called the design of the experiment and the $k \times k$ matrix $\mathbf{M} = \mathbf{X}^T\mathbf{X}$ is called its information matrix, where ^{T} indicates the transpose.

The covariance matrix of the linear least squares estimator $\hat{\mathbf{b}}$ of $\mathbf{b}$ is $\sigma^2 \mathbf{M}^{-1}$ and in constructing optimal designs we minimize a given function of $\mathbf{M}$. In this paper we are interested in minimizing the generalized variance, i.e., maximizing the determinant of $\mathbf{M}$ over all possible designs $\mathbf{X}$. This is called D-optimality.

In the particular case where we have k factors, each at two levels, then $x_{ij} = +1$ or -1 whenever in the ith observation factor j enters at the high level or at the low level. A similar situation arises in weighing experiments with a chemical balance when the objects are put on the right or the left scale-pan.

There exist algorithms, [Fedorov (1972, p. 97), Wynn (1970)] for constructing D-optimal continuous designs where the optimal observation points and the optimal proportion of observations at these points are found. However, if we want these observations to be multiples of $1/n$, where n is the number of observations, the algorithms do not always converge to the global maximum due to the existence of many local optima.

In this paper we only consider experiments where $x_{ij} = \pm 1$, then if the number of observations is a multiple of 4, i.e., $n \equiv 0 \mod 4$, det $\mathbf{M}$ is maximized when the columns of the design $\mathbf{X}$ are orthogonal to each other. If $k = n$ the design is called saturated and the D-optimal design is an Hadamard matrix of order n, whenever $n \equiv 0 \mod 4$.

So the construction of Hadamard matrices of order n gives D-optimal n-observation designs. Plackett and Burman (1946), in studying optimal multifactorial experiments, constructed Hadamard matrices of order $n \leq 100$, $n \equiv 0 \mod 4$.

A great deal of research has been devoted in constructing Hadamard matrices and studying their properties.

In this paper we give some new constructions of Hadamard matrices by extending the results of Turyn (1972). These results are summarized in the three tables given in the Appendix.

2. CONSTRUCTION OF HADAMARD MATRICES

An Hadamard matrix $\mathbf{H}$ of order n is a $(1, -1)$ $n \times n$ matrix with orthogonal rows and thus columns, i.e., $\mathbf{H}^T\mathbf{H} = n\mathbf{I}_n$. A necessary condition for their existence is $n = 2$, or $n \equiv 0 \bmod 4$ and the smallest value of n for which n has not been constructed is $n = 428$.

Here we construct Hadamard matrices by means of Baumert-Hall arrays.

DEFINITION. *The quadruple of circulant and symmetric* $(+1, -1)$ *matrices* $\mathbf{A}, \mathbf{B}, \mathbf{C}, \mathbf{D}$ *of order* m *is called of the Williamson type if*

$$\mathbf{A}^2 + \mathbf{B}^2 + \mathbf{C}^2 + \mathbf{D}^2 = 4m\mathbf{I}_m . \tag{1}$$

DEFINITION. *A Baumert-Hall array of order* t *denoted by* $BH(4t)$ *is a* $4t \times 4t$ *array with entries* $\pm\mathbf{A}, \pm\mathbf{B}, \pm\mathbf{C}, \pm\mathbf{D}$ *such that:*

(i) *In every row of the array there is the same number* t *of* $\pm\mathbf{A}, \pm\mathbf{B}, \pm\mathbf{C}, \pm\mathbf{D}$.

(ii) *Any two rows of the array are orthogonal.*

If we have a $BH(4t)$ and a quadruple of Williamson type matrices of order m , then an Hadamard matrix of order $4mt$ can be constructed [Cooper and Wallis (1972)]. Since Hadamard matrices of order $2m$ are constructed from those of order m, we need only consider the case where m, t are odd.

The $BH(4t)$ we construct in this paper are of the Goethals-Seidel type. This idea is due to Wallis (1973) and employs the notion of T-matrices.

DEFINITION. *The four circulant* $(0, 1, -1)$ *matrices* $\mathbf{X}, \mathbf{Y}, \mathbf{Z}, \mathbf{W}$ *of order* t *form a quadruple of* T*-matrices if:*

(i) $\mathbf{U} * \mathbf{V} = \mathbf{0} \quad \forall\ \mathbf{U}, \mathbf{V} \in \{\mathbf{X}, \mathbf{Y}, \mathbf{Z}, \mathbf{W}\}, \quad \mathbf{U} \neq \mathbf{V}$, $*$ *being the Hadamard product,*

(ii) $\mathbf{X}\mathbf{X}^{\mathrm{T}} + \mathbf{Y}\mathbf{Y}^{\mathrm{T}} + \mathbf{Z}\mathbf{Z}^{\mathrm{T}} + \mathbf{W}\mathbf{W}^{\mathrm{T}} = t\,\mathbf{I}_t$.

The circulant matrices are defined by their first row and in this form are given here.

If $\mathbf{A} = (a_1, \ldots, a_n)$ is a circulant matrix, then we use the following notation:

$$\begin{cases} N_A(i) = \sum_{s=1}^{n-i} a_s\, a_{s+i}, \\ \\ P_A(i) = N_A(i) + N_A(n-i), \end{cases} \qquad i = 1, \ldots, n-1.$$

THEOREM 1.[Turyn (1972)]. *If* $\mathbf{A} = (a_1, \ldots, a_n)$ *and* $\mathbf{B} = (b_1, \ldots, b_n)$ *are circulant* $(1, -1)$ *matrices of order* n *with*:

$$N_A(i) + N_B(i) = 0, \qquad i = 1, \ldots, n-1. \tag{2}$$

Then

(i) $\mathbf{A}\mathbf{A}^{\mathrm{T}} + \mathbf{B}\mathbf{B}^{\mathrm{T}} = 2n\mathbf{I}_n$,

(ii) $\mathbf{X} = (\mathbf{0}_n)$, $\mathbf{Y} = ((\mathbf{A}+\mathbf{B})/2)$, $\mathbf{Z} = ((\mathbf{A}-\mathbf{B})/2)$, $\mathbf{W} = (\mathbf{0}_n)$ *are* T*-matrices of order* n,

(iii) $\mathbf{X} = (1, \mathbf{0}_n)$; $\mathbf{Y} = (0, (\mathbf{A}+\mathbf{B})/2)$, $\mathbf{Z} = (0, (\mathbf{A}-\mathbf{B})/2)$, $\mathbf{W} = (\mathbf{0}_{n+1})$ *are* T*-matrices of order* $n+1$,

where $\mathbf{0}_s$ *is an* $1 \times s$ *row vector of zeros.*

This idea was exploited by Turyn (1972), who constructed an infinite class of Golay sequences and $BH(4t)$ for $t \in \{i\colon i = 1 + 2^a \cdot 10^b \cdot 26^c\}$, a, b, c being non-negative integers.

From (i) we have that n is even and

$$n = (p-q)^2 + (n-p-q)^2, \tag{3}$$

where p, q are the number of -1's in every row (column) of $\mathbf{A}$ and $\mathbf{B}$ respectively. It is known [Geramita and Seberry (1979, p. 136)] that even if (3) is satisfied, the matrices $\mathbf{A}, \mathbf{B}$ satisfying (2) do not always exist as is the case for $n = 18$. The only known values of n for which $\mathbf{A}, \mathbf{B}$ exist, are $n = 2^a \cdot 10^b \cdot 26^c$. However, relation (2) is only sufficient for the construction of the T-matrices of Theorem 1 (iii).

Here we find the necessary conditions and then construct the T-matrices of Theorem 1 (iii). In this way we constructed T-matrices of order 19 in three different ways although the method of Turyn does not give a result. Our method is also different from that of Cooper and Wallis (1972).

THEOREM 2. *If* $\mathbf{A} = (a_1,\dots, a_n)$ *and* $\mathbf{B} = (b_1,\dots, b_n)$ *are circulant* $(1,-1)$ *matrices of order* n *, such that*

$$\mathbf{AA}^{\mathrm{T}} + \mathbf{BB}^{\mathrm{T}} = (2n, c_1,\dots, c_{n-1}) , \tag{4}$$

then the quadruple

$$\mathbf{X} = (1, \mathbf{0}_n) , \ \mathbf{Y} = (0,(\mathbf{A}+\mathbf{B})/2), \ \mathbf{Z} = (0,(\mathbf{A}-\mathbf{B})/2), \ \mathbf{W} = (\mathbf{0}_{n+1})$$

gives T-matrices of order $n+1$ *if and only if*

$$\sum_{i=1}^{n-1} c_i = 0, \tag{5}$$

$$N_A(n-i) + N_B(n-i) = c_1 + \dots + c_i \qquad i = 1,\dots, n-1. \tag{6}$$

Proof. (i) From this construction we have

$$\mathbf{X}_1 * \mathbf{X}_2 = \mathbf{0} \ \ \forall \ \mathbf{X}_1 , \mathbf{X}_2 \in \{\mathbf{X}, \mathbf{Y}, \mathbf{Z}, \mathbf{W}\} \quad \mathbf{X}_1 \neq \mathbf{X}_2 .$$

(ii) If $\mathbf{Q}_s$ is the circulant matrix $(0, 1, 0,\dots, 0)$ of order s, then

$$\mathbf{X} = \mathbf{I}_{n+1}, \mathbf{Y} = \tfrac{1}{2} \sum_{i=1}^{n} (a_i + b_i)\mathbf{Q}^i_{n+1}, \mathbf{Z} = \tfrac{1}{2} \sum_{i=1}^{n} (a_i - b_i)\mathbf{Q}^i_{n+1}, \mathbf{W} = \mathbf{0}_{n+1}.$$

From the requirement

$$\mathbf{XX}^{\mathrm{T}} + \mathbf{YY}^{\mathrm{T}} + \mathbf{ZZ}^{\mathrm{T}} + \mathbf{WW}^{\mathrm{T}} = (n+1)\, \mathbf{I}_{n+1}$$

we obtain the relations (5) and (6).

From (4) and (5) we have

$$n = (p-q)^2 + (n-p-q)^2 ,$$

where p, q are the number of -1's in every row (column) of $\mathbf{A}$ and $\mathbf{B}$ respectively, and also

$$c_i \equiv \begin{cases} 0 \bmod 4 & \text{if } n \text{ is even} \\ 2 \bmod 4 & \text{if } n \text{ is odd.} \end{cases}$$

Note that $c_i = c_{n-i}$, $i = 1,\dots, n-1$, from the symmetry of $\mathbf{AA}^T + \mathbf{BB}^T$.

Since we are interested in T-matrices of odd order $n+1$, then take n to be even. Before we give our algorithm for constructing T-matrices, observe from (6) that:

$$c_1 + \dots + c_i = \begin{cases} 0, \pm 4,\dots, \pm 4(i/2) \text{ if } i \text{ is even} \\ 0, \pm 4,\dots, \pm 4(i-1)/2 \text{ if } i \text{ is odd} \end{cases} \quad i = 1,\dots, n-1. \quad (7)$$

We now give the algorithm for constructing $\mathbf{A}$ and $\mathbf{B}$ and hence T-matrices and Baumert-Hall arrays.

2.1 The algorithm

(i) Find integers p, q, such that

$$(p-q)^2 + (n-p-q)^2 = n, \; 0 \le p, \; q \le n/2. \quad (8)$$

If such a decomposition does not exist, then $\mathbf{A}$ and $\mathbf{B}$ cannot be constructed.

(ii) Take positive integers $k_1,\dots, k_p$ and $m_1,\dots, m_q$ such that:

$$k_1 + \dots + k_p = n \qquad m_1 + \dots + m_q = n$$

and

$$r_1 = 1, \; r_2 = r_1 + k_1,\dots, r_p = r_{p-1} + k_{p-1},$$

$$s_1 = 1, \; s_2 = s_1 + m_1,\dots, s_q = s_{q-1} + m_{q-1}.$$

Then construct $\mathbf{A} = (a_1,\dots, a_n)$, $\mathbf{B} = (b_1,\dots, b_n)$ having p and q -1's respectively in the positions $r_1,\dots, r_p$ and $s_1,\dots, s_q$.

(iii) Calculate $\mathbf{AA}^T + \mathbf{BB}^T = (2n, c_1,\dots, c_{n-1})$ and examine whether the conditions (7), i.e.,

$$c_1 + \dots + c_i = \begin{cases} 0, \pm 4,\dots, \pm 4(i/2) \text{ if } i \text{ is even} \\ 0, \pm 4,\dots, \pm 4(i-1)/2 \text{ if } i \text{ is odd} \end{cases} \quad i = 1,\dots, n-1,$$

are satisfied. If not, go to step (ii) and take another pair of sets $(k_1,\dots, k_p)$, $(m_1,\dots, m_q)$.

(iv) If all conditions (7) are satisfied, examine whether conditions (6) are satisfied, i.e.,

$$N_A(n-i)+N_B(n-i) = c_1+\ldots+c_i\,, \qquad i=1,\ldots,n-1.$$

If all these conditions are satisfied then **A**, **B** have been found, stop.

(v) If not, take $\mathbf{A} = (a_1,\ldots,a_n)$, $\mathbf{B} = (b_n, b_1,\ldots, b_{n-1})$ and go to step (iv).

(vi) Continue until all n cyclic permutations of $\mathbf{A} = (a_1,\ldots,a_n)$ and $\mathbf{B} = (b_1,\ldots,b_n)$ have been exhausted or until conditions (6) are satisfied.

(vii) If step (vi) leads to no solution for **A** and **B**, take another pair of sets $(k_1,\ldots,k_p)$, $(m_1,\ldots,m_q)$ and continue until a solution has been found or all possible pairs of sets $(k_1,\ldots,k_p)$, $(m_1,\ldots,m_q)$ have been exhausted.

Applying the above algorithm for different values of n we obtain the matrices **A**, **B** given in Table 1 of the Appendix. Note that for $n = 18$ we find **A**, **B** although Golay sequences of this order do not exist.

However, if n is not the sum of two integer squares we cannot proceed as before, as in the case for $n = 6, 14$. To overcome this difficulty we give another construction which resolves some of these cases.

THEOREM 3. *If* $\mathbf{A} = (a_1,\ldots,a_n)$, $\mathbf{B} = (b_1,\ldots,b_n)$ *are* (1,–1) *circulant matrices of odd order* n *and*

$$\mathbf{A}\mathbf{A}^{\mathrm{T}}+\mathbf{B}\mathbf{B}^{\mathrm{T}} = (2n, c_1,\ldots,c_{n-1})\,, \tag{9}$$

then the quadruple:

$$\mathbf{X} = (1,0,\mathbf{0}_n),\ \mathbf{Y} = (0,0,(\mathbf{A}+\mathbf{B})/2),\ \mathbf{Z} = (0,0,(\mathbf{A}-\mathbf{B})/2),\ \mathbf{W} = (0,1,\mathbf{0}_n)$$

forms T-matrices of order $n+2$ *if and only if*

$$c_1+c_3+\ldots+c_{n-2} = 0,$$

$$c_2+c_4+\ldots+c_{n-1} = 0,$$

$$N_A(1) + N_B(1) = 0,$$

$$N_A(2) + N_B(2) = 0,$$

$$\begin{cases} N_A(2i+1) + N_B(2i+1) = -(c_1 + c_3 + \ldots + c_{2i-1}) \\ N_A(2i) + N_B(2i) = -(c_2 + c_4 + \ldots + c_{2(i-1)}) \end{cases} \quad i = 1, \ldots, (n-1)/2.$$

Proof. The proof goes as in Theorem 2.

An algorithm similar to the one described, but appropriately modified, was applied and the results are given in Table 2 of the Appendix.

In this way we obtain T-matrices of order 7, 15 which could not be obtained with the previous algorithm.

This type of construction can be extended to form T-matrices of order $n+3$, $n+4$ etc. This is the case for T-matrices of order 13, in this case the two previous algorithms do not apply because $n-1 = 12$, $n-2 = 11$ are not the sum of two integer squares. Applying a similar algorithm for T-matrices of length $n+3$ we find the T-matrices of order 13 as given in Table 3.

APPENDIX

In Table 1 the matrices **A**, **B** of order n are given, which are then used to construct T-matrices of order $n+1$ as in Theorem 2. In Table 2 the matrices **A**, **B** of order n are given, which are then used to construct T-matrices of order $n+2$ as in Theorem 3.

In all cases $n = a^2 + b^2$ with $b = p - q$, $a = n - p - q$ where p, q are the number of -1's in each row (column) of **A**, **B** respectively.

TABLE 1

Giving **A, B** for constructing T-matrices of order $n+1$. (+ stands for 1 and - for −1)

n	a	b	p	q	**A**	**B**
2	1	1	1	0	(-+)	(++)
4	2	0	1	1	(-+++)	(+-++)
8	2	2	4	2	(++--+-+-)	(+--+++++)
10	3	1	4	3	(-+++--+-++)	(+++++--+-+)
16	4	0	6	6	(++++----++-+-+++)	(+-+++--++-+++-+--)
18	3	3	9	6	(++-+--++---++-+-+-)	(++++++-++-+++----+)
20	4	2	9	7	(+++-++--+-++-+-+-+--)	(+++--++----++++++-++)

TABLE 2

Giving **A, B** for constructing T-matrices of order $n + 2$ (+ stands for 1 and - for −1)

n	a	b	p	q	**A**	**B**
1	1	0	0	0	(+)	(+)
5	2	1	2	1	(−+++−)	(+−+++)
9	3	0	3	3	(++++−−+−+)	(−+++−+++−)
13	3	2	6	4	(+−+−−−++−+−++)	(+++++−−−++−++)
17	4	1	7	6	(−++++−++−+++−+−−−)	(++−++−+−+++−−−+++)

TABLE 3

T-matrices of order t. The numbers (without the sign) inside the parentheses indicate the position of ± 1's, the negative sign means that a −1 is there, the remaining elements are 0.

t	**X**	**Y**	**Z**	**W**
3	(1)	(3)	(2)	ϕ
5	(1)	(4,5)	(−2,3)	ϕ
7	(1)	(5,6)	(3,−4,7)	(2)
	(1)	(2,7)	(6)	(−3,4,5)
	(1)	(3,5,6,−7)	(4)	(2)
9	(1)	(2,−4,6,8)	(3,−5,−7,−9)	ϕ
11	(1)	(3,4,5,−7,11)	(2,6,−8,9,−10)	ϕ
	(1)	(4,5,6,−7,9)	(3,−8,−10,11)	(2)
	(1)	(4,8,−9)	(5,−6,−7,10)	(2,3,11)
13	(1)	(5,−6,−7,8,10,12)	(3,4,−9,11)	(2,13)
15	(1)	(3,5,−8,12,−13,14,15)	(4,6,7,−9,−10,11)	(2)
17	(1)	(2,4,5,−7,−8,10,13,16)	(3,−6,−9,11,−12,−14,15,17)	ϕ
	(1)	(−5,6,7,−8,−10,14)	(−9,−11,12,13,−15,16)	(2,3,4,17)
19	(1)	(2,3,5,9,−11,13,14,−15,−17)	(4,6,7,−8,10,12,−16,−18,19)	ϕ
	(1)	(4,6,7,−8,9,12,13,−15)	(3,−5,−10,11,−14,−16,17,18,19)	(2)
	(1)	(4,−5,6,−8,9,10,11)	(7,−12,13,−14,−15,−16,17,18)	(2,3,19)
21	(1)	(2,3,4,−5,7,−9,−11,13,15,17)	(−6,8,−10,−12,14,16,18,−19,20,21)	ϕ

ACKNOWLEDGEMENTS

We thank the referee and the editors for their comments and corrections.

REFERENCES

Cooper, J. and Wallis, J. (1972). A construction for Hadamard arrays. *Bull. Austral. Math. Soc.*, 7, 269-277.

Cooper, J. (1974). A note on Hadamard arrays. *Bull. Austral. Math. Soc.*, 10, 15-21.

Fedorov, V.V. (1972). *Theory of Optimal Experiments*. Academic Press, New York.

Geramita, A.N. and Seberry, J. (1979). *Orthogonal Designs*. Marcel Dekker, New York.

Plackett, R.L. and Burman, J.P. (1946). The design of optimum multifactorial experiments. *Biometrika*, 33, 305-325.

Turyn, R. J. (1972). An infinite class of Williamson matrices. *J. Combinatorial Theory, Ser. A*, 12, 319-321.

Turyn, R. J. (1984). A special class of Williamson matrices and difference sets. *J. Combinatorial Theory, Ser. A*, 36, 111-115.

Wallis, W.D., Street, A.P., Wallis, J.S. (1972). *Combinatorics: Room squares, sum-free sets, Hadamard matrices*. Lecture Notes in Mathematics, Vol. 292. Springer-Verlag.

Wallis, Jennifer (1973). Hadamard matrices of order 28m, 36m, and 44m. *J. Combinatorial Theory, Ser. A*, 15, 323-328.

Wynn, H.P. (1970). The sequential generation of D-optimum experimental designs. *Ann. Math. Statist.*, 41, 1655-1664.

Received 8 January 1987

Department of Mathematics
Aristotle University of Thessaloniki
54006 Thessaloniki
Greece

Proc. Second International Tampere Conference in Statistics
(Tampere, Finland, 1-4 June 1987)
Tarmo Pukkila and Simo Puntanen, *Editors*

pp. 541 - 553

Equalities and inequalities for the canonical correlations associated with some partitioned generalized inverses of a covariance matrix

Dominique LATOUR, Simo PUNTANEN and George P. H. STYAN
North Carolina State University, Raleigh, NC, USA, University of Tampere, Finland, and McGill University, Montreal, QC, Canada

Key words and phrases: Inequalities, unit canonical correlations, symmetric reflexive generalized inverse, Banachiewicz-Schur form, Schur complements, partitioned matrices, eigenvalue inequalities.

ABSTRACT

We consider the positive canonical correlations, ρ_i, $i = 1, \ldots, m$, associated with a possibly singular covariance matrix $\boldsymbol{\Sigma}$, and the positive canonical correlations, $\rho_j^{\#}$, $j = 1, \ldots, m^{\#}$, associated with a symmetric reflexive generalized inverse $\boldsymbol{\Sigma}^{\#}$, partitioned similarly to $\boldsymbol{\Sigma}$ and in Banachiewicz-Schur form. We establish various equalities and inequalities, giving special attention to the numbers u and $u^{\#}$, respectively, of unit canonical correlations. In particular we show that $u^{\#} = 0$, and that the two sets of non-unit canonical correlations ρ_i and $\rho_j^{\#}$ coincide for all such $\boldsymbol{\Sigma}^{\#}$ if and only if $u = 0$.

1. INTRODUCTION AND SUMMARY

Jewell and Bloomfield (1983) showed that two random vectors $\mathbf{x}_1$, $\mathbf{x}_2$ with joint positive definite covariance matrix $\boldsymbol{\Sigma}$, say, have precisely the same canonical correlations as the two random vectors $\mathbf{x}_1^*$, $\mathbf{x}_2^*$ with covariance matrix $\boldsymbol{\Sigma}^{-1}$, where $(\mathbf{x}_1^*, \mathbf{x}_2^*)$ is partitioned similarly to $(\mathbf{x}_1, \mathbf{x}_2)$.

In this paper we extend this result to the situation where the joint covariance matrix $\mathbf{\Sigma}$ of $\mathbf{x}_1$, $\mathbf{x}_2$ is singular; we study the canonical correlations between the sets of random variables with covariance matrix $\mathbf{\Sigma}^{\#}$, belonging to a certain class of symmetric reflexive generalized inverses of $\mathbf{\Sigma}$.

Consider the $1 \times p$ random vector

$$\mathbf{x} = (\mathbf{x}_1, \mathbf{x}_2), \tag{1.1}$$

where $\mathbf{x}_1$ is $1 \times p_1$, and $\mathbf{x}_2$ is $1 \times p_2$, and $p_1 + p_2 = p$, with nonnegative definite (possibly singular) covariance or dispersion matrix partitioned as follows

$$var(\mathbf{x}) = var(\mathbf{x}_1, \mathbf{x}_2) = \mathbf{\Sigma} = \begin{pmatrix} \mathbf{\Sigma}_{11} & \mathbf{\Sigma}_{12} \\ \mathbf{\Sigma}_{21} & \mathbf{\Sigma}_{22} \end{pmatrix}, \tag{1.2}$$

where $\mathbf{\Sigma}_{12}$ is $p_1 \times p_2$.

The positive canonical correlations ρ_i, say, associated with $\mathbf{\Sigma}$ as partitioned in (1.2), satisfy

$$\rho_i^2 = \mathrm{ch}_i(\mathbf{\Sigma}_{11}^- \mathbf{\Sigma}_{12} \mathbf{\Sigma}_{22}^- \mathbf{\Sigma}_{21}); \quad i = 1, \ldots, m, \tag{1.3}$$

where $m = \mathrm{rank}(\mathbf{\Sigma}_{12})$ and any generalized inverses of $\mathbf{\Sigma}_{11}$ and $\mathbf{\Sigma}_{22}$ may be chosen, cf. Styan (1985, Theorem 2.1). In (1.3) ch_i denotes the ith largest characteristic root or eigenvalue. We will write

$$t = \mathrm{num}\{i : 0 < \rho_i < 1\} \quad \text{and} \quad u = \mathrm{num}\{i : \rho_i = 1\}, \tag{1.4}$$

so that t is the number of nonzero canonical correlations strictly less than 1, while u is the number of those equal to 1. Then

$$m = t + u = \mathrm{rank}(\mathbf{\Sigma}_{12}) \leq \min(p_1, p_2). \tag{1.5}$$

Following Puntanen (1987, pp. 43-46) we consider the $p \times p$ matrix

$$\mathbf{\Sigma}^{\#} = \begin{pmatrix} (\mathbf{\Sigma}_{11.2})_{rs}^- & -(\mathbf{\Sigma}_{11.2})_{rs}^- \mathbf{\Sigma}_{12} (\mathbf{\Sigma}_{22})_{rs}^- \\ -(\mathbf{\Sigma}_{22})_{rs}^- \mathbf{\Sigma}_{21} (\mathbf{\Sigma}_{11.2})_{rs}^- & (\mathbf{\Sigma}_{22})_{rs}^- + (\mathbf{\Sigma}_{22})_{rs}^- \mathbf{\Sigma}_{21} (\mathbf{\Sigma}_{11.2})_{rs}^- \mathbf{\Sigma}_{12} (\mathbf{\Sigma}_{22})_{rs}^- \end{pmatrix}$$

$$= \begin{pmatrix} \mathbf{\Sigma}_{11}^{\#} & \mathbf{\Sigma}_{12}^{\#} \\ \mathbf{\Sigma}_{21}^{\#} & \mathbf{\Sigma}_{22}^{\#} \end{pmatrix}, \tag{1.6}$$

say, where the Schur complement

$$\Sigma_{11.2} = \Sigma_{11} - \Sigma_{12}\Sigma_{22}^{-}\Sigma_{21} \tag{1.7}$$

is independent of the choice of generalized inverse of Σ_{22}.

The matrix $\Sigma^{\#}$, defined by (1.6), is a symmetric reflexive generalized inverse of Σ for any choices of symmetric reflexive generalized inverses $(\Sigma_{11.2})^{-}_{rs}$ and $(\Sigma_{22})^{-}_{rs}$, cf. Marsaglia and Styan (1974, p. 439), Ouellette (1981, pp. 232-3). We will say that a symmetric reflexive generalized inverse $\Sigma^{\#}$, defined by (1.6), is in Banachiewicz-Schur form; according to Ouellette (1981, p. 201) Banachiewicz (1937) "appears to have been the first author to study the inverse of a partitioned matrix [using Schur complements]". We note that not all symmetric reflexive generalized inverses need be in Banachiewicz-Schur form and that, in general, the matrix $\Sigma^{\#}$ is not unique (but $\Sigma^{\#}$ is always nonnegative definite since $\Sigma^{\#} = \Sigma^{\#}\Sigma\Sigma^{\#}$ from the reflexive property of the generalized inverse $\Sigma^{\#}$).

Using notation similar to (1.4), let

$$t^{\#} = \text{num}\{0 < \rho_j^{\#} < 1\} \quad \text{and} \quad u^{\#} = \text{num}\{\rho_j^{\#} = 1\}; \tag{1.8}$$

here $\rho_j^{\#}$ denotes the jth largest canonical correlation associated with the matrix $\Sigma^{\#}$, so that $t^{\#}$ is the number of nonzero canonical correlations less than 1, while $u^{\#}$ is the number equal to 1. We note that the number $m^{\#}$ of positive canonical correlations $\rho_j^{\#}$ is

$$m^{\#} = t^{\#} + u^{\#} = \text{rank}(\Sigma_{12}^{\#}) \leq m \leq \min(p_1, p_2). \tag{1.9}$$

In this paper we show (Theorem 1) that the number, $u^{\#}$, of unit canonical correlations associated with $\Sigma^{\#}$ is zero, while the number, $t^{\#}$, of canonical correlations associated with $\Sigma^{\#}$ that are less than 1 is greater than or equal to the number, t, of canonical correlations associated with Σ that are less than 1; necessary and sufficient conditions for equality of t and $t^{\#}$ are obtained.

In Theorem 2 we compare the two sets of non-unit canonical correlations, ρ_{u+h}, $h = 1, \ldots, t$, and $\rho_j^{\#}$, $j = 1, \ldots, t^{\#}$, that are less than one. We show that then

$$0 < \rho_{u+h} \leq \rho_h^{\#} < 1; \quad h = 1, \ldots, t \leq t^{\#}; \tag{1.10}$$

necessary and sufficient conditions for equality in the middle of (1.10) are obtained.

2. SOME INTERMEDIATE RESULTS

Let the $p \times p$ covariance matrix $\mathbf{\Sigma}$ have rank $r \leq p_1 + p_2 = p$. Following Seshadri and Styan (1980, p. 334) we may write

$$\mathbf{\Sigma} = \mathbf{X}\mathbf{X}' = \begin{pmatrix} \mathbf{X}_1' \\ \mathbf{X}_2' \end{pmatrix} (\mathbf{X}_1, \mathbf{X}_2) = \begin{pmatrix} \mathbf{X}_1'\mathbf{X}_1 & \mathbf{X}_1'\mathbf{X}_2 \\ \mathbf{X}_2'\mathbf{X}_1 & \mathbf{X}_2'\mathbf{X}_2 \end{pmatrix}, \tag{2.1}$$

where $\mathbf{X}$ is a $p \times r$ matrix of rank r, $\mathbf{X}_1$ is an $r \times p_1$ matrix of rank r_1 and $\mathbf{X}_2$ is an $r \times p_2$ matrix of rank r_2; and so, cf. (1.2),

$$\mathbf{\Sigma}_{ij} = \mathbf{X}_i'\mathbf{X}_j; \quad i, j = 1, 2. \tag{2.2}$$

Then we may write

$$\mathbf{X}_i = \mathbf{U}_i\mathbf{R}_i'; \quad i = 1, 2, \tag{2.3}$$

where the $r \times r_i$ matrix $\mathbf{U}_i$ satisfies $\mathbf{U}_i'\mathbf{U}_i = \mathbf{I}_{r_i}$ and the $p_i \times r_i$ matrix $\mathbf{R}_i$ has full column rank r_i ($i = 1, 2$). Then, cf. (1.5),

$$m = t + u = \text{rank}(\mathbf{\Sigma}_{12}) \leq \min(r_1, r_2) \leq \min(p_1, p_2). \tag{2.4}$$

The positive canonical correlations, ρ_i, associated with $\mathbf{\Sigma}$ satisfy, cf. (1.3),

$$\begin{aligned} \rho_i^2 &= \text{ch}_i[(\mathbf{X}_1'\mathbf{X}_1)^-\mathbf{X}_1'\mathbf{X}_2(\mathbf{X}_2'\mathbf{X}_2)^-\mathbf{X}_2'\mathbf{X}_1] \\ &= \text{ch}_i[\mathbf{X}_1(\mathbf{X}_1'\mathbf{X}_1)^-\mathbf{X}_1'\mathbf{X}_2(\mathbf{X}_2'\mathbf{X}_2)^-\mathbf{X}_2'] \\ &= \text{ch}_i(\mathbf{U}_1\mathbf{U}_1'\mathbf{U}_2\mathbf{U}_2') \\ &= \text{ch}_i(\mathbf{U}_1'\mathbf{U}_2\mathbf{U}_2'\mathbf{U}_1) = \text{sg}_i^2(\mathbf{U}_1'\mathbf{U}_2); \quad i = 1, \ldots, m, \end{aligned} \tag{2.5}$$

where sg_i denotes the ith largest singular value. The ρ_i are, therefore, the m positive singular values of $\mathbf{U}_1'\mathbf{U}_2$. Let us write the singular value decomposition

$$\mathbf{U}_1'\mathbf{U}_2 = \mathbf{P} \begin{pmatrix} \mathbf{I}_u & & \\ & \mathbf{D}_t & \\ & & \mathbf{0} \end{pmatrix} \mathbf{Q}', \tag{2.6}$$

where $\mathbf{I}_u$ is the $u \times u$ identity matrix and the $t \times t$ diagonal matrix

$$\mathbf{D}_t = \text{diag}(\rho_{u+1}, \rho_{u+2}, \ldots, \rho_{u+t}) \tag{2.7}$$

with the zero matrix $\mathbf{0}$ in (2.6) being $(r_1 - m)\times(r_2 - m)$, where $m = u+t$; the matrices $\mathbf{P}$ and $\mathbf{Q}$ are orthogonal, respectively $r_1\times r_1$ and $r_2\times r_2$. The diagonal elements of $\mathbf{D}_t$ are the positive canonical correlations that are less than 1.

Consider again the $p\times p$ matrix

$$\Sigma^{\#} = \begin{pmatrix} (\Sigma_{11.2})^-_{rs} & -(\Sigma_{11.2})^-_{rs}\Sigma_{12}(\Sigma_{22})^-_{rs} \\ -(\Sigma_{22})^-_{rs}\Sigma_{21}(\Sigma_{11.2})^-_{rs} & (\Sigma_{22})^-_{rs} + (\Sigma_{22})^-_{rs}\Sigma_{21}(\Sigma_{11.2})^-_{rs}\Sigma_{12}(\Sigma_{22})^-_{rs} \end{pmatrix}$$

$$= \begin{pmatrix} \Sigma^{\#}_{11} & \Sigma^{\#}_{12} \\ \Sigma^{\#}_{21} & \Sigma^{\#}_{22} \end{pmatrix}, \tag{2.8}$$

as defined in (1.6). Then $\Sigma^{\#}$ is a symmetric reflexive generalized inverse of Σ in Banachiewicz-Schur form. Substituting (2.2) into (2.8) yields

$$\Sigma^{\#}_{11} = (\mathbf{X}_1{}'\mathbf{M}_2\mathbf{X}_1)^-_{rs}$$

$$\Sigma^{\#}_{12} = -(\mathbf{X}_1{}'\mathbf{M}_2\mathbf{X}_1)^-_{rs}\mathbf{X}_1{}'\mathbf{X}_2(\mathbf{X}_2{}'\mathbf{X}_2)^-_{rs}$$

$$\Sigma^{\#}_{21} = -(\mathbf{X}_2{}'\mathbf{X}_2)^-_{rs}\mathbf{X}_2{}'\mathbf{X}_1(\mathbf{X}_1{}'\mathbf{M}_2\mathbf{X}_1)^-_{rs} = (\Sigma^{\#}_{12})'$$

$$\Sigma^{\#}_{22} = (\mathbf{X}_2{}'\mathbf{X}_2)^-_{rs} + (\mathbf{X}_2{}'\mathbf{X}_2)^-_{rs}\mathbf{X}_2{}'\mathbf{X}_1(\mathbf{X}_1{}'\mathbf{M}_2\mathbf{X}_1)^-_{rs}\mathbf{X}_1{}'\mathbf{X}_2(\mathbf{X}_2{}'\mathbf{X}_2)^-_{rs}, \tag{2.9}$$

where

$$\mathbf{M}_2 = \mathbf{I} - \mathbf{X}_2(\mathbf{X}_2{}'\mathbf{X}_2)^-\mathbf{X}_2{}'. \tag{2.10}$$

The canonical correlations $\rho_j^{\#}$ associated with $\Sigma^{\#}$ are the positive square roots of the positive proper generalized eigenvalues of $\Sigma_{21}{}^{\#}(\Sigma_{11}{}^{\#})^-\Sigma_{12}{}^{\#}$ with respect to $\Sigma_{22}{}^{\#}$, i.e., the positive square roots of the nonzero "generalized" roots of

$$|\rho^2\Sigma^{\#}_{22} - \Sigma^{\#}_{21}(\Sigma^{\#}_{11})^-\Sigma^{\#}_{12}| = 0, \tag{2.11}$$

cf. Puntanen, (1987, pp.17-21), de Leeuw (1982, p. 90). Substituting (2.9) into (2.11) yields

$$|\rho^2(\mathbf{X}_2'\mathbf{X}_2)^-_{rs} - (1-\rho^2)(\mathbf{X}_2'\mathbf{X}_2)^-_{rs}\mathbf{X}_2'\mathbf{X}_1(\mathbf{X}_1'\mathbf{M}_2\mathbf{X}_1)^-_{rs}\mathbf{X}_1'\mathbf{X}_2(\mathbf{X}_2'\mathbf{X}_2)^-_{rs}|$$

$$= |(\mathbf{X}_2'\mathbf{X}_2)^-_{rs}| \cdot |\rho^2\mathbf{I}_{p_2} - (1-\rho^2)\mathbf{X}_2'\mathbf{X}_1(\mathbf{X}_1'\mathbf{M}_2\mathbf{X}_1)^-_{rs}\mathbf{X}_1'\mathbf{X}_2(\mathbf{X}_2'\mathbf{X}_2)^-_{rs}|$$

$$= |(\mathbf{X}_2'\mathbf{X}_2)^-_{rs}| \cdot (\rho^2)^{p_2-p_1} \cdot |\rho^2\mathbf{I}_{p_1} - (1-\rho^2)\mathbf{X}_1'\mathbf{H}_2\mathbf{X}_1(\mathbf{X}_1'\mathbf{M}_2\mathbf{X}_1)^-_{rs}| = 0, \tag{2.12}$$

cf. e.g., Styan (1985, p. 41), where

$$\mathbf{H}_2 = \mathbf{X}_2(\mathbf{X}_2'\mathbf{X}_2)^-\mathbf{X}_2' = \mathbf{I} - \mathbf{M}_2; \tag{2.13}$$

ignoring the leading null determinant and the factor $(\rho^2)^{p_2-p_1}$ in the last row of (2.12) we are led to the nonzero roots ρ^2 of

$$|\rho^2\mathbf{I}_{p_1} - (1-\rho^2)\mathbf{X}_1'\mathbf{H}_2\mathbf{X}_1(\mathbf{X}_1'\mathbf{M}_2\mathbf{X}_1)^-_{rs}| = 0. \tag{2.14}$$

Substituting (2.3) into (2.14) and noting that $\mathbf{H}_2 = \mathbf{U}_2\mathbf{U}_2'$, we obtain

$$|\rho^2\mathbf{I}_{p_1} - (1-\rho^2)\mathbf{R}_1\mathbf{U}_1'\mathbf{U}_2\mathbf{U}_2'\mathbf{U}_1\mathbf{R}_1'[\mathbf{R}_1(\mathbf{I}_{r_1} - \mathbf{U}_1'\mathbf{U}_2\mathbf{U}_2'\mathbf{U}_1)\mathbf{R}_1']^-_{rs}| = 0, \tag{2.15}$$

which has the same nonzero roots as

$$|\rho^2\mathbf{I}_{r_2} - (1-\rho^2)\mathbf{Q}'\mathbf{U}_2'\mathbf{U}_1\mathbf{B}\mathbf{U}_1'\mathbf{U}_2\mathbf{Q}| = 0, \tag{2.16}$$

where the $r_2 \times r_2$ orthogonal matrix $\mathbf{Q}$ is as defined in the singular value decomposition (2.6) and $\mathbf{B} = \mathbf{R}_1'[\mathbf{R}_1(\mathbf{I} - \mathbf{U}_1'\mathbf{U}_2\mathbf{U}_2'\mathbf{U}_1)\mathbf{R}_1']^-_{rs}\mathbf{R}_1$; we now see that $\mathbf{B} = (\mathbf{I} - \mathbf{U}_1'\mathbf{U}_2\mathbf{U}_2'\mathbf{U}_1)^-_{rs}$ in view of the following:

LEMMA 1. *Let the $p \times r$ matrix $\mathbf{R}$ have full column rank r and let the $r \times r$ matrix $\mathbf{T}$ be symmetric. Then the $r \times r$ matrix $\mathbf{V}$ is a symmetric reflexive generalized inverse of $\mathbf{T}$ if and only if it is of the form $\mathbf{V} = \mathbf{R}'(\mathbf{RTR}')^-_{rs}\mathbf{R}$.*

We omit the proof which is straightforward; details are given by Latour (1987, Lemma A.3.4); see also Puntanen (1987, Lemma 1.3, p. 7).

We may, therefore, write (2.16) as

$$|\rho^2\mathbf{I}_{r_2} - (1-\rho^2)\mathbf{A}| = |\rho^2(\mathbf{I}_{r_2} + \mathbf{A}) - \mathbf{A}| = 0, \tag{2.17}$$

where the $r_2 \times r_2$ nonnegative definite matrix

$$\mathbf{A} = \mathbf{Q}'\mathbf{U}_2'\mathbf{U}_1(\mathbf{I}_{r_1} - \mathbf{U}_1'\mathbf{U}_2\mathbf{U}_2'\mathbf{U}_1)^-_{rs}\mathbf{U}_1'\mathbf{U}_2\mathbf{Q}. \tag{2.18}$$

The nonzero roots ρ^2 of (2.17) are, therefore, the nonzero eigenvalues of $(\mathbf{I}+\mathbf{A})^{-1}\mathbf{A}$, which are positive and strictly less than 1 since $\mathbf{A}$ is nonnegative definite. In view of (2.6) we have the general representation

$$(\mathbf{I}-\mathbf{U}_1'\mathbf{U}_2\mathbf{U}_2'\mathbf{U}_1)^-_{rs} = \mathbf{P}\begin{pmatrix} \mathbf{C}_1(\mathbf{I}-\mathbf{D}_t^2)\mathbf{C}_1' + \mathbf{C}_2\mathbf{C}_2' & \mathbf{C}_1 & \mathbf{C}_2 \\ \mathbf{C}_1' & (\mathbf{I}-\mathbf{D}_t^2)^{-1} & \mathbf{0} \\ \mathbf{C}_2' & \mathbf{0} & \mathbf{I}_{r_1-m} \end{pmatrix}\mathbf{P}', \tag{2.19}$$

where the two matrices $\mathbf{C}_1$ and $\mathbf{C}_2$ are arbitrary, respectively $u \times t$ and $u \times (r_1 - m)$; the $r_1 \times r_1$ matrix $\mathbf{P}$ is orthogonal, cf. (2.8). Substituting (2.19) and (2.6) into (2.18) yields

$$\mathbf{A} = \begin{pmatrix} \mathbf{C}_1(\mathbf{I}-\mathbf{D}_t^2)\mathbf{C}_1' + \mathbf{C}_2\mathbf{C}_2' & \mathbf{C}_1\mathbf{D}_t & \mathbf{0} \\ \mathbf{D}_t\mathbf{C}_1' & \mathbf{D}_t(\mathbf{I}-\mathbf{D}_t^2)^{-1}\mathbf{D}_t & \mathbf{0} \\ \mathbf{0} & \mathbf{0} & \mathbf{0} \end{pmatrix} \tag{2.20}$$

and so the last $r_2 - (u+t)$ rows and the last $r_2 - (u+t)$ columns are all zero. Since rank is additive on the Schur complement, it follows that

$$\operatorname{rank}(\mathbf{A}) = t + \operatorname{rank}(\mathbf{C}_2). \tag{2.21}$$

We will also need the following:

LEMMA 2. *Let the $t \times t$ matrix $\mathbf{H}$ be positive definite and let the matrix $\mathbf{F}$ be $s \times t$. Then*

$$\lambda_i = \operatorname{ch}_i\begin{pmatrix} \mathbf{F}\mathbf{H}^{-1}\mathbf{F}' & \mathbf{F} \\ \mathbf{F}' & \mathbf{H} \end{pmatrix} \geq \operatorname{ch}_i(\mathbf{H}); \quad i = 1, \ldots, t. \tag{2.22}$$

Equality holds in (2.22) *for all* $i = 1, \ldots, t$ *if and only if* $\mathbf{F} = \mathbf{0}$.

As pointed out by Marshall and Olkin (1979, pp. 225-226) the first part of Lemma 2 is a simple consequence of a result due to Fan (1954). We present the following proof of Lemma 2, however, in view of its brevity and elegance.

Proof. We note that for $i = 1, ..., t$,

$$\lambda_i = \text{ch}_i\left[\begin{pmatrix} \mathbf{FH}^{-1} \\ \mathbf{I}_t \end{pmatrix}\mathbf{H}(\mathbf{H}^{-1}\mathbf{F}', \mathbf{I}_t)\right]$$

$$= \text{ch}_i[\mathbf{H}^{\frac{1}{2}}(\mathbf{I}_t + \mathbf{H}^{-1}\mathbf{F}'\mathbf{F}\mathbf{H}^{-1})\mathbf{H}^{\frac{1}{2}}]$$

$$= \text{ch}_i(\mathbf{H} + \mathbf{H}^{-\frac{1}{2}}\mathbf{F}'\mathbf{F}\mathbf{H}^{-\frac{1}{2}}) \geq \text{ch}_i(\mathbf{H}). \tag{2.23}$$

If equality holds in (2.22) for all $i = 1, ..., t$ then adding yields

$$\text{tr}(\mathbf{FH}^{-1}\mathbf{F}') + \text{tr}\mathbf{H} = \text{tr}\mathbf{H}, \tag{2.24}$$

since the s other eigenvalues $\lambda_{t+1} = ... = \lambda_{t+s} = 0$. Hence $\mathbf{F} = \mathbf{0}$. □

3. THE MAIN RESULTS

THEOREM 1. *Consider the positive canonical correlations ρ_i, $i = 1, ..., m$, associated with the covariance matrix Σ, partitioned as in* (1.2), *and the positive canonical correlations $\rho_j^{\#}$, $j = 1, ..., m^{\#}$, associated with any symmetric reflexive generalized inverse $\Sigma^{\#}$ in the Banachiewicz-Schur form* (1.6). *Let t, $t^{\#}$ denote, respectively, the numbers of these ρ_i, $\rho_j^{\#}$ that are strictly less than* 1, *and let u, $u^{\#}$ denote the numbers equal to* 1, *cf.* (1.4) *and* (1.8). *Then*

$$u^{\#} = 0 \tag{3.1}$$

and

$$m^{\#} = t^{\#} \geq t. \tag{3.2}$$

Equality holds on the right of (3.2) *for all $\Sigma^{\#}$ of the form* (1.6) *if and only if either $u = 0$ or $m = \text{rank}(\Sigma_{11})$.*

Proof. From (2.17) we recall that the positive canonical correlations $\rho^{\#}$ are the positive square roots of the nonzero eigenvalues of $(\mathbf{I}+\mathbf{A})^{-1}\mathbf{A}$, where $\mathbf{A}$ is defined by (2.18) and (2.20). Since these eigenvalues are all strictly less than 1 it follows immediately that $u^{\#} = 0$. That $m^{\#} = t^{\#} \geq t$ follows at once from (2.21); equality holds if and only if in (2.19) $\mathbf{C}_2$ is absent [or equal to the $u \times (r_1 - m)$ zero matrix], where $r_1 = \text{rank}(\Sigma_{11})$, and so it follows that $t^{\#} = t$ for all $\Sigma^{\#}$ of the form (1.6) if and only if either $u = 0$ or $r_1 - m = 0$. □

We note that the canonical correlations associated with $\mathbf{\Sigma}^{\#}$ do not depend upon the choice of $(\mathbf{X}_2'\mathbf{X}_2)^-_{rs}$, but only upon the choice of $(\mathbf{X}_1'\mathbf{M}_2\mathbf{X}_1)^-_{rs}$.

If $\mathbf{\Sigma}^{\#}$ were defined by the parallel Banachiewicz-Schur form using the Schur complement $\mathbf{\Sigma}_{22.1}$, so that $\mathbf{\Sigma}_{22}^{\#} = (\mathbf{\Sigma}_{22.1})^-_{rs}$, and so on, then Theorem 1 would remain valid except that the condition $m = \operatorname{rank}(\mathbf{\Sigma}_{11})$ for equality on the right of (3.2) would now be $m = \operatorname{rank}(\mathbf{\Sigma}_{22})$.

THEOREM 2. *Let the positive canonical correlations ρ_i and $\rho_j^{\#}$ be defined as in Theorem 1. If ρ_{u+h}, $\rho_h^{\#}$ denote, respectively, the $(u+h)$-th, h-th largest then*

$$0 < \rho_{u+h} \leq \rho_h^{\#} < 1; \quad h = 1, \ldots, t \leq t^{\#}. \tag{3.3}$$

Equality holds in the middle of (3.3) for all $h = 1, \ldots, t \leq t^{\#}$ and for all $\mathbf{\Sigma}^{\#}$ of the form (1.6) if and only if $u = 0$, i.e., there are no unit canonical correlations associated with $\mathbf{\Sigma}$.

Proof. Since $(\rho_h^{\#})^2$ is the hth largest eigenvalue of $(\mathbf{I} + \mathbf{A})^{-1}\mathbf{A}$, where $\mathbf{A}$ is as defined in (2.20), it follows at once that

$$\begin{aligned}
\frac{(\rho_h^{\#})^2}{1-(\rho_h^{\#})^2} &= \operatorname{ch}_h(\mathbf{A}) \\
&\geq \operatorname{ch}_h\begin{pmatrix} \mathbf{C}_1(\mathbf{I}-\mathbf{D}_t^2)\mathbf{C}_1' & \mathbf{C}_1\mathbf{D}_t \\ \mathbf{D}_t\mathbf{C}_1' & \mathbf{D}_t(\mathbf{I}-\mathbf{D}_t^2)^{-1}\mathbf{D}_t \end{pmatrix} \\
&\geq \operatorname{ch}_h[\mathbf{D}_t(\mathbf{I}-\mathbf{D}_t^2)^{-1}\mathbf{D}_t] \\
&= \frac{\rho_{u+h}^2}{1-\rho_{u+h}^2}; \qquad h = 1, \ldots, t \leq t^{\#}.
\end{aligned} \tag{3.4}$$

The second inequality follows from Lemma 2. This establishes (3.3). Equality holds in the middle of (3.3) for all $h = 1, \ldots, t \leq t^{\#}$ if and only if equality holds in both inequalities in (3.4). This is possible if and only if in (2.19) both the $u \times t$ matrix $\mathbf{C}_1$ and the $u \times (r_1 - m)$ matrix $\mathbf{C}_2$ are absent (or equal to zero matrices), where $r_1 = \operatorname{rank}(\mathbf{\Sigma}_{11})$. Hence $\rho_{u+h} = \rho_h^{\#}$, $h = 1, \ldots, t \leq t^{\#}$, for all $\mathbf{\Sigma}^{\#}$ of the form (1.6) if and only if $u = 0$. □

The conditions for equality in Theorem 1 and in Theorem 2 were derived considering all $\Sigma^{\#}$ of the Banachiewicz-Schur form (1.6). Our proofs, however, involve particular choices of the matrix $\mathbf{A}$, as defined in (2.18), for which the desired equalities hold when they are not valid for all such matrices $\mathbf{A}$. This suggests a question concerning the possibility of strengthening Theorems 1 and 2: Is it possible to characterize all those $\Sigma^{\#}$ for which the corresponding $\mathbf{A}$ have $\mathbf{C}_2=\mathbf{0}$ or both $\mathbf{C}_1=\mathbf{0}$ and $\mathbf{C}_2=\mathbf{0}$?

As we will see from the numerical example in Section 4 below, $\rho_h^{\#}$ may assume a value arbitrarily close to 1; it follows, therefore, that the bounds (3.3) for $\rho_h^{\#}$ are "best possible".

When $u = 0$ and only then the Moore-Penrose inverse of Σ may be written, cf. Marsaglia and Styan (1974, Corollary 2), in the Banachiewicz-Schur form

$$\Sigma^{+} = \begin{pmatrix} \Sigma_{11.2}^{+} & -\Sigma_{11.2}^{+}\Sigma_{12}\Sigma_{22}^{+} \\ -\Sigma_{22}^{+}\Sigma_{21}\Sigma_{11.2}^{+} & \Sigma_{22}^{+}+\Sigma_{22}^{+}\Sigma_{21}\Sigma_{11.2}^{+}\Sigma_{12}\Sigma_{22}^{+} \end{pmatrix}, \tag{3.5}$$

since $u = \operatorname{rank}(\Sigma_{11}) + \operatorname{rank}(\Sigma_{22}) - \operatorname{rank}(\Sigma) = 0$, cf. Styan (1985, p. 51); it then follows directly from Theorem 2 that when $u= 0$ the canonical correlations associated with Σ^{+} coincide with those associated with Σ. It is an open question, however, as to what happens in general when $u \neq 0$.

When Σ is positive definite we may write, cf. Ouellette (1981, p. 202),

$$\Sigma^{-1} = \begin{pmatrix} \Sigma_{11.2}^{-1} & -\Sigma_{11}^{-1}\Sigma_{12}\Sigma_{22.1}^{-1} \\ -\Sigma_{22}^{-1}\Sigma_{21}\Sigma_{11.2}^{-1} & \Sigma_{22.1}^{-1} \end{pmatrix}; \tag{3.6}$$

this formula (3.6), using the two Schur complements $\Sigma_{11.2}$ and $\Sigma_{22.1}$, combines the two Banachiewicz-Schur forms and is due to Hotelling (1943). [A similar formula for the Moore-Penrose inverse also holds when $u = 0$.] It is immediate from (3.6) that the canonical correlations associated with Σ^{-1} coincide with those associated with Σ, cf. Jewell and Bloomfield (1983).

4. A NUMERICAL EXAMPLE

We illustrate our results with a numerical example based on the two-way layout example considered by Latour and Styan (1985, Section 3). Let Σ, partitioned as in (1.2), denote the covariance matrix of the row and column totals of the responses in the binary layout with 3×4 incidence matrix

$$\mathbf{N} = \begin{pmatrix} 0 & 1 & 1 & 1 \\ 1 & 1 & 1 & 1 \\ 1 & 1 & 1 & 1 \end{pmatrix} \tag{4.1}$$

and white noise (with $\sigma^2 = 1$). Then

$$\mathbf{\Sigma}_{12} = \mathbf{N} = \mathbf{\Sigma}_{21}', \quad \mathbf{\Sigma}_{11} = \text{diag}(3, 4, 4), \quad \mathbf{\Sigma}_{22} = \text{diag}(2, 3, 3, 3). \tag{4.2}$$

We see that $u = 1$ since the layout is connected; $m = \text{rank}(\mathbf{N}) = 2 = u+t = 1 + t$ and so $t = 1$. From Latour and Styan (1985, p. 237) we obtain $\rho_2^2 = 1/12$ (and $\rho_1 = 1$). The matrix

$$\mathbf{A}_1 = \begin{pmatrix} \frac{11}{12}c_1^2 + c_2^2 & c_1/\sqrt{12} \\ c_1/\sqrt{12} & 1/11 \end{pmatrix} \tag{4.3}$$

is the matrix $\mathbf{A}$, as defined in (2.20), without its last $r_2 - m = 2$ rows and columns (which are all zero). Thus

$$(\rho_h^\#)^2 = \omega_h/(1+\omega_h); \qquad h = 1, 2, \tag{4.4}$$

cf. (3.4), where $\omega_h = \text{ch}_h(\mathbf{A}_1)$. When $c_1 = 0$ the eigenvalues of $\mathbf{A}_1$ are c_2^2 and $1/11$; thus $(\rho_h^\#)^2 = c_2^2/(1+c_2^2)$ and $1/12$. When $c_2 = 0$ the matrix $\mathbf{A}_1$ has rank 1, and so $\omega_1 = (x+12)/132$ and $(\rho^\#)^2 = (x+12)/(x+144) \geq 1/12$, where $x = 121c_1^2$. When neither c_1 nor c_2 is zero the matrix $\mathbf{A}_1$ has rank 2 and we obtain the values for $\rho_1^\#$ and $\rho_2^\#$ shown in the following table:

	$\rho_1^\#$					$\rho_2^\#$				
c_1	$c_2 = 1$	2	3	4	5	1	2	3	4	5
1	0.814	0.912	0.953	0.972	0.981	0.210	0.262	0.276	0.281	0.284
2	0.909	0.941	0.963	0.976	0.983	0.137	0.212	0.246	0.262	0.271
3	0.950	0.962	0.972	0.980	0.985	0.098	0.169	0.213	0.238	0.253
4	0.970	0.974	0.980	0.984	0.988	0.076	0.138	0.183	0.213	0.233
5	0.980	0.982	0.985	0.987	0.990	0.061	0.115	0.158	0.190	0.213

ACKNOWLEDGEMENTS

This research was begun while the first and second authors were visiting McGill University in June 1986, and completed while the first and third authors were visiting the University of Tampere in June 1987. We are most grateful for the very generous provision of facilities both at McGill and in Tampere. We also thank Tapio Nummi for help with the computations associated with our numerical example, and Jerzy K. Baksalary for several helpful suggestions. This research was supported in part by the Academy of Finland, the Natural Sciences and Engineering Research Council of Canada and by the Fonds pour la Formation de Chercheurs et l'Aide à la Recherche du Gouvernement du Québec.

REFERENCES

Banachiewicz, T. (1937). Zur Berechnung der Determinanten, wie auch der Inversen, und zur darauf basierten Auflösung der Systeme linearer Gleichungen. *Acta Astronom. Sér. C*, 3, 41-67.

de Leeuw, Jan (1982). Generalized eigenvalue problems with positive semidefinite matrices. *Psychometrika*, 47, 87-93.

Fan, K. (1954). Inequalities for eigenvalues of Hermitian matrices. *National Bureau of Standards Applied Math. Ser.*, 39, 131-139.

Hotelling, H. (1943). Some new methods in matrix calculations. *Annals of Mathematical Statistics*, 14, 1-34.

Jewell, Nicholas P. and Bloomfield, Peter (1983). Canonical correlations of past and future for time series: definitions and theory. *Annals of Statistics*, 11, 837-847.

Latour, Dominique (1987). Equalities and inequalities for canonical correlations coefficients, with special emphasis on the two-way layout of experimental design. M. Sc. thesis, Dept. of Mathematics and Statistics, McGill University.

Latour, Dominique and Styan, George P. H. (1985). Canonical correlations in the two-way layout. *Proc. First International Tampere Seminar on Linear Statistical Models and their Applications*. Dept. of Mathematical Sciences, University of Tampere, Finland, 225-243.

Marsaglia, George and Styan, George P.H. (1974). Rank conditions for generalized inverses of partitioned matrices. *Sankhyā, Ser. A*, 36, 437-442.

Marshall, Albert W. and Olkin, Ingram (1979). *Inequalities: Theory of Majorization and Its Applications*. Academic Press, New York.

Ouellette, Diane Valérie (1981). Schur complements and statistics. *Linear Algebra and Its Applications*, 36, 187-295.

Puntanen, Simo (1987). On the relative goodness of ordinary least squares and on some associated canonical correlations. *Acta Universitatis Tamperensis, Ser. A*, 216, 1-77 (Paper [1]).

Seshadri, V. and Styan, George P.H. (1980). Canonical correlations, rank additivity and characterizations of multivariate normality. *Colloquia Math. Soc. János Bolyai, vol. 21: Analytic Function Methods in Probability Theory (Debrecen, Hungary, August 1977)*. János Bolyai, Budapest, and North-Holland, Amsterdam, 331-344.

Styan, George P.H. (1985). Schur complements and linear statistical models. *Proc. First International Tampere Seminar on Linear Statistical Models and their Applications*. Dept. of Mathematical Sciences, University of Tampere, Finland, 37-75.

Received 1 February 1987
Revised 5 October 1987

Department of Statistics
North Carolina State University
P. O. Box 8203
Raleigh, NC 27695-8203
U.S.A.

Department of Mathematical Sciences
University of Tampere
P. O. Box 607
SF-33101 Tampere
Finland

Department of Mathematics and Statistics
McGill University, Burnside Hall
805 ouest, rue Sherbrooke Street West
Montréal, Québec
Canada H3A 2K6

Proc. Second International Tampere Conference in Statistics
(Tampere, Finland, 1-4 June 1987)
Tarmo Pukkila and Simo Puntanen, *Editors*

pp. 555 - 566

Identifying influential data in a growth curves model

Erkki P. Liski
University of Tampere, Finland

Key words and phrases: Measuring influence, influence matrix, inference stage measure, design stage measure.

ABSTRACT

In this paper we consider measures for identifying influential measurements with respect to one or several parametric functions in a growth curves model. We examine the problem of influential data both at the design and at the inference stage and correspondingly we propose two kinds of influence measures. We study also the interrelationship between these two measures. We conclude with an example using data from a growth experiment for bulls.

1. INTRODUCTION

We study the generalized multivariate analysis of variance model suggested by Potthoff and Roy (1964). The model is given by

$$\mathbf{Y} = \mathbf{XBT}' + \mathbf{E}, \tag{1.1}$$

where $\mathbf{Y}$ is an $n \times q$ matrix of responses, $\mathbf{X}$ is an $n \times m$ matrix of known constants and $\mathbf{T}$ is a $q \times p$ regression matrix, both $\mathbf{X}$ and $\mathbf{T}$ are of full column rank. $\mathbf{B}$ is an $m \times p$ parameter matrix and the rows of $\mathbf{E}$ are independently distributed with mean vector $\mathbf{0}$ and covariance matrix $\boldsymbol{\Sigma}$. Model (1.1) was originally developed to analyze data obtained from a growth curve experiment, and hence it is also called as the growth curve model.

A large number of statistical quantities have been proposed to study outliers and the influence of individual observations in regression analysis, see for example Cook (1977), Cook and Weisberg (1980, 1982) and Belsey, Kuh and Welsch (1980). However, in the growth curve model (1.1) the problem of influential observations is more complicated than in the ordinary regression model. In this paper we define a design stage measure and an inference stage influence measure, and we show also certain relationships between these two quantities.

As an application we investigate the growth of bulls tested at an experimental station in Finland in the years 1965 and 1966. These are a part of data on 2712 bulls tested between the years 1965 and 1977. One of the objectives in studying this data was to estimate growth curves for these three breeds. In the years 1965-1969 the testing covered the interval 30-365 days of age, and bulls were weighed every 30th day (12 time-points). For more details concerning these data we refer to Lindström and Maijala (1970) and Liski (1987). In these experiments the ages for weighing were chosen on purely practical and economical grounds. Suppose now that our aim is to estimate growth curves for bulls on the basis of these data. If the number of weighing times must be reduced, for example on economical grounds, we should consider the influence of different time-points with respect to a chosen family of curves and with respect to the parametric functions which are of interest. The methods we develop in this paper serve as a means for handling such problems, for example.

2. INFLUENCE OF DELETING MEASUREMENTS

Khatri (1966) and Rao (1965 and 1966) independently developed the maximum likelihood estimator of $\mathbf{B}$:

$$\tilde{\mathbf{B}} = (\mathbf{X}'\mathbf{X})^{-1}\,\mathbf{X}'\mathbf{Y}\mathbf{S}^{-1}\mathbf{T}(\mathbf{T}'\mathbf{S}^{-1}\mathbf{T})^{-1}\,, \tag{2.1}$$

where $\mathbf{S} = \mathbf{Y}'[\mathbf{I} - \mathbf{X}(\mathbf{X}'\mathbf{X})^{-1}\mathbf{X}']\mathbf{Y}$. If we denote $\mathbf{Y}' = (\mathbf{y}_{(1)}, \mathbf{y}_{(2)},\ldots, \mathbf{y}_{(n)})$ and suppose that $cov(\mathbf{y}_{(i)}) = \boldsymbol{\Sigma}$ for every $i = 1, 2,\ldots, n$ and $\boldsymbol{\Sigma}$ is known, then the minimum variance unbiased estimator of $\mathbf{B}$ is

$$\dot{\mathbf{B}} = (\mathbf{X}'\mathbf{X})^{-1}\,\mathbf{X}'\mathbf{Y}\boldsymbol{\Sigma}^{-1}\mathbf{T}(\mathbf{T}'\boldsymbol{\Sigma}^{-1}\mathbf{T})^{-1}\,. \tag{2.2}$$

Rao (1967) showed that in some important subclasses of the growth curve model (1.1) the minimum variance estimator of **B** is as follows:

$$\hat{\mathbf{B}} = (\mathbf{X}'\mathbf{X})^{-1}\mathbf{X}'\mathbf{Y}\mathbf{T}(\mathbf{T}'\mathbf{T})^{-1}, \qquad (2.3)$$

although $\boldsymbol{\Sigma}$ is unknown and $\boldsymbol{\Sigma} \neq \mathbf{I}$. For the sake of simplicity we first investigate the estimators (2.2) and (2.3), and then consider the more complicated case (2.1).

We write now the growth curves model (1.1) as a linear model. We use the vec operation, which rearranges the columns of a matrix underneath each other. Thus, for example, vec $\mathbf{Y}' = (\mathbf{y}_{(1)}{}', \mathbf{y}_{(2)}{}', \ldots, \mathbf{y}_{(n)}{}')'$ is an $nq \times 1$ vector and vec $\mathbf{B}' = (\boldsymbol{\beta}_{(1)}{}', \boldsymbol{\beta}_{(2)}{}', \ldots, \boldsymbol{\beta}_{(m)}{}')'$ is a $pm \times 1$ vector. Since we want to put the rows of **Y** underneath each other, we write first $E(\mathbf{Y}') = \mathbf{T}\mathbf{B}'\mathbf{X}'$ and note that $\mathrm{vec}(\mathbf{T}\mathbf{B}'\mathbf{X}') = (\mathbf{X}\otimes\mathbf{T})\mathrm{vec}\mathbf{B}'$, where $\mathbf{X}\otimes\mathbf{T}$ is the Kronecker product of **X** and **T**. This mode of notation yields the generalized model

$$E\,(\mathrm{vec}\mathbf{Y}') = (\mathbf{X}\otimes\mathbf{T})\,\mathrm{vec}\mathbf{B}' \qquad (2.4)$$

and

$$cov\,(\mathrm{vec}\mathbf{Y}') = \mathbf{I}\otimes\boldsymbol{\Sigma}. \qquad (2.5)$$

Then the BLUE (best linear unbiased estimate) for vec**B**′ is

$$\mathrm{vec}\dot{\mathbf{B}}' = (\mathbf{X}'\mathbf{X}\otimes\mathbf{T}'\boldsymbol{\Sigma}^{-1}\mathbf{T})^{-1}\,(\mathbf{X}\otimes\mathbf{T}\boldsymbol{\Sigma}^{-1})'\,\mathrm{vec}\mathbf{Y}', \qquad (2.6)$$

which is equivalent to the estimator (2.2). Now we may again employ the results derived for linear regression.

We look now at an important special case, where all the deleted measurements come from the same observations. Measurements of the last 10 observations deleted at the first and at the last time-point, and deleting a single measurement are examples of this special case. Explicit expressions for the change in parameter estimates, when general subsets of data are dropped out, can be derived by applying the generalized growth curve model by Kleinbaum (1973). Derivations are rather straightforward but contain an appreciable amount of linear algebra. Therefore, development of these general formulas are not presented here.

By definition a case may be jugded influential if important features of the analysis are substantially altered when it is deleted from the data. To determine the degree of influence of a given subset of observations we compute the estimate of **B** with the given subset deleted. Let $\mathbf{Y}_{\mathrm{I.}}$ denote the set of observations from which we delete measurements and $\mathbf{Y}_{(\mathrm{I.})}$

contains the rest of observations, where I is a vector of indices specifying the cases to be deleted. Further, we partition $\mathbf{Y}_{I.}$ such that $\mathbf{Y}_{IJ}$ contains the deleted measurements and $\mathbf{Y}_{I(J)}$ the other measurements in $\mathbf{Y}_{I.}$, where the index vector J specifies the time points at which measurements are deleted. This representation of incomplete data proves sufficiently general for most practical purposes.

Let $\hat{\mathbf{B}}_{(IJ)}$ and $\dot{\mathbf{B}}_{(IJ)}$ denote the estimates of $\mathbf{B}$ without the measurements $\mathbf{Y}_{IJ}$. Correspondingly $\hat{\mathbf{B}}_{(I.)}$ denotes the estimate of $\mathbf{B}$ with observations indexed by I deleted and $\hat{\mathbf{B}}_{(.J)}$ denote the estimate with the time points J deleted. The matrices $\mathbf{X}_I$ and $\mathbf{T}_J$ consist of the rows of $\mathbf{X}$ and $\mathbf{T}$, correspondingly, as indicated by the sets of indices I and J, whereas $\mathbf{X}_{(I)}$ and $\mathbf{T}_{(J)}$ denote the matrices obtained from $\mathbf{X}$ and $\mathbf{T}$ by deleting the rows indexed by I and J respectively. Denoting $\text{vec}\mathbf{Y}_{(I)}' = \mathbf{y}_1$, $\text{vec}\mathbf{Y}_{I(J)}' = \mathbf{y}_2$ and $\text{vec}\mathbf{Y}_{IJ}' = \mathbf{y}_3$ we may write model (2.4) as follows

$$\begin{pmatrix} \mathbf{y}_1 \\ \mathbf{y}_2 \\ \mathbf{y}_3 \end{pmatrix} = \begin{pmatrix} \mathbf{X}_{(I)} \otimes \mathbf{T} \\ \mathbf{X}_I \otimes \mathbf{T}_{(J)} \\ \mathbf{X}_I \otimes \mathbf{T}_J \end{pmatrix} \text{vec}\mathbf{B}' + \text{vec}\mathbf{E}'. \tag{2.7}$$

If $\mathbf{y}_3$ is deleted from the data and $\mathbf{\Sigma} = \mathbf{I}$, we obtain

$$\text{vec}\hat{\mathbf{B}}_{(IJ)}' = [(\mathbf{X}_{(I)}'\mathbf{X}_{(I)} \otimes \mathbf{T}'\mathbf{T} + \mathbf{X}_I'\mathbf{X}_I \otimes \mathbf{T}_{(J)}'\mathbf{T}_{(J)}]^{-1}[(\mathbf{X}_{(I)} \otimes \mathbf{T})'\mathbf{y}_1 + (\mathbf{X}_1 \otimes \mathbf{T}_{(J)})'\mathbf{y}_2] \tag{2.8}$$

and correspondingly

$$\text{vec}\hat{\mathbf{B}} = (\mathbf{X}'\mathbf{X} \otimes \mathbf{T}'\mathbf{T})^{-1} [(\mathbf{X}_{(I)} \otimes \mathbf{T})'\mathbf{y}_1 + (\mathbf{X}_I \otimes \mathbf{T}_{(J)})'\mathbf{y}_2] + (\mathbf{X}'\mathbf{X} \otimes \mathbf{T}'\mathbf{T})^{-1} (\mathbf{X}_I \otimes \mathbf{T}_J)'\mathbf{y}_3. \tag{2.9}$$

After some algebra we can write

$$\text{vec}\hat{\mathbf{B}}' - \text{vec}\hat{\mathbf{B}}_{(IJ)}' = (\mathbf{X}'\mathbf{X} \otimes \mathbf{T}'\mathbf{T})^{-1} (\mathbf{X}_I \otimes \mathbf{T}_J)' (\mathbf{I} - \mathbf{H}_{IJ})^{-1} \mathbf{r}_{IJ}, \tag{2.10}$$

where $\mathbf{H}_{IJ} = \mathbf{V}_I \otimes \mathbf{Q}_J$, $\mathbf{V}_I = \mathbf{X}_I(\mathbf{X}'\mathbf{X})^{-1}\mathbf{X}_I'$, $\mathbf{Q}_J = \mathbf{T}_J(\mathbf{T}'\mathbf{T})^{-1}\mathbf{T}_J'$ and $\mathbf{r}_{IJ} = \mathbf{y}_3 - (\mathbf{X}_I \otimes \mathbf{T})\text{vec}\hat{\mathbf{B}}'$. The algebra needed here resembles very much derivations in Cook and Weisberg (1980). If some of the eigenvalues of $\mathbf{H}_{IJ}$ are equal to 1, then the inverse in (2.10) does not exist and $\hat{\mathbf{B}}_{(IJ)}$ is not unique.

When $\mathbf{\Sigma}$ is known, we can always consider the transformed model

$$\mathbf{Y\Sigma}^{-1/2} = \mathbf{XBT'\Sigma}^{-1/2} + \mathbf{E\Sigma}^{-1/2}, \tag{2.11}$$

where $\mathbf{\Sigma}^{-1/2}$ is the symmetric square root of $\mathbf{\Sigma}$. Now $cov(\mathbf{\Sigma}^{-1/2}\mathbf{y}_{(i)}) = \mathbf{I}$ for all $i=1, 2, \ldots, n$, and therefore assuming $\mathbf{\Sigma} = \mathbf{I}$ is no restriction. As mentioned before, also in some special growth curve models the estimator $\hat{\mathbf{B}} = (\mathbf{X'X})^{-1}\mathbf{X'YT}(\mathbf{T'T})^{-1}$ is the minimum variance estimator, although $\mathbf{\Sigma} \neq \mathbf{I}$. It proves difficult to derive a convenient formula for the empirical influence function of $\tilde{\mathbf{B}}$. An approximation to the empirical influence function of $\tilde{\mathbf{B}}$ is obtained from (2.10) by substituting $\mathbf{S}/(n-m)$ into (2.11) in place of $\mathbf{\Sigma}$ and using the transformed model (2.11) instead of (1.1).

3. INFLUENTIAL OBSERVATION DIAGNOSTICS

First we develop a measure for detecting influential measurements at the design stage. Suppose now that the covariance matrix of observations is $\mathbf{\Sigma}$. Let us consider the estimation of estimable linear functions of the form $\mathbf{CBD}$, where $\mathbf{C}$ and $\mathbf{D}$ are known matrices of order $g \times m$ and $p \times v$ respectively. This function can be written in vector form as vec $(\mathbf{D'B'C'})$ $= (\mathbf{C}\otimes\mathbf{D'})$ vec $\mathbf{B'}$. Let $V(\mathbf{C\dot{B}D})$ denote the covariance matrix of $(\mathbf{C}\otimes\mathbf{D'})$ vec$\dot{\mathbf{B}}$. It can easily be shown that

$$\begin{aligned} V(\mathbf{C\dot{B}D}) &= (\mathbf{C}\otimes\mathbf{D'})\,[\mathbf{X'X}\otimes\mathbf{T'\Sigma}^{-1}\mathbf{T}]^{-1}\,(\mathbf{C'}\otimes\mathbf{D}) \\ &= \mathbf{C}(\mathbf{X'X})^{-1}\mathbf{C'}\otimes\mathbf{D'}(\mathbf{T'\Sigma}^{-1}\mathbf{T})^{-1}\mathbf{D} \end{aligned} \tag{3.1}$$

and covariance matrix of $\mathbf{C\dot{B}}_{(\mathrm{IJ})}\mathbf{D}$ is

$$\begin{aligned} V(\mathbf{C\dot{B}}_{(\mathrm{IJ})}\mathbf{D}) = (\mathbf{C}\otimes\mathbf{D'})[&\mathbf{X'}_{(\mathrm{I})}\mathbf{X}_{(\mathrm{I})}\otimes\mathbf{T'\Sigma}^{-1}\mathbf{T} \\ &+ \mathbf{X}_{\mathrm{I}}'\mathbf{X}_{\mathrm{I}} \otimes \mathbf{T'}_{(\mathrm{J})}\mathbf{\Sigma}_{(\mathrm{J})}^{-1}\mathbf{T}_{(\mathrm{J})}]^{-1}(\mathbf{C}\otimes\mathbf{D'})', \end{aligned} \tag{3.2}$$

where $\mathbf{\Sigma}_{(\mathrm{J})}$ is obtained from $\mathbf{\Sigma}$ by deleting the rows and columns indexed by I. It is common in the theory of experimental design to discuss the location of observations on the independent variable so that a given regression coefficient, say, has minimal variance. The location of time points in the growth curve model has a bearing on the design of experiments. Discussion about a design stage diagnostic addresses the associated question of how deleting measurements influences the covariance matrix of location parameters.

3.1 A diagnostic of design influence

One possible design criterion is $\mathrm{tr}(\mathbf{C}\dot{\mathbf{B}}\mathbf{D})$. This is a particular case of a linear criterion function and corresponds to the concept of c-optimality (cf. Silvey (1980), p. 12). To identify measurements with large vs. small design influence compare traces of covariance matrices. The design stage diagnostic measure $DS_{(\mathbf{IJ})}$ is defined as

$$DS_{(\mathbf{IJ})}(\mathbf{CBD}) = [\mathrm{tr}\, V(\mathbf{C}\dot{\mathbf{B}}_{(\mathbf{IJ})}\mathbf{D}) - \mathrm{tr}\, V(\mathbf{C}\dot{\mathbf{B}}\mathbf{D})]\,/\mathrm{tr}\, V(\mathbf{C}\dot{\mathbf{B}}\mathbf{D}). \tag{3.3}$$

Note that $\mathbf{CBD} = \{\mathbf{c}_{(i)}'\mathbf{Bd}_j\}$, where $\mathbf{C}' = (\mathbf{c}_{(1)}, \mathbf{c}_{(2)}, \ldots, \mathbf{c}_{(g)})$ and $\mathbf{D} = (\mathbf{d}_1, \mathbf{d}_2, \ldots, \mathbf{d}_v)$. Flexible and simple approach is to calculate influence $DS_{(\mathbf{IJ})}(\mathbf{c}_{(i)}'\mathbf{Bd}_j)$ on every $\mathbf{c}_{(i)}'\mathbf{Bd}_j$ according to the formula (3.3). Ghosh (1983) considered a similar kind of measure in the context of ordinary regression.

If all elements of **CBD** are of equal interest, then the arithmetic mean

$$\overline{DS}_{(\mathbf{IJ})}(\mathbf{CBD}) = \sum_{i=1}^{g} \sum_{j=1}^{v} DS_{(\mathbf{IJ})}(\mathbf{c}_{(i)}'\mathbf{Bd}_j)/(gv) \tag{3.4}$$

would be an appropriate measure.

In practice $\boldsymbol{\Sigma}$ is usually unknown. But if we have before the experiment an estimate of $\boldsymbol{\Sigma}$ from similar earlier experiments, then using such a priori value may yield an useful approximation for $DS_{(\mathbf{IJ})}$. This is the case in our example discussed in Section 4. A simple but crude way of choosing $\boldsymbol{\Sigma}$ is to set $\boldsymbol{\Sigma} = \mathbf{I}$. For certain covariance structures, say for the first order autoregressive disturbances, measure (3.3) can be investigated as a function of one parameter (autocorrelation). Since the choice of the value of $\boldsymbol{\Sigma}$ affects the measure (3.3), it is illustrative to calculate $DS_{(\mathbf{IJ})}$ for many values of $\boldsymbol{\Sigma}$.

3.2 Influence at the inference stage

Perhaps the most popular influence measure in linear regression is the distance measure proposed by Cook (1977). No similar measure can be used straightforwardly in a growth curves model. In order to derive a measure for influence suitable in multivariate situations, consider the following χ^2 statistic:

$$\chi^2 = (\mathbf{G}'\mathrm{vec}\dot{\mathbf{B}}')'\{\mathbf{G}'[\mathbf{X}'\mathbf{X}\otimes\mathbf{T}'\boldsymbol{\Sigma}^{-1}\mathbf{T}]^{-1}\mathbf{G}\}^{-1}(\mathbf{G}'\mathrm{vec}\dot{\mathbf{B}}')$$
$$= \mathrm{tr}\,\{\mathbf{S}_H[\mathbf{D}'(\mathbf{T}'\boldsymbol{\Sigma}^{-1}\mathbf{T})^{-1}\mathbf{D}]^{-1}\}, \tag{3.6}$$

where $\mathbf{G}' = \mathbf{C}\otimes\mathbf{D}'$ and

$$\mathbf{S}_H = (\mathbf{C}\dot{\mathbf{B}}\mathbf{D})'[\mathbf{C}(\mathbf{X}'\mathbf{X})^{-1}\mathbf{C}']^{-1}(\mathbf{C}\dot{\mathbf{B}}\mathbf{D}). \tag{3.7}$$

Substituting $\mathbf{S}/(n-m)$ into (3.6) in place of $\boldsymbol{\Sigma}$ yields the Lawley-Hotelling trace statistic T^2. We will use the Lawley-Hotelling trace statistics as a basis for our influence measure, since this statistics can be easily interpreted as a distance measure and it has a simple connection with the influence at the design stage. Comparing (3.7) with the likelihood method proposed by Khatri (1966) and Rao (1966) reveals that each $\mathbf{S}_H$ is different for both procedures. However, the procedures are asymptotically equivalent. Little information is available about the relative power of each technique for small samples. Here we have chosen the computationally simpler method.

To determine the degree of influence we suggest the measure defined by

$$D_{(\mathrm{IJ})}(\mathbf{CBD}) = \mathrm{tr}\,(\mathbf{S}_H\mathbf{S}_{\mathrm{E}}^{-1}), \tag{3.8}$$

where

$$\mathbf{S} = \mathbf{D}'(\mathbf{T}'\mathbf{S}^{-1}\mathbf{T})^{-1}\mathbf{D} \tag{3.9a}$$

and

$$\mathbf{S}_H = [\mathbf{C}(\tilde{\mathbf{B}}-\tilde{\mathbf{B}}_{(\mathrm{IJ})})\mathbf{D}]'[\mathbf{C}(\mathbf{X}'\mathbf{X})^{-1}\mathbf{C}']^{-1}[\mathbf{C}(\tilde{\mathbf{B}}-\tilde{\mathbf{B}}_{(\mathrm{IJ})})\mathbf{D}], \tag{3.9b}$$

where $\tilde{\mathbf{B}}-\tilde{\mathbf{B}}_{(\mathrm{IJ})}$ is an approximation obtained from (2.10) by using $\mathbf{S}/(n-m)$ in place of $\boldsymbol{\Sigma}$ in model (2.11). The magnitude of $(n-m)D_{(\mathrm{IJ})}(\mathbf{CBD})$ may be assessed by comparing it to the probability points of the corresponding Lawley-Hotelling statistics.

The $1-a$ confidence set for $\mathbf{B}$ is given by the set of all matrices $\mathbf{B}$ satisfying

$$(n-m)\,\mathrm{tr}(\mathbf{S}_H\mathbf{S}_E^{-1}) \le T_a^2, \tag{3.10}$$

where T_a^2 is the upper $1-a$ percentage point of the corresponding Lawley-Hotelling distribution. The comparison of $D_{(\mathrm{IJ})}(\mathbf{B})$ with probability points T_a^2 is not a test of significance, but only a monotonic transformation of $\tilde{\mathbf{B}}-\tilde{\mathbf{B}}_{(\mathrm{IJ})}$ to a more familiar scale.

Let $\mathbf{c}$ be an $m\times 1$ vector and $\mathbf{d}$ a $p\times 1$ vector. It follows from (3.9) that

$$D_{(\mathrm{IJ})}(\mathbf{c}'\mathbf{Bd}) = [\mathbf{c}'(\tilde{\mathbf{B}}-\tilde{\mathbf{B}}_{(\mathrm{IJ})})\mathbf{d}]^2/[\mathbf{c}'(\mathbf{X}'\mathbf{X})^{-1}\mathbf{c}\cdot\mathbf{d}'(\mathbf{T}'\mathbf{S}^{-1}\mathbf{T})^{-1}\mathbf{d}]. \tag{3.11}$$

When $\boldsymbol{\Sigma}$ is known, (2.10) yields $E[\mathbf{c}'(\dot{\mathbf{B}} - \dot{\mathbf{B}}_{(\mathbf{IJ})})\mathbf{d}]^2 = V(\mathbf{c}'\dot{\mathbf{B}}_{(\mathbf{IJ})}\mathbf{d}) - V(\mathbf{c}'\dot{\mathbf{B}}\mathbf{d})$. Therefore we have

$$D_{(\mathrm{IJ})}(\mathbf{c}'\mathbf{B}\mathbf{d}) = DS_{(\mathrm{IJ})}(\mathbf{c}'\mathbf{B}\mathbf{d}) \cdot \frac{[\mathbf{c}'(\mathbf{B} - \mathbf{B}_{(\mathrm{IJ})}\mathbf{d}]^2}{V[\mathbf{c}'(\mathbf{B} - \mathbf{B}_{(\mathrm{IJ})}\mathbf{d}]}, \tag{3.12}$$

which implies $E[D_{(\mathbf{IJ})}(\mathbf{c}'\mathbf{B}\mathbf{d})] = DS_{(\mathbf{IJ})}(\mathbf{c}'\mathbf{B}\mathbf{d})$. Clearly $DS_{(\mathbf{IJ})}(\mathbf{c}'\mathbf{B}\mathbf{d})$ measures the relative sensitivity of the estimate, $\mathbf{c}'\dot{\mathbf{B}}\mathbf{d}$, to potential influential measurements $\mathbf{Y}_{\mathbf{IJ}}$. A small value of $DS_{(\mathbf{IJ})}$ indicates that the associated 'design point' has not much weight in the determination of $\mathbf{c}'\dot{\mathbf{B}}\mathbf{d}$. The two factors combine in (3.12) to produce a measure of the overall impact of the measurements $\mathbf{Y}_{\mathbf{IJ}}$.

If the influence on single elements, β_{ij}, of $\mathbf{B}$ is of the special interest, we compare the influence matrices

$$\mathbf{MD}_{(\mathbf{IJ})} = \{D_{(\mathbf{IJ})}(\beta_{ij})\}_{m\times p} \quad \text{and} \quad \mathbf{MDS}_{(\mathbf{IJ})} = \{DS_{(\mathbf{IJ})}(\beta_{ij})\}_{m\times p}, \tag{3.13}$$

where the influence on every element of $\mathbf{B}$ separately can be seen. When all elements of $\mathbf{B}$ are of equal interest, the average measures

$$\overline{DS}_{(\mathrm{IJ})} = \sum_{i=1}^{m}\sum_{j=1}^{q} DS_{(\mathrm{IJ})}(\beta_{ij})/mq \quad \text{and} \quad \overline{D}_{(\mathrm{IJ})} = \sum_{i=1}^{m}\sum_{j=1}^{q} D_{(\mathrm{IJ})}(\beta_{ij})/mq \tag{3.14}$$

prove useful. It follows from (3.12) that $E[\overline{D}_{(\mathbf{IJ})}] = \overline{DS}_{(\mathbf{IJ})}$. Although the above relations have been derived under the assumption that $\boldsymbol{\Sigma}$ is known, these intuitively meaningful forms can serve as a rule of thumb also when $\boldsymbol{\Sigma}$ is unknown. In any case, at the design stage there is available only a priori guess of $\boldsymbol{\Sigma}$ or an estimate of $\boldsymbol{\Sigma}$ stochastically independent of $\mathbf{Y}$. Therefore, before the experiment, inference is conditional and depends upon the value $\boldsymbol{\Sigma} = \boldsymbol{\Sigma}_0$ we have chosen.

In an influence analysis we calculate first appropriate values of a design measure for a given $\boldsymbol{\Sigma}$ (or for many values of $\boldsymbol{\Sigma}$). After running the experiment we may recalculate the design measure with estimated $\boldsymbol{\Sigma}$ and compare all these results with the values of an inference measure. The importance of such an analysis is in doing better current inference and also in future planning of similar experiments.

4. AN EXAMPLE

Now we investigate the influence of deleting measurements at different time-points, when data on 208 bulls born in 1966 are under consideration. The data were described in the introduction. A polynomial of third degree was fitted to them. In Figure 1 the values of the design measure $\overline{DS}_{(IJ)}$ for different ages are given.

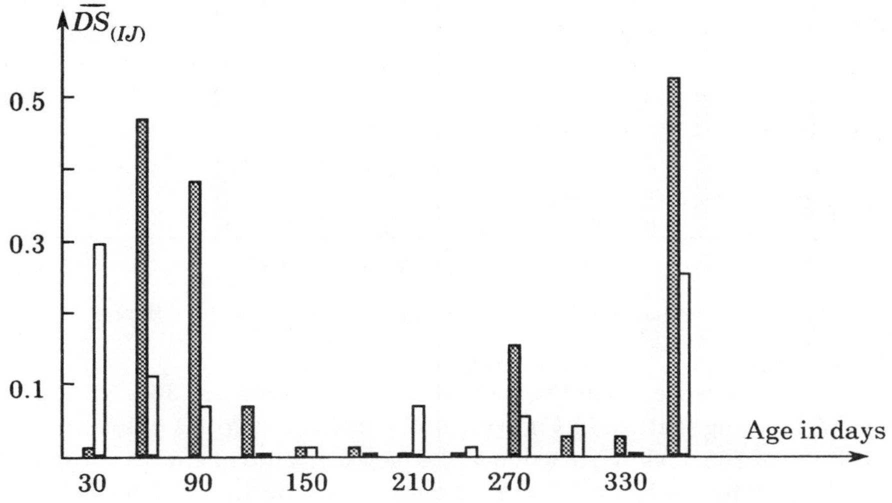

Figure 1. The values of the design measure $\overline{DS}_{(IJ)}$ for different time-points when the measurements at the corresponding point are deleted. Data concern bulls born in 1966. As covariance matrix was used the estimate of Σ (= $\hat{\Sigma}_{65}$) for bulls born in 1965 (▒) and the estimate from the year 1966 (□).

The most influential time-points at the design stage are 365, 60 and 90 days of age, when $\hat{\Sigma}_{1965}$ is used in place of Σ. Correspondingly the least influential points are 210, 30, 150, 180 and 240 days of age, which are approximatively equally uninfluential.

The covariance matrix $\hat{\Sigma}_{66}$ gives the least influential points 180, 120 and 240 days of age. A chosen covariance matrix has a considerable effect on values of the design stage measure. In our data the two different covariance matrices $\hat{\Sigma}_{65}$ and $\hat{\Sigma}_{66}$ lead totally different results only at the age of 30 days.

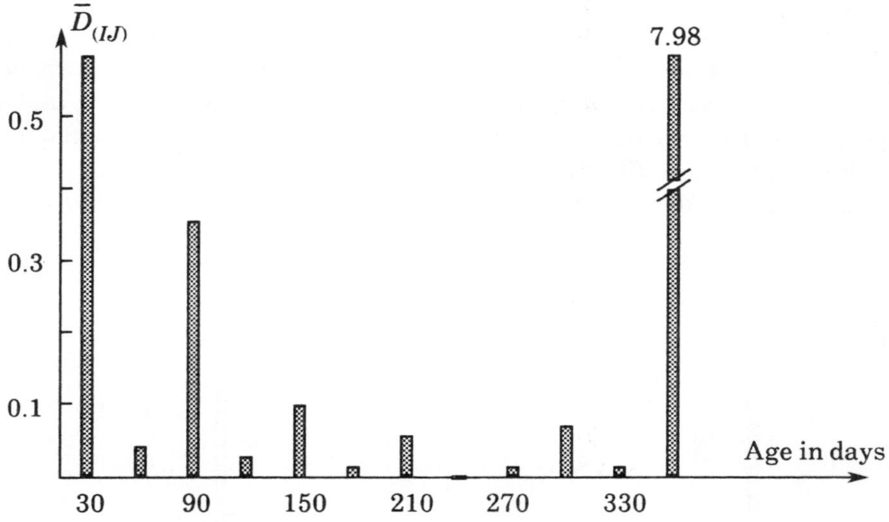

Figure 2. The values of the inference stage measure $\bar{D}_{(IJ)}$ for different time-points.

At the inference stage the most influential points are 365, 30 and 90 days. A striking feature in Figure 2 is the extremely great value of $\bar{D}_{(IJ)}$ at the age of 365 days. This is due to the fact that growth curves estimated in the age interval [30, 330] are no more adequate in the interval (330, 365]. The least influential points are 240, 180, 270 and 330 days of age. It is now interesting to compare the values of the design and inference stage measures. The age of 30 days is influential at the inference stage for same reasons as the age 365 days. However, the age 30 days was not influential at the design stage and therefore the inference stage influence at this point is not great compared with influence at the point 365 days. Examining the influence matrices (3.13) reveals that the design stage influence is regularly greatest on the third degree coefficients of **B**, whereas at the inference stage no such simple influence pattern can be observed.

The main problem in this application was as follows: which points to delete if we must? An ideal point to delete is one, which has little influence both at design and inference stages. Such a point in our data is 240 days, which has the least inference influence. Another candidate is the age 180 days, which has the least design influence when using covariance matrix $\hat{\Sigma}_{66}$. A given set of observations may be influential at the inference stage but not at the design stage, and vice versa. This is no contradiction, however. The measure $\bar{D}_{(IJ)}$ measures the discrepancy between the growth curve coefficients estimated from complete data and the estimate $\tilde{\mathbf{B}}_{(IJ)}$ from incomplete data. The measure $\overline{DS}_{(IJ)}$ is the expectation of $\bar{D}_{(IJ)}$, when $\mathbf{\Sigma}$ is known. When using an estimate in place of $\mathbf{\Sigma}$, $\overline{DS}_{(IJ)}$ is only an estimate of the expectation of $\bar{D}_{(IJ)}$. Thus $\overline{DS}_{(IJ)}$ can also be viewed as a measure of robustness of a design.

ACKNOWLEDGEMENTS

This work was initiated when I was a Senior Researcher in the Academy of Finland, and it was completed while I was working as a Humboldt Research Fellow at the University of Dortmund. I am greatly indebted to Professor G. Trenkler for his hospitality during my stay in Dortmund. My thanks go also to Professor Matti Ojala for making available the data analysed in this paper, and to Mr. Tapio Nummi for computational assistance.

REFERENCES

Belsley, D. A., Kuh, E. and Welsch, R.E. (1980). *Regression Diagnostics: Identifying Influential Data and Sources of Collinearity.* John Wiley, New York.

Cook, R. D. (1977). Detection of influential observations in linear regression. *Technometrics*, 19, 15-18.

Cook, R.D. and Weisberg, S. (1980). Characterizations of an empirical influence function for detecting influential cases in regression. *Technometrics*, 22, 495-508.

Cook, R.D. and Weisberg, S. (1982). *Residuals and Influence in Regression.* Chapman and Hall, New York.

Ghosh, S. (1983). Influential observations in view of design and inference. *Communications in Statistics A - Theory and Methods*, 12, 1675-1683.

Khatri, C.G. (1966). A note on a MANOVA model applied to problems in growth curves. *Annals of the Institute of Statistical Mathematics*, 18, 75-86.

Kleinbaum, D.G. (1973). A generalization of the growth curve model which allows missing data. *Journal of Multivariate Analysis*, 3, 117-124.

Liski, E.P. (1987). A growth curve analysis for bulls tested at station. *Biometrical Journal*, 29, 331-343.

Lindström, U. and Maijala, K. (1970). Evaluation of performance test results for A.I. bulls. *Acta Agriculturae Scandinavica*, 10, 207-217.

Polasek, W. (1984). Regression diagnostics for general linear regression models. *Journal of the American Statistical Association*, 79, 336-340.

Potthoff, R.F. and Roy, S.N. (1964). A generalized multivariate analysis of variance model useful especially for growth curve problems. *Biometrika*, 51, 313-326.

Rao, C.R. (1965). The theory of least squares when the parameters are stochastic and its application to the analysis of growth curves. *Biometrika*, 52, 447-458.

Rao, C.R. (1966). Covariance adjustment and related problems in multivariate analysis. *Multivariate Analysis I* (P.R. Krishnaiah, *ed.*). Academic Press, New York, 87-103.

Rao, C.R. (1967). Least squares theory using an estimated dispersion matrix and its application to measurement of signals. *Proceedings of the Fifth Berkeley Symposium on Mathematical Statistics and Probability*, 1, 355-372.

Silvey, S.D. (1980). *Optimal Design.* Chapman and Hall, London.

Received 1 December 1986
Revised 3 September 1987

Department of Mathematical Sciences
University of Tampere
P.O. Box 607
SF - 33101 Tampere
Finland

Proc. Second International Tampere Conference in Statistics
(Tampere, Finland, 1-4 June 1987)
Tarmo Pukkila and Simo Puntanen, *Editors*

pp. 567 - 573

A randomization alternative to the Bonferroni inequality with multiple F tests

Bryan F.J. MANLY and Lynn McALEVEY
University of Otago, Dunedin, New Zealand

Key words and phrases: Significance levels, analysis of variance, multivariate analysis, comparison of means.

ABSTRACT

An alternative to a multivariate analysis of variance for comparing the mean values of several variables for samples from several populations involves carrying out univariate F tests on each of the variables, using a significance level designed to control the overall probability of a type I error. Here a randomization procedure for determining the significance level and carrying out the F tests is discussed. This procedure takes account of the correlations between variables and accommodates missing values and non-normal data without difficulty. Two applications are discussed. Simulation results are presented to show how the significance levels determined for individual F tests depend upon the number of tests and the correlation between variables.

1. INTRODUCTION

The comparison of the means of samples from several multivariate populations is often complicated by missing observations and non-normal distributions of variables. The first of these problems complicates the use of standard multivariate tests since (unless missing values are replaced with estimates) a missing value on one variable will mean that all the other available measurements for the individual concerned will have to be excluded from the analysis. The second problem makes the use of most parametric univariate and multivariate tests somewhat suspect, and indicates that a randomization method for testing for significant mean differences should be used (Edgington, 1980, Chapter 1; Romesburg, 1985).

Another matter that often has to be considered is the need to know which particular variables, if any, show significant mean differences between samples. In medical research, for example, the knowledge that there is a significant difference between two groups of patients on a battery of tests is often not useful if it is not possible to pinpoint which tests are causing the significance. For this reason, a common practice is to test each variable separately using a parametric or nonparametric test with a significance level that is chosen to ensure that the overall probability of a type I error is realistic. The simplest approach is to use Bonferroni's inequality, which says that if m tests are all made at the $(100\alpha/m)\%$ level, then the probability of any significant results by chance alone is less than α.

In practice Bonferroni's inequality may produce a type I error probability that is much lower than α since it makes no allowance for the correlation between variables. Tests will then lack power and significance may not even be possible in some instances (O'Brien, 1984). In fact, the appropriate significance level to use for m individual tests must lie somewhere between the Bonferroni value of $(100\alpha/m)\%$, which is correct for uncorrelated variables, and $100\alpha\%$, which is correct for perfectly correlated variables.

In a recent paper (Manly *et al.*, 1986) we proposed a randomization method that overcomes the three problems of missing values, non-normal data, and multiple testing without using Bonferroni's inequality, for the case of comparing means of two groups on several variables. We also briefly discussed in that paper the extension of the procedure for the comparison of several groups. Here we enlarge on the use of the procedure with more than two groups. We also give the results of a simulation study which indicates how the appropriate level of significance for m tests depends upon the correlation between variables.

Of course, if missing values and non-normality are not problems, and other necessary assumptions can be made, then there is no need to resort to randomization to test the significance of individual variables. Seber (1984, p. 437) reviews methods based upon the normal distribution that are available in that case.

2. A RANDOMIZATION PROCEDURE

Suppose that there are s samples, with sizes $n_1, n_2, \ldots, n_s$ and m variables. The proposed randomization procedure then involves the following steps:

(a) The $n = \Sigma n_i$ cases are allocated at random to the s samples so that sample i receives n_i cases. An analysis of variance is carried out separately on each variable to get a set of values $\mathbf{F} = (F_1, F_2, \ldots, F_m)$. This process is repeated a large number, R, times to generate a good approximation to the multivariate randomization distribution of $\mathbf{F}$.

(b) For each variable, each generated F_i value is replaced by its randomization significance level (the proportion of F values for that variable that are as large or larger than F_i).

(c) The most significant (i.e. smallest) significance level is found for each of the R randomized sets of data. These are ranked in order from the smallest to the largest in a list $(V_1, V_2, \ldots, V_R)$.

The value of V that is $100\alpha\%$ from the start of the list generated in (c) is the appropriate level of significance to use for randomization F tests on each of the m variables if it is desired that the probability of finding any result significant by chance alone should be approximately α.

See our earlier paper (Manly *et al.*, 1986) for a more detailed description of this procedure for two groups, using t tests rather than F tests. This earlier paper also explains why a minimum of $100m$ randomizations should be made for data with m variables.

We are not aware of any previous description of this randomization procedure. However, it is fairly obvious approach to multiple testing so we accept that it may have been used before. If this is the case then the versatility of the approach makes it worthy of further discussion.

3. EXAMPLES

Two examples indicate how the proposed randomization procedure works in practice. The first is the two sample, ten variable example discussed by Manly *et al.* (1986). Here one sample consisted of 21 patients with multiple sclerosis and the other consisted of 22 normal individuals.

The ten variables were scores on semi-automated computer tests. Missing values and non-normal distributions were features of the data. Bonferroni's inequality suggests that using a significance level of (5/10)% = 0.5% with each individual variable will produce a test with an overall significance level of about 5%. The randomization procedure shows that in fact 0.7% is a more appropriate level for the individual tests to obtain the overall 5% level. However, using either the 0.5 % or the 0.7 % level gives the same outcome in terms of which variables have significantly different means.

The second example involves 14 samples and 16 variables. The samples are obsidian artefacts from different sources in New Zealand and the Pacific. There were 55 specimens in total, so sample sizes are small, with an average of only about four specimens per source. The variables are concentration values for 16 elements determined by neutron activation analysis. Missing values and non-normal distributions of values for specimens from the same source are usual problems with data obtained in this way. The data being considered are part of a larger set of element concentration values discussed by Leach and Manly (1982).

Table 1 shows a summary of the analysis of the data. Randomization testing using the Bonferroni inequality suggests that for an overall test at the 5% level of significance the observed F ratios for the 16 variables (column 2 of the table) should be compared to the (5/16)% = 0.31% critical values of their respective randomization distributions (column 4 of the table). Thirteen of the 16 observed F ratios are significant. On the other hand, the randomization method proposed in Section 2 indicates that the appropriate level of significance for individual F tests is 0.45%. This gives the critical values shown in column 3 of the table that are slightly lower than the Bonferroni levels. However, the significant variables are the same using the Bonferroni critical values and the proposed alternative critical values. The very large critical values for variables 9 and 12 are an indication of their extremely non-normal distributions.

The two examples indicate that our proposed randomization method may not give results that are much different from what is obtained using the Bonferroni inequality. This raises the question of how large the correlations have to be between variables in order to produce substantial differences. For both examples the correlations between variables covered a wide range with some values between 0.8 and 0.9, and many values close to zero. It appears to be the case that a few fairly large correlations do not have the effect of making large differences between the Bonferroni and the alternative critical values.

TABLE 1

Two randomization methods for determing critical values for an overall 5% test applied to data on 16 variables measured on 55 specimens of obsidian from 14 sources. Columns 3 and 4 were determined by randomizing the data 2000 times.

Element	Observed F value	Proposed alternative method critical value (0.45 %)	Bonferroni critical value (0.31 %)
1	137.89	2.88	2.94
2	11.13	2.88	2.97
3	326.83	2.98	3.16
4	2.75	3.38	3.68
5	10.07	2.76	2.84
6	31.02	2.89	3.12
7	13.18	2.95	3.09
8	46.61	3.21	3.72
9	2.11	45.30	49.03
10	8.59	2.76	2.89
11	8.28	3.17	3.39
12	1.83	11.87	12.51
13	189.33	2.83	2.94
14	123.72	2.85	2.95
15	94.48	3.12	3.20
16	24.67	3.09	3.45

4. SIMULATIONS

The examples discussed in the previous section suggest that for many sets of data the Bonferroni inequality will give a good approximation to the significance level that is obtained by our proposed procedure. However, as was mentioned in Section 1, it will give conservative significance levels for highly correlated variables. We have therefore carried out a small simulation study to determine how the level of significance found by our procedure depends upon the correlation between variables, and other factors.

A factorial experimental design was used with the factors being (1) the number of variables (2,5 or 10); (2) the number of samples (2 or 5); (3) the sample size (5 or 20); and (4) the correlation between variables (0.1,

0.5, 0.9 or 0.95). Five replicate results were obtained for each combination of factor levels.

Multivariate normally distributed data with mean values of zero were generated on a computer and arbitrarily divided into samples to produce artificial data. The randomization procedure described above, with 1000 randomizations, was then applied in order to determine the level of significance required for tests on individual variables in order to fix the overall probability of a type I error at 0.05.

Analyses of variance have been carried out on the derived significance levels, using a reciprocal transformation to stabilize variances. A separate analysis was done for the number of variables equal to 2, 5 and 10. For 2 and 10 variables the level of correlation had an effect that was significant at the 0.1 % level. For 5 variables the group size had a highly significant effect as well as the correlation. No other effects were significant. Table 2 shows the results obtained after averaging over the non-significant factor of the number of samples. It is clear that the correlation between variables has to be rather high before the Bonferroni approximation becomes very conservative, particularly with only a few variables.

TABLE 2

Significance levels appropriate for tests on individual variables in order to give a 5 % level of significance for all variables taken together. The values given in the body of the table for correlations of 0.1, 0.5, 0.9 and 0.95 are average values obtained by doing 1000 randomizations on each of five replicate sets of data with two samples, and each of five replicate sets of data with five samples. The values for correlations of 0.0 are what is expected from the Bonferroni inequality. With a correlation of 1.0, individual tests should be given the same 5 % significance level.

Number of variables	Number of samples	Sample sizes	Correlation between variables 0.0	0.1	0.5	0.9	0.95	1.0
2	2 or 5	5	2.5	2.5	2.7	3.1	3.9	5.0
		20	2.5	2.5	2.7	3.5	3.7	5.0
5	2 or 5	5	1.0	1.1	1.1	2.0	2.5	5.0
		20	1.0	1.1	1.2	2.3	2.7	5.0
10	2 or 5	5	0.5	0.6	0.7	1.6	2.2	5.0
		20	0.5	0.6	0.7	1.6	2.3	5.0

5. DISCUSSION

It must be stressed that the main advantage of using the randomization procedure described in Section 2 is that it can accommodate missing values and non-normal data. Part (c) of the procedure, for determining the significance level to use for individual F tests, is simple to carry out when the other parts of the procedure have been done. From this point of view, the fact that the Bonferroni inequality gives a good approximation to this significance level for variables with low or moderate correlations is not particularly relevant.

If it is considered desirable, the proposed multiple testing procedure can be augmented by carrying out an overall test for a difference between the means of all the sample being compared. The statistic E proposed by Romesburg (1985) is very suitable for this purpose.

REFERENCES

Edgington, E.S. (1980). *Randomization Tests.* Marcel Dekker, New York.

Leach, F. and Manly, B.F.J. (1982). Minimum Mahalanobis distance functions and lithic source characterisation by multi-element analysis. *New Zealand Journal of Archaeology*, 4, 77-109.

Manly, B.F.J., McAlevey, L. and Stevens, D. (1986). A randomization procedure for comparing group means on multiple measurements. *British Journal of Mathematical and Statistical Psychology*, 39, 183-189.

O'Brien, P.C. (1984). Procedures for comparing samples with multiple end points. *Biometrics*, 40, 1079-1087.

Romesburg, H.C. (1985). Exploring, confirming, and randomization tests. *Computers and Geosciences*, 22, 19-37.

Seber, G.A.F. (1984). *Multivariate Observations.* Wiley, New York.

Received 5 March 1987

Department of Mathematics and Statistics
University of Otago
P.O. Box 56, Dunedin
New Zealand

Department of Quantitative and Computer Studies
University of Otago
P.O. Box 56, Dunedin
New Zealand

Proc. Second International Tampere Conference in Statistics
(Tampere, Finland, 1-4 June 1987)
Tarmo Pukkila and Simo Puntanen, *Editors*

pp. 575 - 584

Experiments in incomplete split-plot designs

Stanislaw MEJZA
Academy of Agriculture, Poznań, Poland

Key words and phrases: Model building, block design, analysis of variance.

ABSTRACT

Some problems connected with incomplete split-plot design are considered. The designs considered in this paper can be incomplete with regard to the wholeplot treatments or with regard to subplot treatments, or with regard to both. The starting points of paper are considerations connected with the procedure of randomization in design and model building for the analysis. In the next part of the paper an analysis of the linear mixed model obtained before is given. In the analysis of the linear model the approach given by Nelder is adopted.

1. INTRODUCTION

The split-plot design is very often used in agricultural experiments. But there may be situations in which, due to the structure of experimental materials, it is impossible to plan an experiment in the classical complete split-plot design. In such cases it may be desirable to plan an experiment using some incomplete design based on the split-plot structure. But then we have some new problems resulting from incompleteness. The designs considered in this paper can be incomplete with regard to the wholeplot treatments or with regard to the subplot treatments, or with regard to both.

We start by considering the procedure of randomization in an incomplete split-plot design and model building for the analysis. The mixed model we obtain is based on two assumptions. Firstly, we assume that the observed yield is a sum of three components: a "zero yield" connected with experimental units, a pure effect of treatment combinations and "technical errors" connected with measurements. Secondly, we assume that additivity holds among these components.

The model building will be based on the ideas given by Neyman *et al.* (1935), Wilk and Kempthorne (1956); see also Kempthorne (1952), Kempthorne *et al.* (1961).

In the next part of the paper we give an analysis of the linear mixed model obtained. The analysis is based on the approach given by Nelder for multistrata designed experiments.

2. MODEL BUILDING

Let us consider a set of experimental units with a nested structure, i.e., let the experimental material be divided into b blocks and let each block be divided into k_1 wholeplots, and let each wholeplot be divided into k_2 subplots. It is assumed that the following relations hold: $k_1 \leq S$, $k_2 \leq T$, where S denotes the number of levels of factor A and T denotes the number of levels of factor B. If the equality holds then the resulting design will be a complete split-plot design.

From the practical point of view the most interesting case is that of $k_1 < S$ or $k_2 < T$ or both $k_1 < S$ and $k_2 < T$. In the last case we must decide which levels of factor A and which levels of factor B will occur on wholeplots and on the subplots in each block, i.e., we must design an experiment.

Let Ω be a (theoretical) plan of our experiment. It will be assumed that the plan Ω is chosen not at random but it is so chosen that all suggestions of experimenter are satisfied. This means that the experimenter can decide which comparisons are the most interesting and they should be estimated with full efficiency. The loss in efficiency can be suffered by less interesting comparisons or in comparisons for which the significance is evident before the experiment.

Let us consider the components of observed yield. In the paper it will be assumed that the observed yield is a sum of so called "zero yield" (conceptual response), a pure effect due to the treatment combination and a "technical error".

Note that every unit possesses some kind of fertility which gives some yield both in the case when treatments do not occur on a unit and in the case in which no treatments have an effect on yields. This yield will be called the zero yield.

By the pure effect due to the treatment combination we mean the increase (or decrease) in the zero yield due to the treatment combination used on an experimental unit. Usually the sum of zero yield and the pure effect due to treatment is called the pure yield and it is often the base of statistical analysis. There are many experiments in which the pure yield is changed by measurements. The increase (or decrease) in the pure yield by measurements is called the technical error. Assuming that additivity holds among the zero yield, the pure effect due to the treatment combination and the technical error, we can regard the observed yield (in short, yield) as a sum of the above three components.

Let A_h denote the h-th level of factor A and B_j denote the j-th level of factor B. Then the combination A_hB_j $(h = 1, 2,..., S,\ j = 1, 2,..., T)$ will be called the treatment combination. The levels of factors A and B will be arranged on our experimental material in the following way. From the experimental material we choose at random, one block for r-th block of plan Ω. According to the chosen plan Ω we know which levels of factor A occur in r-th block. Hence in the block obtained we choose randomly a wholeplot for A_h. Likewise, we choose a subplot for B_j for the wholeplot obtained. All of the random choices are independent and they are repeated for all r in Ω, for all h within r-th block and for all j within A_h.

Let m_{wst} denote zero yield of unit (w, s, t), where $w = 1, 2,..., b$, $s = 1, 2,..., k_1$, $t = 1, 2,..., k_2$. Then the zero yield, i.e., the yield disregarding the influence of the treatment combination, can be written as

$$Y_{rhj} = \sum_w l_{rw} \sum_s f_{hs}{}^{w} \sum_t g_{jt}{}^{ws}\, m_{wst}, \tag{1}$$

where random variables $l_{rw}, f_{hs}{}^{w}, g_{jt}{}^{ws}$ are such that

$$l_{rw} = \begin{cases} 1, & \text{if } w\text{-th experimental block is assigned to the } r\text{-th block of plan } \Omega, \\ 0, & \text{otherwise,} \end{cases}$$

$$f_{hs}{}^{w} = \begin{cases} 1, & \text{if } s\text{-th wholeplot of } w\text{-th block is assigned to } A_h, \\ 0, & \text{otherwise,} \end{cases}$$

$$g_{jt}{}^{ws} = \begin{cases} 1, & \text{if } t\text{-th subplot of } s\text{-th wholeplot in } w\text{-th block is assigned to } B_j, \\ 0, & \text{otherwise.} \end{cases}$$

After some operations, which are due to Kempthorne and were adapted to the design here considered by Mejza (1986), the linear model for zero yield has the form

$$Y_{rhj} = \mu + \rho_r + \eta_{rh} + \varepsilon_{rhj}, \tag{2}$$

$r = 1, 2,..., b$, $h = 1, 2,..., S$, $j = 1, 2,..., T$, where μ is a general mean of the experiment, ρ_r, η_{rh}, ε_{rhj} are random variables which represent: ρ_r – the effect of the r-th block, η_{rh} – the effect of the h-th wholeplot in r-th block, ε_{rhj} – the effect of the j-th subplot within h-th wholeplot and within r-th block.

Some of the statistical properties of these effects are as follows: $E(\rho_r) = E(\eta_{rh}) = E(\varepsilon_{rhj}) = 0$,

$$\mathrm{Cov}(\rho_r, \rho_{r'}) = \begin{cases} (b-1)b^{-1}\sigma_\rho^2, & r = r', \\ (-1/b)\,\sigma_\rho^2, & r \neq r', \end{cases} \tag{3}$$

$$\mathrm{Cov}(\eta_{rh}, \eta_{r'h'}) = \begin{cases} (k_1-1)k_1^{-1}\sigma_\eta^2, & r = r', h = h', \\ (-1/k_1)\,\sigma_\eta^2, & r = r', h \neq h', \\ 0, & \text{otherwise}, \end{cases} \tag{4}$$

$$\mathrm{Cov}(\varepsilon_{rhj}, \varepsilon_{r'h'j'}) = \begin{cases} (k_2-1)k_2^{-1}\sigma_\varepsilon^2, & r = r', h = h', j = j', \\ (-1/k_2)\,\sigma_\varepsilon^2, & r = r', h = h', j \neq j', \\ 0, & \text{otherwise}, \end{cases} \tag{5}$$

where σ_ρ^2, σ_η^2, σ_ε^2 denote the variances of ρ_r, η_{rh}, ε_{rhj} respectively. Moreover, random variables ρ_r, η_{rh}, ε_{rhj} are mutually independent.

Let μ_{hj} denote the mean effect due to treatment combination A_hB_j and let $\tau_{hj}^* = \mu_{hj} - \mu$ denote the pure effect due to the above treatment combination. Then we have equality $\mu_{hj} - \mu = (\mu_{h\cdot} - \mu) + (\mu_{\cdot j} - \mu) + (\mu_{hj} - \mu_{h\cdot} - \mu_{\cdot j} + \mu)$ or in a new notation

$$\tau_{hj}^* = a_h + \beta_j + (a\beta)_{hj}, \tag{6}$$

where a_h is the effect of the h-th level of A, β_j is the effect of the j-th level of B and $(a\beta)_{hj}$ is the effect of the interaction of the h-th level of A and the j-th level of B.

In the paper it is assumed that additivity holds between treatment combination A_hB_j and the experimental unit assigned to A_hB_j. Hence, the linear model of pure yield is of the form $Y_{rhj}^* = Y_{rhj} + \tau_{hj}^*$.

Let e_{rhj} denote the random variable connected with measurements made on unit (r, h, j). It will be assumed that $E(e_{rhj}) = 0$ and

$$\mathrm{Cov}(e_{rhj}, e_{r'h'j'}) = \begin{cases} \sigma_e^2, & \text{if } r = r', h = h', j = j', \\ 0, & \text{otherwise.} \end{cases}$$

Summarizing, the observed yield (yield) obtained on the (r, h, j) -th unit has the linear model

$$y_{rhj} = \mu + \rho_r + a_h + \eta_{rh} + \beta_j + (a\beta)_{hj} + \varepsilon_{rhj} + e_{rhj}, \qquad (7)$$
$$r = 1, 2,\ldots, b,\ h = 1, 2,\ldots, S,\ j = 1, 2,\ldots, T,$$

with $E(y_{rhj}) = \mu + \tau_{hj}^*$ and variance-covariance structure as follows:

$\mathrm{Cov}(y_{rhj}, y_{r'h'j'}) =$

$$\begin{cases} (b-1)b^{-1}\sigma_\rho^2 + (k_1-1)k_1^{-1}\sigma_\eta^2 + (k_2-1)k_2^{-1}\sigma_\varepsilon^2 + \sigma_e^2, & r = r', h = h', j = j', \\ (b-1)b^{-1}\sigma_\rho^2 + (k_1-1)k_1^{-1}\sigma_\eta^2 - (1/k_2)\sigma_\varepsilon^2, & r = r', h = h', j \neq j', \\ (b-1)b^{-1}\sigma_\rho^2 - (1/k_1)\sigma_\eta^2, & r = r', h \neq h', \\ (-1/b)\,\sigma_\rho^2, & r \neq r'. \end{cases} \qquad (8)$$

Let $\tau_i (= \tau_{hj}^*) = a_h + \beta_j + (a\beta)_{hj}$ denote the effect of i-th treatment (treatment combination) with $i = (h-1)T + j$, $h = 1, 2,\ldots, S$, $j = 1, 2,\ldots, T$.

Matrix notation will now be used. Hence, model (7) in matrix notation has a form

$$\mathbf{y} = \mathbf{1}_n\mu + \mathbf{D}'\boldsymbol{\rho} + \boldsymbol{\Delta}'\boldsymbol{\tau} + \mathbf{G}'\boldsymbol{\eta} + \boldsymbol{\varepsilon} + \mathbf{e}, \qquad (9)$$

where $\mathbf{y}$ is a vector of lexicographically ordered observations (with respect to blocks, wholeplots, subplots), $\mathbf{1}_n$ is an $n\times 1$ vector of ones, $n = bk_1k_2$, $\mathbf{D}'$ is an $n\times b$ design matrix for blocks, $\boldsymbol{\rho}$ is a $b\times 1$ vector of block effects, $\boldsymbol{\Delta}'$ is an $n\times v$ design matrix for treatments, $\boldsymbol{\tau}$ is a $v\times 1$ vector of treatment effects, $\mathbf{G}'$ is an $n\times bk_1$ design matrix for wholeplots, $\boldsymbol{\eta}$ is a $bk_1\times 1$ vector of wholeplot errors, $\boldsymbol{\varepsilon}$ is an $n\times 1$ vector of subplot errors, $\mathbf{e}$ is an $n\times 1$ vector of technical errors.

The covariance structure of (8) in matrix notation has a form

$$\mathrm{Cov}(\mathbf{y}) = \mathbf{V} = \mathbf{D}'\mathbf{V}_\rho\mathbf{D} + \mathbf{G}'\mathbf{V}_\eta\mathbf{G} + \mathbf{V}_\varepsilon + \sigma_e^2\mathbf{I}_n, \qquad (10)$$

where $\mathbf{V}_\rho = \mathrm{Cov}(\boldsymbol{\rho})$, $\mathbf{V}_\eta = \mathrm{Cov}(\boldsymbol{\eta})$, $\mathbf{V}_\varepsilon = \mathrm{Cov}(\boldsymbol{\varepsilon})$, $E(\boldsymbol{\rho}\boldsymbol{\eta}') = \mathbf{0}$, $E(\boldsymbol{\rho}\boldsymbol{\varepsilon}') = E(\boldsymbol{\rho}\mathbf{e}') = \mathbf{0}$, $E(\boldsymbol{\eta}\boldsymbol{\varepsilon}') = E(\boldsymbol{\eta}\mathbf{e}') = \mathbf{0}$, $E(\boldsymbol{\varepsilon}\mathbf{e}') = \mathbf{0}$, $\mathbf{I}_n$ denotes identity matrix.

According to our lexicographic ordering we have $\mathbf{D}' = \mathbf{I}_b \otimes \mathbf{1}_{k_1} \otimes \mathbf{1}_{k_2}$, $\mathbf{G}' = \mathbf{I}_b \otimes \mathbf{I}_{k_1} \otimes \mathbf{1}_{k_2}$. Note that $\mathbf{V}_\rho = \sigma_\rho^2(\mathbf{I}_b - b^{-1}\mathbf{J}_b)$, $\mathbf{V}_\eta = \sigma_\eta^2\mathbf{I}_b \otimes (\mathbf{I}_{k_1} - k_1^{-1}\mathbf{J}_{k_1})$, $\mathbf{V}_\varepsilon = \sigma_\varepsilon^2\mathbf{I}_{bk_1} \otimes (\mathbf{I}_{k_2} - k_2^{-1}\mathbf{J}_{k_2})$, $\mathbf{J}_t = \mathbf{1}_t\mathbf{1}_t'$, where $\otimes$ denotes Kronecker product of matrices.

3. ANALYSIS OF VARIANCE

In this paper, the approach proposed by Nelder (1965) to the analysis of multistrata experiments satisfying general balance will be adopted. It will be assumed that the incomplete split-splot design here considered is generally balanced in the sense of Houtman and Speed (1983). Hence we are interested in finding a suitable set of positive semi-definite matrices. These matrices will play an important role in the statistical analysis of the experiment.

Let $\mathbf{P}_0 = n^{-1}\mathbf{J}_n$, $\mathbf{P}_1 = k^{-1}\mathbf{D}'\mathbf{D} - n^{-1}\mathbf{J}_n$, $\mathbf{P}_2 = k_2^{-1}\mathbf{G}'\mathbf{G} - k^{-1}\mathbf{D}'\mathbf{D}$, $\mathbf{P}_3 = \mathbf{I}_n - k_2^{-1}\mathbf{G}'\mathbf{G}$, $k = k_1k_2$.

It can be proved that the matrices $\mathbf{P}_t$, $t = 0, 1, 2, 3$, are symmetric, idempotent, pairwise orthogonal and satisfy the relations: $\mathbf{P}_0 + \mathbf{P}_1 + \mathbf{P}_2 + \mathbf{P}_3 = \mathbf{I}_n$, $\mathbf{P}_0\mathbf{1} = \mathbf{1}$, $\mathbf{P}_t\mathbf{1} = \mathbf{0}$, $t = 1, 2, 3$, $r(\mathbf{P}_0) = 1$, $r(\mathbf{P}_1) = b - 1$, $r(\mathbf{P}_2) = bk_1 - b$, $r(\mathbf{P}_3) = n - bk_1$.

The matrix $\mathbf{V}$ of (10) can be written as

$$\mathbf{V} = \gamma_0\mathbf{P}_0 + \gamma_1\mathbf{P}_1 + \gamma_2\mathbf{P}_2 + \gamma_3\mathbf{P}_3,$$

where $\gamma_0 = \sigma_e^2$, $\gamma_1 = k\sigma_\rho^2 + \sigma_e^2$, $\gamma_2 = k_2\sigma_\eta^2 + \sigma_e^2$, $\gamma_3 = \sigma_\varepsilon^2 + \sigma_e^2$.

From the properties of matrices $\mathbf{P}_t$ it follows that $\mathbf{V}\mathbf{P}_t = \gamma_t\mathbf{P}_t$, $\mathbf{V}^{-1} = \gamma_0^{-1}\mathbf{P}_0 + \ldots + \gamma_3^{-1}\mathbf{P}_3$. Assuming that strata variances γ_t, $t = 0, 1, 2, 3$, are known – the overall analysis can be based on the (reduced) normal equation for $\boldsymbol{\tau}$ obtained after eliminating μ,

$$\mathbf{C}\boldsymbol{\tau}^0 = \mathbf{Q}, \tag{11}$$

where

$$\mathbf{C} = \gamma_1^{-1}\mathbf{C}_1 + \gamma_2^{-1}\mathbf{C}_2 + \gamma_3^{-1}\mathbf{C}_3,\ \mathbf{C}_t = \boldsymbol{\Delta}\mathbf{P}_t\boldsymbol{\Delta}',\ t = 1, 2, 3,$$
$$\mathbf{Q} = \gamma_1^{-1}\mathbf{Q}_1 + \gamma_2^{-1}\mathbf{Q}_2 + \gamma_3^{-1}\mathbf{Q}_3,\ \mathbf{Q}_t = \boldsymbol{\Delta}\mathbf{P}_t\mathbf{y},\ t = 1, 2, 3. \tag{12}$$

In this paper, only estimable linear functions $\mathbf{c}'\boldsymbol{\tau}$ will be examined. From the fact that $\mathbf{C1} = \mathbf{0}$ follows that if the linear function $\mathbf{c}'\boldsymbol{\tau}$ is estimable then it must be a contrast, i.e., $\mathbf{c}'\mathbf{1} = 0$.

It can be proved that in model (9) each contrast is estimable $[r(\mathbf{C}) = ST - 1]$. The BLUE of some contrast $\mathbf{c}'\boldsymbol{\tau}$ is given by $\widehat{\mathbf{c}'\boldsymbol{\tau}} = \mathbf{C}^{-}\mathbf{Q}$, where $\mathbf{C}^{-}$ denotes a g-inverse of matrix $\mathbf{C}$.

The above procedure is appropriate as long as the strata variances γ_t, $t = 0, 1, 2, 3$, are known. The most interesting case, from the theoretical and practical points of view, is that in which the variances γ_t are unknown. In that case the within strata analysis can be helpful. This analysis is based on the model

$$\mathbf{y}_t = \mathbf{P}_t\mathbf{y},\ \mathrm{Cov}(\mathbf{y}_t) = \gamma_t\mathbf{P}_t,\ t = 1, 2, 3. \tag{13}$$

Model (13) is linear model with singular covariance matrix. However, from the structure of the model (13) it follows [cf. Baksalary and Kala (1983)] that BLUE obtained in the model for the t-th stratum (13) is equal to the BLUE obtained by a simple least squares method, i.e., from model $\mathbf{y}_t^* = \mathbf{P}_t\mathbf{y}$, $\mathrm{Cov}(\mathbf{y}_t^*) = \gamma_t\mathbf{I}_n$, $t = 1, 2, 3$.

Applying the simple least squares method to model (13), the estimator of $\boldsymbol{\tau}$ in t-th stratum can be obtained from the normal equation

$$\mathbf{C}_t\boldsymbol{\tau}_t^0 = \mathbf{Q}_t,\ t = 1, 2, 3. \tag{14}$$

Note that from $\mathbf{C}_t\mathbf{1} = \mathbf{0}$ it follows that only some treatment contrasts can be estimated in t-th stratum.

It can be shown that if a contrast is estimable only in one of the strata then it is the BLUE in general model (9). However, there may also be a situation in which the same contrasts are estimable in two or three strata. Then the stratum BLUE can be used in combining the estimators from all the strata. In that case the general balance plays the important role. It was shown by Houtman and Speed (1983) that estimates obtained from combining individual stratum estimates are BLUEs if and only if the design is generally balanced.

Now let us consider the estimation of variance components γ_t, $t = 1, 2, 3$, and the possibility of constructing some tests of hypotheses. This may be done by the within strata analysis of variance, which can be presented as in Table 1.

TABLE 1. Analysis of variance for stratum t

Source of variation	d.f.	S.S.	E.S.S.
Treatments t	υ_{Tt}	SST_t	$\upsilon_{Tt}\gamma_t + \boldsymbol{\tau}'C_t\boldsymbol{\tau}$
Error t	υ_{Et}	SSE_t	$\upsilon_{Et}\gamma_t$
Total t	υ_t	SSY_t	$\upsilon_t\gamma_t + \boldsymbol{\tau}'C_t\boldsymbol{\tau}$

The undefined symbols in Table 1 denote: $\upsilon_{Tt} = r(\mathbf{C}_t)$, $\upsilon_t = r(\mathbf{P}_t)$, $\upsilon_{Et} = \upsilon_t - \upsilon_{Tt}$, $SSY_t = \mathbf{y}'\mathbf{P}_t\mathbf{y}$, $SST_t = \mathbf{y}'\mathbf{P}_t\boldsymbol{\Delta}'\mathbf{C}_t^{-}\boldsymbol{\Delta}\mathbf{P}_t\mathbf{y}$, $SSE_t = SSY_t - SST_t$.

If the normality of the random variable of model (9) is assumed then it is easy to construct an exact test (F-test) of the hypothesis H_{0t}: $\boldsymbol{\tau}'\mathbf{C}_t\boldsymbol{\tau} = 0$, $t = 1, 2, 3$. Note that the $\mathbf{C}_t$ are usually different and in general, these are different hypotheses. But there may be such contrasts which are testable in two or three strata. Then if we are interested in testing some subhypothesis H_0: $\mathbf{c}'\boldsymbol{\tau} = 0$, we can use the F-test in every stratum (if it is possible) or we can use a combined test. The paper Littel and Folks (1973) can be helpful in choosing the combined test. It is worth noting that stratum F-tests are independent and this property is very helpful in the theory of combining tests.

4. PLANNING OF EXPERIMENTS

The allocation of treatments to blocks and to wholeplots can be described by two incidence matrices $\mathbf{N}_1$ and $\mathbf{N}_2$. Let $\mathbf{N}_1 = \boldsymbol{\Delta}\mathbf{D}'$ and let $\mathbf{N}_2 = \boldsymbol{\Delta}\mathbf{G}'$. The elements (i, w), $i = 1, 2, \ldots, ST$, $w = 1, 2, \ldots, b$; of matrix $\mathbf{N}_1$ show the number of occurences of the i-th treatment in the w-th block, whilst the element of the matrix $\mathbf{N}_2$ show the number of times each treatment occurs in each wholeplot.

Note that $\mathbf{C}_1 = k^{-1}\mathbf{N}_1\mathbf{N}_1' - n^{-1}\mathbf{r}\mathbf{r}'$, $\mathbf{C}_2 = k_2^{-1}\mathbf{N}_2\mathbf{N}_2' - k^{-1}\mathbf{N}_1\mathbf{N}_1'$, $\mathbf{C}_3 = \mathbf{R} - k_2^{-1}\mathbf{N}_2\mathbf{N}_2'$, where $\mathbf{r}$ is the vector of treatment replications, $\mathbf{R} = \boldsymbol{\Delta}\boldsymbol{\Delta}'$ is diagonal matrix with diagonal elements equal to elements of $\mathbf{r}$. Hence in the planning of experiments special importance should be attached to the incidence matrices $\mathbf{N}_1$ and $\mathbf{N}_2$.

In the paper there is no place to consider the details connected with some particular cases of matrices $\mathbf{N}_1$ and $\mathbf{N}_2$. This will be considered in another paper of the author. But we will briefly discuss two examples.

EXAMPLE 1. Let $S = 3$, $T = 2$, $k_1 = 2$, $k_2 = 1$, $b = 12$, and let $\mathbf{N}_1 = \mathbf{N}_A \otimes \mathbf{N}_B$, where

$$\mathbf{N}_A = \begin{pmatrix} 1 & 1 & 0 & 1 & 1 & 0 \\ 1 & 0 & 1 & 0 & 1 & 1 \\ 0 & 1 & 1 & 1 & 0 & 1 \end{pmatrix}, \quad \mathbf{N}_B = \begin{pmatrix} 1 & 0 \\ 0 & 1 \end{pmatrix}.$$

This means that only one from the levels of factor B occurs on wholeplot. In such a design the contrast among the levels of factor A and the interaction contrasts are estimated with efficiency 1/4 in the first stratum and with efficiency 3/4 in the second stratum. The contrast between levels of factor B is estimated with efficiency equal to 1 (full efficiency) in the first stratum.

EXAMPLE 2. Let us consider the design which is complete with respect to the levels of factor A ($k_1 = S$) but is incomplete with respect to the levels of factor B ($k_2 < T$). Let the incidence matrix $\mathbf{N}_1$ be of the form $\mathbf{N}_1 = [\mathbf{N}_{11}', \mathbf{N}_{12}', \ldots, \mathbf{N}_{1S}']'$, where $\mathbf{N}_{1h} = \mathbf{N}$, $h = 1, 2, \ldots, S$, and $\mathbf{N}$ is the incidence matrix of a BIB design for T treatments occurring in b blocks of size k_2 each. Then: (i) all contrasts among the levels of factor A are estimated with full efficiency in the second stratum; (ii) all constrasts among the levels of factor B are estimated with efficiency equal to $(T - k_2) / k_2(T - 1)$ in the first stratum and with efficiency equal to $T(k_2 - 1) / k_2(T - 1)$ in the third stratum; (iii) all interaction constrasts are estimated with efficiency equal to $(T - k_2) / k_2(T - 1)$ in the second stratum and with efficiency equal to $T(k_2 - 1) / k_2(T - 1)$ in the third one.

Finally, it is worth noting that some results given by Rees (1969), Robinson (1970), Bhargava and Shah (1975) are particular cases of results presented in this paper.

ACKNOWLEDGEMENT

The author thanks the referees for their constructive comments.

REFERENCES

Baksalary, J.K. and Kala, R. (1983). On equalities between BLUEs, WLSEs and SLSEs. *The Canadian Journal of Statistics*, 11, 119–123.

Bhargava, R.P. and Shah, K.R. (1975). Analysis of some mixed models forblock and split-plot designs. *Annals of the Institute of Statistical Mathematics*, 27, 365–375.

Houtman, A.M. and Speed, T.P. (1983). Balance in designed experiments with orthogonal block structure. *Annals of Statistics*, 11, 1069-1085.

Kempthorne, O. (1952). *The Design and Analysis of Experiments*. Wiley, New York.

Kempthorne, O., Zyskind, G., Addelman, S., Throckmorton, T.N. and White, R.F. (1961). Analysis of variance procedures. Technical Report 149, Wright-Patterson Air Force Base, Aeronautical Research Laboratory, Ohio.

Littel, R.C. and Folks, J.L. (1973). Asymptotic optimality of Fisher´s method of combining independent tests. *Journal of the American Statistical Association*, 68, 193–194.

Mejza, S. (1986). Doświadczenia w ukladach blokowych niekompletnych o jednostkach rozszczepionych. (Polish) *Roczniki AR w Poznaniu, Rozprawy Naukowe*, z 150, 1–87.

Nelder, J.A. (1965). The analysis of randomized experiments with orthogonal structure. *Proceedings of the Royal Society (London), Ser. A*, 283, 147–178.

Neyman, J., Iwaszkiewicz, K. and Kolodziejczyk, St. (1935). Statistical problems in agricultural experimentation. *Journal of the Royal Statistical Society Supp.*, 2, 107-154.

Rees, D.H. (1969). The analysis of variance of some nonorthogonal designs with split-plots. *Biometrika*, 56, 43–54.

Robinson, J. (1970). Blocking in incomplete split-plot designs. *Biometrika*, 57, 347–350.

Wilk, M.B. and Kempthorne, O. (1956). Some aspects of the analysis of factorial experiments in a completely randomized design. *Annals of Mathematical Statistics*, 27, 950-985.

Received 6 January 1987
Revised 18 September 1987

Dept. of Mathematical and Statistical Methods
Academy of Agriculture in Poznań
Wojska Polskiego 28
PL-60-637 Poznań
Poland

Proc. Second International Tampere Conference in Statistics
(Tampere, Finland, 1-4 June 1987)
Tarmo Pukkila and Simo Puntanen, *Editors*

pp. 585 - 590

Underestimating the largest eigenvalue of a covariance matrix

Jorma Kaarlo MERIKOSKI
University of Tampere, Finland

Key words and phrases: Covariance matrix, eigenvalues, row sums.

ABSTRACT

Let A be a complex $n \times n$ covariance matrix (i.e. a Hermitian nonnegative definite matrix) with row sums $R_1,\ldots,R_n$. The ratio

$$\beta = \frac{|R_1|^2 + \ldots + |R_n|^2}{R_1 + \ldots + R_n},$$

which is a lower bound for the largest eigenvalue of A, is studied and compared with certain other lower bounds.

1. INTRODUCTION

Let A be a complex Hermitian $n \times n$ matrix with ordered eigenvalues $\lambda_1 = \lambda_1(A),\ldots,\lambda_n = \lambda_n(A)$ ($\lambda_1 \geq \ldots \geq \lambda_n$) and row sums $R_1 = R_1(A),\ldots, R_n = R_n(A)$. We denote $\operatorname{su} A = R_1 + \ldots + R_n$ and

$$a = a(A) = \frac{\operatorname{su} A}{n} = \frac{R_1 + \ldots + R_n}{n},$$

$$\beta = \beta(A) = \frac{\operatorname{su} A^2}{\operatorname{su} A} = \frac{|R_1|^2 + \ldots + |R_n|^2}{R_1 + \ldots + R_n},$$

$$\gamma = \gamma(A) = \prod_{j=1}^{n} R_j^{R_j/\Sigma_k R_k}.$$

Then

$\lambda_1 \geq \alpha$ if A is Hermitian; (1)

$\lambda_1 \geq \beta$ if A is a covariance matrix, i.e. Hermitian nonnegative definite (denoted A nnd); (2)

$\lambda_1 \geq \gamma$ if A is symmetric and (elementwise) nonnegative (denoted A snn). (3)

[If A is nnd and su $A = 0$, we define $\beta = 0$. Moreover, we define $\gamma(0) = 0$ and $0^0 = 1$ if $A \neq 0$, $R_j = 0$.] From the well-known property of the Rayleigh quotient

$$\lambda_1 \geq \frac{X^{\mathrm{H}} A X}{X^{\mathrm{H}} X} \quad (X \in \mathbb{C}^n,\ X \neq 0) \tag{4}$$

we obtain (1) by taking $X = E = (1,\ldots, 1)^{\mathrm{T}}$ and (2) by $X = A^{\frac{1}{2}} E$. (3) is due to Deutsch [1].

The bound α often appears to be very good if A is snn [3], and γ is in this case still better [1]. Our main interest is now in β, which we will compare with certain other bounds both theoretically and by examples.

A more general discussion about bounds obtained from (4) can be found in [2].

2. THE BOUNDS

For A nnd, we compare α, β and

$$\tilde{\alpha} = \frac{|R_1| + \ldots + |R_n|}{n}, \quad \tilde{\gamma} = \prod_{j=1}^{n} |R_j|^{|R_j|/\Sigma_k |R_k|}$$

(γ is not necessarily defined).

PROPOSITION 1. $\lambda_1 \geq \beta \geq \tilde{\gamma} \geq \tilde{a} \geq a$ *if A is nnd.*

Proof. Since A is nnd, we have su $A = E^T AE \geq 0$. If su $A = 0$, then, by diagonalizing, it is easy to see that also su $A^2 = 0$, and because su $A^2 = |R_1|^2 + \dots + |R_n|^2$, necessarily $R_1 = \dots = R_n = 0$. Therefore the proof is trivial in the case su $A = 0$, and we can assume su $A > 0$.

$\lambda_1 \geq \beta$. We know it already (2).

$\beta \geq \tilde{\gamma}$. Applying the arithmetic–geometric mean inequality we have

$$\beta = \sum_j \frac{|R_j|^2}{\sum_k R_k} = \sum_j \frac{|R_j|}{\sum_k R_k} |R_j| \geq \sum_j \frac{|R_j|}{\sum_k |R_k|} |R_j| \geq \prod_j |R_j|^{|R_j| / \Sigma_k |R_k|} = \tilde{\gamma}.$$

$\tilde{\gamma} \geq \tilde{a}$. [1].
$\tilde{a} \geq a$. Trivial.

A stronger inequality

$$\lambda_1 \geq \prod_{j=1}^{n} |R_j|^{|R_j| / \Sigma_k R_k}$$

(for A nnd) does not hold. If, for example

$$A = \begin{pmatrix} 9 & -5 \\ -5 & 3 \end{pmatrix},$$

then $\lambda_1 = 11.8$ but the right-hand side $= 32$.
We demonstrate the sharpness of β by a couple of examples.

EXAMPLE 1. Let A be the symmetric matrix given by the upper triangle

$$\begin{pmatrix} 4.9810 & 3.8063 & 4.7740 \\ & 3.0680 & 3.7183 \\ & & 4.8264 \end{pmatrix}.$$

This matrix was studied e.g. in [4] to illustrate the Principal Component Analysis. We have $\lambda_1 = 12.6372$, $\beta = 12.6359$, $\tilde{\gamma} = \gamma = 12.5656$, $\tilde{a} = a = 12.4909$. This good result is not very surprising, since A is close to a rank-one matrix ($\lambda_2 = 0.1386$, $\lambda_3 = 0.0977$) and all our bounds are exact if rank $A = 1$.

EXAMPLE 2. Several tests by random 4×4 nnd matrices showed that β is often good if the real parts of all the elements of A are nonnegative. For a typical example, let the upper triangle of (Hermitian) A be

$$\begin{pmatrix} 2.0 & 3.3+0.2i & 2.5-0.6i & 0.9-0.1i \\ & 7.2 & 1.9+1.1i & 0.4-0.1i \\ & & 10.8 & 3.6+i \\ & & & 2.2 \end{pmatrix}.$$

Then $\lambda_1 = 14.059, \beta = 13.611, \tilde{\gamma} = 12.731, \tilde{a} = 11.873, a = 11.850$. Thus β is good although A is far from rank-one matrices ($\lambda_2 = 7.327, \lambda_3 = 0.777$, $\lambda_4 = 0.037$).

3. IMPROVING THE BOUNDS

A natural attempt to try to improve β, $\tilde{\gamma}$ and $\tilde{a}$ and, in fact, to generalize them for A Hermitian, is to study such real numbers t that

$$A - tI \text{ is nnd.} \tag{5}$$

Then

$$\lambda_1 = \lambda_1(A) = \lambda_1(A - tI) + t$$

and $\lambda_1(A - tI)$ can be underestimated by

$$b(t) = \beta(A - tI) + t,$$

$$g(t) = \tilde{\gamma}(A - tI) + t,$$

$$a(t) = \tilde{a}(A - tI) + t.$$

[It is not worth studying $a(A - tI) + t$, since it is identically $a(A)$.] We study the behaviour of these functions when t increases. First consider b:

$$b(t) = \frac{\operatorname{su}(A - tI)^2}{\operatorname{su}(A - tI)} + t = \frac{\operatorname{su} A^2 - 2t \operatorname{su} A + nt^2}{\operatorname{su} A - tn} + t = \frac{\operatorname{su} A^2 - t \operatorname{su} A}{\operatorname{su} A - tn},$$

$$b'(t) = \frac{(\mathrm{su}\,A - tn)(-\mathrm{su}\,A) - (\mathrm{su}\,A^2 - t\,\mathrm{su}\,A)(-n)}{(\mathrm{su}\,A - tn)^2} = \frac{n\,\mathrm{su}\,A^2 - (\mathrm{su}\,A)^2}{(\mathrm{su}\,A - tn)^2}.$$

It is easy to see that $n\,\mathrm{su}\,A^2 - (\mathrm{su}\,A)^2 \geq 0$, and so

$$b'(t) \geq 0 \quad \text{for } t < \mathrm{su}\,A/n \ \ (\text{or } t > \mathrm{su}\,A/n).$$

Next, let us study a. Since

$$a(t) = \sum_{j=1}^{n} \frac{|R_j - t| + t}{n}$$

and, for $t_1 \leq t_2$,

$$|R_j - t_1| + t_1 = |R_j - t_2 + (t_2 - t_1)| + t_1 \leq |R_j - t_2| + t_2 - t_1 + t_1$$

$$= |R_j - t_2| + t_2,$$

it follows $a(t_1) \leq a(t_2)$. We have now proved the following:

PROPOSITION 2. *For any $t \neq \mathrm{su}\,A/n$, b is increasing. For any real t, a is increasing.*

Also g seems often to be increasing, but not always. If, for example,

$$A = \begin{pmatrix} 2 & -1 \\ -1 & 0.9 \end{pmatrix},$$

then, for $-0.1 \leq t \leq 1$, we namely have

$$g(t) = (1 - t)^{\frac{1-t}{1.1}} (0.1 + t)^{\frac{0.1+t}{1.1}} + t,$$

$g(0.01) = 0.8047$, $g(0) = 0.8111$.

To obtain best possible bounds using $b(t)$ and $a(t)$, t must by Prop. 2 be as large as possible, subject to (5); i.e. $t = \lambda_n$. If we do not know λ_n, we must use some lower bound of it. Thus, if A is Hermitian and t is any lower bound for λ_n (obtained e.g. from the literature), then $b(t)$ and $a(t)$ are lower bounds for λ_1.

REFERENCES

1. E. Deutsch, Lower bounds for the Perron root of a non-negative irreducible matrix. *Math. Proc. Camb. Phil. Soc.* 92 (1982), 49–54.
2. S. Hyyrö, J.K.Merikoski and A.Virtanen, Improving certain simple eigenvalue bounds. *Math. Proc. Camb. Phil. Soc.* 99 (1986), 507–518.
3. J.K. Merikoski, On a lower bound for the Perron eigenvalue. *BIT* 19 (1979), 39–42.
4. D.F. Morrison, *Multivariate Statistical Methods*. Second Edition. McGraw–Hill, Tokyo, 1976.

Received 16 January 1987

Department of Mathematical Sciences
University of Tampere
P.O. Box 607
SF-33101 Tampere
Finland

Proc. Second International Tampere Conference in Statistics
(Tampere, Finland, 1-4 June 1987)
Tarmo Pukkila and Simo Puntanen, *Editors*

pp. 591 - 601

Comparison of some influence measures in nonlinear regression

Jiři MILITKÝ
Research Institute
for Textile Finishing
Dvůr Králové nad Labem
Czechoslovakia

Karel KVĚTOŇ
Czech Technical University
Prague
Czechoslovakia

Jaroslav ČÁP
Applied Cybernetics Centre
Slušovice
Czechoslovakia

Key words and phrases: Influential measures, identification of influential points, nonlinear models, linearized and quadratic expansion, Michelis-Menten model.

ABSTRACT

In the paper basic procedures for identification of influential points in nonlinear regression models are presented. The techniques based on linear or quadratic expansion of the model function as well as procedures making use of influence curves are discussed.

1. INTRODUCTION

One of the problems frequently tackled with practical application of regression tasks is the presence of influential points. Under the term such points or their groups are hidden that considerably affect the results or regression (estimates, their statistical characteristics, prediction, etc.). In the linear least squares with the design matrix $\mathbf{X}$ practically all influence measures depend upon residuals $\hat{\mathbf{e}} = \mathbf{y} - \mathbf{X}\hat{\mathbf{a}}$ and the elements h_{ij} of the hat matrix $\mathbf{H} = \mathbf{X}(\mathbf{X}^T\mathbf{X})^{-1}\mathbf{X}^T$. Most of the techniques for identification of influential points start from $\hat{\mathbf{a}}_{(i)}$ estimates obtained by omitting the i-th point. Considering the regression model linearity, the estimates $\hat{\mathbf{a}}_{(i)}$ may be directly, assessed from $\mathbf{X}$ and $\hat{\mathbf{a}}$ without the need of regression recomputing.

With nonlinear regression models the situation is much more complicated. One consequence of nonlinearity is that the parameter estimates, residuals and the hat matrix are no longer linear functions of observations. Therefore, several approaches are employed to determine the influence measures.

2. FORMULATION OF NONLINEAR REGRESSION PROBLEM

One of the principal phases in constructing nonlinear regression models $f(\mathbf{x}, \mathbf{a})$ is to estimate the model parameters $\mathbf{a} = (a_1, \dots, a_m)^T$. Commonly, it is started from n experimental points $\{\mathbf{x}_i, y_i\}$, $i = 1, \dots, n$, where $\mathbf{x}_i$ are deterministic (explanatory) variables and responses y_i are realizations of random variables.

In solving this problem the following is usually assumed

a) Additive measurement model is valid

$$y_i = f(\mathbf{x}_i, \mathbf{a}) + \varepsilon_i. \tag{1}$$

b) Random errors ε_i have zero mean values $E(\varepsilon_i) = 0$ and the diagonal covariance matrix $D(\boldsymbol{\varepsilon}\boldsymbol{\varepsilon}^T) = \sigma^2\mathbf{E}$ where $\mathbf{E}$ is the diagonal unit matrix of order n.

c) Error distribution is characterized by continuous twice differentiable density $p(\varepsilon)$.

d) Model function $f(\mathbf{x}, \mathbf{a})$ is twice differentiable.

Provided that these assumptions are valid, it is possible to obtain maximum likelihood estimates by maximizing the logarithm of the likelihood function

$$L(\mathbf{a}) = \sum_{i=1}^{n} \ln p[y_i - f(\mathbf{x}_i, \mathbf{a})].. \tag{2}$$

Here, we are restricted to the most frequent case where ε_i are i.i.d. random variables having normal distribution $N(0, \sigma^2)$. After substituting into equation (2) the well-known least squares criterion results

$$S(\mathbf{a}) = \sum_{i=1}^{n} [y_i - f(\mathbf{x}_i, \mathbf{a})]^2. \tag{3}$$

By minimizing equation (3) the sought estimates $\hat{\mathbf{a}}$ may be obtained.

Among the most popular ones the Gauss-Newton method should be mentioned where the increment vector in the i-th iteration has the form

$$\mathbf{\Delta a} = (\mathbf{J}^T\mathbf{J})^{-1}\mathbf{J}^T\mathbf{e}, \tag{4}$$

where $\mathbf{\Delta a} = \mathbf{a}_{i+1} - \mathbf{a}_i$, $\mathbf{e}$ is the vector of residuals for $\mathbf{a}_i$ estimates having the elements

$$e_j = y_j - f(\mathbf{a}_i, \mathbf{x}_j) \tag{5}$$

and $\mathbf{J}$ is the Jacobian matrix ($n \times m$) with the components

$$J_{j,h} = \partial f(\mathbf{x}_j, \mathbf{a}) / \partial a_h \Big|_{\mathbf{a} = \mathbf{a}_i}. \tag{6}$$

In the MINOPT program which has been applied for numerical computations in this work, the so-called double dog-leg strategy of Dennis and Mei (1979) is made use of. A detailed description of the program structure is stated in the work - Militký and Čáp (1987).

It is known that the least squares estimates $\hat{\mathbf{a}}$ are asymptotically unbiased and normally distributed [see e.g. Jennrich (1969)]. However, for finite sample sizes these estimates are biased. Also other quantities being useful in analyzing nonlinear regression models can be characterized in the same way.

In order to assess the size of bias $\mathbf{b}$ of $\hat{\mathbf{a}}$ estimates the Taylor expansion of the model function into quadratic terms around the point $\hat{\mathbf{a}}$ is employed:

$$f(\mathbf{x}_i, \mathbf{a}) = f(\mathbf{x}_i, \hat{\mathbf{a}}) + \mathbf{J}_i^T \cdot \mathbf{\Delta a} + \tfrac{1}{2}\mathbf{\Delta a}^T\mathbf{W}_i\mathbf{\Delta a} + \ldots \tag{7}$$

In equation (7) $\mathbf{\Delta a} = \mathbf{a} - \hat{\mathbf{a}}$ and $\mathbf{J}_i^T$ is the i-th row of the Jacobian matrix. The matrix $\mathbf{W}_i$ of second derivatives of the model function has the elements

$$W_{j,h} = \partial^2 f(\mathbf{x}_i, \mathbf{a}) / \partial a_j \partial a_h. \tag{8}$$

According to Cook and Tsai (1984) the quadratic approximation of bias vector $\mathbf{b} \approx E(\hat{\mathbf{a}}) - \mathbf{a}$ can be evaluated from the equation

$$\mathbf{b} = (\mathbf{J}^{\mathrm{T}}\mathbf{J})^{-1}\mathbf{J}^{\mathrm{T}}\mathbf{d}. \tag{9}$$

In equation (9) $\mathbf{d}$ is the expected difference between the linear and quadratic expansion of the model function. This vector has the components

$$d_i = -\sigma^2 \,\mathrm{tr}[(\mathbf{J}^{\mathrm{T}}\mathbf{J})^{-1}\mathbf{W}_i]/2, \qquad i = 1,\dots,n. \tag{10}$$

In practical cases instead of b_j the relative bias values $b_{Rj} = (b_j / \hat{a}_j)\cdot 100$ are computed. Provided that all b_{Rj} are sufficiently small (usually less than 3 or 1) the bias is considered negligible and the model function is satisfactorily approximated by using the linear approximation

$$f(\mathbf{x}_i, \mathbf{a}) \simeq f(\mathbf{x}_i, \hat{\mathbf{a}}) + \mathbf{J}_i^{\mathrm{T}}\Delta\mathbf{a}. \tag{11}$$

Statistical analysis of estimates may then be performed in the same way as in the linear regression models.

In strongly nonlinear models (having high parameter biases b_{Rj}) other methods of statistical analysis of estimates should be introduced [see e.g. Simonoff and Tsai (1986)].

3. INFLUENCE MEASURES BASED ON LINEARIZATION

On the basis of equation (4) and equation (11) it is possible to assess one step approximation of the estimates $\hat{\mathbf{a}}_{(i)}$ involving all but i-th point $(y_i, \mathbf{x}_i)$ from the equation

$$\hat{\mathbf{a}}_{(i)}^{\,1} = \hat{\mathbf{a}} - (\mathbf{J}^{\mathrm{T}}\mathbf{J})^{-1}\mathbf{J}_i \cdot \hat{e}_i / (1 - h_{ii}), \tag{12}$$

where h_{ii} is the i-th diagonal element of the approximated (tangential) hat matrix $\mathbf{H}_A = \mathbf{J}(\mathbf{J}^{\mathrm{T}}\mathbf{J})^{-1}\mathbf{J}^{\mathrm{T}}$ and $\hat{e}_i = y_i - f(\mathbf{x}_i, \hat{\mathbf{a}})$ is the i-th residual.

Linearization of the model at the least squares estimate $\hat{\mathbf{a}}$ can thus provide one step approximation to various influential measures applied with linear models.

For one step approximation of Jackknife residuals it follows

$$RS_i = \frac{\hat{e}_i}{\hat{s}_{(i)} \cdot (1 - h_{ii})^{\frac{1}{2}}} \tag{13}$$

where

$$\hat{s}_{(i)}^{\,2} = [S(\hat{\mathbf{a}}) - \hat{e}_i^{\,2} / (1 - h_{ii})] / (n - m - 1) \tag{14}$$

is an independent estimate of the variance obtained from all points but the i-th one. Approximately, it may be stated that if the i-th point is not an outlier RS_i possess the Student distribution with $n - m - 1$ degrees of freedom.

Analogically, the relationships for approximating measures of influence, described e.g. by Belsey, Kuh and Welsch (1980) or Cook and Weisberg (1982), may easily be derived. Moreover, the measures based on the influence curve illustrated in Section 5 may be simplified by using the one step approximation $\hat{\mathbf{a}}_{(i)}^{\,1}$.

4. INFLUENCE MEASURES BASED ON QUADRATIC EXPANSION

In order to calculate the $\mathbf{b}_{(i)}$ vector which determines the bias of parameters from all points except of the i-th one Cook and Tsai (1984) derived the relation

$$\mathbf{b}_{(i)} = \mathbf{b}_\delta + \mathbf{b} - \frac{(\mathbf{J}^{\mathrm{T}}\mathbf{J})^{-1}\mathbf{J}_i \cdot a_i}{1 - h_{ii}}, \tag{15}$$

where $\mathbf{b}_\delta = (\mathbf{J}^{\mathrm{T}}\mathbf{J})^{-1}\mathbf{J}^{\mathrm{T}} \cdot \boldsymbol{\delta}_i$ is the vector of coefficients from the ordinary least squares regression of $\boldsymbol{\delta}_i$ on $\mathbf{J}$ and a_i is the i-th residual from regression $(\mathbf{d} + \boldsymbol{\delta}_i)$ on $\mathbf{J}$. The vector $\boldsymbol{\delta}_i$ reflects the change in the expected difference between linear and quadratic approximation of the model function due to omitting the i-th point. It holds that its elements δ_{ij} are given in the form

$$\delta_{ij} = -\frac{\sigma^2}{2} \cdot \frac{\mathbf{J}_i^{\mathrm{T}}(\mathbf{J}^{\mathrm{T}}\mathbf{J})^{-1}\mathbf{W}_j(\mathbf{J}^{\mathrm{T}}\mathbf{J})^{-1}\mathbf{J}_i}{1 - h_{ii}}, \qquad j = 1,\ldots, n. \tag{16}$$

Since $\mathbf{b}_{(i)}$ is a vector of $(n \times 1)$ dimension, it is useful to consider certain norms [see Cook and Weisberg (1982)] or scaled quantities as a KV_i defined by relation

$$KV_i = \mathbf{b}_{(i)}^{\mathrm{T}}(\mathbf{J}^{\mathrm{T}}\mathbf{J})\mathbf{b}_{(i)} / \sigma^2. \tag{17}$$

5. NONLINEAR INFLUENCE MEASURES

Cook and Weisberg (1982) discuss different variants of expressing the influence curves in linear regression models. Here, we are restricted to an empirical influence curve $\mathbf{EIF}_i$ and the sample influence $\mathbf{SIC}_i$. The empirical influence curve $\mathbf{EIF}_i$ connected with the i-th point $\{x_i, y_i\}$ is defined by Militký and Čáp (1985a) as the expression

$$\mathbf{EIF}_i \approx \hat{\mathbf{a}}(\boldsymbol{\varepsilon}) - \hat{\mathbf{a}}(\boldsymbol{\varepsilon}_i), \tag{18}$$

where $\boldsymbol{\varepsilon}$ is the vector of error terms and $\boldsymbol{\varepsilon}_i$ differs from $\boldsymbol{\varepsilon}$ by having the i-th component equal to zero. From linear expansion of $\hat{\mathbf{a}}(\boldsymbol{\varepsilon})$ may easily be derived that

$$\mathbf{EIF}_i \approx \hat{e}_i \cdot (\mathbf{J}^{\mathrm{T}}\mathbf{J})^{-1} \cdot \mathbf{J}_i. \tag{19}$$

Instead of the $\mathbf{EIF}_i$ vector the effect of i-th point may be summed by means of an approximate elliptical norm CD_i which is defined by the relation

$$CD_i = \frac{\hat{e}_i^2 \mathbf{J}_i^{\mathrm{T}}(\mathbf{J}^{\mathrm{T}}\mathbf{J})^{-1}\mathbf{J}_i}{m\, \hat{\sigma}^2}. \tag{20}$$

In order to identify the highly influential points CD_i / m is compared with quantiles $F_a(m, n-m)$ of the Fisher distribution [see Militký and Čáp (1985)].

Sample influence function $\mathbf{SIC}_i$ can be defined as a difference

$$\mathbf{SIC}_i \approx \hat{\mathbf{a}} - \hat{\mathbf{a}}_{(i)}. \tag{21}$$

For the nonlinear models Cook and Weisberg (1982) recommend application of some norm of $\mathbf{SIC}_i$ or a generalized measure derived from the contours of the log likelihood function (called the likelihood distance).

On the basis of alternative expression of $\mathbf{SIC}_i$ it is possible to use the norm of change in the vector of fitted values [see Cook and Weisberg (1982)]

$$FDT_i = \sum_{j=1}^{n} [f(\mathbf{x}_j, \hat{\mathbf{a}}) - f(\mathbf{x}_j, \hat{\mathbf{a}}_{(i)})]^2 / (m\, \hat{\sigma}^2). \tag{22}$$

Provided that in equation (22) the fitted values $f(\mathbf{x}_j, \hat{\mathbf{a}}_{(i)})$ are replaced by their one-step approximations an FDA_i one-step approximation is obtained. Cook and Weisberg (1982) define the likelihood distance LD_i as the difference

$$LD_i = 2[L(\hat{\mathbf{a}}) - L(\hat{\mathbf{a}}_{(i)})]. \tag{23}$$

LD_i can be calibrated by comparing it with the upper point of chi-squared distribution having m degrees of freedom $\chi^2(m)$. For a special case of nonlinear least squares where also the variance σ^2 is estimated, equation (23) may be expressed also in the form, viz.

$$LDT_i = n \ln \left\{ \frac{n}{n-1} \cdot \frac{S(\hat{\mathbf{a}}_{(i)}) - [y_i - f(\mathbf{x}_i, \hat{\mathbf{a}}_{(i)})]^2}{S(\hat{\mathbf{a}}_{(i)})} \right.$$

$$\left. - n + (n-1) \cdot \frac{S(\hat{\mathbf{a}}_{(i)})}{S(\hat{\mathbf{a}}_{(i)}) - [y - f(\mathbf{x}_i, \hat{\mathbf{a}}_{(i)})]^2} \right\}. \tag{24}$$

By substituting its one step approximation for the $\hat{\mathbf{a}}_{(i)}$ estimate in equation (24) we obtain the approximative likelihood distance LDA_i.

6. SIMULATED RESULTS

In order to compare influence measures of individual points or to estimate the masking effect of two aberrant points a simple Michelis-Menten model has been chosen

$$f(x_i, \mathbf{a}) = a_1 x_i / (a_2 + x_i). \tag{25}$$

Experimental points have been obtained by simulation. It has been stated that $a_1 = a_2 = 1$ and into this dependence random errors $\varepsilon_i \sim N(0, \sigma^2)$ for

$\sigma = 0.05$ have been introduced. The resulting points for $n = 7$ are outlined in Table I.

TABLE I

Simulated experimental data

x_i	0.25	0.5	0.75	1	1.25	1.5	1.75
y_i	0.204	0.424	0.429	0.504	0.556	0.652	0.643

TABLE II

Parameter estimates and their relative biases

Modif. point No: h	Added quantity	$\hat{\mathbf{a}}_1$	$\hat{\mathbf{a}}_2$	$\mathbf{b}_{R1}$ [%]	$\mathbf{b}_{R2}$ [%]
unmod.	0	0.939	0.786	1.53	3.926
4	$0.5y_h$	0.947	0.67	7.84	22.06
4	y_h	0.976	0.607	24.9	74.4
7	$0.5y_h$	2.725	3.986	87.18	116.6
7	y_h	$2.61 \cdot 10^4$	$4.55 \cdot 10^4$	$-5 \cdot 10^{11}$	dtto
4	$0.5y_h$	1.709	1.829	35.09	601
7	$0.5y_h$				
4	y_h	4.166	5.23	543.4	684.6
7	y_h				

In the first row of Table II corresponding parameter estimates $\hat{\mathbf{a}}$ are given together with their relative biases.

Influential points were simulated for $h = 4$ and 7 (or both) by adding the quantity of $0.5y_h$ and y_h, respectively, to the one given in Table I.

Computation results are summarized in Tables II, III, IV. In Table II nonlinear least squares estimators $\hat{\mathbf{a}}$ and their relative biases $\mathbf{b}_R$ are given.

TABLE III
Linearized and nonlinear influence measures

Modif. point No: h	Added quantity	RS_h	CD_h	FDT_h	FDA_h	$LDST_h$	$LDSA_h$
4	$0.5y_h$	1.97_M	1.56_M	0.42_M	0.19_M	5.51_M	4.10_M
4	y_h	3.31_M	0.81_M	0.46_M	0.21_M	31.1_M	17.0_M
7	$0.5y_h$	1.59	3.74_M	2.91_M	0.14_M	11.6_M	0.35_M
7	y_h	2.42	2.57_M	3.06_M	$2 \cdot 10^{-9}$	53.6_M	0.14_M
4	$0.5y_h$	2.06	0.92_M	0.29_M	0.07	0.23_M	0.09_M
7		1.71	1.07_M	0.96_M	0.08_M	0.74_M	0.07_M
4	y_h	4.52_M	0.61	0.25	0.04	0.33_M	0.12_M
7		2.81_M	1.15_M	1.06_M	0.13_M	0.78_M	0.10_M

TABLE IV
Influence measures based on quadratic approximation

Modif. point No: k	Added quantity	KV_h	$\mathbf{b}_{(h)1}$	$\mathbf{b}_{(h)2}$
4	$0.5y_h$	0.13	0.075	0.15
4	y_h	0.43	0.243	0.46
7	$0.5y_h$	3.44_M	4.37_M	8.28_M
7	y_h	$3 \cdot 10^{16}{}_M$	$3 \cdot 10^{16}{}_M$	$4.9 \cdot 10^{16}{}_M$
4	$0.5y_h$	0.53	0.65	1.81
7		1.16_M	1.02_M	1.77_M
4	y_h	7.91	25.9	41.3
7		22.74_M	42.63_M	65.6_M

In Table III one-step approximations of Jackknife residuals RS_h are given and the influence measures based upon the influence curve, viz. CD_h, FDT_h, FDA_h, $LDST_h = LDT_h/\chi_{.95}^2(2)$ and $LDSA_h = LDA_h/\chi_{.95}^2(2)$. Finally, Table IV presents vector components $\mathbf{b}_{(h)}$ and measures KV_i based on the bias of the parameters.

Occurrence of the M subscript with a number in Tables II, III, IV indicates that absolute value of a number represents the highest element (or one of the two in modifications with couples of points) from all corresponding characteristics for $i = 1, \ldots, n$.

7. CONCLUSION

On the strength of data from Tables II, III, IV the following conclusions may be postulated for the case of the Michelis-Menten model:

(i) The position of influential points strongly affects both parameter estimates and their relative bias.
(ii) Approximate Jackknife residuals RS_i do not identify always the outliers correctly.
(iii) Influence measures based on bias are not always correct for this model.
(iv) Relatively best results were obtained by using influence measures based upon the influence function.
(v) For qualitative estimation of most influential points both normed empirical and normed influence sample function may be employed.

Provided that more influential points are presented nor the procedures based on correct likelihood distance do not provide the data for quantitative evaluation of influence measure of individual points (LDST_i values are less than one).

On the basis of the above one-step approximation of likelihood distance LDSA_i may be recommended for qualitative identification of single influential points.

REFERENCES

Beckmann, R.J. and Cook, R.D. (1983). Outliers. *Technometrics*, 25, 119-149.

Belsey, D.A., Kuh, E. and Welsch, R.E. (1980). *Regression Diagnostics*. Wiley, New York.

Cook, R.D. and Weisberg, S. (1982). *Residuals and Influence in Regression*. Chapman and Hall, New York.

Cook, R.D. and Tsai, C.L. (1984). Bias in nonlinear regression. Technical Summary Report 2645, University of Wisconsin, Mathematical Research Center.

Dennis, J.E. and Mei, H.W. (1979). Two new unconstrained optimization algorithms. *Journal of the Optimization Theory and Applications*, 28, 457-477.

Gray, J.B. and Ling, R.F. (1984). K-clustering as a detection tool for influential subsets in regression. *Technometrics*, 26, 305-330.

Jennrich, R.I. (1969). Asymptotic properties of nonlinear least squares estimators. *The Annals of Mathematical Statistics*, 40, 633-648.

Militký, J. and Čáp, J. (1985). Detection of influential points in general nonlinear problems. *Proceedings of the 2nd International Symposium on System Analysis and Simulation*, Berlin.

Militký, J. and Čáp, J. (1987). User's oriented MINOPT program for nonlinear regression. *Proc. Conf. Chemical Engineering Fundamentals*, Taormina.

Simonoff, J.S. and Tsai, C.L. (1986). Jackknife-based estimators and confidence regions in nonlinear regression. *Technometrics*, 28, 103-112.

Received 30 January 1987
Revised 2 September 1987

Research Institute for Textile Finishing
Nejedlého 770
54428 Dvur Králové nad Labem
Czechoslovakia

Department of Physics
Faculty of Mechanical Engineering
Czech Technical University
Suchbátarova 4
16607 Prague 6
Czechoslovakia

JZD AK Slušovice
Applied Cybernetics Centre
76315 Slušovice
Czechoslovakia

Proc. Second International Tampere Conference in Statistics
(Tampere, Finland, 1-4 June 1987)
Tarmo Pukkila and Simo Puntanen, *Editors*

pp. 603 - 614

Algebra of subspaces with applications to problems in statistics

K. NORDSTRÖM and D. von ROSEN
University of Helsinki, Helsinki, Finland
and University of Stockholm, Stockholm, Sweden

Key words and phrases: Lattice of subspaces, commuting subspaces, decomposition of vector spaces, orthogonal projector, singular Gauss-Markov model, growth curve model.

ABSTRACT

This paper is concerned with certain fundamental laws as well as basic principles of the algebra of linear subspaces pertaining to the decomposition of finite-dimensional vector spaces. The properties of intersections and sums of subspaces appear to be best understood by observing that the totality of subspaces forms a complete modular lattice with respect to set-inclusion. Treating subspaces as elements of this lattice, a systematic and unified account of the most important relations is given. The basic properties of commuting subspaces are given, and the implications of commutativity on the orthogonal decomposition of a vector space are indicated. Decompositions of vector spaces are discussed and applied to the singular Gauss-Markov model as well as to the growth curve model with linear constraints on the parameters.

1. INTRODUCTION

The starting point for this paper is the search for a general setting that would allow a systematic and unified account of relations between subspaces in a finite-dimensional vector space. The lattice-theoretic approach adopted here seems to provide such a setting. This is perhaps known to mathematicians working in lattice theory and functional analysis, but does not appear to have been exploited by statisticians. It is felt that the theory of lattices and, in particular, the theory of orthomodular lattices contain many interesting concepts and results that may be useful in statistics. As an example one could mention the concept of commuting subspaces, which seems to be helpful for a more thorough understanding of the underlying geometry of many results in the theory of linear models as well as multivariate analysis. In addition this concept provides a suitable framework for a systematic presentation and comparison of numerous results involving e.g. orthogonal subspace-decompositions and projectors that appear scattered in the statistical literature as unrelated "lemmas". (A detailed study of this will be made elsewhere.)

Due to lack of space most of the results are stated without proof. In addition to the references mentioned in the text, detailed proofs may be found in Nordström & von Rosen (1987). As to general references, good expositions of the theory of finite-dimensional vector spaces can be found in Jacobson (1953), Shephard (1966) and Halmos (1974), the first mentioned stressing connections with abstract algebra. For concepts and results in general lattice theory the reader is referred to Birkhoff (1967) and Grätzer (1978). An easily accessible account of orthomodular lattice theory as well as its relationship with von Neumann algebras and continuous geometry is contained in the expository paper by Holland (1970), whereas general treatments of orthomodular lattice theory can be found in Kalmbach (1983) and Beran (1985). Finally we may note that the extension of the results of this paper to infinite-dimensional Hilbert spaces requires certain non-trivial modifications [e.g. Holland (1970)].

2. LATTICE OF SUBSPACES

We consider a finite-dimensional vector space V, and denote its subspaces by $A, B, C,\ldots$ (possibly indexed) and the totality of subspaces of V by Λ. It is obvious that Λ is partially ordered with respect to set-inclusion. Given arbitrary subspaces A and B, the intersection $A \cap B$ and sum $A + B$ act as greatest lower bound (g.l.b.) and least upper bound (l.u.b.) of $\{A, B\}$, respectively. Consequently Λ [or strictly speaking the ordered pair $(\Lambda, \subset)$] forms a *lattice*. Indeed, if $\{A_i\}$ is any subset of Λ, the subspaces $\cap A_i$ and ΣA_i act as g.l.b. and l.u.b. of $\{A_i\}$, respectively, showing the lattice Λ to be *complete*. Moreover, given an arbitrary subspace A there always exists at least one *direct complement* of A, i.e., a subspace B such that $A \cap B = \{0\}$ and $A + B = V$, and hence Λ is *(directly) complemented.* In the following lemma we have brought together some of the most basic algebraic laws for sums and intersections of subspaces.

LEMMA 2.1. *Let A, B and C be arbitrary elements of Λ. Then*

(i) $A \cap A = A, \quad A + A = A,$ *idempotent laws;*

(ii) $A \cap B = B \cap A, \quad A + B = B + A,$ *commutative laws;*

(iii) $A \cap (B \cap C) = (A \cap B) \cap C, \quad A + (B + C) = (A + B) + C,$ *associative laws;*

(iv) $A \cap (A + B) = A + (A \cap B) = A,$ *absorptive laws;*

(v) $A \subset B \Leftrightarrow A \cap B = A,$
$A \subset B \Leftrightarrow A + B = B,$ *consistency laws;*

(vi) $A \subset B \Rightarrow A \cap C \subset B \cap C,$
$A \subset B \Rightarrow A + C \subset B + C,$ *isotonicity of compositions;*

(vii) $A \cap (B + C) \supset (A \cap B) + (A \cap C),$
$A + (B \cap C) \subset (A + B) \cap (A + C),$ *distributive inequalities;*

(viii) $A \supset C \Rightarrow A \cap (B + C) = (A \cap B) + C,$
$A \subset C \Rightarrow A + (B \cap C) = (A + B) \cap C,$ *modular laws.* (2.1)

A proof of (viii) is given e.g. by Köthe (1969, p.58).

By virtue of (2.1), Λ is found to be a *modular lattice.* The properties (vii) and (viii) suggest the occurrence of dual pairs of relations, one being obtainable from the other by inverting the relation of set-inclusion and interchanging the compositions $\cap$ and $+$. To avoid redundancy we state in the sequel only one of a pair of dual relations, the other relation holding automatically by virtue of the *Principle of Duality* [e.g. Grätzer (1978, p.6)]. Lemma 2.1 can be used e.g. to establish relations between *polynomials of subspaces.* In the following corollary we have collected a few easily verified identities between polynomials formed from three subspaces.

COROLLARY 2.1. *Let A, B and C be arbitrary elements of* Λ. *Then*

(i) $$A \cap (B + C) = A \cap [B + (C \cap D)] \qquad \forall D \supset A + B,$$

(ii) $$[A \cap (B + C)] + B = [(A + B) \cap C] + B,$$

(iii) $$[A \cap (B + C)] + (B \cap C) = [A + (B \cap C)] \cap (B + C),$$

(iv) $$(A \cap B) + (A \cap C) = A \cap [B + (A \cap C)],$$

(v) $$(A \cap B) + (A \cap C) + (B \cap C) = [A + (B \cap C)] \cap [B + (A \cap C)]$$
$$= \{[A \cap (B + C)] + (B \cap C)\} \cap \{[B \cap (A + C)] + (A \cap C)\},$$

(vi) $$[(A \cap B) + (A \cap C)] \cap [(A \cap B) + (B \cap C)] = A \cap B.$$

We may note that relations like those exhibited in Lemma 2.1 and Corollary 2.1 can be useful in proving various results in linear algebra as well as in linear model theory. As an example one could mention the problem of idempotency and commutativity of the product of two projectors, and the related "reverse order law" -problem for generalized inverses [Shinozaki & Sibuya (1974 a, b)]. The above relations can also be used to give simple proofs of some less well-known conditions for disjointness of a finite set of subspaces [von Rosen & Nordström (1987)].

As a different kind of consequence of the modularity of Λ we obtain the following.

COROLLARY 2.2. *Let* B *and* C *be* comparable *elements of* Λ, i.e., *subspaces such that* $B \subset C$ or $C \subset B$ *holds. Then*

$$A \cap B = A \cap C,\ A + B = A + C \Rightarrow B = C, \quad \textit{cancellation law.}$$

Considering subspace polynomials formed from a finite set of subspaces, the situation is, in general, much more complex. A special case of this more general situation, which is important in statistical applications, is described in Section 3.

3. ORTHOGONALITY AND COMMUTATIVITY OF SUBSPACES

Let the finite-dimensional vector space V be furnished with a non-degenerate inner product. To every subspace A of V there corresponds a unique perpendicular subspace $A^{\perp}$, the *orthocomplement* of A. Replacing $+$ by $\boxplus$ if the sum consists of orthogonal subspaces, we have the following well-known properties of orthocomplementation.

LEMMA 3.1. *Let A and B be arbitrary elements of* Λ. *Then*

(i) $V = A \boxplus A^{\perp}$, $(A^{\perp})^{\perp} = A$, *projection theorem;*

(ii) $A \subset B \Rightarrow A^{\perp} \supset B^{\perp}$, *antitonicity of orthocomplementation;*

(iii) $A \subset B \Rightarrow B = A \boxplus (A^{\perp} \cap B)$, *orthomodular law;*

(iv) $(A \cap B)^{\perp} = A^{\perp} + B^{\perp}$, $(A + B)^{\perp} = A^{\perp} \cap B^{\perp}$, *de Morgan's laws.*

By virtue of Lemma 3.1, Λ satisfies the axioms defining an abstract orthomodular lattice [Birkhoff (1967, pp. 52 - 53)]. This is an important observation, since it puts concepts and results of the theory of abstract orthomodular lattices at our disposal.

It is interesting to note that the orthomodular lattice structure of Λ carries over in an obvious way to the set of idempotent and self-adjoint linear operators defined on V [Holland (1970, pp. 46 - 49)]. This provides a general setting for studying e.g. combinations and commutativity of orthogonal projectors.

The rest of this section is devoted to the important concept of commuting subspaces, which appears to have been introduced in a more general context by Birkhoff and von Neumann (1936).

DEFINITION 3.1. *Let* $\{A_i\}$ *be a finite set of subspaces of* V.

(i) *The subspaces* $\{A_i\}$ *are said to be* commutative *iff* $A_i = (A_i \cap A_j) \boxplus (A_i \cap A_j^{\perp})$ *(denoted* $A_i | A_j$*) holds for all* i, j.

(ii) *Given subspaces* A_i *and* A_j *we define the* commutator *of* A_i, A_j *as*

$$\mathrm{com}(A_i, A_j) = (A_i \cap A_j) \boxplus (A_i \cap A_j^{\perp}) \boxplus (A_i^{\perp} \cap A_j) \boxplus (A_i^{\perp} \cap A_j^{\perp}). \tag{3.1}$$

The symmetry of the notation introduced in (i) is justified in Corollary 3.1. We may also note that some authors refer to (the orthocomplement of) the r.h.s. of (3.1) as the (upper) lower commutator of A_i, A_j [e.g. Beran (1985, p. 86)].

The following theorem provides an alternative condition for commutativity of subspaces as well as a useful characterization of commutativity in terms of orthogonal projectors. It should be mentioned that the equivalence of conditions (i), (iii), (iv) and (v) has been proved by Rao and Yanai (1979, p. 9). [The equivalence of conditions (i) and (iii) has also been noted by Afriat (1957, p. 801).] Let P_i, P_{ij}, Q_{ij} and P_{i+j} denote the orthogonal projectors onto A_i, $A_i \cap A_j$, $A_i \cap A_j^{\perp}$ and $A_i + A_j$, respectively.

THEOREM 3.1. *The subspaces* $\{A_i\}$ *are commutative iff any of the following equivalent conditions holds:*

(i) $$A_i \cap (A_i \cap A_j)^{\perp} \perp A_j \cap (A_i \cap A_j)^{\perp} \quad \textit{for all } i,j, \tag{3.2}$$

(ii) $$P_i = P_{ij} + Q_{ij} \quad \textit{for all } i,j,$$

(iii) $$P_i P_j = P_j P_i \quad \textit{for all } i,j, \tag{3.3}$$

(iv) $$P_i P_j = P_{ij} \quad \textit{for all } i,j,$$

(v) $$P_{i+j} = P_i + P_j - P_{ij} \quad \textit{for all } i,j.$$

Proof: Obviously $A_i \mid A_j \Leftrightarrow P_i = P_{ij} + Q_{ij}$, and hence it suffices to prove (ii) $\Leftrightarrow$ (iii).

Assume $P_i = P_{ij} + Q_{ij}$ whence $P_i P_j = P_{ij} P_j + Q_{ij} P_j = P_{ij} P_j = P_{ij} = P_j P_{ij} = P_j P_{ij} + P_j Q_{ij} = P_j P_i$. Conversely, $P_i P_j = P_j P_i$ implies $P_i(I - P_j) = (I - P_j)P_i$. Since (iii) implies (iv), we have $P_i P_j = P_{ij}$ and $P_i(I - P_j) = Q_{ij}$ yielding $P_i = P_i P_j + P_i(I - P_j) = P_{ij} + Q_{ij}$. □

Condition (3.2) expresses orthogonality of A_i and A_j modulo $A_i \cap A_j$. If A_i and A_j are subspaces corresponding to factors i and j in an analysis of variance model, then (3.2) is the usual condition for orthogonality of i and j [e.g. Tjur (1984, p. 40)]. We may also note that a pair of subspaces A_i and A_j satisfying (3.2) is, among others, referred to in the literature as "orthogonally incident" [Afriat (1957, p. 801)] or "geometrically orthogonal" [Tjur (1984, p. 40)].

Employing e.g. condition (3.3) we obtain immediately the following.

COROLLARY 3.1. *Let* A_i *and* A_j *be subspaces of* V. *Then*

(i) $A_i \subset A_j \Rightarrow A_i \mid A_j$,

(ii) $A_i \perp A_j \Rightarrow A_i \mid A_j$,

(iii) $A_i \mid A_j \Rightarrow A_i \mid A_j^{\perp}$,

(iv) $A_i \mid A_j \Rightarrow A_j \mid A_i$.

To clarify the implications of commutativity, we make use of the following version of a general result mentioned, but not proved, in Holland (1970, p. 80).

THEOREM 3.2. *Let* $\{A_i\}$ *be a finite set of subspaces of* V *and let* B *be a subspace of* V *that commutes with each of the subspaces* A_i. *Then*

(i) $B \mid \Sigma A_i$, $B \cap (\Sigma A_i) = \Sigma\ (B \cap A_i)$,

(ii) $B \mid \cap A_i$, $B + (\cap A_i) = \cap\ (B + A_i)$.

An elementary proof of Theorem 3.2 is given in Nordström & von Rosen (1987).

COROLLARY 3.2. *Let* $\{A_i\}$ *and* $\{B_j\}$ *be finite sets of subspaces of* V *such that* A_i *commutes with* B_j *for all* i, j. *Then*

(i) $(\Sigma A_i) \cap (\Sigma B_j) = \Sigma\ \Sigma\ (A_i \cap B_j)$,

(ii) $(\cap A_i) + (\cap B_j) = \cap\ \cap (A_i + B_j)$.

COROLLARY 3.3. *Let* $\{A_i\}$ $(1 \leq i \leq n)$ *be a finite set of orthogonal subspaces of* V *and let* B *be a subspace of* V. *Then the following conditions are equivalent:*

(i) $B \mid A_i$ *for all i,*

(ii) $B = (B \cap A_1) \boxplus \ldots \boxplus (B \cap A_n) \boxplus (B \cap A_1^{\perp} \cap \ldots \cap A_n^{\perp}),$

(iii) $V = (B \cap A_1) \boxplus (B^{\perp} \cap A_1) \boxplus \ldots \boxplus (B \cap A_n) \boxplus (B^{\perp} \cap A_n) \boxplus$
$(B \cap A_1^{\perp} \cap \ldots \cap A_n^{\perp}) \boxplus (B^{\perp} \cap A_1^{\perp} \cap \ldots \cap A_n^{\perp}).$

These two corollaries clearly spell out the implications of commutativity of subspaces. From Corollary 3.2 we see explicitly that distributivity holds under commutativity, whereas Corollary 3.3 exhibits simple orthogonal decompositions of a vector space that are valid under commutativity.

The following easily verified result expresses the relationship between commutativity and commutators.

THEOREM 3.3. *The subspaces* $\{A_i\}$ *are commutative iff* $\mathrm{com}(A_i, A_j) = V$ *for all* i, j.

We end this section with a result relating commutativity with the concept of an invariant subspace. In the sequel P stands for the orthogonal projector onto the subspace appearing as a subscript.

THEOREM 3.4. *Let* A *be a subspace of* V *and let* $T : V \rightarrow V$ *be a normal operator with range* R. *If* A *is* T*-invariant (i.e.* $TA \subset A$*), then* A *commutes with* R.

Proof: Let T^* and T^+ denote the adjoint and Moore-Penrose inverse of T, respectively. By virtue of normality T is reduced by the pair $(A, A^{\perp})$, whence $TP_A = P_A T$ [Halmos (1974, pp. 77, 162)]. Using basic properties of the Moore-Penrose inverse [e.g. Rao & Mitra (1971, p. 67)], we have $TT^+ = T^+T$ and $(T^+)^*T^+ = (TT^*)^+ = (T^*T)^+ = T^+(T^+)^*$. Since T and T^+ are commuting normal operators their eigenvectors coincide, implying that T^+ is also reduced by $(A, A^{\perp})$. Hence $T^+P_A = P_A T^+$, yielding $P_A P_R = P_A TT^+ = TP_A T^+ = TT^+P_A = P_R P_A$. The result now follows from Theorem 3.1. □

COROLLARY 3.4. *Let* A *be a subspace of* V *and let* $T : V \rightarrow V$ *be a non-negative definite self-adjoint operator with range* R. *If* A *is* T*-invariant, then* A *commutes with* R.

Consider the coordinate-free Gauss-Markov set-up [see Kruskal (1961) or Drygas (1970)]

$$\{y : y = \mu + e, \mu \in M\}, \; E(e) = 0, \quad C(e) = \Sigma$$

with singular covariance operator Σ having range R. Now assume $P_M P_R = P_R P_M$ holds. Then Theorems 3.1 and 3.3 together with the fundamental direct sum decomposition $M+R = M \oplus \Sigma M^{\perp}$ given by Rao (1974) yield the following decomposition of the observation space

$$M + R = [(M \cap R) \boxplus (M \cap N)] \oplus \Sigma M^{\perp}. \tag{3.4}$$

Comparing (3.4) with the corresponding decomposition given by Nordström (1985, p. 238), we see that under the assumption of commutativity the regression subspace M may be decomposed in a rather simple manner to yield the stochastic and deterministic components of y in M. We may note that, by virtue of Corollary 3.4, commutativity holds if M is Σ-invariant. [This is the well-known necessary and sufficient condition for coincidence of the simple least squares estimator of μ with the BLUE of μ given by Zyskind (1967) and Kruskal (1968).] In this case (3.4) is an orthogonal decomposition, since $\Sigma M^{\perp} = M^{\perp} \cap R$.

Perhaps more important areas of application of the concept of commuting subspaces include analysis of variance models with orthogonal factors (compare the remarks following Theorem 3.1) and variance components models. A study of these models will be made elsewhere.

4. APPLICATION

In this section we consider the growth curve model with linear restrictions on the parameters describing the mean. The model is defined in the following way. Let $\mathbf{X} : p \times n$ and $E(\mathbf{X}) = \mathbf{ABC}$, where $\mathbf{A}: p \times q$ and $\mathbf{C}: k \times n$ are known matrices and $\mathbf{B}: q \times k$ is the matrix of parameters. The columns of $\mathbf{X}$ are assumed to be independently normally distributed with an unknown dispersion matrix $\mathbf{\Sigma}$. Moreover, we assume the restrictions

$$\mathbf{D}_1\mathbf{B}\mathbf{E}_1 = \mathbf{0}, \; \mathbf{D}_2\mathbf{B}\mathbf{E}_2 = \mathbf{0}, \tag{4.1}$$

where the $\mathbf{D}$'s and $\mathbf{E}$'s are known matrices. In the following we indicate how to obtain maximum likelihood estimators for $\mathbf{B}$ and $\mathbf{\Sigma}$ when $R(\mathbf{E}_1)$

$\subset R(\mathbf{E}_2)$ holds. [$R(\cdot)$ denotes the range space of a matrix argument.] We will see that the restricted model can be reduced to the model

$$E(\mathbf{X}) = \mathbf{A}_1\mathbf{B}_1\mathbf{C}_1 + \mathbf{A}_2\mathbf{B}_2\mathbf{C}_2 + \mathbf{A}_3\mathbf{B}_3\mathbf{C}_3,\ R(\mathbf{C}_3') \subset R(\mathbf{C}_2') \subset R(\mathbf{C}_1'), \tag{4.2}$$

where the $\mathbf{A}$'s and $\mathbf{C}$'s are known, the $\mathbf{B}$'s are unknown and $\mathbf{X}$ is as above. Within the model (4.2) maximum likelihood estimators of the $\mathbf{B}$'s and $\boldsymbol{\Sigma}$ have been derived by von Rosen (1984).

Before we proceed we need to state some consequences of Lemma 3.1 (iii).

LEMMA 4.1. *Let A and B be arbitrary elements of Λ. Then*

$$V = \{[A^{\perp} \cap (A + B)] \oplus [B^{\perp} \cap (A + B)]\} \boxplus (A + B)^{\perp} \boxplus (A \cap B).$$

For the purposes of this section we need an extension of Lemma 4.1 to tensor spaces. Let V and W be inner product spaces, and let the inner product on the tensor space $V \otimes W$ be defined in the usual manner [e.g. Greub (1978)].

THEOREM 4.1. *Let the A's and B's be arbitrary subspaces of V and W, respectively. Then*

(i) $$\begin{aligned} V \otimes W = {} & [(A_1 \otimes B_1) + (A_2 \otimes B_2)] \boxplus (\{A_2^{\perp} \otimes [(B_1 + B_2) \cap B_1^{\perp}]\} \\ & \oplus \{A_1^{\perp} \otimes [(B_1 + B_2) \cap B_2^{\perp}]\}) \boxplus [(A_1 + A_2)^{\perp} \otimes (B_1 \cap B_2)] \\ & \boxplus [V \otimes (B_1 + B_2)^{\perp}], \end{aligned}$$

(ii) $A_1 \subset A_2 \Rightarrow$
$$\begin{aligned} V \otimes W = {} & [(A_1 \otimes B_1) + (A_2 \otimes B_2)] \boxplus \{A_1^{\perp} \otimes [(B_1 + B_2) \cap B_2^{\perp}]\} \\ & \boxplus (A_2^{\perp} \otimes B_2) \boxplus [V \otimes (B_1 + B_2)^{\perp}]. \end{aligned}$$

Treating (4.1) as a linear equation in $\mathbf{B}$ and using a vectorized form of this system of equations, we immediately obtain from Theorem 4.1 a representation of solution. Thus when $R(\mathbf{E}_1) \subset R(\mathbf{E}_2)$, Theorem 4.1 (ii) implies that a solution to (4.1) is given by

$$\mathbf{B} = \mathbf{T}_1\boldsymbol{\theta}_1\mathbf{E}_1^{\circ\prime} + \mathbf{D}_2'\boldsymbol{\theta}_2\mathbf{E}_2^{\circ\prime} + (\mathbf{D}_1' : \mathbf{D}_2')^{\circ}\boldsymbol{\theta}_3 \tag{4.3}$$

where $'$ denotes the transpose of a matrix, the $\boldsymbol{\theta}$'s are arbitrary matrices, $\mathbf{E}_1^{\circ}$, $\mathbf{E}_2^{\circ}$ and $(\mathbf{D}_1' : \mathbf{D}_2')^{\circ}$ are any matrices spanning the orthocomplements of $R(\mathbf{E}_1)$, $R(\mathbf{E}_2)$ and $R(\mathbf{D}_1') + R(\mathbf{D}_2')$, respectively, and $\mathbf{T}_1$ is any matrix spanning $[R(\mathbf{D}_1')+R(\mathbf{D}_2')] \cap R(\mathbf{D}_2')^{\perp}$. Plugging (4.3) into the ordinary growth curve model yields

$$E(\mathbf{X})= \mathbf{A}\mathbf{T}_1\boldsymbol{\theta}_1 \mathbf{E}_1^{\circ\prime}\mathbf{C}+\mathbf{A}\mathbf{D}_2'\boldsymbol{\theta}_2\mathbf{E}_2^{\circ\prime}\mathbf{C}+ \mathbf{A}(\mathbf{D}_1' : \mathbf{D}_2')^{\circ}\boldsymbol{\theta}_3\mathbf{C}. \qquad (4.4)$$

Since, by assumption, $R(\mathbf{C}'\,\mathbf{E}_1^{\circ}) \subset R(\mathbf{C}'\,\mathbf{E}_2^{\circ}) \subset R(\mathbf{C}')$ (4.4) is of the form (4.2). Hence maximum likelihood estimators of the $\boldsymbol{\theta}$'s are available and therefore, using (4.3), the maximum likelihood estimator of $\mathbf{B}$ can be found.

ACKNOWLEDGEMENTS

The authors would like to thank the referees for constructive criticism which helped to improve the presentation. The work of the first author was done while on leave of absence at Department of Mathematics and Statistics, University of Pittsburgh, U.S.A. Financial support from the Academy of Finland is gratefully acknowledged.

REFERENCES

Afriat, S.N. (1957). Orthogonal and oblique projectors and the characteristics of pairs of vector spaces. *Proceedings of the Cambridge Philosophical Society*, 53, 800 - 816.

Beran, L. (1985). *Orthomodular Lattices. Algebraic Approach.* D. Reidel Publishing Company, Dordrecht.

Birkhoff, G. (1967). *Lattice Theory.* Third Edition. American Mathematical Society, Providence, R.I.

Birkhoff, G. & von Neumann, J. (1936). The logic of quantum mechanics. *Annals of Mathematics*, 37, 823 - 842.

Drygas, H. (1970). *The Coordinate-free Approach to Gauss-Markov estimation.* Springer, Berlin.

Greub, W.H. (1978). *Multilinear Algebra.* Second Edition. Springer, New York.

Grätzer, G. (1978). *General Lattice Theory.* Birkhäuser Verlag, Basel.

Halmos, P.R. (1974). *Finite-Dimensional Vector Spaces.* Second Edition. Springer, New York.

Holland Jr., S.S. (1970). The current interest in orthomodular lattices. *Trends in Lattice Theory*, 41-126, (J.C. Abbott, Editor). van Nostrand, New York.

Jacobson, N. (1953). *Lectures in Abstract Algebra.* van Nostrand, New York.

Kalmbach, G. (1983). *Orthomodular Lattices.* Academic Press, London.

Kruskal, W.H. (1961). The coordinate-free approach to Gauss-Markov estimation and its application to missing and extra observations. *Proceedings of the Fourth Berkeley Symposium*, I, 435-451.

Kruskal, W.H. (1968). When are Gauss-Markov and least squares estimators identical? A coordinate-free approach. *Annals of Mathematical Statistics*, 39, 70-75.

Köthe, G. (1969). *Topological Vector Spaces I.* Springer, New York.

Nordström, K. (1985). On a decomposition of the singular Gauss-Markov model. *Lecture Notes in Statistics*, 35, 231-245. Springer, New York.

Nordström, K. & von Rosen, D. (1987). Algebra of subspaces with applications to problems in statistics. Research report No. 61, Department of Statistics, University of Helsinki.

Rao, C.R. (1974). Projectors, generalized inverses and the BLUEs. *Journal of the Royal Statistical Society B*, 36, 442-448.

Rao, C.R. & Mitra, S.K. (1971). *Generalized Inverse of Matrices and its Applications.* Wiley, New York.

Rao, C.R. & Yanai, H. (1979). General definition and decomposition of projectors and some applications to statistical problems. *Journal of Statistical Planning Inference*, 3, 1 - 17.

von Rosen, D. (1984). Maximum likelihood estimators in multivariate linear normal models with special references to the growth curve model. Research Report No. 135, Institute of Actuar. Mathematics and Mathematical Statistics, University of Stockholm.

von Rosen, D. & Nordström, K. (1987). On tools and techniques for decomposing vector spaces with applications to statistical problems. Unpublished manuscript.

Shephard, G.C. (1966). *Vector Spaces of Finite Dimension.* Oliver & Boyd, London.

Shinozaki, N. & Sibuya, M. (1974a). The reverse order law $(AB)^- = B^-A^-$. *Linear Algebra and its Applications*, 9, 29-40.

Shinozaki, N. & Sibuya, M. (1974b). Product of projectors. Scientific center report, IBM Japan, Tokyo.

Tjur, T. (1984). Analysis of variance models in orthogonal designs. *International Statistical Review*, 52, 33 - 81.

Zyskind, G. (1967). On canonical forms, nonnegative covariance matrices and best and simple least squares linear estimators in linear models. *Annals of Mathematical Statistics*, 38, 1092-1109.

Received 5 January 1987
Revised 4 June 1987

Department of Statistics
University of Helsinki
Aleksanterinkatu 7
SF-00100 Helsinki
Finland

Department of Mathematical Statistics
University of Stockholm, Box 6701
S-113 85 Stockholm
Sweden

Proc. Second International Tampere Conference in Statistics
(Tampere, Finland, 1-4 June 1987)
Tarmo Pukkila and Simo Puntanen, *Editors*

pp 615 - 626

Robust regression for the heavy contaminated sample

G.A. OSOSKOV

Joint Institute for Nuclear Research, Dubna, U.S.S.R.

Key words and phrases: Robust estimate, location and scale parameters, Studentizing.

ABSTRACT

Robust estimates of regression parameters are studied for samples heavy contaminated by uniformly distributed noise. The maximum likelihood approach is used to define an optimal weight function and joint estimates of location and scale parameters. Special scaling of residuals is proposed to overcome the estimate non-stability due to so-called leverage points.

1. INTRODUCTION

The robust estimations of regression parameters of some models arising in particular in the charged particle track recognition problems in high energy physics is considered. In the case of automatic measurements the experimental data obtained from track detectors consist of a useful part related to track to be found as well of signals of background tracks, fiducials and other noise points. The noise points are usually uniformly distributed. The error distribution of experimental data can be described as the gross-error model

$$f(x) = (1-\varepsilon)\phi(x) + \varepsilon h(x) \tag{1}$$

with $\phi(x) = [\sigma(2\pi)^{\frac{1}{2}}]^{-1} \exp(-x^2/2\sigma^2)$ and some long-tailed noise distribution $h(x)$. Because of reasons mentioned above $h(x)$ is supposed to be uniform, i.e. $h(x) = h_0$ in sufficiently large interval of the length $1/h_0 >> \sigma$ and $\varepsilon > 1/2$ (even close to 1).

These models are usually explored by the pattern recognition or clustering methods. The robust estimates are also applicable in these cases, but with certain necessary modifications or auxiliary means due to the heavy contamination of data and to the influence of the so-called leverage points (outlying point having too large residuals after some gaps in the factor space).

The related problems considered in this paper are:

- optimal choice of the weight function in the weighted least square procedure used to get the M-estimates of regression parameters;
- joint estimates of location and scale parameters;
- overcoming the influence of the leverage points.

Approaches are proposed below to solve these problems and some results of their exploring by Monte Carlo simulations. Such approaches may be useful for the robust estimation theory itself.

2. THE CHOICE OF THE WEIGHT FUNCTION FOR M-ESTIMATION

It is convenient to begin with a one-parameter model of estimating the location parameter $a = Ex$ from a sample $x_1, x_2, \ldots, x_n$, where $x \sim a + (1-\varepsilon)\rho(x) + \varepsilon h(x)$, the functions ρ, h are described above. We use Huber's M-estimates (Huber 1981):

$$L(a,\sigma) = \sum_{i=1}^{n} \rho\left(\frac{x_i - a}{\sigma}\right) \to \inf_a \tag{2}$$

or

$$a = \sum_i w_i x_i / \sum_i w_i \tag{3}$$

with the weights $w_i = w[(x_i - a)/\sigma]$, where $w(t) = \rho'(t)/t$ is the weight function of the estimator. The usual requirements on (2) are: the function $\rho(t)$ must be even, C^2-smooth, not decreasing for $t>0$, $\rho(0) = 0$, $\rho(t) \sim t^2/2$ as $t \to 0$ (i.e. $w(t) \to 1$ as $t \to 0$), the function $w(t) \geq 0$ and does not increase for

$t > 0$, and the estimate (2) must be shift- and scale-invariant, as well as the estimate of σ if its value is unknown.

The problem of the choice of the function $\rho(t)$ [or $w(t)$] is widely discussed in the literature on robust statistics. Unbounded convex functions $\rho(t)$ provide the uniqueness of the estimate (2), its consistency, asymptotic normality in some models and a certain minimax efficiency (Huber 1981, Yohai 1979). But these estimates are practically unsuitable for heavy contaminated data models with $\varepsilon > 1/2$ and asymmetric, not unimodal function $h(t)$.

The M-estimators with bounded functions $\rho(t)$ are very robust in these cases, but there are many difficulties in their use. The first one is that there is almost no theoretical foundation for the use of such functions. In particular, they all are obtained by their authors heuristically. In any case there are certain objections against their application (see Huber 1981).

We shall demonstrate that the maximum likelihood estimation in the framework of our model straightforwardly leads to a bounded function $\rho(t)$ in (2). Evaluating the corresponding likelihood equation

$$\frac{\partial}{\partial a}\sum \ln\left(\frac{1-\varepsilon}{\sigma(2\pi)^{1/2}}\, e^{-\frac{1}{2}(x_i - a)^2/\sigma^2} + \varepsilon h_0\right) = 0$$

we obtain $a = \Sigma w_i x_i/(\Sigma w_i)$, where $w_i = w((x_i - a)/\sigma)$ with the weight function

$$w(t) = w_U(t) = \frac{1+c}{1+ce^{t^2/2}} \qquad (4)$$

with $c = (2\pi)^{\frac{1}{2}}\sigma h_0 \varepsilon/(1-\varepsilon)$ (the factor $1 + c$ is introduced in (4) to fulfil $w(0) = 1$). The weight function (4) corresponds to the bounded function

$$\rho_u(t) = (1 + c)\ln[(1 + c)/(c + e^{-t^2/2})]. \qquad (5)$$

The only parameter c is the ratio of the mean number of noise observations within an interval of the length $\sigma(2\pi)^{\frac{1}{2}}$ to the mean number of useful observations in the sample. Thus it is determined by the contamination of data not in the whole range of the sample but within its essential part where all useful oservations are practically concentrated. The value of c is often approximately known in experimental models.

The upper bound of (5) $(1+c)\ln(1+1/c)$ increases without limit as $c \rightarrow 0$ (with the noise diminishing). Hence the boundedness of this function is significant only for $c > 0.1$ which corresponds to heavy contamination. Fig. 1 shows the function (4) compared to Tukey's bi-square weight (Tukey 1974)

$$w_T(t) = \{[1-(t/c_T)^2]^2 \text{ for } |t| < c_T \text{ and } 0 \text{ for } |t| \leq c_T\}. \quad (6)$$

These functions are close to each other, but (6) is more preferable due to their faster computation, so it is used further instead of (4) whenever that is important.

FIGURE 1

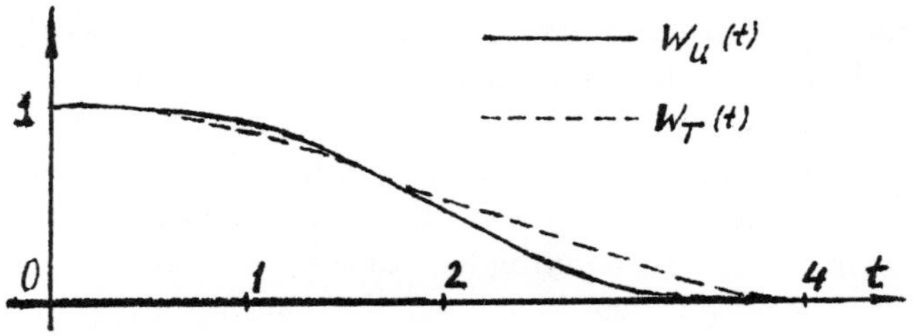

3. ESTIMATION OF σ

Another problem caused by the use of bounded functions $\rho(t)$ is connected with the non-uniqueness of the estimate (2). The function $L(a,\sigma)$ often has several minima. It is difficult to find them all and to choose one of them for estimating a. The number of minima and their location depends on the value of σ, i.e., the problems of M-estimating a and σ are closely related. Ramsay (1977) also notes that separate procedures for estimating a and σ are unwise. Our approach is based upon the joint M-estimates of a and σ.

It is a difficult problem to estimate the parameter σ in our model. The common robust estimate $\hat{\sigma} = \text{const}\cdot\text{med}\,\{|x_i - a|\}$ is unavailable for $\varepsilon > 1/2$ as well as the other estimates based on the order statistics. Differentiating (2) with respect to σ one obtains the likelihood equation for σ, which gives

$$\hat{\sigma} = [\sum_i w_i(x_i - a)^2] / \sum_i w_i \tag{7}$$

with w_i defined in (4). This estimate is applicable for $\varepsilon > 1/2$, too. It was proposed by Ramsay (1977), Aivazyan *et al.* (1985) with different functions $w(t)$. It also satisfies Huber's definition of M-estimate of σ through the solution of the following equation (Huber 1981)

$$\sum_i \chi\left(\frac{x_i - a}{\sigma}\right) = 0 \tag{8}$$

with an even function $\chi(t)$. The estimate (7) corresponds to (8) when $\chi(t) = t^2\rho'(t) - w(t)$. Therefore it is shift- and scale-invariant. Using the same weight function in (3) and (7) one can consider a- and σ-estimating as a single problem.

Let us consider some geometric properties of $L(a, \sigma)$ in (2) as the function of two parameters. The set of local conditional minima of $L(a, \sigma)$ for all fixed $\sigma > 0$ forms a finite collection of smooth curves in the semi-plane $\{(a, \sigma), \sigma > 0\}$. Denote them $\gamma_1, \gamma_2, \ldots, \gamma_m$. There is sufficiently large $\sigma_1 > 0$ such that the semi-plane $\{(a, \sigma), \sigma > \sigma_1\}$ contains only one of these curves which is unbounded and has the asymptote $a_0 = (x_1 + x_2 + \ldots + x_n)/n$ as $\sigma \to \infty$ (we denote this curve by γ_1).

In Fig. 2 there are two examples of the surface $z = L(a, \sigma)$ for the samples each containing 20 points grouped into three and two clusters, correspondingly. On these surfaces one can see "ravines", the number of which increases as $\sigma \to 0$. Fig. 3 represents the curves $\gamma_1, \gamma_2, \ldots, \gamma_m$ which are the "bottoms" of these ravines (the corresponding samples are marked with asterisks on the axis a).

FIGURE 2

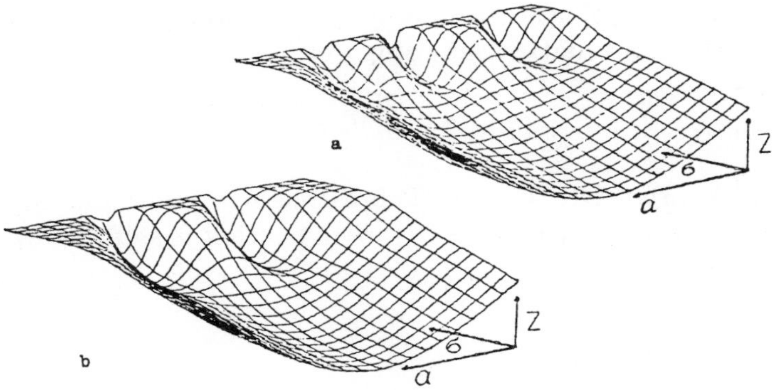

FIGURE 3

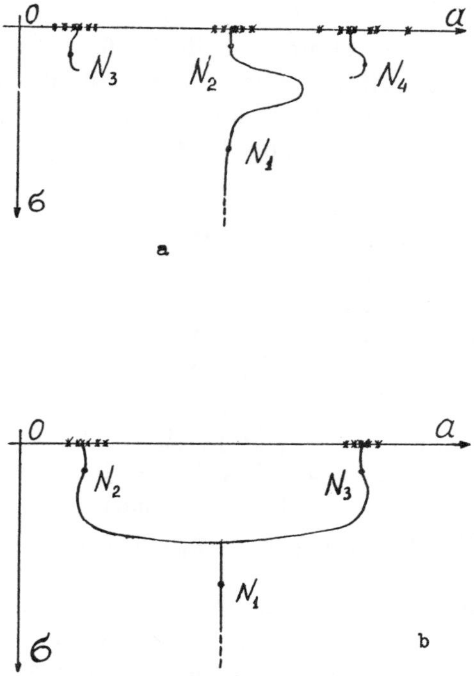

These curves are always disconnected except in the special case of artificial symmetric samples like in Fig. 3b, which leads to branching γ_1 into two different curves.

The detailed study of positions on the curves $\gamma_1, \ldots, \gamma_m$ of points corresponding to the estimates (3) and (7) was made by Chernov and Ososkov (1986). Some annoying cases have been found with the absence of such points on certain curves. After considering different attempts to modify the function $L(a, \sigma)$ in such a way that its minima would be joint M-estimates of a, σ Chernov and Ososkov proposed to replace (7) by

$$\hat{\sigma}^2 = (\Sigma w_i (x_i - a)^2)/\Sigma H_i \tag{9}$$

where $H_i = H((x_i - a)/\sigma)$, $H(t) = \rho''(t)$. This estimate is robust and satisfies the definition of (8), where $\chi(t) = t^2 w(t) - H(t)$, hence it is shift- and scale-invariant.

The estimate (9) is more suitable for our purposes than (7) because, as can be proved, on the curve γ_1 it is at least one point corresponding to (9), i.e. satisfying the equation $M(a, \sigma) = \sigma^2 \Sigma H - \Sigma w_i (x_i - a)^2 = 0$ with the necessary condition $\partial M/\partial\sigma > 0$.

The obtained joint estimate (2), (9) can be defined by the equation system containing the function $L(a, \sigma)$ only:

$$\begin{cases} \partial L/\partial a = 0; \\ \partial^2 L/\partial a^2 + \sigma^{-1}\partial L/\partial\sigma = 0 \end{cases} \tag{10}$$

with the following condition: $\partial^2 L/\partial a^2 > 0$, $\partial M/\partial\sigma > 0$.

Note, that (10) defines just the maximum likelihood estimate in the classical case of $\varepsilon = 0$ and $\rho = t^2/2$.

The estimates obtained from (10) are denoted by $N_1, N_2, \ldots$ in Fig.3. Studying many different samples by computer we noticed the cases with simultaneous estimates (9) on several curves $\gamma_1, \gamma_2, \ldots$ The multiplicity of solutions (10) corresponds to the existence of clusters in the sample $x_1, x_2, \ldots, x_n$.

The existence of at least one estimate (10) on the curve γ_1 leads to the following algorithm. Starting from some point (a_0, σ_0) on the curve γ_1 with sufficiently large σ_0 and $a_0 = (x_1 + \ldots + x_n)/n$ move along γ_1 decreasing σ and looking through all solutions of (10). The solution closest to the axis $\sigma = 0$ must be chosen as the M-estimate. In our model $\sigma >> 1/h_0$, i.e., σ is much less than the range of the sample. Therefore one can simplify this algorithm by moving along γ_1 without looking for solutions of (10) but just stopping when one will reach a small threshold $\sigma = \sigma_{min}$.

The generalization of the one-dimensional case to the many-dimensional linear regression is not a difficult problem. In the space of parameters $\Theta = (\theta_1, \theta_2, \ldots, \theta_m)$ the function

$$L(\theta_1, \ldots, \theta_m) = \sum_{i=1}^{n} \rho\left(\frac{y_i - \theta_1 x_{i1} - \ldots - \theta_{m-1} x_{i,m-1} - \theta_m}{\sigma}\right) \tag{11}$$

has the conditional minima for fixed $\sigma > 0$ which satisfy the equations $\partial L/\partial\theta j = 0$; $j = 1, \ldots, m$ and lie on the finite number of curves. One of these curves (namely γ_1) is unbounded and converges to $\hat{\Theta}$ which is the least square estimate of Θ as $\sigma \to \infty$. By analogy with (10) the system

$$\begin{cases} \partial L/\partial\theta_j = 0; \; j = 1, \ldots, m \\ \partial^2 L/\partial\theta_m{}^2 + \sigma^{-1}\partial L/\partial\sigma = 0 \end{cases} \tag{12}$$

has at least one solution on the unbounded curve γ_1.

4. ESTIMATE COMPUTING PROCEDURE

Estimate computing procedure is iterative. The solution of the system (12) is obtained similar to (3) and (9) with the weights expressed via parameter estimates from previous iteration

$$w_i^{(k)} = w\left(\frac{y_i - \theta_1^{(k-1)}x_{i1} - \ldots - \theta_m^{(k-1)}}{\hat{\sigma}^{(k-1)}}\right). \tag{13}$$

Denoting by $\mathbf{W}^{(k)}$ the weight matrix $n \times n$ with the weights on its diagonal, by $\mathbf{Y}$ - the vector of observations $(y_1,\ldots, y_n)$ and by $\mathbf{X} = ((X_{ij}); i = 1,\ldots, n; j = 1,\ldots, m)$ - the factor matrix, one can obtain the estimate of $\mathbf{\Theta}$ on the k-th iteration as

$$\hat{\mathbf{\Theta}}^{(k)} = (\mathbf{X}^T\mathbf{W}^{(k)}\mathbf{X})^{-1}\mathbf{X}^T\mathbf{W}^{(k)}\mathbf{Y}$$

(here ^{T} is the transposition symbol). If there are no a prior initial values the estimate $\hat{\mathbf{\Theta}}^{(1)}$ on the first iteration can be obtained as the least square estimate with the unit weight matrix $\mathbf{W}^{(1)} = \mathbf{I}$.

Concerning $\hat{\sigma}$ in the denominator of (13) the proposition was made, in correspondence to the algorithm given above, to take as $\hat{\sigma}^{(1)}$ the sample range, i.e. $\hat{\sigma}^{(1)} = y_n - y_1$ and decrease it on the every iteration by the rule: $\hat{\sigma}^{(k)} = \lambda\hat{\sigma}^{(k-1)}$ $(0 < \lambda < 1)$. For iteration number reducing it is desirable to keep λ close to 1, but not much to prevent loosing the curve γ_1. In our calculations we choose $\lambda \sim 0.9$.

5. STUDENTIZING

The σ-estimating does not solve completely the problem of the non-uniqueness of function (2) or (12) minima for bounded functions $\rho(t)$. The phenomenon of the local minimum appearence harms especially in cases of so-called leverage points. In such a point the observation $y_1 = \Sigma x_{ij}\theta_j + e_i$ has not only a big error e_i but the factor values $\bar{\mathbf{x}}_i = (x_{i1},\ldots, x_{im})$ lie in the space $\mathbb{R}^m$ apart from other factors (see example in Fig. 4b for $m = 2$).

FIGURE 4

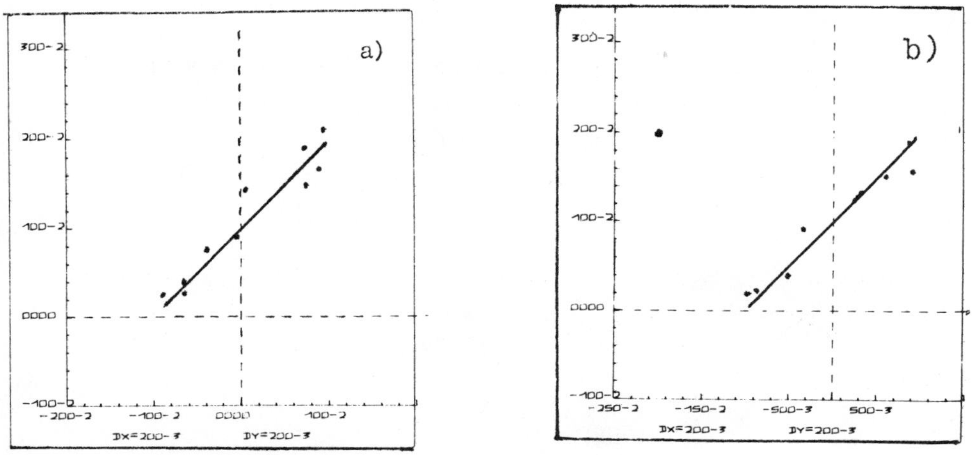

For such cases the M-estimate modification is proposed (Kunyaev *et al.* 1984), which consists in scaling residuals $\hat{e}_i = y_i - \Sigma x_{ij}\hat{\theta}_j$; $i = 1,...,n$ in (13) not by $\hat{\sigma}$ but by the quantity d_i, where d_i^2 is the individual variance of $\hat{e}_i$. This variance can be evaluated as

$$d_i^2 = \sigma^2(1 + \mathbf{H}_i\mathbf{H}_i^T - 2\mathbf{H}_i\mathbf{A}_i), \tag{14}$$

where $H_i = \bar{\mathbf{x}}_i(\mathbf{X}^T\mathbf{W}\mathbf{X})^{-1}\mathbf{X}^T\mathbf{W}$, $\mathbf{A}_i = (0,..., 0, 1, 0,..., 0)$ is the i-th basic vector in $\mathbb{R}^n$. The innovation consists here in recalculating the values of $\hat{e}_i$ and d_i on the every iteration. Due to such scaling studentized residuals $\hat{e}_i/d_i$ would have zero mean and unit covariance independently of the point location in the factor space. That would effectively suppress the harmful influence of leverage points with large $|\hat{e}_i|$ without loosing information for small ones.

A qualitative picture of how this studentizing works is given in Fig. 5, where for $m = 2$ the surfaces $z = -L(\theta_1, \theta_2)$ are shown (a) without leverage points (see Fig. 4a); (b) when such a point presents (see Fig. 4b) (c) when scaling $\hat{e}_i/d_i$ is made. One can see the studentizing effect as reducing the area of the horizontal projection of the extraneous peak. That decreases the probability, when the calculated estimate would occur in this area.

As it has been proven (Astapov *et al.* 1985), the ratio $\hat{e}_i^{(k)}/d_i^{(k)}$ for the k-th iteration does not depend on the weight $w_i^{(k)}$, and it guarantees no risk iteration convergence inversing.

FIGURE 5

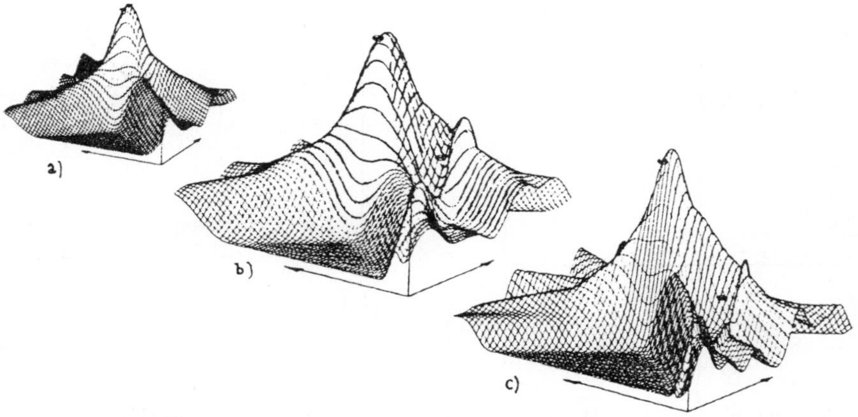

6. SUMMARY OF MONTE CARLO RESULTS

The above mentioned modifications of M-estimates were studied for different distributions $h(x)$ and contamination levels ε in the model (1). Monte Carlo samples have been simulated for uniform density $h(x)$ (symmetric and non-symmetric with respect to a regression line) as well as for normal and Cauchy distributions. Calculations have been also performed for our estimators with the weight function (4) and some others like those of Forsythe (1972), Huber (1981), Ramsay (1977), Andrews (1974) and Hampel [found in Huber (1981)].

The results (see Astapov *et al.* 1985, Kunyaev *et al.* 1984, Chernov and Ososkov 1986) can be summarized as follows:

1. If there is no contaminations ($\varepsilon = 0$) all robust methods listed above have mean error only 10 - 15 % greater than for least square method (LSM). When the contamination grows up to $\varepsilon = 0.1$, LSM loses accuracy in the order of magnitude. For $\varepsilon = 0.25$ or Cauchy distribution methods with the unbounded function $\rho(t)$ become 2 - 3 times less precise than methods with the bounded $\rho(t)$.

2. M-estimators with weight function (4) show rather high efficiency for uniformly distributed contamination and ε between 0.5 and 0.9 (note, the last value corresponds to the signal/noise ratio equal 0.1 only). Thus for $m = 2$ and ε from the above denoted limits a set of samples, each sample containing 60 points, was simulated. The relative number of cases was found when this estimate of Θ appeared to be closer to its mean value than 10^{-3}. This number changed from 94% to 80% when ε increased from 0.5 to 0.9.

3. Residual studentizing givinig a rather small gain in accuracy (12 - 17% for $\varepsilon = 0.25$ and 30% for Cauchy distribution) is very effective in cases of the leverage point occurence.

Besides that, it was found very fruitful to apply the modified M-estimators for data processing in experimental high energy physics. In particular, it increases significantly the efficiency and the speed of procedures of linear and non-linear (circular) fitting, which are intensively used for track finding in many actual data handling systems.

REFERENCES

Aivasyan, S.A., Yenyukov, I.S., Meshalkin, L.D. (1985). Applied statistics. Study of relationships. *Finansy i Statistika*, Moscow. (In Russian).

Andrews, D.F. (1974). A robust method for multiple linear regression. *Technometrics*, 16, 523-531.

Astapov, A.A. Borodyuk, V.P., Kunyaev, S.V., Ososkov, G.A. and Chernov, N.I. (1985). Numerical analysis of robust regression methods. Communication of Joint Institute for Nuclear Research P5-85-492. Dubna (in Russian).

Chernov, N.I. and Ososkov, G.A. (1986). Joint robust estimates of location and scale parameters. Preprint of Joint Institute for Nuclear Research E10-86-282, Dubna.

Forsythe, A.B. (1972). Robust estimation of straight line regression coefficients by minimizing p-th power deviations. *Technometrics*, 14, 159-166.

Huber, P.J. (1981). *Robust Statistics*. Wiley, New York.

Kunyaev, S.A., Ososkov, G.A. and Chernov, N.I. (1984). Statistical recognition methods and robust estimations of track parameters. Communication of Joint Institute for Nuclear Research P10-84-553, Dubna (in Russian).

Ramsay, J.O. (1977). A comparative study of several robust estimates of slope, intercept and scale in linear regression. *Journal of the American Statistical Association*, 72, 608-615.

Tukey, J.W. (1974). Introduction to today's data analysis. *Critical Evaluation of Chemical and Physical Structural Information*. National Academy of Science, Washington, 3-14.

Yohai, V.J. and Maronna, R.A. (1979). Asymptotic behaviour of M-estimators for the linear model. *Annals of Statistics*, 7, 258-268.

Received 15 January 1987

Joint Institute for Nuclear Research
Head Post Office Box 79
101000 Moscow
U.S.S.R.

Proc. Second International Tampere Conference in Statistics
(Tampere, Finland, 1-4 June 1987)
Tarmo Pukkila and Simo Puntanen, *Editors*

pp. 627 - 636

Selection of variables in nonlinear discriminant analysis by information criteria

Seppo PYNNÖNEN
University of Vaasa, Vaasa, Finland

Key words and phrases: Additional information, sufficiency, multivariate normal distribution, classsification.

ABSTRACT

In this paper variable selection criteria based on criterion functions, like Akaike's (1973) AIC or Schwarz's (1978) BIC, for normal-theory nonlinear discriminant analysis is proposed. The criteria are derived by using Rao's (1970) additional information principle to evaluate the independent information supplied by variables not in the model given the information of the variables in the model. The behaviour of the criteria are illustrated by simulation experiments.

1. INTRODUCTION

The problem of discarding superfluous variables from the model has been extensively studied in multiple group linear discriminant analysis, where the populations are normally distributed with a common covariance matrix. Several methods for variable selection have been suggested. Useful summaries of methods based upon test theoretic approaches can be found in McKay and Campbell (1982a and 1982b). Also graphical aids for selecting suitable variables have been suggested (e.g. McKay 1978). Fujikoshi (1983, 1985) has derived a criterion for variable selection based upon Akaike's (1973) AIC. Furthermore, he has derived some asymptotical results for the behaviour of his criterion. Also Pynnönen (1986) has considered the use of criteria like AIC in selecting variables in linear discriminant analysis. Bernardo and Bermudez (1985) have given a general decision theoretic procedure, where the selection criterion depends on the utility function that is used.

In this paper we suggest an approach to evaluate the importance of variables in nonlinear discriminant analysis of normally distributed populations. This gives an extension of the variable selection method in Fujikoshi (1983) for the case of unequal covariance matrices. The approach is based upon Rao's (1970) additional information or sufficiency principle. As the criterion to evaluate the additional information of the variables we use the AIC (Akaike 1973) and BIC (Schwarz 1978, Rissanen 1978) criteria. The behaviour of these criteria are illustrated by simulation experiments.

2. DEFINITIONS AND PRELIMINARY RESULTS

Let $\boldsymbol{x} = (x_1, \dots, x_K)'$ be a K-dimensional observation vector on an individual to be classified in one of the multivariate normal populations π_g: $N_K(\boldsymbol{\mu}^{(g)}, \boldsymbol{\Sigma}^{(g)})$ $g = 1, \dots, q$. The random vector associated to $\boldsymbol{x}$ is denoted by $\boldsymbol{X}' = (X_1, \dots, X_K)$. The mean vectors $\boldsymbol{\mu}^{(g)}$ $(K \times 1)$ and the covariance matrices $\boldsymbol{\Sigma}^{(g)}$ $(K \times K)$ of full ranks are unknown.

Let us identify variables $X_{j_1}, \dots, X_{j_k}$ by the corresponding subset, $J_k := \{j_1, \dots, j_k\}$, $(1 \le j_i \le K,\ i = 1, \dots, k)$ of the index set $J := \{1, \dots, K\}$. Define $J_k = \emptyset$, if $k = 0$. Furthermore, call J_k the model containing variables X_{i_1}, $\dots, X_{i_k}$. For the sake of simplicity, we assume from now on that $J_k = \{1, \dots, k\}$, $k = 0, 1, \dots, K$. With this convention partition X according to J_k and $J - J_k$ as $\boldsymbol{X}' = (\boldsymbol{X}_1', \boldsymbol{X}_2')$, where $\boldsymbol{X}_1 = (X_1, \dots, X_k)'$, and $\boldsymbol{X}_2 = (X_{k+1}, \dots, X_K)'$. Suppose, furthermore, that there are N_g $(> K)$ observations $\boldsymbol{x}_{ig}' = (\boldsymbol{x}_{1ig}', \boldsymbol{x}_{2ig}')$ available from each population π_g, $i = 1, \dots, N_g$, $g = 1, \dots, q$.

As in Rao(1970) we say that $\boldsymbol{X}_1$ is *sufficient* for discrimination between the populations $\pi_1, \dots, \pi_q$, if the conditional distributions of $\boldsymbol{X}$ given $\boldsymbol{X}_1$ are the same in each population. This definition may also be termed as the absence of additional information contained in $\boldsymbol{X}_2$, when the information supplied by $\boldsymbol{X}_1$ is already available (Rao 1970).

Partition the $\boldsymbol{\mu}^{(g)}$'s and $\boldsymbol{\Sigma}^{(g)}$'s corresponding to $\boldsymbol{X}' = (\boldsymbol{X}_1', \boldsymbol{X}_2')$ as

$$\boldsymbol{\mu}^{(g)} = \begin{pmatrix} \boldsymbol{\mu}_1^{(g)} \\ \boldsymbol{\mu}_2^{(g)} \end{pmatrix}, \quad \text{and} \quad \boldsymbol{\Sigma}^{(g)} = \begin{pmatrix} \boldsymbol{\Sigma}_{11}^{(g)} & \boldsymbol{\Sigma}_{12}^{(g)} \\ \boldsymbol{\Sigma}_{21}^{(g)} & \boldsymbol{\Sigma}_{22}^{(g)} \end{pmatrix}$$

$g = 1, \dots, q$. Then for (nonsingular) multivariate normal populations, we have the following result

THEOREM. *Let* $\boldsymbol{X}' = (\boldsymbol{X}_1', \boldsymbol{X}_2') \sim N(\boldsymbol{\mu}^{(g)}, \boldsymbol{\Sigma}^{(g)})$ *in population* π_g, $g = 1, \ldots, q$. *Then* $\boldsymbol{X}_1$ *is sufficient for discrimination between populations, if and only if*

$$\boldsymbol{\mu}_2^{(1)} - \boldsymbol{\Sigma}_{21}^{(1)}\boldsymbol{\Sigma}_{11}^{(1)-1}\boldsymbol{\mu}_1^{(1)} = \cdots = \boldsymbol{\mu}_2^{(q)} - \boldsymbol{\Sigma}_{21}^{(q)}\boldsymbol{\Sigma}_{11}^{(q)-1}\boldsymbol{\mu}_1^{(q)}, \tag{2.1}$$

$$\boldsymbol{\Sigma}_{21}^{(1)}\boldsymbol{\Sigma}_{11}^{(1)-1} = \cdots = \boldsymbol{\Sigma}_{21}^{(q)}\boldsymbol{\Sigma}_{11}^{(q)-1}, \tag{2.2}$$

$$\boldsymbol{\Sigma}_{22.1}^{(1)} = \cdots = \boldsymbol{\Sigma}_{22.1}^{(q)}, \tag{2.3}$$

where

$$\boldsymbol{\Sigma}_{22.1}^{(g)} = \boldsymbol{\Sigma}_{22}^{(g)} - \boldsymbol{\Sigma}_{21}^{(g)}\boldsymbol{\Sigma}_{11}^{(g)-1}\boldsymbol{\Sigma}_{12}^{(g)}, \tag{2.4}$$

$g = 1, \ldots, q$.

Proof. The conditional distributions of $\boldsymbol{X}$, given $\boldsymbol{X}_1 = \boldsymbol{x}_1$, are identical in all populations, if and only if

$$\mathcal{E}^{(g)}(\boldsymbol{X}_2 | \boldsymbol{X}_1 = \boldsymbol{x}_1) = \mathcal{E}^{(h)}(\boldsymbol{X}_2 | \boldsymbol{X}_1 = \boldsymbol{x}_1),$$

and

$$\text{cov}^{(g)}(\boldsymbol{X}_2 | \boldsymbol{X}_1 = \boldsymbol{x}_1) = \text{cov}^{(h)}(\boldsymbol{X}_2 | \boldsymbol{X}_1 = \boldsymbol{x}_1)$$

for all $g, h = 1, \ldots, q$. That is, if and only if

$$\boldsymbol{\mu}_2^{(g)} + \boldsymbol{\Sigma}_{21}^{(g)}\boldsymbol{\Sigma}_{11}^{(g)-1}(\boldsymbol{x}_1 - \boldsymbol{\mu}_1^{(g)}) = \boldsymbol{\mu}_2^{(h)} + \boldsymbol{\Sigma}_{21}^{(h)}\boldsymbol{\Sigma}_{11}^{(h)-1}(\boldsymbol{x}_1 - \boldsymbol{\mu}_1^{(h)}) \tag{2.5}$$

and

$$\boldsymbol{\Sigma}_{22.1}^{(g)} = \boldsymbol{\Sigma}_{22.1}^{(h)}, \tag{2.6}$$

for all $g, h = 1, \ldots, q$.

The equality (2.6) implies (2.3) of the theorem. From (2.5) we obtain

$$[\boldsymbol{\Sigma}_{21}^{(g)}\boldsymbol{\Sigma}_{11}^{(g)-1} - \boldsymbol{\Sigma}_{21}^{(h)}\boldsymbol{\Sigma}_{11}^{(h)-1}]\boldsymbol{x}_1 = \boldsymbol{c}_{gh}, \tag{2.7}$$

where

$$\boldsymbol{c}_{gh} = \boldsymbol{\mu}_2^{(h)} - \boldsymbol{\Sigma}_{21}^{(h)}\boldsymbol{\Sigma}_{11}^{(h)-1}\boldsymbol{\mu}_1^{(h)} - [\boldsymbol{\mu}_2^{(g)} - \boldsymbol{\Sigma}_{21}^{(g)}\boldsymbol{\Sigma}_{11}^{(g)-1}\boldsymbol{\mu}_1^{(g)}]. \tag{2.8}$$

Now (2.1) and (2.2) follow from (2.7) and (2.8), respectively, since $\boldsymbol{c}_{gh} = \boldsymbol{0}$ for all $g, h = 1, \ldots, q$. In fact, since the relation (2.7) must hold for all $\boldsymbol{x}_1$, and hence also for $\boldsymbol{x}_1 = \boldsymbol{0}$, the equalities $\boldsymbol{c}_{gh} = \boldsymbol{0}$, $g, h = 1, \ldots, q$ follow. This completes the proof.

Define

$$\boldsymbol{\Gamma}^{(g)} := (\boldsymbol{\Gamma}_1^{(g)}, \boldsymbol{\Gamma}_2^{(g)}) := (\boldsymbol{\mu}_2^{(g)} - \boldsymbol{\Gamma}_2^{(g)}\boldsymbol{\mu}_1^{(g)}, \boldsymbol{\Sigma}_{21}^{(g)}\boldsymbol{\Sigma}_{11}^{(g)-1}), \tag{2.9}$$

$(K-k) \times (k+1)$ matrix ($g = 1, \ldots, q$). Then using the above theorem we define as our *Additional Information* (AI) hypothesis

$$H : \begin{cases} \boldsymbol{\Gamma}^{(1)} & = \cdots = \boldsymbol{\Gamma}^{(q)} =: \boldsymbol{\Gamma} =: (\boldsymbol{\Gamma}_1, \boldsymbol{\Gamma}_2) \\ \boldsymbol{\Sigma}_{22.1}^{(1)} & = \cdots = \boldsymbol{\Sigma}_{22.1}^{(q)} =: \boldsymbol{\Sigma}_{22.1} \end{cases} \tag{2.10}$$

Hence, we observe that the additional information hypothesis becomes just the hypothesis of testing total homogeneity of populations in multivariate regression. As a special case, when no variable contains discrimination information, the hypothesis becomes the usual hypothesis of total homogeneity of multivariate populations.

Let $p_g > 0$ such that $p_1 + p_2 + \cdots + p_q = 1$. Furthermore, let

$$\boldsymbol{\Sigma} = \sum_{g=1}^{q} p_g \boldsymbol{\Sigma}^{(g)} = \begin{pmatrix} \boldsymbol{\Sigma}_{11} & \boldsymbol{\Sigma}_{12} \\ \boldsymbol{\Sigma}_{21} & \boldsymbol{\Sigma}_{22} \end{pmatrix}, \tag{2.11}$$

$$\boldsymbol{\Omega} = \sum_{g=1}^{q} p_g \left(\boldsymbol{\mu}^{(g)} - \bar{\boldsymbol{\mu}}\right)\left(\boldsymbol{\mu}^{(g)} - \bar{\boldsymbol{\mu}}\right)' = \begin{pmatrix} \boldsymbol{\Omega}_{11} & \boldsymbol{\Omega}_{12} \\ \boldsymbol{\Omega}_{21} & \boldsymbol{\Omega}_{22} \end{pmatrix}, \tag{2.12}$$

and

$$\boldsymbol{\Phi} = \boldsymbol{\Sigma} + \boldsymbol{\Omega} = \begin{pmatrix} \boldsymbol{\Phi}_{11} & \boldsymbol{\Phi}_{12} \\ \boldsymbol{\Phi}_{21} & \boldsymbol{\Phi}_{22} \end{pmatrix}, \tag{2.13}$$

where

$$\bar{\boldsymbol{\mu}} = \sum_{g=1}^{q} p_g \boldsymbol{\mu}^{(g)} = \begin{pmatrix} \bar{\boldsymbol{\mu}}_1 \\ \bar{\boldsymbol{\mu}}_2 \end{pmatrix}, \tag{2.14}$$

and the partitions correspond to partitioning of $\boldsymbol{X}$. Then the relationship between the parameters under the hypothesis, H, is as follows:

LEMMA. *In* (2.10)

$$\boldsymbol{\Gamma}_1 = \bar{\boldsymbol{\mu}}_2 - \boldsymbol{\Gamma}_2\bar{\boldsymbol{\mu}}_1, \tag{2.15}$$

$$\boldsymbol{\Gamma}_2 = \boldsymbol{\Phi}_{21}\boldsymbol{\Phi}_{11}^{-1}, \tag{2.16}$$

and

$$\boldsymbol{\Sigma}_{22.1} = \boldsymbol{\Sigma}_{22} - \boldsymbol{\Sigma}_{21}\boldsymbol{\Sigma}_{11}^{-1}\boldsymbol{\Sigma}_{12} = \boldsymbol{\Phi}_{22} - \boldsymbol{\Phi}_{21}\boldsymbol{\Phi}_{11}^{-1}\boldsymbol{\Phi}_{12} = \boldsymbol{\Phi}_{22.1}. \tag{2.17}$$

Proof. The equality (2.15) is obvious. In (2.16) we have

$$\begin{aligned}
\boldsymbol{\Sigma}_{21}\boldsymbol{\Sigma}_{11}^{-1} &= \left(\sum_{g=1}^{q} p_g \boldsymbol{\Sigma}_{21}^{(g)}\right)\left(\sum_{g=1}^{q} p_g \boldsymbol{\Sigma}_{11}^{(g)-1}\right)^{-1} \\
&= \left(\sum_{g=1}^{q} p_g \boldsymbol{\Sigma}_{21}^{(g)}\boldsymbol{\Sigma}_{11}^{(g)-1}\boldsymbol{\Sigma}_{11}^{(g)}\right)\left(\sum_{g=1}^{q} p_g \boldsymbol{\Sigma}_{11}^{(g)}\right)^{-1} \\
&= \left(\sum_{g=1}^{q} p_g \boldsymbol{\Gamma}_2 \boldsymbol{\Sigma}_{11}^{(g)}\right)\left(\sum_{g=1}^{q} p_g \boldsymbol{\Sigma}_{11}^{(g)}\right)^{-1} \\
&= \boldsymbol{\Gamma}_2.
\end{aligned}$$

Further

$$\boldsymbol{\Omega}_{21} = \sum_{g=1}^{q} p_g\left(\boldsymbol{\mu}_2^{(g)} - \bar{\boldsymbol{\mu}}_2\right)\left(\boldsymbol{\mu}_2^{(g)} - \bar{\boldsymbol{\mu}}_2\right)' = \boldsymbol{\Gamma}_2\boldsymbol{\Omega}_{11},$$

where we have used the equality

$$\boldsymbol{\mu}_2^{(g)} - \bar{\boldsymbol{\mu}}_2 = \boldsymbol{\Gamma}_2\left(\boldsymbol{\mu}_1^{(g)} - \bar{\boldsymbol{\mu}}_1\right). \tag{2.18}$$

But then

$$\begin{aligned}
\boldsymbol{\Gamma}_2 = \boldsymbol{\Sigma}_{21}\boldsymbol{\Sigma}_{11}^{-1} &= \boldsymbol{\Sigma}_{21}\left(\mathbf{I} + \boldsymbol{\Sigma}_{11}^{-1}\boldsymbol{\Omega}_{11}\right)\left(\mathbf{I} + \boldsymbol{\Sigma}_{11}^{-1}\boldsymbol{\Omega}_{11}\right)^{-1}\boldsymbol{\Sigma}_{11}^{-1} \\
&= \left(\boldsymbol{\Sigma}_{21} + \boldsymbol{\Sigma}_{21}\boldsymbol{\Sigma}_{11}^{-1}\boldsymbol{\Omega}_{11}\right)\left(\boldsymbol{\Sigma}_{11} + \boldsymbol{\Omega}_{11}\right)^{-1} \\
&= \left(\boldsymbol{\Sigma}_{21} + \boldsymbol{\Gamma}_2\boldsymbol{\Omega}_{11}\right)\left(\boldsymbol{\Sigma}_{11} + \boldsymbol{\Omega}_{11}\right)^{-1} \\
&= \left(\boldsymbol{\Sigma}_{21} + \boldsymbol{\Omega}_{21}\right)\left(\boldsymbol{\Sigma}_{11} + \boldsymbol{\Omega}_{11}\right)^{-1} \\
&= \boldsymbol{\Phi}_{21}\boldsymbol{\Phi}_{11}^{-1},
\end{aligned}$$

which proves (2.16).

To see that $\boldsymbol{\Sigma}_{22.1} = \boldsymbol{\Phi}_{22.1}$, we use (2.18), and $\boldsymbol{\Sigma}_{21}\boldsymbol{\Sigma}_{11}^{-1} = \boldsymbol{\Phi}_{21}\boldsymbol{\Phi}_{11}^{-1}$ to obtain

$$\begin{aligned}
\boldsymbol{\Omega}_{22} &= \sum_{g=1}^{q} p_g\left(\boldsymbol{\mu}_2^{(g)} - \bar{\boldsymbol{\mu}}_2\right)\left(\boldsymbol{\mu}_2^{(g)} - \bar{\boldsymbol{\mu}}_2\right)' \\
&= \boldsymbol{\Gamma}_2\boldsymbol{\Omega}_{11}\boldsymbol{\Gamma}_2' = \boldsymbol{\Phi}_{21}\boldsymbol{\Phi}_{11}^{-1}\boldsymbol{\Omega}_{11}\boldsymbol{\Phi}_{11}^{-1}\boldsymbol{\Phi}_{12} \\
&= \boldsymbol{\Phi}_{21}\boldsymbol{\Phi}_{11}^{-1}\left(\boldsymbol{\Phi}_{11} - \boldsymbol{\Sigma}_{11}\right)\boldsymbol{\Phi}_{11}^{-1}\boldsymbol{\Phi}_{12} \\
&= \boldsymbol{\Phi}_{21}\boldsymbol{\Phi}_{11}^{-1}\boldsymbol{\Phi}_{12} - \boldsymbol{\Sigma}_{21}\boldsymbol{\Sigma}_{11}^{-1}\boldsymbol{\Sigma}_{12}.
\end{aligned}$$

Thus

$$\boldsymbol{\Sigma}_{22} = \boldsymbol{\Phi}_{22} - \boldsymbol{\Omega}_{22} = \boldsymbol{\Phi}_{22} - \boldsymbol{\Phi}_{21}\boldsymbol{\Phi}_{11}^{-1}\boldsymbol{\Phi}_{12} + \boldsymbol{\Sigma}_{21}\boldsymbol{\Sigma}_{11}^{-1}\boldsymbol{\Sigma}_{12}.$$

Hence by inserting, we have

$$\boldsymbol{\Sigma}_{22.1} = \boldsymbol{\Sigma}_{22} - \boldsymbol{\Sigma}_{21}\boldsymbol{\Sigma}_{11}^{-1}\boldsymbol{\Sigma}_{12} = \boldsymbol{\Phi}_{22} - \boldsymbol{\Phi}_{21}\boldsymbol{\Phi}_{11}^{-1}\boldsymbol{\Phi}_{12} = \boldsymbol{\Phi}_{22.1},$$

completing the proof.

Define

$$\bar{\boldsymbol{x}}^{(g)} = \frac{1}{N_g}\sum_{i=1}^{N_g}\boldsymbol{x}_{ig} = \begin{pmatrix}\bar{\boldsymbol{x}}_1^{(g)}\\ \bar{\boldsymbol{x}}_2^{(g)}\end{pmatrix}, \text{ and } \bar{\boldsymbol{x}} = \frac{1}{N}\sum_{g=1}^{q}N_g\bar{\boldsymbol{x}}^{(g)} = \begin{pmatrix}\bar{\boldsymbol{x}}_1\\ \bar{\boldsymbol{x}}_2\end{pmatrix}, \tag{2.19}$$

$$\boldsymbol{W}^{(g)} = \sum_{i=1}^{N_g}\left(\boldsymbol{x}_{ig} - \bar{\boldsymbol{x}}^{(g)}\right)\left(\boldsymbol{x}_{ig} - \bar{\boldsymbol{x}}^{(g)}\right)' = \begin{pmatrix}\boldsymbol{W}_{11}^{(g)} & \boldsymbol{W}_{12}^{(g)}\\ \boldsymbol{W}_{21}^{(g)} & \boldsymbol{W}_{22}^{(g)}\end{pmatrix}, \tag{2.20}$$

$$\boldsymbol{T} = \sum_{g=1}^{q}\sum_{i=1}^{N_g}\left(\boldsymbol{x}_{ig} - \bar{\boldsymbol{x}}\right)\left(\boldsymbol{x}_{ig} - \bar{\boldsymbol{x}}\right)' = \begin{pmatrix}\boldsymbol{T}_{11} & \boldsymbol{T}_{12}\\ \boldsymbol{T}_{21} & \boldsymbol{T}_{22}\end{pmatrix}. \tag{2.21}$$

Now we use these definitions together with the notations given in the lemma above, with $p_g = N_g/N$, $g = 1, \ldots, q$. Furthermore, we apply some familiar identities in least squares regression, and the conditionality representation for normal distributions (e.g. Anderson 1984, Sections 2.5 and 8.2), so that we can write the likelihood function under the AI-hypothesis

$$\begin{aligned} L = (2\pi)^{-\frac{1}{2}NK}&\left[\prod_{g=1}^{q}\left|\boldsymbol{\Sigma}_{11}^{(g)}\right|^{-\frac{1}{2}N_g}\right]\left|\boldsymbol{\Phi}_{22.1}\right|^{-\frac{1}{2}N} \\ &\times \exp\left\{-\frac{1}{2}\operatorname{tr}\sum_{g=1}^{q}\boldsymbol{\Sigma}_{11}^{(g)-1}\left[\boldsymbol{W}_{11}^{(g)} + (\bar{\boldsymbol{x}}_1^{(g)} - \boldsymbol{\mu}_1^{(g)})(\bar{\boldsymbol{x}}_1^{(g)} - \boldsymbol{\mu}_1^{(g)})'\right]\right\} \\ &\qquad \times \exp\left\{-\frac{1}{2}\operatorname{tr}\boldsymbol{\Phi}_{22.1}^{-1}\left[\boldsymbol{T}_{22.1} + (\hat{\boldsymbol{\Gamma}} - \boldsymbol{\Gamma})\boldsymbol{A}(\hat{\boldsymbol{\Gamma}} - \boldsymbol{\Gamma})'\right]\right\}, \end{aligned} \tag{2.22}$$

where

$$\hat{\boldsymbol{\Gamma}} = (\hat{\boldsymbol{\Gamma}}_1, \hat{\boldsymbol{\Gamma}}_2) \text{ with } \hat{\boldsymbol{\Gamma}}_1 = \bar{\boldsymbol{x}}_2 - \hat{\boldsymbol{\Gamma}}_2\bar{\boldsymbol{x}}_1 \text{ and } \hat{\boldsymbol{\Gamma}}_2 = \boldsymbol{T}_{21}\boldsymbol{T}_{11}^{-1},$$

and

$$\boldsymbol{A} = \sum_{g=1}^{q}\sum_{i=1}^{N_g}\boldsymbol{a}_{ig}\boldsymbol{a}_{ig}' \quad \text{with} \quad \boldsymbol{a}_{ig} = \begin{pmatrix}1\\ \boldsymbol{x}_{1ig}\end{pmatrix}.$$

3. SELECTION CRITERIA

Akaike (1973) suggested a criterion for model selection, namely to choose that model J_k which minimizes

$$\mathrm{AIC}(J_k) := -2\log L\left[\hat{\theta}(J_k)\right] + 2m(J_k), \tag{3.1}$$

where $\hat{\boldsymbol{\theta}}(J_k)$ is the maximum likelihood (ML) estimator of the parameter $\boldsymbol{\theta}$ under the restrictions of model J_k, and $m(J_k)$ is the number of independently estimated parameters.

Another well known criterion, derived from Bayesian point of view due to Schwarz (1978), and from non-Bayesian point of view due to Rissanen (1978), is

$$\mathrm{BIC}(J_k) := -2\log L[\hat{\boldsymbol{\theta}}(J_k)] + m(J_k)\log N, \tag{3.2}$$

where N is the total sample size.

The maximum of the likelihood (2.22) becomes

$$\max L = (2\pi)^{-\frac{1}{2}NK}\left|\frac{1}{N}\boldsymbol{T}_{22.1}\right|^{-\frac{1}{2}N}\left[\prod_{g=1}^{q}\left|\frac{1}{N_g}\boldsymbol{W}_{11}^{(g)}\right|^{-\frac{1}{2}N_g}\right]e^{-\frac{1}{2}NK}. \tag{3.3}$$

Hence, the maximum log-likelihood can be written as

$$\log L[\hat{\boldsymbol{\theta}}(J_k)] = -\frac{1}{2}\sum_{g=1}^{q} N_g \log[\Lambda_g(J_k)] + \text{ constant}, \tag{3.4}$$

where we have used the identity of the form $|\boldsymbol{T}| = |\boldsymbol{T}_{11}||\boldsymbol{T}_{22.1}|$ to obtain

$$\Lambda_g(J_k) := \frac{\left|\frac{1}{N_g}\boldsymbol{W}_k^{(g)}\right|}{\left|\frac{1}{N}\boldsymbol{T}_k\right|}. \tag{3.5}$$

Here $\Lambda_g(\emptyset) = 1$, and $\boldsymbol{W}_k^{(g)}$ and $\boldsymbol{T}_k$ are $k\times k$ submatrices of $\boldsymbol{W}^{(g)}$ and $\boldsymbol{T}$, respectively, corresponding to the k variables in the current model.

The number of independently estimated parameters consist of qk group means in $\boldsymbol{\mu}_1^{(g)}$'s, $K-k$ grand means in $\bar{\boldsymbol{\mu}}_2$, $qk(k+1)/2$ and $(K-k)k+(K-k)(K-k+1)/2$ variance-covariance parameters in $\boldsymbol{\Sigma}_{11}^{(g)}$'s, and $\boldsymbol{\Phi}_{22.1}$ respectively. Hence, all told,

$$m(J_k) = \frac{1}{2}(q-1)k(k+3) + \frac{1}{2}K(K+3). \tag{3.6}$$

By dropping constants that remain the same for each modelling alternative, AIC and BIC become equivalent to

$$A(J_k) := \sum_{g=1}^{q} N_g \log[\Lambda_g(J_k)] + (q-1)k(k+3), \tag{3.7}$$

and

$$B(J_k) := \sum_{g=1}^{q} N_g \log[\Lambda_g(J_k)] + \frac{1}{2}(q-1)k(k+3)\log N, \tag{3.8}$$

respectively, where $A(\emptyset) = B(\emptyset) = 0$.

4. EXAMPLES

As examples we consider simulation experiments, where the observations arise equally likely from the following two normally distributed populations, π_1: $\boldsymbol{\mu}^{(1)} = (0,0,0,0,0,0)'$, $\boldsymbol{\Sigma}^{(1)} = \mathbf{I}$, and π_2: $\boldsymbol{\mu}^{(2)} = (1.5,0,0,0,0,0)'$, $\boldsymbol{\Sigma}^{(2)} = \mathrm{diag}(1/2,1,1,1,1,1)$. Hence, only the first variable has discrimination power. Using the likelihood ratio as the classification rule and assuming the misclassification costs equal, the *true error rate* (Hand 1981, p. 9) becomes 0.185. The classification function in this case is of the form

$$(\boldsymbol{x}-\boldsymbol{\mu}^{(1)})'\boldsymbol{\Sigma}^{(1)-1}(\boldsymbol{x}-\boldsymbol{\mu}^{(1)}) - (\boldsymbol{x}-\boldsymbol{\mu}^{(2)})'\boldsymbol{\Sigma}^{(2)-1}(\boldsymbol{x}-\boldsymbol{\mu}^{(2)}) + \log\frac{|\boldsymbol{\Sigma}^{(1)}|}{|\boldsymbol{\Sigma}^{(2)}|}, \qquad (4.1)$$

where the rule is to classify into π_1 if (4.1) is negative, and into π_2 otherwise (cf. Srivastava and Khatri 1979, p. 261).

To see how the two selection criteria, considered above, behave in this situation, we generated 200 samples of sizes 20, 50, and 100 from each of the populations, and performed the model selections. The results are given in Tables 4.1 a and b. According to these results, AIC seems to over-estimate the true dimension in about fifty percent of cases. BIC, on the other hand, seems to give order estimates that tend to concentrate around the true dimension. In one case it has underestimated the model when $N_1 = N_2 = 20$. We may note that asymptotically BIC yields the true model with probability one, and AIC yields asymptotically a model that contains the true model with probability one, but it is not necessarily the most parsimonious one.

TABLE 4.1a
Distributions of the estimated order (AIC)

	Order						
$N_1 = N_2$	0	1	2	3	4	5	6
20	0.000	0.405	0.305	0.205	0.080	0.005	0.000
50	0.000	0.485	0.375	0.125	0.015	0.000	0.000
100	0.000	0.530	0.335	0.120	0.010	0.005	0.000

TABLE 4.1b
Dsitributions of the estimated order (BIC)

	Order						
$N_1 = N_2$	0	1	2	3	4	5	6
20	0.005	0.895	0.080	0.020	0.000	0.000	0.000
50	0.000	0.975	0.025	0.000	0.000	0.000	0.000
100	0.000	0.995	0.005	0.000	0.000	0.000	0.000

In Table 4.2 we have reported the *apparent error rates* obtained by reclassifying the samples at each simulation round after estimating the parameters. Furthermore, we have reported in that table the error rates (called *simulated error rates*) obtained by generating on each simulation round 100 new observations from each population and classifying them by the estimated classification functions. In the table it is seen the known fact that the apparent error rate tends to under-estimate the true error rate. Similarly, especially in small samples, the full model tends to yield higher misclassification rates than the more parsimonious models. On the other hand, the apparent error rate of the full model gives much too optimistic a feeling about the behaviour of the classifier in small samples.

TABLE 4.2
Apparent and simulated error rates

$N_1 =$	$N_2 =$	20	50	100
AIC	apparent	0.160	0.181	0.181
	simulated	0.219	0.196	0.186
BIC	apparent	0.177	0.186	0.184
	simulated	0.197	0.191	0.183
Full	apparent	0.119	0.164	0.172
	simulated	0.273	0.223	0.202

True error rate = 0.185

ACKNOWLEDGEMENTS

This paper was written while the author was a research fellow at the Academy of Finland. The author is grateful to the referees for their kind suggestions that greatly improved the presentation. The author is also grateful to Mr. Arto Laakkonen for his help in automatic word processing during the preparation of the manuscript.

REFERENCES

Akaike, H. (1973). Information theory and an extension of the maximum likelihood principle. In *Proceedings of the 2nd International Symposium on Information Theory*, eds. B.N. Petrov and F. Czaki, 267–81, Budapest, Akademia Kiado.

Anderson, T.W. (1984). *An Introduction to Multivariate Statistical Analysis*, 2nd ed. Wiley: New York.

Bernardo, J.M. and Bermudez, J.D. (1985). The choice of variables in probabilistic classification. *Bayesian Statistics*, 2, 67–82.

Fujikoshi, Y. (1983). A criterion for variable selection in multiple discriminant analysis. *Hiroshima Mathematical Journal*, 13, 203–14.

Fujikoshi, Y. (1985). Selection of variables in two-group discriminant analysis by error rate and Akaike's information criterion. *Journal of Multivariate Analysis*, 17, 27–37.

Hand, D.J. (1981). *Discrimination and Classification*. Wiley: New York.

McKay, R.J. (1978). A graphical aid to selection of variables in two group discriminant analysis. *Applied Statistics*, 27, 259–263.

McKay R.J. and Campbell, N.A. (1982a). Variable selection techniques in discriminant analysis I. Description. *British Journal of Mathematical and Statistical Psychology*, 35, pp. 1-29.

McKay R.J. and Campbell, N.A. (1982b). Variable selection techniques in discriminant analysis II. Allocation. *British Journal of Mathematical and Statistical Psychology*, 35, 30–41.

Pynnönen, S. (1986). Selection of variables in linear discriminant analysis by information criteria. *Proceedings of the University of Vaasa, Discussion Papers 81*.

Rao, C. Radhakrishna (1970). Inference on discriminant function coefficients. In *Essays in Probability and Statistics*. Eds. R.C. Bose, I.M. Chakravarti, P.C. Mahalanobis, C.R. Rao and K.J.C. Smith, 587–602.

Rissanen, J. (1978). Modeling by shortest data description. *Automatica*, 14, 465–471.

Schwarz, G. (1978). Estimating the dimension of a model. *Annals of Statistics*, 6, 461–464.

Srivastava, M.S. and Khatri, C.G. (1979). *An Introduction to Multivariate Statistics*. Elsevier North Holland, Inc.: New York.

Received 31 December 1986
Revised 7 May 1987

University of Vaasa
Raastuvankatu 31
SF-65100 Vaasa
Finland

Proc. Second International Tampere Conference in Statistics
(Tampere, Finland, 1-4 June 1987)
Tarmo Pukkila and Simo Puntanen, *Editors*

pp. 637 - 646

On the determination of the amplitude of quasi-stationary phenomena

A.M. SALEM and S.A. WAHAB
Ain Shams University, Abbassia, Cairo, Egypt
and Sadat Academy of Management Sciences, Cairo, Egypt

Key words and phrases: Fourier analysis, power spectrum analysis, cosmic ray data, amplitude.

ABSTRACT

The determination of the amplitudes of periodical phenomena masked by superimposed statistical fluctuations was investigated. The study was carried out applying both Fourier series expansion and power spectrum technique. The data used in the analysis are the hourly counting rates of cosmic rays recorded at five worldwide cosmic ray stations, namely Deep River, Leeds, Kiel, Inuvik and Oulu. The amplitudes of the different components of daily variation phenomena in the cosmic ray intensity were computed. The result of analysis shows that the computed values for all five stations by the power spectrum are in high agreement with those obtained by Fourier analysis. Such result confirms the validity of the power spectrum technique in the amplitude determination beside its use in periodicity detection of quasi-stationary phenomena.

1. INTRODUCTION

In fact, all of the periodic and transient time variations in the cosmic ray intensity represent anisotropies arising from solar phenomena. Thus, cosmic rays serve as an effective space probes for investigating the electromagnetic conditions in space and in the earth's immediate environment. The study of the solar anisotropies has been much advanced since the appearance of super-neutron monitores and large area multi-directional muon telescopes. The discovery of tridiurnal phenomenon (periodic variation of 3 cycles per day (cpd)), in addition to the confirmation of the solar diurnal (1 cpd) and semidiurnal (2 cpd) variations, is one of the results thus obtained. These phenomena are studied by relating time variations in the intensity observed at stations distributed over the earth's surface to their asymptotic directions of viewing in space. In this respect, the characteristics of the responsible anisotropies are usually estimated by using the minimum variance analysis between the observed amplitudes and the theoretical ones of the studied phenomenon. The characteristics of the diurnal, semidiurnal and tridiurnal anisotropies have been determined by a number of workers. The diurnal anisotropy, over the range from a few GV (gigavolt) to about 100 GV, is essentially independent of rigidity. The mean free space diurnal amplitude and direction are found to be 0.4% and 90° east of the sun-earth line respectively (Pomerantz and Duggal, 1971). The semidiurnal anisotropy has a positive spectral exponent ≈ 1 and an amplitude $\approx 0.1\%$ at 10 GV with the maximum flux incident from a direction perpendicular to the mean direction of the interplanetary magnetic field, i.e. 135° west of the sun-earth line (Fujii *et al.*, 1970). Moreover, Ahluwalia (1975) showed the existence of the tridiurnal wave in data of detectors having response to primary rigidities in the range of several gigavolt to 100 GV, thus indicating its extraterrestrial origin.

On the other hand, the study of the daily variation phenomena (diurnal, semidiurnal and tridiurnal) has made a substantial contribution to our present knowledge of the interplanetary medium, in particular the characteristics of the interplanetary magnetic field and the interaction of the galactic cosmic rays and solar particles with this field. In general, the observed amplitudes of the different components of this variation (diurnal, semidiurnal, and tridiurnal) are compared with those theoretically predicted from the appropriate cosmic ray diffusion coefficients, where the diffusion coefficients are derived from power spectra of interplanetary magnetic field fluctuation (e.g. Sari *et al.* 1975). Therefore, accurate determination of the amplitudes of these components is important for an understanding of the solar modulation mechanism of the daily variation in interplanetary space.

The amplitudes of the diurnal, semidiurnal and tridiurnal waves are generally derived by Fourier analysis (FA) of the recorded data. The details of this procedure have been discussed in the literature (e.g. Sandstrom 1965). Moreover, several investigators (e.g. Sari *et al.* 1975; Girgis *et al.* 1977; Salem *et al.* 1984) have used the power spectrum technique in the determination of the amplitude of the daily variation components. This is because it serves as a good measure for the contribution of the concerned periodicity relative to the whole frequence spectrum of the input data. Moreover, the FA is valid if the phenomenon under investigation is characterized by constant periodic length. This constricts the application of the FA to quasi-stationary phenomena. Sari *et al.* (1975) determined the amplitude of the diurnal component of neutron monitor counting rates from Sulphur Mountain station (cutoff rigidity 1 GV) on the solar rotation basis (mid-December 1965 - April 1966) by using both Fourier and power spectrum analysis. Their results showed that the amplitude value obtained from the power spectrum technique is larger. Girgis *et al.* (1977) put forward an improved method, through the power spectrum, of computing the low amplitude periodical phenomena masked by superimposed statistical fluctuations. They have used this method in the detection of the tridiurnal component as well as in the determination of the amplitude of this component.

The present work reports a comparison of the magnitudes of the amplitudes of the daily variation components (diurnal, semidiurnal, and tridiurnal) as computed from power spectrum analysis and from the conventional Fourier analysis. The investigation was carried out on five cosmic ray stations (Deep River, Leeds, Oulu, Inuvik and Kiel) over the solar activity period 1965-1968.

2. DATA USED IN THE PRESENT WORK

Cosmic ray data selected for this analysis were the hourly counting rates recorded at Deep River, Kiel, Oulu, Leeds and Inuvik cosmic ray stations, covering the period from 1965 to 1968. The physical characteristics of these stations, given in Table 1, prompt the following remarks:

TABLE 1

Physical characteristics of the cosmic ray stations used in the present analysis

Station Name	Geogr. Long. (Deg.)	Geogr. Lat. (Deg.)	Altitude (m)	Vertical cut-off rigidity GV
Deep River (Canada)	77.50 W	46.10 N	145	1.02
Leeds (England)	1.50 W	53.80 N	72	2.20
Kiel (W. Germany)	10.10 E	54.30 N	54	2.29
Oulu (Finland)	25.50 E	55.02 N	15	0.81
Inuvik (Canada)	133.72 W	68.35 N	21	0.18

(1) The available data cover a period characterized by an increasing phase of solar activity.
(2) Three of these stations (Deep River, Kiel and Leeds) lie in the middle latitude region, while the other two (Oulu and Inuvik) are located in the high latitude region.
(3) The vertical cutoff regidities applicable to the detectors of these stations lie in the range 0.18 GV to 2.29 GV. Such low cutoff rigidities permit the observation of variations in the low energy range of cosmic ray primaries.
(4) The asymptotic cones of acceptance for these stations are quite narrow; thus they exhibit great response in detecting the daily variation wave with its diurnal, semidiurnal, and tridiurnal components. The mean cone broadening for these stations lie in the range 36 degree at Deep River to 47 degree at Inuvik (Girgis *et al.* 1977).

3. COMPUTATIONAL ALGORITHM USING FOURIER TECHNIQUE

In determining the mean amplitudes of the diurnal, semidiurnal, and tridiurnal components over the given period, the following algorithm of analysis was carried out:

Step 1: The data for the period concerned (1965-1968) were first arranged in a two dimensional array $X(i, t)$, where i represents the day number and takes the values 1, 2,..., M and M is the total number of days, t represents the hour and takes the values 1,2,..., 24.

Step 2: The mean hourly values $N(t)$ were then calculated for each particular hour in order to obtain a mean time series of 24 points. This time series may be represented in the following form,

$$N(t) = [\sum_{i=1}^{M} X(i,t)]/M,$$

where M is the number of days during the investigated period.

Step 3: The percentage deviation $\Delta N(t)$ corresponding to each hour in the new series $N(t)$ is then computed using the expression

$$\Delta N(t) = [(N(t) - \bar{N}) / \bar{N}]*100$$

where $\bar{N}$ is the mean of the total hourly counting rates for the whole period.

Step 4: The Fourier coefficients and the corresponding amplitudes of the first, second and third harmonic components for the produced time series $\Delta N(t)$ were computed using the Fourier analysis (FA) formulae (Sandstrom, 1965).

Step 5: The error (ε) in the amplitude was calculated according to Dorman's estimation (1963), given as

$$\varepsilon = 0.29(\frac{1}{24}\sum_{t=1}^{24} (d_t - \bar{d})^2)^{1/2},$$

where d_t is the deviation of the mean hourly value from the mean value of the time series $N(t)$ and $\bar{d}$ is the mean of these deviations.

4. COMPUTATIONAL ALGORITHM USING POWER SPECTRUM TECHNIQUE

The basic procedure of analysis used in this treatment is given by Salem (1977). This method can be summarized in the following steps:

Step 1: Averaging of the observed time series data in order to reduce the noise effects. In this respect, the hourly counting rates are arranged in trains of length L days and the average train $Y(1)$, $Y(2)$,..., $Y(24L)$ is obtained, where

$$Y(h) = [\sum_{j=1}^{T} \Delta D(j, h)]/T,$$

where j is the train order, T is the total number of trains and $\Delta D(j, h)$ is the percentage deviation from the mean counting rate of train j.

Step 2: The resultant train of data $Y(h)$ is then analyzed utilizing the power spectrum technique with hanning window following the formulae of Blackman and Tukey (1959).

Step 3: For low noise level and significant peaks in the smoothed power spectrum, the amplitude $A(f)$ of the wave of frequently f is calculated according to Martinic *et al.*, (1971) from the relation

$$A(f) = 2(P(f)/\tau_0)^{\frac{1}{2}} \tag{I}$$

where $P(f)$ is the density corresponding to the frequency f and τ_0 is the maximum lag.

The error in the density, at any frequency, was computed according to Martinic *et al.* (1971). Salem (1977) proved that the smoothing process through hanning or hamming windows decreased the power spectrum density to its half value. This should be considered in case of amplitude determination from the power spectrum analysis.

The effect of the averaging process, the smoothing principle, the train length (L), the maximum lag (τ_0), and the number of trains (T) on the interaction of noise with existing waves in the power spectrum was comprehensively studied by Girgis *et al.* (1976).

5. RESULTS OF ANALYSIS

In the present study the above two analytical procedures were applied to determine the amplitudes of the different components (diurnal, semidiurnal, and tridiurnal) of the daily variation for the data concerned. In the power spectrum analysis (PSA), the power spectrum density distribution (PSDD) for a given station was recomputed for different maximum lags (τ_0) ranging from 20% n to 45% n in steps of 5% n, where n is the number of points in the average train. Also, in all computations the train length was taken to be 10 days (Salem, 1977). Fig. 1 shows the power spectrum density as a function of the maximum lag at the tridiurnal frequency 3 cycles/day. The fitted least square straight lines in the figure are in high agreement with the theoretical predictions of equation I. Fig. 2 represents samples of the PSDD obtained, where the PSDD is characterized by highly significant peaks at frequencies 1, 2, 3 cpd. The amplitudes of the three components computed by the two procedures are listed in tables 2, 3 and 4 for the first, second, and third harmonics respectively.

From tables 2, 3 and 4 it can be seen that the amplitudes calculated by PSA for the three harmonics for all stations are in high agreement with that obtained using FA. This result confirms the validity of the power spectrum technique in amplitude determination, besides its use in periodicity detection of quasi-stationary phenomena. On the other hand, the Fourier analysis is preferred to power spectrum technique when dealing with phenomena characterized by phase determination.

TABLE 2

First Harmonic Amplitudes

Station	FA	PSA
Deep River	0.2682 ± 0.0010	0.2740 ± 0.0614
Leeds	0.2749 ± 0.0026	0.2769 ± 0.0620
Oulu	0.2632 ± 0.0064	0.2697 ± 0.0605
Kiel	0.2613 ± 0.0068	0.2576 ± 0.0560
Inuvik	0.2141 ± 0.0050	0.2173 ± 0.0653

TABLE 3

Second Harmonic Amplitudes

Station	FA	PSA
Deep River	0.0443 ± 0.0032	0.0459 ± 0.0103
Leeds	0.0341 ± 0.0019	0.0300 ± 0.0070
Oulu	0.0255 ± 0.0036	0.0246 ± 0.0055
Kiel	0.0309 ± 0.0021	0.0326 ± 0.0073
Inuvik	0.0235 ± 0.0050	0.0279 ± 0.0056

TABLE 4

Third Harmonic Amplitudes

Station	FA	PSA
Deep River	0.0145 ± 0.0010	0.0138 ± 0.0031
Leeds	0.0041 ± 0.0017	0.0038 ± 0.0008
Oulu	0.0096 ± 0.0030	0.0104 ± 0.0023
Kiel	0.0073 ± 0.0015	0.0067 ± 0.0015
Inuvik	0.0035 ± 0.0010	0.0017 ± 0.0026

FA : Fourier Analysis.
PSA : Power Spectrum Analysis.

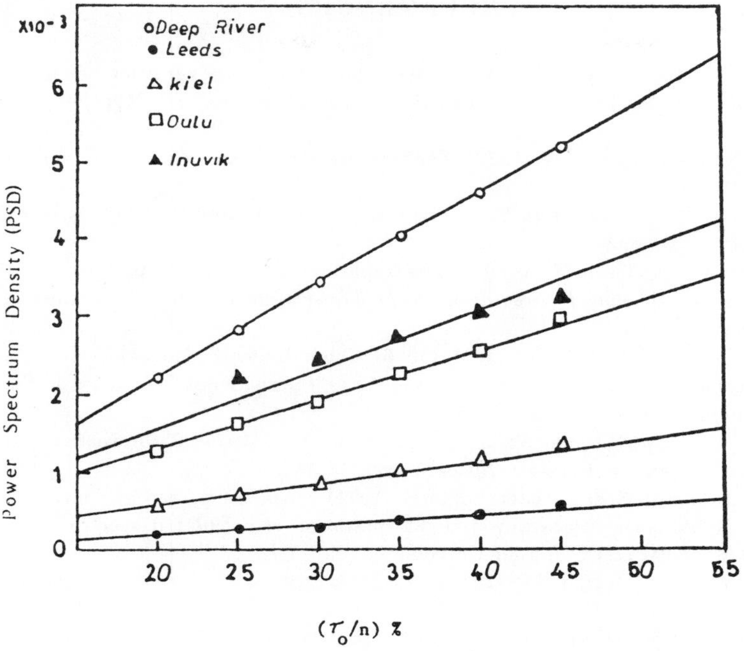

FIGURE 1. The relation between the PSD and $(\tau_0/n)\%$ for the stations concerned at the tridiurnal wave.

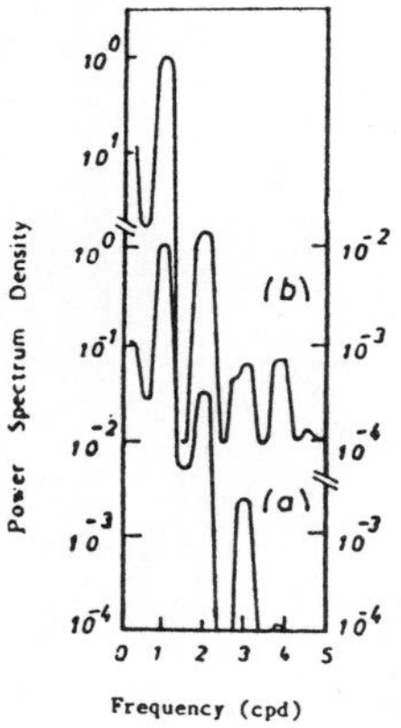

FIGURE 2. PSDD for (a) Deep River station, (b) Kiel station.

REFERENCES

Ahluwalia, H.S. (1975). Some characteristics of solar tridiurnal variation of cosmic rays. *Proceedings of the 14th International Conference on Cosmic Rays*, München, 12, 4207-4212.

Blackman, R.B. and Tukey, J.W. (1959). *The Measurement of Power Spectra*. Dover Publications, New York.

Dorman, L.T. (1963). *Progress in Elementary Particles and Cosmic Ray Physics, VII*. North Holland, Amsterdam.

Fujii, Z., Fujimoto, K., Ueno, H., Kondo, I. and Nagashima, K. (1970). Semidiurnal anisotropy of cosmic radiation. *Proceedings of the 11th International Conference on Cosmic Rays*, Budapest, 2, 83-88.

Girgis, A.H., Tolba, M.F. and Salem, A.M. (1976). Computer method for detecting low amplitude periodical variations. *Proceedings of the 1st International Conference on Statistics and Computer Science*, Cairo, III, 15-26.

Girgis, A.H., Tolba, M.F., Abdel-Wahab, S. and Salem, A.M. (1977). Tridiurnal variations in cosmic ray intensity. *Planetary Space Science*, 25, 39-50.

Martinic,M.,Lindgren, S. and Abdel-Wahab,S. (1971). On power spectrum analysis of time series, signal-noise interaction and interpretation of calculated spectra. Scientific Report, UU/CRG 72-12, Uppsala University.

Pomerantz, M.A. and Duggal, S.P. (1971). The cosmic ray solar diurnal anisotropy. *Space Science Reviews*, 12, 75-130.

Sandstrom, A.E. (1965). *Cosmic Ray Physics*. North Holland, Amsterdam.

Salem, A.M. (1977). Ph.D. Thesis, "A Study of the Solar Periodic Variations in Cosmic-Ray Primaries", Faculty of Science, Ain Shams University, Cairo, Egypt.

Salem, A.M., Wahab, S.A. and Abdel-Aziz, A.H. (1984). Application of the digital filtering technique to study the cosmic ray diurnal variation. *Proceedings of the 9th International Conference on Statistics and Computer Science*, Cairo, 4, 27-46.

Sari, J.M., Lanzerotti, L.J., Macelennan, G.G. and Venkatesan, D. (1975). The cosmic ray diurnal variations and fluctuations of the interplanetary magnetic field. *Proceedings of the 14th International Conference on Cosmic Rays*, München, 4, 1150-1155.

Received 5 January 1987
Revised 23 May 1987

Physics Department
Faculty of Science
Ain Shams University
Abbassia, Cairo
Egypt

Computer Science Department
Faculty of Management
Sadat Academy for Management Sciences
Cairo
Egypt

Proc. Second International Tampere Conference in Statistics
(Tampere, Finland, 1-4 June 1987)
Tarmo Pukkila and Simo Puntanen, *Editors*

pp. 647 - 664

Less sensitive tests by introducing stochastic linear hypotheses

Burkhard SCHAFFRIN
Stuttgart University, F.R. Germany

Key words and phrases: Mixed regression, Helmert's knack, Mixed Linear Model, inhomBLIP versus homBLUP, weaker test criteria.

ABSTRACT

In order to test hypotheses within the theory of linear models, the relative increase of the variance estimate when the respective hypothesis is introduced as additional restriction, classically serves as a test criterion. However, decisions based on such tests may turn out to be *too sensitive* for practical applications. Therefore *stochastic linear hypotheses* have been introduced which can be treated by means of H. Theil's "*Mixed Regression*" or, equivalently, in the context of a "*Mixed Linear Model*" by considering inhomBLIP (*B*est *inhom*ogeneously *LI*near Prediction) after applying "*Helmert's knack*". Here we further extend these results to the case of homBLUP (*B*est *hom*ogeneously *L*inear weakly *U*nbiased *P*rediction) which readily proves to be not only *robust against non-sample misinformation* but which thereby also leads to even *less sensitive tests* as frequently required in practice.

1. INTRODUCTION

In geodesy as in many other engineering and applied sciences, the expectation of some *observation increments* taken with respect to some fixed approximation and collected in the $n \times 1$ vector $\mathbf{y}$ can be modeled by a specific *linear* function (the *Jacobian* $\mathbf{X}$) of a $m\times 1$ vector $\boldsymbol{\beta}$ which contains the *unknown parameter increments*. Thus we are led to the *Linearized Gauss-Markov model*

$$E\{\mathbf{y}\} = \mathbf{y}-\mathbf{u} = \mathbf{X}\boldsymbol{\beta}, \quad \text{rk}\,\mathbf{X} = m, \quad D\{\mathbf{y}\} = D\{\mathbf{u}\} = \sigma^2\mathbf{P}^{-1}, \tag{1.1}$$

where $\sigma^2\mathbf{P}^{-1}$ is the *positive-definite* $n\times n$ dispersion matrix and

$$\mathbf{u} \sim N(\mathbf{0}, \sigma^2\mathbf{P}^{-1}) \tag{1.2}$$

denotes a *normally distributed* random $n\times 1$ vector of *inconsistency*, the so-called "*observational error vector*". It is well known that, by the method of *weighted least squares*, we firstly obtain the *Best Linear Uniformly Unbiased Estimate (BLUUE)* of $\boldsymbol{\beta}$ as solution of the "*normal equations*"

$$\mathbf{N}\hat{\boldsymbol{\beta}} = \mathbf{z}, \ \mathbf{N} := \mathbf{X}^T\mathbf{P}\mathbf{X}, \ \mathbf{z} := \mathbf{X}^T\mathbf{P}\mathbf{y}, \tag{1.3a}$$

with

$$D\{\hat{\boldsymbol{\beta}}\} = \sigma^2\mathbf{N}^{-1} = \sigma^2(\mathbf{X}^T\mathbf{P}\mathbf{X})^{-1} \tag{1.3b}$$

as dispersion matrix, and secondly the *Best Invariant Quadratic Uniformly Unbiased Estimate (BIQUUE)* of σ^2 through

$$\hat{\sigma}^2 = (n-m)^{-1}\Omega \tag{1.4a}$$

with the quadratic form Ω of the *residual vector*

$$\tilde{\mathbf{u}} := \mathbf{y}-\mathbf{X}\hat{\boldsymbol{\beta}} \sim \mathrm{N}(\mathbf{0}, \sigma^2[\mathbf{P}^{-1}-\mathbf{X}\mathbf{N}^{-1}\mathbf{X}^T]) \tag{1.4b}$$

being defined by

$$\Omega := \tilde{\mathbf{u}}^T\mathbf{P}\tilde{\mathbf{u}} = \mathbf{y}^T\mathbf{P}\tilde{\mathbf{u}} = \mathbf{y}^T\mathbf{P}[\mathbf{P}^{-1}-\mathbf{X}\mathbf{N}^{-1}\mathbf{X}^T]\mathbf{P}\mathbf{y} \tag{1.4c}$$

and therefore *distributed as*

$$\Omega/\sigma^2 \sim \chi^2(n-m) \tag{1.5a}$$

with *expectation*

$$E\{\Omega/\sigma^2\} = \text{tr}(\mathbf{I}_n - \mathbf{P}\mathbf{X}\mathbf{N}^{-1}\mathbf{X}^T)\mathbf{P}(\mathbf{P}^{-1} + \mathbf{X}\boldsymbol{\beta}\sigma^{-2}\boldsymbol{\beta}^T\mathbf{X}^T) = n-m. \tag{1.5b}$$

Let us now introduce r different *stochastic linear restrictions* the validity of which we wish to test. Thus, in matrix notation, we have to constrain the $r\times 1$ *random* vector $\mathbf{g}$ with (unknown) expectation $\boldsymbol{\gamma}$ according to

$$\boldsymbol{\gamma} := E\{\mathbf{g}\} = \mathbf{g}-\mathbf{e} \doteq \mathbf{G}\boldsymbol{\beta}, \ \text{rk}\,\mathbf{G} = r, \tag{1.6a}$$

with the *positive-definite* $r\times r$ dispersion matrix

$$D\{\mathbf{g}\} = D\{\mathbf{e}\} = \sigma^2\mathbf{M}^{-1}, \quad \mathbf{M} := \rho^{-1}(\mathbf{G}\bar{\mathbf{P}}_e^{-1}\mathbf{G}^T)^{-1}, \quad \rho := \sigma_e^2/\sigma^2, \tag{1.6b}$$

where ρ is supposed to be given and both,

$$\bar{\mathbf{e}} \sim N(\bar{\boldsymbol{\gamma}} - \boldsymbol{\beta}, \sigma_e^2\bar{\mathbf{P}}_e^{-1}) \quad \textit{and} \quad \mathbf{e} := \mathbf{G}\bar{\mathbf{e}} \sim N(\boldsymbol{\gamma} - \mathbf{G}\boldsymbol{\beta}, \sigma^2\mathbf{M}^{-1}), \quad \boldsymbol{\gamma} = \mathbf{G}\bar{\boldsymbol{\gamma}}, \tag{1.6c}$$

denote the respective $m \times 1$ and $r \times 1$ *normally distributed* random vectors of *inconsistency*, the so-called "*hypothetical errors*", for which we assume

$$C\{\bar{\mathbf{e}}, \mathbf{u}\} = \mathbf{0} \quad \textit{or} \quad C\{\mathbf{e}, \mathbf{u}\} = \mathbf{0}, \quad \text{respectively.} \tag{1.6d}$$

Note that $E\{\mathbf{e}\} = \boldsymbol{\gamma} - \mathbf{G}\boldsymbol{\beta} = \mathbf{0}$ holds true *under the null hypothesis* (1.6a, b). Thus by combining (1.6a, b) with (1.1) we arrive at the model of "*Mixed Regression* " due to H. Theil / A.S. Goldberger (1961) and H. Theil (1963) which then reads as

$$\begin{bmatrix} \mathbf{y} \\ \mathbf{g} \end{bmatrix} = \begin{bmatrix} \mathbf{X} \\ \mathbf{G} \end{bmatrix} \boldsymbol{\beta} + \begin{bmatrix} \mathbf{u} \\ \mathbf{e} \end{bmatrix}, \tag{1.7a}$$

$$\begin{bmatrix} \mathbf{u} \\ \mathbf{e} \end{bmatrix} \sim \left(\begin{bmatrix} \mathbf{0} \\ \mathbf{0} \end{bmatrix}, \sigma^2 \begin{bmatrix} \mathbf{P}^{-1} & \mathbf{0} \\ \mathbf{0} & \mathbf{M}^{-1} \end{bmatrix} \right). \tag{1.7b}$$

Following G.G. Judge/M.E. Bock (1978, Chapter 2.5) or H.D. Vinod/A. Ullah (1981, Chapter 3.4), e.g., the method of *weighted least squares* again yields the BLUUE of $\boldsymbol{\beta}$ as solution of the "*normal equation*"

$$\begin{bmatrix} \mathbf{N} & \mathbf{G}^T \\ \mathbf{G} & -\mathbf{M}^{-1} \end{bmatrix} \begin{bmatrix} \hat{\boldsymbol{\beta}}_R \\ \hat{\boldsymbol{\lambda}}_R \end{bmatrix} = \begin{bmatrix} \mathbf{z} \\ \mathbf{g} \end{bmatrix}, \quad [\mathbf{N}, \mathbf{z}] := \mathbf{X}^T\mathbf{P}[\mathbf{X}, \mathbf{y}], \tag{1.8}$$

with the $r \times 1$ vector $\boldsymbol{\lambda}$ of "*Lagrange multipliers*", which can be represented by

$$\begin{aligned} \hat{\boldsymbol{\beta}}_R &= (\mathbf{N} + \mathbf{G}^T\mathbf{M}\mathbf{G})^{-1}(\mathbf{z} + \mathbf{G}^T\mathbf{M}\mathbf{g}) = \hat{\boldsymbol{\beta}} - (\mathbf{N} + \mathbf{G}^T\mathbf{M}\mathbf{G})^{-1}\mathbf{G}^T\mathbf{M}(\mathbf{G}\hat{\boldsymbol{\beta}} - \mathbf{g}) \\ &= \hat{\boldsymbol{\beta}} + \mathbf{N}^{-1}\mathbf{G}^T(\mathbf{M}^{-1} + \mathbf{G}\mathbf{N}^{-1}\mathbf{G}^T)^{-1}(\mathbf{g} - \mathbf{G}\hat{\boldsymbol{\beta}}) \end{aligned} \tag{1.9a}$$

with

$$\begin{aligned} D\{\hat{\boldsymbol{\beta}}_R\} &= \sigma^2(\mathbf{N} + \mathbf{G}^T\mathbf{M}\mathbf{G})^{-1} = \sigma^2[\mathbf{N}^{-1} - (\mathbf{N} + \mathbf{G}^T\mathbf{M}\mathbf{G})^{-1}\mathbf{G}^T\mathbf{M}\mathbf{G}\mathbf{N}^{-1}] \\ &= D\{\hat{\boldsymbol{\beta}}\} - \sigma^2\mathbf{N}^{-1}\mathbf{G}^T(\mathbf{M}^{-1} + \mathbf{G}\mathbf{N}^{-1}\mathbf{G}^T)^{-1}\mathbf{G}\mathbf{N}^{-1} \end{aligned} \tag{1.9b}$$

as dispersion matrix. Furthermore, we easily calculate the estimated vector of "Lagrange multipliers" by

$$\hat{\boldsymbol\lambda}_R = \mathbf{M}(\mathbf{G}\hat{\boldsymbol\beta}_R - \mathbf{g}) = [\mathbf{M} - \mathbf{G}\mathbf{N}^{-1}\mathbf{G}^{\mathrm T}(\mathbf{M}^{-1} + \mathbf{G}\mathbf{N}^{-1}\mathbf{G}^{\mathrm T})^{-1}](\mathbf{G}\hat{\boldsymbol\beta} - \mathbf{g})$$
$$= -(\mathbf{M}^{-1} + \mathbf{G}\mathbf{N}^{-1}\mathbf{G}^{\mathrm T})^{-1}(\mathbf{g} - \mathbf{G}\hat{\boldsymbol\beta}). \tag{1.9c}$$

On the other hand, the BIQUUE of σ^2 is readily given by

$$\hat\sigma_R{}^2 = (n - m + r)^{-1}\Omega_R \tag{1.10a}$$

with the quadratic form Ω_R of the *extended residual vector*

$$\begin{bmatrix}\tilde{\mathbf{u}}_R\\ \tilde{\mathbf{e}}_R\end{bmatrix} := \begin{bmatrix}\mathbf{y}\\ \mathbf{g}\end{bmatrix} - \begin{bmatrix}\mathbf{X}\\ \mathbf{G}\end{bmatrix}\hat{\boldsymbol\beta}_R = \begin{bmatrix}\tilde{\mathbf{u}}\\ 0\end{bmatrix} + \begin{bmatrix}\mathbf{X}\mathbf{N}^{-1}\mathbf{G}^{\mathrm T}\\ -\mathbf{M}^{-1}\end{bmatrix}\hat{\boldsymbol\lambda}_R$$
$$= \begin{bmatrix}\tilde{\mathbf{u}}\\ \mathbf{g} - \mathbf{G}\hat{\boldsymbol\beta}\end{bmatrix} + \begin{bmatrix}\mathbf{X}\\ \mathbf{G}\end{bmatrix}\mathbf{N}^{-1}\mathbf{G}^{\mathrm T}\hat{\boldsymbol\lambda}_R \tag{1.10b}$$

being defined by

$$\Omega_R := \tilde{\mathbf{u}}_R{}^{\mathrm T}\mathbf{P}\tilde{\mathbf{u}}_R + \tilde{\mathbf{e}}_R{}^{\mathrm T}\mathbf{M}\tilde{\mathbf{e}}_R = \tilde{\mathbf{u}}_R{}^{\mathrm T}\mathbf{P}\mathbf{y} + \tilde{\mathbf{e}}_R{}^{\mathrm T}\mathbf{M}\mathbf{g}$$
$$= \tilde{\mathbf{u}}^{\mathrm T}\mathbf{P}\mathbf{y} - \hat{\boldsymbol\lambda}_R{}^{\mathrm T}(\mathbf{g} - \mathbf{G}\hat{\boldsymbol\beta}). \tag{1.10c}$$

Consequently, the *relative increase*

$$R/\sigma^2 := \sigma^{-2}(\Omega_R - \Omega)$$
$$= (\mathbf{g} - \mathbf{G}\hat{\boldsymbol\beta})^{\mathrm T}(\mathbf{M}^{-1} + \mathbf{G}\mathbf{N}^{-1}\mathbf{G}^{\mathrm T})^{-1}(\mathbf{g} - \mathbf{G}\hat{\boldsymbol\beta})/\sigma^2, \tag{1.11a}$$

due to

$$\sigma^{-1}(\mathbf{g} - \mathbf{G}\hat{\boldsymbol\beta}) \sim N(\sigma^{-1}(\boldsymbol\gamma - \mathbf{G}\boldsymbol\beta), \mathbf{M}^{-1} + \mathbf{G}\mathbf{N}^{-1}\mathbf{G}^{\mathrm T}), \tag{1.11b}$$

is *distributed as*

$$R/\sigma^2 \sim \chi'^2(r; \theta_R) \tag{1.12a}$$

with *non-centrality parameter*

$$\theta_R := (\boldsymbol\gamma - \mathbf{G}\boldsymbol\beta)^{\mathrm T}(\mathbf{M}^{-1} + \mathbf{G}\mathbf{N}^{-1}\mathbf{G}^{\mathrm T})^{-1}(\boldsymbol\gamma - \mathbf{G}\boldsymbol\beta)/2\sigma^2 \tag{1.12b}$$

and *expectation*

$$E\{R/\sigma^2\} = \mathrm{tr}(\mathbf{M}^{-1} + \mathbf{G}\mathbf{N}^{-1}\mathbf{G}^{\mathrm T})^{-1}(D\{\mathbf{g} - \mathbf{G}\hat{\boldsymbol\beta}\} + E\{\mathbf{g} - \mathbf{G}\hat{\boldsymbol\beta}\}E\{\mathbf{g} - \mathbf{G}\hat{\boldsymbol\beta}\}^{\mathrm T})/\sigma^2$$
$$= r + 2\theta_R. \tag{1.12c}$$

Since the *independence* of Ω and $R = \Omega_R - \Omega$ is readily established we finally conclude along the usual line of reasoning that our *test statistic*

$$T_R := \frac{(n-m)R}{r\Omega} = \frac{(n-m+r)\hat{\sigma}_R^2 - (n-m)\hat{\sigma}^2}{r\hat{\sigma}^2} \tag{1.13a}$$

is *distributed as*

$$T_R = F'(r, n-m; \theta_R) \tag{1.13b}$$

with $\theta_R = 0$ under the null hypothesis. Note, in particular, that in this case the resulting *central* F-distribution is *not* influenced by the choice of the "weight matrix" $\mathbf{M}$. Hence for $\mathbf{M}^{-1} \to \mathbf{0}$ we obtain the *traditional* F*-test* of a *non-stochastic linear hypothesis* while $\mathbf{M} \to \mathbf{0}$ leads to $T_R \to 0$ and therefore, in the limit, to the obvious acceptance of the hypothesis. As a consequence we are able to introduce an *indicator for the sensitivity* by looking at the functions $\theta_R(\mathbf{M})$ and $R(\mathbf{M})$, respectively, of the variable "weight matrix" $\mathbf{M}$ finding the important *inequalities*

$$\begin{aligned}\theta_R(\infty) - \theta_R(\mathbf{M}) &= (2\sigma^2)^{-1}(\boldsymbol{\gamma} - \mathbf{G}\boldsymbol{\beta})^{\mathrm{T}}[(\mathbf{G}\mathbf{N}^{-1}\mathbf{G}^{\mathrm{T}})^{-1} - (\mathbf{M}^{-1} + \mathbf{G}\mathbf{N}^{-1}\mathbf{G}^{\mathrm{T}})^{-1}](\boldsymbol{\gamma} - \mathbf{G}\boldsymbol{\beta}) \\ &= (2\sigma^2)^{-1}(\boldsymbol{\gamma} - \mathbf{G}\boldsymbol{\beta})^{\mathrm{T}}(\mathbf{G}\mathbf{N}^{-1}\mathbf{G}^{\mathrm{T}})^{-1}(\mathbf{I}_r + \mathbf{G}\mathbf{N}^{-1}\mathbf{G}^{\mathrm{T}}\mathbf{M})^{-1}(\boldsymbol{\gamma} - \mathbf{G}\boldsymbol{\beta}) \\ &= (2\sigma^2)^{-1}(\boldsymbol{\gamma} - \mathbf{G}\boldsymbol{\beta})^{\mathrm{T}}[\mathbf{G}\mathbf{N}^{-1}(\mathbf{N} + \mathbf{G}^{\mathrm{T}}\mathbf{M}\mathbf{G})\mathbf{N}^{-1}\mathbf{G}^{\mathrm{T}}]^{-1}(\boldsymbol{\gamma} - \mathbf{G}\boldsymbol{\beta}) \geq 0\end{aligned} \tag{1.14a}$$

and

$$\begin{aligned}R(\infty) - R(\mathbf{M}) &= (\mathbf{g} - \mathbf{G}\hat{\boldsymbol{\beta}})^{\mathrm{T}}[(\mathbf{G}\mathbf{N}^{-1}\mathbf{G})^{-1} - (\mathbf{M}^{-1} + \mathbf{G}\mathbf{N}^{-1}\mathbf{G}^{\mathrm{T}})^{-1}](\mathbf{g} - \mathbf{G}\hat{\boldsymbol{\beta}}) \\ &= (\mathbf{g} - \mathbf{G}\hat{\boldsymbol{\beta}})^{\mathrm{T}}(\mathbf{G}\mathbf{N}^{-1}\mathbf{G}^{\mathrm{T}})^{-1}(\mathbf{I}_r + \mathbf{G}\mathbf{N}^{-1}\mathbf{G}^{\mathrm{T}}\mathbf{M})^{-1}(\mathbf{g} - \mathbf{G}\hat{\boldsymbol{\beta}}) \\ &= (\mathbf{g} - \mathbf{G}\hat{\boldsymbol{\beta}})^{\mathrm{T}}[\mathbf{G}\mathbf{N}^{-1}(\mathbf{N} + \mathbf{G}^{\mathrm{T}}\mathbf{M}\mathbf{G})\mathbf{N}^{-1}\mathbf{G}^{\mathrm{T}}]^{-1}(\mathbf{g} - \mathbf{G}\hat{\boldsymbol{\beta}}) \geq 0\end{aligned} \tag{1.14b}$$

with the *special limits*

$$\theta_R(\mathbf{M} \to \mathbf{0}) = (2\sigma^2)^{-1}\lim(\boldsymbol{\gamma} - \mathbf{G}\boldsymbol{\beta})^{\mathrm{T}}\mathbf{M}(\mathbf{I}_r + \mathbf{G}\mathbf{N}^{-1}\mathbf{G}^{\mathrm{T}}\mathbf{M})^{-1}(\boldsymbol{\gamma} - \mathbf{G}\boldsymbol{\beta}) = 0 \tag{1.15a}$$

and

$$R(\mathbf{M} \to \mathbf{0}) = \lim(\mathbf{g} - \mathbf{G}\hat{\boldsymbol{\beta}})^{\mathrm{T}}\mathbf{M}(\mathbf{I}_r + \mathbf{G}\mathbf{N}^{-1}\mathbf{G}^{\mathrm{T}}\mathbf{M})^{-1}(\mathbf{g} - \mathbf{G}\hat{\boldsymbol{\beta}}) = 0. \tag{1.15b}$$

In the following *chapter 2* we shall present an *equivalence theorem* which allows us to derive the same results as above within the *Mixed Linear Model* by considering *B*est *inhom*ogeneously *Li*near *P*rediction (inhomBLIP) after applying "*Helmert's knack*". Even *less sensitive tests* are then constructed in *chapter 3* by extending these results to a *robust* alternative like *B*est *hom*ogeneously *L*inear (weakly) *U*nbiased *P*rediction (homBLUP) which has been developed in B. Schaffrin (1983; 1985; 1987) for the *Mixed Linear Model* in various forms, but cannot easily be interpreted in the context of "*Mixed Regression*"; such less sensitive tests are frequently required in practice.

2. THE MIXED LINEAR MODEL AND AN EQUIVALENCE THEOREM

Taking into account that the $r \times m$ matrix $\mathbf{G}$ of the restrictions (1.6a, b) can be partitioned as

$$\mathbf{G} = [\mathbf{G}_1, \mathbf{G}_2] = \mathbf{G}_1[\mathbf{I}_r, \mathbf{G}_1^{-1}\mathbf{G}_2], \ \ \mathbf{G}_1 \text{ regular,} \tag{2.1}$$

we obtain the appropriate *reparametrized model equation* from (1.1a) as

$$\begin{aligned} E\{\mathbf{y}\} &= \mathbf{y} - \mathbf{u} = \mathbf{X}\boldsymbol{\beta} = \mathbf{X}_1\boldsymbol{\beta}_1 + \mathbf{X}_2\boldsymbol{\beta}_2 \\ &= (\mathbf{X}_1\mathbf{G}_1^{-1})(\mathbf{G}\boldsymbol{\beta}) + (\mathbf{X}_2 - \mathbf{X}_1\mathbf{G}_1^{-1}\mathbf{G}_2)\boldsymbol{\beta}_2 = \bar{\mathbf{X}}_1\bar{\boldsymbol{\beta}}_1 + \bar{\mathbf{X}}_2\boldsymbol{\beta}_2 \end{aligned} \tag{2.2a}$$

with

$$[\bar{\mathbf{X}}_1, \bar{\mathbf{X}}_2] := [\mathbf{X}_1\mathbf{G}_1^{-1}, \mathbf{X}_2 - \mathbf{X}_1\mathbf{G}_1^{-1}\mathbf{G}_2], \quad \bar{\boldsymbol{\beta}}_1 := \mathbf{G}\boldsymbol{\beta}, \tag{2.2b}$$

so that (1.6a) turns over to

$$\boldsymbol{\gamma} = E\{\mathbf{g}\} = \mathbf{g} - \mathbf{e} \doteq \mathbf{G}\boldsymbol{\beta} = [\mathbf{I}_r, \mathbf{0}][\bar{\boldsymbol{\beta}}_1^{\mathrm{T}}, \boldsymbol{\beta}_2^{\mathrm{T}}]^{\mathrm{T}}. \tag{2.3}$$

Thus there is *no loss in generality* if we assume from the beginning

$$\mathbf{X} = [\bar{\mathbf{X}}_1, \bar{\mathbf{X}}_2], \ \ \mathbf{G} = [\mathbf{I}_r, \mathbf{0}], \ \ \mathbf{P}_e := \rho\mathbf{M}, \ \ \boldsymbol{\beta}_1 = \bar{\boldsymbol{\beta}}_1 \doteq \mathbf{g} - \mathbf{e}, \tag{2.4}$$

instead of (1.6a, b). Now we apply "*Helmert's knack*" as implicitly proposed already in F.R. Helmert (1872), and *interchange the stochastic nature* of the vectors $\mathbf{g}$ and $\boldsymbol{\beta}_1$ by formally adding a $r \times 1$ "*random zero vector*" $\underline{\mathbf{0}}$ such that (2.4) is replaced by

$$\boldsymbol{\gamma}_1 := \mathbf{g} + \underline{\mathbf{0}} \doteq \mathbf{b}_1 + \underline{\mathbf{0}}, \ \ \mathbf{b}_1 := \boldsymbol{\beta}_1 + \mathbf{e}, \tag{2.5a}$$

with the constraining *non-random* $r \times 1$ vector $\boldsymbol{\gamma}_1$ being *numerically identical* with $\mathbf{g}$ (but *not* stochastically) and, moreover,

$$\underline{\mathbf{0}} \sim N(\boldsymbol{\gamma}_1 - \boldsymbol{\beta}_1 = \mathbf{0}, \sigma^2\mathbf{P}_1^{-1}), \ \ \mathbf{P}_1^{-1} := \sigma^{-2}D\{\mathbf{e}\} = \rho\mathbf{P}_e^{-1}, \tag{2.5b}$$

under the *null hypothesis*

$$\mathbf{b}_1 \sim N(\gamma_1, \sigma^2 \mathbf{P}_1^{-1}), \quad \gamma_1 \text{ and } \mathbf{P}_1 \text{ given,} \tag{2.6}$$

opposed to the alternative $\mathbf{b}_1 \sim N(\gamma_1, \infty)$. Combining (2.6) with (1.1) yields the *Mixed Linear Model*

$$\mathbf{y}_* := \mathbf{y} - \mathbf{X}_1 \underline{\mathbf{0}} = \mathbf{X}_1 \mathbf{b}_1 + \mathbf{X}_2 \boldsymbol{\beta}_2 + \mathbf{u}, \quad \operatorname{rk} \mathbf{X}_2 = m - r, \tag{2.7a}$$

$$E\{\mathbf{b}_1\} = \mathbf{b}_1 - \mathbf{e} = \boldsymbol{\beta}_1 \doteq \gamma_1, \quad \gamma_1 \textit{ given}, \tag{2.7b}$$

$$\begin{bmatrix} \mathbf{u} \\ \mathbf{e} \end{bmatrix} \sim \left(\begin{bmatrix} \mathbf{0} \\ \mathbf{0} \end{bmatrix}, \sigma^2 \begin{bmatrix} \mathbf{P}^{-1} & \mathbf{0} \\ \mathbf{0} & \mathbf{P}_1^{-1} \end{bmatrix} \right), \tag{2.7c}$$

in case null hypothesis holds true, and with γ_1 *unknown* (or equivalently $\mathbf{P}_1 \to \mathbf{0}$) otherwise. Note that there is *no numerical difference* between the observational vectors $\mathbf{y}$ and $\mathbf{y}_*$ though they differ *stochastically* from each other due to the relation

$$\begin{aligned} D\{\mathbf{y}_*\} &= D\{\mathbf{y}\} + \mathbf{X}_1 D\{\mathbf{b}_1\} \mathbf{X}_1^{\mathrm{T}} \\ &= D\{\mathbf{u}\} + \mathbf{X}_1 D\{\mathbf{e}\} \mathbf{X}_1^{\mathrm{T}} = \sigma^2 (\mathbf{P}^{-1} + \mathbf{X}_1 \mathbf{P}_1^{-1} \mathbf{X}_1^{\mathrm{T}}). \end{aligned} \tag{2.8}$$

According to H. Läuter (1970), H. Toutenburg (1970), S.R. Searle (1974), C.R. Rao (1975), D.A. Harville (1976) and several other authors, the *Best inhomogeneously Linear Prediction (inhomBLIP)* of $\mathbf{b}_1$ turns out to be *weakly unbiased* and can, together with the BLUUE of $\boldsymbol{\beta}_2$, be calculated from the "*normal equations*"

$$\begin{bmatrix} \mathbf{N}_{11} & \mathbf{N}_{12} & \mathbf{I}_r \\ \mathbf{N}_{21} & \mathbf{N}_{22} & \mathbf{0} \\ \mathbf{I}_r & \mathbf{0} & -\mathbf{P}_1^{-1} \end{bmatrix} \begin{bmatrix} \tilde{\mathbf{b}}_{1I} \\ \hat{\boldsymbol{\beta}}_{2I} \\ \hat{\boldsymbol{\lambda}}_I \end{bmatrix} = \begin{bmatrix} \mathbf{z}_1 \\ \mathbf{z}_2 \\ \gamma_1 \end{bmatrix}, \tag{2.9a}$$

with

$$\begin{bmatrix} \mathbf{N}_{11} & \mathbf{N}_{12} & \mathbf{z}_1 \\ \mathbf{N}_{21} & \mathbf{N}_{22} & \mathbf{z}_2 \end{bmatrix} := \begin{bmatrix} \mathbf{X}_1^{\mathrm{T}} \\ \mathbf{X}_2^{\mathrm{T}} \end{bmatrix} \mathbf{P}[\mathbf{X}_1, \mathbf{X}_2, \mathbf{y}_*] \tag{2.9b}$$

and $\boldsymbol{\lambda}$ as $r \times 1$ vector of "*Lagrange multipliers*". The equivalent and computationally most appealing form

$$\begin{bmatrix} \mathbf{N}_{11} + \mathbf{P}_1 & \mathbf{N}_{12} \\ \mathbf{N}_{21} & \mathbf{N}_{22} \end{bmatrix} \begin{bmatrix} \tilde{\mathbf{b}}_{1I} \\ \hat{\boldsymbol{\beta}}_{2I} \end{bmatrix} = \begin{bmatrix} \mathbf{z}_1 + \mathbf{P}_1 \gamma_1 \\ \mathbf{z}_2 \end{bmatrix} \tag{2.10}$$

of the "normal equations" is now easily be derived and leads to the solution

$$\begin{bmatrix} \tilde{\mathbf{b}}_{1I} \\ \hat{\boldsymbol{\beta}}_{2I} \end{bmatrix} = \begin{bmatrix} (\mathbf{S}_1 + \mathbf{P}_1)^{-1}(\mathbf{z}_1 - \mathbf{N}_{12}\mathbf{N}_{22}^{-1}\mathbf{z}_2 + \mathbf{P}_1\boldsymbol{\gamma}_1) \\ \mathbf{N}_{22}^{-1}(\mathbf{z}_2 - \mathbf{N}_{21}\tilde{\mathbf{b}}_{1I}) \end{bmatrix} \tag{2.11a}$$

with the *mean square error matrix*

$$D\{\begin{bmatrix} \tilde{\mathbf{b}}_{1I} - \mathbf{b}_1 \\ \hat{\boldsymbol{\beta}}_{2I} \end{bmatrix}\} = \sigma^2 \begin{bmatrix} \mathbf{N}_{11} + \mathbf{P}_1 & \mathbf{N}_{12} \\ \mathbf{N}_{21} & \mathbf{N}_{22} \end{bmatrix}^{-1}$$

$$= \sigma^2 \begin{bmatrix} \bar{\mathbf{S}}_1^{-1} & -\bar{\mathbf{S}}_1^{-1}\mathbf{N}_{12}\mathbf{N}_{22}^{-1} \\ -\mathbf{N}_{22}^{-1}\mathbf{N}_{21}\bar{S}_1^{-1} & \mathbf{N}_{22}^{-1} + \mathbf{N}_{22}^{-1}\mathbf{N}_{21}\bar{S}_1^{-1}\mathbf{N}_{12}\mathbf{N}_{22}^{-1} \end{bmatrix} \tag{2.11b}$$

in *numerical* analogy to (1.9a, b) if $\mathbf{P}_1 \neq \mathbf{0}$, and in *complete* analogy to (1.3a, b) if $\mathbf{P}_1 = \mathbf{0}$. Similarly, the estimated vector of "Lagrange multipliers" is given by

$$\hat{\boldsymbol{\lambda}}_I = \mathbf{P}_1(\tilde{\mathbf{b}}_{1I} - \boldsymbol{\gamma}_1) = \mathbf{P}_1(\mathbf{S}_1 + \mathbf{P}_1)^{-1}(\mathbf{X}_1^T - \mathbf{N}_{12}\mathbf{N}_{22}^{-1}\mathbf{X}_2^T)\mathbf{P}(\mathbf{y}_* - \mathbf{X}_1\boldsymbol{\gamma}_1)$$
$$= \mathbf{S}_1(\mathbf{S}_1 + \mathbf{P}_1)^{-1}\mathbf{P}_1(\tilde{\mathbf{b}}_1 - \boldsymbol{\gamma}_1) = -(\mathbf{S}_1^{-1} + \mathbf{P}_1^{-1})^{-1}(\boldsymbol{\gamma}_1 - \tilde{\mathbf{b}}_1). \tag{2.11c}$$

Here the $r \times r$ matrix

$$\mathbf{S}_1 := \mathbf{N}_{11} - \mathbf{N}_{12}\mathbf{N}_{22}^{-1}\mathbf{N}_{21} \tag{2.12a}$$

denotes the first "*Schur complement*" such that

$$\bar{\mathbf{S}}_1^{-1} := (\mathbf{S}_1 + \mathbf{P}_1)^{-1} = \mathbf{S}_1^{-1} - (\mathbf{S}_1 + \mathbf{P}_1)^{-1}\mathbf{P}_1\mathbf{S}_1^{-1}$$
$$= \mathbf{S}_1^{-1} - \mathbf{S}_1^{-1}(\mathbf{S}_1^{-1} + \mathbf{P}_1^{-1})^{-1}\mathbf{S}_1^{-1} \tag{2.12b}$$

holds true and (2.11a) can be transformed into

$$\begin{bmatrix} \tilde{\mathbf{b}}_{1I} \\ \hat{\boldsymbol{\beta}}_{2I} \end{bmatrix} = \begin{bmatrix} \tilde{\mathbf{b}}_1 \\ \hat{\boldsymbol{\beta}}_2 \end{bmatrix} + \begin{bmatrix} -\mathbf{I}_r \\ \mathbf{N}_{22}^{-1}\mathbf{N}_{21} \end{bmatrix}(\mathbf{S}_1 + \mathbf{P}_1)^{-1}\mathbf{P}_1(\tilde{\mathbf{b}}_1 - \boldsymbol{\gamma}_1)$$

$$= \begin{bmatrix} \tilde{\mathbf{b}}_1 \\ \hat{\boldsymbol{\beta}}_2 \end{bmatrix} - \begin{bmatrix} -\mathbf{I}_r \\ \mathbf{N}_{22}^{-1}\mathbf{N}_{21} \end{bmatrix}\mathbf{S}_1^{-1}(\mathbf{S}_1^{-1} + \mathbf{P}_1^{-1})^{-1}(\boldsymbol{\gamma}_1 - \tilde{\mathbf{b}}_1) \tag{2.13a}$$

with

$$\begin{bmatrix} \tilde{\mathbf{b}}_1 \\ \hat{\boldsymbol{\beta}}_2 \end{bmatrix} = \begin{bmatrix} \mathbf{S}_1^{-1}(\mathbf{z}_1 - \mathbf{N}_{12}\mathbf{N}_{22}^{-1}\mathbf{z}_2) \\ \mathbf{N}_{22}^{-1}(\mathbf{z}_2 - \mathbf{N}_{21}\tilde{\mathbf{b}}_1) \end{bmatrix} \tag{2.13b}$$

as "*unconstrained solution*" corresponding to $\hat{\boldsymbol{\beta}}$ in formula (1.3a). Furthermore, the BIQUUE of σ^2 is readily obtained from

$$\hat{\sigma}_I^2 = (n - m + r)^{-1}\Omega_I \tag{2.14a}$$

with the quadratic form Ω_I of the appropriate *residual vector*

$$\begin{bmatrix} \tilde{\mathbf{u}}_I \\ \tilde{\mathbf{e}}_I \end{bmatrix} := \begin{bmatrix} \mathbf{y}_* \\ -\mathbf{y}_1 \end{bmatrix} - \begin{bmatrix} \mathbf{X}_1 & \mathbf{X}_2 \\ -\mathbf{I}_r & \mathbf{0} \end{bmatrix} \begin{bmatrix} \tilde{\mathbf{b}}_{1I} \\ \hat{\boldsymbol{\beta}}_{2I} \end{bmatrix}$$

$$= \begin{bmatrix} \mathbf{y}_* - \mathbf{X}_1\tilde{\mathbf{b}}_1 - \mathbf{X}_2\hat{\boldsymbol{\beta}}_2 \\ -(\mathbf{y}_1 - \tilde{\mathbf{b}}_1) \end{bmatrix} + \begin{bmatrix} \mathbf{X}_1 - \mathbf{X}_2\mathbf{N}_{22}^{-1}\mathbf{N}_{21} \\ -\mathbf{I}_r \end{bmatrix} \mathbf{S}_1^{-1}\hat{\boldsymbol{\lambda}}_I$$

$$= \begin{bmatrix} \tilde{\mathbf{u}} \\ \mathbf{0} \end{bmatrix} + \begin{bmatrix} (\mathbf{X}_1 - \mathbf{X}_2\mathbf{N}_{22}^{-1}\mathbf{N}_{21})\mathbf{S}_1^{-1} \\ \mathbf{P}_1^{-1} \end{bmatrix} \hat{\boldsymbol{\lambda}}_I \tag{2.14b}$$

being defined by

$$\Omega_I := \tilde{\mathbf{u}}_I^{\mathrm{T}}\mathbf{P}\tilde{\mathbf{u}}_I + \tilde{\mathbf{e}}_I^{\mathrm{T}}\mathbf{P}_1\tilde{\mathbf{e}}_I = \tilde{\mathbf{u}}_I^{\mathrm{T}}\mathbf{P}\mathbf{y} - \tilde{\mathbf{e}}_I^{\mathrm{T}}\mathbf{P}_1\mathbf{y}_1 = \tilde{\mathbf{u}}^{\mathrm{T}}\mathbf{P}\mathbf{y}_* - \hat{\boldsymbol{\lambda}}_I^{\mathrm{T}}(\mathbf{y}_1 - \tilde{\mathbf{b}}_1)$$
$$= \tilde{\mathbf{u}}^{\mathrm{T}}\mathbf{P}\tilde{\mathbf{u}} + (\mathbf{y}_1 - \tilde{\mathbf{b}}_1)^{\mathrm{T}}(\mathbf{S}_1^{-1} + \mathbf{P}_1^{-1})^{-1}(\mathbf{y}_1 - \tilde{\mathbf{b}}_1). \tag{2.14c}$$

Taking into account that in the "*unconstrained case*" for

$$\tilde{\mathbf{u}} := \mathbf{y}_* - \mathbf{X}_1\tilde{\mathbf{b}}_1 - \mathbf{X}_2\hat{\boldsymbol{\beta}}_2 \sim N(\mathbf{0}, \sigma^2(\mathbf{P}^{-1} - [\mathbf{X}_1, \mathbf{X}_2] \begin{bmatrix} \mathbf{N}_{11} & \mathbf{N}_{12} \\ \mathbf{N}_{21} & \mathbf{N}_{22} \end{bmatrix}^{-1} \begin{bmatrix} \mathbf{X}_1^{\mathrm{T}} \\ \mathbf{X}_2^{\mathrm{T}} \end{bmatrix})) \tag{2.15a}$$

we simply find

$$\Omega := \tilde{\mathbf{u}}^{\mathrm{T}}\mathbf{P}\tilde{\mathbf{u}} = \tilde{\mathbf{u}}^{\mathrm{T}}\mathbf{P}\mathbf{y}_* = \mathbf{y}_*^{\mathrm{T}}\mathbf{P}(\mathbf{P}^{-1} - [\mathbf{X}_1, \mathbf{X}_2] \begin{bmatrix} \mathbf{N}_{11} & \mathbf{N}_{12} \\ \mathbf{N}_{21} & \mathbf{N}_{22} \end{bmatrix}^{-1} \begin{bmatrix} \mathbf{X}_1^{\mathrm{T}} \\ \mathbf{X}_2^{\mathrm{T}} \end{bmatrix})\mathbf{P}\mathbf{y}_* \tag{2.15b}$$

to be *distributed* again as

$$\Omega / \sigma^2 \sim \chi^2(n - m) \tag{2.16a}$$

with *expectation*

$$E\{\Omega/\sigma^2\} = \mathrm{tr}(\mathbf{I}_n - \mathbf{P}[\mathbf{X}_1, \mathbf{X}_2]\begin{bmatrix}\mathbf{N}_{11} & \mathbf{N}_{12}\\ \mathbf{N}_{21} & \mathbf{N}_{22}\end{bmatrix}^{-1}\begin{bmatrix}\mathbf{X}_1^{\mathrm{T}}\\ \mathbf{X}_2^{\mathrm{T}}\end{bmatrix})\mathbf{P}E\{\mathbf{y}_*\mathbf{y}_*^{\mathrm{T}}\}/\sigma^2 = n-m, \tag{2.16b}$$

we may well conclude in the "*constrained case*" that the *relative increase*

$$R_I/\sigma^2 := \sigma^{-2}(\Omega_I - \Omega) = (\boldsymbol{\gamma}_1 - \tilde{\mathbf{b}}_1)^{\mathrm{T}}(\mathbf{S}_1^{-1} + \mathbf{P}_1^{-1})^{-1}(\boldsymbol{\gamma}_1 - \tilde{\mathbf{b}}_1)/\sigma^2, \tag{2.17a}$$

due to

$$\sigma^{-1}(\boldsymbol{\gamma}_1 - \tilde{\mathbf{b}}_1) \sim N(\sigma^{-1}(\boldsymbol{\gamma}_1 - \boldsymbol{\beta}_1), \mathbf{S}_1^{-1} + \mathbf{P}_1^{-1}), \tag{2.17b}$$

now is *distributed as*

$$R_I/\sigma^2 \sim \chi'^2(r; \theta_I) \tag{2.18a}$$

with *non-centrality parameter*

$$\theta_I := (\boldsymbol{\gamma}_1 - \boldsymbol{\beta}_1)^{\mathrm{T}}(\mathbf{S}_1^{-1} + \mathbf{P}_1^{-1})^{-1}(\boldsymbol{\gamma}_1 - \boldsymbol{\beta}_1)/2\sigma^2 \tag{2.18b}$$

and *expectation*

$$E\{R_I/\sigma^2\} = \mathrm{tr}(\mathbf{S}_1^{-1} + \mathbf{P}_1^{-1})^{-1}(D\{\boldsymbol{\gamma}_1 - \tilde{\mathbf{b}}_1\} + E\{\boldsymbol{\gamma}_1 - \tilde{\mathbf{b}}_1\}E\{\boldsymbol{\gamma}_1 - \tilde{\mathbf{b}}_1\}^{\mathrm{T}})/\sigma^2 = r + 2\theta_I. \tag{2.18c}$$

If we express Ω and R_I as quadratic forms in the *common* random vector

$$\mathbf{y}_* - \mathbf{X}_1\boldsymbol{\gamma}_1 = \mathbf{y}_* - [\mathbf{X}_1, \mathbf{X}_2]\begin{bmatrix}\boldsymbol{\gamma}_1\\ \mathbf{0}\end{bmatrix}$$

$$\sim N(\mathbf{X}_1(\boldsymbol{\beta}_1 - \boldsymbol{\gamma}_1) + \mathbf{X}_2\boldsymbol{\beta}_2, \sigma^2(\mathbf{P}^{-1} + \mathbf{X}_1\mathbf{P}_1^{-1}\mathbf{X}_1^{\mathrm{T}})) \tag{2.19a}$$

which yields

$$\Omega = (\mathbf{y}_* - \mathbf{X}_1\boldsymbol{\gamma}_1)^{\mathrm{T}}(\mathbf{P} - \mathbf{P}[\mathbf{X}_1, \mathbf{X}_2]\begin{bmatrix}\mathbf{N}_{11} & \mathbf{N}_{12}\\ \mathbf{N}_{21} & \mathbf{N}_{22}\end{bmatrix}^{-1}\begin{bmatrix}\mathbf{X}_1^{\mathrm{T}}\\ \mathbf{X}_2^{\mathrm{T}}\end{bmatrix}\mathbf{P})(\mathbf{y}_* - \mathbf{X}_1\boldsymbol{\gamma}_1) \tag{2.19b}$$

and

$$R_I = (\mathbf{y}_* - \mathbf{X}_1\boldsymbol{\gamma}_1)^{\mathrm{T}}\mathbf{P}(\mathbf{X}_1 - \mathbf{X}_2\mathbf{N}_{22}^{-1}\mathbf{N}_{21})\mathbf{S}_1^{-1}(\mathbf{S}_1^{-1} + \mathbf{P}_1^{-1})^{-1} \cdot \mathbf{S}_1^{-1}(\mathbf{X}_1^{\mathrm{T}} - \mathbf{N}_{12}\mathbf{N}_{22}^{-1}\mathbf{X}_2^{\mathrm{T}})\mathbf{P}(\mathbf{y}_* - \mathbf{X}_1\boldsymbol{\gamma}_1), \tag{2.19c}$$

the *independence* of both, Ω and R_I, immediately follows such that the *test statistic*

$$T_I := \frac{(n-m)R_I}{r\Omega} = \frac{(n-m+r)\hat{\sigma}_I^2 - (n-m)\hat{\sigma}^2}{r\hat{\sigma}^2} \tag{2.20a}$$

becomes *distributed as*

$$T_I \sim F'(r, n-m; \theta_I) \qquad (2.20b)$$

with $\theta_I = 0$ under the null hypothesis. Again the resulting *central F-distribution* is *not* influenced by the choice of the "weight matrix" $\mathbf{P}_1$. Comparing our result (2.20a, b) with the previously derived result (1.13a, b) we summarize these, together with a sensitivity statement, in the following.

EQUIVALENCE THEOREM:

(i) *Instead of testing a stochastic linear hypothesis* (1.6) *in the context of the Mixed Regression Model* (1.7) *by applying* (1.13), *the hypothesis* (2.6) *can* equivalently *be tested within the Mixed Linear Model* (2.7) *by exploiting* (2.20).

(ii) *The* sensitivity of the test *according to* (2.20) *is governed by the "weight matrix"* $\mathbf{P}_1$*; if* $\mathbf{P}_1 \to \mathbf{0}$ *the sensitivity is going to be lost, while for* $\mathbf{P}_1$ *growing the sensitivity is increasing, too, until the traditional F-test is achieved in the limit case* $\mathbf{P}_1 \to \infty$ *or* $\mathbf{P}_1^{-1} \to \mathbf{0}$*, respectively.*

The second part of the *Equivalence Theorem* follows the same reasoning as demonstrated at the end of *chapter 1*. For, looking at the functions $\theta_I(\mathbf{P}_1)$ and $R_I(\mathbf{P}_1)$, respectively, of the variable "weight matrix" $\mathbf{P}_1$ we again find the important *inequalities*

$$\begin{aligned}\theta_I(\infty) - \theta_I(\mathbf{P}_1) &= (2\sigma^2)^{-1}(\boldsymbol{\gamma}_1 - \boldsymbol{\beta}_1)^T[\mathbf{S}_1 - (\mathbf{S}_1^{-1} + \mathbf{P}_1^{-1})^{-1}](\boldsymbol{\gamma}_1 - \boldsymbol{\beta}_1)\\ &= (2\sigma^2)^{-1}(\boldsymbol{\gamma}_1 - \boldsymbol{\beta}_1)^T\mathbf{S}_1(\mathbf{S}_1 + \mathbf{P}_1)^{-1}\mathbf{S}_1(\boldsymbol{\gamma}_1 - \boldsymbol{\beta}_1) \geq 0 \qquad (2.21a)\end{aligned}$$

and

$$\begin{aligned}R_I(\infty) - R_I(\mathbf{P}_1) &= (\boldsymbol{\gamma}_1 - \tilde{\mathbf{b}}_1)^T[\mathbf{S}_1 - (\mathbf{S}_1^{-1} + \mathbf{P}_1^{-1})^{-1}](\boldsymbol{\gamma}_1 - \tilde{\mathbf{b}}_1)\\ &= (\boldsymbol{\gamma}_1 - \tilde{\mathbf{b}}_1)^T\mathbf{S}_1(\mathbf{S}_1 + \mathbf{P}_1)^{-1}\mathbf{S}_1(\boldsymbol{\gamma}_1 - \tilde{\mathbf{b}}_1) \geq 0 \qquad (2.21b)\end{aligned}$$

with the *special limits*

$$\theta_I(\mathbf{P}_1 \to \mathbf{0}) = (2\sigma^2)^{-1}\lim(\boldsymbol{\gamma}_1 - \boldsymbol{\beta}_1)^T\mathbf{P}_1(\mathbf{I}_r + \mathbf{S}_1^{-1}\mathbf{P}_1)^{-1}(\boldsymbol{\gamma}_1 - \boldsymbol{\beta}_1) = 0 \qquad (2.22a)$$

and

$$R_I(\mathbf{P}_1 \to \mathbf{0}) = \lim(\boldsymbol{\gamma}_1 - \tilde{\mathbf{b}}_1)^T\mathbf{P}_1(\mathbf{I}_r + \mathbf{S}_1^{-1}\mathbf{P}_1)^{-1}(\boldsymbol{\gamma}_1 - \tilde{\mathbf{b}}_1) = 0. \qquad (2.22b)$$

In the following *chapter 3* we shall introduce *even less sensitive tests*, not by varying the "weight matrix" $\mathbf{P}_1$, but by considering a *robust alternative* for inhomBLIP $\tilde{\mathbf{b}}_{1I}$ in combination with BLUUE $\hat{\boldsymbol{\beta}}_{1I}$ within the *Mixed Linear Model.*

3. EVEN LESS SENSITIVE TESTS WITHIN THE MIXED LINEAR MODEL

In the following we further assume the *Mixed Linear Model* (2.7) for testing the *null hypothesis* (2.6) against its alternative. However, instead of inhomBLIP $\tilde{\mathbf{b}}_{1I}$ and BLUUE $\hat{\boldsymbol{\beta}}_{1I}$ we consider the *Best homogeneously Linear (weakly) Unbiased Prediction (homBLUP)* of $\mathbf{b}_1$ together with the *appropriate uniformly unbiased estimate* of $\boldsymbol{\beta}_2$ which turn out to be *robust* against errors in the prior information. Therefore, we expect *even less sensitive tests* from their application.

As shown in full detail by B. Schaffrin (1983; 1985; 1987), for this purpose we have to solve basically the same "*normal equations*" as in (2.10), namely

$$\begin{bmatrix} \mathbf{N}_{11}+\mathbf{P}_1 & \mathbf{N}_{12} \\ \mathbf{N}_{21} & \mathbf{N}_{22} \end{bmatrix} \begin{bmatrix} \mathbf{c}_1, \boldsymbol{\eta}_1 \\ \mathbf{c}_2, \boldsymbol{\eta}_2 \end{bmatrix} = \begin{bmatrix} \mathbf{z}_1, \mathbf{N}_{11}\mathbf{y}_1 \\ \mathbf{z}_2, \mathbf{N}_{21}\mathbf{y}_1 \end{bmatrix} \tag{3.1a}$$

which gives

$$\begin{bmatrix} \mathbf{c}_1, \boldsymbol{\eta}_1 \\ \mathbf{c}_2, \boldsymbol{\eta}_2 \end{bmatrix} = \begin{bmatrix} \mathbf{N}_{11}+\mathbf{P}_1 & \mathbf{N}_{12} \\ \mathbf{N}_{21} & \mathbf{N}_{22} \end{bmatrix}^{-1} \begin{bmatrix} \mathbf{z}_1, \mathbf{N}_{11}\mathbf{y}_1 \\ \mathbf{z}_2, \mathbf{N}_{21}\mathbf{y}_1 \end{bmatrix}$$

$$= \begin{bmatrix} (\mathbf{S}_1+\mathbf{P}_1)^{-1}(\mathbf{z}_1-\mathbf{N}_{12}\mathbf{N}_{22}^{-1}\mathbf{z}_2) & , (\mathbf{S}_1+\mathbf{P}_1)^{-1}\mathbf{S}_1\mathbf{y}_1 \\ \mathbf{N}_{22}^{-1}(\mathbf{z}_2-\mathbf{N}_{21}\mathbf{c}_1) \; , & \mathbf{N}_{22}^{-1}\mathbf{N}_{21}(\mathbf{S}_1+\mathbf{P}_1)^{-1}\mathbf{P}_1\mathbf{y}_1 \end{bmatrix}. \tag{3.1b}$$

From (3.1b) we are able to derive *not only* inhomBLIP $\tilde{\mathbf{b}}_{1I}$ with BLUUE $\hat{\boldsymbol{\beta}}_{2I}$ by calculating

$$\begin{bmatrix} \tilde{\mathbf{b}}_{1I} \\ \hat{\boldsymbol{\beta}}_{2I} \end{bmatrix} = \begin{bmatrix} \mathbf{y}_1 \\ \mathbf{0} \end{bmatrix} + \begin{bmatrix} \mathbf{c}_1-\boldsymbol{\eta}_1 \\ \mathbf{c}_2-\boldsymbol{\eta}_2 \end{bmatrix} \tag{3.2}$$

in accordance with (2.11a), *but also* homBLUP $\tilde{\mathbf{b}}_{1H}$ with its "companion estimate" $\hat{\boldsymbol{\beta}}_{2H}$ by calculating

$$\begin{bmatrix} \tilde{\mathbf{b}}_{1H} \\ \hat{\boldsymbol{\beta}}_{2H} \end{bmatrix} = \begin{bmatrix} a\boldsymbol{\gamma}_1 \\ \mathbf{0} \end{bmatrix} + \begin{bmatrix} \mathbf{c}_1 - a\boldsymbol{\eta}_1 \\ \mathbf{c}_2 - a\boldsymbol{\eta}_2 \end{bmatrix} \tag{3.3a}$$

where the *random number* a is defined through

$$\begin{aligned} a := & (\boldsymbol{\gamma}_1^{\mathrm{T}}\mathbf{P}_1\boldsymbol{\eta}_1)^{-1}(\boldsymbol{\gamma}_1^{\mathrm{T}}\mathbf{P}_1\mathbf{c}_1) \\ = & [\boldsymbol{\gamma}_1^{\mathrm{T}}(\mathbf{S}_1^{-1} + \mathbf{P}_1^{-1})^{-1}\boldsymbol{\gamma}_1]^{-1}[\boldsymbol{\gamma}_1^{\mathrm{T}}(\mathbf{S}_1^{-1} + \mathbf{P}_1^{-1})^{-1}\mathbf{S}_1^{-1}(\mathbf{z}_1 - \mathbf{N}_{12}\mathbf{N}_{22}^{-1}\mathbf{z}_2)]. \end{aligned} \tag{3.3b}$$

Similarly, the appropriate *mean square error matrix* is readily obtained by a "*rank-1 modification*" of (2.11b) from

$$D\{\begin{bmatrix} \tilde{\mathbf{b}}_{1H} - \mathbf{b}_1 \\ \hat{\boldsymbol{\beta}}_{2H} \end{bmatrix}\} = \sigma^2(\begin{bmatrix} \mathbf{N}_{11} + \mathbf{P}_1 & \mathbf{N}_{12} \\ \mathbf{N}_{21} & \mathbf{N}_{22} \end{bmatrix}^{-1} + \alpha \begin{bmatrix} \boldsymbol{\eta}_1 - \boldsymbol{\gamma}_1 \\ \boldsymbol{\eta}_2 \end{bmatrix} [(\boldsymbol{\eta}_1 - \boldsymbol{\gamma}_1)^{\mathrm{T}}, \boldsymbol{\eta}_2^{\mathrm{T}}]) \tag{3.4a}$$

with the *non-random number*

$$\alpha := (\boldsymbol{\gamma}_1^{\mathrm{T}}\mathbf{P}_1\boldsymbol{\eta}_1)^{-1} = [\boldsymbol{\gamma}_1^{\mathrm{T}}(\mathbf{S}_1^{-1} + \mathbf{P}_1^{-1})^{-1}\boldsymbol{\gamma}_1]^{-1} \tag{3.4b}$$

such that

$$a \sim N(\alpha(\boldsymbol{\beta}_1^{\mathrm{T}}\mathbf{P}_1\boldsymbol{\eta}_1) = 1, \alpha\sigma^2) \tag{3.5}$$

holds true under the *null hypothesis*

$$\mathbf{b}_1 \sim N(\boldsymbol{\gamma}_1, \sigma^2\mathbf{P}_1^{-1}),\ \boldsymbol{\gamma}_1 \text{ and } \mathbf{P}_1 \text{ given,} \tag{3.6}$$

opposed to the alternative $\mathbf{b}_1 \sim N(\boldsymbol{\gamma}_1, \infty)$. Of course, also (3.3a) can be given a form corresponding to (2.13a), namely

$$\begin{aligned} \begin{bmatrix} \tilde{\mathbf{b}}_{1H} \\ \hat{\boldsymbol{\beta}}_{2H} \end{bmatrix} & = \begin{bmatrix} \tilde{\mathbf{b}}_1 \\ \hat{\boldsymbol{\beta}}_2 \end{bmatrix} + \begin{bmatrix} -\mathbf{I}_r \\ \mathbf{N}_{22}^{-1}\mathbf{N}_{21} \end{bmatrix} (\mathbf{S}_1 + \mathbf{P}_1)^{-1}\mathbf{P}_1(\tilde{\mathbf{b}}_1 - a\boldsymbol{\gamma}_1) \\ & = \begin{bmatrix} \tilde{\mathbf{b}}_1 \\ \hat{\boldsymbol{\beta}}_2 \end{bmatrix} - \begin{bmatrix} -\mathbf{I}_r \\ \mathbf{N}_{22}^{-1}\mathbf{N}_{21} \end{bmatrix} \mathbf{S}_1^{-1}(\mathbf{S}_1^{-1} + \mathbf{P}_1^{-1})^{-1}(a\boldsymbol{\gamma}_1 - \tilde{\mathbf{b}}_1) \end{aligned} \tag{3.7a}$$

where again

$$\begin{bmatrix} \tilde{\mathbf{b}}_1 \\ \hat{\boldsymbol{\beta}}_2 \end{bmatrix} = \begin{bmatrix} \mathbf{S}_1^{-1}(\mathbf{z}_1 - \mathbf{N}_{12}\mathbf{N}_{22}^{-1}\mathbf{z}_2) \\ \mathbf{N}_{22}^{-1}(\mathbf{z}_2 - \mathbf{N}_{21}\tilde{\mathbf{b}}_1) \end{bmatrix} = \begin{bmatrix} \mathbf{c}_1 + \mathbf{S}_1^{-1}\mathbf{P}_1\mathbf{c}_1 \\ \mathbf{c}_2 - \mathbf{N}_{22}^{-1}\mathbf{N}_{21}\mathbf{S}_1^{-1}\mathbf{P}_1\mathbf{c}_1 \end{bmatrix} \tag{3.7b}$$

represents the "*unconstrained solution*" as in (2.13b). Moreover, we obtain the proper IQUUE of σ^2 simply from

$$\hat{\sigma}_H^{\,2} = (n - m + r - 1)^{-1}\Omega_H \tag{3.8a}$$

with the quadratic form Ω_H of the appropriate *residual vector*

$$\begin{aligned}\begin{bmatrix}\tilde{\mathbf{u}}_H \\ \tilde{\mathbf{e}}_H\end{bmatrix} &:= \begin{bmatrix}\mathbf{y}_* \\ -a\boldsymbol{\gamma}_1\end{bmatrix} - \begin{bmatrix}\mathbf{X}_1 & \mathbf{X}_2 \\ -\mathbf{I}_r & \mathbf{0}\end{bmatrix}\begin{bmatrix}\tilde{\mathbf{b}}_{1H} \\ \hat{\boldsymbol{\beta}}_{2H}\end{bmatrix} \\ &= \begin{bmatrix}\mathbf{y}_* - \mathbf{X}_1\tilde{\mathbf{b}}_1 - \mathbf{X}_2\hat{\boldsymbol{\beta}}_2 \\ -(a\boldsymbol{\gamma}_1 - \tilde{\mathbf{b}}_1)\end{bmatrix} + \begin{bmatrix}\mathbf{X}_1 - \mathbf{X}_2\mathbf{N}_{22}^{-1}\mathbf{N}_{21} \\ -\mathbf{I}_r\end{bmatrix}\mathbf{S}_1^{-1}\hat{\boldsymbol{\lambda}}_H \\ &= \begin{bmatrix}\tilde{\mathbf{u}} \\ \mathbf{0}\end{bmatrix} + \begin{bmatrix}(\mathbf{X}_1 - \mathbf{X}_2\mathbf{N}_{22}^{-1}\mathbf{N}_{21})\mathbf{S}_1^{-1} \\ \mathbf{P}_1^{-1}\end{bmatrix}\hat{\boldsymbol{\lambda}}_H \end{aligned} \tag{3.8b}$$

being defined by

$$\begin{aligned}\Omega_H &:= \tilde{\mathbf{u}}_H^{\mathrm{T}}\mathbf{P}\tilde{\mathbf{u}}_H + \tilde{\mathbf{e}}_H^{\mathrm{T}}\mathbf{P}_1\tilde{\mathbf{e}}_H = \tilde{\mathbf{u}}_H^{\mathrm{T}}\mathbf{P}\mathbf{y} - a\tilde{\mathbf{e}}_H^{\mathrm{T}}\mathbf{P}_1\boldsymbol{\gamma}_1 \\ &= \tilde{\mathbf{u}}^{\mathrm{T}}\mathbf{P}\mathbf{y}_* - \hat{\boldsymbol{\lambda}}_H^{\mathrm{T}}(a\boldsymbol{\gamma}_1 - \tilde{\mathbf{b}}_1) \\ &= \tilde{\mathbf{u}}^{\mathrm{T}}\mathbf{P}\tilde{\mathbf{u}} + (a\boldsymbol{\gamma}_1 - \tilde{\mathbf{b}}_1)^{\mathrm{T}}(\mathbf{S}_1^{-1} + \mathbf{P}_1^{-1})^{-1}(a\boldsymbol{\gamma}_1 - \tilde{\mathbf{b}}_1). \end{aligned} \tag{3.8c}$$

Here we introduced the *auxiliary* $r\times 1$ vector $\hat{\boldsymbol{\lambda}}_H$ in formal analogy to (2.11c) as

$$\begin{aligned}\hat{\boldsymbol{\lambda}}_H &:= \mathbf{P}_1(\tilde{\mathbf{b}}_{1H} - a\boldsymbol{\gamma}_1) = \mathbf{P}_1(\mathbf{c}_1 - a\boldsymbol{\eta}_1) \\ &= \mathbf{P}_1(\mathbf{S}_1 + \mathbf{P}_1)^{-1}(\boldsymbol{\eta}_1^{\mathrm{T}} - \mathbf{N}_{12}\mathbf{N}_{22}^{-1}\boldsymbol{\eta}_2^{\mathrm{T}})\mathbf{P}(\mathbf{y}_* - a\mathbf{X}_1\boldsymbol{\gamma}_1) \\ &= \mathbf{S}_1(\mathbf{S}_1 + \mathbf{P}_1)^{-1}\mathbf{P}_1(\tilde{\mathbf{b}}_1 - a\boldsymbol{\gamma}_1) = -(\mathbf{S}_1^{-1} + \mathbf{P}_1^{-1})^{-1}(a\boldsymbol{\gamma}_1 - \tilde{\mathbf{b}}_1). \end{aligned} \tag{3.9}$$

Taking into account that in the "*unconstrained case*" for

$$\tilde{\mathbf{u}} := \mathbf{y}_* - \mathbf{X}_1\tilde{\mathbf{b}}_1 - \mathbf{X}_2\hat{\boldsymbol{\beta}}_2 \sim N(\mathbf{0}, \sigma^2(\mathbf{P}^{-1} - [\mathbf{X}_1, \mathbf{X}_2]\begin{bmatrix}\mathbf{N}_{11} & \mathbf{N}_{12} \\ \mathbf{N}_{21} & \mathbf{N}_{22}\end{bmatrix}^{-1}\begin{bmatrix}\mathbf{X}_1^{\mathrm{T}} \\ \mathbf{X}_2^{\mathrm{T}}\end{bmatrix})) \tag{3.10a}$$

we simply find

$$\Omega := \tilde{\mathbf{u}}^{\mathrm{T}}\mathbf{P}\tilde{\mathbf{u}} = \tilde{\mathbf{u}}^{\mathrm{T}}\mathbf{P}\mathbf{y}_* = \mathbf{y}_*^{\mathrm{T}}\mathbf{P}(\mathbf{P}^{-1} - [\mathbf{X}_1, \mathbf{X}_2]\begin{bmatrix}\mathbf{N}_{11} & \mathbf{N}_{12} \\ \mathbf{N}_{21} & \mathbf{N}_{22}\end{bmatrix}^{-1}\begin{bmatrix}\mathbf{X}_1^{\mathrm{T}} \\ \mathbf{X}_2^{\mathrm{T}}\end{bmatrix})\mathbf{P}\mathbf{y}_* \tag{3.10b}$$

to be *distributed* again as

$$\Omega / \sigma^2 \sim \chi^2(n-m) \tag{3.11a}$$

with *expectation*

$$E\{\Omega / \sigma^2\} = \mathrm{tr}(\mathbf{I}_n - \mathbf{P}[\mathbf{X}_1, \mathbf{X}_2]\begin{bmatrix}\mathbf{N}_{11} & \mathbf{N}_{12}\\ \mathbf{N}_{21} & \mathbf{N}_{22}\end{bmatrix}^{-1}\begin{bmatrix}\mathbf{X}_1{}^{\mathrm{T}}\\ \mathbf{X}_2{}^{\mathrm{T}}\end{bmatrix})\mathbf{P}E\{\mathbf{y}_*\mathbf{y}_*{}^{\mathrm{T}}\} / \sigma^2 = n-m, \tag{3.11b}$$

we now conclude in the *"constrained case"* that the *relative increase*

$$R_H / \sigma^2 := \sigma^{-2}(\Omega_H - \Omega) = (a\boldsymbol{\gamma}_1 - \tilde{\mathbf{b}}_1)^{\mathrm{T}}(\mathbf{S}_1{}^{-1} + \mathbf{P}_1{}^{-1})^{-1}(a\boldsymbol{\gamma}_1 - \tilde{\mathbf{b}}_1)/\sigma^2 \tag{3.12a}$$

due to

$$\sigma^{-1}(a\boldsymbol{\gamma}_1 - \tilde{\mathbf{b}}_1) \sim N(\sigma^{-1}a(\boldsymbol{\gamma}_1\boldsymbol{\beta}_1{}^{\mathrm{T}} - \boldsymbol{\beta}_1\boldsymbol{\gamma}_1{}^{\mathrm{T}})(\mathbf{S}_1{}^{-1} + \mathbf{P}_1{}^{-1})^{-1}\boldsymbol{\gamma}_1, (\mathbf{S}_1{}^{-1} + \mathbf{P}_1{}^{-1}) - a\boldsymbol{\gamma}_1\boldsymbol{\gamma}_1{}^{\mathrm{T}}), \tag{3.12b}$$

is obviously *distributed as*

$$R_H / \sigma^2 \sim \chi'^2(r-1; \theta_H) \tag{3.13a}$$

with *non-centrality parameter*

$$\theta_H := a\boldsymbol{\beta}_1{}^{\mathrm{T}}(\mathbf{S}_1{}^{-1} + \mathbf{P}_1{}^{-1})^{-1}(\boldsymbol{\beta}_1\boldsymbol{\gamma}_1{}^{\mathrm{T}} - \boldsymbol{\gamma}_1\boldsymbol{\beta}_1{}^{\mathrm{T}})(\mathbf{S}_1{}^{-1} + \mathbf{P}_1{}^{-1})^{-1}\boldsymbol{\gamma}_1 / 2\sigma^2 \tag{3.13b}$$

and *expectation*

$$\begin{aligned} E\{R_H / \sigma^2\} &= \mathrm{tr}(\mathbf{S}_1{}^{-1} + \mathbf{P}_1{}^{-1})^{-1}(D\{a\boldsymbol{\gamma}_1 - \tilde{\mathbf{b}}_1\} + E\{a\boldsymbol{\gamma}_1 - \tilde{\mathbf{b}}_1\}E\{a\boldsymbol{\gamma}_1 - \tilde{\mathbf{b}}_1\}^{\mathrm{T}}) / \sigma^2 \\ &= \mathrm{tr}(\mathbf{I}_r - a(\mathbf{S}_1{}^{-1} + \mathbf{P}_1{}^{-1})^{-1}\boldsymbol{\gamma}_1\boldsymbol{\gamma}_1{}^{\mathrm{T}}) + E\{a\boldsymbol{\gamma}_1 - \tilde{\mathbf{b}}_1\}^{\mathrm{T}}(\mathbf{S}_1{}^{-1} + \mathbf{P}_1{}^{-1})^{-1}E\{a\boldsymbol{\gamma}_1 - \tilde{\mathbf{b}}_1\} \\ &= (r-1) + 2\theta_H. \end{aligned} \tag{3.13c}$$

If we further express Ω and R_H as quadratic forms in the *common* random vector

$$\begin{aligned} \mathbf{y}_* - a\mathbf{X}_1\boldsymbol{\gamma}_1 &= \mathbf{y}_* - [\mathbf{X}_1, \mathbf{X}_2][a\boldsymbol{\gamma}_1{}^{\mathrm{T}}, \mathbf{0}^{\mathrm{T}}]^{\mathrm{T}} \\ &\sim N(a\mathbf{X}_1(\boldsymbol{\beta}_1\boldsymbol{\gamma}_1{}^{\mathrm{T}} - \boldsymbol{\gamma}_1\boldsymbol{\beta}_1{}^{\mathrm{T}})(\mathbf{S}_1{}^{-1} + \mathbf{P}_1{}^{-1})^{-1}\boldsymbol{\gamma}_1 + \mathbf{X}_2\boldsymbol{\beta}_2, \\ &\qquad \sigma^2(\mathbf{P}^{-1} + \mathbf{X}_1\mathbf{P}_1{}^{-1}\mathbf{X}_1{}^{\mathrm{T}}) + \sigma^2 a(\boldsymbol{\eta}_1 - \boldsymbol{\gamma}_1)(\boldsymbol{\eta}_1 - \boldsymbol{\gamma}_1)^{\mathrm{T}} - \sigma^2 a\boldsymbol{\eta}_1\boldsymbol{\eta}_1{}^{\mathrm{T}}) \end{aligned} \tag{3.14a}$$

which yields

$$\Omega = (\mathbf{y}_* - a\mathbf{X}_1\boldsymbol{\gamma}_1)^{\mathrm{T}}(\mathbf{P} - \mathbf{P}[\mathbf{X}_1, \mathbf{X}_2]\begin{bmatrix}\mathbf{N}_{11} & \mathbf{N}_{12}\\ \mathbf{N}_{21} & \mathbf{N}_{22}\end{bmatrix}^{-1}\begin{bmatrix}\mathbf{X}_1{}^{\mathrm{T}}\\ \mathbf{X}_2{}^{\mathrm{T}}\end{bmatrix}\mathbf{P})(\mathbf{y}_* - a\mathbf{X}_1\boldsymbol{\gamma}_1) \tag{3.14b}$$

and

$$R_H = (\mathbf{y}_* - a\mathbf{X}_1\gamma_1)^{\mathrm{T}}\mathbf{P}(\mathbf{X}_1 - \mathbf{X}_2\mathbf{N}_{22}^{-1}\mathbf{N}_{21})\mathbf{S}_1^{-1}(\mathbf{S}_1^{-1} + \mathbf{P}_1^{-1})^{-1}\mathbf{S}_1^{-1} \cdot(\mathbf{X}_1^{\mathrm{T}} - \mathbf{N}_{12}\mathbf{N}_{22}^{-1}\mathbf{X}_2^{\mathrm{T}})\mathbf{P}(\mathbf{y}_* - a\mathbf{X}_1\gamma_1), \quad (3.14c)$$

the *independence* of both, Ω and R_H, immediately follows such that the *test statistic*

$$T_H := \frac{(n-m)R_H}{(r-1)\Omega} = \frac{(n-m+r-1)\hat{\sigma}_H^2 - (n-m)\hat{\sigma}^2}{(r-1)\hat{\sigma}^2} \quad (3.15a)$$

becomes *distributed as*

$$T_H \sim F'(r-1, n-m; \theta_H) \quad (3.15b)$$

with $\theta_H = 0$ under the null hypothesis. Again the resulting *central* F-distribution is *not* influenced by the choice of the "weight matrix" $\mathbf{P}_1$. Note, in particular, that for its *fractiles* at any significance level always holds

$$F_{1-\alpha;\, r-1,\, n-m} \geq F_{1-\alpha;\, r,\, n-m} \quad \text{if } n-m \geq 3. \quad (3.15c)$$

Comparing (3.15a, b, c) with our previous result (2.20a, b) and assuming $r \geq R_I(\mathbf{P}_1)[R_I(\mathbf{P}_1) - R_H(\mathbf{P}_1)]^{-1}$ we may state the following

SENSITIVITY THEOREM: *A stochastic linear hypothesis of type* (2.6) *can also be tested within the Mixed Linear Model* (2.7) *by applying* (3.15) *instead of* (2.20). *This test turns out to be* less sensitive *in the relevant case* $n-m \geq 3$ *with basically the* same *limit behaviour for* $\mathbf{P}_1 \to \mathbf{0}$, *but* not *for* $\mathbf{P}_1 \to \infty$ *or* $\mathbf{P}_1^{-1} \to \mathbf{0}$, *respectively*.

Following a similar reasoning as above (e.g. at the end of *chapter 1*), we first look at the functions $\theta_H(\mathbf{P}_1)$ and $R_H(\mathbf{P}_1)$, respectively, of the variable "weight matrix" $\mathbf{P}_1$ and obtain the important *inequalities*

$$\begin{aligned}\theta_I(\mathbf{P}_1) - \theta_H(\mathbf{P}_1) &= (2\sigma^2)^{-1}[(\gamma_1 - \boldsymbol{\beta}_1)^{\mathrm{T}}(\mathbf{S}_1^{-1} + \mathbf{P}_1^{-1})^{-1}(\gamma_1 - \boldsymbol{\beta}_1) \\ &\quad - \boldsymbol{\beta}_1^{\mathrm{T}}(\mathbf{S}_1^{-1} + \mathbf{P}_1^{-1})^{-1}\boldsymbol{\beta}_1 + \alpha(\boldsymbol{\beta}_1^{\mathrm{T}}(\mathbf{S}_1^{-1} + \mathbf{P}_1^{-1})^{-1}\gamma_1)^2] \\ &= (2\sigma^2)^{-1}\alpha[(\gamma_1 - \boldsymbol{\beta}_1)^{\mathrm{T}}(\mathbf{S}_1^{-1} + \mathbf{P}_1^{-1})^{-1}\gamma_1]^2 \geq 0 \quad (3.16a)\end{aligned}$$

and

$$\begin{aligned}R_I(\mathbf{P}_1) - R_H(\mathbf{P}_1) &= (\gamma_1 - \tilde{\mathbf{b}}_1)^{\mathrm{T}}(\mathbf{S}_1^{-1} + \mathbf{P}_1^{-1})^{-1}(\gamma_1 - \boldsymbol{\beta}_1) \\ &\quad - (a\gamma_1 - \tilde{\mathbf{b}}_1)^{\mathrm{T}}(\mathbf{S}_1^{-1} + \mathbf{P}_1^{-1})^{-1}(a\gamma_1 - \tilde{\mathbf{b}}_1) \\ &= \alpha[(\gamma_1 - \tilde{\mathbf{b}}_1)^{\mathrm{T}}(\mathbf{S}_1^{-1} + \mathbf{P}_1^{-1})^{-1}\gamma_1]^2 \geq 0 \quad (3.16b)\end{aligned}$$

because of the *identity*

$$a = a(\gamma_1{}^T\mathbf{P}_1\mathbf{c}_1) = a\gamma_1{}^T\mathbf{P}_1(\mathbf{S}_1 + \mathbf{P}_1)^{-1}\mathbf{S}_1\tilde{\mathbf{b}}_1 = a\tilde{\mathbf{b}}_1{}^T(\mathbf{S}_1{}^{-1} + \mathbf{P}_1{}^{-1})^{-1}\gamma_1, \quad (3.17)$$

which clearly show the *amount of lost sensitivity* for any arbitrarily chosen "weight matrix" $\mathbf{P}_1$ if we apply (3.15) instead of (2.20). Of course, in the *limit case* $\mathbf{P}_1 \rightarrow \mathbf{0}$ we easily find from (3.16a, b) using (2.22a, b)

$$\theta_H(\mathbf{P}_1 \rightarrow \mathbf{0}) = (2\sigma^2)^{-1} \lim a\boldsymbol{\beta}^T\mathbf{P}_1(\mathbf{I}_r + \mathbf{S}_1{}^{-1}\mathbf{P}_1)^{-1}(\boldsymbol{\beta}_1\gamma_1{}^T - \gamma_1\boldsymbol{\beta}_1{}^T)\mathbf{P}_1 \cdot \cdot(\mathbf{I}_r + \mathbf{S}_1{}^{-1}\mathbf{P}_1)^{-1}\gamma_1 = 0 \quad (3.18a)$$

although $a \rightarrow \infty$, and

$$R_H(\mathbf{P}_1 \rightarrow \mathbf{0}) = -\lim(a\gamma_1 - \tilde{\mathbf{b}}_1)^T\mathbf{P}_1(\mathbf{I}_r + \mathbf{S}_1{}^{-1}\mathbf{P}_1)^{-1}\tilde{\mathbf{b}}_1 = 0 \quad (3.18b)$$

since a is bounded. *In contrast,* for $\mathbf{P}_1 \rightarrow \infty$ using *Cauchy-Schwarz' inequality"* we quickly arrive at

$$\theta_H(\infty) = (2\sigma^2)^{-1}(\gamma_1{}^T\mathbf{S}_1\gamma_1)^{-1}\boldsymbol{\beta}_1{}^T\mathbf{S}_1(\boldsymbol{\beta}_1\gamma_1{}^T - \gamma_1\boldsymbol{\beta}_1{}^T)\mathbf{S}_1\gamma_1 \geq 0 \quad (3.19a)$$

and

$$R_H(\infty) = (\gamma_1{}^T\mathbf{S}_1\gamma_1)^{-1}\tilde{\mathbf{b}}_1{}^T\mathbf{S}_1(\tilde{\mathbf{b}}_1\gamma_1{}^T - \gamma_1\tilde{\mathbf{b}}_1{}^T)\mathbf{S}_1\gamma_1 \geq 0 \quad (3.19b)$$

which leads to a *less sensitive test* even if the considered hypothesis is introduced as *non-stochastic.*

REMARK: *The above method of designing less sensitive tests can also be extended to the case of T.D. Wallace (1972) and T.A. Yancey et al. (1974) where any* improvement of the predictors *due to – or inspite of perhaps* incorrect – *(stochastic) prior information is tested along similar lines; however, this extension is beyond the scope of the present paper and will be published elsewhere.*

ACKNOWLEDGEMENT

This paper was partially funded by the *Stiftung Volkswagenwerk,* Projekt-No. I/61 793; this support is gratefully acknowledged.

REFERENCES

Harville, D.A. (1976). Extension of the Gauss-Markov theorem to include the estimation of random effects. *Annals of Statistics*, 4, 384-395.

Helmert, F.R. (1872). *Die Ausgleichungsrechnung nach der Methode der kleinsten Quadrate.* Teubner, Leipzig.

Judge, G.G. and Bock, M.E. (1978). *The Statistical Implications of Pre-Test and Stein-Rule Estimators in Econometrics.* North-Holland.

Läuter, H. (1970). Optimale Vorhersage und Schätzung in regulären und singulären Regressionsmodellen. *Mathematische Operationsforschung & Statistik*, 1, 229-243.

Rao, C.R. (1975). Simultaneous estimation of parameters in different linear models and applications to biometric problems. *Biometrics*, 31, 545-554.

Schaffrin, B. (1983). A note on linear prediction within a Gauss-Markov model linearized with respect to a random approximation. *Proceedings of the First Tampere Seminar on Linear Statistical Models and their Applications* (T. Pukkila / S. Puntanen, *eds.*). Department of Mathematical Sciences, University of Tampere, Finland, 285-300.

Schaffrin, B. (1985). *Das geodätische Datum mit stochastischer Vorinformation.* Deutsche Geodätische Kommission, Bayr. Akad. der Wiss., Publ. C-313, Munich.

Schaffrin, B. and Grafarend, E. (1987). A unified computational scheme for traditional and robust prediction of random effects with some applications in geodesy. *The First International Conference on Statistical Computing*, Izmir, Turkey, forthcoming.

Searle, S.R. (1974). Prediction, mixed models, and variance components. *Reliability and Biometry* (F. Proschan / R.J. Serfling, eds.). SIAM, Philadelphia, 229-266.

Theil, H. and Goldberger, A.S. (1961). On pure and mixed statistical estimation in economics. *International Economics Review*, 2, 65-78.

Theil, H. (1963). On the use of incomplete prior information in regression analysis. *Journal of the American Statistical Association*, 58, 401-414.

Toutenburg, H. (1970). Probleme linearer Vorhersagen im allgemeinen linearen Regressionsmodell. *Biometrische Zeitschrift*, 12, 242-252.

Vinod, H.D. and Ullah, A. (1981). *Recent Advances in Regression Methods.* Marcel Dekker, New York-Basel.

Wallace, T.D. (1972). Weaker criteria and tests for linear restrictions in regression. *Econometrica*, 40, 689-698.

Yancey, T.A., Judge, G.G. and Bock, M.E. (1974). A mean square error test when stochastic restrictions are used in regression. *Communications in Statistics*, 3, 755-768.

Received 14 January 1987

Department of Geodetic Science
Stuttgart University
Keplerstrasse 11
D-7000 Stuttgart 1
F.R. Germany

Proc. Second International Tampere Conference in Statistics
(Tampere, Finland, 1-4 June 1987)
Tarmo Pukkila and Simo Puntanen, *Editors*

pp. 665 - 678

On outliers and influence in the general multivariate normal linear model

Robert SCHALL and Timothy T. DUNNE

Messerschmitt Bolkow Blohm GmbH, München, F.R. Germany;
Rice University, Houston, TX, U.S.A. and University of Cape Town, South Africa

Key words and phrases: Arbitrary variance-covariance, missing data.

ABSTRACT

Previous work has lead to the definition of three types of outlier which can be handled by the use of additional regressor variables in the univariate linear model with normality assumption, but correlated error terms with known variance-covariance structure $\sigma^2\mathbf{V}$.

A formal development for the multivariate linear model with regressors is presented here, but under the additional complications of simultaneously estimating $\mathbf{V}$ from the data, and adjusting estimates for the presence of some outliers. Relationships to measures of influence and to missing data approaches such as the EM algorithm are sketched.

1. INTRODUCTION

1.1 Review of the literature

Outliers and influence have been extensively examined in such texts as those of Hawkins (1980), and Cook and Weisberg (1981). In the main the interest has focused on the univariate case, though techniques for multivariate outliers were suggested by Gnanadesikan and Kettenring (1972), and are discussed in Barnett and Lewis (1984), amongst others.

Rao (1971), and elsewhere, has provided a unified approach to estimation of the regression mean vector $\mathbf{X}\beta$ and scalar variance parameter σ^2, in a model for a univariate response vector Y with arbitrary known variance-covariance structure $\sigma^2\mathbf{V}$. Using that development, and outlier formulations due to Gentleman and Wilk (1975), and to John and Draper (1978), Dunne

(1982) and, more extensively, Schall (1986) proposed the distinction of three outlier types: additive shift (A-), transformational (T-) and distributional (D-) outliers. Outlier tests involve hypotheses H_0: $\mathbf{A}_{22}\lambda = 0$ in the model

$$\begin{bmatrix} y_1 \\ y_2 \end{bmatrix} = \begin{bmatrix} \mathbf{X}_1 & \mathbf{A}_{12} \\ \mathbf{X}_2 & \mathbf{A}_{22} \end{bmatrix} \begin{bmatrix} \beta \\ \lambda \end{bmatrix} + \begin{bmatrix} e_1 \\ e_2 \end{bmatrix} \tag{1.1}$$

where the conformable partitioning reflects the prior view that the observations y_2 should be examined for possible anomalies. Here $\mathbf{A} = \begin{bmatrix} \mathbf{A}_{11} & \mathbf{A}_{12} \\ \mathbf{A}_{21} & \mathbf{A}_{22} \end{bmatrix}$ is a square matrix with $\mathbf{A}_{11}$ and $\mathbf{A}_{22}$ square. For A-outliers, T-outliers, and D-outliers we take $\mathbf{A} = \mathbf{V}^0 = \mathbf{I}_n$, $\mathbf{A} = \mathbf{V}^{1/2}$ and $\mathbf{A} = \mathbf{V}^1$ respectively. We write H_0: $\lambda = 0$ as a loose convenient description, in each of these cases.

1.2 Multivariate regression

We consider the multivariate linear regression model

$$\underset{(m \times n)}{\mathbf{Y}} = \underset{(m \times p)}{\mathbf{X}} \cdot \underset{(p \times n)}{\mathbf{B}} + \underset{(m \times n)}{\mathbf{E}} \tag{1.2}$$

where $\mathbf{Y}$ is the matrix of m observations, on an n-dimensional response vector Y, $\mathbf{X}$ is the known design or regressor matrix, $\mathbf{B}$ is the matrix of regression parameters. The matrix $\mathbf{E}$ of unobservable error components has rows which are assumed to be stochastically independent and identically distributed as $N_n(0, \mathbf{V})$, where $\mathbf{V}$ is the unknown $n \times n$ variance-covariance matrix of the responses within observations.

The superscript in $\mathbf{A}^-$ will indicate a generalized inverse of $\mathbf{A}$, defined by $\mathbf{A}\mathbf{A}^-\mathbf{A} = \mathbf{A}$. Projections $\mathbf{M}$ and $\mathbf{N}$ yield the maximum likelihood estimates (MLE's) $\mathbf{X}\hat{\mathbf{B}}$ for $\mathbf{XB}$, and $\hat{\mathbf{V}}$ for $\mathbf{V}$, under model (1.2), and we require $n \leq m - rank(\mathbf{X})$, so $\hat{\mathbf{V}}^{-1}$ exists almost surely. Thus

$$\mathbf{X}\hat{\mathbf{B}} = \mathbf{X}(\mathbf{X}'\mathbf{X})^-\mathbf{X}'\mathbf{Y} = \mathbf{MY} \tag{1.3}$$

$$\hat{\mathbf{E}} = (\mathbf{I} - \mathbf{X}(\mathbf{X}'\mathbf{X})^-\mathbf{X}')\mathbf{Y} = (\mathbf{I} - \mathbf{M})\mathbf{Y} = \mathbf{NY} \tag{1.4}$$

$$m \cdot \hat{\mathbf{V}} = \hat{\mathbf{E}}'\hat{\mathbf{E}} = \mathbf{Y}'\mathbf{NY}\,. \tag{1.5}$$

2. OUTLIERS AND INFLUENCE

No conceptual difficulties arise in extending the outlier testing methods for the general linear model to corresponding types of outlier in the multivariate regression model. If J, an arbitrary subset of data points y_{ij} from the data matrix $\mathbf{Y} = (y_{ij})$, is the set of observations suspected to be outlying, then the adjusted model (1.2) is

$$\mathbf{Y} = \mathbf{XB} + \mathbf{\Theta V} + \mathbf{E} \tag{2.1}$$

if J is assumed to be a set of D-outliers, and

$$\mathbf{Y} = \mathbf{XB} + \mathbf{\Theta I}_n + \mathbf{E} \tag{2.2}$$

if J is assumed to be a set of A-outliers. The parameter matrix $\mathbf{\Theta} = (\theta_{ij})$ is of order $m \times n$, and θ_{ij} is *a priori* specified to be zero if and only if $y_{ij} \notin J$.

Let $\mathbf{V} = \mathbf{P\Delta P}'$ be the singular value decomposition (SVD) of $\mathbf{V}$, then the (uncentered) principal components (PC's) $\mathbf{Y}^*$ of $\mathbf{Y}$ are given by

$$\mathbf{Y}^* = \mathbf{YP} \,. \tag{2.3}$$

If J^*, an arbitrary subset of PC's y_{ij}^* of $\mathbf{Y}^* = (y_{ij}^*)$ is the set of PC's suspected to outlying, then the model (1.2) adjusted for the T-outliers J^* is

$$\mathbf{Y} = \mathbf{XB} + \mathbf{\Theta P}' + \mathbf{E} \,. \tag{2.4}$$

As above, the parameter matrix $\mathbf{\Theta} = (\theta_{ij})$ is of order $m \times n$, and θ_{ij} is *a priori* specified to be zero if and only if $y_{ij}^* \notin J^*$. If $\mathbf{V}$ is unknown, then $\mathbf{P}$ and the dummy variables fitted in models (2.1) and (2.4) are unknown.

2.1 Distributional outliers

We consider the adjusted model (2.1) and, initially, assume that the $k \times l$-submatrix $\mathbf{Y}_{22}$ consists of the set J of suspect data points, with $\mathbf{Y}$ and the model partitioned as

$$\mathbf{Y} = \begin{bmatrix} \mathbf{Y}_{11} : \mathbf{Y}_{12} \\ \mathbf{Y}_{21} : \mathbf{Y}_{22} \end{bmatrix} = \begin{bmatrix} \mathbf{Y}_1 : \mathbf{Y}_2 \end{bmatrix} \text{, and} \tag{2.5}$$

$$\begin{bmatrix} \mathbf{Y}_{11} : \mathbf{Y}_{12} \\ \mathbf{Y}_{21} : \mathbf{Y}_{22} \end{bmatrix} = \begin{bmatrix} \mathbf{X}_1 \\ \mathbf{X}_2 \end{bmatrix} \begin{bmatrix} \mathbf{B}_1 : \mathbf{B}_2 \end{bmatrix} + \begin{bmatrix} 0 \\ \mathbf{\Theta} \end{bmatrix} \begin{bmatrix} \mathbf{V}_{21} : \mathbf{V}_{22} \end{bmatrix} + \begin{bmatrix} \mathbf{E}_{11} : \mathbf{E}_{12} \\ \mathbf{E}_{21} : \mathbf{E}_{22} \end{bmatrix} \tag{2.6}$$

where $\mathbf{X} = \begin{bmatrix} \mathbf{X}_1' : \mathbf{X}_2' \end{bmatrix}'$ and $\mathbf{B} = \begin{bmatrix} \mathbf{B}_1 : \mathbf{B}_2 \end{bmatrix}$ are conformably partitioned, i.e. $\mathbf{X}_2$ is $k \times p$ and $\mathbf{B}_2$ is $p \times l$. The parameter matrix $\mathbf{\Theta}$ is here $k \times l$, and $\mathbf{V}_{22}$ is a $l \times l$-submatrix of $\mathbf{V}$. For $\mathbf{X}_2\mathbf{B}_2$ to be estimable under the adjusted model (2.6) we require, as in the univariate case, that $R(\mathbf{X}_2) \subset R(\mathbf{X}_1)$.

The ML-estimates $\mathbf{X\hat{B}}$, $\mathbf{\hat{E}}$ and $\mathbf{\hat{V}}$ of the unknowns $\mathbf{XB}$, $\mathbf{E}$ and $\mathbf{V}$ in model (2.1) are given by equations (1.3) through (1.5). Now let $\mathbf{X\tilde{B}}$, $\mathbf{\tilde{E}}$, $\mathbf{\tilde{\Theta}}$ and $\mathbf{\tilde{V}}$ denote the ML-estimates of $\mathbf{XB}$, $\mathbf{E}$, $\mathbf{\Theta}$ and $\mathbf{V}$ in the adjusted model (2.6). If $l = n$, then the submatrices $\mathbf{Y}_{11}$, $\mathbf{Y}_{21}$, $\mathbf{E}_{11}$, $\mathbf{E}_{21}$ and $\mathbf{B}_1$ vanish and the model (2.6) is obtained as

$$\begin{bmatrix} \mathbf{Y}_{12} \\ \mathbf{Y}_{22} \end{bmatrix} = \begin{bmatrix} \mathbf{X}_1 \\ \mathbf{X}_2 \end{bmatrix} \mathbf{B}_2 + \begin{bmatrix} 0 \\ \mathbf{\Theta} \end{bmatrix} \mathbf{V} + \begin{bmatrix} \mathbf{E}_{12} \\ \mathbf{E}_{22} \end{bmatrix} \tag{2.7}$$

Clearly, the ML-estimates $\mathbf{X}\tilde{\mathbf{B}}_2 = \mathbf{X}\tilde{\mathbf{B}}$, $\tilde{\mathbf{E}}$ and $\tilde{\mathbf{V}}$ for $\mathbf{XB}$, $\mathbf{E}$ and $\mathbf{V}$ under the model (2.7) are obtained in the reduced model

$$\mathbf{Y}_{12} = \mathbf{X}_1\mathbf{B}_2 + \mathbf{E}_{12} \tag{2.8}$$

using the formulae (1.3) through (1.5), replacing $\mathbf{Y}$ by $\mathbf{Y}_{12}$ and $\mathbf{X}$ by $\mathbf{X}_1$. Then $\mathbf{\Theta}$ can be estimated by $\tilde{\mathbf{\Theta}} = (\mathbf{Y}_{22} - \mathbf{X}_2\tilde{\mathbf{B}}_2)\tilde{\mathbf{V}}^{-1}$.

In the following we assume that $l < n$. Transforming the model (2.6) from the right by the transformation

$$\mathbf{T} = \begin{bmatrix} \mathbf{I} & : & 0 \\ -\mathbf{V}_{22}^{-1}\mathbf{V}_{21} & : & \mathbf{I} \end{bmatrix} \tag{2.9}$$

we observe that the MLE $\mathbf{X}_1\tilde{\mathbf{B}}_2$ for $\mathbf{X}_1\mathbf{B}_2$ is given by

$$\mathbf{X}_1\tilde{\mathbf{B}}_2 = \mathbf{X}_1(\mathbf{X}_1'\mathbf{X}_1)^{-}\mathbf{X}_1'\mathbf{Y}_{12}, \tag{2.10}$$

i.e. $\mathbf{X}_1\mathbf{B}_2$ is estimated from uncontaminated data alone, and similarly

$$\tilde{\mathbf{E}}_{22} = 0. \tag{2.11}$$

But clearly, using (2.11) we have equations

$$\mathbf{Y}_{22} - \mathbf{X}_2\tilde{\mathbf{B}}_2 = \tilde{\mathbf{\Theta}}\tilde{\mathbf{V}}_{22} \tag{2.12}$$

$$\tilde{\mathbf{\Theta}} = (\mathbf{Y}_{22} - \mathbf{X}_2\tilde{\mathbf{B}}_2)\tilde{\mathbf{V}}_{22}^{-1} \tag{2.13}$$

so the MLE $\tilde{\mathbf{V}}_{22}$ for $\mathbf{V}_{22}$ is obtained in terms of the uncontaminated data alone, through (2.10) and (2.11), from

$$m \cdot \tilde{\mathbf{V}}_{22} = \begin{bmatrix} \tilde{\mathbf{E}}_{12}' : \tilde{\mathbf{E}}_{22}' \end{bmatrix} \begin{bmatrix} \tilde{\mathbf{E}}_{12} \\ \tilde{\mathbf{E}}_{22} \end{bmatrix} = \mathbf{Y}_{12}'(\mathbf{I} - \mathbf{X}_1(\mathbf{X}_1'\mathbf{X}_1)^{-}\mathbf{X}_1')\mathbf{Y}_{12}. \tag{2.14}$$

Similarly to (2.14), the MLE $\tilde{\mathbf{V}}_{21}$ for $\mathbf{V}_{21}$ is obtained by expansion and collection of terms in

$$m \cdot \tilde{\mathbf{V}}_{21} = \begin{bmatrix} \tilde{\mathbf{E}}_{12}' : \tilde{\mathbf{E}}_{22}' \end{bmatrix} \begin{bmatrix} \tilde{\mathbf{E}}_{11} \\ \tilde{\mathbf{E}}_{21} \end{bmatrix} = \tilde{\mathbf{E}}_{12}'\mathbf{Y}_{11} \tag{2.15}$$

We note that as a consequence of (2.15) we have that

$$\tilde{\mathbf{E}}_{12}'\mathbf{Y}_{11} = \tilde{\mathbf{E}}_{12}'\tilde{\mathbf{E}}_{11} = \tilde{\mathbf{E}}_{12}'\hat{\mathbf{E}}_{11}. \tag{2.16}$$

Now, knowing $\tilde{\Theta}$, $\tilde{V}_{22}$ and $\tilde{V}_{21}$, we can compute $X\tilde{B}_1$ as

$$X\tilde{B}_1 = X(X'X)^- X' \begin{bmatrix} Y_{11} \\ Y_{21} - \tilde{\Theta}\tilde{V}_{21} \end{bmatrix} \tag{2.17}$$

Now the MLE $\tilde{V}_{11}$ for V_{11} is obtained form equations

$$\tilde{E}_{11} = Y_{11} - X_1\tilde{B}_1, \tag{2.18}$$

$$\tilde{E}_{21} = Y_{21} - X_2\tilde{B}_1 - \tilde{\Theta}\tilde{V}_{21}, \text{ and} \tag{2.19}$$

$$m \cdot \tilde{V}_{11} = \begin{bmatrix} \tilde{E}'_{11} : \tilde{E}'_{21} \end{bmatrix} \begin{bmatrix} \tilde{E}_{11} \\ \tilde{E}_{21} \end{bmatrix}. \tag{2.20}$$

Thus the ML-estimates $X\tilde{B}$, $\tilde{E}$, $\tilde{\Theta}$ and $\tilde{V}$ for XB, E, Θ and V in the adjusted model (2.6) can be written in closed form. As in the univariate case we have that $\tilde{E}_{22} = 0$ in the adjusted model (2.6). That is, the fitted value for the mean of Y_{22} in model (2.6) is the observed Y_{22} itself.

Alternatively the ML-estimates $X\tilde{B}$, $\tilde{E}$, $\tilde{\Theta}$ and $\tilde{V}$ can be given in terms of the ML-estimates $X\hat{B}$, $\hat{E}$ and $\hat{V}$ under the original model (2.1) to facilitate the computation of the adjusted estimates under model (2.6). Since $\tilde{E}_{22} = 0$ from (2.11) and using N as in (1.4), we can write

$$0 = \tilde{E}_{22} = \begin{bmatrix} 0 : I \end{bmatrix} \begin{bmatrix} N_{11} & : & N_{12} \\ N_{21} & : & N_{22} \end{bmatrix} \begin{bmatrix} Y_{12} \\ Y_{22} - \tilde{\Theta}\tilde{V}_{22} \end{bmatrix} = \hat{E}_{22} - N_{22}\tilde{\Theta}\tilde{V}_{22}. \tag{2.21}$$

The nonsingularity of N_{22} follows from $R(X_2) \subset R(X_1)$. This implies

$$\tilde{\Theta}\tilde{V}_{22} = N_{22}^{-1}\hat{E}_{22} \tag{2.22}$$

Consequently $X\tilde{B}_2$ can be written as

$$X\tilde{B}_2 = X(X'X)^- X' \begin{bmatrix} Y_{12} \\ Y_{22} - N_{22}^{-1}\hat{E}_{22} \end{bmatrix} = X\hat{B}_2 - M_2 N_{22}^{-1}\hat{E}_{22}. \tag{2.23}$$

Further, using (2.23), after extensive algebraic substitutions, we obtain

Theorem 2.1

The ML-estimates $X\tilde{B}$, $\tilde{\Theta}$ and $\tilde{V}$ for XB, Θ, and V in the adjusted model (2.6) are given by

(i) $X\begin{bmatrix} \tilde{B}_1 : \tilde{B}_2 \end{bmatrix} = X\begin{bmatrix} \hat{B}_1 : \hat{B}_2 \end{bmatrix} - M_2\tilde{\Theta}\begin{bmatrix} \tilde{V}_{21} : \tilde{V}_{22} \end{bmatrix}$

(ii) $\tilde{\Theta} = N_{22}^{-1}\hat{E}_{22}\tilde{V}_{22}^{-1} = m \cdot N_{22}^{-1}\hat{E}_{22}(\hat{E}_2'\hat{E}_2 - \hat{E}_{22}'N_{22}^{-1}\hat{E}_{22})^{-1}$

(iii) $m \cdot \tilde{V} = \hat{E}'\hat{E} - \begin{bmatrix} \hat{E}_{21}'\tilde{\Theta}\tilde{V}_{21} + \tilde{V}_{12}\tilde{\Theta}'[\hat{E}_{21} - N_{22}\tilde{\Theta}\tilde{V}_{21}] & : & \hat{E}_{22}'N_{22}^{-1}\hat{E}_{21} \\ \hat{E}_{21}'N_{22}^{-1}\hat{E}_{22} & : & \hat{E}_{22}'N_{22}^{-1}\hat{E}_{22} \end{bmatrix}$

where the original ML-estimates $X\hat{B}$, $\hat{E}$ and $\hat{V}$ for XB, E and V in (2.1), generate all the adjusted estimates. •

We note that because of the removal of the error term E_{22} from the estimation of V in the adjusted model (2.6), the estimate $\tilde{V}$ for V as given by Theorem 2.1 is biased. A bias-corrected estimate $\ddot{V}$ for V is obtained as

$$\ddot{V} = \begin{bmatrix} \tilde{V}_{11} + \frac{k}{m-k} \cdot \tilde{V}_{12}\tilde{V}_{22}^{-1}\tilde{V}_{21} & : & \frac{m}{m-k} \cdot \tilde{V}_{12} \\ \frac{m}{m-k} \cdot \tilde{V}_{21} & : & \frac{m}{m-k} \cdot \tilde{V}_{22} \end{bmatrix} . \tag{2.24}$$

The hypothesis that Y_{22} is not a D-outlier in (2.6), is examined by

Corollary 2.1.1

The LRT-statistic for the hypothesis $H_0 : \Theta = 0$ in (2.6) is

$$\chi^2 = m \cdot \ln(|\hat{V}| \,/\, |\tilde{V}|) \tag{2.25}$$

where $\hat{V}$ and $\tilde{V}$ are respectively the ML-estimates for V under the models (2.1) and (2.6). By the general theory of likelihood ratio tests, as presented by Rao (1973), the statistic χ^2 is asymptotically ($m \to \infty$) distributed chi-squared with $k \cdot l$ degrees of freedom. •

2.2 Outliers by additive shifts

As in the previous section we consider the adjusted model (2.2) assuming initially that the set J of possible A-outliers forms a $k \times l$-submatrix Y_{22} of Y as given in (2.5). The adjusted model (2.2) can then be written as

$$\begin{bmatrix} Y_{11} & : & Y_{12} \\ Y_{21} & : & Y_{22} \end{bmatrix} = \begin{bmatrix} X_1 \\ X_2 \end{bmatrix} \begin{bmatrix} B_1 : B_2 \end{bmatrix} + \begin{bmatrix} 0 \\ \Theta \end{bmatrix} \begin{bmatrix} 0 : I \end{bmatrix} + \begin{bmatrix} E_{11} : E_{12} \\ E_{21} : E_{22} \end{bmatrix} \tag{2.26}$$

where the parameter matrix Θ is $k \times l$. Again we require $R(X_2) \subset R(X_1)$ for X_2B_2 to be estimable in the model (2.26).

If $l = n$, the model (2.26) is equivalent to the model (2.7) under reparametrization. When a set of complete observational vectors is outlying, A- and D-outliers cannot be distinguished. We proceed assuming that $l < n$.

The ML-estimate $\mathbf{X}\tilde{\mathbf{B}}$ for $\mathbf{XB}$ under the adjusted model (2.26) is obtained from $\tilde{\boldsymbol{\Theta}}$ the MLE for $\boldsymbol{\Theta}$ by

$$\mathbf{X}\tilde{\mathbf{B}} = \mathbf{M}\left(\begin{bmatrix}\mathbf{Y}_{11}:\mathbf{Y}_{12}\\ \mathbf{Y}_{21}:\mathbf{Y}_{22}\end{bmatrix} - \begin{bmatrix}0\\ \boldsymbol{\Theta}\end{bmatrix}\begin{bmatrix}0:\mathbf{I}\end{bmatrix}\right) = \mathbf{X}\begin{bmatrix}\hat{\mathbf{B}}_1:\hat{\mathbf{B}}_2-\mathbf{M}_2\tilde{\boldsymbol{\Theta}}\end{bmatrix} \tag{2.27}$$

These results are summarized and an explicit formulation is given by the following theorem, which is essentially due to Anderson (1957). However, we represent here the adjusted estimates $\mathbf{X}\tilde{\mathbf{B}}$, $\tilde{\mathbf{V}}$ and $\tilde{\boldsymbol{\Theta}}$ in terms of the estimates $\mathbf{X}\hat{\mathbf{B}}$, $\hat{\mathbf{V}}$ and $\hat{\mathbf{E}}$ under the original model (2.1).

Theorem 2.2 (Anderson, 1957)

The ML-estimates $\tilde{\boldsymbol{\Theta}}$, $\mathbf{X}\tilde{\mathbf{B}}$ and $\tilde{\mathbf{V}}$ for $\boldsymbol{\Theta}$, $\mathbf{XB}$ and $\mathbf{V}$ in the adjusted model (2.26) are obtained from

(i) $\tilde{\boldsymbol{\Theta}} = (\mathbf{N}_{22}-\hat{\mathbf{E}}_{21}\hat{\mathbf{V}}_{11}^{-1}\hat{\mathbf{E}}_{21}'\cdot m^{-1})^{-1}(\hat{\mathbf{E}}_{22}-\hat{\mathbf{E}}_{21}\hat{\mathbf{V}}_{11}^{-1}\hat{\mathbf{V}}_{12})$.

(ii) $\mathbf{X}\tilde{\mathbf{B}} = \mathbf{X}\begin{bmatrix}\hat{\mathbf{B}}_1:\hat{\mathbf{B}}_2-\mathbf{M}_2\tilde{\boldsymbol{\Theta}}\end{bmatrix} = \mathbf{X}\begin{bmatrix}\hat{\mathbf{B}}_1:\hat{\mathbf{B}}_2-(\mathbf{X}'\mathbf{X})^{-1}\mathbf{X}_2'\tilde{\boldsymbol{\Theta}}\end{bmatrix}$

(iii)
$$\begin{aligned} m\cdot\tilde{\mathbf{V}} &= (\mathbf{Y}-\mathbf{X}\tilde{\mathbf{B}} - \begin{bmatrix}0\\ \boldsymbol{\Theta}\end{bmatrix}\begin{bmatrix}0:\mathbf{I}\end{bmatrix})'(\mathbf{Y}-\mathbf{X}\tilde{\mathbf{B}} - \begin{bmatrix}0\\ \boldsymbol{\Theta}\end{bmatrix}\begin{bmatrix}0:\mathbf{I}\end{bmatrix}) \\ &= (\begin{bmatrix}\hat{\mathbf{E}}_1:\hat{\mathbf{E}}_2\end{bmatrix} - \begin{bmatrix}0:\mathbf{N}_2\tilde{\boldsymbol{\Theta}}\end{bmatrix})'(\begin{bmatrix}\hat{\mathbf{E}}_1:\hat{\mathbf{E}}_2\end{bmatrix} - \begin{bmatrix}0:\mathbf{N}_2\tilde{\boldsymbol{\Theta}}\end{bmatrix}) \\ &= m\cdot\hat{\mathbf{V}} - \begin{bmatrix}0 & : & \hat{\mathbf{E}}_{21}'\tilde{\boldsymbol{\Theta}} \\ \tilde{\boldsymbol{\Theta}}'\hat{\mathbf{E}}_{21} & : & \hat{\mathbf{E}}_{22}'\tilde{\boldsymbol{\Theta}} + \tilde{\boldsymbol{\Theta}}'\hat{\mathbf{E}}_{22} - \tilde{\boldsymbol{\Theta}}'\hat{\mathbf{N}}_{22}\tilde{\boldsymbol{\Theta}}\end{bmatrix} . \end{aligned}$$
•

We note that the estimate $\tilde{\mathbf{V}}$ for $\mathbf{V}$ in the adjusted model (2.26) as given by Theorem 2.2 is biased, due to the missing data $\mathbf{Y}_{22}$. Orchard and Woodbury (1972) give a bias-corrected estimate $\ddot{\mathbf{V}}$ for $\mathbf{V}$ as

$$\ddot{\mathbf{V}} = \begin{bmatrix}\tilde{\mathbf{V}}_{11} & : & \tilde{\mathbf{V}}_{12} \\ \tilde{\mathbf{V}}_{21} & : & \dfrac{m}{m-k}\cdot\tilde{\mathbf{V}}_{22} - \dfrac{k}{m-k}\cdot\tilde{\mathbf{V}}_{21}\tilde{\mathbf{V}}_{11}^{-1}\tilde{\mathbf{V}}_{12}\end{bmatrix} . \tag{2.28}$$

Corollary 2.2.1

The LRT-statistic for the hypothesis $H_0\colon \boldsymbol{\Theta}=0$ in the adjusted model (2.26), that $\mathbf{Y}_{22}$ is not an A-outlier, is given by $\chi^2 = m\cdot\ln(|\hat{\mathbf{V}}|/|\tilde{\mathbf{V}}|)$, which is asymptotically ($m\rightarrow\infty$) distributed chi-squared with $k\cdot l$ degrees of freedom. •

The results of this section also apply when $\mathbf{Y}_{22}$ is not a set of possible outliers but a set of arbitrary values (e.g. $\mathbf{Y}_{22}=0$) for missing data points. Then $\tilde{\mathbf{Y}}_{22} = \mathbf{X}_2\tilde{\mathbf{B}}_2$ under model (2.26) is the missing data estimate for $\mathbf{Y}_{22}$.

2.3 Mean shifts and outliers in principal components

As a third type of outlier in the multivariate regression model (2.1), we examine in this section transformational (T–) outliers or outliers in principal components (PC's). Gnanadesikan and Kettenring (1972) and Hawkins (1974, 1980) used principal components to detect outliers in multivariate data, but the development here is different in that it represents outliers in principal components through one type of model from a unified family.

When a subset of l (say) principal components corresponding to the principal axes $\mathbf{P}_2$ (say) is specified as outlying for all observations belonging to a subset $\mathbf{Y}_2$ (say) of the observations $\mathbf{Y}$ in the model (2.1), then, possibly after some rearrangement of the rows and columns of $\mathbf{Y}$, the adjusted model (2.4) can be written as

$$\begin{bmatrix}\mathbf{Y}_1\\ \mathbf{Y}_2\end{bmatrix} = \begin{bmatrix}\mathbf{X}_1\\ \mathbf{X}_2\end{bmatrix}\mathbf{B} + \begin{bmatrix}0\\ \boldsymbol{\Theta}\end{bmatrix}\mathbf{P}_2' + \begin{bmatrix}\mathbf{E}_1\\ \mathbf{E}_2\end{bmatrix} . \tag{2.29}$$

The parameter matrix $\boldsymbol{\Theta}$ is here $k \times l$, that is, l PC's corresponding to the principal axes $\mathbf{P}_2$ are specified as outlying, for k observations $\mathbf{Y}_2$. The matrix $\mathbf{Y}_2$ is thus $k \times n$ and $\mathbf{X}_2$ is $k \times p$. Note that $\mathbf{Y}$ and $\mathbf{E}$ are partitioned differently from the partitioning in the previous sections, and thus $\mathbf{Y}_1, \mathbf{Y}_2$, and $\mathbf{E}_1, \mathbf{E}_2$ now denote different submatrices of $\mathbf{Y}$ and $\mathbf{E}$ respectively. The matrix $\mathbf{P}_2$ is a $n \times l$-submatrix of $\mathbf{P} = \begin{bmatrix}\mathbf{P}_1 : \mathbf{P}_2\end{bmatrix}$, from the SVD:

$$\mathbf{V} = \mathbf{P}\,\Delta\mathbf{P}' = \begin{bmatrix}\mathbf{P}_1 : \mathbf{P}_2\end{bmatrix}\begin{bmatrix}\Delta_1 & \\ & \Delta_2\end{bmatrix}\begin{bmatrix}\mathbf{P}_1'\\ \mathbf{P}_2'\end{bmatrix} . \tag{2.30}$$

When $l = n$, the model (2.29) is a reparametrization of models (2.7) and (2.26). Thus, D-, A- and T-outliers cannot be distinguished for complete observations in (1.2). In the following we assume that $l < n$.

A model which is more general than (2.29) is the model

$$\begin{bmatrix}\mathbf{Y}_1\\ \mathbf{Y}_2\end{bmatrix} = \begin{bmatrix}\mathbf{X}_1\\ \mathbf{X}_2\end{bmatrix}\mathbf{B} + \begin{bmatrix}0\\ \mathbf{A}\boldsymbol{\Theta}\end{bmatrix}\mathbf{P}_2' + \begin{bmatrix}\mathbf{E}_1\\ \mathbf{E}_2\end{bmatrix} \tag{2.31}$$

where $\mathbf{A}$ is an arbitrary known $k \times a$–matrix, so that $\boldsymbol{\Theta}$ is here $a \times l$. The model (2.31) allows for an arbitrary mean shift in a subset of l principal components of $\mathbf{Y}_2$. Taking $\mathbf{A} = \mathbf{I}_k$ we obtain the model (2.29) as a special case. Without loss of generality we take $a \leq k$.

If the adjusted model (2.31) is transformed from the right by $\mathbf{P}$ we obtain a mean shift $\mathbf{A}\boldsymbol{\Theta}$ in the PC's $\mathbf{Y}_2\mathbf{P}_2$ through

$$\begin{bmatrix} \mathbf{Y}_1\mathbf{P}_1 : \mathbf{Y}_1\mathbf{P}_2 \\ \mathbf{Y}_2\mathbf{P}_1 : \mathbf{Y}_2\mathbf{P}_2 \end{bmatrix} = \begin{bmatrix} \mathbf{X}_1 \\ \mathbf{X}_2 \end{bmatrix} \mathbf{BP} + \begin{bmatrix} 0 \\ \mathbf{A\Theta} \end{bmatrix} \begin{bmatrix} 0 : \mathbf{I} \end{bmatrix} + \begin{bmatrix} \mathbf{E}_1\mathbf{P}_1 : \mathbf{E}_1\mathbf{P}_2 \\ \mathbf{E}_2\mathbf{P}_1 : \mathbf{E}_2\mathbf{P}_2 \end{bmatrix} . \tag{2.32}$$

The maximum likelihood estimation of the unknowns in the adjusted model (2.31) can be performed in closed form. The ML-estimates $\widetilde{\mathbf{XB}}$, $\tilde{\mathbf{\Theta}}$ and $\tilde{\mathbf{V}}$ for $\mathbf{XB}$, $\mathbf{\Theta}$ and $\mathbf{V}$ in the model (2.29) are then obtained by taking $\mathbf{A} = \mathbf{I}_k$ in the corresponding formulae for the ML-estimates $\widetilde{\mathbf{XB}}$, $\tilde{\mathbf{\Theta}}$ and $\tilde{\mathbf{V}}$ in the model (2.31).

The likelihood ratio test statistic for the hypothesis $H_0 : \mathbf{A\Theta} = 0$ is obtained by comparing the maximum likelihood under the original model (2.1) with the maximum likelihood under the adjusted model (2.31).

Corollary 2.3.1

The LRT-statistic for the hypothesis $H_0 : \mathbf{A\Theta} = 0$ under the adjusted model (2.31), which is the hypothesis that no mean shift is present in the PC's, is given by

$$\chi^2 = m \cdot \ln(\,|\tilde{\mathbf{V}}\,| \,/\, |\hat{\mathbf{V}}|) = m \cdot \ln(\,|\tilde{\Delta}\,| \,/\, |\hat{\mathbf{V}}|) \tag{2.33}$$

which is asymptotically $(m \rightarrow \infty)$ distributed chi-squared with $rank(\mathbf{A}) \cdot l$ degrees of freedom. •

As noted above, taking $\mathbf{A} = \mathbf{I}_k$ we obtain the model (2.29) which is the model (2.1) adjusted for outliers in PC's or T-outliers. The LRT-statistic for testing the hypothesis $H_0 : \mathbf{\Theta} = 0$ in the adjusted model (2.29) is obtained in a manner similar to Corollary 2.3.1.

2.4 Nested outlier patterns

More general than the outlier pattern formed a submatrix $\mathbf{Y}_{22}$ of the data matrix $\mathbf{Y}$, is the nested outlier pattern. In this case we assume that the data matrix $\mathbf{Y}$ of the multivariate regression model (2.1) can be partitioned as

$$\mathbf{Y} = \begin{bmatrix} \mathbf{Y}_{11} & \cdots & \mathbf{Y}_{1q} \\ \vdots & & \vdots \\ \mathbf{Y}_{q1} & \cdots & \mathbf{Y}_{qq} \end{bmatrix} \tag{2.34}$$

such that the rectangular submatrices

$$\{\mathbf{Y}_{ij} \mid j > q - i + 1\} \tag{2.35}$$

contain the data points suspected to be outlying. The submatrices $\mathbf{Y}_{ij}$ of $\mathbf{Y}$ given by (2.35) are the submatrices below the contragredient block diagonal of $\mathbf{Y}$. Naturally, the partitioning of $\mathbf{Y}$ as in (2.34) and (2.35) may have been obtained after a suitable rearrangement of the rows and column of $\mathbf{Y}$, and a corresponding rearrangement of the rows and columns of $\mathbf{X}$ and $\mathbf{B}$ in model (2.1). We may adjust for A-outliers occurring in a nested pattern by

$$\begin{bmatrix} \mathbf{Y}_{11} & \cdots & \mathbf{Y}_{1q} \\ \vdots & & \vdots \\ \mathbf{Y}_{q1} & \cdots & \mathbf{Y}_{qq} \end{bmatrix} = \begin{bmatrix} \mathbf{X}_1 \\ \vdots \\ \mathbf{X}_q \end{bmatrix} \begin{bmatrix} \mathbf{B}_1 : & \cdots & : \mathbf{B}_q \end{bmatrix} + \\ + \begin{bmatrix} 0 & \cdots & 0 \\ \vdots & & \mathbf{\Theta}_{2q} \\ \vdots & & \vdots \\ 0 & \mathbf{\Theta}_{q2} \cdots & \mathbf{\Theta}_{qq} \end{bmatrix} + \begin{bmatrix} \mathbf{E}_{11} & \cdots & \mathbf{E}_{1q} \\ \vdots & & \mathbf{E}_{2q} \\ \vdots & & \vdots \\ \mathbf{E}_{q1} & \cdots & \mathbf{E}_{qq} \end{bmatrix} . \tag{2.36}$$

Anderson (1957) and Rubin (1974) treated a model of this type in the context of missing data estimation where the set of submatrices (2.35) of the data matrix $\mathbf{Y}$ represents missing data points rather than A-outliers. But the problem is equivalent to the A-outlier problem, as far as ML-estimation of the unknowns in model (2.36) is concerned.

2.5 Extensions

In the general case, when outliers occur in an arbitrary pattern, ML-estimation of the parameters in the adjusted models (2.1), (2.2) and (2.4) is not possible in closed form. Iterative methods will be required. For A-outliers there exist several well-known techniques to estimate the parameters in model (2.2). Dempster, Laird and Rubin (1977) give an extensive treatment of the EM-algorithm, and the respective merits of the EM, scoring, and Newton-Raphson methods are discussed. Wu (1983) corrects an error in the EM theory and obtains several convergence results for the EM-algorithm. These methods extend to D-outliers and T-outliers.

Schall (1986) uses the foregoing theory to generalize Cook's (squared) distance and the statistic of Andrews and Pregibon (1978) for multivariate multiple regression situations.

3. APPLICATION

The well-known iris data of Anderson (1935), examined by Fisher (1936) by linear discriminant methods, was subjected to the outlier tests envisaged in the preceding section. The data consists of three sets of fifty observations on four variables: petal and sepal length and width. The three sets are from the species *versicolor, virginica* and *setosa*. Particular elements of the data sets are referred to by case and variable number in the original listing of Anderson (1935). No subsets of any data set were designated prior to the analysis as being of particular interest. Within each set, outlier statistics were calculated for each of the 200 data elements. All elements with large individual test-statistic values were aggregated into sets whose joint test-statistic values were obtained. This approach is not immune to masking or swamping effects, and was regarded as exploratory. We established that the test-statistics of each method focus attention on different elements of the data. We were not successful in locating major data features that had previously escaped detection in the many studies of these data sets.

The computation of 1800 individual χ^2-values to locate A- and D-outliers in the three samples using an iterative algorithm like the EM-algorithm takes a large amount of computing time. Alternatively, the χ^2-values could be computed in closed form using the formulae given in Sections 2.1 and 2.2. A third possibility, which involves the least number of computations, is to compute for each data point y_{ij} the statistic $\hat{Q}_{ij}$, that is the sample outlier sum of squares corresponding to the observation y_{ij}. In the multivariate location model, using the statistic $\hat{Q}_{ij}^D$ is equivalent to using the scaled residuals $\hat{e}_{ij}^*/\sqrt{\hat{\mathbf{V}}_{jj}}$, and similarly using $\hat{Q}_{ij}^T$ is equivalent to using the scaled PC-residuals $\hat{e}_{ij}^*/\sqrt{\hat{\Delta}_j}$. Finally, using $\hat{Q}_{ij}^A$ is equivalent to using the scaled residuals in the model transformed by $\hat{\mathbf{V}}^{-1}$. Thus we have

$$
\begin{aligned}
\hat{Q}_{ij}^D &= \frac{\hat{e}_{ij}^2}{\hat{\mathbf{V}}_{jj}} \cdot \frac{m}{m-1}\,, \\
\hat{Q}_{ij}^A &= \frac{\hat{\epsilon}_{ij}^2}{(\hat{\mathbf{V}}^{-1})_{jj}} \cdot \frac{m}{m-1}\,, \quad \text{where } \hat{\mathbf{E}}\hat{\mathbf{V}}^{-1} = (\hat{\epsilon}_{ij})\,, \text{ and} \qquad (3.1) \\
\hat{Q}_{ij}^T &= \frac{(\hat{e}_{ij}^*)^2}{\hat{\Delta}_j} \cdot \frac{m}{m-1}\,, \quad \text{where } \hat{\mathbf{E}}\hat{\mathbf{P}} = (\hat{e}_{ij}^*)\,.
\end{aligned}
$$

To locate T-outliers in the data, the scaled PC-residuals $\hat{e}_{ij}^*/\sqrt{\hat{\Delta}_j}$ were computed. The scaled PC-residuals are approximately distributed as $N(0,1)$. Elements were declared to be in a set of interest, for a particular outlier type, if they generated individual values of the corresponding statistic which achieved significance levels 0.005, 0.025 or 0.050. These sets satisfy an obvious inclusion relation, and were defined for each of the three samples, and with respect to D-, A- and T-outliers. Unfortunately it is difficult to determine a level of significance at which to reject the null-hypothesis that a constructed set is non-outlying. Ideally the chi-square value for k outliers would have to be compared with the $(1-\alpha)$-fractile of the distribution of the maximum of $c_k = \binom{n \cdot m}{k}$ possible non-independent χ^2-statistics from the data. Since this distribution is generally intractable, the $(1-\alpha/c_k)$-fractile of the χ_k^2-distribution is used as an approximation, resulting from the first Bonferroni inequality. Hawkins (1980) notes that this approximation is conservative but generally very good for $k=1$. For $k>1$, however, the approximation can be extremely conservative. We keep this in mind while screening the data outliers.

The $(1-0.05/c_k)$-fractiles of the χ_k^2-distribution for $n \cdot m = 200$ and $k=1,\ldots,7$ were computed using the routine MDCH of the IMSLIB (1985)-Library in single precision. They are

k	1	2	3	4	5	6	7
$\chi_k^2(1-0.05/c_k)$	13.5	25.8	37.4	48.4	59.0	69.2	78.8

The $(1-0.025/c_1)=(1-0.025/200)$-fractile of $N(0,1)$ is 3.66, and for the distribution of the maximum of 200 independent $N(0,1)$-variates we obtain a $(1-0.025)$-fractile of 3.02. Those two fractiles can be used as values against which the scaled PC-residuals may be compared.

For the samples of *versicolor* and *virginica* there is little agreement between the location and the number of those data points which yield high χ^2-values when specified respectively as D- and A-outlier.

The sample of *versicolor* has a largest χ_1^2-value 6.44 for D-outliers, much smaller than 13.5, and thus the hypothesis that D-outliers are in the sample is discarded. Similarly we discard the hypothesis of T-outliers, because the absolute value of the largest scaled PC-residual is 2.51, compared with 3.66 or even 3.02. The largest χ_1^2-value for A-outliers, however, is 13.06, with the second largest being 8.10 for observations (19,2) and (49,3). These warrant more investigation in view of the conservative nature of the Bonferroni-approximation. Their χ_2^2-value for is 21.19, compared with $\chi_2^2(1-0.05/c_2)=25.8$. It appears that they are possibly A-outliers.

In the *virginica* sample, the largest χ_1^2-value for D-outliers is 6.73, the largest χ_1^2-value for A-outliers is 8.42 and the largest absolute value of the PC-residuals is 2.48, all well below Bonferroni significance. We accept the hypothesis that no outliers are present.

The sample of *setosa* is different in that it shows substantial agreement between possible D-, A- and T-outliers. This may arise from the fact that the sample variance-covariance matrix $\hat{\mathbf{V}}$ for *setosa* is closest to being diagonal of all three sample covariance matrices. The largest χ_1^2-values for D- and A-outliers are respectively 11.95 and 12.96, both for the data point (44,4). The largest absolute PC-residual is 3.28, also appearing in the observational vector $\mathbf{y}_{44}$. All those values are close to significance, but the hypothesis of A-outliers in the data appears best supported by the data. Observations (42,2) and (44,4) have the maximal χ_2^2-values 19.21 and 22.77 for D- and A-outliers respectively, compared with $\chi_2^2(1-0.05/c_2)=25.8$. Similarly the maximal χ_3^2-values are 24.29, 32.41 and 26.26 for D-, A- and T-outliers respectively, compared with $\chi_3^2(1-0.05/c_3)=37.3$. It appears that at most three A-outliers are present, data points (25,3), (42,2) and (44,4).

In summary it appears to be worthwhile to consider the three different types of outlier, distributional, additive, and transformational, since there may generally be little agreement between the sets of suspicious data points corresponding to the respective types. It is generally a single component which signals a certain observational vector to be possibly outlying. Thus it is instructive to search the individual data elements for outliers, as well as complete observational vectors.

REFERENCES

Anderson, E. (1935). The irises of the Gaspe peninsula. *Bulletin of the American Iris Society,* **59**, 2-5.

Anderson, T.W. (1957). Maximum likelihood estimates for a multivariate normal distribution when some observations are missing. *Journal of the American Statistical Association,* **52**, 200-203.

Andrews, D.F. and Pregibon, D. (1978). Finding the outliers that matter. *Journal of the Royal Statistical Society, Ser. B,* **40**, 85-93.

Barnett, V. and Lewis, T. (1984). *Outliers in Statistical Data* (2nd ed.), John Wiley & Sons, Chichester.

Cook, R.D. and Weisberg, S. (1982). *Residuals and Influence in Regression.* Chapman and Hall, New York.

Dempster, A.P., Laird, N.M. and Rubin, D.B. (1977). Maximum likelihood from incomplete data via the EM-algorithm (with discussion). *Journal of the Royal Statistical Society, Ser. B,* **39**, 1-38.

Draper, N.R. and John, J.A. (1981). Influential observations and outliers in regression. *Technometrics,* **23**, 21-26.

Dunne, T.T. (1982). *Contributions to the theory of generalized inverses, the linear model and outliers.* Unpublished PhD thesis. University of Cape Town, South Africa.

Fisher, R.A. (1936). The use of multiple measurements in taxonomic problems. *Annals of Eugenics,* **7**, 179-188.
Gentleman, J.F. and Wilk, M.B. (1975). Detecting outliers. II. Supplementing the direct analysis of residuals. *Biometrics,* **31**, 387-410.
Gnanadesikan R., and Kettenring, J.R. (1972). Robust estimates, residuals and outlier detection with multiresponse data. *Biometrics,* **28**, 81-124.
Hawkins, D.M. (1974). The detection of errors in multivariate data using principal components. *Journal of the American Statistical Association,* **69**, 340-344.
Hawkins, D.M. (1980). *Identification of Outliers.* Chapman and Hall, London.
IMSLIB (1985). International Mathematical and Statistical Library, Users Manual. IMSL Inc., Houston, Texas
John, J.A. and Draper, N.R. (1978). On testing for two outliers or one outlier in two-way tables. *Technometrics,* **20**, 69-78.
Orchard, T. and Woodbury, M.A. (1972). A missing information principle; theory and applications. *Proceedings of the 6th Berkeley Symposium on Mathematical Statistics and Probability,* , 697-715.
Rao, C.R. (1971). Unified theory of linear estimation. *Sankhyā, Ser. A,* **33**, 371-394.
Rao, C.R. (1973). *Linear Statistical Inference and its Applications.* Wiley, New York.
Rubin, D.B (1974). Characterizing the estimation of parameters in incomplete-data problems. *Journal of the American Statistical Association,* **69**, 467-474.
Schall, R. (1986). *Outliers and influence under arbitrary variance.* Unpublished PhD thesis. University of Cape Town, South Africa.
Wu, C.F.J. (1983). On the convergence properties of the EM-algorithm. *Annals of Statistics,* **11**, 95-103.

Received 13 January 1987
Revised 12 May 1987

Friedhofstr. 13
6725 Roemerberg 1
F.R. Germany

Department of Mathematical Sciences
Rice University
P.O. Box 1892
Houston, TX 77251-1892
U.S.A.

Proc. Second International Tampere Conference in Statistics
(Tampere, Finland, 1-4 June 1987)
Tarmo Pukkila and Simo Puntanen, *Editors*

pp. 679 - 690

Iterative improvements of a partial minimax estimator in regression analysis

G. TRENKLER
University of Dortmund, F.R.G.

P. STAHLECKER
University of Mainz, F.R.G.

B. SCHIPP
University of Dortmund, F.R.G.

F. HERING
University of Dortmund, F.R.G.

Key words and phrases: Minimax estimation, linear restrictions, ellipsoid restrictions, iterative improvements.

ABSTRACT

The general minimax estimator of the linear regression model is applicable when the whole parameter vector is restricted to an ellipsoid. In many applications, however, it is more realistic to assume that only a part of the parameter vector is constrained. In this paper a partial minimax estimator is investigated where the unrestricted part is estimated by the OLS-estimator in the first step. An iterative procedure leads to an estimator with good statistical properties.

1. INTRODUCTION

Consider the linear regression model

$$\mathbf{y} = \mathbf{X}\boldsymbol{\beta} + \mathbf{u},\ E[\mathbf{u}] = \mathbf{0},\ Cov[\mathbf{u}] = \sigma^2\mathbf{V} \tag{1.1}$$

where $\mathbf{y}$ is an $(n\times 1)$ -vector of observations on the dependent variable, $\mathbf{X}$ is an $(n\times p)$ -matrix of observations on p explanatory variables, $\mathbf{u}$ is an

$(n \times 1)$ -vector of unobservable disturbances and $\boldsymbol{\beta}$ is an unknown $(p \times 1)$ -vector of parameters. The positive definite (p.d.) matrix $\mathbf{V}$ is known, and the scalar $\sigma^2 > 0$ is also assumed to be given for the present.

In many practical regression situations there is some prior information about the parameter vector $\boldsymbol{\beta}$. For instance, one main problem of monetary economics consists of measuring the interrelated effects of wealth, income, interest rates and prices on the demand for assets and liabilities. Empirical investigations on this subject "should not only embody many such interrelated influences, but also satisfy balance-sheet identities in both equilibrum and dynamic models" (Saïto 1977, p. 1).

As Saito pointed out, individually specified regression equations (flow of funds equations) usually will not add up to meet the balance-sheet restrictions. Furthermore, the equations contain many parameters to be estimated. This will lead to serious multicollinearity problems when using the traditional least squares procedure. To overcome these difficulties Saito applies the linear expenditure system (LES) to the demand for assets and liabilities, which from our point of view also serves as an interesting example for using the minimax approach.

By maximizing the expected utility of holding assets Saito (1977, p. 3) derives the following static set of demand functions:

$$H_i = c_i q_i + b_i(W - \sum_{k=1}^{n} c_k q_k), \qquad i = 1,\dots, n, \tag{1.2}$$

where H_i = real value of end-of-period holdings of the financial asset (nominal value deflated by consumer price index); $q_i = 1/(1 + r_i)$, where r_i is the current interest rate of the ith asset; W = real end-of-period net worth of financial assets, c_i and b_i are parameters.

The first order conditions for a maximal expected utility imply that

$$b_i = a_i/(a_1 + \dots + a_n), \tag{1.3}$$

where $a_i > 0$, $i = 1,\dots, n$, are parameters of the expected utility function. Thus from (1.3) several restrictions on b_i may be obtained:

$$b_i \in [0, 1] \quad \text{and} \quad b_1 + \dots + b_n = 1\,. \tag{1.4}$$

With the restriction on the sum of b_i the wealth constraint

$$\sum_{i=1}^{n} H_i = \sum_{i=1}^{n} c_i q_i + \sum_{i=1}^{n} b_i(W - \sum_{k=1}^{n} c_k q_k) = W \tag{1.5}$$

will hold, which is a particular property of LES.

Since (1.2) is nonlinear in the parameters, Saito (1977, pp.7-9) combines cross-section with time-series data in a two-step estimation procedure. In the first step cross-section data are used such that the q_i, $i = 1,\dots, n$, may be regarded as constants.

Furthermore, model (1.2) is modified to incorporate income effects on the demand of assets. This yields the following final form of the demand system:

$$H_i = b_i W + h_i Y + g_i + u_i, \quad i = 1,\dots, n, \tag{1.6}$$

where Y represents income; u_i the stochastic term and h_i, g_i are parameters.

The system (1.6) of equations may be estimated with cross-section data. The corresponding regression equations are given by

$$H_{ij} = b_i W_j + h_i Y_j + g_i + u_{ij}, \quad j = 1,\dots, m, \tag{1.7}$$

where m is the number of observations (households).

Setting $\boldsymbol{\beta}_i = (b_i, h_i, g_i)'$, $\mathbf{y}_i = (H_{i1},\dots, H_{im})'$, $\mathbf{u}_i = (u_{i1},\dots, u_{im})'$, $i = 1,\dots, n$, and

$$\bar{\mathbf{X}} = \begin{pmatrix} W_1 & Y_1 & 1 \\ \vdots & \vdots & \vdots \\ W_m & Y_m & 1 \end{pmatrix}$$

we see that (1.7) can be written as

$$\begin{pmatrix} \mathbf{y}_1 \\ \cdot \\ \cdot \\ \cdot \\ \mathbf{y}_n \end{pmatrix} = \begin{pmatrix} \bar{\mathbf{X}} & \mathbf{0} & \dots & \mathbf{0} \\ \mathbf{0} & \bar{\mathbf{X}} & \dots & \mathbf{0} \\ \cdot & \cdot & & \cdot \\ \cdot & \cdot & & \cdot \\ \mathbf{0} & \mathbf{0} & \dots & \bar{\mathbf{X}} \end{pmatrix} \begin{pmatrix} \boldsymbol{\beta}_1 \\ \cdot \\ \cdot \\ \cdot \\ \boldsymbol{\beta}_n \end{pmatrix} + \begin{pmatrix} \mathbf{u}_1 \\ \cdot \\ \cdot \\ \cdot \\ \mathbf{u}_n \end{pmatrix} \tag{1.8}$$

i.e.

$$\mathbf{y} = \mathbf{X}\boldsymbol{\beta} + \mathbf{u}. \tag{1.9}$$

The restrictions (1.4) on the b_i may be viewed as a combination of linear and ellipsoidal restrictions on $\boldsymbol{\beta}$ where, however, the ellipsoid constraints only approximate the exact inequalities (cf. Toutenburg 1982, pp. 80-82).

Setting $\mathbf{R} = (1, 0, 0, 1, 0, 0,\dots, 1, 0, 0)$ and $r = 1$ we see that $\mathbf{R}\boldsymbol{\beta} = \mathbf{r}$ corresponds to $\Sigma b_i = 1$. By Toutenburg's procedure we derive

$$\mathbf{T} = (4/n)\,\mathrm{diag}\,(1, 0, 0, 1, 0, 0,\dots, 1, 0, 0)$$

and with $\beta_0 = 0$ we obtain the ellipsoid restriction

$$\boldsymbol{\beta}'\mathbf{T}\boldsymbol{\beta} \leq 1 \tag{1.10}$$

approximating the information $b_i \in [0, 1]$, $i = 1, \ldots, n$.

In a recent paper Pilz (1986) derived minimax estimators based on interval information using connections to Bayesian experimental design. These estimators need the computation of the second order moment matrix of a least favourable prior distribution of $\boldsymbol{\beta}$ on the interval region. This task can be done only by an interative search algorithm over the corner points which is difficult to implement especially, when the number of regressors is large. Therefore we propose to accept the loss caused by weakening the pure interval restrictions.

In this paper we assume that the parameter vector $\boldsymbol{\beta}$ is simultaneously constrained by ellipsoidal and inhomogenous linear restrictions. The ellipsoid restriction is characterized by the set

$$E = \{\boldsymbol{\beta} \in \mathbb{R}^p \,|\, (\boldsymbol{\beta} - \boldsymbol{\beta}_0)'\mathbf{T}(\boldsymbol{\beta} - \boldsymbol{\beta}_0) \leq c\}, \tag{1.11}$$

where $\mathbf{T}$ is a known p.d. matrix of order $p \times p$. The center of the ellipsoid $\boldsymbol{\beta}_0$ is also known, and without loss of generality the constant c may be set to 1.

The linear restrictions are given by the set

$$C = \{\boldsymbol{\beta} \in \mathbb{R}^p \,|\, \mathbf{R}\boldsymbol{\beta} = \mathbf{r}\}, \tag{1.12}$$

where the matrix $\mathbf{R}$ is of type $m \times p$ and the vector $\mathbf{r}$ is a member of $\mathbb{R}^m$, both $\mathbf{R}$ and $\mathbf{r}$ are known. To avoid trivialities assume that the set

$$B := E \cap C \tag{1.13}$$

is nonempty.

2. MINIMAX ESTIMATION

In the following we are interested in estimating the quantity $\mathbf{B}\boldsymbol{\beta}$ where $\mathbf{B}$ is a given $(q \times p)$ -matrix. Simultaneously we wish to incorporate the information $\boldsymbol{\beta} \in B$ into the estimation process. For this consider the class of heterogeneous linear estimators.

$$D = \{\mathbf{d} \mid \mathbf{d} = \mathbf{B}[\mathbf{d}_0 + \mathbf{C}(\mathbf{y} - \mathbf{X}\mathbf{d}_0)],\ \mathbf{C} \in \mathbb{R}^{p\times n}\}, \tag{2.1}$$

where $\mathbf{d}_0 \in \mathbb{R}^p$ is a constant initial guess of $\boldsymbol{\beta}$ to be specified independently of $\mathbf{y}$ in a suitable way. To measure the performance of an estimator $\mathbf{d}$ we use the weighted risk function

$$R_A(\mathbf{d}, \mathbf{B}\boldsymbol{\beta}) = E[(\mathbf{d} - \mathbf{B}\boldsymbol{\beta})'\mathbf{A}(\mathbf{d} - \mathbf{B}\boldsymbol{\beta})], \tag{2.2}$$

where $\mathbf{A}$ denotes a $q \times q$ nonnegative definite (n.n.d.) matrix of weights.

Our objective is to identify an optimal matrix $\mathbf{C}_* \in \mathbb{R}^{p\times n}$ which minimizes $R_A(\mathbf{d}, \mathbf{B}\boldsymbol{\beta})$ in the worst situation with respect to $\boldsymbol{\beta}$.

As a matter of direct derivations we obtain for $\mathbf{d} \in D$:

$$E[\mathbf{d}] = \mathbf{B}[\mathbf{d}_0 + \mathbf{C}\mathbf{X}(\boldsymbol{\beta} - \mathbf{d}_0)], \tag{2.3}$$

$$Cov[\mathbf{d}] = \sigma^2\mathbf{B}\mathbf{C}\mathbf{V}\mathbf{C}'\mathbf{B}', \tag{2.4}$$

$$Bias[\mathbf{d}] = E[\mathbf{d}] - \mathbf{B}\boldsymbol{\beta} = \mathbf{B}(\mathbf{C}\mathbf{X} - \mathbf{I})(\boldsymbol{\beta} - \mathbf{d}_0), \tag{2.5}$$

$$R_A(\mathbf{d}, \mathbf{B}\boldsymbol{\beta}) = \sigma^2\mathrm{tr}\{\mathbf{V}\mathbf{C}'\mathbf{G}\mathbf{C}\} + (\boldsymbol{\beta} - \mathbf{d}_0)'(\mathbf{C}\mathbf{X} - \mathbf{I})'\mathbf{G}(\mathbf{C}\mathbf{X} - \mathbf{I})(\boldsymbol{\beta} - \mathbf{d}_0), \tag{2.6}$$

where $\mathbf{G} = \mathbf{B}'\mathbf{A}\mathbf{B}$.

Let us first explore the case where the matrix $\mathbf{T}$ from (1.1) is positive definite (p.d.). Put

$$\mathbf{S} = \mathbf{R}\mathbf{T}^{-\frac{1}{2}},\ \ \mathbf{s} = \mathbf{r} - \mathbf{R}\boldsymbol{\beta}_0,\ \ \mathbf{t}_* = \mathbf{S}^+\mathbf{s},\ \ \mathbf{P} = \mathbf{I} - \mathbf{S}^+\mathbf{S}, \tag{2.7}$$

$$Q(\mathbf{C}) = \mathbf{P}\mathbf{T}^{-\frac{1}{2}}(\mathbf{C}\mathbf{X} - \mathbf{I})'\mathbf{G}(\mathbf{C}\mathbf{X} - \mathbf{I})\mathbf{T}^{-\frac{1}{2}}\mathbf{P},\ \ \mathbf{d}_0 = \boldsymbol{\beta}_0 + \mathbf{T}^{-\frac{1}{2}}\mathbf{t}_*. \tag{2.8}$$

The symbol $\mathbf{S}^+$ denotes the Moore-Penrose inverse of $\mathbf{S}$.

Then we may state

LEMMA 1. *If* $B \neq \varnothing$, *then for all* $\mathbf{d} \in D$ *we have*

$$\max_{\boldsymbol{\beta}\in B} R_A(\mathbf{d}, \mathbf{B}\boldsymbol{\beta}) = \sigma^2 tr\{\mathbf{V}\mathbf{C}'\mathbf{G}\mathbf{C}\} + \lambda_{max}[Q(\mathbf{C})](1 - \mathbf{t}_*'\mathbf{t}_*), \tag{2.9}$$

where $\lambda_{max}[Q(\mathbf{C})]$ *denotes the maximal eigenvalue of* $Q(\mathbf{C})$.

Proof. See Stahlecker/Trenkler (1987). For the case of pure ellipsoid restrictions we refer to Kuks/Ol'man (1972).

Formula (2.9) gives an expression for the maximal risk which occurs in the worst situation. Our next aim is to minimize (2.9) subject to **C** to obtain the minimax estimator. However, due to the nonlinear dependence of $\lambda_{max}[Q(\mathbf{C})]$ on the matrices **P**, **T**, **G** and **C** explicit minimization of $\lambda_{max}[Q(\mathbf{C})]$ with respect to **C** fails if rank(**G**) > 1.

To circumvent this difficulty observe that

$$(1/p)\mathrm{tr}[Q(\mathbf{C})] \le \lambda_{max}[Q(\mathbf{C})] \le \mathrm{tr}[Q(\mathbf{C})] \tag{2.10}$$

(see Humak 1977, p. 484). Thus it appears reasonable to solve the subsidiary problem

$$F_\gamma(\mathbf{C}) \to \min_{\mathbf{C}},$$

where

$$F_\gamma(\mathbf{C}) = \sigma^2\mathrm{tr}\{\mathbf{VC'GC}\} + a\gamma\mathrm{tr}[Q(\mathbf{C})] \tag{2.11}$$

with

$$a = 1 - \mathbf{t}_*'\mathbf{t}_*, \quad \gamma = 1 \text{ or } \gamma = 1/p.$$

If we minimize $F_\gamma(\mathbf{C})$ with respect to **C** the resulting estimators ($\gamma = 1$ or $\gamma = 1/p$) are not minimax, but their risks provide a lower and an upper bound for the risk of the general minimax estimator. Let now

$$\mathbf{U} = \mathbf{T}^{-\frac{1}{2}}\mathbf{P}. \tag{2.12}$$

Then we derive

THEOREM 1. *The following statements are valid* ($\gamma = 1$ or $\gamma = 1/p$):

(i) *All matrices* $\mathbf{C}_\gamma \in \mathbb{R}^{p\times n}$ *minimizing* $F_\gamma(\mathbf{C})$ *are given by the equation*

$$\mathbf{GC}_\gamma(\sigma^2\mathbf{V} + a\gamma\mathbf{XUT}^{-\frac{1}{2}}\mathbf{X}') = a\gamma\mathbf{GUT}^{-\frac{1}{2}}\mathbf{X}'. \tag{2.13}$$

(ii) *A special solution of* (2.13) *not depending on* **G** *is given by*

$$\mathbf{C}_\gamma = \mathbf{U}[(a\gamma)^{-1}\sigma^2\mathbf{T}^{\frac{1}{2}} + \mathbf{X}'\mathbf{V}^{-1}\mathbf{XU}]^{-1}\mathbf{X}'\mathbf{V}^{-1}. \tag{2.14}$$

(iii) *If* $\mathbf{G} = \mathbf{gg}'$ (i.e. rank(**G**) = 1) *then there exists an equality restricted minimax estimator*

$$\mathbf{d}_1 = \mathbf{B}[\mathbf{d}_0 + \mathbf{C}_1(\mathbf{y} - \mathbf{Xd}_0)], \tag{2.15}$$

where the minimax initial guess is given by $\mathbf{d}_0 = \boldsymbol{\beta}_0 + \mathbf{T}^{-\frac{1}{2}}\mathbf{t}_*$.

(iv) *The minimal value of* $F_\gamma(\mathbf{C})$ *is given by*

$$F_\gamma(\mathbf{C}_\gamma) = \sigma^2 tr\mathbf{G}[(\alpha\gamma)^{-1}\sigma^2\mathbf{I} + \mathbf{U}\mathbf{T}^{-\frac{1}{2}}\mathbf{X}'\mathbf{V}^{-1}\mathbf{X}]^{-1}\mathbf{U}\mathbf{U}'. \tag{2.16}$$

This minimum represents a lower bound ($\gamma = 1/p$) *and an upper bound* ($\gamma = 1$) *for the minimax risk in the general case of restricted minimax estimation.*

Proof. (i) Since F_γ is convex, we only need to differentiate $F_\gamma(\mathbf{C})$ with respect to $\mathbf{C}$. Then $(\partial F_\gamma)/(\partial\mathbf{C})(\mathbf{C}_\gamma) = \mathbf{0}$ is equivalent to

$$\sigma^2\mathbf{G}\mathbf{C}_\gamma\mathbf{V} + \alpha\gamma\mathbf{G}(\mathbf{C}_\gamma\mathbf{X} - \mathbf{I})\mathbf{U}\mathbf{U}'\mathbf{X}' = \mathbf{0} \tag{2.17}$$

which readily yields (2.13).

(ii) Omitted, follows by direct calculations.

(iii) Since $\mathbf{G}$ has rank one, $Q(\mathbf{C})$ has also rank one for all $\mathbf{C} \in \mathbb{R}^{p\times n}$. Hence there can be only one eigenvalue of $Q(\mathbf{C})$ different from zero. Consequently we have

$$\mathrm{tr}\{Q(\mathbf{C})\} = \lambda_{max}[Q(\mathbf{C})]. \tag{2.18}$$

By a symmetry argument analogous to Läuter (1975) we see that $\mathbf{d}_0 = \boldsymbol{\beta}_0 + \mathbf{T}^{-\frac{1}{2}}\mathbf{t}_*$ is the minimax initial guess. Then proceed as in (i).

(iv) It is easy to establish that

$$F_\gamma(\mathbf{C}_\gamma) = \alpha\gamma\mathrm{tr}\{\mathbf{G}(\mathbf{I} - \mathbf{C}_\gamma\mathbf{X})\mathbf{U}\mathbf{U}'\} \tag{2.19}$$

which straightforwardly leads to (2.16). □

The estimator (2.15) clearly is important also in the case when rank($\mathbf{G}$) > 1, and will be called upper restricted quasi-minimax estimator (URQME) correspondingly,

$$\mathbf{d}_{1/p} = \mathbf{B}[\mathbf{d}_0 + \mathbf{C}_{1/p}(\mathbf{y} - \mathbf{X}\mathbf{d}_0)] \tag{2.20}$$

is lower restricted quasi-minimax estimator (LRQME), where $\mathbf{G}$ may be of arbitrary rank. While the estimators from (2.15) and (2.20) do not necessarily lie within the set E it is easy to show that they satisfy the linear restrictions given by L at least.

It remains an open question, whether URQME or LRQME possess the matrix risk minimax property, c.f. Bunke (1975).

3. PARTIAL MINIMAX ESTIMATION

Let us turn our attention to the case, as in our example, when only a subvector of $\boldsymbol{\beta}$ is restricted to an ellipsoid. For this let

$$\boldsymbol{\beta} = (\boldsymbol{\beta}_1', \boldsymbol{\beta}_2')' \tag{3.1}$$

and suppose that $\boldsymbol{\beta}_2$ is a member of the ellipsoid

$$E_2 = \{\mathbf{b} \in \mathbb{R}^r \mid (\mathbf{b}-\mathbf{b}_2)'\mathbf{T}_2(\mathbf{b}-\mathbf{b}_2) \leq c_2\} \tag{3.2}$$

where $0 < r \leq p$. The $(r \times r)$ -matrix $\mathbf{T}_2$ is known and p.d., $\mathbf{b}_2 \in \mathbb{R}^r$ is the known center of E_2, and the positive scalar c_2 is also given. Partitioning $\boldsymbol{\beta}$ this way is always possible by rewriting the regression equation. For $r < p$ the subvector $\boldsymbol{\beta}_1$ contains the unrestricted parameters of the model. Without loss of generality we may choose $c_2 = 1$.

Suppose we wish to combine (3.2) with the linear restrictions $\mathbf{R}\boldsymbol{\beta} = \mathbf{r}$. Then the basis set of our estimation procedure is given by

$$L_2 = \{\boldsymbol{\beta} \mid \boldsymbol{\beta} \in \mathbb{R}^p \mid \boldsymbol{\beta} = (\boldsymbol{\beta}_1', \boldsymbol{\beta}_2')',\ \boldsymbol{\beta}_2 \in E,\ \mathbf{R}\boldsymbol{\beta} = \mathbf{r}\}. \tag{3.3}$$

Unfortunately we cannot expect that L_2 is a compact set. Therefore the maximal risk may not exist. To tackle the problem of an infinite risk, Drygas (1985) proposed to calculate minimax estimators by imposing additional restrictions on the class of linear estimators. Subsequently we develop a "partial minimax procedure", which is much more easier to handle.

Suppose for a moment that the quantity

$$\eta = (1 + \boldsymbol{\beta}_1'\boldsymbol{\beta}_1)^{-1} \tag{3.4}$$

is known. After setting

$$\mathbf{T} = \begin{pmatrix} \mathbf{I} & \mathbf{0} \\ \mathbf{0} & \mathbf{T}_2 \end{pmatrix},\quad \boldsymbol{\beta}_0 = (\mathbf{0}', \mathbf{b}_2')',\quad \mathbf{T}_\eta = \eta\mathbf{T}. \tag{3.5}$$

we obtain for all $\boldsymbol{\beta} \in L_2$

$$(\boldsymbol{\beta}-\boldsymbol{\beta}_0)'\mathbf{T}_\eta(\boldsymbol{\beta}-\boldsymbol{\beta}_0) = \eta[\boldsymbol{\beta}_1'\boldsymbol{\beta}_1 + (\boldsymbol{\beta}_2-\mathbf{b}_2)'\mathbf{T}_2(\boldsymbol{\beta}_2-\mathbf{b}_2)] \leq 1. \tag{3.6}$$

Since $\mathbf{T}_\eta$ is p.d. we may consider the set

$$B_\eta = \{\boldsymbol{\beta} \in \mathbb{R}^p \mid (\boldsymbol{\beta} - \boldsymbol{\beta}_0)'\mathbf{T}_\eta(\boldsymbol{\beta} - \boldsymbol{\beta}_0) \leq 1, \quad \mathbf{R}\boldsymbol{\beta} = \mathbf{r}\} \tag{3.7}$$

and (formally) apply Theorem 1 after replacing $\mathbf{T}$ by $\mathbf{T}_\eta$. Then we obtain lower and upper partial restricted quasi-minimax estimators determined by

$$\mathbf{C}_\gamma^{\text{part}} = \mathbf{U}[(\bar{a}\gamma)^{-1}\sigma^2\eta\mathbf{T}^{\frac{1}{2}} + \mathbf{X}'\mathbf{V}^{-1}\mathbf{X}\mathbf{U}]\mathbf{X}'\mathbf{V}^{-1}, \tag{3.8}$$

where $\bar{a} = 1 - \eta\mathbf{t}_*'\mathbf{t}_*$. The corresponding partial minimax guess is

$$\mathbf{d}_0^{\text{part}} = \boldsymbol{\beta}_0 + \mathbf{T}^{-\frac{1}{2}}\mathbf{t}_* \tag{3.9}$$

where $\mathbf{T}^{-\frac{1}{2}}$ and $\mathbf{t}_*$ are computed from the matrix $\mathbf{T}$ given by (3.5). Thus our partial restricted quasi-minimax estimators are

$$\mathbf{d}_\gamma^{\text{part}} = \mathbf{B}[\mathbf{d}_0^{\text{part}} + \mathbf{C}_\gamma^{\text{part}}(\mathbf{y} - \mathbf{X}\mathbf{d}_0^{\text{part}})].$$

Clearly $\mathbf{C}_\gamma^{\text{part}}$ depends on the unknown quantities σ^2 and $\boldsymbol{\beta}_1'\boldsymbol{\beta}_1$ (via η). To make our estimators feasible we propose the following strategy:

(I) Replace σ^2 by a reliable substitute (e.g. s^2, its MSE-minimal estimate).

(II) Replace η by $\hat{\eta} = (1 + \hat{\boldsymbol{\beta}}_1'\hat{\boldsymbol{\beta}}_1)^{-1}$, where $\hat{\boldsymbol{\beta}}_1$ is taken from the least squares or the restricted least squares estimator. Compute $\mathbf{d}_\gamma^{\text{part}}$ with $\mathbf{B} = \mathbf{I}$ and $\gamma = 1$. That part of $\mathbf{d}_\gamma^{\text{part}}$ which corresponds to $\boldsymbol{\beta}_1$ is chosen as our approximation for $\boldsymbol{\beta}_1$. Repeat this procedure until convergence is sufficient.

4. A SIMULATION STUDY

Since the insertion strategies proposed in the previous section yield nonlinear estimators whose statistical properties cannot be analyzed analytically we performed a simulation study to get some hints at their behaviour under varying model parameters. We chose a model with 5 regressors, where the parameter vector $\boldsymbol{\beta}$ was fixed at the surface of the unit ball $\boldsymbol{\beta}'\boldsymbol{\beta} \leq 1$. The disturbances were generated as pseudorandom numbers with varying variance σ^2 according to

$\sigma^2 =$	0.001	0.01	0.1	0.2	1.0
	very low	low	medium	high	very high

On principle, the regressor matrix was fixed, but we incorporated different degrees of multicollinearity.

Setting

$$\mathbf{X} = (1 - \phi^2)^{\frac{1}{2}} \begin{pmatrix} (\mathbf{G}^{-1})' \\ \mathbf{0} \end{pmatrix} \tag{4.1}$$

and

$$\mathbf{G}_{p\times p} = \begin{pmatrix} (1-\phi^2)^{\frac{1}{2}} & 0 & 0 & \dots & 0 \\ -\phi & 1 & 0 & \dots & 0 \\ 0 & -\phi & 1 & \dots & 0 \\ \cdot & \cdot & \cdot & & \cdot \\ \cdot & \cdot & \cdot & & \cdot \\ \cdot & \cdot & \cdot & & \cdot \\ 0 & 0 & 0 & \dots & 1 \end{pmatrix} \tag{4.2}$$

leads to

$$\mathbf{X}'\mathbf{X} = (1-\phi^2)^{\frac{1}{2}}\mathbf{G}^{-1}(\mathbf{G}^{-1})' = \begin{pmatrix} 1 & & \phi^{p-1} \\ & \ddots & \\ \phi^{p-1} & & 1 \end{pmatrix}. \tag{4.3}$$

Varying ϕ over the interval [0, 1] yields different degrees of multicollinearity. For $\phi \to 1$ we obtain a singular matrix $\mathbf{X}'\mathbf{X}$ consisting of ones, if $\phi \to 1$ we get an orthogonal design.

In the study we chose the following values of ϕ:

$\phi =$	0.99	0.75	0.5	0.1	0.0
yielding	very high	high	medium	low	very low

degree of multicollinearity.

The additional linear information is represented by five different restriction matrices R:

"low information" $R1 = \begin{bmatrix} 1 & 0 & 1 & -1 & 1 \end{bmatrix}$

"medium information" $R2 = \begin{bmatrix} 1 & -1 & 0 & 1 & -1 \\ 0 & 1 & -1 & -1 & 1 \end{bmatrix}$

"high information" $R3 = \begin{bmatrix} 1 & 0 & 1 & -1 & -1 \\ 0 & 1 & -1 & 1 & -1 \\ 1 & 1 & -1 & -1 & 0 \end{bmatrix}$

"very high information" $R4 = \begin{bmatrix} 1 & 0 & 1 & -1 & -1 \\ 0 & 1 & -1 & 1 & -1 \\ 1 & 1 & -1 & -1 & 0 \\ -1 & -1 & 0 & 1 & 1 \end{bmatrix}$

"incorrect information" $R5 = \begin{bmatrix} 1 & 1 & 1 & 1 & 1 \end{bmatrix}$, where from R1 to R4 the hyperspace information increases since the rows of R are independent. By R5 the case of incorrect prior information is considered, to get some insight into the behaviour of the restricted estimators when the linear restriction is misspecified.

In every of 100 runs we computed the empirical mean square error of

- ordinary least squares estimator
- restricted least squares estimator
- URQME
- LRQME.

For the last two estimators we included also their partial counterparts, where either $\hat{\boldsymbol{\beta}}_1$ was inserted for $\boldsymbol{\beta}_1$ or we repeated insertion until convergence according to (II) was observed.

RESULTS

- All partial estimators exhibited distinct advantages over ordinary and restricted least squares estimator. The gain was impressive even in the case when only one component of $\boldsymbol{\beta}$ was restricted to an interval.

- When the number of ellipsoid restrictions concerning the components of $\boldsymbol{\beta}$ was increased at the same time the MSE-values improved.

- URQME and LRQME could never be outperformed by their partial counterparts, but the loss was surprisingly small even in the case of one component restricted by an interval.

- The proposed iterative procedure for insertion of the unknown $\boldsymbol{\beta}_1$ converged in all cases. Moreover, the resulting estimator lead to better MSE-properties than the noniterative estimator using $\hat{\boldsymbol{\beta}}_1$ from LS-estimator only.

5. CONCLUDING REMARKS

The lower and upper restricted quasi-minimax estimators and their partial counterparts can be improved by observing that there are sharper bounds for the maximal eigenvalue of $Q(\mathbf{C})$. When minimizing these bounds we can achieve results analogous to those derived by Stahlecker (1987) and Stahlecker/Lauterbach (1987). However the optimal statistics have to be determined by an extensive numerical procedure.

There still remains to show theoretically that the insertion strategy proposed in situation 3 and applied in Section 4 always leads to convergence. Preliminary studies indicated that this will be the case even if we start from an arbitrary $\boldsymbol{\beta}_1$-value.

We note in passing that the preceding investigations are also possible without linear restrictions on $\boldsymbol{\beta}$. Some related results may be found in Hering/Stahlecker/Trenkler (1987).

REFERENCES

Bunke, O. (1975). Minimax linear, ridge and shrunken estimators for linear parameters. *Mathematische Operationsforschung und Statistik*, 6, 697-701.

Drygas, H. (1985). Minimax prediction in linear models. *Lecture notes in statistics. Linear statistical inference.* Ed. D. Brillinger, 48-60.

Hering, F., Stahlecker, P. and Trenkler, G. (1987). Partial minimax estimation in regression analysis. *Statistica Neerlandica*, 41, 111-128.

Humak, K.M.S. (1977). *Statistische Methoden der Modellbildung.* Akademie Verlag, Berlin.

Läuter, H. (1975). A minimax linear estimator for linear parameters under restrictions in form of inequalities. *Mathematische Operationsforschung und Statistik*, 6, 689-695.

Saito, M. (1977). Household flow-of-funds equations. *The Journal of Money Credit and Banking*, 9, 1-20.

Stahlecker, P. (1987). A priori Information und Minimax-Schätzung im Linearen Regressionsmodell. Verlag Athenäum. *Mathematical Systems in Economics*, 108.

Stahlecker, P. and Lauterbach, J. (1987). Approximate minimax estimation in linear regression: Theoretical results. *Commun. Statist.* A16, 4 (forthcoming).

Stahlecker, P. and Trenkler, G. (1987). Full and partial minimax estimation in regression analysis with additional linear constraints. Submitted to *Linear Algebra and its Applications*.

Toutenburg, H. (1982). *Prior Information in Linear Models.* Wiley, New York.

Received 3 March 1987
Revised 17 August

Department of Statistics
University of Dortmund
Vogelpothsweg
Postfach 500500
4600 Dortmund 50
F.R. Germany

Department of Economics
Johannes Gutenberg University of Mainz
Saarstrasse 21
6500 Mainz
F.R. Germany

Proc. Second International Tampere Conference in Statistics
(Tampere, Finland, 1-4 June 1987)
Tarmo Pukkila and Simo Puntanen, *Editors*

pp. 691 - 699

On classification strategies in medical functional diagnostics

K.-D. WERNECKE — Charité Eye Clinic
G. KALB — Charité Eye Clinic
E. STÜRZEBECHER — Charité-ENT-Clinic
The Humboldt-University of Berlin, G.D.R.

Key words and phrases: Feature reduction and selection, time series classifiers, coupling of classifiers, mixed model, hearing loss.

ABSTRACT

In otological and ophthalmological functional diagnostics it is typical to record for a certain individual several similar biopotentials under different conditions of stimulation and/or to collect various potentials for the same individual. Moreover, most diagnoses are done after using further information about the individual, e.g. from some clinical basic investigations. That means we have to cope with various data sets of the same type of but different structure (e.g. time series) and furthermore with data of different type of scaling. In the paper a classification scheme is given applicable to different kinds of continuous data as well as to categorial ones by coupling the single classifiers.

1. INTRODUCTION

In otological and ophthalmological functional diagnostics several similar biopotentials are recorded under different conditions of stimulation and/or potentials of various kinds are gathered for the same patient. Moreover, it is important to use in addition further diagnostic findings such as audiometric data and others in order to gain the desired diagnosis. In previous papers [Wernecke and Kalb (1987), Wernecke (1987)] we could show that the most often used linear discriminant analysis (LDA) as classifier should not be applied when $n_i/p < k$, $i = 1(1)K$ (n_i class sizes, p number of features, $1 < k < 3$). Therefore LDA is not appropriate to the classification of time series like the given biopotentials.

In the paper various classifiers for time series are represented in connection with a special method of feature reduction. Last but not least, a coupling procedure is described combining logically the different classification rules to get a new (common) classifier.

2. CLASSIFICATION RULES

The allocation of the individuals is principally to be performed according to Bayes' decision rule. Therefore an observation vector $\mathbf{x} = (x_1,\dots, x_p)'$ is classified into the class (population) A_j for which

$$V_j(\mathbf{x}) = \sum_{k=1}^{K} a_k r_{kj} f_k(\mathbf{x}) \Rightarrow \min!$$

[$f_k(\mathbf{x})$ are probability densities, a_k prior probabilities of the class A_k, r_{kj} the loss incurred when an element originated from class A_k is allocated to A_j, $r_{kk} = 0$, $k, j = 1(1)K$]. That leads to the minimum of the expected (total) loss

$$V = \sum_{k,j=1}^{K} a_k r_{kj} \int_{R_j} f_k(\mathbf{x})\, d\mathbf{x}$$

$[R_j = \{\mathbf{x} \mid \mathbf{x} \Rightarrow A_j\} = \{\mathbf{x} \Rightarrow A_j \mid V_j(\mathbf{x}) = \min_{k=1(1)K} V_k(\mathbf{x})\}]$.

2.1 Discriminant analysis

Under the supposition

$$f_k(\mathbf{x}) = f_k(x_1,\dots, x_p) = (2\pi)^{-p/2}|\boldsymbol{\Sigma}_k|^{-\frac{1}{2}}\exp\{-\tfrac{1}{2}(\mathbf{x}-\boldsymbol{\mu}_k)'\boldsymbol{\Sigma}_k^{-1}(\mathbf{x}-\boldsymbol{\mu}_k)\}$$

($\boldsymbol{\mu}_k$ mean vector, $\boldsymbol{\Sigma}_k$ covariance matrix of population A_k) and $a_k r_{kj} = 1 - \delta_{kj}$ (δ_{kj} Kronecker's symbol) we gain with $\boldsymbol{\Sigma}_k = \boldsymbol{\Sigma}\ \forall\ k$ the special classifier (LDA)

$$L_j = \{\mathbf{x} \Rightarrow A_j \mid \hat{f_j}(\mathbf{x}) = \max_{k=1(1)K} \hat{f_k}(\mathbf{x})\}$$
$$= \{\mathbf{x} \Rightarrow A_j \mid d_j(\mathbf{x}) = \min_{k=1(1)K}(\mathbf{x} - \bar{\mathbf{x}}_k)'\mathbf{S}^{-1}(\mathbf{x} - \bar{\mathbf{x}}_k)\}$$

[$\bar{\mathbf{x}}_k$, $\mathbf{S}$ maximum-likelihood-estimates (MLE) of $\boldsymbol{\mu}_k$, $\boldsymbol{\Sigma}$ from the given sample. $L = (L_1,..., L_K)$ splits the feature space into K disjoint subsets].

If $\boldsymbol{\Sigma}_k = \boldsymbol{\Sigma}\ \forall k$ is not to maintain Bayes' rule yields

$$Q_j = \{\mathbf{x} \Rightarrow A_j \mid d_j'(\mathbf{x}) = \min[(\mathbf{x} - \bar{\mathbf{x}}_k)'\mathbf{S}_k^{-1}(\mathbf{x} - \bar{\mathbf{x}}_k) + \ln|\mathbf{S}_k|]\}$$

the so-called quadratic discriminant analysis (QDA).

2.2 Discriminant analysis in the complex (frequency) domain

The biopotentials used in ophthalmological and otological diagnostics are always time functions $x(t)$ containing some background activity (EEG-noise) $n(t)$ and a certain evoked response $\mu(t)$ $(-\infty < t < \infty)$. Writing these functions in terms of their sample points we assume for a periodic and W-band-limited signal

$$x(t) = x(t + T), \qquad T > 0, \quad -\infty < t < \infty,$$

the model

$$\mathbf{x} = \left(x(0),\ x\left(\frac{1}{2W}\right),..., x\left(\frac{2WT-1}{2W}\right)\right)' = (x_1,..., x_p)' = \boldsymbol{\mu}_j + \mathbf{n}_j$$

where $\boldsymbol{\mu}_j$ denotes a fixed signal and $\mathbf{n}_j$ represents some stochastic process with respect to class A_j ($j = 1(1)K$). In the analysis of EEG-recordings $\mathbf{n}_j(t) = \mathbf{n}_j$ is very often supposed as a zero mean (weakly) stationary process, which implies $E(\mathbf{n}_j) = \mathbf{0}$, i.e. $E(\mathbf{x}) = \boldsymbol{\mu}_j$ and a stationary covariance matrix $\boldsymbol{\Sigma}_j$

$$cov(\mathbf{x}, \mathbf{x}') = \boldsymbol{\Sigma}_j = cov(\mathbf{n}_j, \mathbf{n}_j') = \{\sigma_j(t-u);\ t, u = 0(1)\,2WT-1\},$$

where $\sigma_j(t - u)$ denotes the autocorrelation function of the stochastic process $n_j(t)$:

$$\sigma_j(t-u) = \sigma_j(\tau) = \int_{-\infty}^{\infty} e^{i\tau\lambda} p_j(\lambda) d\lambda$$

[$p_j(\lambda)$ the power spectrum of $n_j(t)$, $\tau = t-u$, $\lambda = 2\pi/T$, i imaginary unit].

Using the discrete Fourier-transform

$$DFT\{x(t)\} = DFT\left\{x\left(\frac{k}{2W}\right)\right\} = \frac{1}{2WT}\sum_{m=0}^{2WT-1} x\left(\frac{m}{2W}\right) e^{-\frac{ikm\pi}{WT}} = c(k)$$

[k = 0(1) 2WT − 1] we apply the following spectral approximations:

(i) Replace $\mathbf{\Sigma}_j = \{\sigma_j(t-u)\}$ by the matrix

$$\mathring{\mathbf{\Sigma}}_j = \{\mathring{\sigma}_j(\tau)\} = \left\{\frac{1}{2WT}\sum_{k=0}^{2WT-1} p_j\left(\frac{k\pi}{WT}\right) e^{\frac{ik\pi}{WT}\tau}\right\}$$

[which is practically the replacement of the inverse continuous Fourier-transform by the discrete one – Wahba (1968)].

(ii) Replace $\mathbf{\Sigma}_j^{-1}$ by

$$\mathring{\mathbf{\Sigma}}_j^{-1} = \{\mathring{\sigma}_j^{-1}(\tau)\} = \left\{\frac{1}{2WT}\sum_{k=0}^{2WT-1} p_j^{-1}\left(\frac{k\pi}{WT}\right) e^{\frac{ik\pi}{WT}\tau}\right\}$$

[which is practically the inverse DFT $\left\{p_j^{-1}\left(\frac{k\pi}{WT}\right)\right\}$ – Ligett (1971)].

(iii) Replace the spectra $p_j\left(\frac{k\pi}{WT}\right)$ by $V[c_j(k)]$ with

$$\hat{V}[c_j(k)] = \frac{1}{n_j - 1}\sum_{\ell=1}^{n_j} \left| c_{j\ell}(k) - \frac{1}{n_j}\sum_{\ell=1}^{n_j} c_{j\ell}(k)\right|^2 = \hat{p}_j\left(\frac{k\pi}{2W}\right)$$

for some training sample $\mathbf{x}_{j\ell}$ [$j = 1(1)K$, $\ell = 1(1)n_j$, $k = 0(1)2\,WT-1$] [Shumway (1982)].

With these simple spectral approximations we perform LDA and QDA such as described above [cf. Shumway (1982)] but avoiding complicated expressions (involving matrix inverses) and critical relations between the number of features (sample points) and the size of the training sample [cf. Wernecke (1987)].

2.3 Feature selection

An important question in connection with practical applications is the selection of such subsets of features that are most essential for classification.

Choosing discriminant analysis as classifier the often-used selection procedure for LDA (as well its complex variant) is based on the F-criterion.

Another generally applicable criterion is the minimization of an estimation of the true error rate

$$V_T = \sum_{k,j=1}^{K} a_k r_{kj} \int_{\hat{R}_j} f_k(\mathbf{x})\, d\mathbf{x}$$

($\bigcup_{j=1}^{K} \hat{R}_j = (\hat{R}_1, \hat{R}_2, \ldots, \hat{R}_K)$ splits the feature space according to the classifier) for this desired subset of features. Appropriate estimations $\hat{V}_T$ are available [cf. Wernecke *et al.* (1980), (1983)].

2.4 Feature reduction

Another possibility in handling data sets with a large number of features is a feature reduction by means of spectral analysis (before the classification process, because of the admissible relation between number of features and sample size - see above).

Among others we use the discrete Fourier-transform in the sense:

Find a number $N' = 2W'T - 1$ of frequencies which yields a given amount M of the power

$$P\{x(t)\} = \sum_{k=0}^{N} |c(k)|^2 \qquad (N = 2WT - 1),$$

in other words,

$$(2W'T - 1 =) N' < N (= 2WT - 1) \text{ under the condition } \sum_{k=0}^{N'} |c(k)|^2 = M$$

and take the N' sample points

$$x\left(\frac{k}{2W'}\right) = \mathrm{DFT}^{-1}\{c(k)\} = \sum_{m=0}^{2W'T-1} c(m)e^{\frac{ikm\pi}{W'T}} \qquad (k = 0(1)N')$$

as new features (where $\mathrm{DFT}^{-1}\{...\}$ is the inverse DFT).

3. COUPLING OF CLASSIFIERS

3.1 Simple coupling

To simplify matters we describe the procedure with only two different classifiers and note the classification results obtained using the π-error estimation [cf. Wernecke *et al.* (1980) e.g.].

The classifiers $KL1$, $KL2$ may provide the allocation vectors

$$KL1:\ \mathbf{k} = (k_1,..., k_n)' \qquad k_j, \ell_j \in \{1,..., K\}$$
$$KL2:\ \boldsymbol{\ell} = (\ell_1,..., \ell_n)'$$

($n = (n_1 + ... + n_K)$; object j is allocated to class $k_j(\ell_j)$, $j = 1(1)n$).

From these two allocation vectors we construct for each class A_k(k = 1(1)K) a matrix $M(k)$ with the frequencies $h(k, \ell, j)$ which are defined by the number of objects from class A_k, being allocated to class ℓ with the classifier $KL1$, and to class j with the classifier $KL2$ ($\ell, j, k = 1(1)K$). If the given classes, based on the samples drawn, are sufficiently well separable, and the classifiers $KL1$, $KL2$ reflect this, then the (three-dimensional) main diagonal elements $h(k, \ell, j)$ with $k = \ell = j$ will be maximal over all matrices $M(k)$.

Accordingly we can increase the safety of assignment to class $K_1,...,$ K_K with different classification results from $KL1$ and $KL2$, respectively, if we allocate to the class where $h(k, \ell, j)$ will be maximal among all matrices $M(k)$ for a given combination $\ell \neq j$. Based on these considerations we define a matrix $M = M_{\max}$, whose elements m_{lj} are determined according to

$$m_{\ell j} = k'(\ell, j) \qquad (\ell, j = 1(1)K)$$

with $k' = k'(\ell, j)$ from $\max\limits_{k=1(1)K} h(k, \ell, j) = h(k', \ell, j)$, i.e., $m_{\ell j} = k'(\ell, j)$ is the class index k' with the maximal $h(k, \ell, j)$ over all matrices $M(k)$ for a given combination ℓ, j.

With this matrix we define a new allocation rule in the sense that an object from any given class, which was allocated to class ℓ with the classifier $KL1$ and to class j with the classifier $KL2$, is now assigned to class $m_{\ell j} = k'(\ell, j)$. In case of $h(k', \ell, j) = h(k'', \ell, j)$, $k' \neq k''$, i.e. with equal frequency numbers, assignment can be made arbitrarily.

This classification from the combination of two classifiers $KL1$, $KL2$ can be only an improvement upon the single classifications, and will be judged by a certain estimation of the true error rate (cf. 2.3). The case $K > 2$ is very easy to realize.

3.2 Modified coupling

The safety of decision may be increased still further by modification of the allocations from M made on the basis of just a few cases, i.e. small frequencies $h(k', \ell, j)$. For this purpose we determine in the matrix $M_H = ((h(k', \ell, j)))$ $(\ell, j = 1(1)K)$ the frequencies $h(k', \ell', j')$, lying below a limit S to be specified, and calculate for these frequencies confidence intervals $p(k', \ell', j')_\ell$, $p(k', \ell', j')_u$ by which means we obtain a modified (new) classifier for the fields regarded of matrices M and M_H, respectively. For further details see Wernecke *et al.* (1986).

The essential advantage of these coupling procedures is the possibility to combine total different classifiers to a new classification rule because the method needs only the allocation vectors of each classifier.

4. EXAMPLE

In order to discriminate between patients suffering from noise-induced hearing loss and cochlear hearing loss of other genesis we investigated two corresponding groups of patients in 93 and 165 subjects, respectively. Evaluated were so-called auditory evoked potentials (AEP) of two different kinds, namely auditory brain stem potentials and slow cortical potentials under different conditions of stimulation. In addition, we used the findings of hearing threshold measurements at four different frequencies.

The results are given in the next tables.

SINGLE CLASSIFICATION RESULTS

data[1]	classifier	number of features (sample points)	error %[4]
E 21	complex (LDA + QDA)	18[2]	31,40
E 24	complex (LDA + QDA)	2[2]	40,31
B 2	complex QDA	78[2]	37,21
AU 2	LDA	3[3]	24,81

[1] E 21: slow cortical potential, evoked by 1000 Hz
E 24: slow cortical potential, evoked by 4000 Hz
B 2: auditory brain stem response
AU 2: audiometric threshold

[2] After digital filtering according to 2.4

[3] After feature selection according to 2.3

[4] Estimation of the true error rate cf. 2.3

Error after coupling of classifiers:

E 21 ∧ E 24 ∧ B2 : 29,84%
E 21 ∧ E 24 ∧ B2 ∧ AU2 : 24,81%

Only in the cases of the slow cortical responses LDA could be applied because of a sufficient high amount of power after digital filtering (cf. 2.4).

SINGLE CLASSIFICATION RESULTS

data[1]	classifier	number of features (sample points)	error %[4]
E 21	LDA	8[2), 3)]	27,37
E 24	LDA	2[2), 3)]	38,37
B 2	complex QDA	78[2]	37,21
AU 2	LDA	3[3]	24,81

[1), 2), 3), 4)] see above!

Error after coupling of classifiers:

E 21 ∧ E 24 ∧ B2 : 27,13%
E 21 ∧ E 24 ∧ B2 ∧ AU2 : 21,32%

REFERENCES

Ligett, W.S. (1971). On the asymptotic optimality of spectral analysis for testing hypothesis about time series. *Ann. Math. Statist.*, 42, 1348-1358.

Shumway, R.H. (1982). Discriminant analysis for time series. *Handbook of Statistics* 2 (P.R. Krishnaiah and L.N. Kanal, *eds.*), 1-46.

Wahba, G. (1968). On the distribution of some statistics useful in the analysis of jointly stationary time series. *Ann. Math. Statist.*, 38, 1849-1862.

Wernecke, K.-D., Kalb, G. and Stürzebecher, E (1980). Comparison of various procedures for estimation of the classification error in discriminance analysis, *Biom. J.*, 25, 247-258.

Wernecke, K.-D., Unger, S. and Kalb, G. (1986). The use of combined classifiers in medical functional diagnostics. *Biom. J.*, 28, 81-88.

Wernecke, K.-D. and Kalb, G. (1987). Untersuchungen zur Merkmalsauswahl bei Bayes'schen Klassifikatoren. *EDV in Med. u. Biol.* in press.

Wernecke, K.-D. (1987). Some results on the feature selection in discriminance analysis. *Proceedings of the DIANA II - conference in Liblice, CSSR.*

Received 3 March 1987

Charité Eye Clinic
The Humboldt University of Berlin
Schumannstrasse 20/21
1040 Berlin
G.D.R.

Charité-ENT-Clinic
The Humboldt University of Berlin
Schumannstrasse 20/21
1040 Berlin
G.D.R.

LIST OF REFEREES

ALALOUF, I.S.
AMEMIYA Yasuo
ANDERSON, T.W.
ANDREWS, Fred C.
ANSLEY, Craig F.
ARJAS, Elja
ARNOLD, Steven F.
ATKINSON, A.C.
BAKSALARY, Jerzy K.
BARNETT, Vic
BECKMAN, Richard J.
BENTLER, Peter M.
BIRKES, David S.
BOCK, Mary Ellen
BOLLERSLEV, Tim
BONDAR, James V.
BROWNE, M.W.
BUNKE, O.
CARMICHAEL, Jean-Pierre
CHALONER, Kathryn M.
CHANDA, Kamal C.
CHIKUSE, Yasuko
COOK, Dennis R.
CORNELL, John A.
CSÖRGÖ, S.
DAWID, A. Philip
DELECROIX, Michel
DENTENEER, Dee
DEUTSCH, Emeric
DRYGAS, Hilmar
DUNNETT, Charles W.
EDGINGTON, Eugene S.
EPPS, T.W.
ERIKSSON, Tor
FAREBROTHER, R.W.
FELLMAN, Johan
FIENBERG, Stephen E.
FINDLEY, David F.
FUJIKOSHI, Yasunori
GAFFKE, Norbert
GENTLEMAN, Jane F.
de GOOIJER, Jan G.
GRANGER, Clive W.J.
GREEN, P.J.
HABERMAN, Shelby J.
HAN, Chien-Pai
HANNAN, E.J.
HARVEY, Andrew C.
HAVRÁNEK, Tomas
HAWKINS, Douglas M.
HENDERSON, Charles R.
HJORT, Nils Lid
HOAGLIN, David C.
HUBER, Peter
KALA, Radoslav
KANTO, Antti J.
KARIYA, Takeaki
KEMPTHORNE, Oscar
KHATRI, C.G.
KHURI, André
KIIVERI, Harri T.
KLEFFE, Jürgen
KNOTT, Martin
KOREISHA, Sergio G.
KORHONEN, Pekka J.
KRÄMER, Walter
KRUSKAL, William H.
KSHIRSAGAR, Anant M.
KULLBACK, Solomon
LEHMANN, Erich L.
LüTKEPOHL, Helmut
MÄKELÄINEN, Timo
MATHEW, Thomas
McCABE, George
McLEOD, Ian
MELLIN, Ilkka
MILHØJ, Anders
MORRIS, Carl N.
MUKERJEE, Rahul
NAEVE, Peter
NELDER, J.A.
NERLOVE, Marc
NEUDECKER, Heinz
NEWMAN, Morris
NIEMI, Hannu
NIEMINEN, Mauri
NYBLOM, Jukka
NYQUIST, Hans
O´BRIEN, Peter C.
OJA, Hannu
OMAN, Samuel D.
PALM, Franz
PARZEN, Emanuel
POLASEK, Wolfgang
PRAKASA RAO, B.L.S.
PUKELSHEIM, Friedrich
PYNNÖNEN, Seppo
RAHIALA, Markku
RAMIREZ, Jose
RAMSAY, James O.
RAO, C. Radhakrishna
RÉVÉSZ, Pál
RIPLEY, Brian D.
RISSANEN, Jorma
ROBINSON, Enders A.
ROUSSEEUW, Peter J.
SAIKKONEN, Pentti
SÄRNDAL, Carl-Erik
SCHAFFRIN, Burkhard
SESHADRI, V.
SHAPIRO, Alexander
SHIBATA, Ritei
SHUMWAY, Robert H.
SIMONOFF, Jeffrey S.
SINHA, Bikas K.
SPEED, T.P.
SUBRAMANYAM, Kasala
SUGIURA, Nariaki
TANIGUCHI, Masanobu
TATSUOKA, Maurice M.
THISTED, Ronald A.
THOMPSON, Robin
TITTERINGTON, D.M.
TOUTENBURG, Helge
TRENKLER, Götz
TRUAX, Donald
TSAY, Ruey S.
TUKEY, John W.
VALKEILA, Esko
WATSON, Geoffrey S.
WEGMAN, Edward J.
WEISBERG, Sanford
WELSCH, Roy E.
WERMUTH, Nanny
WHITTLE, Peter
WOLD, Herman
WORSLEY, Keith J.
YANAI, Haruo
ZACKRISSON, Uno
ZELLNER, Arnold

PHOTOGRAPHS [by Heimo Mäkinen]

T.W. Anderson

George E.P. Box

Jerzy K. Baksalary

C.G. Khatri
Michael D. Perlma
Bimal K. Sinha
Takeaki Kariya
Gunnar Rosenqvist
Kurt Brännäs

Andrzej Zielinski
Sergio G. Koreisha
Friedrich Pukelshe
John W. Pratt
Daryl Pregibon
Manfred Deistler
A.C. Atkinson
Marc Nerlove

Merja Koivula
Pirjo Larima
Elisa Lahtinen
Virpi Mäntylä

To the sauna

Tarmo Pukkila
Dominique Latour
George P.H. Styan
Knut Conradsen,
Heikki Hella
E.J. Hannan

Alastair J. Scott Margaret Scott Dominique Latour Simo Puntanen

Jorma Rissanen

Knut Conradsen E.J. Hannan Alexander Novikov

Andrzej Matuszewski Stanislaw Mejza Kenneth Nordström
Friedrich Pukelsheim

Stratis Kounias Alexander Novikov V.D. Konakov Peter Naeve

Seppo Mustonen

Terry Speed